Structure and Dynamics of Elementary Matter

NATO Science Series

A Series presenting the results of scientific meetings supported under the NATO Science Programme.

The Series is published by IOS Press, Amsterdam, and Kluwer Academic Publishers in conjunction with the NATO Scientific Affairs Division

Sub-Series

I. Life and Behavioural Sciences	IOS Press
II. Mathematics, Physics and Chemistry	Kluwer Academic Publishers
III. Computer and Systems Science	IOS Press
IV. Earth and Environmental Sciences	Kluwer Academic Publishers
V. Science and Technology Policy	IOS Press

The NATO Science Series continues the series of books published formerly as the NATO ASI Series.

The NATO Science Programme offers support for collaboration in civil science between scientists of countries of the Euro-Atlantic Partnership Council. The types of scientific meeting generally supported are "Advanced Study Institutes" and "Advanced Research Workshops", although other types of meeting are supported from time to time. The NATO Science Series collects together the results of these meetings. The meetings are co-organized bij scientists from NATO countries and scientists from NATO's Partner countries – countries of the CIS and Central and Eastern Europe.

Advanced Study Institutes are high-level tutorial courses offering in-depth study of latest advances in a field.
Advanced Research Workshops are expert meetings aimed at critical assessment of a field, and identification of directions for future action.

As a consequence of the restructuring of the NATO Science Programme in 1999, the NATO Science Series has been re-organised and there are currently Five Sub-series as noted above. Please consult the following web sites for information on previous volumes published in the Series, as well as details of earlier Sub-series.

http://www.nato.int/science
http://www.wkap.nl
http://www.iospress.nl
http://www.wtv-books.de/nato-pco.htm

Series II: Mathematics, Physics and Chemistry – Vol. 166

Structure and Dynamics of Elementary Matter

edited by

Walter Greiner
Johann Wolfgang Goethe University,
Institute for Theoretical Physics,
Frankfurt, Germany

Mikhail G. Itkis
Joint Institute of Nuclear Research,
Dubna, Russia

Joachim Reinhardt
Johann Wolfgang Goethe University,
Institute for Theoretical Physics,
Frankfurt, Germany

and

Mehmet Cem Güçlü
Istanbul Technical University,
Turkey

Kluwer Academic Publishers

Dordrecht / Boston / London

Published in cooperation with NATO Scientific Affairs Division

Proceedings of the NATO Advanced Study Institute on
Structure and Dynamics of Elementary Matter
Camyuva-Kemer (Antalya), Turkey
22 September–2 October 2003

A C.I.P. Catalogue record for this book is available from the Library of Congress.

ISBN 1-4020-2446-0 (PB)
ISBN 1-4020-2445-2 (HB)
ISBN 1-4020-2705-2 (e-book)

Published by Kluwer Academic Publishers,
P.O. Box 17, 3300 AA Dordrecht, The Netherlands.

Sold and distributed in North, Central and South America
by Kluwer Academic Publishers,
101 Philip Drive, Norwell, MA 02061, U.S.A.

In all other countries, sold and distributed
by Kluwer Academic Publishers,
P.O. Box 322, 3300 AH Dordrecht, The Netherlands.

Printed on acid-free paper

Printed in the Netherlands.

TABLE OF CONTENTS

Relativistic Heavy Ion Physics – Theory

Astroparticle and Nuclear Astrophysics

Superheavy Elements

Exotic Nuclei and Nuclear Structure

Posters

PREFACE

Ladies and gentlemen,
dear colleagues,
welcome to Kemer to the NATO Advanced Study Institute

Structure and Dynamics of Elementary Matter.

We have chosen Kemer as the place of our NASI because it is located in a beautiful and hospitable surrounding. This part of the Mediterranean at the Turkish Riviera is a historic region where many cultures meet (e.g., the Oriental and the Greek and Roman European cultures) and where you find numerous places which played a role in ancient science and in early Christianity. Moreover, with the hotel Ceylan Inter-Continental we have found a most excellent meeting place, directly located at the beach, equipped with wonderful swimming pools and restaurants – an absolutely first-class location.

Our NASI will deal with the most recent developments in high-energy heavy ion physics and in the search for superheavy nuclei – two rather distinct areas of research. Indeed, we want to bring two very active communities of nuclear and high-energy physics into close contact. The meeting is both a school and has also the character of a conference: A school because there are many advanced students, many of which are themselves already top researchers and who are contributing with their own research in seminars and posters. It is also a conference because new results in the exciting and wonderful fields of low- and high-energy heavy ion physics will be presented. We are mainly focussing on the topics of superheavy elements and of hot and dense nuclear matter.

Here we investigate, how elementary matter may have been formed and what its properties were just fractions of a second after the big bang. We try to reveal how our world was formed, how baryosynthesis and nucleosynthesis may have proceeded, how ordinary or exotic stars (e.g., neutron or quark stars) are built, how supernova explosions occur, and many similar questions. It is believed that hot and/or dense elementary matter displays completely new phases, most notably the famous Quark-Gluon-Plasma predicted by Quantum Chromodynamics (QCD) which most probably is a cluster plasma where strong correlations of hadronic type survive. The most recent experimental data from RHIC indicate that we are close to the discovery of this new phase of elementary matter. Another intriguing phenomenon are the phases of Color Superconductivity which may exist in the interior of collapsed stars. Other new forms of matter involving strangeness (strangelets, multi-Λ-hypernuclei, multiply strange exotic mesonic objects (Memos)) may be formed in nuclear collisions, as well as clusters of matter and antimatter. The extension of our periodic system into the new directions of strange matter and antimatter leads to exciting prospects. The matter-antimatter clusters, furthermore, open the door for a new form of cold compression of elementary matter. This would stand in contrast to the compression by nuclear shock waves which automatically also heats up the matter (Rankine-Hugoniot equations). We want to create dense nuclear matter and this up to now was achieved through the shock wave mechanism in its various forms. When nuclei collide at high energies they compress and heat up tremendously. On the other hand, in matter-antimatter clusters the heating should be much less (Mishustin, Satarov, ...).

Another tremendously rich and exciting field is the search for superheavy nuclei. The theoretical predictions by the Lund and Frankfurt groups now are 35 years old. Experiments at GSI (S. Hofmann, Münzenberg, and collaborators) and at Dubna (Oganessian and his group) have brought great progress. Short-living superheavy nuclei along the proton drip line were found and their α-decay has been observed. The fusion of these superheavy nuclei discovered so far occurs if the projectile and target are chosen such that they approach through a *cold valley* in the potential energy surface. In this case no push and also no 'extra push' is necessary. The fused elements are produced rather cold.

Also quasifission processes (Itkis) of various nuclear systems are seen. The theory has been better founded over the years, using, e.g., the framework of meson field theory, and new magic numbers occur. The extension of the periodic system is a challenge!

We want to bring both research communities together. It is important to understand each other, to talk to each other and to learn from each other! This is why in this NASI we have mixed the talks evenly, so that there is little possibility to escape each other.

We will have plenty of discussions, I hope, both here in the lecture hall as well as outside in the surroundings and at the beach side. The wonderful hotel Ceylan will provide an ideal setting for our conference and we will spend nearly two weeks in a most agreeable confinement.

As it turned out, we had far too many applications, many of which had to be turned down, and also the number of accepted participants lies somewhat above the normal quota for a NATO Advanced Studies Institute. Not surprisingly, quite a number of spouses and companions have joined their husbands and friends.

Let me finally make a number of remarks concerning the organization of the meeting.

The idea for such a NASI down here in Turkey has been suggested to me by A. V. Ramayya (Vanderbilt University) and I took it up with pleasure. My colleagues and friends Dr. Joachim Reinhardt from Goethe University, Frankfurt, and Prof. Çem Güçlü from Istanbul Technical University were helping tremendously in the organization. I thank them cordially. NATO's scientific program has been very generous in providing the main support for this school, deserving our warmest thanks. The German Ministry for Science (BMBF), in particular Dr. D. Müller, and the Gesellschaft für Schwerionenforschung (GSI, Prof. Walter Henning) have also been very helpful with their support. The same is true for the Institut National de Physique Nucléaires et de Physique Particules (IN2P3), for Vanderbilt University, Nashville TN (Prof. J. Hamilton) and for the Turkish Atomic Energy Agency, Ankara. All of them deserve our "thank you" for their support, without which this meeting could not take place in this form. Travel support for students was granted by the Scientific and Technical Research Council of Turkey (Tübitak) and by the National Science Foundation, Washington.

Special acknowledgments go once more to Dr. Joachim Reinhardt who keeps contact with all the speakers and will be organizing most of the editorial work for the Proceedings volume.

The next eleven days will be time filled with hard work, but I am confident that it will also be a fruitful, memorable, and enjoyable period.

I wish us all a very pleasant, interesting and successful NASI. Let us get to work now!

Walter Greiner

NATO ADVANCED STUDIES INSTITUTE

"Structure and Dynamics of Elementary Matter"

Kemer/Turkey, Sept. 22 - Oct. 2, 2003

List of Participants

Abe, Yasuhisa
Yukawa Institute for Theoretical Physics
Kyoto University
Kyoto 606-8502
Japan
abey@yukawa.kyoto-u.ac.jp

Aichelin, Jörg
SUBATECH
Ecole des Mines, B.P. 20722
4, Rue Alfred Kastler
F-44307 Nantes, Cedex 03
France
Aichelin@subatech.in2p3.fr

Aritomo, Yoshihiro
Flerov Lab of Nuclear Reactions
Joint Institute of Nuclear Research
141908 Dubna,
Moscow Region, Russia
Aritomo@nt.phys.s.u-tokyo.ac.jp

Bass, Steffen
Duke University
Department of Physics
Physics Bldg., Science Dr.
Box 90305
Durham, NC 27708
USA
bass@phy.duke.edu

Baudot, Jerome
Institut de Recherches Subatomiques (IReS)
23, rue du Loess
BP 28
67037 Strasbourg Cedex 2
France
baudot@in2p3.fr

Bauer, Wolfgang
Department of Physics and Astronomy
Michigan State University
East Lansing, MI 48824-1116
USA
Bauer@pa.msu.edu

Bezverkhny, Betty
Yale University
A.W. Wright Nuclear Structure Laboratory
272 Whitney Avenue
New Haven, CT 06520-8124
USA
betty.bezverkhny@yale.edu

Biró, Tamas
Theoretical Department
Research Institute for Particle and Nuclear Physics
P.O.B. 49
1525 Budapest
Hungary
tsbiro@sunserv.kfki.hu

Blume, Christoph
Institut für Kernphysik
Johann Wolfgang Goethe-Universität
August-Euler-Straße 6
60486 Frankfurt
Germany
blume@ikf.uni-frankfurt.de

Bürvenich, Thomas
Los Alamos National Laboratory
T16, MS B283
Los Alamos NM 87545
USA
tbuerven@lanl.gov

Büyükcizmeci, Nihal
Department of Physics
Faculty of Science and Arts
Selcuk University
42079, Konya
Turkey
nihal@selcuk.edu.tr

Catu, Oana
Yale University
New Haven CT
USA
oana.catu@yale.edu

Cerutti, Francesco
Universita' degli Studi di Milano
Dipartimento di Fisica
via Celoria, 16
20133 Milano
Italy
Francesco.Cerutti@mi.infn.it

Coffin, Jean Pierre
Institut de Recherches Subatomiques
23 Rue du Loess BP28
F 67037 Strasbourg Cedex 2

France
coffin@in2p3.fr

Csernai, Laszlo P.
Section for Theoretical and Computational Physics
University of Bergen
Allegaten 55
N 5007 Bergen
Norway
csernai@fi.uib.no

d'Enterria, David G.
Nevis Laboratories
Columbia University, NY
and
Brookhaven National Laboratory
Phys. Dept. Bldg 510C
Upton, 11973, NY
USA
denterria@nevis.columbia.edu

Diaz-Torres, Alexis
Institut für Theoretische Physik
Johann Wolfgang Goethe-Universität
Robert Mayer Str. 10
60054 Frankfurt am Main
Germany
diaz@th.physik.uni-frankfurt.de

Dikmen, Erdal
Suleyman Demirel Universitesi
Fen-Edebiyat Fakultesi, Fizik Bolumu
Isparta
Turkey
and
Drexel University
Physics Department
Philadelphia,PA, 19151
USA
erdal_s_dikmen@yahoo.com

Eremeev, Vitalie
Institute of Applied Physics,
Academy of Sciences of Moldova
Academiei str 5
Chisinau MD-2028
Moldova
eremeev@cc.acad.md

Eren, Nuretdin
Selcuk University
Arts and Science Faculty
Department of Physics
42079 Kampus/Konya
Turkey
eren@selcuk.edu.tr

Fong, Dennis
Department of Physics and Astronomy
Vanderbilt University
Nashville, TN 37235
USA
d.fong@vanderbilt.edu

Gainutdinova, Razia
Institute of Physics
National Academy of Sciences
Chui Prosp., 265-A
Bishkek 720071
Kyrgyz Republic
epfk@rambler.ru

Gales, Sydney
Institute de Physique Nucléaire
15, rue Georges Clémenceau
F-91406 Orsay cedex
France
gales@ipno.in2p3.fr

Gelir, Ali
Istanbul Technical University
Fizik Bolumu, Ayazaga kampusu
Maslak
Istanbul
Turkey
gelira@itu.edu.tr

Gherghescu, Radu
Horia Hulubei National Institute of Physics and Nuclear Engineering
Str. Atomistilor 407
P. O. Box MG-6
RO-76900 Bucharest-Magurele
Romania
rgherg@ifin.nipne.ro

Gorenstein, Mark
Bogolyubov Institute for Theoretical Physics
14b Metrologichna str.
252143 Kiev
Ukraine
mark@mgor.kiev.ua

Greene, Victoria
Department of Physics and Astronomy
Vanderbilt University
Box 1807, Station B
Nashville, TN 37235
USA
senta.v.greene@vanderbilt.edu

Greiner, Carsten
Institut für Theoretische Physik
Justus-Liebig-Universität
Heinrich-Buff-Ring 16
35392 Giessen
Germany
Carsten.Greiner@theo.physik.uni-giessen.de

Greiner, Walter
Institut für Theoretische Physik
Johann Wolfgang Goethe-Universität
Robert Mayer Str. 10
60054 Frankfurt am Main
Germany
Greiner@th.physik.uni-frankfurt.de

Gridnev, Konstantin
St. Petersburg State University
Institute of Physics
Uljanova 1, Peterhof
St.Petersburg, 198504
Russia
gridnev@nuclpc1.phys.spbu.ru

Güçlü, Cem
Istanbul Technical University
Department of Physics
Maslak 80626 Istanbul
Turkey
guclu@itu.edu.tr

Gutbrod, Hans
Gesellschaft für Schwerionenforschung
Postfach 110552
64220 Darmstadt
Germany
H.Gutbrod@gsi.de

Gyulassy, Miklos
Columbia University
Physics Department
538 West 120th Street
New York, NY 10027
USA
gyulassy@nt3.phys.columbia.edu

Hagen, Gaute
Fysisk institutt
Universitetet i Bergen
Allégt. 55
5007 Bergen
Norway
Gaute.Hagen@fi.uib.no

Harris, John
Physics Department
Wright Nuclear Structure Laboratory
Yale University
272 Whitney Avenue
New Haven, CT 06520-8124
USA
John.Harris@yale.edu

Hofmann, Sigurd
Gesellschaft für Schwerionenforschung
Postfach 110552
64220 Darmstadt
Germany
S.Hofmann@gsi.de

Holmes, Martha
Department of Physics and Astronomy
Vanderbilt University
Box 1807, Station B
Nashville, TN 37235
USA
martha.j.holmes@vanderbilt.edu

Hossenfelder, Sabine
Institut für Theoretische Physik
Johann Wolfgang Goethe-Universität
Robert Mayer Str. 10
60054 Frankfurt am Main
Germany
hossi@th.physik.uni-frankfurt.de

Hostler, Jacquelyn
Berea College
Department of Physics
CPO 2191
Berea, KY 40404
USA
Jaki_Hostler@berea.edu

Insolia, Antonio
Dipartimento di Fisica e Astronomia & INFN
Universita` di Catania
Via S. Sofia, 64
I-95123 Catania
Italy
Antonio.Insolia@ct.infn.it

Itkis, Mikhail
Flerov Laboratory of Nuclear Reactions
Joint Institute for Nuclear Research
141980 Dubna, Moscow region
Russian Federation
itkis@flnr.jinr.ru

Karimov, Kazimir
Institute of Physics
National Academy of Sciences
Chui Prosp., 265-A
Bishkek 720071
Kyrgyz Republic
kazimir@aknet.kg

Keil, Christoph
Institut für Theoretische Physik
Justus-Liebig-Universität
Heinrich-Buff-Ring 16
35392 Giessen
Germany
christoph.m.keil@theo.physik.uni-giessen.de

Kezzar, Khalid
Gesellschaft für Schwerionenforschung
Postfach 110552

64220 Darmstadt
Germany
k.kezzar@gsi.de

Kleinert, Hagen
Institut für Theoretische Physik
Freie Universität Berlin
Arnimallee 14
14195 Berlin
Germany
Hagen.kleinert@physik.fu-berlin.de

Kostyuk, Andriy
Bogolyubov Institute for Theoretical Physics
14b Metrologichna str.
252143 Kiev
Ukraine
kostyuk@th.physik.uni-frankfurt.de

Kozlov, Michael
HEP department
School of Physics
Tel Aviv University
Ramat Aviv, Tel Aviv 69978
Israel
bmk007@motorola.com

Kuhn, Christian
Institut de Recherches Subatomiques
23 Rue du Loess - BP 28
67037 Strasbourg Cedex 2
France
Christian.kuhn@ires.in2p3.fr

Lokhtin, Igor
D.V. Skobeltsyn Institute of Nuclear Physics
Moscow State University
119992, Vorobievy Gory
Moscow
Russia
igor@lav01.sinp.msu.ru

Ludolphs, Wilrid
Physikalisches Institut
Philosophenweg 12
69120 Heidelberg
Germany
ludolphs@physi.uni-heidelberg.de

Magas, Volodymyr
Center for Physics of Fundamental Interactions (CFIF)
Instituto Superior Tecnico
Av. Rovisco Pais,
1049-001 Lisboa,
Portugal
vladimir@cfif.ist.utl.pt

Markert, Christina
Physics Department
Wright Nuclear Structure Laboratory
Yale University
272 Whitney Avenue
New Haven, CT 06520-8124
USA
christina.markert@yale.edu

McLerran, Larry
Brookhaven National Laboratory
Physics Department/Nuclear Theory Group
P.O.Box 5000
Upton, NY 11973-5000
USA
mclerran@bnl.gov

Mishustin, Igor
Russian Research Center
Kurchatov Institute
Kurchatov sq. 1
Moscow, 123182
Russia
mish@mbslab.kiae.ru

Mişicu, Şerban
Department of Theoretical Physics
Institute of Atomic Physics
Magurele-Bucharest, MG-6
Romania
misicu@theor1.theory.nipne.ro

Molnar, Etele
University of Bergen
Department of Physics
Allegaten 55
5075 Bergen
Norway
etele@fi.uib.no

Muntian, Igor
Soltan Institute for Nuclear Studies,
Hoza 69
PL-00-681 Warsaw
Poland
muntian@fuw.edu.pl

Münzenberg, Gottfried
Gesellschaft für Schwerionenforschung
Postfach 110552
64220 Darmstadt
Germany
G.Muenzenberg@gsi.de

Oganessian, Yuri
Flerov Laboratory of Nuclear Reactions
Joint Institute for Nuclear Research
141980 Dubna, Moscow region
Russia
oganessian@sungraph.jinr.ru

Ogul, Riza

Selcuk University
Department of Physics
42079 Kampus
Konya
Turkey
rogul@selcuk.edu.tr

Onbasli, Ulker
University of Marmara
Department of Physics
Ridvanpasa cad. 2. sok. 85/12
Göztepe/Istanbul
Turkey

Özcan, Aynur
Gazi University
Physics Department
06500 Teknikokullar
Ankara
Turkey
aynur@gazi.edu.tr

Özel, Banu
Department of Physics
Faculty of Arts and Sciences
Cukurova University
01330, Adana
Turkey
banuozel@cu.edu.tr

Paar, Nils
Physik-Department T30
Technische Universität München
James-Franck-Strasse 1
85748 Garching
Germany
npaar@ph.tum.de

Parkhomenko, Alexandr
Soltan Institute for Nuclear Studies,
Hoza 69
PL-00-681 Warsaw
Poland
shura@fuw.edu.pl

Petrushanko, Sergey
Hadronic Interactions Laboratory,
Skobeltsyn Institute of Nuclear Physics,
Moscow State University,
Moscow 119992
Russia
sergant@lav01.sinp.msu.ru

Plostinar, Dan-Ciprian
Babes-Bolyai University
Romania
pcipri@yahoo.com

Poenaru, Dorin
Horia Hulubei National Institute of Physics
and Nuclear Engineering
Str. Atomistilor 407
P. O. Box MG-6
RO-76900 Bucharest-Magurele
Romania
poenaru@ifin.nipne.ro

Pretzl, Klaus
Laboratorium für Hochenergiephysik
Universität Bern
Sidlerstr. 5
CH 3012 Bern
Schweiz
Klaus.Pretzl@lhep.unibe.ch

Ramayya, Akunuri V.
Department of Physics and Astronomy
Vanderbilt University
VU Station B 1807
Nashville, TN 37235
USA
a.v.ramayya@vanderbilt.edu

Rami, Fouad
Institut de Recherches Subatomiques - IRES
23 Rue du Loess - BP 28
67037 Strasbourg Cedex 2
France
fouad.rami@ires.in2p3.fr

Reinhardt, Joachim
Institut für Theoretische Physik
Johann Wolfgang Goethe-Universität
Robert Mayer Str. 10
60054 Frankfurt am Main
Germany
jr@th.physik.uni-frankfurt.de

Reisdorf, Willibrord
Gesellschaft für Schwerionenforschung
Postfach 110552
64220 Darmstadt
Germany
W.Reisdorf@gsi.de

Renfordt, Rainer
Institut für Kernphysik
Johann Wolfgang Goethe-Universität
August-Euler-Straße 6
60486 Frankfurt
Germany
renfordt@ikf.uni-frankfurt.de

Reuter, Philipp
Institut für Theoretische Physik
Johann Wolfgang Goethe-Universität
Robert Mayer Str. 10
60054 Frankfurt am Main
Germany

preuter@th.physik.uni-frankfurt.de

Ring, Peter
Physik-Department T30
Technische Universität München
James-Franck-Strasse 1
85748 Garching
Germany
Ring@ph.tum.de

Rischke, Dirk
Institut für Theoretische Physik
Johann Wolfgang Goethe-Universität
Robert Mayer Str. 10
60054 Frankfurt am Main
Germany
drischke@th.physik.uni-frankfurt.de

Ruppert, Jörg
Institut für Theoretische Physik
Johann Wolfgang Goethe-Universität
Robert Mayer Str. 10
60054 Frankfurt am Main
Germany
ruppert@th.physik.uni-frankfurt.de

Sa'd, Basil
Physics Dept.
Yarmouk University
Irbid 21163
Jordan
basil.sad@mail.com

Salomov, Uktam
Heat Physics Department
of the Uzbek Academy of Sciences
28 Katartal St.
700135, Tashkent
Uzbekistan
lps@mnet.uz

Salur, Sevil
Yale University
POB 204285
New Haven, CT, 06520
USA
sevil.salur@yale.edu

Schaffner-Bielich, Jürgen
Institut für Theoretische Physik
Johann Wolfgang Goethe-Universität
Robert Mayer Str. 10
60054 Frankfurt am Main
Germany
schaffner@th.physik.uni-frankfurt.de

Scheid, Werner
Institut für Theoretische Physik
Justus-Liebig-Universität
Heinrich-Buff-Ring 16
35392 Giessen
Germany
werner.scheid@theo.physik.uni-giessen.de

Schmitt, Andreas
Institut für Theoretische Physik
Johann Wolfgang Goethe-Universität
Robert Mayer Str. 10
60054 Frankfurt am Main
Germany
aschmitt@th.physik.uni-frankfurt.de

Schramm, Stefan
Institut für Theoretische Physik
Johann Wolfgang Goethe-Universität
Robert Mayer Str. 10
60054 Frankfurt am Main
Germany
schramm@th.physik.uni-frankfurt.de

Shovkovy, Igor
Institut für Theoretische Physik
Johann Wolfgang Goethe-Universität
Robert Mayer Str. 10
60054 Frankfurt am Main
Germany
shovkovy@th.physik.uni-frankfurt.de

Sobiczewski, Adam
Soltan Institute for Nuclear Studies,
Hoza 69
PL-00-681 Warsaw
Poland
Adam.Sobiczewski@fuw.edu.pl

Solov'yov, Andrey
A.F. Ioffe Physical Technical Institute
Russian Academy of Sciences
Polytechnicheskaja str. 26
St. Petersburg 194021
Russia
E-mail: solovyov@rpro.ioffe.rssi.ru

Solov'yov, Ilya
A.F. Ioffe Physical Technical Institute
Russian Academy of Sciences
Polytechnicheskaja str. 26
St. Petersburg 194021
Russia
E-mail: ilia@th.physik.uni-frankfurt.de

Stock, Reinhard
Institut für Kernphysik
Johann Wolfgang Goethe-Universität
August-Euler-Straße 6
60486 Frankfurt
Germany
stock@ikf.uni-frankfurt.de

Stöcker, Horst

Institut für Theoretische Physik
Johann Wolfgang Goethe-Universität
Robert Mayer Str. 10
60054 Frankfurt am Main
Germany
stoecker@th.physik.uni-frankfurt.de

Strottman, Dan
Theoretical Division
MS B210
Los Alamos National Laboratory
Los Alamos, NM 87545
USA
dds@lanl.gov

Sucu, Yusuf
Akdeniz University
Faculty of Sciences and Arts
Department of Physics
P.B.510, 07058, Antalya
Turkey

Svirikin, Alexander
Flerov Laboratory of Nuclear Reactions
Joint Institute for Nuclear Research
Joliot-Curie 6
141980 Dubna, Moscow region
Russia
sasha@sunvas.jinr.ru

Ter-Akopian, Gurgen
Joint Institute for Nuclear Research
141980 Dubna
Russia
Gurgen.TerAkopian@jinr.ru

Türkan, Nurettin
Erciyes University
Yozgat Fenedebiyat Fakültesi
66100 Yozgat
Turkey
nurettin_turkan@yahoo.com

Turkoz, Semsettin
Ankara University
Department of Physics
06100 Tandogan Ankara
Turkey
turkoz@science.ankara.edu.tr

Udovyk, Oleg
ITP, Ukrainian Academy of Sciences
3, Dobrohotov St.
Kyiv 03142
Ukraine
oleg_udovyk@hotmail.com

Umar, Sait
Department of Physics and Astronomy
Vanderbilt University
Box 1807, Station B
Nashville, TN 37235
USA
Umar@compsci.cas.vanderbilt.edu

Ünal, Nuri
Akdeniz University
Faculty of Sciences and Arts
Department of Physics
P.B.510, 07058, Antalya
Turkey
nuriunal@akdeniz.edu.tr

Vaagen, Jan
Fysisk institutt
Universitetet i Bergen
Allégt. 55
5007 Bergen
Norway
JanS.Vaagen@fi.uib.no

Voinov, Alexei
Flerov Laboratory of Nuclear Reactions
Joint Institute for Nuclear Research
Joliot-Curie St., 6
Dubna 141980
Russia
voinov_2000@mail.ru

Westfall, Gary
National Superconducting Cyclotron
Laboratory
Department of Physics and Astronomy
Michigan State University
East Lansing, Michigan 48824-1321
USA
Westfall@nscl.msu.edu

Wheaton, Spencer
University of Cape Town
Department of Physics
Rondebosch 7701
South Africa
wheaton@hostess.phy.uct.ac.za

Yücel, Melike Behiye
Akdeniz University
Faculty of Sciences and Arts
Department of Physics
P.B.510, 07058, Antalya
Turkey
myucel@akdeniz.edu.t

CREATING BULK QCD MATTER AT RHIC

John W. Harris
Physics Department, Yale University,
P.O. Box 208124, 272 Whitney Avenue, New Haven CT, U.S.A. 06520-8124
John.Harris@Yale.edu

Abstract The combined results from RHIC depict formation of a dense system at an energy density greater than 5GeV/fm^3, well above that where hadrons exist. Statistical models reproduce the RHIC produced particle ratios indicating a temperature and baryo-chemical potential at the deconfinement phase transition boundary predicted by lattice QCD calculations. The anisotropic collective flow observed in collisions at RHIC provides direct evidence of the existence of strong pressure gradients in the highly interacting early stages of the dense medium. Hard-scattering, suppression of hadrons at large transverse momentum, and quenching of di-jets are observed in central Au + Au collisions and not in d + Au collisions, providing evidence for extreme energy loss of partons traversing the dense QCD medium created at RHIC.

Keywords: RHIC, QCD matter, QGP, collective flow, high transverse momentum, suppression, jet quenching

Introduction

Collisions at the Relativistic Heavy Ion Collider (RHIC) are expected to melt the QCD vacuum, allowing quarks and gluons to interact over a large volume thus forming bulk QCD matter. Lattice QCD calculations indicate a deconfinement phase transition (and accompanying chiral phase transition) both at high temperatures and at high densities. The equation of state of QCD and properties of the medium depend critically on the number of flavors and on the quark masses that are used in the lattice calculations. This is presently being investigated intensively [1, 2] (also using improved actions). As an example, the results of calculations on the lattice for the energy density as a function of temperature are displayed in Fig. 1. A relatively sharp deconfinement transition occurs at a temperature of approximately 175 MeV in 2-flavor QCD and at about 20 MeV lower temperature for 3-flavors (both in the chiral limit). There has

W. Greiner et al. (eds.), Structure and Dynamics of Elementary Matter, 1–18.

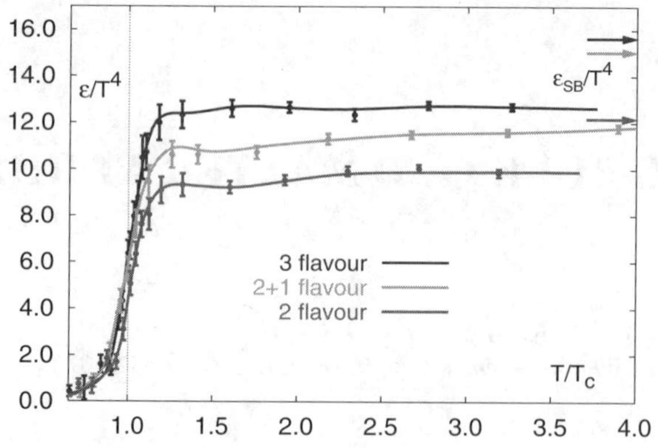

Figure 1. Lattice QCD predictions for the energy density $(\epsilon)/T^4$ as a function of the temperature T relative to the critical temperature T_c [1]. The flavor dependence is shown by the three curves, depicting results for 3 light quarks (u,d,s), 2 light quarks (u,d) plus 1 heavier (s) quark, and 2 light quarks (u,d) as indicated above. The Stephan-Boltzmann values of ϵ_{SB}/T^4 are depicted by the arrows on the right.

also been recent success in implementation of techniques to calculate on the lattice at small but finite baryon density [3].

With recent predictions of new phases of matter, the phase diagram of QCD has become considerably more complicated in recent years. Displayed in Fig. 2 is an illustration of a conceptual phase diagram for QCD matter [4] that incorporates present thinking [5]. RHIC and the LHC heavy ion program will explore the high temperature regime, as depicted in the diagram. A first order phase transition is expected along the curve up to a critical point shown in Fig. 2. At higher temperatures a quark-gluon plasma (QGP) phase is expected. The early Universe cooled down from higher temperatures near the $\mu_{baryon} = 0$ axis as shown. The region of high baryo-chemical potential μ_{baryon} appears rich with structure and is presently an area of intensive theoretical investigation [5]. Understanding the nature of these phase transitions has implications for nuclear physics, astrophysics, cosmology and particle physics.

RHIC was constructed to create the QGP in the laboratory and study its properties. This presentation will describe the experimental search for the QGP at RHIC. It will be divided into two parts: 1) creating and establishing the properties of the high density matter at RHIC through the study of soft (low transverse momentum) particle production, and 2) probing the matter with hard (high transverse momentum) probes to determine its response.

1. Creating High Density Matter

Many aspects of the matter created at RHIC, such as its bulk properties and collective effects, can be understood by studying the particles most abundantly

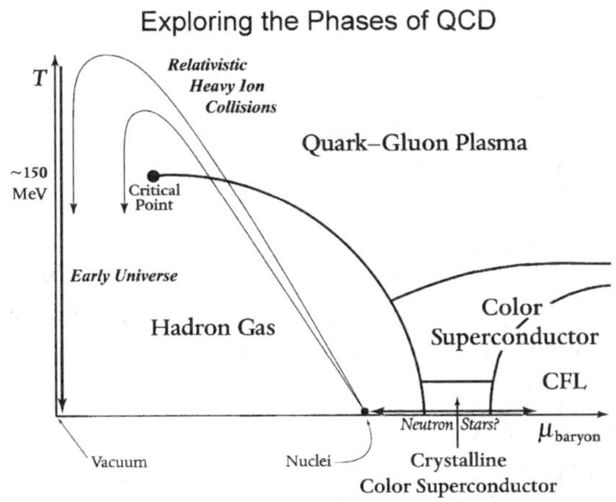

Figure 2. Schematic phase diagram of QCD matter as a function of temperature T and baryochemical potential μ_{baryon}.

produced in these collisions, i.e. those with transverse momentum less than 2 GeV/c. A brief overview of the results on this soft physics will be presented in this section.

RHIC accelerates and collides ions from protons to the heaviest nuclei over a range of energies, up to 250 GeV for protons and 100 A-GeV for Au nuclei. Since 2000 in the three production runs for physics, RHIC has collided Au + Au at center-of-mass energies $\sqrt{s_{NN}}$ = 19.6, 130, and 200 GeV, d + Au at $\sqrt{s_{NN}}$ = 200 GeV, and p + p at $\sqrt{s}$ = 200 GeV. RHIC has also begun to collide polarized protons for studies of the spin content of the proton. RHIC and its four experiments (BRAHMS, PHENIX, PHOBOS, and STAR) are described comprehensively in [6].

1.1 Particle Production

A large number of particles is produced in a central collision at RHIC. The measured charged hadron multiplicity density at midrapidity is $dn_{ch}/d\eta\ |_{\eta=0}=$ $700{\pm}27(syst)$ [7] for the 3% most central collisions of Au + Au at $\sqrt{s_{NN}}=200$ GeV. This corresponds to a hadron multiplicity density $dn_{total}/d\eta\ |_{\eta=0}=1050$ and a total hadron multiplicity in the most central events of ~ 7000 [7, 8]. In terms of the number of quarks, consider the case where all observed hadrons in the final state are mesons, each containing a quark and anti-quark. Thus 7,000 hadrons correspond to 14,000 quarks and anti-quarks. A lower limit for the number of produced quarks and anti-quarks can be obtained by subtracting off the valence quarks that originally enter the collision in the incident nuclei. The number of valence quarks in a head-on collision of two Au nuclei is 2 (Au nuclei) x 197 nucleons/Au x 3 quarks/nucleon $\sim$ 1200. Thus, more than 90% of the more than 14,000 quarks and anti-quarks in the final state are produced in the collision. In addition, there are a large number of gluons present that have not been considered in this simple estimate.

1.2 Energy Density

Large energy densities are reached in a central collision at RHIC. Measurements of the transverse energy per unit pseudorapidity $dE_T/d\eta$ [9] or the total particle multiplicity density and mean transverse momentum per particle [10] can be used to estimate the energy density produced in a Bjorken longitudinal expansion scenario [11].

It has been shown in Au + Au collisions at $\sqrt{s_{NN}}$ = 130 GeV that there is longitudinal boost invariance in a region around mid-rapidity at RHIC [10]. Thus, the Bjorken scenario applies and the energy density can be estimated by $\epsilon_{Bj} = \frac{1}{\tau_o \pi R^2} \times \frac{dE_T}{dy}$, with dE_T/dy the transverse energy per unit rapidity, R the transverse radius of the system, and τ_o the formation time. Assuming a maximum value for the formation time τ_o = 1 fm/c, the minimum energy density for central collisions of Au + Au at $\sqrt{s_{NN}}$ = 130 GeV is 4.6 GeV/fm^3 for the 2% most central Au + Au collisions [9] and 4.3 GeV/fm^3 for the 5% most central Au + Au collisions [10]. This is a 60% increase over the value 2.9 GeV/fm^3 measured for Pb + Pb at $\sqrt{s_{NN}}$ = 17.2 GeV at the CERN SPS. Note that the value for τ_o is expected to be significantly smaller at RHIC than at the SPS, thus the energy density derived for RHIC from this formula is a minimum and may be significantly ($\sim$ ten times) larger.

1.3 Net Baryon Density

The anti-baryon/baryon ratios have been measured at the SPS and at two RHIC energies. Displayed in Fig. 3 are the $\overline{p}/p$, $\overline{\Lambda}/\Lambda$, $\overline{\Xi}^+/\Xi^-$, and $\overline{\Omega}^+/\Omega^-$ ratios in the left panel for Pb + Pb collisions at the SPS and Au + Au collisions at RHIC and in the right panel for p + p collisions at the SPS and RHIC. The anti-baryon/baryon ratios increase with the strange quark content of the observed anti-baryon/baryon pair. In going from the SPS to RHIC energies, there is a large increase in the ratios with energy, approaching unity for the $\overline{\Omega}^+/\Omega^-$ at RHIC. Of significance are the new measurements in the p + p system [12] that give ratios similar to those found in A + A collisions. These results indicate that at RHIC the baryon production is dominated by pair production both in p + p and Au + Au, although the slight excess of baryons over anti-baryons (anti-baryon/baryon ratio $<$ 1) at mid-rapidity indicates that a fraction of the net baryon number in the incident nuclei is present in the final system and observed in the anti-baryon/baryon ratios. The ratios at the SPS energies are significantly less than unity and therefore the net baryon densities are significantly larger

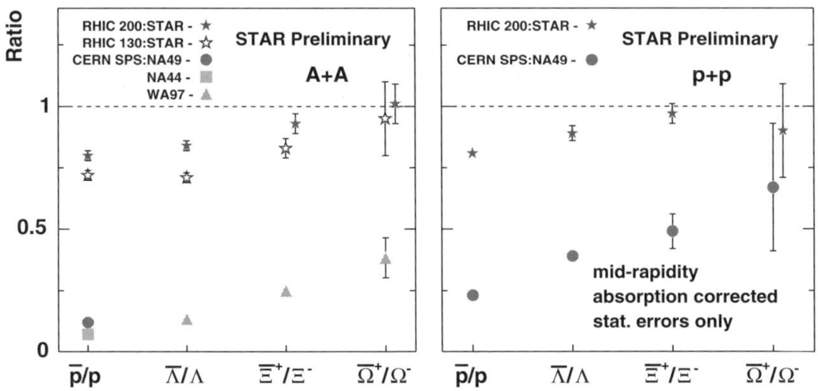

Figure 3. The $\overline{p}/p$, $\overline{\Lambda}/\Lambda$, $\overline{\Xi}^+/\Xi^-$, and $\overline{\Omega}^+/\Omega^-$ anti-baryon to baryon ratios plotted by increasing number of strange quarks. The data (specified in legend) are measured at mid-rapidity at the SPS and at RHIC . Left panel is for A + A collisions and right panel for p + p collisions.

and play a prominent role in particle production. The RHIC ratios indicate a nearly net-baryon-free region at mid-rapidity, while those at the SPS reflect a net-baryon-rich mid-rapidity region.

The anti-baryon/baryon ratios in A + A systems can be described in terms of quark coalescence [13] or by a statistical-thermal model [14] without distinction between quark and hadron degrees of freedom. The latter provides us with the baryo-chemical potential and temperature at chemical freeze-out (discussed below) at the time when particle ratios are established. The anti-baryon/baryon ratios particularly constrain the baryo-chemical potential, whereas non-identical particle ratios establish the chemical freeze-out temperature. In elementary p + p collisions, string fragmentation models reproduce the anti-baryon/baryon ratios at RHIC, but typically overpredict these ratios at the SPS.

2. Production of Strangeness

An enhancement in the production of rare strange hadrons was the first signature that was predicted to determine whether a deconfined quark-gluon plasma state is formed in relativistic heavy ion collisions. [15] To determine whether strangeness is enhanced in heavy ion collisions, it is important to study systematically the production of strange anti-particles relative to non-strange anti-particles, and the production of strange particles relative to non-strange particles as a function of collision energy.

Displayed in Fig. 4 (top panel) are the ratios of K^+/π^+ and K^-/π^- and (bottom panel) the ratios of Λ/π and $\overline{\Lambda}/\pi$. It is obvious that the production of strange mesons (K^+) and strange baryons (Λ) relative to non-strange particle production follow the same trend which is distinctly different from that of their anti-particles (K^- and $\overline{\Lambda}$). The production of strange particles peaks sharply at

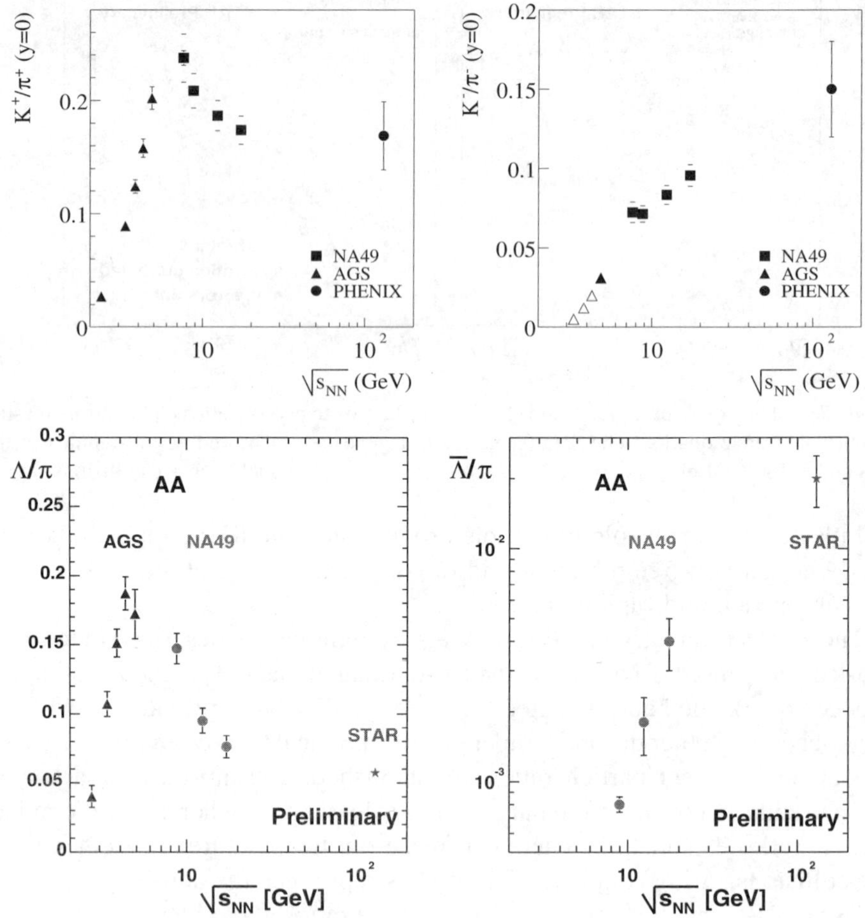

Figure 4. Top panel: K^+/π^+ (left) and K^-/π^- (right) ratios at midrapidity as a function of center-of-mass energy per nucleon pair for Au + Au collisions at the AGS and RHIC and for Pb + Pb collisions at the SPS. Bottom panel: Λ/π (left) and $\overline{\Lambda}/\pi$ (right) ratios.

an energy intermediate between the AGS and SPS, while strange anti-particle production increases continuously with increasing energy.

In order to understand strange particle production, it is important to consider the Wroblewski factor [16] $\lambda_s = 2 < s\overline{s} > /(< d\overline{d} > + < u\overline{u} >)$. This factor gives the ratio of produced strange quarks to produced non-strange quarks. It is dominated at these energies by the production of kaons, the lightest strange particle. The Wroblewski factor depends on the energy available for particle production (or temperature in a thermal model) and the density of baryons (baryo-chemical potential in a thermal model) of the system. The value of λ_s as a function of $\sqrt{s}$ is shown in Fig. 5 for all produced strange particles (solid

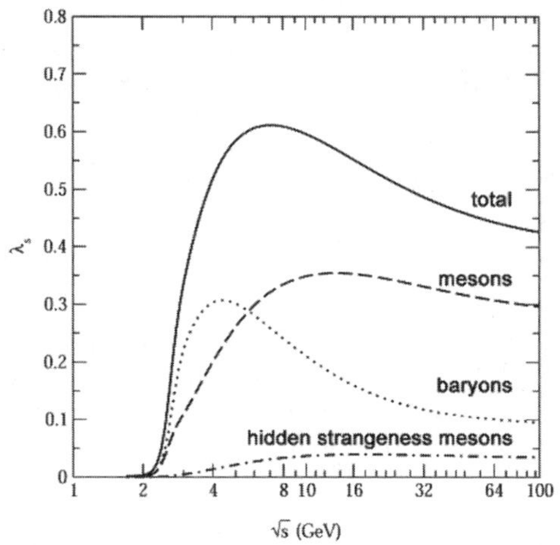

Figure 5. The Wroblewski factor λ_s as a function of $\sqrt{s}$ for produced strange particles (solid curve), strange mesons (dashed curve), strange baryons (dotted curve), and particles with hidden strangeness (dot-dashed curve). [17]

curve), produced strange mesons (dashed curve), produced strange baryons (dotted curve), and for particles with hidden strangeness (dot-dashed curve). A sharp maximum in the Wroblewski factor occurs at $\sqrt{s} \sim 4$ GeV for strange baryons. The measured ratios in Fig. 4 (left panels) for K^+/π^+ and Λ/π peak in the same energy range predicted by the Wroblewski factor for strange baryons. This can be attributed to the large baryon density in this energy regime and its effect of enhancing strange baryon and K^+ production relative to light quark production. On the other hand, the anti-particle ratios in Fig. 4 (right panels) for K^-/π^- and $\overline{\Lambda}/\pi$ exhibit a continual rise as a function of $\sqrt{s}$ up to RHIC energies indicating that creation of these anti-particles depends critically on the energy available for particle pair creation rather than on baryon density. Note that the Wroblewski factor for strange mesons has the same behavior as a function of c.m. energy as observed in the sum of the K^+/π^+ and K^-/π^- ratios (top left and top right panels of Fig. 4).

In order to pursue the quark-gluon plasma signature of enhanced strangeness production [15], the WA97 and NA57 experiments have investigated strange baryon and anti-baryon production at the SPS. An enhancement in the production of strange particles in A + A collisions relative to elementary collisions scaled by the number of wounded nucleons has been measured in NA57 [18]. The enhancement is observed for various strange baryons at mid-rapidity and increases with the centrality of the collision for a given A + A system. NA57 defines this enhancement factor as the yield per wounded-nucleon in Pb + Pb collisions relative to p + Be interactions (to approximate p + p) after extrapolation over all transverse momenta and one unit of rapidity centered at mid-rapidity. A factor of 20 enhancement is observed for the $\Omega^- + \overline{\Omega}^+$ in Pb + Pb

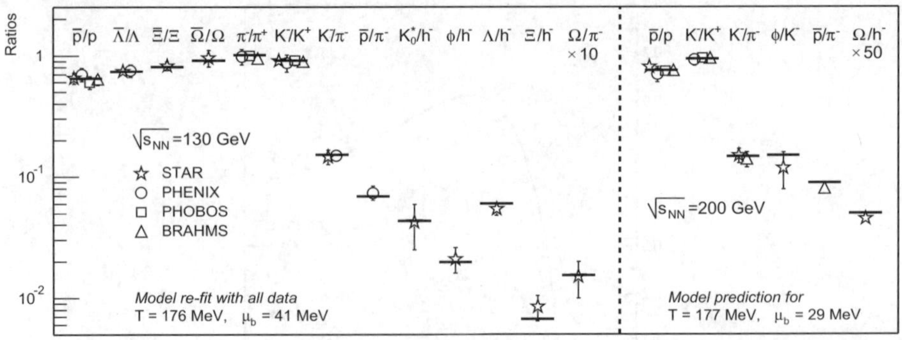

Figure 6. Particle ratios measured in RHIC experiments denoted by symbols in the legend for central collisions of Au + Au at $\sqrt{s_{NN}} = 130$ GeV (left panel) and 200 GeV (right panel). Results from a statistical-thermal model [23] fit to the entire set of data are shown as a horizontal line for each ratio. Parameters (μ_B, T) for the best fit at each energy are shown at the bottom of each panel and discussed in the text.

relative to the p + Be and p + Pb systems. The presently observed enhancement appears to correlate with the inverse of the strangeness suppression factor or approximately twice the inverse of the Wroblewski factor at the SPS, when the nonzero quark chemical potential is taken into account. With the exception of the $\overline{\Lambda}$, the ratios can be interpreted to result from the change in the strangeness suppression factor. [19–21] Similar measurements are underway in STAR at RHIC. It will be necessary to determine experimentally the evolution of the suppression of strangeness as a function of centrality or system size (to distinguish applicability of a canonical approach versus a grand canonical approach), in order to ascertain whether there is in fact an enhancement of strangeness at the SPS or at RHIC.

3. Chemical Freezeout

The particle yields are established at chemical freeze-out and can provide information on this relatively early stage prior to hadronic expansion, cooling and thermal freeze-out. Statistical thermal models have been able to reproduce the measured ratios using as parameters the chemical freeze-out temperature and the baryo-chemical potential. These models employ hadronic degrees of freedom in a grand canonical ensemble. See [22] for a recent review. Displayed in Fig. 6 are the particle ratios measured at RHIC along with results of a statistical-thermal model fit. [23, 24] The particle ratios observed at RHIC for $\sqrt{s_{NN}} = 130$ GeV Au + Au can be fit with the parameters T = 176 MeV and $\mu_B = 41$ MeV. For $\sqrt{s_{NN}} = 200$ GeV Au + Au T = 177 MeV and $\mu_B = 29$ MeV are required. The statistical-thermal model fits reproduce the data surprisingly well. When the same approach is used for the SPS Pb + Pb data reasonable fits are found with T = 164 MeV and μ_B = 274 MeV. [25, 26] When a similar approach [27] is applied to the new particle production data from NA49 in Pb + Pb collisions, the best-fit parameters are T = 148 MeV and μ_B = 377 MeV at E_{lab} = 40 A-GeV and T = 154 MeV and μ_B = 293 MeV at E_{lab} = 80 A-GeV. In both these cases an additional strangeness saturation parameter [1] of $\gamma_s = 0.75$ and 0.72 is required, respectively, for the two lowest energy cases. It is clear that the chemical freeze-out follows a curve in (μ_B, T) space as depicted in Fig. 7 where μ_B decreases from 293 MeV at the lowest SPS energy to 29 MeV at the highest RHIC energy, and the chemical freeze-out temperature increases gradually from 154 MeV to 176 MeV. When drawn on a (μ_B, T) plane, the values of (μ_B, T) from statistical model fits to the experimental data approach, at RHIC energies, the deconfinement phase transition boundary predicted by Lattice QCD [29].

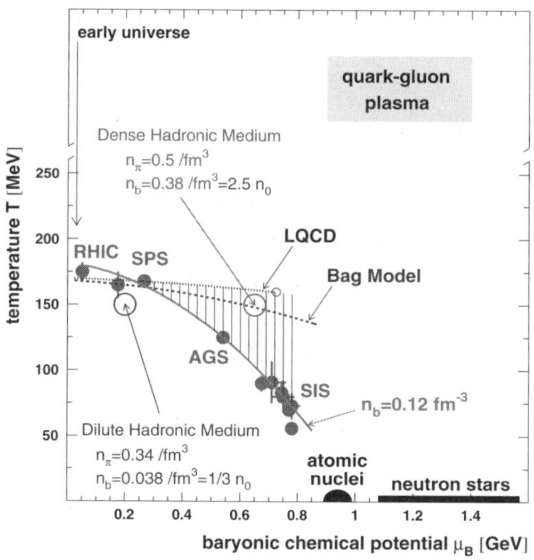

Figure 7. Nuclear phase diagram extracted from a thermal model analysis of experimental results at various energies and heavy ion facilities (SIS, AGS, SPS, and RHIC) [28]. The dotted curve labeled "LQCD" corresponds to the deconfinement phase boundary predicted by lattice QCD [29]. The solid curve corresponds to freeze-out at a constant baryon density.

4. Thermal Freezeout

The ability of the statistical-thermal model to fit the particle abundances lends credence to equilibration at the time of chemical freeze-out. Further study specifically of the particle spectra can provide additional information on the dynamics of the expansion and the temperature of the system at kinetic decoupling or thermal freeze-out, when particles cease interacting. Ideally, one would like to be able to determine from the particle spectra the extent to which the system maintains equilibrium during expansion. But, this is a difficult task as the spectra are considerably more complicated as we shall see below.

The particle spectra reflect various contributions from different times to the dynamics of expansion. Most of the particles (> 99%) have momenta less

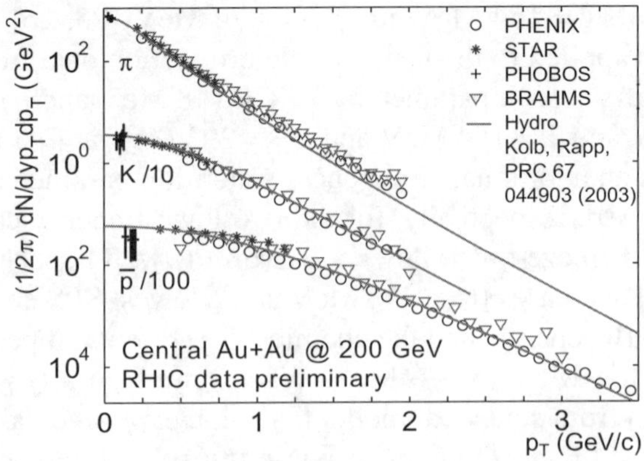

Figure 8. Transverse momentum spectra of charged pions, kaons and anti-protons measured in $\sqrt{s_{NN}}$ = 200 GeV Au + Au central collisions. Shown as the curves are the results of a hydrodynamics calculation [31].

than 2 GeV/c, and give information on the bulk properties of the system at thermal freeze-out. Higher momentum particles usually reflect earlier times of emission, and are often created in perturbative QCD processes such as hard parton scattering and fragmentation. These can be used to probe the early stages of the collision. We first discuss the low momentum or thermal component of the particle spectra.

Thermal Freeze-out and Transverse Flow. The transverse momentum spectra of hadrons provide information about the system at the time of thermal freeze-out, when interactions cease, while also containing a history of the expansion process and any collective effects. Displayed in Fig. 8 are the transverse momentum spectra measured for charged pions, kaons and anti-protons in $\sqrt{s_{NN}}$ = 200 GeV Au + Au central collisions. Also, shown are the results of a hydrodynamics calculation [31] that incorporates a quark-gluon to hadron phase transition at T_c = 165 MeV, and subsequent hydrodynamic expansion in a hadronic phase in chemical equilibrium including effective meson and baryon chemical potentials. In this calculation, pre-equilibrium collective flow at the time of thermalization is required to fit the spectra of the various particles at the top RHIC energy.

Elliptic Flow. Non-zero impact parameter collisions have an inherent initial spatial asymmetry in the plane perpendicular to the beam axis due to the impact parameter. The larger the impact parameter the larger the asymmetry. If the initial spatial asymmetry is transferred into a momentum-space anisotropy during the brief traversal time of the incident nuclei ($\sim$ 1 fm/c at these energies), the resulting azimuthal asymmetry is evidence for the presence of large pressure gradients early in the collision process and requires a dense, highly interacting medium.

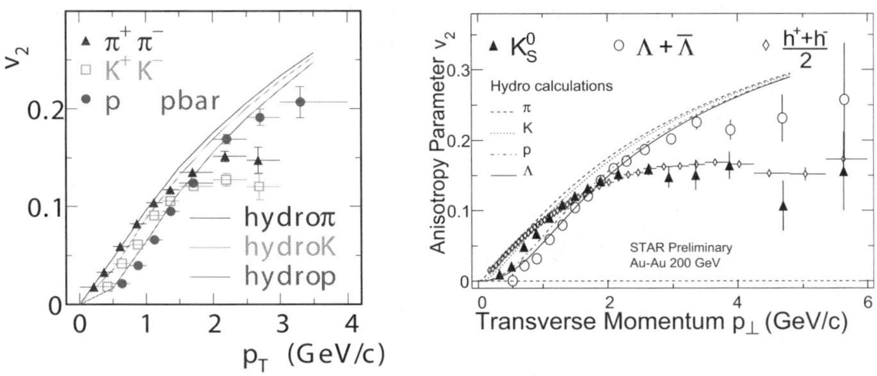

Figure 9. Measured elliptic flow v_2 for $\sqrt{s_{NN}}$ = 200 GeV Au + Au minimum bias collisions as a function of p_T. Left: $\pi^\pm$, $K^\pm$, and $p^\pm$ from PHENIX [32]. Right: $K^o{}_s$, $\Lambda + \overline{\Lambda}$, and $(h^+ + h^-)/2$ from STAR. [33] Curves are hydrodynamics predictions. [34, 35]

A way to measure the azimuthal asymmetry is to construct the second Fourier harmonic component of the azimuthal distribution of particles in momentum space with respect to the reaction plane, $v_2 = \langle\cos(2\phi)\rangle$ where $\phi = \mathrm{atan}\,(p_y/p_x)$. This v_2 is called the elliptic flow. Displayed in the left (right) panel of Fig. 9 is v_2 measured by PHENIX [32] (STAR [33]) for $\pi^\pm$, $K^\pm$, and $p^\pm$ ($K^o{}_s$, $\Lambda + \overline{\Lambda}$,$(h^+ + h^-)/2$) as a function of transverse momentum in $\sqrt{s_{NN}}$ = 200 GeV Au + Au minimum bias collisions. Also shown are predictions from hydrodynamics. The elliptic flow is well described at low transverse momenta by hydrodynamical models incorporating a softening of the equation of state due to quark and gluon degrees of freedom [34, 35]. However, at transverse momenta $>$ 1.5 GeV/c the lighter particles (π, K) deviate from the predictions of hydrodynamics as do the heavier particles (p, Λ) at slightly higher transverse momenta. The baryons continue to have higher values of v_2 than the mesons at the largest transverse momenta measured. The saturation of v_2 for larger momenta reflects effects of the energy loss due to the large gluon densities of the evolving medium [36].

Larger values of v_2 for baryons than mesons at larger transverse momenta may result from the particles being created in soft processes and promoted to higher p_T by a collective boost. Alternatively or in addition, coalescence of quarks to form composite particles may occur. The v_2 for baryons and mesons at large transverse momenta are consistent with a quark coalescence picture for strange hadrons at quark $p_T > 0.5$ GeV/c [37].

5. Hard Probes of the Medium

For the first time in relativistic heavy ion collisions, hard scattering has been used at RHIC to probe the medium through which the hard-scattered partons propagate. The radiation energy loss of a parton traversing a dense medium is predicted to be significant and is sensitive to the gluon density of the medium [38–40]. In order to investigate parton energy loss in the medium, the RHIC experiments have measured hadron spectra and azimuthal correlations of particles with large transverse momentum.

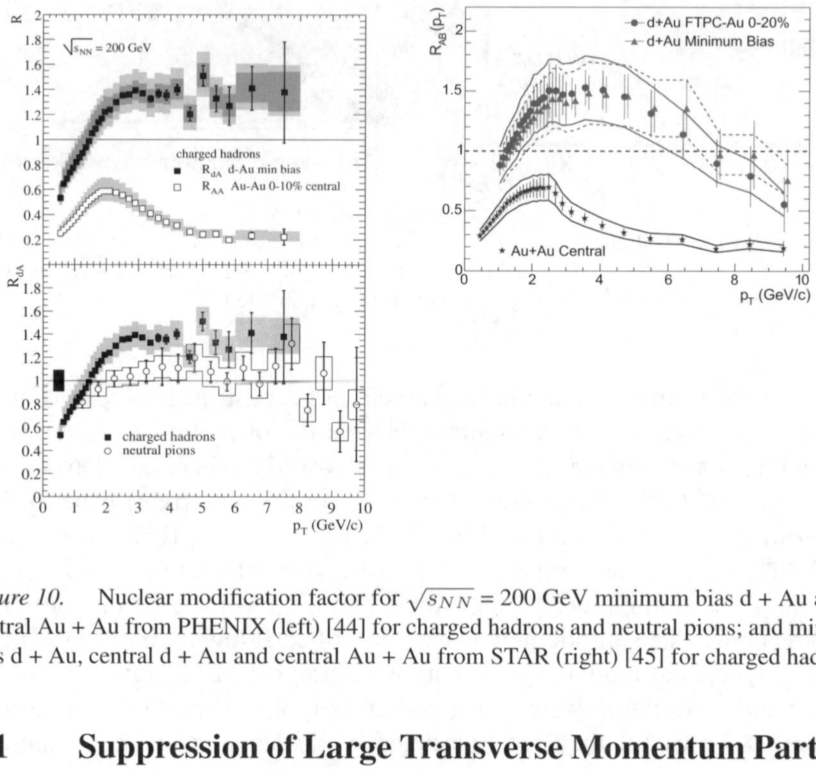

Figure 10. Nuclear modification factor for $\sqrt{s_{NN}}$ = 200 GeV minimum bias d + Au and for central Au + Au from PHENIX (left) [44] for charged hadrons and neutral pions; and minimum bias d + Au, central d + Au and central Au + Au from STAR (right) [45] for charged hadrons.

5.1 Suppression of Large Transverse Momentum Particles

In order to make comparisons of results from collisions of nuclei with those of elementary particles, such as p + p, a nuclear modification factor can be defined as $R_{AA}(p_T) = \frac{dN^2/dp_T dy\,(AA)}{\langle N_{binary}\rangle dN^2/dp_T dy\,(NN)}$. $\langle N_{binary}\rangle$ is the number of binary collisions in a geometrical model used to scale from elementary nucleon-nucleon (NN) collisions to nucleus-nucleus (AA) collisions. R_{AA} = 1 when AA is an incoherent superposition of NN collisions corresponding to scaling with the number of binary collisions (binary scaling).

Suppression of the hadron spectra at large transverse momenta $p_T > 2$ GeV/c has been observed in the nuclear modification factors measured in central Au + Au collisions at RHIC [41],[42],[43]. Displayed in Fig. 10 is the nuclear modification factor as a function of transverse momentum for central Au + Au collision data and for d + Au collisions from PHENIX (left panel) [44] and STAR (right panel)[45]. The ratios are taken relative to p + p collisions scaled by the number of binary collisions. These data exhibit a clear suppression by a factor 4-5 in the central Au + Au case. The central Au + Au ratio remains rather flat up to the statistical limits of the data at 10 GeV/c transverse momentum, where no sign of reaching the pQCD limit of binary scaling is observed. The peripheral collision data (not shown) exhibit no nuclear modification, i.e. R_{AA} = 1, within errors. The Au + Au relative to p + p yields are reproduced by pQCD calculations incorporating parton energy loss in dense matter [46] and by a model incorporating initial-state gluon saturation [47]. The ratio $R_{dAu}(p_T) >$ 1 in d + Au reflects a Cronin enhancement (initial state multiple scattering) for $1 < p_T < 7$ GeV/c and no suppression, rather a slight enhancement. The d + Au charged hadron data when compared to scaled p + p data [44],[45],[48],[49] rule out initial-state gluon saturation leaving only final state effects as a cause of the suppression in the Au + Au data at mid-rapidity at RHIC.

Displayed in the left panel of Fig. 11 are the ratios of $K^o{}_s$, $\Lambda + \overline{\Lambda}$, and $(h^+ + h^-)/2$ produced in central relative to binary-scaled peripheral collisions. [50] The $K^o{}_s$ and $(h^+ + h^-)/2$ ratio of central to peripheral peak around $p_T \sim 2$ GeV/c at ratios considerably below 1 and continue to be suppressed up to $p_T \sim$ 6 GeV/c. The ratios of central to peripheral yields of $\Lambda + \overline{\Lambda}$ peak near unity and are suppressed relative to binary-scaling only above around 4 GeV/c. For comparison, the PHENIX data on π^o and $p + \overline{p}$ are shown in the right panel of Fig. 11 [51]. The π^o are suppressed by a factor of $\sim$ 4 - 5, while the protons do not deviate from binary scaling up to transverse momentum 4 GeV/c.

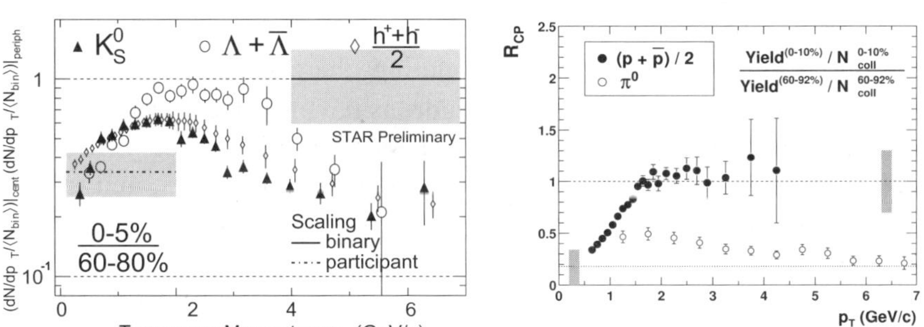

Figure 11. $\sqrt{s_{NN}}$ = 200 GeV Au + Au collision data for (left panel) ratios of $K^o{}_s$, Λ, $\overline{\Lambda}$, and $(h^+ + h^-)$ / 2 from STAR [50] and π^o and $(p + \overline{p})/2$ from PHENIX [51].

While the baryons (p, Λ) exhibit a larger elliptic flow compared to the mesons (π, K), they are not as suppressed as the mesons. If energy loss is the basis for elliptic flow at high p_T, one would expect that particles exhibiting large suppression in the high p_T spectra (e.g. π, K) will also have larger elliptic flow. But this is not the case, presenting a possible dilemma.

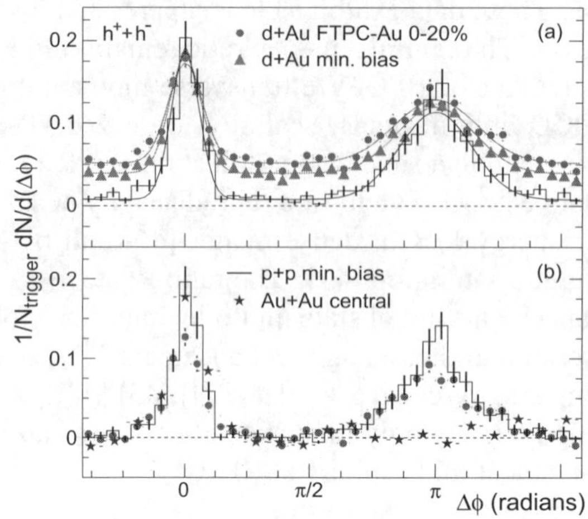

Figure 12. a) Correlation functions in relative azimuth for p + p interactions, and minimum bias and central collisions of d + Au at $\sqrt{s_{NN}} = 200$ GeV [53]. b) Comparison of correlations for p + p, central d + Au, and central Au + Au collisions with pedestals subtracted.

5.2 Jet Quenching of Azimuthal Correlations

Due to the complexities of identifying and determining precisely the energies of jets in the large background of soft particles in relativistic heavy ion collisions, two-particle correlations of large transverse momentum hadrons can be used to identify the hard scattering component on a statistical basis.

The fragmentation of partons should lead to jets, a correlation of hadrons in a small angular region of relative azimuthal angle ($\Delta\phi$) and pseudorapidity ($\Delta\eta$). As seen in Fig. 12 an enhancement is observed at $\Delta\phi = 0$ in azimuthal correlations near midrapidity upon correlating charged hadrons having $4 < p_T^{trig} < 6$ GeV/c with all other charged hadrons having 2 GeV/c $< p_T < p_T^{trig}$ in the same event. This has been seen in collisions of Au + Au at $\sqrt{s_{NN}} = 130$ and 200 GeV ranging from minimum bias to central collisions, and in p + p and d + Au at $\sqrt{s}$ = 200 GeV [52, 53]. The peak at small relative azimuthal angle is indicative of short range correlations and suggests contributions from jets and/or resonance decays. The p + p, d + Au, and peripheral Au + Au data exhibit a distinct back-to-back correlation, a characteristic of hard-parton scattering, with peaks at $\Delta\phi$ = 0 and π, indicating di-jets. This is shown in Fig. 12 for the p + p and d + Au reactions, whereas for the most central Au + Au collisions the backward ($\Delta\phi = \pi$) correlation disappears. These results indicate that strong dissipative effects of the medium created in central Au + Au collisions quench high momentum partons or their fragmentation products, and provide further evidence that the suppression at large p_T in the spectra is not a product of initial state gluon saturation.

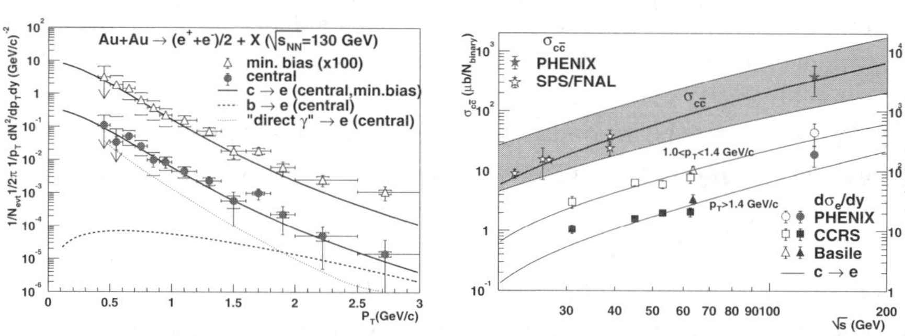

Figure 13. Left: electron spectra after background subtraction for $\sqrt{s_{NN}}$ = 130 GeV Au + Au minimum bias and central collisions from PHENIX [59]. Various contributions from decays and conversions are also shown, as described in the text. Right: Single electron cross sections derived from these spectra and for measurements at other c.m. energies for all charm and for the two momentum ranges shown. The curve for $\sigma_{c\bar{c}}$ and the shaded band correspond to predictions of PYTHIA [61] and a next-to-leading-order pQCD calculation [62], respectively.

6. Production of Heavy Flavors

The production of charm is expected to be dominated by interactions in the early stage of nucleus-nucleus collisions. Charm yields are sensitive to gluon shadowing in the incident nuclei [54] and the gluon densities reached [55, 56] in the collision. An early prediction of a deconfinement signature was the suppression of J/ψ production in a deconfined medium [57]. J/ψ suppression has been observed in central nucleus-nucleus collisions at the SPS [58]. It is imperative to also measure charm production to be able to understand the di-lepton intermediate mass spectrum as well as the J/ψ suppression.

6.1 Open Charm

PHENIX has measured the spectra of single electrons for transverse momenta $p_T < 3$ GeV/c in $\sqrt{s_{NN}} = 130$ GeV Au + Au minimum bias and central collisions after subtraction of backgrounds from π^o-Dalitz decays and γ-conversion using measured pion spectra [59]. Shown in Fig. 13 (left panel) are the background-subtracted electron spectra. Also shown are curves for contributions from open charm decays, bottom decays, and conversion electrons from direct photons. The latter two are shown for central collisions only. The measured electron spectra are consistent with what is expected primarily from charm decays, and the centrality-selected data scale approximately as the number of binary collisions. This binary scaling is different from that seen for lighter non-charm particles, where a large suppression (by factor ~ 4) is seen as discussed above. There is a prediction that charm quarks propagating through the medium should experience less energy loss due to the "dead cone" effect [60].

Assuming that the electron spectra are from charm decays, PHENIX has derived the charm production cross section at RHIC. This is shown in Fig.

13 (right panel) along with the c.m. energy dependence as measured in other experiments. PHENIX finds that the charm cross section is consistent with predictions from PYTHIA [61] and a next-to-leading-order pQCD calculation [62]. This appears to be in contrast to the increased charm production associated with the intermediate mass range for the di-muon measurements at the SPS [63].

6.2 J/ψ

A suppression of J/ψ production has been observed in $\sqrt{s_{NN}}$ = 17.3 GeV Pb + Pb collisions at the SPS [64] and attributed to a deconfinement phase transition as predicted by [57]. Alternative models based on interactions with hadronic co-movers have been proposed, but have problems reproducing the suppression as a function of centrality. Expectations at RHIC vary between the different models. If the suppression at the SPS is due to deconfinement, then the J/ψ should also be suppressed at RHIC energies. On the other hand, coalescence of $c\bar{c}$ pairs [65] and statistical-thermal models [66, 67] predict significant J/ψ production at RHIC. First measurements of J/ψ have been reported by PHENIX at RHIC in $\sqrt{s_{NN}}$ = 200 GeV p + p and Au + Au collisions [68]. These results disfavor enhancement due to $c\bar{c}$ coalescence [65] and thermal $c\bar{c}$ production [67], while still not yet being able to distinguish between models for suppression relative to binary scaling. Future high statistics measurements from PHENIX for J/$\psi \rightarrow$ e^+e^- at mid-rapidity and J/$\psi \rightarrow \mu^+\mu^-$ at forward rapidities should provide data to adequately address the J/ψ production and evolution mechanisms.

7. Conclusions and Outlook

Results from RHIC depict the formation of a dense system at an energy density greater than 5 GeV/fm^3, well above that where hadrons can exist. As the system evolves it can be described by the coalescence of quarks and subsequently by statistical models that indicate a temperature and baryo-chemical potential at the deconfinement phase transition boundary predicted by lattice QCD calculations. Collective flow is observed and provides direct evidence of strong pressure gradients in a highly interacting dense medium. Hard-scattering, suppression of hadrons at large transverse momentum, and quenching of di-jets are observed providing evidence for extreme energy loss of partons or their fragmentation products propagating in the hot matter created at RHIC. This is supported by the measured ratio $R_{dAu}(p_T) > 1$ in d + Au that exhibits a Cronin enhancement (initial state multiple scattering) for $1 < p_T < 7$ GeV/c and no suppression. The d + Au charged hadron data when compared to scaled p + p data rule out initial-state gluon saturation leaving only final state effects as a cause of the suppression in the Au + Au data at mid-rapidity at RHIC.

Notes

1. A strangeness saturation parameter is necessary in a canonical approach or when additional dynamical effects are present affecting extraction of true chemical equilibrium values.

References

[1] see F. Karsch, Nuc. Phys. A698 (2002) 199c.

[2] see K. Kanaya, Nuc. Phys. A715 (2003) 233c.

[3] for review of recent results at finite temperature and density see Z. Fodor, Nuc. Phys. A715 (2003) 319c.

[4] T.S. Ullrich, private communication.

[5] see for example M. Alford J. Phys. G: Nucl. Part. Phys. 30 (2004) S441.

[6] Nuc. Inst. Meth. A499 (2003) p. 235-824.

[7] B. Back et al (PHOBOS), Phys. Rev. C65 (2002) 061901R.

[8] B. Back et al (PHOBOS), Phys. Rev. Lett. 87 (2001) 102303.

[9] K. Adcox et al. (PHENIX) Phys. Rev. Lett. 87 (2001) 052301.

[10] C. Adler et al. (STAR), Phys. Rev. Lett. 87 (2001) 112303.

[11] J.D. Bjorken, Phys. Rev. D27 (1983) 140.

[12] A.E.M. Billmeier et al (STAR), J. Phys. G: Nucl. Part. Phys. 30 (2004) S363.

[13] A. Bialas, Phys. Let. B442 (1998) 449; J. Zimanyi et al, Phys. Lett. B472 (2000) 243.

[14] J. Letessier, A. Tounsi and J. Rafelski, Phys. Lett. B292 (1992) 417.

[15] J. Rafelski and B. Muller, Phys. Rev. Lett. 48 (1982) 1066; ibid., Phys. Rev. Lett. 56 (1986) 2334.

[16] A. Wroblewski, Acta Physica Polonica B16 (1985) 379.

[17] P. Braun-Munzinger, J. Cleymans, H. Oeschler, K. Redlich, Nucl. Phys. A697 (2002) 902.

[18] F. Antinori et al WA97, J. Phys. G: Nucl. Part. Phys. 30 (2004) S129.

[19] J. Rafelski, Proceedings of this Conference.

[20] J. Stachel, Proceedings of this Conference.

[21] A. Tounsi, A. Mischke and K. Redlich, Nucl. Phys. A715 (2003) 565c.

[22] P. Braun-Munzinger, K. Redlich, and J. Stachel, preprint nucl-th/0304013 v1 (2003).

[23] P. Braun-Munzinger, D. Magestro, K. Redlich, and J. Stachel, Phys. Lett. B518 (2001) 41 and D. Magestro (private communication).

[24] for another statistical-thermal approach see F. Becattini, J. Phys. G28 (2002) 1553.

[25] P. Braun-Munzinger, I. Heppe, and J. Stachel, Phys. Lett. B465 (1999) 15.

[26] see also F. Becattini et al., Phys. Rev. C64 (2001) 024901.

[27] F. Beccatini, M. Gazdzicki, and J. Sollfrank, Eur. J. Phys. C5 (1998) 143.

[28] P. Braun-Munzinger and J. Stachel, nucl-th/0112051 (2001).

[29] Z. Fodor and S.D. Katz, preprint hep-lat/0106002 (2001).

[30] T.S. Biro, P. Levai and J. Zimanyi, Phys. Lett. B347 (1995) 6.

[31] P. Kolb and R. Rapp, Phys. Rev. C 67 (2003) 044903.

[32] S.S. Adler et al (PHENIX), nucl-ex/0305013.

[33] J. Adams et al (STAR), nucl-exp/0306007.
[34] P.F. Kolb et al., Phys. Lett. B500 (2001) 232; P.F. Kolb et al., Nucl. Phys. A696 (2001) 197.
[35] P. Huovinen et al., Phys. Lett. B 503 (2001) 58.
[36] M. Gyulassy, I. Vitev, and X.N. Wang, Phys. Rev. Lett. 86 (2001) 2537.
[37] C. Nonaka, R. Fries, and S. Bass, nucl-th/0308051.
[38] M. Gyulassy and M. Plümer, Phys. Lett. B243 (1990) 432.
[39] X. N. Wang and M. Gyulassy, Phys. Rev. Lett. 68 (1992) 1480.
[40] R. Baier, D. Schiff, and B. G. Zakharov, Annu. Rev. Part. Sci. 50 (2000) 37.
[41] K. Adcox et al (PHENIX), Phys. Rev. Lett. 88 (2002) 022301.
[42] C. Adler et al (STAR), Phys. Rev. Lett. 89 (2002) 202301.
[43] B.B. Back et al. (PHOBOS), nucl-exp/0302015v1.
[44] K. Adcox et al (PHENIX), Phys. Rev. Lett. 91 (2003) 072303.
[45] J. Adams et al (STAR), Phys. Rev. Lett. 91 (2003) 072304.
[46] I. Vitev and M. Gyulassy, Phys. Rev. Lett. 89 (2002) 252301.
[47] D. Kharzeev, E. Levin and L. McLerran, Phys. Lett. 561 (2003) 93.
[48] B.B. Back et al. (PHOBOS), Phys. Rev. Lett. 91 (2003) 072302.
[49] I. Arsene et al. (BRAHMS), Phys. Rev. Lett. 91 (2003) 072305.
[50] J. Adams et al (STAR), nucl-ex/0306007.
[51] S.S. Adler et al (PHENIX), nucl-ex/0307022.
[52] C. Adler et al. (STAR) Phys. Rev. Lett. 90 (2003) 032301.
[53] C. Adler et al. (STAR) Phys. Rev. Lett. 89 (2002) 202301.
[54] Z. Lin and M. Gyulassy, Phys. Rev. Lett. 77 (1996) 1222.
[55] J.A. Appel, Ann, Rev. Nucl. Part. Sci. 42 (1992) 367.
[56] B. Muller and X.N. Wang, Phys. Rev. Lett. 68 (1992) 2437.
[57] T. Matsui and H. Satz, Phys. Lett. B 178 (1986) 416.
[58] M.C. Abreu et al (NA50), Phys. Lett. B 477 (2000) 28; ibid. Phys. Lett. B 521 (2001) 195.
[59] K. Adcox et al (PHENIX), Phys. Rev. Lett. 88 (2002) 192303.
[60] Y.L. Dokshitzer and D.E. Kharzeev, Phys. Lett. B519 (2001) 199.
[61] T. Sjostrand, Comp. Phys. Commun. 82 (1994) 74.
[62] M. Mangano, P. Nason, and G. Ridolfi, Nucl. Phys. B405 (1993) 507.
[63] M.C. Abreu et al (NA50), Eur. Phys. J. C14 (2000) 443.
[64] L. Ramello et al (NA50), Nucl. Phys. A715 (2003) 243c.
[65] R.L. Thews, M. Schroedter, and J. Rafelski, Phys. Rev. C 63 (2001) 054905.
[66] L. Grandchamp and R. Rapp, Nucl. Phys. A709 (2002) 415.
[67] A. Andronic et al, nucl-th/0303036.
[68] S.S. Adler et al (PHENIX), nucl-ex/0305030 v1.

THE STRANGENESS SIGNAL IN HADRON PRODUCTION AT RELATIVISTIC ENERGY

Reinhard Stock
Johann Wolfgang Goethe-University, Frankfurt am Main, Germany.

Abstract Strangeness enhancement is discussed as a feature specific to relativistic nuclear collisions which create a "fireball" of strongly interacting matter at high energy density. At very high energy this is suggested to be partonic matter, but at lower energy it should consist of yet unknown hadronic degrees of freedom. The freeze-out of this high density state to a hadron gas can tell us about properties of fireball matter. The hadron gas at the instant of its formation captures conditions directly at the QCD phase boundary at top SPS and RHIC energy, chiefly the critical temperature and energy density.

1. Introduction

Lattice-QCD theory predicts the disappearance of the hadronic phase of matter once the energy density exceeds a critical value of about 1 GeV per fm^3 [1], giving rise to a continuous, deconfined QCD state that is governed by the elementary interaction of quarks and gluons. To recreate this phase in the laboratory one collides heavy nuclei at relativistic energy with the goal of ascertaining the QCD predictions, and to pin down the decay point from the partonic to the hadronic phase by obtaining estimates for the transition temperature and energy density. The CERN SPS Lead (^{208}Pb) beam facility provides for a top energy of 158 GeV per projectile nucleon, corresponding to a total internal CM-system energy of about 3.5 TeV, to heat and compress the primordial reaction volume. A tenfold higher $\sqrt{s}$ is reached at the BNL heavy ion collider RHIC. In fact calorimetric data [2] show that the average transverse energy density exceeds about 2.5 GeV/fm^3 already at the SPS in central Pb+Pb collisions. Moreover the study of J/Ψ production [3] demonstrates a suppression of the yield in such collisions, characteristic of the QCD "Debye" screening mechanism expected in a deconfined partonic medium [4].

This lecture will deal with physics observables that could tell us about that medium. As an introduction I will look at the reaction dynamics in AA central collisions and then turn to my main topic, bulk hadron production and the location of the QCD phase boundary. We shall see that the production rates

W. Greiner et al. (eds.), Structure and Dynamics of Elementary Matter, 19–43.

of all strange hadronic species play a prominent role, as a signal that helps to understand dense QCD matter, and its decay.

1.1 Fireball Matter Dynamics

The initial Woods-Saxon nucleon density distributions, with average energy density of about 0.16 GeV per cubic fermi, impinge onto each other leading, at first, to nucleon-nucleon collisions occuring concurrently at the microscopic level as enveloped by the overall impact geometry of the target/projectile nuclei. This first generation of binary nucleon-nucleon encounters will involve essentially all the incident 2 A nucleons if we consider head-on collisions of nuclei with mass number A. In Pb+Pb collisions we thus have about 200 primary such nucleon-nucleon collisions of first generation. If we could now somehow stop the reaction dynamics, letting the reaction products escape to a detector system that identifies them, we would expect to register a final multiparticle state with a composition closely resembling A-times the (well known) outcome of a nucleon-nucleon collision at similar center-of-mass energy $\sqrt{s}$: nothing new!

However what makes relativistic nuclear collisions interesting is the fact that, in reality, the first generation set of A microscopic nucleon-nucleon interaction systems will immediately re-interact while still in statu nascendi of its "pending" output of asymptotically distinguishable, "on shell" reaction hadrons which would consist of about 10-20 produced mesons and baryon-antibaryon pairs. However, before being fully formed they run into a secondary generation cycle of subsequent collisions within the nuclear density distributions of the heavy nuclei. In fact there may be up to 6 secondary, subsequent collision generations of such pre-exited (not finally formed) microscopic collision volumes in a central Pb+Pb collision. Moreover they will occur in successive time steps spaced by fractions of a fm/c only - due to the relativistic Lorenz contraction of the nuclear density profiles of the Pb nuclei in beam direction ($\gamma_{CM} = 9$ at the SPS).

Something new! As we have no experimental knowledge of such secondary collisions of partly incompletely formed ("off shell"), partly decomposed hadrons we have to withstand the temptation to capture the overall reaction dynamics in a classical billiard ball cascade of subsequent generations of an inelastically multiplying gas of known hadrons. Into such a model we would insert the known, vacuum elementary cross sections at each microscopic binary encounter and proceed via Monte Carlo probability sampling methods. Such approaches are called microscopic hadron transport models [5]. From the above we would be surprised if they could give a correct description of the final outcome of a highly relativistic heavy ion collision because their set of microscopic degrees of freedom (isolated binary collisions of "on shell" hadrons) fails to capture the unknown nature of secondary, tertiary etc. encounters in the dense medium. These encounters, moreover, occur unresolved in time which should lead to quantum mechanical coherence, which might render the entire picture of isolated, sequential microscopic cascade-sub-processes obsolete. Let us, therefore, conclude that the overall large interaction volume of a central Pb+Pb collision will (after about 2-3 fm/c of interpenetration and reaction time have elapsed) be composed of a hitherto unknown state of strongly interacting matter that, however, contains all the quantum numbers and relative center of mass energy of the initial nuclear projectiles. We thus suspect that it will, quite generally, feature a high spatial density and, similarly, a high energy density. We therefore sometimes call this short-lived object a "fireball". The state of strongly interacting matter inside it is the topic of modern QCD theory [6, 7] and of our experiments.

Now there is something simple about this picture of dense matter in a fireball that experiments can check quickly. A new "state" of "matter" supposedly has been formed by fusing spheres of target and projectile cold nuclear matter which are initially located at opposite ends of longitudinal momentum space in the CM-system. By the symmetry of Pb+Pb collisions, a fused fireball should occupy a common momentum space volume centered in the CM-system at zero longitudinal and transverse momentum. Fig.1 shows the distribution in longitudinal phase space (measured here by the rapidity variable $y = 0.5ln[(1+\beta_L)/(1-\beta_L)]$ for the negative and neutral K-mesons, and for the anti-strange hyperon $\overline{\Lambda}$ [8]: all are Gaussians well centered at $y_{CM} = 0$. Note that the K^- and $\overline{\Lambda}$ consist only of newly created quarks that were not brought into the fireball as initial nucleon valence quarks. For an ideal, spherically symmetric fireball we would also require an isotropic momentum distribution. Looking simultaneously at the transverse momentum (not shown here) and at the y-distributions of these particles we see that this is not strictly the case (the fireball is longitudinally stretched and looks like a fire-football of US vintage) - but closely enough [9].

Furthermore we can experimentally gather almost **all** hadrons created in a central Pb+Pb collision at top SPS energy: it is about 2500 of them! Superimposing the total transverse energy in the football, carried by the various hadronic species we can estimate [10] its total average transverse energy density **in space**. This is the 2.5 GeV/fm^3 that we mentioned initially [2]. The fireball should thus initially exceed the QCD phase boundary between hadrons and quark-gluon matter, which is in the vicinity of 1 GeV/fm^3! Then it will expand explosively, back to hadrons via the phase transformation.

1.2 Deconfinement and Phase Boundary: Signals

How do we get experimental signals that elucidate the state of matter in the fireball maximum density stage? Like with supernova analysis, we may

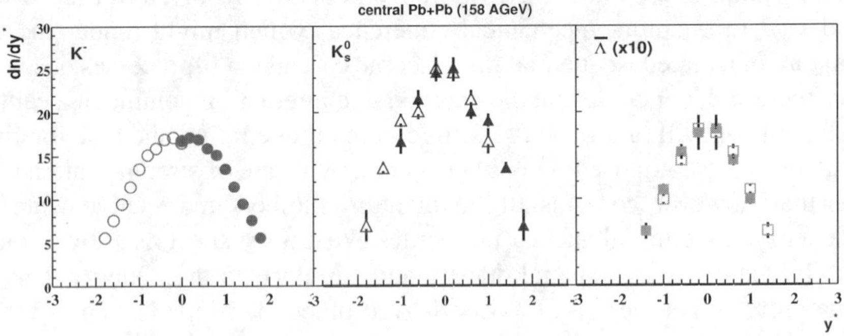

Figure 1. Rapidity distributions of K^-, K^0 and $\overline{\Lambda}$ produced in central Pb+Pb collisions at top SPS energy (158 GeV per nucleon). From Ref.[8].

try two ways: observe the primordial "light curve" (i.e. electromagnetic and neutrino radiation) or look at the bulk explosion material (e.g. expansion modes, element composition etc.). For the nuclear fireball the neutrino signal is out but we are left with "black body radiation": directly emitted photons and lepton pairs which leave the fireball while it is hot and dense. The corresponding experiments [11] are very demanding but successful. However I will not cover them here for lack of space.

Focusing on experiments that analyze the fireball material after expansion and cooling I briefly mention the suppression of the J/Ψ yield confirmed experimentally [3] to occur in central Pb+Pb collisions at top SPS energy, 158 GeV per nucleon. Still discussed vigorously in the community, we may take such observations to be, at least, consistent with creation of a fireball energy density well in excess of 1 GeV/fm^3 at which QCD predicts deconfinement thus dissolving the J/Ψ. This observation agrees with our above estimate of the transverse energy density as derived from the total final phase space density of the produced hadrons.

If one tentatively takes for granted such indications of a deconfinement state at top SPS energy one expects, likewise, to receive signals of the parton to hadron phase transformation bound to occur once the primordial high density state expands and cools toward the critical temperature and energy density. This brings me to the principal topic of the following chapters, in which I shall develop this observable in broader detail: could it be imagined that the detailed composition of the fireball decay products, in terms of abundances of the various hadronic species, captures the parton to hadron QCD phase transformation period of the dynamical fireball evolution? Or, more generally speaking: we expect to create, at least, a novel state of strongly interacting matter in the short lived fireball which must then decay. We infer from the above considerations of microscopic fireball dynamics that this high density state should feature novel degrees of freedom at the microscopic level. These may be partons (at top SPS energy and beyond), or shadows of the familiar hadron spectrum as modified by interaction with their dense, surrounding fireball medium (as might occur at lower collisions energies), and by the microscopic reaction cycles during the interpenetration stage of the nuclear density distributions. In any case it is the **decay** of this unknown coherent quantum mechanical state to a quasi classical state with familiar degrees of freedom, in our case a hadron and resonance gas, that we have to consider. The composition of that hadron gas is detectable: our signal.

Let me make a few remarks referring to QCD folklore. Almost trivially, all conceivable states of strongly interacting matter are falling under QCD governance. Thus in a decay of a state higher in energy density and temperature, upon fireball expansion, to the lower density hadron gas it is just the set of proper QCD degrees of freedom that is changing - we are thus dealing with a QCD phase transformation. At the hadron gas level, QCD resides in the spectrum of mass, spin, isospin, flavour, width etc. of the vast array of hadronic species. Thus in a QCD parton to hadron phase transformation, an initial fireball ensemble of flavoured plus coloured quarks and antiquarks, and of colour-anticolour carrying gluons is, sloppily speaking, "looking down" at the QCD realization below: the hadron spectrum. The ensuing, colour neutralizing "condensation" process of partons leads to hadronic objects featuring a spectrum of specific flavour-colour-spin-momentum internal compositions that absorb the initial partonic degrees of freedom. Physics experience suggests that such a process, occurring in an extended volume, should be governed by statistics. In the partonic view "from above" a light ($u\overline{d}$ positive) pion will be certainly the easiest way to hadronize in a predominant u, $\overline{u}$, d, $\overline{d}$ population (likewise for the other pions). In strong contrast, a heavy (sss) Omega hyperon will be a highly unlikely hadronization outcome, in view of competing K^- and K^0 mesons that could take care of the s-quarks combining them with the more abundant $\overline{u}$ and $\overline{d}$ quarks. Net result of all of this: the hadronization transition populates the hadronic spectrum in order of the relative statistical weight of the hadrons. We call such a process "phase space (statistical weight) dominated". It will thus create a hadron gas ensemble of maximum entropy, a decoherent, classical system [12] which exhibits a characteristic ordering pattern concerning the relative abundance of each hadronic species. It is, thus, not surprising that the abundance spectrum (expressed in terms of the various average hadronic multiplicities per collision event) obeys a statistical Gibbs ensemble [13], either in the so-called "canonical" or in the "grand-canonical" realization.

The statistical analysis of the composition of multihadronic final states was pioneered by Hagedorn in the 1960's [14]. It was revived in the last decade, applying it to the hadron populations observed in relativistic nucleus-nucleus collisions as well as in *pp* and e^+e^- annihilation [13]. In this approach one captures the temperature and energy density prevailing **at birth** of the multi-hadronic final state, i.e. the point in the fireball dynamics where it decouples, by decoherence, from the novel state of high energy density/temperature created in the early phase of the dynamics. This is clearly an interesting signal!

2. Hadron Multiplicity and Strangeness Enhancement

The first SPS experiments with ^{32}S-beams at 200 GeV/A showed an enhancement of various strange particle multiplicities, chiefly K^+, Λ and $\overline{\Lambda}$, relative to pion multiplicities, in going from peripheral to central S + (S, Ag, Au) collisions [15]. This observation appeared to be in-line with the pioneering analysis of Rafelski and Müller [16] who first linked strangeness enhancement to the advent of transition from the hadronic to a partonic phase. This offered lower effective $s\overline{s}$ threshold, shorter dynamical relaxation time toward flavour equilibrium, and an additional, nontrivial effect of relatively high net baryon number or baryochemical potential: the light quark Fermi energy levels move up, perhaps even to the s-quark mass at high μ_B, and the Boltzmann penalty factor for the higher mass $s\overline{s}$ pair creation might be removed. This latter aspect was mostly ignored in the late 1980s but receives fresh significance as we become increasingly aware of the crucial role of μ_B.

In this section, and in sections 3 and 4, I will present a sketch of our recent progress, both in gathering far superior data and in the understanding of the statistical model that was rudimentarily anticipated in such early strangeness enhancement speculations.

Fig.2 shows a modern version of the typical strangeness enhancement phenomena. ϕ-meson to positive pion multiplicity ratios obtained by NA49 [17, 20] in Pb+Pb SPS collisions at 158 GeV/A (corresponding to $\sqrt{s}$=17.3 GeV) are plotted for a sequence of collision centrality conditions from peripheral to central. At the peripheral end the minimum bias p+p point [18] matches with the trend. The centrality scale employed here is, at first, the number of participating nucleons (left hand side of Fig.2). The raw data centrality bins are ordered in NA49 data by decreasing projectile spectator energy as recorded in a zero degree calorimeter. This information is converted to mean participant nucleon number by a VENUS calculation from which one also derives the mean impact parameter b, corresponding to each centrality bin. Neither of these scales turn out to be satisfactory [17, 19, 20] in merging data from central *light* nuclei collisions such as C+C with the various centralities of the Pb+Pb collisions. For example, a central C+C collision has $b \approx 2$ fm and $N_{part} \approx 18$ but on a b scale the ϕ/π^+ value is about 50% lower than the $b = 2$ result for the much heavier Pb+Pb system. Inversely on the N_{part} scale: N_{part}=18 corresponds

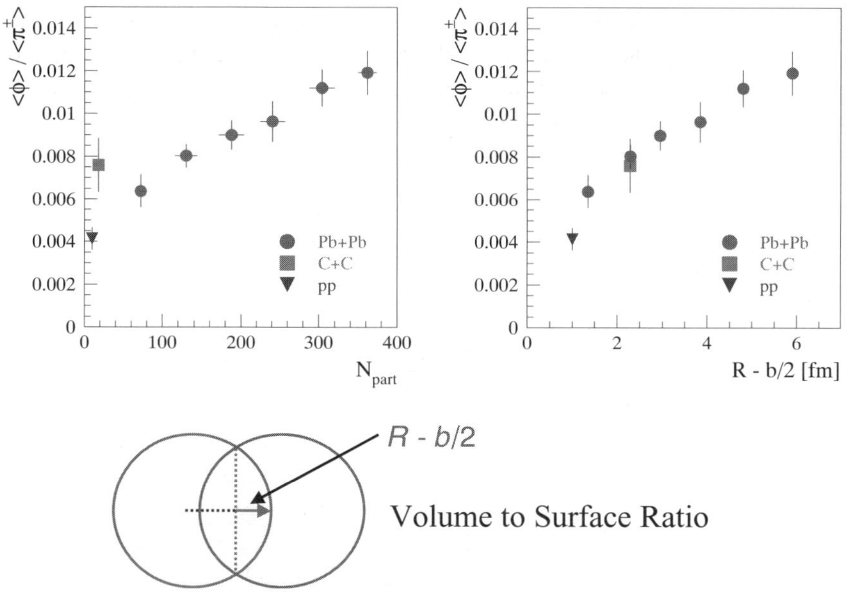

Figure 2. Phi meson to pion multiplicity ratios for Pb+Pb at 158 GeV/A, as a function of collision centrality given by the number of participating nucleons (left), and by R-b/2 where R is the nuclear radius. Data for min. bias pp and for central light nucleus collisions are also given, Ref.17 and 20.

to *very* peripheral Pb+Pb and the central C+C result is about 50% higher than the Pb+Pb curve. A central collision of a relatively light nuclear pair thus behaves quite differently from a very peripheral heavy nuclear collision where only the dilute Woods-Saxon density tails interact! The scale of the right hand side of Fig.2 is an intuitive guess [19] to represent the relative compactness, or volume-to-surface ratio of the primordial interaction volume, by the variable R-*b*/2, where R is the radius of the colliding nuclear species. It might be connected with the energy or collision density reached in the primordial collision volume (see chapter 5). We see that the central light nuclear collision data now merge with the Pb+Pb centrality scale.

The "strangeness enhancement factor" is given oftentimes as the production ratio of AA central/(pp min. bias times 0.5 N_{part}). In the case of Fig.2 it would be roughly 2.5. We also have a systematic study of strangeness enhancement in Au+Au collisions at the BNL AGS energy of 11 GeV per nucleon where this factor is about three [21]. Multistrange hyperons [22, 23] show factors between 4 and 15 at top SPS energy.

Bulk strangeness enhancement in central collisions is a nuclear feature, absent in pp collisions. Of course we lack a detailed picture about "centrality" in pp collisions but we could still employ e.g. the total charged particle multiplicity to select more or less "violent" collisions. Fig.3 shows the K^+/π ratio of pp at 158 GeV versus charged particle multiplicity to be essentially flat [24]. Similar findings are made up to Tevatron $\sqrt{s} = 1.8$ TeV $p\overline{p}$ collisions: the K^+/π ratio is 50% higher here but also almost independent of N_{ch} [25].

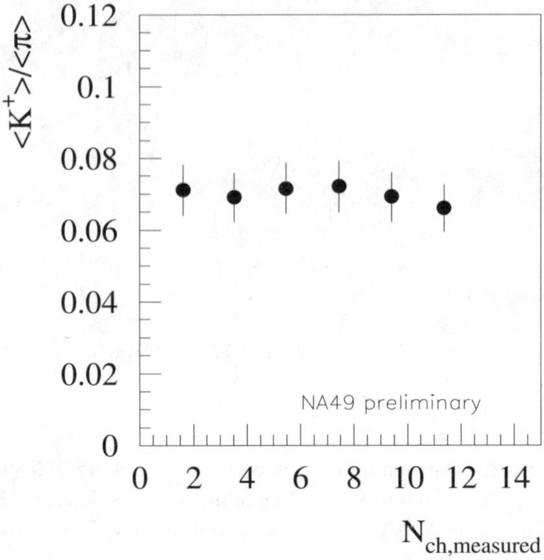

Figure 3. The multiplicity ratio of positive kaons to pions in pp collisions at 158 GeV, as a function of charged particle multiplicity, Ref.24.

A picture emerges in which strangeness enhancement, or more generally speaking the yield order in the overall bulk hadron population is connected with "sequentiality" of interactions at the microscopic level, i.e. with the fact that a number of successive collisions occurs in narrow intervals of time if one may employ a naive Glauber picture. This number reflects the size and density of the primordial interaction zone. It thus seems that a certain (extended) size **and** a certain interaction rate (and thus energy density), typical of central nuclear collisions, conspire in strangeness enhancement. Unfortunately we do not really know how to describe a second, third etc. collision of a hadron, within fractions of a fm/c space-time distance. We would then have probably resolved the key issue: does it dissolve into a parton cascade from which the final hadrons are reconstituted? Proton-nucleus collisions should hold a key to this question but nobody has succeeded in isolating the second, third, n'th successive collision of the projectile, as of yet [26–28].

At the moment we thus forgo pA as an intermediate step although it certainly also features changes in the hadronic production ratios [26–28] and base the analysis on comparing pp to AA. Fig.4 shows the hadronic multiplicities,

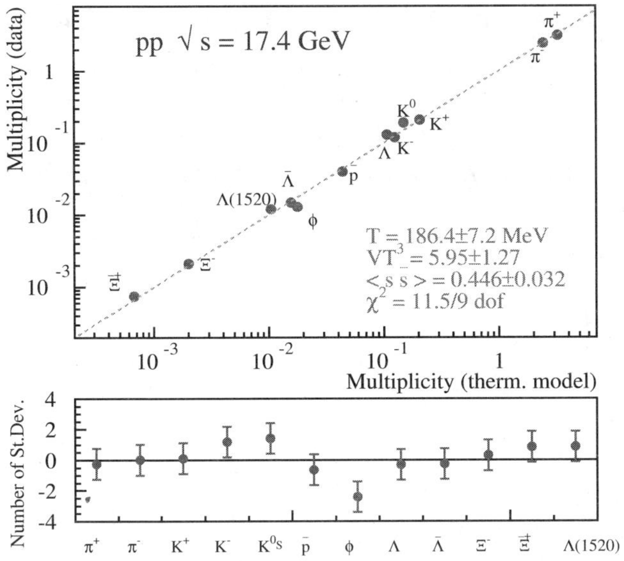

Figure 4. NA49 data for hadron multiplicities in pp collisions at 158 GeV confronted with the canonical model of Becattini, Ref.30.

from pion to cascade hyperon, obtained by NA49 for min. bias pp at $\sqrt{s}$=17.3 GeV [29]. The data are confronted with the Hagedorn statistical model in its canonical Gibbs ensemble form as employed by F. Becattini [30], leading to very good agreement (as it is well known also for other elementary collisions and energies [31]). The three parameters are T=186±7 MeV, a reaction volume of about 7 fm^3, and a total of about 0.5 $s\bar{s}$ pairs. The apparent validity of a statistical weight-dominated picture of phase-space filling has been considered a puzzle already since Hagedorn's time. It is clear, however, that the apparent canonical "hadrochemical equilibrium" pattern can *not* result from "rescattering" of produced hadrons: there is none. In Hagedorn's view [32] a creation "from above" must hold the key to the apparent maximum entropy state, i.e. the QCD process of hadronization [33]. In it a medium of unknown composition (the "fireball" or "string" or "cluster") undergoes a quantum mechanical decay which is dominated by the phase space weight distribution that corresponds to the spectrum of hadrons and resonances [12, 34]. The canonical abundance pattern and the T-value are a fingerprint of QCD hadronization - do AA collision data at high $\sqrt{s}$ also confirm this picture (they must, of course, also result from a hadronization process)?

3. AA collisions in the Grand Canonical Model

Fig.5 shows the grand canonical fit by Becattini to the NA49 data from central Pb+Pb at 158 GeV/A [30]. The temperature is 160 ± 5 MeV and μ_B=240 MeV; besides, this model employs the much discussed strangeness undersaturation factor γ_s=0.8. Leaving the second order concern about γ_s to the theoretical

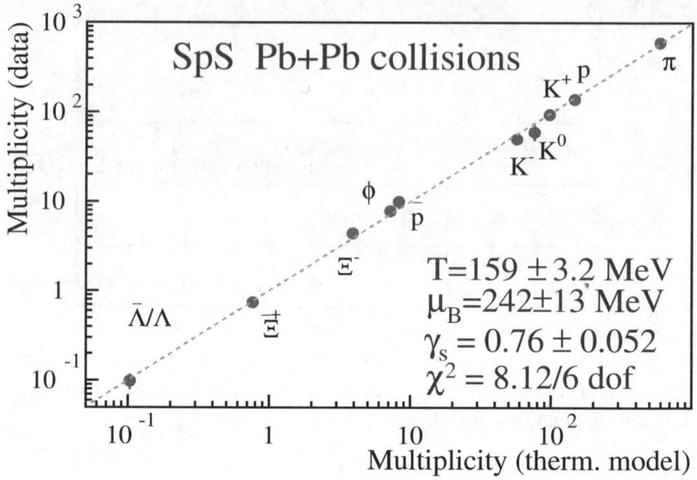

Figure 5. Hadron multiplicities for central Pb+Pb collisions at 158 GeV/A from NA49 confronted with the grand canonical statistical model, Ref.30.

community I note here that Braun-Munzinger et al. [35] fit a set of data at the same top SPS energy without introducing a γ_s; they report T=170 $\pm$5 MeV, at μ_B=270 MeV, close enough. There are also studies of the new RHIC STAR data [36] at $\sqrt{s}$=130 GeV by this model [37] and by Kaneta and Xu [38], averaging at 175 $\pm$ 5 MeV and μ_B=48 MeV. And the new data of NA49 [39, 40] at 80 and 40 GeV/A have resulted in fit values of T=155 MeV, μ_B=270 MeV and T=150 MeV, μ_B=395 MeV, respectively [41]. I will return shortly to a further discussion of the grand canonical approach but wish to, first of all, show an overall impression from these analyses which are confronted in Fig.6 with the new lattice QCD calculations at finite μ_B by Fodor and Katz [42].

The latter predict the T, μ dependence of the QCD phase transformation which in this model consists of a crossover for all $\mu < 650$ MeV, i.e. in the SPS-RHIC domain. Note that physics observables can change rapidly in a crossover, too: the familiar steep rise of e.g. lattice ϵ/T^4 at T_c does not, by itself, reveal the order of the phase transformation [6]. Anyhow: the hadronization points from grand canonical ensemble analysis merge with the phase transformation site of lattice QCD at top SPS and RHIC energy. Quite a remarkable result, but also a plausible one [33, 43] if we recall that Ellis and Geiger did already

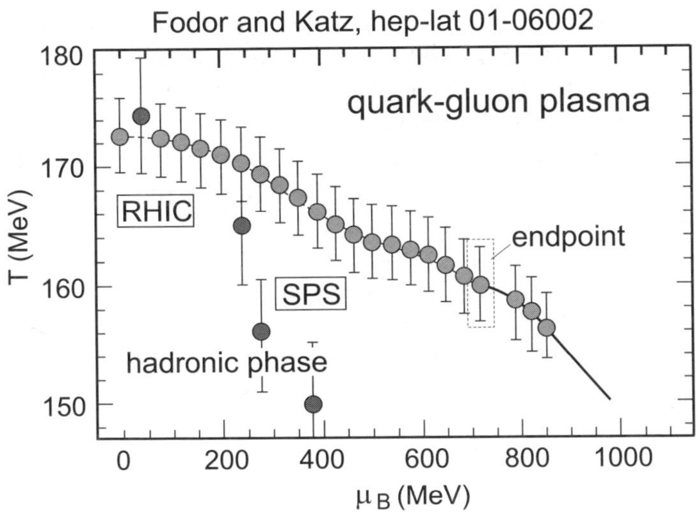

Figure 6. The lattice QCD phase boundary in the plane of T vs. μ_B, Ref.42. The hadronization points captured by grand canonical analysis for SPS and RHIC energies are also shown.

point out in 1996 that hadronic phase space weight dominance appears to result from the colour-flavour-spin-momentum "coalescence" of partons that occurs at hadronization [12, 44]. Unfortunately a rigorous QCD treatment of the parton to hadron transition is still missing.

At this point the following objection is always raised: if the same basical model describes hadronic yield ratios in pp, e^+e^- and in central AA collisions, Figs. 4 and 5, what is special about AA, as you will not tell us now that a QGP is also formed in pp?! Answer: on the one hand both collision systems reveal the QCD hadronization process which features, furthermore, the Hagedorn limiting hadronic temperature T_H. At top SPS and at RHIC energy T (hadronic ensemble) $\approx T_H \approx T_c$ (QCD), *this* is the common feature; it should not be a coincidence. On the other hand hadronization appears to occur under distincly different conditions in AA collisions, as captured in the transition from a canonical to a grand canonical description. Inspection of Fig.4 and 5 shows that the hadronic population ratios are quite different: the falloff from pions to strangeness-two cascade hyperons in the former case is about four orders of magnitude whereas it reduces to three in the grand canonical situation: strangeness enhancement! In the canonical case the small reaction fireball volume is strongly constrained by local conservation of baryon number, strangeness neutrality and isospin whereas these constraints fade away in the grand canonical ensemble which represents a situation in which, remarkably, these conservation laws act only *on the average*, over a rather large volume, as captured by a collective chemical potential μ_B. This leaves one global quantity μ_B essentially in charge of all the conservation tasks. Note that the statistical model does *predict* nothing, it *merely captures* this most remarkable feature of the hadron gas emerging after hadronic freeze-out. Its observed success implies some kind of long range collective behaviour in the hadronizing source, the origin of which is yet unknown, but must be specific to central AA collisions. We shall address this topic in the following Sections but note, for now, that **Strangeness enhancement is the fading away of small volume canonical constraints, in the terminology of the statistical model [45]**.

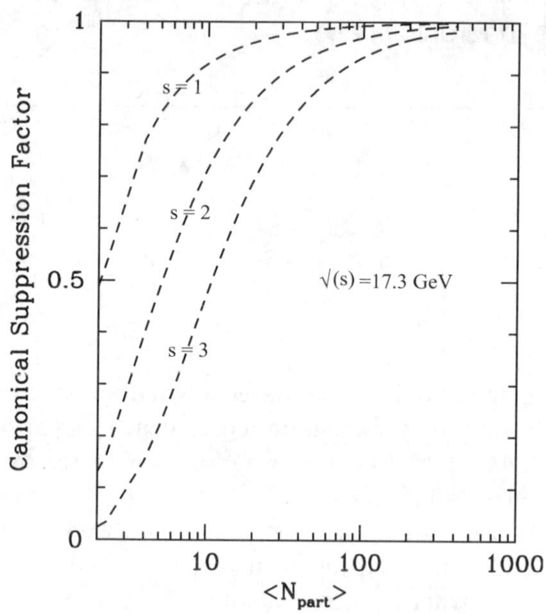

Figure 7. The canonical to grand-canonical transition as reflected in the canonical suppression factor which is the inverse of strangeness enhancement, shown for strange hadron species with s=1,2,3 at top SPS energy, Ref.46.

This aspect has been recently studied in all detail by Cleymans, Redlich, Tounsi and collaborators [45, 46]. Fig.7 illustrates their results concerning the transition from canonical to grand canonical behaviour with increasing number of participants, i.e. overall "source" size. It is intuitively clear that it should occur, first, in singly strange hadrons, the increase occuring with offset (but having a larger specific effect on the yields per participant) in S=2,3 hyperons.

A further, appropriate critical question: how can we understand the other aspect of Fig.6, i.e. the steep falloff from the QCD transition domain occuring at the lower SPS energies? We even have a further GC analysis, at top AGS energy, by Stachel [47], for central Si+Au collisions at 14.6 GeV/A, shown in

Fig.8. The result is T=125 MeV, μ_B=540 MeV, far below the T scale of Fig.6.

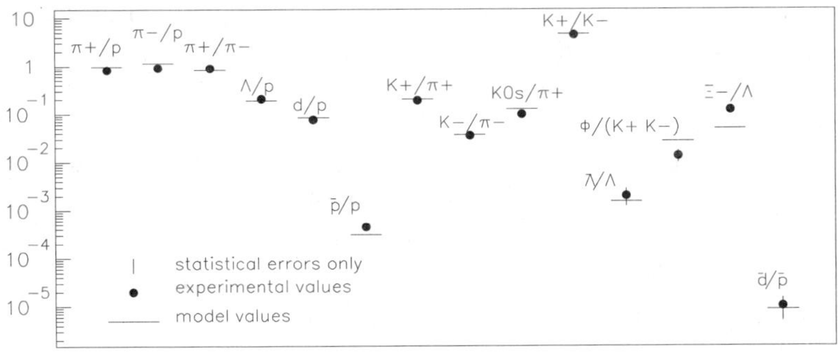

Figure 8. Hadron yield ratios at top AGS energy, in central Si+Au collisions at 14.8 GeV/A as fitted with the grand canonical statistical model, Ref.47.

The picture of a direct parton to hadron transition is intuitively inapplicable at these lower energies. Still the overall dynamical trajectory that ends in hadronic chemical (abundance) freezeout should arrive there "from above" as hydrodynamical models [48] show. The hadronic system is initially very dense along its trajectory, it is not a hadron gas, and it is thus a quantum mechanical coherent state composed of excited and in-medium modified hadrons that decays to the finally observed classical hadron gas ensemble. This decay might also ignore the classical concept of a relaxation time. Recall the nucleus, also still a dense system: we do not invoke relaxation time in a transition within such a quantal medium, such as β-decay. And yet "Fermis Golden Rule" asserts that the transition strength depends "only" on the squared matrix element times final state phase space volume weight plus global conservation laws. And we know that the phase space factor oftentimes far overrides the matrix element, in the net decay strength. High density *hadronic* matter behaviour is essentially unknown: an old and new research paradigma but at present we state that it is **not** a classical hadron-resonance gas and that it may decay quantum mechanically, the decay products well described by a grand canonical ensemble of hadrons and resonances. At top SPS and at RHIC energy, in turn, the increasingly "explosive" nature of partonic and hadronic expansion may almost instantaneously dilute the hadronizing source toward chemical freezeout, as indicated by T(GC)$\approx$T (Hagedorn) $\approx T_c$ (QCD). We may, thus, directly look here at the QCD parton to hadron phase boundary, located at T $\approx$ 170 MeV.

4. Working with the Grand Canonical Ensemble

4.1 The Hagedorn Fireball

The previous sections have described our recent attempts to understand the order of multiplicity per collision in which the various hadronic species are populated. We have referred to the statistical Hagedorn model [32] in its modern versions. It has turned out that elementary collisions, such as pp, $p\overline{p}$, $e^+e^- \rightarrow$ hadrons, are well described in the canonical version of the statistical model, whereas central nucleus-nucleus collision hadronic final states appear to obey grand canonical statistical order. Let me emphasize again that in this model we consider the finally observed multihadron state to result from the *decay* of a quantum mechanically coherent fireball stage that resides *"above"* the finally observed, "frozen-out" classical hadron gas, in terms of energy density and temperature. Its decay occurs at a certain, late stage in the overall dynamical evolution, imbedded into an overall cycle of initial interpenetration, fireball stage of maximum energy density, and expansion dilution and cooling. The composition and structure of matter in the intermediate fireball stage is the object of foremost interest. Its direct radiation output signals form one set of appropriate observable signals [11]. The other signals are derived from the fireball freeze-out decay into hadrons.

The freeze-out product of the fireball is sufficiently dilute to be quantum mechanically incoherent, thus being describable as a classical hadron gas. The multiplicities, and multiplicity ratios of the various hadronic species in this gas form an image of the instant of decay: we thus study the *decay properties* of the unknown state of matter in the high density fireball: the conditions prevailing "at birth" of the frozen-out hadron gas, common to all hadrons emerging from the fireball. These common conditions are, chiefly, the temperature, energy density and net baryon number density. They are captured by the statistic ensemble analysis. Its success in describing the composition of the frozen out hadron gas (residing in the multiplicity ratios of dozens of hadronic species, from pions to multiply strange heavy hyperons) shows that the entire, at first sight exceedingly complicated final state, of up to several thousands of hadrons, can be understood from a few common macroscopic parameters. To summarize: it is the underlying assumption of Hagedorn analysis that the "decay from above" is dominated primarily by the energy density (and conserved quantum numbers) at the instant of decay, and by the statistical weight distribution offered by the hadronic spectrum into which the decay occurs. Immediately after decay the hadronic gas thus exhibits a statistical equilibrium population, as captured by a Gibbs ensemble of mixed hadronic species. At the risk of overdue repetition, I stress again that this apparent "thermal" equilibrium is a result of the decay process, the nature of which lies well beyond the statistical model which "merely" captures the apparent statistical order, prevailing right after decay. The hadrons and their weight distribution are "born into" this state [32, 33]! The observed equilibrium is, thus, *not* achieved by inelastic transmutation of the various hadronic species densities, in final hadron gas rescattering cascades, i.e. not by hadron rescattering approaching a dynamical equilibrium. The system explodes and cools rapidly after the initial hadron gas formation phase [33], the canonical or grand canonical order staying frozen-in throughout subsequent expansion (while the momentum spectra etc. still get modified by elastic interactions, resonance decays etc.). Hadro-chemical composition freeze-out thus occurs prior to the final decoupling from all strong interaction (spectral freeze-out). In a typical central Pb+Pb collision at top SPS energy, 158 GeV per projectile nucleon, hadro-chemical freeze-out occurs at a temperature of 165 ± 5 MeV (thus capturing the hypothetical QCD phase boundary) but final, spectral decoupling occurs at about 110 MeV [49], as we learn from hadronic spectra and correlations.

I have inserted this somewhat lengthy section to avoid the misunderstandings and controversies that have accompanied the Hagedorn model ever since its inception. In the 1970's particle physicists were seeking an understanding of proton collisions in analogy to the elementary Feynman graphs of QED, thus trying to insolate similarly "elementary" hadronic processes, connecting the initial and final constituent (massive) quarks. However the most probable outcome of a pp collision, namely to go to a ten-hadron final state (of statistically varying microscopic composition) clearly defied such a picture, thus being called "background" reactions, outside of the primary research focus. On the other hand Fermi, Landau and Hagedorn just put the emphasis on this side of hadron collisions, guided by the intuition that an overall process of ever increasing density of potentially coexisting microscopic subprocesses should approach the statistical situation: all subprocesses would feed into a symbolic, intermediate "compound state" (reminiscent of Bohrs excited nuclear compound state)that was called fireball. Its decay would then feed into a frozen-out, statistically ordered hadron gas. The appropriate formulation turned out to be a canonical Gibbs ensemble [32]. We have seen the modern version of this model in Fig.4.

Furthermore, the Hagedorn concept of freeze-out from the symbolic fireball (of microscopically unknown degrees of freedom) to a classical hadron gas of statistical weight dominated composition - that then travels on to final observation unchanged by the subsequent dynamical evolution - took a long time to be comprehended. On the other hand we are cordially familiar with the same picture from explosive Big Bang nucleosynthesis. We know that the cosmic average proton to helium composition ratio froze out to its observed ratio once the inelastic transmutations in the cosmic fireball (among the various light nuclear species) stopped at about 1 MeV temperature - thus presenting to us highly relevant data concerning this dynamical stage. With further expansion this chemical composition travels on unchanged (frozen-in), while the spectral temperature of the cosmic inventory has dropped down to a few 10^{-4} eV. Likewise, hadro-chemical fireball freeze-out creates hadron composition ratios that travel onward throughout further hadronic expansion.

4.2 The Grand Canonical Ensemble

Let us now turn to the formalities of grand canonical ensemble analysis. One starts from the formulation of a partition function which specifies the relative weight Z_i (it counts the sum of possible states) for each particle (or resonance) species i in a multihadronic mixed gas at temperature T:

$$lnZ_i = \frac{g_i V}{6\pi^2 T} \int_0^\infty \frac{k^4 dk}{E_i(k)[exp\{(E_i(k) - \mu_i)/T\} \pm 1]} \qquad (1)$$

where g_i is the statistical Landé factor (the "statistical degeneracy") of species i, V is the total common volume shared by all species, $E_i^2(k) = k^2 + m_i^2$ the total energy square of species i at momentum k, and $\mu_i = \mu_B B_i + \mu_s S_i + \mu_I I_i$ the "chemical potential" of species i. The latter is the typical, unfamiliar ingredient of the grand canonical ensemble. Please consult textbooks describing its occurence in Z_i as a result of adding a so called Lagrange multiplier to the Lagrange density of the system in order to enforce global conservation of certain net quantum numbers that are specific to the total system contained in the total volume V. For our case of "hadrochemistry", these are the net quantum numbers initially carried into the fireball volume by the incident nuclei. Their total baryon number B, total strangeness S, and total isospin (Z-component) I are initially given by the total participating nucleon (=baryon) number, S being zero (nuclei have zero strangeness), and by the net isospin of all participant nucleons: I=(Z-N)/2. These net quantum numbers are conserved by strong interaction, and accompany the collision volume throughout the fireball evolution, during which they will be re-distributed over the entire ensemble of the hadronic species that is being created. Thus the three components of μ_i in Z_i represent the net impact of the overall quantum number conservation on each separate species i. We notice, most significantly, that the conservation laws are thus enforced *on average only*, not locally but over the entire fireball volume. This total volume thus acts somehow coherently, balancing globally the quantum number exchanges that have occured microscopically during fireball dynamics preceding the decoupled hadron gas that is described by this set of Z_i's. From our considerations above, and following Hagedorn's advice [32], we state again that this global coherence is not achieved by inelastic relaxation toward equilibrium within the frozen-out hadron gas ensemble but must be a characteristic feature either of the quantum mechanically coherent high density fireball preceding freeze-out, or by the mechanism of its decay process to the decoupled hadron gas (or by both acting together). After decay the resulting hadron gas is imprinted with these global coherence aspects of its birth process. The grand canonical ensemble description merely "takes note" of the freeze-out product of such a process, analyzing a temporary snapshot of the apparent order in the frozen-out state, ignorant of its dynamical origin.

A short comment concerning the terminology: chemical "potential". You see from equ. (1) that the exponential "penalty factor" in the denominator, that expresses the cost of realizing a certain hadron with total energy $E_i(k)$ within a bath of temperature T (characterizing the average kinetic energy, or energy density), gets modified by μ_i. This modification thus takes into account the "affinity" - to employ the classical terminology - that the medium offers to species i. In modern terms we realize that μ_i in the exponent acts like an average potential for species i in the medium, modifying its vacuum energy $E_i(k)$, to now read $E_i(k) - \mu_i$. The medium thus modifies the cost of realizing species i at momentum k. An "affine" medium (positive μ_i) reduces the penalty factor and thus increases the relative weight Z_i.

From the partition function in (1) we derive the distribution of number densities of the species i in the medium,

$$n_i = \frac{g_i}{2\pi^2} \int \frac{k^2 dk}{exp\{(E_i(k) - \mu_i)/T\} \pm 1} \tag{2}$$

The contribution of species i to the overall energy density ϵ_i is obtained by inserting the factor $E_i(k)$ in the numerator of the integral. From the number density of species i we get its total multiplicity per event by multiplying n_i with the total reaction volume V.

Now we want to analyze experimental data, with the set of N_i hadronic multiplicities supposed to be known. From this we have to determine the basic parameters V, T and μ_B, μ_S and μ_I. It turns out that we can, before proceeding to a fit to equ. (2), reduce the three chemical potentials to one remaining quantity. We do this by exploiting global baryon, strangeness and isospin conservation in a set of coupled equations which, in the end, express μ_s and μ_I in terms of μ_B which thus remains as the only independent potential, expressing all conservation tasks. Three parameters remain: V, T and μ_B. We have shown in Fig.5 the outcome of such a fit procedure as applied to a set of NA49 hadron multiplicity data for central Pb+Pb collisions at top SPS energy. Repeating at various energies [41] we get the systematics of hadronic chemical freeze-out conditions as represented in the T, μ_B plane (Fig.6). That is, basically, how it works.

An easy qualitative understanding of the procedure that determines the two parameters T and μ_B, in a first cycle (which then leads to fixing V in a second step) is obtained from considering the special case of particle to antiparticle multiplicity ratios. For easy approximation we go from Bose or Fermi statistics (resulting in the ± 1 term in the denominator of equ. (2)) to Boltzmann classical statistics dropping that term. We consider the ratio of a certain particle-antiparticle multiplicity,

$$< n_i > / < \overline{n}_i >= n_i V / \overline{n}_i V = n_i / \overline{n}_i \quad (3)$$

in which the common total volume parameter drops out. Consideration of the appropriate ratio of two integrals (2) shows that almost everything drops out because statistics and phase space are identical for particle-antiparticle pairs. What remains is

$$< n_i > / < \overline{n}_i >= exp\{(\mu_i - \overline{\mu}_i)/T\} = exp(2\mu_i/T). \quad (4)$$

If we do this for several such ratios in combination, say K^+/K^-, $p/\overline{p}$, $\Lambda/\overline{\Lambda}$, $\Omega/\overline{\Omega}$ etc., we get a set of equations from which to determine the maximum likelyhood $\{\mu_B, T\}$ combination. In a second step we fix the volume parameter V by fitting equ. (2) to the pion multiplicity $< n_\pi >$ with known μ_B and T.

In the real procedure one has to go through one additional complication. The finally observed hadrons are not really the ones that existed at freeze-out, T=165 MeV or so, in the created hadronic gas. It is composed, at first, of excited hadronic states such as N^*, Δ, K^* etc. and of resonances like ϱ and ω. It is *their* population that obeys the set of equations (2)! After being established at freeze-out this population decays in well known ways to the finally observed π, K, Λ, p, $\overline{p}$ etc.. These latter particles have *never* been in a state that obeys the grand canonical multiplicity order! They serve as an observational input to a grand canonical ensemble fit via an attached procedure that was invented by Wroblewski [50], which relates the final, observed set of hadron multiplicities to a set of excited hadron and resonance multiplicities that forms the initial hadron composition at freeze-out, via the known decay branching ratios. This important backward-transformation was oftentimes ignored in early applications of the statistical model, causing considerable confusion.

5. The Onset of Grand Canonical Order in AA Collisions

We infer from the previous section that grand canonical ensemble analysis captures a snapshot at birth of the classical hadron gas, emerging from central collisions of heavy nuclei. The apparent most interesting feature was the quantum number conservation on average only, over a large volume of microscopic ingredients. We have claimed that this coherence is not a property of the frozen-out hadron gas but should reside in its birth process "from above", referring hypothetically, to the dynamics of fireball evolution preceding decoupling and freeze-out. We have, thus, invoked a prior dynamics: coherence in a large volume, high energy density fireball (of unknown matter composition) which, upon decay, creates grand canonical order. The grand canonical analysis merely takes note of this order; it does not know about our additional, hypothetical speculation concerning its dynamical origin.

On the other hand, as we are interested, in particular, in just this dynamical origin (as it might tell us about the properties of the unknown fireball matter above: its decay temperature and chemical potential), we would now want to inspect a wider variety of fireball hadronization processes. We thus turn to elementary hadron collisions, and to e^+e^--annihilation, in which the interaction volume is very small in comparison to AA fireball volumes. Nevertheless, hadronization must occur under similar QCD dominance here. This leads us to expect that the statistical temperature, which is the decay image of the critical energy density at which any QCD hadronization occurs, must be common to both elementary and AA collisions at $\sqrt{s} \geq 20$ GeV (top SPS to RHIC energy). Because we assume, from all existing observables, that we are witnessing a parton to hadron QCD phase transformation at these high energies. Indeed the temperature parameters all fall within 170 ± 10 MeV, from e^+e^- to Au+ Au [13, 31, 36, 38, 41].

The apparent difference between small and large freeze-out volumes appears, thus, to arise not from the decay temperature nor from the final hadronic level spectrum but from a specific difference in the size of the coherently decaying system. Indeed, in the canonical decay mode the "volume" amounts to a few fm^3 only. The relevant conserved quantities are forced in the canonical case to be conserved locally - if an Omega hyperon is to be produced it must be locally accompanied by an anti-Omega or by an anti-Lambda plus two K^+ and a π^- (charge, strangeness and baryon number conservation), etc.. Of course this requirement is relatively hard to meet locally, and it inflicts a high statistical "punishment factor". We call this effect canonical suppression. In the grand canonical case nothing couples to the Omega locally, we only require three strange quarks and some energy density. The resulting enhancement of the production rate can be understood as an absence of canonical suppression. This statement forms the essence of the analysis [46] of the transition from C to GC behaviour that we showed in Fig.7.

Before looking at the formal expression for this transition, as derived by Tounsi and Redlich [46] let me try a crudely oversimplified qualitative argument for illustration, involving the punishment factor only. Consider Omega formation by decay of a partonic medium. Canonical decay then requires 3 s quarks and 3 $\overline{s}$ quarks jointly participating (plus other factors that I ignore here). The penalty factor arising from local strangeness and baryon number conservation will, in this naive picture, be $exp(-6m_s/T)$ whereas in a GC situation it diminishes to $exp(-3m_s/T)$, the conservation being taken care of elsewhere in the large volume. The enhancement factor then is $exp(3m_s/T)$. Taking m_s=140 MeV and T=165 MeV we get a factor of about 13: rather close to the ratio of about 15 for the Omega production per participant nucleon pair in central Pb+Pb collisions at top SPS energy, relative to the p+p production rate at similar energy [23]. The Omega is, of course, a very rare species at all incident energies. The "bulk" strangeness abundance in the system is more directly reflected in the singly strange K^+ yield. Let us try: the canonical penalty factor would naively be $exp(-2m_s/T)$ for a K^+, but only $exp(-m_s/T)$ in a GC scenario. The GC enhancement factor of bulk strangeness would thus be $exp(m_s/T) = 2.3$: close enough to the experimentally observed [17] kaon enhancement, by a factor of about two.

Let us take the above as an amusing illustration concerning the Boltzmann phase space factor for different hadronic species occuring in the denominator of equ. (1). We have, of course, ignored the effect of the baryochemical potential μ_B; and many other appropriate considerations. An exact treatment can be found in ref. [46]: Tounsi and Redlich show that the suppression factor governing the yields of strangeness 1, 2, 3 hadrons in proceeding from the canonical case (elementary hadron collisions and e^+e^- annihilation to hadrons) to the GC limit (central AA collisions) is given by

$$F = I_s(V)/I_0(V) \tag{5}$$

where the I are modified Bessel functions of ascending order in strangeness $S = 1, 2, 3$ and V is the quantum number coherent fireball volume that feeds downward to the frozen-out hadronic gas volume (which may initially be of equal size to the decaying fireball volume). The Bessel functions essentially depend on the coherent volume; the asymptotic limits for this ratio are 1 for $V \rightarrow \infty$, and $V/2$ for V near zero. The ascend to the "no suppression limit" at large volume is steepest for singly strange hadrons, and slower for multiply strange species. At small volumes, the factor grows about linearly with the reaction volume. Fig.7 refers to the top SPS energy, E/A=158 GeV corresponding to $\sqrt{s}$=17.3 GeV per participant nucleon pair. From this calculation we infer an enhancement factor of two for singly strange hadrons in the large coherent volume limit, and a much larger factor for s=2 and 3 hyperons. This study also shows that the canonical suppression - or grand canonical enhancement factor depends, in more detail, on $\sqrt{s}$: it falls down with $\sqrt{s}$ increasing from SPS to RHIC energies. As we have seen this analysis captures the observations made at top SPS energy [30, 35]. However it also holds at lower SPS energy [41] and further downward to AGS [47] and even SIS [48] energy. Strangeness enhancement (GC behaviour) is, thus, not indicating deconfinement, as proposed by its pioneers [16]. There must be a more general property of the maximum density fireball stage causing GC behaviour: probably large size and coherence. On the other hand let us avoid a misunderstanding: a QCD quark-gluon plasma represents one such case of large size and quantum coherence, and if it decouples directly to a hadron gas one expects to observe GC order [12, 33]. But there ought to be other forms of fireball matter that decouple similarly.

5.1 Fireball Size and Coherence

In order to pin down the origin of GC freeze-out let us first look at the size dependence. At top SPS energy we have data for central C+C, Si+Si, S+S and Pb+Pb collisions from NA35 and NA49 [15,17,51]. Fig.9 shows compilations of these data [51] concerning the average total π^+, K^+, K^- and Φ multiplicities of such central collisions (for Pb+Pb we include the top two centrality windows). The ratio of kaon and Φ multiplicities to those observed for pions is plotted against the projectile nucleon participant number. We see that all these ratios shoot up steeply from the (canonical) pp values that are included for reference [18]. The overall trend resembles the Tounsi-Redlich model [46] illustrated in Fig 7. We see that S+S already exhausts the GC limit to about 80%, with $< N_{part} >\approx 55$. In general the model predicts a transition from C to GC order that is yet steeper than the data. However we have to be precise here. Firstly, the model is run with constant temperature and baryochemical potential - perhaps not exactly true in the data. Second, $< N_{part} >$ has a different meaning in the data and the model: the statistical model has in reality no such $< N_{part} >$ parameter whatsoever, but only a reaction volume in which to enforce either local quantum number conservation or conservation *on the average only* (from the C to the GC situation). I.e. the authors of Fig.7 [46] fit a coherence volume V, starting from pp analysis where they find $V \approx 7fm^3$ in a situation with, undoubtedly, two participants. With this normalization they merely relabel their volume by $N_{part} = 2V/7fm^3$. Comparison with AA data is thus not straight forward. We may read off Fig.7 that for s=1 the grand canonical limit is reached for "N_{part}" ≥ 20 which thus really means $V \geq 70fm^3$. Not unreasonable: the rms radius of ^{32}S is about $3fm$, thus its area $19fm^2$. The rapidity distribution of K^+ has a FWHM of about 2.5 units [40]. Employing the Bjorken-equivalence [10] which equates the unit rapidity width with a spatial cylinder length of $1fm$ we estimate the primordial source volume of a central S+S collision to amount to about $50fm^3$.

As a last, important step in our attempt to understand the onset and origin of grand canonical behaviour we shall demonstrate now that it is **neither** the participant number **nor** the mere volume of the fireball by itself that matters (recall Fig.2). To this end we show in Fig.10 (left column) the data of Fig.9 confronted with the results of minimum bias PbPb collisions at similar energy. They exhibit a strikingly different, smooth ascend with N_{part} which thus can not be the right scaling with variable.

We are now prepared to formulate the essential hypothesis that we have repeatedly hinted at in sections 2-4. We propose that the crucial GC coherence effect resides in the space-time density of "successive" collisions at the microscopic level, during fireball formation time. This dynamical quantity is, of

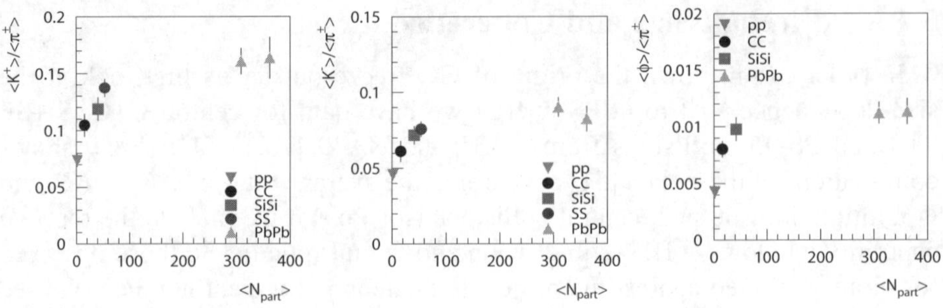

Figure 9. The ratios of K^+, K^- and Φ to positive pion production in full phase space, plotted vs. participant nucleon number, for central C+C, Si+Si, S+S and Pb+Pb collisions at 158 GeV per nucleon [15,17,54]. The values observed for $p+p$ collisions [18] are given for reference.

course, not an ingredient of the statistical model description of hadronization. But it might be the property that justifies a grand canonical description of the system right after hadronization. In order to obtain an estimate of the space-time density of "successive" collisions in the early fireball, we have to consult a model of the microscopic dynamics. The calculation employs the Frankfurt UrQMD model [52].

These considerations are illustrated [51] in the right column of Fig.10. The data of Fig.10 (left column) are replotted here as a function of the average density of successive inelastic collisions that would characterize each of the collisions systems if a quasi classical cascade picture would be applicable. Important: we shall conclude shortly that this picture of sequentiality should *not* be applicable but we consider the implicit transport ansatz nevertheless to be reliable insofar as it captures the consequences of initial impact geometry and density distributions in a symbolic quantity "inelastic collisions per cubic fermi and unit time fm/c", in the emerging fireball. Fig.10 shows a smooth increase of the strangeness enhancement quantities, with saturation occuring in central Pb+Pb. All colliding systems line up smoothly, and we see that saturation, i.e. the GC limit, is reached at $<\varrho>\geq 6$.

Now let us, finally, turn the argument around. A system with 6 separate inelastic collisions concurrent, on average, in each of its space-time unit cells is a strange idea! The uncertainty relation tells us that the energy uncertainty at each "successive" step would exceed 1.2 GeV here. I.e. the average density of energy uncertainty would exceed the critical QCD energy density. Such a system can not possibly be described as a sequence of hadron collisions. Its decay must occur under global quantum mechanical coherence, from interfering

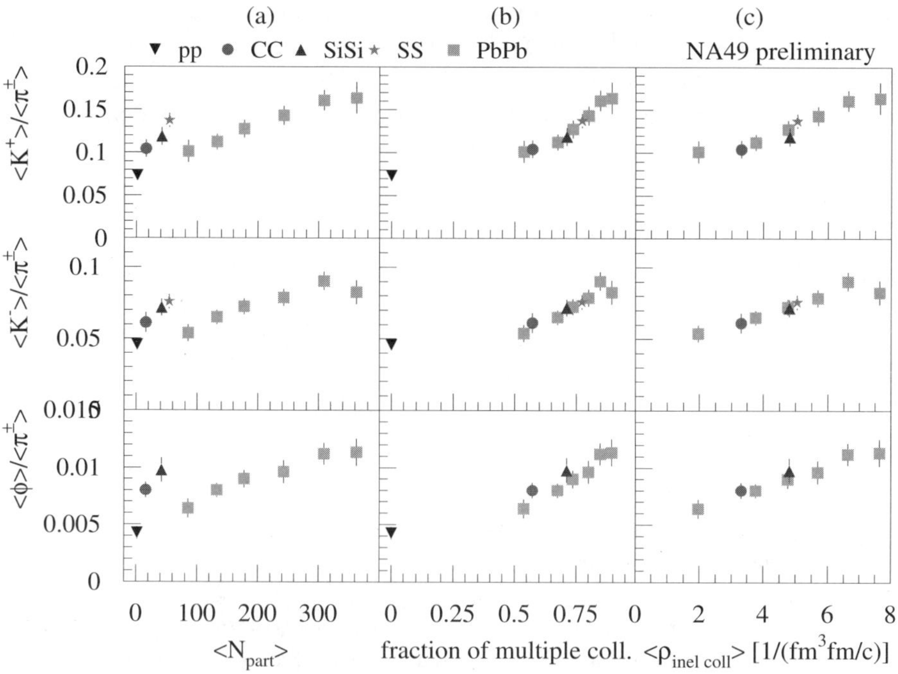

Figure 10. Minimum bias Pb+Pb data confronted with the data shown in Fig.9 (left column). Right: all data plotted versus collision density, ref. 51. Middle: the fraction of nucleons in each collision system undergoing more than one inelastic collision.

local subprocesses (for which we have no microscopic model as of yet). This mechanism should be the origin of the apparent grand canonical coherence. The cascade-transport model itself thus indicates the point at which it becomes invalid (indeed it does not predict the GC multiplicity pattern). The new physics seems to set in with C+C already, and is well established from S+S onward (Fig.9) where the hot core of the fireball may have a volume of about 50 fm^3. Not terribly far from the prediction [46] of Fig.7 for strangeness one.

It would be extremely interesting to check also the "onset curves" for s=2,3 with hyperon data for lighter collision systems. Meanwhile we conclude that the occurence of a global chemical potential in the grand canonical description reflects quantum mechanical coherence over a significant volume at the stage of hadronization decay of the fireball. It is important to repeat that in this view strangeness enhancement is not, by itself, an unambiguous "smoking gun" signal of deconfinement and quark gluon plasma formation. High density hadronic matter (whatever that might be) should also exhibit grand canonical hadron yields. In both cases, however, strangeness enhancement signals the production of hitherto unknown QCD states of matter. This should even apply to the relatively modest effects observed in pA collisions [26–28] that we mentioned in sect. 2. The "tube" of target matter struck by the incident nucleon should also be a coherent (albeit small) system.

References

[1] A. Ukawa, Nucl. Phys. A638 (1998) 339c; E. Laerman, Nucl. Phys. A610 (1996) 1c.

[2] T. Alber et al., NA49 Coll., Phys. Rev. Lett. 75 (1995) 3814.

[3] L. Ramello et al., NA50 Coll., Nucl. Phys. A638 (1998) 26c.

[4] T. Matsui and H. Satz, Phys. Lett. B178 (1986) 416; D. Kharzeev, Nucl. Phys. A638 (1998) 279c.

[5] S. A. Bass, J. Phys. G28 (2002) 1543.

[6] F. Karsch, Nucl. Phys. A698 (2002) 199.

[7] M. Alford, K. Rajagopal and F. Wilczek, Phys.Lett. B422 (1998) 247.

[8] A. Mischke et al., NA49, nucl-ex/0201012.

[9] In order to avoid a misunderstanding I have to add that this picture of a nearly stopped fireball applies only up to SPS energy. At the ten times higher RHIC energy, $\sqrt{s}$=200 GeV per nucleon pair, final phase space already looks like a fire-saussage in beam direction. However a more involved study shows that hydrodynamic equilibrium persists in each slice.

[10] J. D. Bjorken, Phys. Rev. D27 (1983)140.

[11] M. M Aggarwal et al., WA98, Phys. Rev. Lett. 85 (2000) 3595; G. Agakichiev et al., NA45, Phys. Lett. B422 (1998) 405.

[12] J. Ellis and K. Geiger, Phys. Rev. D54 (1996) 1967.

[13] V. Koch, B. Müller and J. Rafelski, Phys. Rev. 142 (1986) 167; F. Becattini, Z. Phys. C69 (1996) 485; J. Letessier, J. Rafelski and A. Tounsi, Nucl. Phys. A590 (1995) 613; R. Stock, Nucl. Phys. A661 (1999) 282; P. Braun-Munzinger, P. Heppe anf J. Stachel, Phys. Lett. B465 (1999) 15; J. Letessier and J. Rafelski, Hadrons and Quark-Gluon Plasma, Cambridge Univ. Press 2002.

[14] R. Hagedorn, Nuovo Cimento 35 (1965) 395.

[15] J. Bartke et al., NA35, Z. Phys. C48 (1990) 191; P. Foka et al., NA35, Nucl. Phys. A583 (1995) 687; T. Alber et al., NA35, Z. Phys. C64 (1994) 195.

[16] B. Müller and J. Rafelski, Phys. Rev. Lett. 48 (1982) 1066;
R. Hagedorn and J. Rafelski, Phys. Lett. 97B (1980) 180.

[17] S. V. Afanasiev et al., NA49, Nucl. Phys. A698 (2002) 104.

[18] C. Höhne et al., NA49, Nucl. Phys. A661 (1999) 485; J. Bächler et al., NA49, ibidem p 45.

[19] F. Sikler, hep-ex/0102004.

[20] F. Friese et al., NA49, Nucl. Phys. A698 (2002) 487.

[21] L. Ahle et al., E802, Phys. Rev.C60 (1999) 044904.

[22] S. V. Afanasiev et al., NA49, Nucl. Phys. A715 (2003) 161.

[23] F. Antinori et al., WA97/NA57, Nucl. Phys. A698(2002) 118.

[24] C. Höhne (NA49), PhD thesis, Marburg University 2003.

[25] T. Alexopoulos et al., E735 Coll., Nucl. Phys. A498 (1989)181

[26] B. Cole, Nucl. Phys. A661 (1999) 366.

[27] T. Susa et al., NA49, Nucl. Phys. A698 (2002) 491.

[28] H. G. Fischer, Nucl. Phys. A715 (2003) 118.

[29] T. Susa et al., NA49, to be published; G. I. Veres et al., NA49, Nucl. Phys. A661 (1999) 383; D. Barna PhD thesis, Budapest 2002.

[30] F. Becattini, J. of Physics G28 (2002) 1553.

[31] F. Becattini, M. Gazdzicki and J. Sollfrank, Nucl. Phys. A638 (1998) 403; J. Cleymans and K. Redlich, Phys. Rev. Lett. 81 (1998) 5284.

[32] R. Hagedorn, Nucl. Phys. B24 (1979) 93.

[33] R. Stock, Phys. Lett. B456 (1999) 277; U. Heinz, Nucl. Phys. A661 (1999) 140.

[34] F. Becattini and L. Ferroni, hep-ph/030706.

[35] P. Braun-Munzinger, I. Heppe and J. Stachel, Phys. Lett. B465 (1999) 15.

[36] L. S. Barnby et al., STAR, J. of Physics G28 (2002) 1535.

[37] P. Braun-Munzinger, D. Magestro and J. Stachel, ibidem p. 1745, and Phys. Lett. B518 (2001) 41.

[38] N. Xu and M. Kaneta, Nucl. Phys. A 698 (2002) 306.

[39] P. Seyboth, Proc. ICPAQGP, Jaipur 2001, to be published by Ind. Acad. Science; T. Kollegger, NA49, J. of Physics G28 (2002) 1689; nucl-ex/0201019.

[40] S. V. Afanasiev et al., NA49 Coll., Phys. Rev. C66 (2002) 054902.

[41] F. Becattini et al., hep-ph/0310049.

[42] Z. Fodor and S. Katz, hep-lat/0106002.

[43] R. Stock, Nucl. Phys. A661 (1999) 282.

[44] K. Geiger and D. Srivastava, Phys. Rev. C56 (1997) 2718.

[45] J. S. Hamieh, K. Redlich and A. Tounsi, Phys. Lett. B486 (2000) 61; J. Cleymans et al., Phys. Rev. C56 (1997) 2747.

[46] A. Tounsi and K. Redlich, hep-ph/0111159, and J. of Physics G28 (2002) 2095.

[47] J. Stachel, Nucl. Phys. A610 (1996) 509.

[48] H. W. Bartz et al., Phys. Rev. D40 (1989) 157.

[49] H. Appelshäuser et al., NA49, Eur. J. Phys. C2 (1998) 661.

[50] A. K. Wroblewski, Acta Phys. Pol. B16 (1985) 379.

[51] C. Höhne et al., NA49 Coll., Nucl. Phys. A715 (2003) 474.

[52] UrQMD version 3.1, H. Weber, Thesis Frankfurt 2002.

MULTISTRANGE BARYONS AT STAR

Jerôme Baudot
Institut de Recherche Subatomique (IN2P3/CNRS - ULP)
23 rue du Loess, F-67037 Strasbourg, France
baudot@in2p3.fr

for the STAR Collaboration

Abstract

We discuss in this contribution the role played by multistrange baryons, Ξ^- and Ω^-, in studying heavy ion collisions at RHIC. The STAR experiment has obtained signals for these particles already in the three collisions system provided by the accelerator. The yields of multistrange baryons fit into a picture where all hadrons are produced from an equilibrated source. Also the transverse momentum spectra of Ξ^- and Ω^- reveals a dynamic different from other hadrons which suggest that they acquire their dynamic properties from a pre-hadronic phase. Finally we conclude on the recent measurement of the elliptic flow for Ξ^- which supports the existence of an equilibrated phase at the earliest time of the collision.

Introduction

The interest for studying strange particles in high energy heavy ion collisions, in the view of evidencing a new state of matter [1] (the Quark Gluon Plasma, QGP), relies on the observation that all strange quarks are newly created during the collision. Hence strangeness production is sensitive to the nuclear medium, possibly a new state of matter, obtained through the collision. Regarding hadrons, the higher the strangeness content, the more sensitive will be the properties of the hadron to the medium. That's why we will concentrate in this contribution on multistrange baryons, that is to say the Ξ^- and the Ω^- as well as their anti-particles which carry respectively two and three strange quarks.

We will first see why the STAR (Solenoid Tracker At RHIC) experiment at the RHIC (Relativistic Heavy Ion Collider) is so well suited to conduct these studies at the highest energy available today for heavy-ion collisions.

W. Greiner et al. (eds.), Structure and Dynamics of Elementary Matter, 45–52.

A discussion on the yields measured will be presented in the light of a long announced prediction that multistrange baryon production should be enhanced if a QGP is formed. We will then have a closer look into their production mechanism using momentum dependant observables and unravelled new hope to probe the dense matter created at RHIC.

1. A powerfull tool for strangeness

Though their production level is very low (at most a few particles per event) experimentalists can use a specific property of multistrange baryons to actually "see" them into the stayhack of the numerous particles produced in the collisions: they are weakly decaying. Specifically they undergo a chain of desintegration, $\Xi^- \rightarrow \Lambda^0\,\pi^-$, $\Omega^- \rightarrow \Lambda^0\,K^-$ followed by $\Lambda^0 \rightarrow K^+\,\pi^-$, which displays two displaced vertex with respect to the primary vertex of the collision. The corresponding $c\tau$ of these decays range from 3 to 8 cm allowing an unambiguous identification once the full chain has been reconstructed. One has to note that this method is not momentum limited as the usual particle identification using the linear energy loss dE/dx. Nevertheless the apparatus is required to have dedicated characterisitics to allow such a reconstruction efficiently: a large acceptance because of the openeing angle between the three daughter particles, a very efficient tracking, vertexing abilities and finally particle identification to distinguish the different flavors of the daughter particles.

The STAR detector [2] provides all these features thanks its large volume Time Projection Chamber which tracks particles in the full azimuthal plane over two units of rapidity. Since year 2002, three layers of silicon drift detectors have been installed to allow a better secondary vertex determination and finding. This Silicon Vertex Tracker has been upgraded with one more layer made of silicon strip detectors which will be completed in 2004.

The RHIC [2] started its operation in 2000 and has provided three run periods so far with following beam-energy (center of mass of the nucleon-nucleon system) specifications: Au+Au @ 130 GeV, Au+Au @ 200 GeV, p+p @ 200 GeV and d+Au @ 200 GeV. Clear signals of Ξ^- and Ω^- (and anti-particles) have been obtained in all four types of collisions allowing a detailed and systematic study of the production mechanism, especially comparing systems where a quark gluon plasma is or not expected.

However, measurements of multistrange baryons for p+p and d+Au are still to be finalized and will not be presented here.

2. Yields

As early as 1982 [4], it was conjectured that different production mechanism for strangeness in a pure hadronic system and in a matter with quasi-free roaming quarks and gluons will lead to a significant enhancement of multistrange baryon yields, even more pronounced for anti-baryons and with increasing strangeness content. This effect has been actually discovered at SPS 18 years later by the WA97 and NA57 collaborations [5] when comparing yields observed in A+A collisions to the ones obtained in p+p or p+A collisions. But new interpretations [8] explain this enhancement without the QGP hypothesis. The situation is not known at RHIC because STAR p+p and d+Au results for strangeness yields are not yet available. One can simply consider the centrality dependence of the Ξs yields in Au+Au collisions, see [9], where no stunning behaviour is exhibited but rather an almost linear dependance with centrality.

Nevertheless the question arises wether it is possible to understand qualitatively the multistrange baryon production by quantitatively reproducing their yields with a phenomenological model. So far, the most succesfull theory-experiment matching has been obtained by statistical models, see J. Harris lecture [3]. Provided in slightly different implementations these models aim at reproducing the many ratio of yields for like and unlike particles both at SPS and RHIC. They use a reduced set of parameters the main ones being the chemical freeze-out temperature T_{ch} and the baryonic chemical potential μ_B (or directly the light quarks chemical potential μ_q). But they assume hadrons are produced from an equilibrated hadronic system. All the fits to the A+A collisions [6] point out to the same conclusions: all the ratio are well reproduced including those involving multistrange (anti-)baryons and $T_{ch} \sim 170 GeV$. We conclude from this result that the hadrons outgoing from the collision come from a system which can be described as thermalized at a temperature equivalent to the one lattice QCD predicts [7] to be the limit for deconfined matter.

We should now turn to additional information to learn more about the production mechanism.

3. Spectra dependant oservables

Instead of averaging the yields over the momentum one can consider their transverse momentum, $p_\perp$, distribution or spectra which is characterized by their mean transverse momentum $< p_\perp >$. This can be obtained at different centralities and for different particle masses as pictured on figure 1. The $< p_\perp >$ increase both with mass and centrality indicates a high level of rescattering among particles which leads to a transverse flow velocity pushing them outward. The Ξ and the Ω spectra show also the influence of such a rescattering, see [9]. But they seem to depart from a simple linear relation between the mass and the $< p_\perp >$ which lighter particles fullfill. It has been proposed [11] that this behaviour may be due to a lower hadronic cross-section of the multistrange baryons explaining a lower level of re-interaction in the hadronic phase.

Another qualitative information comes from the so-called nuclear modification factors, R_{CP} which is the ratio of the yields, normalized by the number of

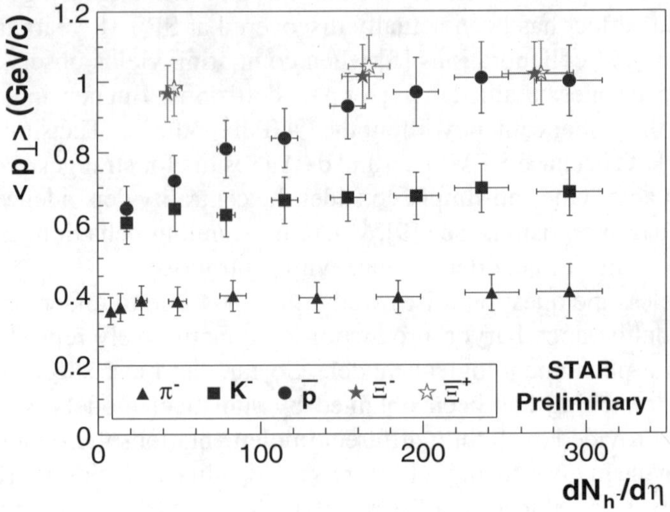

Figure 1. Evolution of the mean transverse momentum, $< p_\perp >$, of different particles with respect to the centrality of the collision given by the negatively charged multiplicity per unit of pseudo-rapidity $dN/d\eta$, as measured by STAR [10] at $\sqrt{S_{NN}} = 130$ GeV. Large $dN/d\eta$ corresponding to central collision and small ones to peripheral collisions.

binary collisions N_{bin}, for a given particle at a given $p_\perp$ between two different centralities. STAR has measured [12] this ratio for the main particles and for the multistrange baryons, see figure 2. At high $p_\perp$, if the yields were scaling with the number of binary collisions, it should be 1. However it has been predicted [13], that high $p_\perp$ particles should be shifted to lower $p_\perp$ due to jet-quenching if particles cross a dense partonic medium, see J. Harris and L. McLerran lectures [3] for details. Grossly, this suppression between central and peripheral collisions is actually observed but its detail pattern for $p_\perp$ around 2-4 GeV/c reveals differences between mesons and baryons. The high statistics sample of K^0 and Λ clearly show that baryons reach their maximum later in $p_\perp$ and at a higher value. Then they fall down also joining the mesons at high $p_\perp$. The Ξ seems to follow this trend to. Several authors [14] have suggested that this may reveal the nature of the production mechanism of hadrons. Indeed, if one assumes that hadrons are mainly created through the recombination out of a partonic medium and that their $p_\perp$ spectrum is exponential in this regime, such difference between mesons and baryons are reproducable.

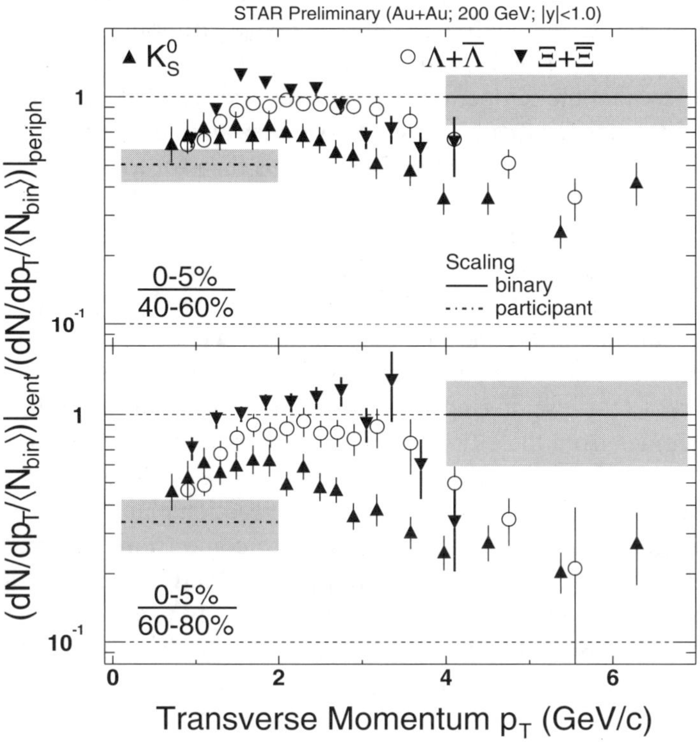

Figure 2. Distribution of the nuclear modification factor R_{CP} with $p_\perp$ for different particles. Spectra from the $0-5\%$ of the most central collisions are compared to mid-peripheral $(40-60\%)$ and peripheral $(60-80\%)$ collisions, as measured by STAR [12].

4. Two crucial tests

Before going further, let us come back on what we have learned so far from multistrange baryons production. First the yields have been shown to fit in a statistical production model which essential assumption is a thermalized source. Then spectra dependant observables $< p_\perp >$ teached us both that a large degree of collectivity is reached in the collision, which we have called transverse flow. Furthemore there is some indication that hadrons are produced from a partonic medium. We have assembled here two essential characteristics of a partonic (quarks and gluons) matter (equilibrated medium). Nevertheless we lack an ingredient that would tell us that the equilibrium is reached during the partonic phase. Multistrange baryons are again helping us to build the timescale for the collision scenario with two important measurements.

The first point brings us back to the previous discussion where it was suggested that Ξs and the Ωs may freeze-out kinetically earlier than other hadrons. Fitting the $p_\perp$ spectra of a given particle with a model inspired by hydrodynamic [15] allows to extract the temperature T_k and the mean radial boost $<\beta_\perp>$ at the kinetic freeze-out for this particle. STAR measurements [9] demonstrate that the temperature T_k is higher for Ξ (189$\pm$29) than for a common fit to π, K, p and Λ whereas the boost $<\beta_\perp>$ is lower (0.42$\pm$0.006 c) though still significant. This trend matches again the suggestion that multistrange baryons decouple from the hadronic medium quite early. But since they still show a significant radial flow this means that their flow takes mainly or partly its origin from the pre-hadronic phase *i.e.* the partonic one, which in turn implies that rescattering is occuring in the partonic phase.

Another proof that rescattering effects are large in the early (probably partonic) phase comes from the elliptic flow measurement represented in figure 3 obtained by STAR [16]. The elliptic flow parameters v_2 reflects the anisotropy of the azymuthal distribution with respect to the reaction plane for a given particle. The origine of this effect is understood to come from a conversion of the space anisotropy of the initial collision zone in mid-central events into a momentum-space anisotropy by rescatering in the early phase of the collision [17]. One actually observes from figure 3 that the Ξ elliptic flow is of the same magnitude than for other hadrons and matches hydrodynamic predictions at low $p_\perp$. This late fact strongly points out that a very high level of equilibrium is reached at the beginning of the collision (the first few fm/c). In addition it is known that such a magnitude of elliptic flow is hardly obtained in a hadronic system.

It is fruitfull to view the elliptic flow measurement in the light of the baryons-mesons difference observed in R_{CP}. Indeed from figure 3 one may infer that Λ and Ξ v_2 departs from K_s one for $p_\perp$ around 2-4 GeV/c. Under the assumption that hadrons are dominantly produced from the coalescence of partons in this regime as considered in [14], it is predicted that the distribution of $v_2(p_\perp/n)/n$, where $p_\perp$ is the transverse momentum of a hadron containing n valence quarks, is the same for any kind of hadrons. And this universality is observed in [16] indicating that the elliptic flow probably mainly builds up in a partonic phase.

A tentative conclusion

The STAR experiment is ideally suited to characterise the production of multistrange baryons at the midrapidity of the colliding system. With just a few years of data taking results on Ξ^- and Ω^- already provide important information on the matter produced by ultra-relativistic heavy ion collisions.

The enhancement factor for multistrange yields is not yet determined at RHIC. Nevertheless, the dynamic of the Ξ^- production provides in itself valu-

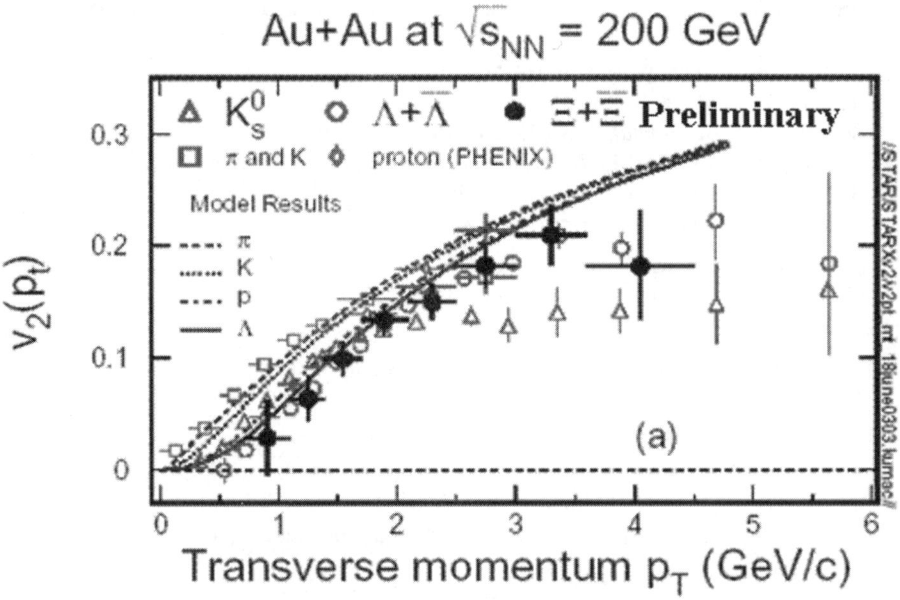

Figure 3. Distribution of the elliptic flow v_2 with $p_\perp$ for diffrent particles for mid-peripheral collisions as measured by STAR [16]. Lines are predictions from hydrodynamical models.

able insight into the collision scenario. It suggests that this baryon do not acquire its flow mainly from hadronic re-interactions in the final phase but rather from an earlier phase which is probably a partonic phase. Moreover it is inferred from the elliptic flow measurement that a high level of equilibirum is reached in this partonic phase.

Such interpretation clearly asks to be confirmed with more data taking which will allow to measure the Ω^- flow dynamic as well. But in any case they indicate that multistrange baryons constitute a very valuable probe of the deconfined phase which is probably produced in the collisions available at RHIC.

References

[1] H. Satz, Nucl. Phys. **A715**(2003)3-19c

[2] RHIC and STAR are described in a dedicated volume: Nucl. Instr. Meth. **A499**(2003)

[3] Other contributions in these proceedings

[4] B. Müller, J.Rafelski, Phys.Rev.Lett. **B178**(1982)1066

[5] F. Antinori *et al.*, Eur. Phys. J. **C11**(1999)79-88
and for the newest data L. Ŝándor for the NA57 collaboration, in Proceedings fo the 7th

International conference on Strangeness in Quark Matter, Atlantic Beach, USA, March 12-17, 2003

[6] D. Zschiesche *et al.*, Phys. Lett. **B547**(2002)7-14,
F. Becattini *et al.*, Phys. Rev. **C64(2001)**024901,
P. Braun-Munzinger *et al.*, Phys. Lett. **B518**(2001)41-46,
J. Letessier and J. Rafelski, Int. J. Mod. Phys. **E9**(2000)107-147 and
P. Braun-Munzinger *et al.*, Phys. Lett. **B465**(1999)15-20

[7] F. Karsch *et al.*, Phys. Lett. **B571**(2003)67-74

[8] A. Tounsi *et al.*, Nucl. Phys. **A715**(2003)565-568,
S. Yacoob, J. Cleymans, Preprint hep-ph/0208246

[9] J. Adams et al., Preprint nucl-ex/0307024

[10] J. Castillo for the STAR Collaboration, in Proceedings fo the 7th International conference on Strangeness in Quark Matter, Atlantic Beach, USA, March 12-17, 2003

[11] H. van Hecke *et al.*, Phys. Rev. Lett. **81**(1998)5764-5767

[12] P.Sorensen for the STAR collaboration, in Proceedings fo the 7th International conference on Strangeness in Quark Matter, Atlantic Beach, USA, March 12-17, 2003

[13] X.N. Wang and M. Gyulassy, Phys. Rev. Lett. **86**(1992)1480-1483
and X.N. Wang, preprint nucl-th/0307036

[14] R.J. Fries *et al.*, Phys. Rev. **C68**(2003)044902
and contribution of S. Bass in these proceedings

[15] E.Schneedermann *et al.*, Phys. Rec. **C48**(1993)2462-2475

[16] J. Castillo for the STAR Collaboration, presented at the conference Topics in Heavy Ion Collisions, Montreal, Canada, June 25-28, 2003

[17] J.Y. Ollitrault, Phys. Rev. **D46**(1992)229-245
and P.F. Kolb *et al.*, Phys. Rev. **C62**(2000)054909

RECENT RESULTS FROM THE NA49 EXPERIMENT

C. Blume
for the NA49 collaboration
Institut für Kernphysik der J. W. Goethe Universität, Frankfurt am Main, Germany

T. Anticic[21], B. Baatar[8],D. Barna[4], J. Bartke[6], M. Behler[13], L. Betev[9], H. Bialkowska[19], A. Billmeier[9], C. Blume[7], B. Boimska[19], M. Botje[1], J. Bracinik[3], R. Bramm[9], R. Brun[10], P. Buncic[9,10], V. Cerny[3], P. Christakoglou[2], O. Chvala[15], J.G. Cramer[17], P. Csato[4], N. Darmenov[18], A. Dimitrov[18], P. Dinkelaker[9], V. Eckardt[14], P. Filip[14], H.G. Fischer[10], D. Flierl[9], Z. Fodor[4], P. Foka[7], P. Freund[14], V. Friese[7], J. Gal[4], M. Gazdzicki[9], G. Georgopoulos[2], E. Gladysz[6], S. Hegyi[4], C. Höhne[13], K. Kadija[21], A. Karev[14], V.I. Kolesnikov[8], T. Kollegger[9], E. Kornas[6], R. Korus[12], M. Kowalski[6], I. Kraus[7], M. Kreps[3], M. van Leeuwen[1], P. Levai[4], L. Litov[18], M. Makariev[18], A.I. Malakhov[8], C. Markert[7], M. Mateev[18], B.W. Mayes[11], G.L. Melkumov[8], C. Meurer[9], A. Mischke[7], M. Mitrovski[9], J. Molnar[4], St. Mrowczynski[12], G. Palla[4], A.D. Panagiotou[2], D. Panayotov[18], K. Perl[20], A. Petridis[2], M. Pikna[3], L. Pinsky[11], F. Pühlhofer[13], J.G. Reid[17], R. Renfordt[9], W. Retyk[20], C. Roland[5], G. Roland[5], M. Rybczynski[12], A. Rybicki[6,10], A. Sandoval[7], H. Sann[7], N. Schmitz[14], P. Seyboth[14], F. Siklér[4], B. Sitar[3], E. Skrzypczak[20], G. Stefanek[12], R. Stock[9], H. Ströbele[9], T. Susa[21], I. Szentpetery[4], J. Sziklai[4], T.A. Trainor[17], D. Varga[4], M. Vassiliou[2], G.I. Veres[4], G. Vesztergombi[4], D. Vranic[7], S. Wenig[10], A. Wetzler[9], Z. Wlodarczyk[12] I.K. Yoo[16], J. Zaranek[9], J. Zimanyi[4]

[1]*NIKHEF, Amsterdam, Netherlands.*

[2]*Department of Physics, University of Athens, Athens, Greece.*

[3]*Comenius University, Bratislava, Slovakia.*

[4]*KFKI Research Institute for Particle and Nuclear Physics, Budapest, Hungary.*

[5]*MIT, Cambridge, USA.*

[6]*Institute of Nuclear Physics, Cracow, Poland.*

[7]*Gesellschaft für Schwerionenforschung (GSI), Darmstadt, Germany.*

W. Greiner et al. (eds.), Structure and Dynamics of Elementary Matter, 53–60.

[8] *Joint Institute for Nuclear Research, Dubna, Russia.*

[9] *Fachbereich Physik der Universität, Frankfurt, Germany.*

[10] *CERN, Geneva, Switzerland.*

[11] *University of Houston, Houston, TX, USA.*

[12] *Institute of Physics Świetokrzyska Academy, Kielce, Poland.*

[13] *Fachbereich Physik der Universität, Marburg, Germany.*

[14] *Max-Planck-Institut für Physik, Munich, Germany.*

[15] *Institute of Particle and Nuclear Physics, Charles University, Prague, Czech Republic.*

[16] *Department of Physics, Pusan National University, Pusan, Republic of Korea.*

[17] *Nuclear Physics Laboratory, University of Washington, Seattle, WA, USA.*

[18] *Atomic Physics Department, Sofia University St. Kliment Ohridski, Sofia, Bulgaria.*

[19] *Institute for Nuclear Studies, Warsaw, Poland.*

[20] *Institute for Experimental Physics, University of Warsaw, Warsaw, Poland.*

[21] *Rudjer Boskovic Institute, Zagreb, Croatia.*

Abstract

Results from the NA49 experiment on the energy dependence of hadronic observables, including new preliminary data at 30 AGeV beam energy, are presented. The transverse mass spectra can generally be described by assuming a thermal source with transverse expansion at all energies. A clear step-like pattern is observed in the energy dependence of the inverse slope parameters for kaons. The K^+/π^+ ratio reveals a distinct maximum at around 30 AGeV. The deuteron coalescence parameter B_2, as well as the elliptic flow parameter v_2 vary smoothly as a function of the beam energy.

1. Introduction

In the recent years the NA49 experiment has collected data at beam energies between 20 to 158 AGeV with the objective to cover the critical region of energy densities where the expected phase transition to a deconfined phase might occur. In this contribution therefore the energy dependence of various observables that characterize the thermal and the chemical freeze-out of the reaction are presented. While the conditions at thermal freeze-out are reflected in the transverse mass spectra, HBT- as well as deuteron coalescence measurements, the particle yields are determined by the chemical freeze-out. NA49 is a fixed target experiment at the CERN SPS and consists of a large acceptance magnetic spectrometer equipped with four TPCs as tracking devices and a forward

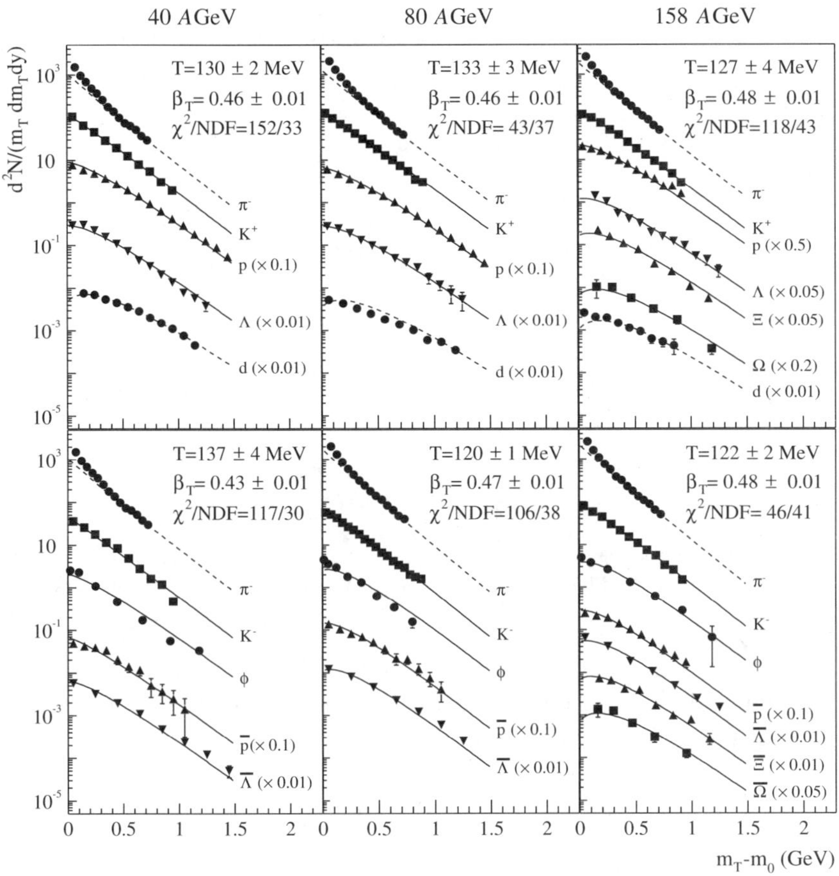

Figure 1. Transverse mass spectra in central Pb+Pb collisions at 40, 80, and 158 *A*GeV. The lines represent the result of a fit, including transverse expansion, as explained in the text.

calorimeter for centrality selection. Details on the experimental setup can be found in [1].

2. Thermal Freeze-Out

Figure 1 summarizes the transverse mass spectra, measured at three different beam energies. All spectra are measured in one unit around midrapidity, except for the ϕ, Ξ, and Ω which correspond to a larger rapidity interval. The paon and pion data are published for all three energies [2], the ϕ and Ξ for 158 *A*GeV [3, 4], while the remaining data is still preliminary. If the differences of the m_T-spectra for particles of different mass are only due to the presence

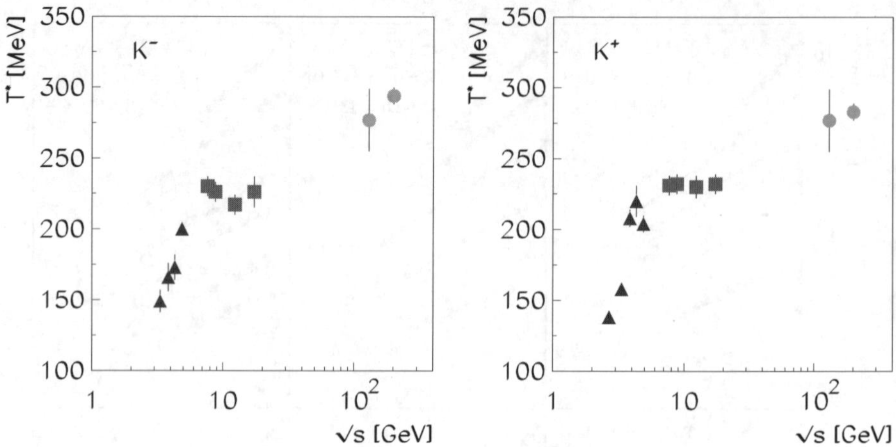

Figure 2. The inverse slope parameters T, derived from transverse mass spectra of K^- (left) and K^+ (right) in central Pb+Pb (Au+Au) collisions, as a function of the center-of-mass energy.

of transverse flow, a common description of all spectra might be possible by a hydro-dynamical model that takes the expansion into account. Here, we fit the spectra with the hydro-dynamic inspired formula [5]:

$$\frac{dN}{m_t dm_t dy} \propto m_t K_1 \left(\frac{m_t \cosh \rho}{T} \right) I_0 \left(\frac{p_t \sinh \rho}{T} \right) \tag{1}$$

The result of the fit is represented by the lines in Fig. 1. The simplified model provides a good fit to all particle species, including the rare Ξ and Ω and the resulting fit parameter for the thermal freeze-out temperature T and the transverse flow velocity β_T ($\rho = \tanh^{-1} \beta_T$) do not show any clear trend in the energy range studied here (pions and deuterons are excluded from the fit [6]).

As can be seen from Fig. 1 the spectra of most particles have a non-exponential shape, due to the expansion and, predominantly in the case of pions, the contribution of resonance decays. The kaon spectra, however, are an exception and can be fitted over the entire measured range by an exponential. Figure 2 displays the energy dependence of the resulting inverse slope parameter T in combination with data from the AGS and RHIC (for reference see [2]). Surprisingly, the kaon slopes exhibit a step like excitation function. While they are strongly increasing at AGS energies, there is no significant change in the SPS energy regime, followed by a slight increase to RHIC energies. This behaviour is very suggestive in terms of latent-heat-like temperature evolution and might point to a change of the equation of state [7].

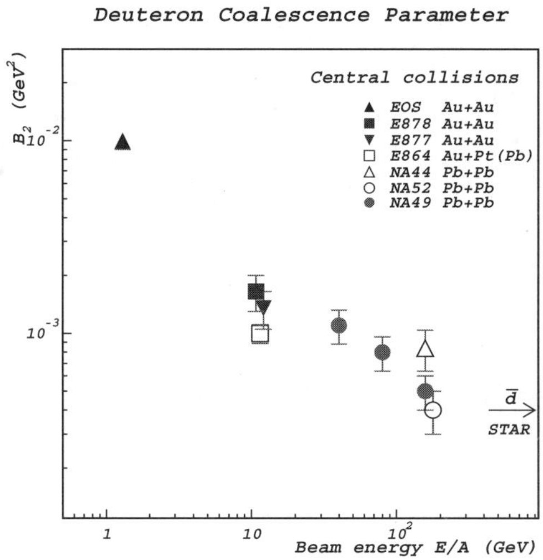

Figure 3. The deuteron coalescence factor B_2 as a function of the beam energy for central nucleus-nucleus reactions.

From the measurement of proton and deuteron spectra (all at mid-rapidity for central reactions[1]), a deuteron coalescence factor B_2 can be derived:

$$E_d \frac{d^3 N_d}{dp_d^3} = B_2 \left(E_p \frac{d^3 N_p}{dp_p^3} \right)^2 , \; p_d = 2p_p \tag{2}$$

Figure 3 displays the extracted B_2 for different beam energies, also including data from the AGS and other CERN SPS experiments. B_2 decreases significantly with increasing beam energy. Even in the SPS energy range, taken alone, there is a change by a factor of 2. This is in contrast to the weak energy dependence of the radii from Bose Einstein correlations. Similarly, also the centrality dependence of B_2 is much more pronounced that the one seen in the HBT-radii [8]. This fact suggests that deuteron coalescence is not directly reflecting the thermal freeze-out conditions in the same way that is seen in Bose-Einstein correlations.

3. Chemical Freeze-Out

The large acceptance of the NA49 spectrometer allows to measure particle spectra over a wide range of the longitudinal phasespace. This allows to determine the total yields with only small extrapolations. The Fig.4 shows the ratio

[1] 40 and 80 *A*GeV: 7% most central, 158 *A*GeV: 5% most central

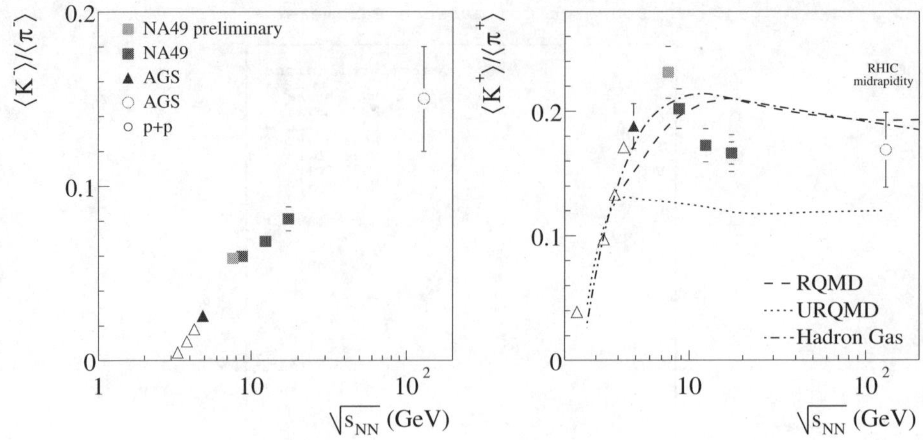

Figure 4. The energy dependence of the $\langle K \rangle / \langle \pi \rangle$ ratios.

of the 4π-yields of K and π. While this ratio for negatively charged particles more or less rises continuously (left hand side of Fig. 4) – except a small indication for a kink at 30 AGeV – a very distinct maximum is observed in the positively charged case (right hand side of Fig. 4). The lines included in the $\langle K^+ \rangle / \langle \pi^+ \rangle$-plot are current predictions of an extended hadron gas model [9] and the transport codes RQMD [10] and URQMD [11]. Even though the hadron gas model and RQMD predict a maximum of the $\langle K^+ \rangle / \langle \pi^+ \rangle$-ratio in the SPS energy range, none of the models can fully describe the sharp feature of its energy dependence. A feature, which is also not present in p+p collisions. On the other hand, a strong non-monotonic energy dependence of the total strangeness to pion ratio was predicted by the Statistical Model of The Early Stage [12], assuming a phase transition from confined matter to a quark-gluon plasma.

The anti-baryon / baryon ratio reflects the change of baryon density due to stopping. As can bee seen from Fig. 5 this ratio at midrapidity generally increases with the beam energy, slowly approaching 1 as expected for a baryon free region [13]. However, the strength of the energy dependence is strongly dependent on the strangeness content of the particle species. While it is very pronounced in the $\bar{p}/p$-case with $S = 0$, it is only relatively weak for the Ξ ($S = 2$) and even weaker for the Ω ($S = 3$).

4. Collective effects

Anisotropic flow allows the measurement of the pressure of the system in an early stage of the reaction. Therefore, it is supposed to be sensitive to changes in the equation of state. NA49 has studied the elliptic flow parameter v_2 in

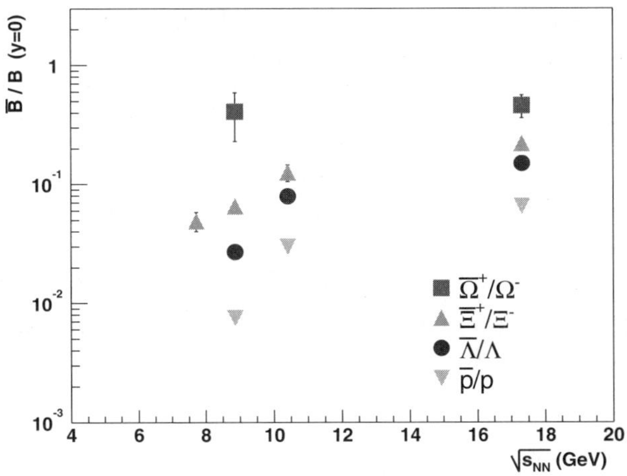

Figure 5. The anti-baryon / baryon ratio at midrapidity for particles with different strangeness content as a function of the center-of-mass energy.

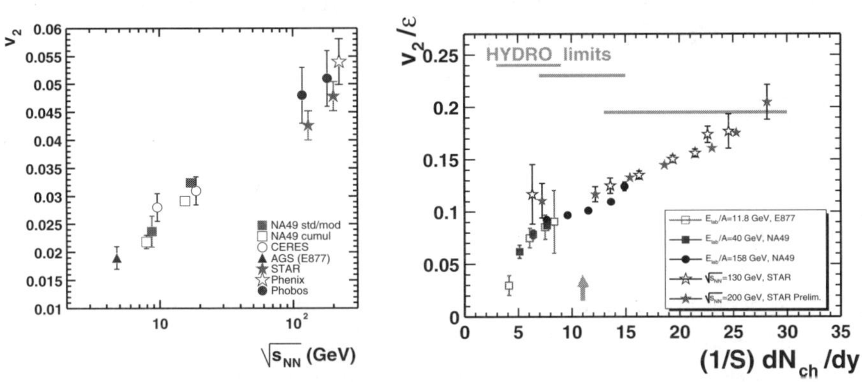

Figure 6. Left: The elliptic flow parameter v_2 for pions at midrapidity for mid-central collisions (approx. 12-34% of the geometrical cross section) as a function of the center-of-mass energy. Right: v_2/ϵ as a function of the particle density.

minimum bias Pb+Pb collisions at 40 and 158 *A*GeV [14]. The plot on the left hand side of Fig. 6 compares the results of these measurements to data of other experiments. A smooth increase of v_2 with beam energy can be observed, reflecting the decreasing effect of spectators and the increasing influence of thermalization. The results for SPS and RHIC energies can be directly compared by dividing v_2 by the initial excentricity of the overlap region ϵ and plotting it as a function of the particle densitity. As can be seen on the right hand side of Fig.6, all data points fall on a common curve, which is approaching the limes of an ideal hydrodynamic fluid at RHIC energies.

5. Summary

The recent study of the excitation functions of hadronic obeservables in the SPS energy range has revealed a number of interesting and unexpected results. This includes a step-like energy dependence of the inverse slope parameter of Kaons and a sharp maximum in the strangeness to pion ratio. This study will be completed in the near future by results from the currently analyzed 20 AGeV data and other observables, whose analysis is not yet completed.

Acknowledgments

Acknowledgements: This work was supported by the US Department of Energy Grant DE-FG03-97ER41020/A000, the Bundesministerium fur Bildung und Forschung, Germany, the Polish State Committee for Scientific Research (2 P03B 130 23, SPB/CERN/P-03/Dz 446/2002-2004, 2 P03B 02418, 2 P03B 04123), the Hungarian Scientific Research Foundation (T032648, T032293, T043514), the IIungarian National Sciencc Foundation, OTKΛ, (F034707), and the Polish-German Foundation.

References

[1] S. V. Afanasiev et al., Nucl. Instrum. Meth. A430 (1999) 210-244.

[2] S. V. Afanasiev et al., Phys. Rev. C66 (2002) 157-166.

[3] S. V. Afanasiev et al., Phys. Lett. B491 (2002) 59-66.

[4] S. V. Afanasiev et al., Phys. Lett. B538 (2002) 275-281.

[5] E. Schnedermann and U. Heinz, Phys. Rev. C50 (1994) 1675.

[6] M. van Leeuwen et al., Nucl. Phys. A715 (2003) 161c-170c.

[7] M. Gazdzicky, M. I. Gorenstein, and K Bugaev, Phys. Lett. B567 (2003) 175-178.

[8] C. Blume et al., Nucl. Phys. A715 (2003) 55c - 64c.

[9] P. Braun-Munzinger, K. Redlich, and H. Oeschler, Nucl. Phys. A697 (2002) 902-912.

[10] H. Sorge, H. Stöcker, and W. Greiner, Nucl. Phys. A498 (1989) 567.

[11] S. A. Bass et al., Prog. Part. Nucl. Phys. 41 (1998) 255-369.

[12] M. Gazdzicky and M. I. Gorenstein, Acta Phys. Polon. B30 (1999) 2705-2735.

[13] M. Mitrovski et al., arXiv:nucl-ex/0305029, to be published in J. Phys. G.

[14] C. Alt et al., Phys. Rev. C68 (2003) 034903.

QCD HARD SCATTERING RESULTS FROM PHENIX AT RHIC

David d'Enterria for the PHENIX Collaboration
Nevis Laboratories, Columbia University
Irvington, NY 10533, and New York, NY 10027, USA
denterria@nevis.columbia.edu

Abstract Data on hadron production at high transverse momentum ($p_T > 2$ GeV/c) in p+p, d+Au, and Au+Au collisions at $\sqrt{s_{NN}} = 200$ GeV from the PHENIX experiment at RHIC are reviewed. The single inclusive spectrum of light hadrons produced in central Au+Au reactions shows significant differences compared to p+p and d+Au collisions, and provides interesting information on the properties of the underlying QCD medium present in heavy-ion reactions at collider energies.

Keywords: Relativistic nucleus-nucleus collisions, QCD, hard scattering, PHENIX, RHIC

1. Introduction

The fundamental degrees of freedom of the theory of the strong interaction (Quantum Chromodynamics, QCD) are colored quarks (fermions) and gluons (gauge bosons). In the real world, however, quarks and gluons are not observed free but confined in color singlet states (hadrons). In addition, the observed effective *constituent* mass of the quarks ($\bar{m}_{u,d} \approx 300$ MeV for the lightest u and d quarks) is much larger than the bare *current* mass of the QCD Lagrangian ($m_{u,d} \approx 2 - 8$ MeV). Theoretically understanding from first principles these two basic properties of QCD: confinement and chiral symmetry breaking, remains in fact one of the most important problems in fundamental physics [1]. Lattice calculations of QCD in the bulk [2] predict that above an energy density of $\epsilon_{crit} \approx 0.7 \pm 0.3$ GeV/fm^3, the (long range part of the) strong color potential becomes screened and, as a result, quarks and gluons can move freely with their bare mass within the deconfined medium. Experimentally accessing and studying the properties of this "Quark-Gluon Plasma" (QGP) phase is one of the main goals of the physics program of the PHENIX experiment at the Relativistic Heavy-Ion Collider (RHIC) at BNL. By colliding heavy nuclei at high energies one expects to produce a thermalized system of quarks and gluons with energy densities above ϵ_{crit}, albeit for very short time scales, $\mathcal{O}(10^{-23}$

W. Greiner et al. (eds.), Structure and Dynamics of Elementary Matter, 61–69.

s) and small volumes, $\mathcal{O}(10^2\ \text{fm}^3)$. In the period 2000-2003, PHENIX collected data from Au+Au, d+Au and p+p collisions at $\sqrt{s_{NN}}$ = (130)200 GeV. The most interesting phenomena observed so far in the data are in the high p_T ($p_T \gtrsim 2$ GeV/c) sector where the production of hadrons in (non-peripheral) Au+Au collisions shows substantial differences compared to more elementary reactions either in free space (p+p, e^+e^-) or in a "cold" nuclear matter environment (d+Au). Hard scattering processes are an excellent experimental probe in heavy-ion collisions inasmuch as: (i) they are the result of violent (large Q^2) short-distance (parton-parton) collisions in the first instants of the reaction ($\tau \sim 1/p_T \lesssim 0.1$ fm/c) and, as such, direct probes of the **partonic** phase(s) of the reaction; (ii) in the absence of medium effects, their cross section in A+A reactions is expected to **scale simply** with that measured in **p+p** collisions times the number of scattering centers (A^2, see discussion below); and (iii) their production yields can be **theoretically calculable** in QCD via standard perturbative (collinear factorization plus Glauber multiscattering [3]) or via "classical-field" ("Color-Glass-Condensate" approach [4]) methods.

2. The PHENIX experiment

The PHENIX detector [5] is specifically designed to measure hard QCD probes such as high p_T hadrons, direct photon radiation, lepton pairs, and heavy flavour production. PHENIX achieves good mass and PID resolution, and small granularity by combining 13 detector subsystems (~350,000 channels) divided into: (i) 2 central arm spectrometers for electron, photon and hadron measurement at mid-rapidity ($|\eta| < 0.38$, $\Delta\phi = \pi/2$); (ii) 2 forward-backward ($|\eta|$ = 1.15 - 2.25, $\Delta\phi = 2\pi$) spectrometers for muon detection; and (iii) 4 global (inner) detectors for trigger and centrality selection. Neutral mesons are reconstructed through invariant mass analysis of decay γ pairs detected in two types of electromagnetic calorimeters (15552 lead scintillator towers with $18X_0$, and 9216 lead glass modules with $14.4X_0$). The trajectories and momenta of charged hadrons in the axial central magnetic field (B_{max} = 1.15 T m) are measured by a drift chamber (DC) and three layers of MWPC's with pad readout (PC). Hadron identification ($\pi^\pm$, $K^\pm$, and $p, \bar{p}$) is achieved by matching the reconstructed tracks to hits in a time-of-flight wall (TOF).

The occurrence of a p+p, d+Au or Au+Au collision (with vertex position $|z| < 60$ cm within the center of the detector) is triggered by a coincidence between the two Beam-Beam Counters (BBC)[1] which cover $|\eta|$ = 3.0 - 3.9. These minimum bias triggers accept respectively (52±4)%, (88±4)%, and (92±3)% of the total inelastic cross-sections ($\sigma_{pp}^{inel} \approx 42$ mb, $\sigma_{dAu}^{inel} \approx 2150$ mb, and $\sigma_{AuAu}^{inel} \approx 6850$ mb). Since hard probes are rare, high luminosities are required.

[1] Plus the Zero Degree Calorimeters, ZDC, at $|\theta| < 2$ mrad in the case of Au+Au.

PHENIX is an efficient high interaction rate experiment with a state-of-the-art data acquisition system capable of recording 120 MB/s to disk with event sizes of $\sim$100 KB, and event rates of $\sim$1-2 KHz. A total of $0.2{\cdot}10^9$, $5.5{\cdot}10^9$, and $4.0{\cdot}10^9$ events have been sampled in Au+Au, d+Au, and p+p collisions respectively at $\sqrt{s_{NN}}$ = 200 GeV.

3. Scaling of yields and cross-sections from p+p to A+B

Insights on the mechanisms of particle production (and "destruction") in nucleus-nucleus (A+B) collisions are obtained from the study of the scaling behavior of their yields with respect to p+p collisions. Depending on the p_T range in which they are produced, two different "scaling laws" are relevant:

• Soft particle production at low p_T ($p_T \lesssim 1$ GeV/c) in A+B reactions is dominated by non-perturbative nucleon-nucleon collisions with small momentum transfers. Since those collisions occur in "parallel" and with large probabilities, they suffer destructive interference which effectively limits the maximal number of processes that can lead to secondary particle production [6]. As a result soft particle yields turn out to be proportional to the average number of participating (or "wounded") nucleons, i.e. nucleons which undergo at least one inelastic collision [7]. Thus, in a A+B reaction at impact parameter b:

$$E\frac{dN^{soft}_{AB}(b)}{d^3p} = \langle N_{part}(b)\rangle \cdot E\frac{dN^{soft}_{pp}}{d^3p}, \tag{1}$$

where $N_{part}(b)$ can be obtained from the nuclear thickness functions $T_{A,B}(b) = \int dz\, \rho_{A,B}(z,b)$ (normalized to A, B resp.) via a Glauber model[2]. Such a "participant scaling" is indeed approximately observed in the total multiplicities measured in heavy-ion collisions [8]. Integrating $N_{part}(b)$ over impact parameter one obtains the average number of participant nucleons in a minimum bias A+B collision: $\langle N_{part}\rangle|_{MB} = (A\,\sigma^{inel}_{pB} + B\,\sigma^{inel}_{pA})/\sigma^{inel}_{AB}$. Using this expression, Eq. (1), and $dN^X_{AA} = d\sigma^X_{AA}/\sigma^{inel}_{AA}$, the soft invariant yields and cross-sections in a symmetric A+A minimum bias collision[3] are related to the corresponding p+p ones via:

$$\left.E\frac{dN^{soft}_{AA}}{d^3p}\right|_{MB} = \frac{2A\,\sigma^{inel}_{pA}}{\sigma^{inel}_{AA}} \cdot E\frac{dN^{soft}_{pp}}{d^3p}, \quad \text{and}$$

$$\left.E\frac{d\sigma^{soft}_{AA}}{d^3p}\right|_{MB} = \frac{2A\,\sigma^{inel}_{pA}}{\sigma^{inel}_{pp}} \cdot E\frac{d\sigma^{soft}_{pp}}{d^3p}. \tag{2}$$

[2]The corresponding formulas can be found e.g. in [7].

[3]Also, for min. bias A+A collisions: $\langle N_{part}\rangle|_{MB} = 2A\sigma^{inel}_{pA}/\sigma^{inel}_{AA} \approx 2A\pi R_A^2/\pi(2R_A)^2 \approx A/2$.

• Hard particle production at high p_T ($p_T \gtrsim 2$ GeV/c) results from incoherent parton-parton scatterings with large Q^2. In this regime, the pQCD "factorization theorem" [9] holds and the inelastic cross-section for the production of a given particle can be separated in the product of long-distance (non-perturbative parton distribution functions, $f_{a/A}$, and fragmentation functions, $D_{c/h}$) and short-distance (parton-parton scattering) contributions:

$$E\frac{d\sigma^{hard}_{AB\to h}}{d^3p} = f_{a/A}(x,Q^2) \otimes f_{b/B}(x,Q^2) \otimes \frac{d\sigma^{hard}_{ab\to c}}{d^3p} \otimes D_{c/h}(z,Q^2). \quad (3)$$

The assumption of incoherent scattering at high p_T entails also that $f_{a/A} = A \cdot f_{a/N}$, i.e. that the density of partons in a nucleus A should be equivalent to the superposition of A independent nucleons. Thus,

$$E\frac{d\sigma^{hard}_{AB\to h}}{d^3p} = A\,B\; \cdot f_{a/p}(x,Q^2) \otimes\; f_{b/p}(x,Q^2) \otimes E\frac{d\sigma^{hard}_{ab\to c}}{d^3p} \otimes D_{c/h}(z,Q^2), \quad (4)$$

and minimum bias hard cross-sections in A+B are expected to scale simply as

$$E\frac{d\sigma^{hard}_{AA}}{d^3p}\bigg|_{MB} = A\,B \cdot E\frac{d\sigma^{hard}_{pp}}{d^3p}. \quad (5)$$

In the general case, for a given A+B reaction with impact parameter b:

$$E\frac{dN^{hard}_{AB}(b)}{d^3p} = \langle T_{AB}(b)\rangle \cdot E\frac{d\sigma^{hard}_{pp}}{d^3p}, \quad (6)$$

where $T_{AB}(b) = \int d^2s\, T_A(s) T_B(b-s)$ (normalized to $A\,B$) is the nuclear overlap function[4] at b. Since T_{AB} is proportional to the number of nucleon-nucleon (NN) collisions: $T_{AB}(b) = N_{coll}(b)/\sigma^{inel}_{pp}$, one alternatively quotes Eq. (6) in the form of "(binary) collision scaling" of invariant yields:

$$E\frac{dN^{hard}_{AB}(b)}{d^3p} = \langle N_{coll}(b)\rangle \cdot E\frac{dN^{hard}_{pp}}{d^3p}. \quad (7)$$

From (5) and (7), it is easy to see that[5] $\langle N_{coll}\rangle|_{MB} = A\,B \cdot \sigma^{inel}_{pp}/\sigma^{inel}_{AB}$.

Following Eqs. (2) and (5), since $\sigma^{inel}_{pA,AA} \propto R_A^2 \propto A^{2/3}$, the atomic number dependence of soft processes in symmetric A+A reactions is of the type

$$E\frac{dN^{soft}_{AA}}{d^3p}\bigg|_{MB} \propto A \cdot E\frac{dN^{soft}_{pp}}{d^3p} \quad \text{and} \quad E\frac{d\sigma^{soft}_{AA}}{d^3p}\bigg|_{MB} \propto A^{5/3} \cdot E\frac{d\sigma^{soft}_{pp}}{d^3p}, \quad (8)$$

[4]The "natural" magnitudes of $T_A(b)$ and $T_{AB}(b)$ are $A/\pi R_A^2$ and $AB/\pi(R_A+R_B)^2$ respectively [10].
[5]Also, for min. bias A+A collisions: $\langle N_{coll}\rangle|_{MB} = A^2\sigma^{inel}_{pp}/\sigma^{inel}_{AA} \approx A^2\sigma^{inel}_{pp}/\pi(2R_A)^2 \approx A^{4/3}/4$.

whereas the A dependence of hard processes is of the form:

$$E\frac{dN_{AA}^{hard}}{d^3p}\bigg|_{MB} \propto A^{4/3}\cdot E\frac{dN_{pp}^{hard}}{d^3p} \quad \text{and} \quad E\frac{d\sigma_{AA}^{hard}}{d^3p}\bigg|_{MB} = A^2\cdot E\frac{d\sigma_{pp}^{hard}}{d^3p}. \tag{9}$$

4. High p_T production: Au+Au vs. p+p

One of the most interesting experimental results at RHIC so far is the *breakdown* of the expected incoherent parton scattering assumption for high p_T production in non-peripheral Au+Au collisions. Fig. 1 shows the comparison of the measured p+p π^0 spectrum [11] to peripheral (left) and central (right) Au+Au spectra [12], and to standard NLO pQCD calculations [13]. Whereas peripheral data is consistent with a simple superposition of individual NN collisions, central data shows a suppression factor of 4 – 5 with respect to this expectation.

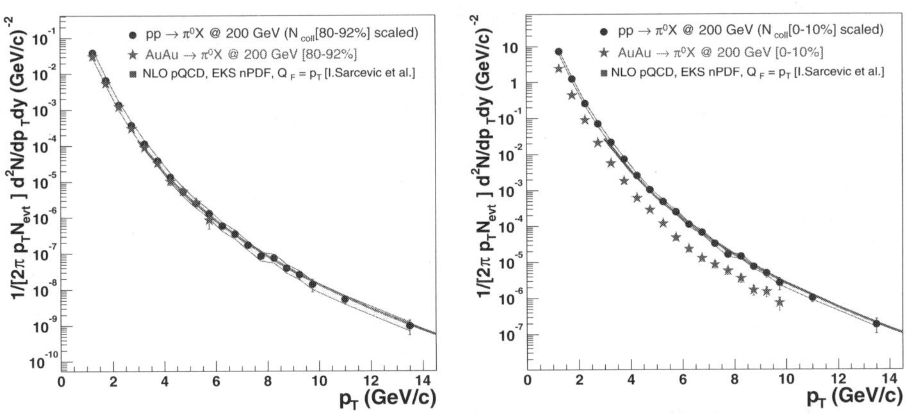

Figure 1. Invariant π^0 yields measured by PHENIX in peripheral (left) and in central (right) Au+Au collisions (stars) [12], compared to the T_{AB} scaled p+p π^0 cross-section (circles) [11] and to a NLO pQCD calculation (gray line) [13]. The yellow band around the scaled p+p points represents the overall normalization uncertainties.

From Eq. (6), it is customary to quantify the medium effects at high p_T using the *nuclear modification factor* given by the ratio of the A+A invariant yields to the p+p cross-sections scaled by the nuclear overlap at impact parameter b:

$$R_{AA}(p_T, b) = \frac{d^2N_{AA}^{\pi^0}(b)/dydp_T}{\langle T_{AB}(b)\rangle \cdot d^2\sigma_{pp}^{\pi^0}/dydp_T}. \tag{10}$$

$R_{AA}(p_T)$ measures the deviation of A+A from an incoherent superposition of NN collisions in terms of suppression (R_{AA} <1) or enhancement (R_{AA} >1).

Figure 2 shows R_{AA} as a function of p_T for several π^0 measurements in high-energy A+A collisions. Much of the excitement at RHIC comes from the fact that the PHENIX R_{AA} values for central collisions at 200 GeV (circles) and 130 GeV (triangles) are noticeably below unity in contrast to the enhanced production ($R_{AA} > 1$) observed at CERN-ISR [14] (stars) and CERN-SPS [15] (squares) energies. This enhanced production, observed first in p+A fixed-target experiments [16] ("Cronin effect"), is interpreted in terms of multiple initial-state soft and semi-hard interactions which broaden the transverse momentum of the colliding partons prior to the hard scattering itself.

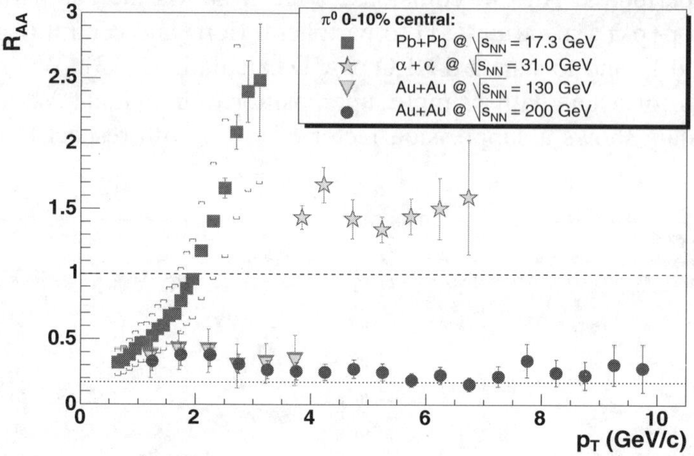

Figure 2. Nuclear modification factor, $R_{AA}(p_T)$, for π^0 measured in central ion-ion reactions at CERN-SPS [15], CERN-ISR [14], and BNL-RHIC [17, 12] energies. The dashed (dotted) line is the expectation of "N_{coll} (N_{part}) scaling" for "hard" ("soft") particle production.

The breakdown of the expectations from collinear factorization for high p_T production in central A+A collisions at RHIC by such a large factor, has been interpreted as due to one (or more) of the following facts:

1 Breakdown of leading-twist QCD collinear factorization itself: The incoherence between long- and short-distance effects in which the factorized product Eq. (3) relies upon, does not hold for A+A collisions.

2 Strong **initial-state** effects: The parton distribution functions in the nuclei are strongly modified: $f_{a/A} << A \cdot f_{a/p}$ in the relevant (x, Q^2) range, resulting in a reduced number of effective partonic scattering centers in the initial-state.

3 Strong **final-state** effects: The parton fragmentation functions (or, more generally, any post hard collision effect on the scattered partons) are strongly modified in the nuclear medium compared to free space.

Explanations 1. and 2. are usually invoked in the context of the "Color-Glass-Condensate" picture [4] which assumes that the kinematical conditions prevailing in the initial-state of an atomic nucleus boosted to RHIC energies (moderate $Q^2 \approx Q_s^2$ = 1–2 GeV$^2/c^2$, and small parton fractional momenta $x \lesssim 10^{-2}$) are such that nonlinear QCD effects ($g + g \rightarrow g$ processes, amplified by a $A^{1/3}$ factor compared to the proton case) are important and lead to a strong saturation of the parton (mostly gluon) densities in the nuclei. In this scenario, one expects N_{part} (instead of N_{coll}) scaling at moderately high p_T's [18], as approximately observed in the data (dotted line in Fig. 2). Explanation 3., on the other hand, relies on the expectations of "jet quenching" in a Quark Gluon Plasma [19] in which the hard scattered partons lose energy by final-state "gluonstrahlung" in the dense partonic system formed in the reaction. After traversing the medium, the partons fragment into high p_T (leading) hadrons with a reduced energy compared to standard fragmentation in the "vacuum". Different jet quenching calculations can reproduce the magnitude of the π^0 suppression assuming the formation of a hot and dense partonic system characterized by different, but closely related, properties: i) large initial gluon densities $dN^g/dy \approx 1000$ [20], ii) large "transport coefficients" $\hat{q}_0 \approx 3.5$ GeV/fm^2 [21], iii) high opacities $L/\lambda \approx 3.5$ [22], iv) effective parton energy losses of the order of $dE/dx \approx 14$ GeV/fm [23], or v) plasma temperatures of $T \approx 0.4$ GeV [24].

5. High p_T production: d+Au vs p+p

In order to disentangle between the two different (QGP and CGC) QCD scenarios, it was of paramount interest to determine experimentally the modification, if any, of high p_T hadron production due to initial-state nuclear effects alone, i.e. for a system in which a hot and dense medium is not produced in the final state. The results of high p_T π^0 production at midrapidity in d+Au collisions at $\sqrt{s_{NN}}$ = 200 GeV [25] do not show any indication of suppression (Fig. 3). On the contrary, π^0 production seems to be slightly enhanced ($R_{dAu} \approx 1.1$) compared to the expectations of collinear factorization.

This result, reminiscent of the "Cronin enhancement", indicates that the observed suppression in Au+Au central collisions at mid-rapidity is not an initial-state effect arising from strong modifications of the gluon distribution functions in nuclei at moderately small values of parton fractional momenta x, but more likely a final-state effect of the produced dense medium.

6. Summary

PHENIX has measured π^0 at mid-rapidity up to $p_T \approx 10$ GeV$/c$ in p+p, d+Au and Au+Au collisions at $\sqrt{s_{NN}}$ = 200 GeV. The spectral shape and invariant yields in peripheral Au+Au reactions are consistent with those of p+p reactions scaled by the number of inelastic NN collisions, in agreement with

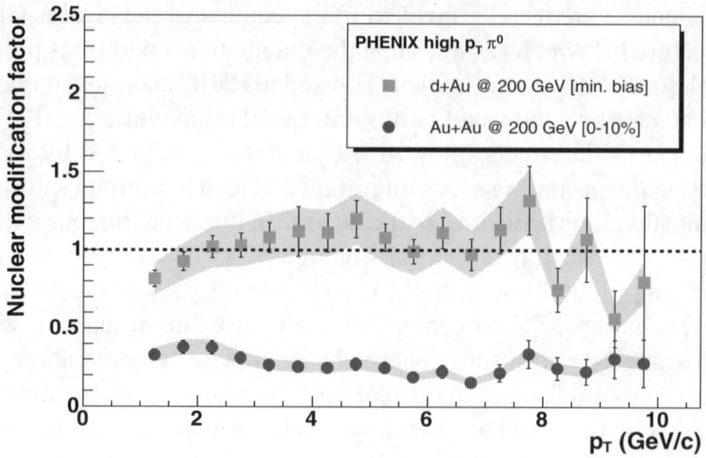

Figure 3. Nuclear modification factor, $R_{AA}(p_T)$, for π^0 measured at $y = 0$ in minimum bias d+Au (squares) [25] and central Au+Au (circles) [12] reactions at $\sqrt{s_{NN}}$ = 200 GeV.

pQCD collinear factorization expectations. In central Au+Au reactions, the π^0 are suppressed by a factor 4–5 with respect to the same expectations. The magnitude of this deficit can be reproduced by pQCD-based parton final-state energy loss calculations in an opaque medium, as well as by initial-state gluon saturation models. The unquenched high p_T π^0 production in minimum bias d+Au collisions, however, seems to rule out any significant "cold" nuclear matter effect as responsible for the observed suppression in Au+Au central reactions.

References

[1] A. M. Jaffe and E. Witten, "Quantum Yang-Mills Theory", Clay Mathematics Institute Millennium Prize problem. `http://www.claymath.org/Millennium_Prize_Problems/`, and D. Gross (2000), Problem # 10. in "Ten Problems in Fundamental Physics", `http://feynman.physics.lsa.umich.edu/strings2000/millennium.html`.

[2] See e.g. F. Karsch, Lect. Notes Phys. **583** (2002) 209.

[3] See e.g. A. Accardi *et al.*, "CERN Yellow Report on Hard Probes in Heavy Ion Collisions at the LHC: Jet Physics", hep-ph/0310274.

[4] See e.g. E. Iancu and R. Venugopalan in Quark Gluon Plasma 3, eds. R.C. Hwa and X.N. Wang, World Scientific, Singapore, hep-ph/0303204, for a recent review.

[5] K. Adcox *et al.* [PHENIX Collaboration], Nucl. Instrum. Meth. **A**499, (2003) 469.

[6] This is a purely heuristic argumentation, a rigorous discussion within the Dual Parton Model, can be found e.g. in A. Capella *et al.*, Phys. Lett. **B**108 (1982) 347.

[7] A. Bialas, M. Bleszynski and W. Czyz, Nucl. Phys. **B111** (1976) 461.

[8] K. Adcox *et al.* [PHENIX Collaboration], Phys. Rev. Lett. **86** (2001) 3500.

[9] J.C. Collins, D.E. Soper and G. Sterman, Nucl. Phys. **B**261 (1985) 104.

[10] A summary of useful analytical results of $T_{A,AB}(b)$ for simple parametrizations of the nuclear geometry can be found e.g. in K.J. Eskola *et al.* Nucl. Phys. **B**323 (1989) 37.

[11] S.S. Adler *et al.* [PHENIX Collaboration], Phys. Rev. Lett. **91** (2003) 241803.

[12] S.S. Adler *et al.* [PHENIX Collaboration], Phys. Rev. Lett. **91** (2003) 072301.

[13] S. Jeon, J. Jalilian-Marian and I. Sarcevic, Phys. Lett. **B**562 (2003) 45.

[14] A.L.S. Angelis *et al.*, Phys. Lett. **B**185 (1987) 213.

[15] M.M. Aggarwal *et al.* [WA98 Collaboration], Eur. Phys. J. **C**23 (2002) 225.

[16] D. Antreasyan *et al.*, Phys. Rev. **D19** (1979) 764.

[17] K. Adcox *et al.* [PHENIX Collaboration], Phys. Rev. Lett. **88** (2002) 022301.

[18] D. Kharzeev, E. Levin and L. McLerran, Phys. Lett. **B**561 (2003) 93.

[19] See e.g. M. Gyulassy *et al.* in Quark Gluon Plasma 3, eds. R.C. Hwa and X.N. Wang, World Scientific, Singapore, nucl-th/0302077, for a recent review.

[20] I. Vitev and M. Gyulassy, Phys. Rev. Lett. **89** (2002) 252301.

[21] F. Arleo, J. High Energy Phys. 11 (2002) 44.

[22] G.G. Barnafoldi, P. Levai, G. Papp, G. Fai, and Y. Zhang, nucl-th/0212111.

[23] X. N. Wang, nucl-th/0305010.

[24] S. Jeon and G.D. Moore, hep-ph/0309332.

[25] S. S. Adler *et al.* [PHENIX Collaboration], Phys. Rev. Lett. **91** (2003) 072303.

A PANORAMA OF PHENIX PHYSICS

S.V.Greene[1]

[1] *Department of Physics and Astronomy,*
Vanderbilt University,
Nashville, TN 37235, USA

Abstract

Looking back on the development of the universe in the context of our present scientific understanding, we believe that the universe began with the big bang and subsequently expanded and cooled. Increasingly more complex forms of matter began to emerge as temperatures dropped sufficiently to allow matter to bind. Early on, the universe consisted of a plasma that included deconfined quarks and gluons. Within microseconds after the big bang, the temperature had decreased sufficiently for the strong force to bind the quarks and gluons into hadrons. With further cooling, nucleosynthesis followed and eventually electrons joined nuclei electromagnetically to form neutral atoms. After this stage, the gravitational force began to dominate the universe. A primary goal of relativistic heavy ion physics is to produce, detect, and describe the quark-gluon plasma (QGP) phase. Towards this end, four experiments have been built at RHIC: BRAHMS [1], PHENIX [2], PHOBOS [3], and STAR [4]. This paper describes experimental results from the first three years of experiments at the Relativistic Heavy Ion Collider with a strong focus on data taken with the PHENIX detector.

1. Introduction

The Relativistic Heavy Ion Collider (RHIC) at Brookhaven National Laboratory can accelerate and collide atomic nuclei ranging from protons to gold. The collisions of these nuclei can reach center-of-mass energies of $\sqrt(s_{\mathrm{NN}}) = 200\mathrm{GeV}$ per nucleon . The PHENIX collaboration has built a versatile detector capable of observing many features of the collisions. Measuring leptons and photons can probe the plasma phase directly, while studying the hadrons produced duriing the collisions gives information on the subsequent hadronization. RHIC also provides the opportunity to study collisions

W. Greiner et al. (eds.), Structure and Dynamics of Elementary Matter, 71–86.

of polarized protons with the purpose of measuring the spin structure of the nucleon [5].

The PHENIX detector comprises several detector technologies in order to accomplish these physics goals. Global detectors characterize the collisions, a pair of central spectrometers at mid rapidity measures electrons, hadrons, and photons, and a pair of forward spectrometers measures muons. Each spectrometer has a large geometric acceptance, excellent energy and momentum resolution, and good particle identifcation.

2. Overview

Quantum Chromodynamics (QCD) predicts that matter at a sufficiently high temperature and energy density such as that produced in ultrarelativistic heavy ion collisions, should undergo a phase transition from ordinary hadronic matter to a deconfined state of quarks and gluons moving freely. In the standard picture, the system proceeds through different stages from the initial collision through deconfinement and, possibly, chiral symmetry restoration. After thermalization, the system expands and cools before undergoing hadronization. PHENIX is designed to study each stage of this evolution because of its ability to study rare processes involving photons, electrons and muons as well as the predominant hadronic production. The experiment has a high rate capability and segmentation combined with excellent momentum, energy and mass resolution. Direct photons and lepton pairs emerging from the collision with minimal final state interaction are sensitive to the full time evolution from the initial state through thermalization.

Many observables can give insight into the state of matter produced in high-energy heavy-ion collisions. This paper focuses on a small subset of possible QGP probes as well as measurements that allow the characterization of the collision geometry. One example of a potential QGP probe is the observation of jets produced during the initial collision stages through the hard scattering of constituent quarks. The jets are thus sensitive to the properties of the medium in which they are formed. An observable modification of jet structure is predicted if a jet forms within the QGP.

Another well-known potential probe of the deconfined state is the systematic study of the J/ψ meson production. The attendant Debye screening of the color force in a QGP is expected to decrease J/ψ production relative to collisions that result in normal hadronic matter. The J/ψ's can be studied by observation of their decays into lepton pairs.

Many of the proposed QGP signals can also be produced in interactions between particles in hot hadronic matter. Therefore, it is essential to understand purely hadronic effects.by studying pp and pA collisions as a baseline.

3. Transverse Energy

The transition from hadronic matter to a quark-gluon plasma is predicted for energy densities on the order of a few GeV/fm^3 [6]. Following Bjorken [7], the spatial energy density (ϵ) in a relativistic collision can be estimated by measuring the transverse energy density in rapidity, dE_T/dy, which is effectively the co-moving energy density in a longitudinal expansion:

$$\epsilon_{Bj} = \frac{dE_T}{dy}\frac{1}{\tau_0\,\pi R^2} \tag{1}$$

where τ_0, the formation time, is usually taken as 1 fm/c, and πR^2 is the effective area of the collision. The transverse energy (E_T) is defined as:

$$E_T = \sum_i E_i \sin\theta_i,\;\; dE_T(\eta)/d\eta = \sin\theta(\eta)\, dE(\eta)/d\eta, \tag{2}$$

where θ is the polar angle, $\eta = -\ln\tan\theta/2$ is the pseudorapidity, E_i is by convention taken as the kinetic energy for nucleons and the total energy for all other particles [8], and the sum is over all particles emitted into a fixed solid angle for each event. E_T measurements, even in limited apertures at mid-rapidity, provide excellent characterization of the nuclear geometry of a reaction on an event-by-event basis and are sensitive to the underlying reaction dynamics [6].

4. Pizero Results

The large center-of-mass energies available in Au+Au collisions at RHIC have resulted in significant production of high transverse momentum hadrons, defined as hadrons with $p_T > 2$ GeV$/c$ [9]. High p_T particle production in hadronic collisions results from quark and gluon fragmentation as they emerge from the initial high Q^2 parton-parton scatterings [10]. Thus, hard processes in nucleus-nucleus (AA) collisions provide direct information on the partonic phases of the reaction. If there are no nuclear medium effects, hard scattering yields in AA reactions should scale like an incoherent superposition of nucleon-nucleon (NN) collisions because of the small probability of hard scattering processes per NN collision.

In AA reactions, the number of NN collisions (N_{coll}) at impact parameter b is proportional to the geometric nuclear overlap function, $T_{AA}(b)$ which can be calculated using a Glauber model [11]. After scaling by the nuclear geometry, spectra of high p_T particles measured in AA reactions can be compared to the baseline $p{+}p$ and $d{+}A$ data, as well as to perturbative [12–15] and classical [16] QCD predictions. Any departure from N_{coll} scaling provides information on the strongly interacting medium in central heavy-ion reactions.

The suppressed yield of moderately high p_T neutral pions ($p_T \approx 1.5$ - 4.0 GeV/c) in central Au+Au at $\sqrt{s_{NN}}$ = 130 GeV with respect to the N_{coll}-scaled $p+p$ and peripheral Au+Au data [17] was a particularly significant observation from the first RHIC run. This result points to strong medium effects present in central Au+Au and has triggered extensive theoretical studies on its origin [12–16, 18–21]. Most of these studies are based on the prediction [22, 23] that a QGP would induce multiple gluon radiations from the scattered fast partons, leading to a subsequent suppression of high p_T hadronic fragmentation products, a process known as jet quenching. Alternative interpretations have been proposed based on initial-state gluon saturation [16] or final-state hadronic interactions [21].

Results of π^0 measurements by PHENIX for Au+Au collisions at $\sqrt{s_{NN}}$ = 200 GeV can be compared to $p + p \rightarrow \pi^0\, X$ data measured by PHENIX at the same center-of-mass energy [25]. During the 2001-2002 RHIC run (Run-2), PHENIX obtained 30 million minimum bias π^0 event measured in the electromagnetic calorimeter (EMCal).

Event centrality is determined by correlating the charge detected in the beam-beam counter (BBC) with the energy measured in the ZDC detectors. A Glauber model Monte Carlo (MC) combined with a simulation of the BBC and ZDC responses [29–31] allows an estimate to be made of the associated number of binary collisions (N_{coll}) and participating nucleons (N_{part}) in each centrality bin. The combined π^0 p_T distributions, as measured using the Pb-scintillator (PbSc) and Pb-glass (PbGl) calorimeters, are shown in Fig. 1 for minimum bias and for 9 centrality bins (scaled by factors of 10 for clarity).

It is useful to quantify the medium effects on high p_T production in AA collisions with the *nuclear modification factor* given by the ratio of the measured AA invariant yields to the NN collision scaled $p + p$ invariant yields:

$$R_{AA}(p_T) = \frac{(1/N^{evt}_{AA})\, d^2N^{\pi^0}_{AA}/dp_T dy}{\langle N_{coll}\rangle/\sigma^{inel}_{pp} \times d^2\sigma^{\pi^0}_{pp}/dp_T dy}, \tag{3}$$

where the $\langle N_{coll}\rangle/\sigma^{inel}_{pp}$ is the average Glauber nuclear overlap function $\langle T_{AuAu}\rangle$ in a given centrality bin $R_{AA}(p_T)$ is a measure of the deviation of AA data from an incoherent superposition of NN collisions. For $p_T \leq 2$ GeV/c, R_{AA} is below one, since the bulk of particle production is comes from soft processes which scale more closely as the number of participant nucleons [29].

Figure 2 shows R_{AA} as a function of p_T for π^0 measured in 0-10% central (closed circles) and 80-92% peripheral (open circles) Au+Au. The PHENIX $p + p \rightarrow \pi^0$ data [25] is used as the reference in the denominator. The R_{AA} values for central collisions are clearly below unity, as found at 130 GeV [17], and in contrast to the enhanced high p_T π^0 production (R_{AA} >1) observed at CERN-SPS energies [32] and interpreted in terms of the Cronin effect, an

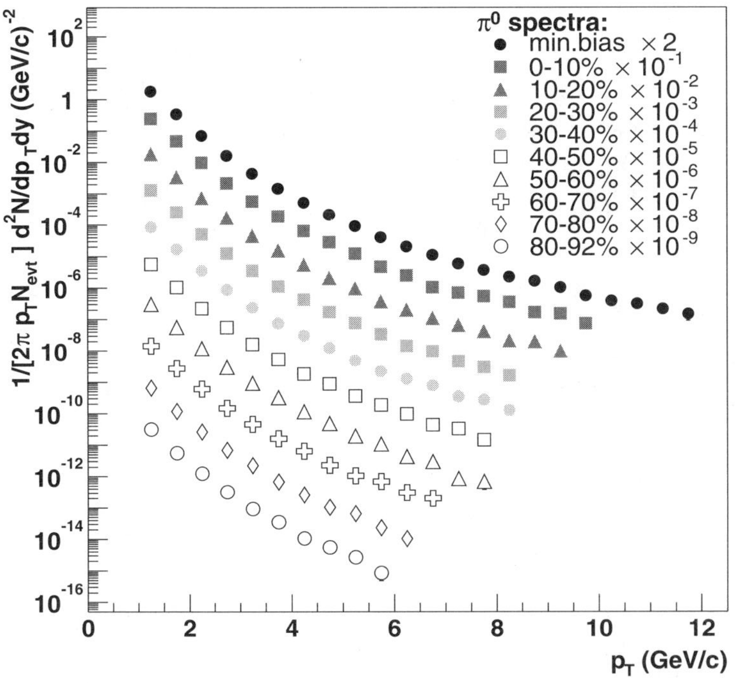

Figure 1. Invariant π^0 yields at mid-rapidity as a function of p_T for minimum bias and 9 centralities in Au+Au at $\sqrt{s_{NN}}$ = 200 GeV [0-10% (80-92%) is most central (peripheral)].

initial-state p_T broadening. [33]). Within errors, high p_T π^0 production in peripheral Au+Au collisions acts as a superposition of $p+p$ collisions ($R_{AA} \approx$ 1). In central collisions, the suppression is smallest at 2 GeV$/c$ and increases to a nearly constant suppression factor of $1/R_{AA} \approx$ 4-5 over the p_T range of 4 - 10 GeV$/c$, ~30% above the expectation from N_{part} scaling (dotted line in Fig. 2).

Fig. 3 shows high p_T π^0 suppression as a function of N_{part}, a measure of centrality. The suppression is described as the ratio of Au+Au over $p+p$ yields integrated above 4 GeV/*c* and normalized using two different scalings. R_{AA} (circles) denotes the N_{coll} scaling as in Eq. (3). R_{AA}^{part} (crosses) describes N_{part} scaling as expected in scenarios dominated by gluon saturation [16] or by surface emission of the quenched jets [18]. Fig. 3 indicates that the transition from the N_{coll} scaling behaviour ($R_{AA} \sim 1$) seen in the most peripheral collisions, to the strong suppression observed in central collisions ($R_{AA} \sim 0.25$) is smooth. Although there is no exact participant scaling ($R_{AA}^{part} > 1$ for all centralities), the π^0 production per participant pair above 4 GeV/*c* is approximately constant over a broad range of centralities. This is in qualitative agreement with the parton saturation model prediction [16].

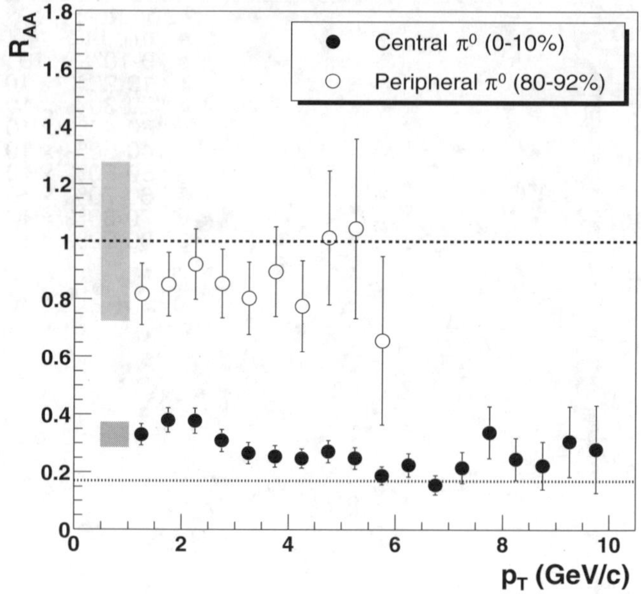

Figure 2. Nuclear modification factor $R_{AA}(p_T)$ for π^0 in central (closed circles) and peripheral (open circles) Au+Au at $\sqrt{s_{NN}}$ = 200 GeV. The error bars include all point-to-point experimental ($p + p$, Au+Au) errors. The shaded bands represent the fractional uncertainties in $\langle T_{AuAu} \rangle$ and in the π^0 yields normalization added in quadrature, which can move all the points up or down together (in the central case the shaded band shown is the fractional error for the first point).

In summary, the spectral shape and invariant yield for peripheral reactions are consistent with those of $p+p$ reactions scaled by the average number of inelastic NN collisions. In contrast, central yields are significantly lower than peripheral Au+Au and $p+p$ scaled yields, as found for collisions at $\sqrt{s_{NN}}$ = 130 GeV. The observed suppression increases slowly with p_T to as much as a factor 4-5 in the 10% most central collisions, and is constant within errors above ~4 GeV/c. This suppression is seen to increase smoothly with centrality. Finally, the magnitude of the observed suppression can be described by calculations of parton energy loss in an opague medium.

Theoretical treatments of the propagation of partons in high density matter suggest that partons will lose a significant fraction of their energy through gluon bremsstrahlung [36, 37]. This energy loss reduces the parton momentum and decreases the yield of high p_T hadrons [42, 12–15, 19]. This is a final state effect that takes place within the medium created in A+A collisions. By contrast, initial state effects include nuclear modifications to the parton structure functions, and soft scatterings of the incoming parton prior to the hard scattering. These should be present in p+A and d+A as well as in A+A collisions.

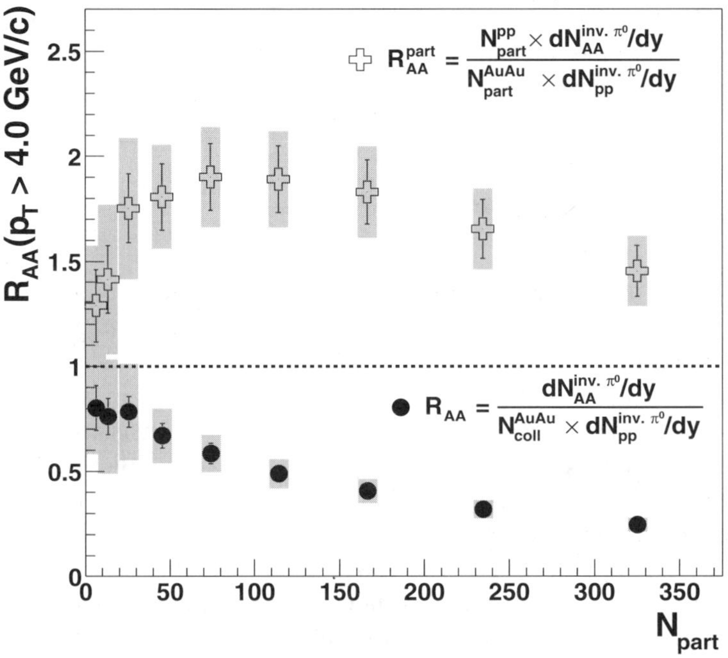

Figure 3. Ratio of Au+Au over $p + p$ π^0 yields integrated above 4 GeV/c and normalized using N_{coll} (circles) and N_{part} (crosses), as a function of centrality given by N_{part}. The errors bands and bars are the same as for Fig. 2.

Interpretations of Au+Au interactions that include either initial-state parton saturation effects [16] or final-state hadronic interactions [21] also predict hadron suppression of hadron at high p_T. Thus it is necessary to determine experimentally what modification of high p_T hadron yields from initial state effects may occur if a hot dense medium is not produced in the final state. PHENIX has measured charged hadron and π^0 production at midrapidity in d+Au collisions at $\sqrt{s_{NN}}$ = 200 GeV [24]. The results are compared to those in p+p and Au+Au [41, 25, 44] at the same nucleon-nucleon center-of-mass energy. Similar d+Au measurements are also reported in [45, 46].

The fully corrected p_T distributions of π^0's and $(h^+ + h^-)/2$ are shown in Fig. 1. Each panel shows the reference spectrum from p+p collisions and the particle spectrum from d+Au. For π^0, the reference is the π^0 spectrum measured in p+p collisions by PHENIX [25], shown on the right panel. For charged hadrons, the reference is obtained by scaling up the π^0 spectrum by the p_T-dependent h/π ratio observed at the ISR [48] at $\sqrt{s} = 23$ to 63 GeV and measured by PHENIX. The value of h/π is constant at 1.6 $\pm$ 0.16 above 1.5 GeV/c, and decreases at lower p_T to a value of $1.25^{+0.12}_{-0.25}$.

As in A+A collisions, nuclear medium effects on high p_T production can be quantified by the *nuclear modification factor*, which is defined for d+Au collisions as the ratio of invariant yield in d+Au to that of p+p, scaled by the number of binary collisions.

$$R_{\text{dA}}(p_T) = \frac{(1/N_{\text{dA}}^{evt})\, d^2N_{\text{dA}}/d\eta dp_T}{\langle N_{coll}\rangle/\sigma_{\text{pp}}^{inel}\, d^2\sigma_{\text{pp}}/d\eta dp_T},$$

where $\langle N_{coll}\rangle$ is the average number of inelastic nucleon-nucleon (NN) collisions per event in the minimum bias collisions, and $\langle N_{coll}\rangle/\sigma_{\text{pp}}^{inel}$ is the nuclear overlap function $\langle T_A(b)\rangle$.

The ratio R_{dA} is plotted separately for π^0 measured in the PbGl and in the PbSc calorimeters in Fig. 2. The two results are consistent within errors. The data are compared to the corresponding nuclear modification factor R_{AA} measured for central Au+Au collisions. The top panel of Fig. 4 shows R_{dA} for inclusive charged particles $(h^+ + h^-)/2$ compared with R_{AA} as observed in central Au+Au collisions. The lower panel compares $(h^+ + h^-)/2$ with π^0. In both Figs. 2 and 4, the uncertainties are plotted as follows: error bars represent the quadrature sum of statistical errors and those systematic errors which vary point-to-point in p_T while systematic errors on the absolute yield and correlated systematic errors are shown as bands. The data show no evidence of suppression of high p_T particles in d+Au collisions. There is, however, an enhancement in inclusive charged particle production at $p_T > 2$ GeV/c which may be attributed to the Cronin effect described above. To facilitate comparison of the Cronin effect in both inclusive charged particles and π^0, the lower part of Fig. 4 shows all systematic uncertainties common to both analyses in the bar on the left. This uncertainty should be added in quadrature with the bands shown for each curve to obtain the 1σ allowed range of R_{dA}. The π^0 data suggest that the enhancement is smaller for pions than for inclusive charged particles at p_T = 2-4 GeV/c. The charged particle spectrum includes baryons and antibaryons, which may have a different nuclear enhancement than the mesons [33].

The observation of an enhancement of high-p_T hadron production in the d+Au collisions indicates that the suppression in central Au+Au collisions is neither an initial state effect, nor is it caused by a modification of the parton structure functions in nuclei. Rather, the data suggest that the suppression of high p_T hadrons in Au+Au is most probably a final state effect of the produced dense medium.

5. Production of J/psi in p + p collisions

Suppression of charmonium from color screening in a deconfined medium has long been considered as a possible indication of QGP production. In addition, we would like to determine the extent to which heavy quarks may lose

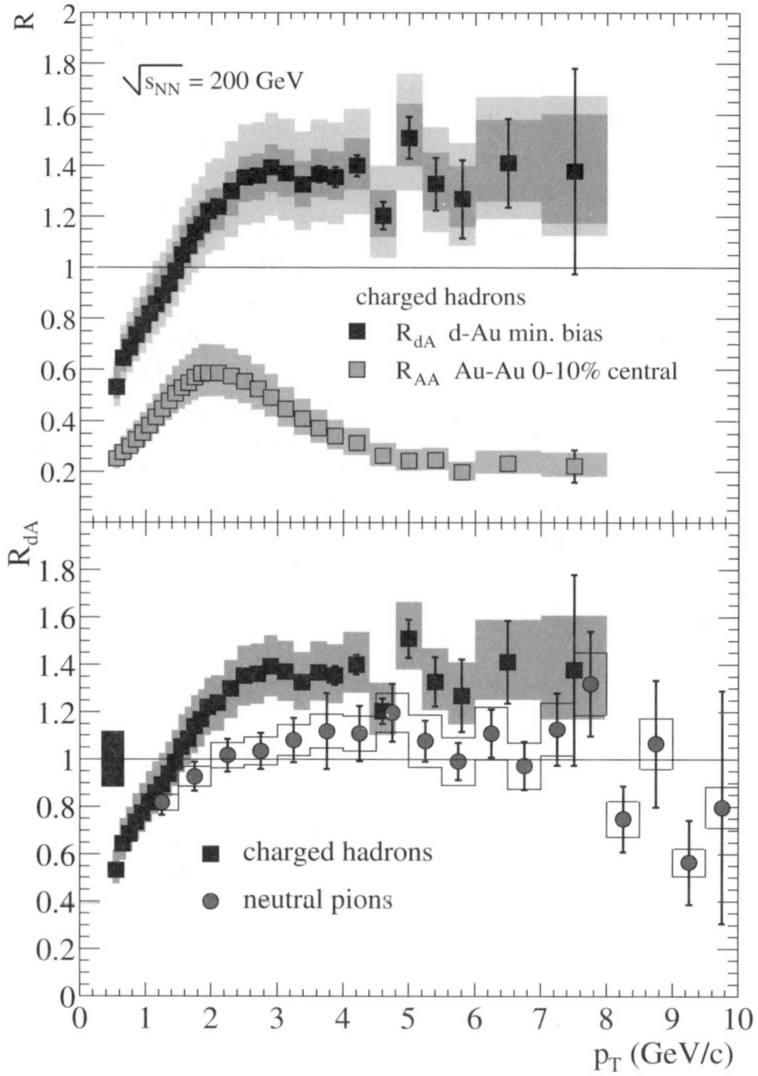

Figure 4. Top: Nuclear modification factor R_{dA} for $(h^+ + h^-)/2$ in minimum bias d+Au compared to R_{AA} in the 10% most central Au+Au collisions. Inner bands show systematic errors which can vary with p_T, and outer bands include also the normalization uncertainty. Bottom: Comparison of R_{dA} for $(h^+ + h^-)/2$ and the average of the π^0 measurements in d+Au. The bar at the left indicates the systematic uncertainty in common for the charged and π^0 measurements.

energy in quark matter. Finally, thermal charm production could lead to possible charm enhancement. For all these measurements, J/ψ yields in proton-proton collisions at RHIC will provide aan essential baseline for studying cold and hot nuclear matter using J/ψ as a probe.

There is also theoretical interest in J/ψ the production mechanism in its own right. The Color Singlet Model (CSM) was found [54] to underpredict the high p_T CDF prompt J/ψ and $\psi(2S)$ cross sections [53]. Models emerged in which color octet $c\bar{c}$ states could contribute to the J/ψ yield. The Color Octet Model (COM), based on the Non-Relativistic QCD model [55], has been successful in reproducing the high p_T CDF prompt J/ψ cross sections. The more phenomenological Color Evaporation Model [56]has also been successful in this respect.

In order to understand J/ψ production, data is needed which spans a large range of collision energies, rapidity and transverse momentum (p_T). Data at lower energies from fixed target hadron experiments give total cross sections and mean p_T ($\langle p_T \rangle$) values in the energy range $\sqrt{s}$ = 7 - 38.8 GeV [51].However, the limited kinematic coverage of collider experiments [51, 52] has precluded measurement of the total cross sections and mean p_T values. The systematic study of J/ψ production at Relativistic Heavy Ion Collider (RHIC) energies with wide p_T and rapidity coverage should provide important tests of various J/ψ production models.

In [69] the PHENIX experiment reported results of the first measurements of $pp \rightarrow J/\psi + X$ at RHIC at $\sqrt{s}$ = 200 GeV by the PHENIX experiment. The data yield the first total cross sections for J/ψ production beyond fixed target energies, and the first measurement of $\langle p_T \rangle$ greater than $\sqrt{s}$ = 63 GeV. They can constrain models at lower p_T, in the region where gluon fusion is expected to dominate.

The PHENIX experiment [58] can detect electrons in the pseudo-rapidity range $|\eta| \leq 0.35$ in two central spectrometer arms that cover $\Delta\phi$ = 90°, and forward rapidity muons in two muon arms covering $\Delta\phi$ = 360°. (Only one muon arm, covering $1.2 < \eta < 2.2$, was used for this data.) Electrons are identified by matching charged particle tracks to energy deposits in the Electromagnetic Calorimeter (EMCal) and to rings in the Ring Imaging Čerenkov Detector (RICH). Muons are identified by matching tracks that penetrate the Muon Identifier (MuID) with tracks in the Muon Tracker (MuTr). The data were recorded during the 2001/2002 pp run at $\sqrt{s}$ = 200 GeV.

J/ψ yields in the central arms were obtained by reconstructing electron-positron pairs and J/ψ yields in the muon arm were obtained by reconstructing $\mu^+\mu^-$ pairs. Unlike-sign pairs and like-sign pairs for backgrounnd subtraction were combined to form invariant mass spectra. Simulations show that the acceptance for like-sign and unlike-sign pairs is the same to within a few percent for invariant masses above 1 GeV/c^2 for electrons. In Fig. 5, unlike-sign and

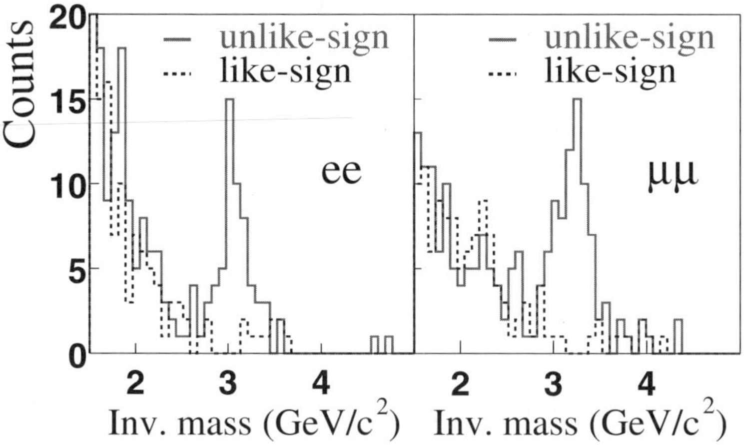

Figure 5. The invariant mass spectra for dielectron and dimuon pairs. Unlike-sign pairs are shown as solid lines, the sum of like-sign pairs as dashed lines.

like-sign invariant mass spectra from the entire pp data set are shown together. For electrons, the net yield inside 2.8 - 3.4 GeV/c^2 is 46, for muons inside 2.71 - 3.67 GeV/c^2 it is 65. For electrons, the peak width is 110 MeV/c^2 and the centroid agrees well with the PDG value [60]. For muons, the width is 160 MeV/c^2. The muon peak centroid is higher than the PDG value by about 3% which is consistent with the uncertainty in the muon magnetic field calibration.

The measured trigger efficiency from high p_T π^0 events was used when calculating the J/ψ cross section, under the assumption that the trigger efficiency is the same for both processes. Good agreement is observed in the $dN_{ch}/d\eta$ distribution between PYTHIA simulations and measurements [63] for events involving high p_T π^0 production.

For the electron analysis, $A\,\epsilon_{rec}$ was determined as a function of p_T using a full GEANT simulation. An estimate of the systematic uncertainty in $A\,\epsilon_{rec}$ due to z vertex dependence of the acceptance, momentum resolution effects, the pair mass cut, and electron identification cuts is given in [69], along with the uncertainty in the yield due to Drell-Yan and correlated charm decay contributions. The efficiency ϵ_{lvl1} of the level-1 J/ψ triggers in the central arms was determined as a function of p_T by analyzing simulated single J/ψ events.

For the muon arm, $A\,\epsilon_{rec}\,\epsilon_{lvl1}$ was using a full GEANT simulation with J/ψ events generated by PYTHIA (with GRV94LO PDFs). The systematic uncertainty includes discrepancies between Monte Carlo and real detector response, run to run variations in the detector state, and uncertainties in the PYTHIA dis-

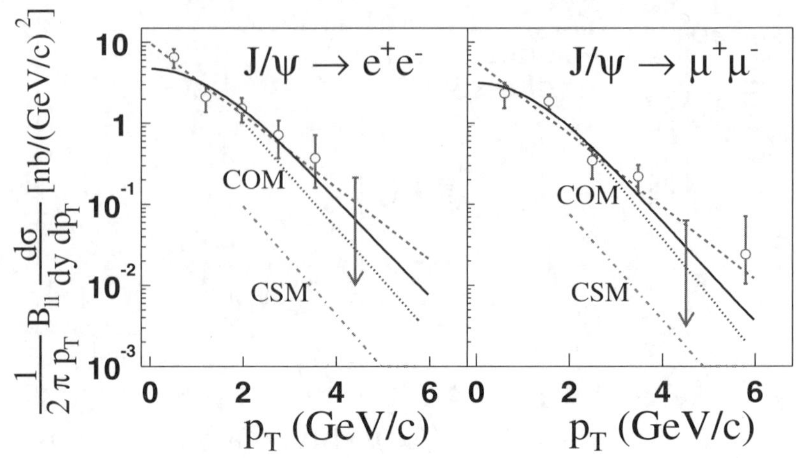

Figure 6. The J/ψ p_T distributions for the dielectron and dimuon measurements, with statistical uncertainties. The solid line is a phenomenological fit of the form $1/(2\pi p_T)\ d\sigma/dp_T = A\ (1 + (p_T/B)^2)^{-6}$. The dashed line is an exponential fit. The CSM (dot-dashed) and COM (dotted) calculations are from [65].

tributions. The results, integrated over the rapidity range of the muon arm, are shown in [69].

The p_T distributions for $J/\psi \to e^+e^-$ and $J/\psi \to \mu^+\mu^-$ are shown in Fig. 6 along with predictions [65] from the Color Octet model (COM). Predictions of the Color Singlet model (CSM) are also shown. The solid lines are a phenomenological fit of a form that has been shown to fit J/ψ data well at fixed target energies [66]. The dashed line is an exponential fit. The phenomenological fits yield $\langle p_T \rangle$ values of 1.85 $\pm$ 0.46(stat) $\pm$ 0.16(sys) GeV/c (central arm) and 1.78 $\pm$ 0.27(stat) $\pm$ 0.16(sys) GeV/c (muon arm), with a combined value of 1.80 $\pm$ 0.23(stat) $\pm$ 0.16(sys) GeV/c. The systematic uncertainties were estimated from the spread in $\langle p_T \rangle$ from a weighted mean of the binned data, the phenomenological fit, and the exponential fit. An additional 3% was assigned to the muon $\langle p_T \rangle$ because of the uncertainty in momentum scale.

The J/ψ rapidity distribution obtained by combining the dielectron and dimuon measurements is shown in Fig. 7, with the muon arm data divided into two rapidity bins. The COM curves are theoretical predictions [59] of the shapes of the distributions normalized to the data to allow for easier comparison. Since gluon fusion is the dominant process in all of the models, the rapidity shape depends primarily on the gluon distribution function rather than the production model. A PYTHIA calculation that reproduces the shape of the data best is also shown in Fig. 7. After normalization, the total cross section was determined to be 3.99 $\pm$ 0.61(stat) $\pm$ 0.58(sys) $\pm$ 0.40(abs) μb. The sys-

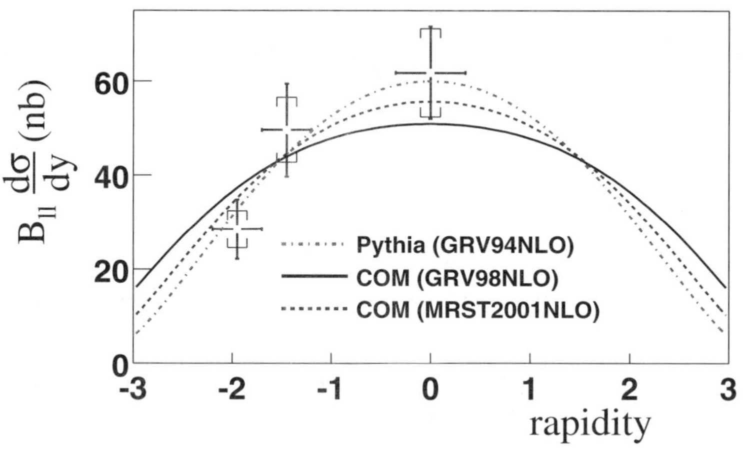

Figure 7. The central rapidity point is from $J/\psi \rightarrow e^+e^-$, the others from $J/\psi \rightarrow \mu^+\mu^-$. The brackets represent systematic uncertainties. All curves have their overall normalization fitted to the data. The PYTHIA shape was used to determine the cross section. There is an overall 10% absolute normalization error not shown.

tematic error of 14% was estimated by setting the measured cross sections to the extremes of their systematic error limitsand observing the change in the cross section.

Fig. 8a compares the present $\langle p_T \rangle$ value with values from previous experiments [51]. The total measured J/ψ cross section is shown in Fig. 8b, along with cross sections determined by lower energy experiments [51] and predictions from the COM [59] using two different PDFs. Since the shape of the PDFs depends on the factorization scale Q,the $\sqrt{s}$ dependence of the cross section is sensitive to this scale. The values of Q (3.1 GeV for GRV98NLO and 2.3 GeV for MRST2001NLO) were chosen to give good agreement with the data. The total cross section normalization was obtained using color octet matrix elements from [67], but there are large theoretical uncertainties. The CEM is also able to describe the total cross section data [56]. All measurements and models include feed-down from the χ_c and the ψ' to the J/ψ. B decay feed-down is estimated [68] to contribute less than 4% to the J/ψ total cross section at $\sqrt{s}$ = 200 GeV. Thus, the transverse momentum distributions above 2 GeV/c are reasonably well described by the COM. Within the available statistics, the rapidity distribution shape is consistent with most of the available PDFs. COM calculations reproduce the $\sqrt{s}$ dependence of the cross section using standard matrix elements found in the literature, with a reasonable choice of QCD parameters.

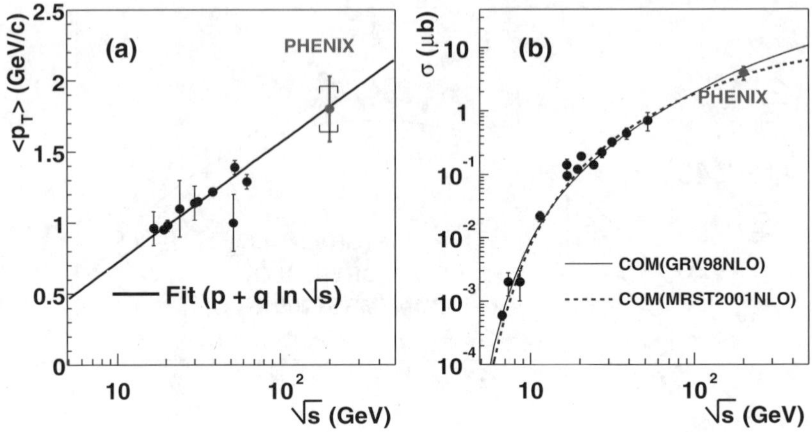

Figure 8. (a) The present J/ψ mean p_T value compared with previous measurements at lower energy. The linear fit parameters are $p = 0.53$, $q = 0.19$. (b) The present J/ψ total cross section compared with previous measurements at other values of $\sqrt{s}$. The curves are discussed in the text.

6. Conclusions

RHIC is expected to have an extended A+A run at $\sqrt{s}$ = 200 and proton-proton runs at $\sqrt{s}$ = 200 and 500 GeV in the near future. The anticipated increased luminosity will improve the statistical precision and p_T reach of the PHENIX data, and will ultimately make it possible to measure the J/ψ production in A + A collisions. This measurement will complement the already intriguing results from measurements of particles produced at high p_T and bring us closer to an understanding of the nature of the matter produces in these collisions.

References

[1] Adamczyk, et al., Nucl. Instr. Meth. A 499, 437 (2003).

[2] K. Adcox, et al., Nucl. Instr. Meth. A 499, 469 (2003).

[3] Back, et al.,Nucl. Instr. Meth. A 499, 603 (2003).

[4] Ackermann et al.,Nucl. Instr. Meth. A 499, 624 (2003).

[5] N. Saito et al., Nucl. Phys. A638, 575c (1998).

[6] For example, see Proceedings Quark Matter 1984, edited by K. Kajantie (Springer, Berlin, 1985); Proceedings Quark Matter 1987, edited by H. Satz, H. J. Specht, and R. Stock [Z. Phys. C 38, 1–370 (1988)].

[7] J. D. Bjorken, Phys. Rev. D 27, 140 (1983).

[8] Helios Collaboration, T. Åkesson et al., Z. Phys. C 38, 383 (1988); 38, 397 (1988).

[9] S.S.Adler et al., Phys. Rev. Lett. 91, 072301 (2003).

[10] J.F. Owens et al., Phys. Rev. D18, 1501 (1978).

[11] R.J. Glauber and G. Matthiae, Nucl. Phys. B 21, 135 (1970).

[12] I. Vitev and M. Gyulassy, Phys. Rev. Lett. 89 (2002).

[13] S. Jeon, J. Jalilian-Marian and I. Sarcevic, Phys. Lett. B562,2003.

[14] G.G. Barnafoldi, P. Levai, G. Papp, G. Fai, and Y. Zhang, nucl-th/0212111.

[15] X. N. Wang, Nucl. Phys. A715, 775 (2003).

[16] D. Kharzeev, E. Levin and L. McLerran Phys.Lett.B561:93,2003.

[17] K. Adcox *et al.* [PHENIX Collaboration], Phys. Rev. Lett. **88**, 022301 (2002).

[18] B. Müller, Phys.Rev.C67:061901,2003

[19] F. Arleo, J. High Energy Phys. 11, 44 (2002).

[20] C. A. Salgado and U. A. Wiedemann, Phys.Rev.D68:014008,2003

[21] K. Gallmeister, C. Greiner and Z. Xu (to appear in Phys. Rev. **C**), hep-ph/0212295.

[22] M. Gyulassy and M. Plümer, Phys. Lett. **B243**, 432 (1990); X.N. Wang and M. Gyulassy, Phys. Rev. Lett. **68**, 1480 (1992).

[23] R. Baier, D. Schiff and B.G. Zakharov, Annu. Rev. Nucl. Part. Sci. **50**, 37 (2000), and references therein.

[24] S. S. Adler *et al.*, Phys. Rev. Lett. 91, 072303 (2003).

[25] S.S. Adler *et al.* [PHENIX Collaboration], submitted to Phys. Rev. Lett.

[26] L. Aphecetche *et al.* [PHENIX Collaboration], Nucl. Instrum. Methods **A**499, 521 (2003).

[27] GEANT 3.21, CERN program library.

[28] X. N. Wang and M. Gyulassy, Phys. Rev. **D44**, 3501 (1991).

[29] K. Adcox *et al.* [PHENIX Collaboration], Phys. Rev. Lett. **86**, 3500 (2001).

[30] The Run-2 minimum bias trigger is inefficient for very peripheral reactions and records $92.2^{+2.5}_{-3.0}\%$ of σ_{AuAu}.

[31] Woods-Saxon Au nuclear radius $R = 6.38^{+0.27}_{-0.13}$ fm, diffusivity $a = 0.53 \pm 0.01$ fm [B. Hahn, D.G. Ravenhall and R. Hofstadter, Phys. Rev. **101**, 1131 (1956)], and nucleon-nucleon cross section $\sigma_{NN}^{inel} = 42 \pm 3$ mb.

[32] M.M. Aggarwal *et al.* [WA98 Collaboration], Eur. Phys. J. **C**23, 225 (2002).

[33] D. Antreasyan *et al.*, Phys. Rev. **D19**, 764 (1979).

[34] K. J. Eskola, V. J. Kolhinen and C. A. Salgado, Eur. Phys. J. C **9**, 61 (1999); S. R. Klein and R. Vogt, Phys. Rev. **C67**, 047901 (2003).

[35] M. Gyulassy, P. Levai and I. Vitev, Phys. Rev. Lett. **85**, 5535 (2000), Nucl. Phys. B **594**, 371 (2001).

[36] M. Gyulassy and M. Plümer, Phys. Lett. **B243**, 432 (1990); X.N. Wang and M. Gyulassy, Phys. Rev. Lett. **68**, 1480 (1992).

[37] R. Baier *et al.*, Phys. Lett. **B345**, 277 (1995).

[38] K. Adcox *et al.*, Phys. Rev. Lett. **88**, 022301 (2002).

[39] K. Adcox *et al.*, Phys. Lett. **B561**, 82 (2003).

[40] C. Adler *et al.*, Phys. Rev. Lett. **89**, 202301 (2002).

[41] S.S. Adler *et al.*, nucl-ex/0304022, sub. to Phys. Rev. Lett.

[42] M. Gyulassy and X.N. Wang, Nucl. Phys. **B420**, 583 (1994); X.N. Wang, Phys. Rev. **C58**, 2321 (1998). nucl-th/0212111; V. Greco, C.M. Ko and P. Lévai, Phys. Rev. Lett. **90**, 202302 (2003).

[43] K. Adcox *et al.*, Nucl. Instrum. Methods **A499**, 469 (2003), and references therein.

[44] J. Jia *et al.*, Nucl. Phys. **A715**, 769c (2003).

[45] J. Adams *et al.*, nucl-ex/0306024, sub. to Phys. Rev. Lett.

[46] B.B. Back *et al.*, nucl-ex/0306025, sub. to Phys. Rev. Lett.

[47] J. T. Mitchell *et al.*, Nucl. Instrum. Methods **A482**, 491 (2002).

[48] B. Alper *et al.*, Nucl. Phys. **B100**, 237 (1975).

[49] In the present work the Woods-Saxon nuclear density parameters radius $R = 6.38$ fm, diffusivity $a = 0.54$ fm and N-N cross section $\sigma_{NN}^{inel} = 42$ mb were used. The deuteron is described by a Hulthén wave function (L. Hulthén and M.Sagawara, Handbuch der Physik **39** (1957).) with α= 0.228 fm^{-1} and β= 1.18 fm^{-1}.

[50] B. Kopeliovich, Phys. Rev. Lett. **88**, 232303 (2002).

[51] R. Vogt, Phys. Repts. **310**, 197 (1999); G A. Schuler, hep-ph/9403387, and references therein.

[52] F. Abe *et al.*, Phys. Rev. Lett. **69**, 3704 (1992); F. Abe *et al.*, Phys. Rev. **D66**, 092001 (2002); S. Abachi *et al.*, Phys. Lett. **B370**, 239 (1996); B. Abbot *et al.*, Phys. Rev. Lett. **82**, 35 (1999).

[53] F. Abe *et al.*, Phys. Rev. Lett. **79**, 572 (1997).

[54] E. Braaten *et al.*, Ann.Rev.Nucl.Part.Sci. **46**, 197 (1996).

[55] G. T. Bodwin *et al.*, Phys. Rev. **D51**, 1125 (1995); erratum Phys. Rev. **D55**, 5853 (1997).

[56] J. F. Amundson *et al.*, Phys. Lett. **B390**, 323 (1997), and private communication.

[57] E. Braaten *et al.*, Phys. Rev. Lett. **71**, 1673 (1993).

[58] K. Adcox *et al.*, Nucl. Instrum. Methods **A499**, 469 (2003).

[59] H. D. Sato, Ph.D Thesis, hep-ph/0305239.

[60] Particle Data Group, Phys. Rev. **D66**, 010001 (2002).

[61] S. S. Adler *et al.*, hep-ex/0304038.

[62] T. Sjöstrand, Comp. Phys. Comm. **135**, 238 (2001).

[63] R. Kephart *et al.*, Phys. Rev. **D14**, 2909 (1976).

[64] T. Affolder *et al.*, Phys. Rev. Lett. **85**, 2886 (2000); E. Braaten *et al.*, Phys. Rev. **D62**, 094005 (2000).

[65] Gouranga C. Nayak *et al.*, hep-ph/0302095.

[66] A. Gribushin *et al.*, Phys. Rev. **D62**, 012001 (2000) and references therein.

[67] M. Beneke and I. Z. Rothstein, Phys. Rev. **D54**, 2005 (1996).

[68] S. S. Adler *et al.*, Phys. Rev. C, Phys. Rev. C 69, 014901 (2004).

[69] S. S. Adler *et al.*, Phys. Rev. **C69**, 014901 (2004).

[70] I. Vitev, Phys. Lett. **B562**, 36 (2003).

THE FUTURE INTERNATIONAL ACCELERATOR FACILITY FOR BEAMS OF IONS AND ANTI-PROTONS AT DARMSTADT

Hans H. Gutbrod
GSI, Planckstr. 1, 64291 Darmstadt, Germany, Email: H.Gutbrod@gsi.de

Abstract A brief description is given of the main research areas, technical aspects and the overall performance characteristics of the future international accelerator facility at the GSI Laboratory.

1. Introduction

The principal goal of the new facility is to provide the European science community with a worldwide unique and technically innovative accelerator system to perform future forefront research in the sciences concerned with the basic structure of matter, and in intersections with other fields. The facility will provide an extensive range of particle beams from protons and their antimatter partners, antiprotons, to ion beams of all chemical elements up to the heaviest one, uranium, with in many respects world record intensities. The new facility builds, and substantially expands, on the present accelerator system at GSI, both in its research goals and its technical possibilities.
These plans for a new facility are based on extensive discussions about future research opportunities between GSI Darmstadt, the universities and the various international user communities. They also build on the priority recommendations made over recent years by various high-level science committees worldwide that have reviewed the areas of research addressed by this facility.

Various national and international committees have assessed the GSI plans for a future facility, e.g. the German "Komitee für Hadronen- und Kernphysik (KHUK)" [2] and NuPECC. The proposal received very positive evaluations and strong support by all these boards.
Following a request of the Federal Ministry of Education and Research, the Wissenschaftsrat (German Science Council), the highest-ranking Scientific Advisory Committee of the German Government, performed an in-depth assessment of nine projects for new large-scale facilities for basic research – amongst them the GSI future plans. The results were publicized in July 2002 [3]. Again, the GSI project received a very favorable evaluation. The Wissenschaftsrat classified it among the five projects that got a very supportive recommendation.

Finally, on February 5th, 2003, the German government decided to build the International Accelerator Facility for Beams of Ions and Anti-Protons at Darmstadt. The government requested a Technical Report within two years and a participation of at least 25% of the costs from the international community. The construction of the Facility is foreseen to take 8 years.

W. Greiner et al. (eds.), Structure and Dynamics of Elementary Matter, 87–94.

This paper presents an overview of the future facility planned for GSI Darmstadt. More detailed information is contained in the Conceptual Design Report (CDR), which is available from the GSI Website and as a CD [1].

2. Facility Concept and Performance

The research to be performed with this future facility covers, in the broadest sense, studies with energetic primary and secondary beams of ions of highest intensity and quality, including an "antimatter beam" of antiprotons. Technical challenges that arise from the requirements of highest beam intensity include new rapidly cycling

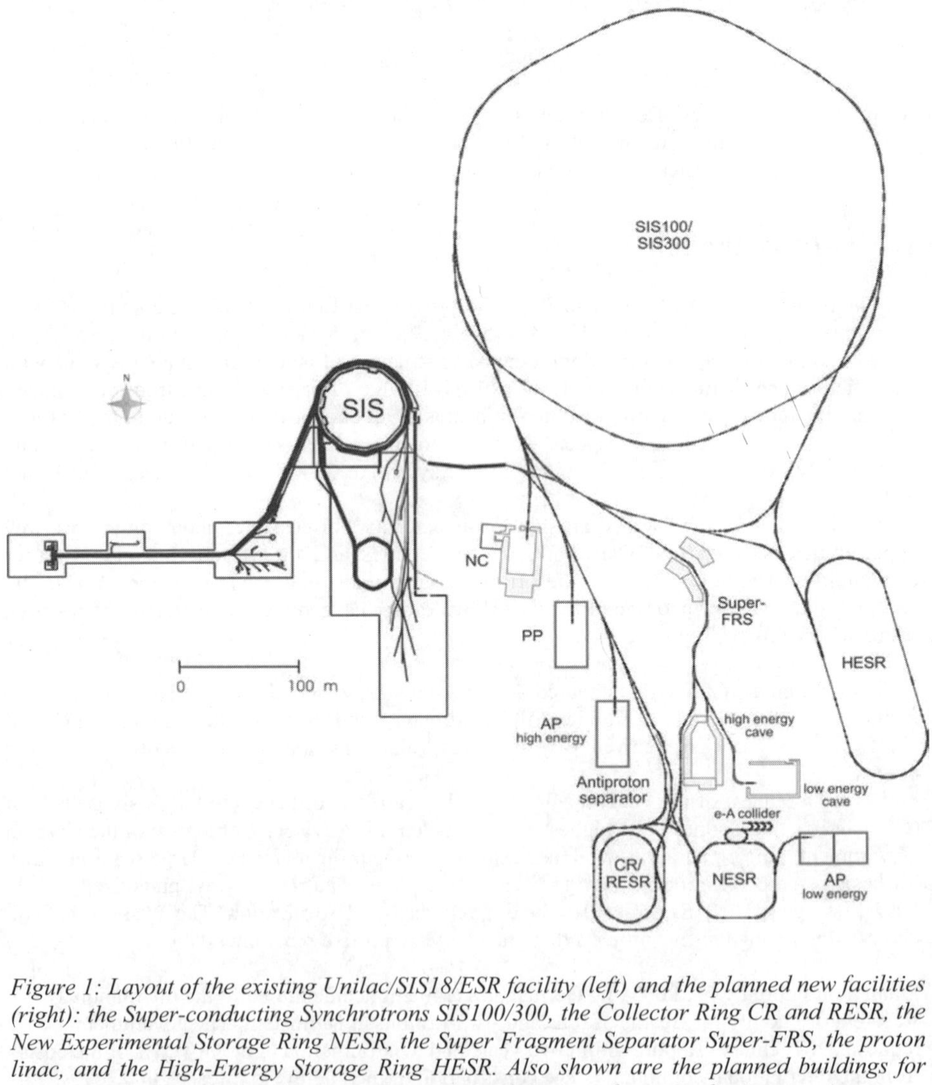

Figure 1: Layout of the existing Unilac/SIS18/ESR facility (left) and the planned new facilities (right): the Super-conducting Synchrotrons SIS100/300, the Collector Ring CR and RESR, the New Experimental Storage Ring NESR, the Super Fragment Separator Super-FRS, the proton linac, and the High-Energy Storage Ring HESR. Also shown are the planned buildings for plasma physics, nuclear collisions, radioactive ion beams, and atomic physics experiments.

superconducting magnets, both of the iron-core and of the cosine-theta types, and stringent vacuum requirements. High beam quality will be achieved in collector and storage rings with beam cooling, in particular high-energy electron-beam cooling of ion and antiproton beams.

The proposed layout of the technical facility is shown above in Figure 1. Key aspects of this facility are:

Full range of ion beam species: The proposed facility will accelerate all ions from protons (and antiprotons) up to the heaviest element, uranium.

Highest beam intensities: Intensities of primary heavy-ion beams will increase by a factor 100, secondary radioactive beams by a factor of up to 10,000 over the present GSI facility.

Substantial increase in beam energy: an increase in energy by at least a factor of 20 is planned for beams as heavy as uranium. In this energy regime nucleus-nucleus collisions are believed to exhibit maximum stopping and to generate hadronic matter at the highest baryon densities. It also enables the study of heavy quarks, i. e. charm production in highly compressed matter.

Precision beams: as already discussed, these will be achieved through sophisticated beam handling, such as stochastic and electron cooling of the ion beams. Together with the statistical accuracy and sensitivity that result from high beam intensities and high interaction rates, these beams will allow entering new areas of precision studies in the science fields envisioned.

Rings as accelerator structures of choice: for ion beams, from protons to uranium, synchrotrons provide the most economical solution to generate high beam energies. But more importantly, accelerator rings are unique in their capability to store, cool, bunch, and stretch beams. The choice of a ring system for the new facility also matches to the existing accelerator to be used as the injector.

The concept and layout of the new facility has evolved from the science requirements as follows: substantially higher intensities are achieved, compared to the present system, through faster cycling and, for heavy ions, lower charge state which enters quadratically into the space charge limit. The reduced charge state, and still a desired energy of up to 1.5 AGeV for radioactive beam production, requires a larger magnetic bending power. The SIS100 synchrotron fulfills these aspects. It also generates intense beams of energetic protons, up to 30 GeV, and from these antiprotons.

Heavy ion beams of high energy, i.e. up to 33 AGeV Uranium, are generated using ions in a high charge state plus the additional, somewhat slower but still rapidly cycling SIS300 synchrotron ring. This is possible because of the 3 times stronger magnets of SIS300, which are going up to 6T magnetic field. The intensity required for these beams allows for long spills of up to 100s. Similarly, the SIS300 can be used as a stretcher for radioactive beams. Both, primary and secondary beams can be injected, cooled and stored in a system of rings with internal targets and in-ring experimentation. Rings may be shared for uses with different beams. Based on the developments and excellent experiences with cooled beams at the present GSI facility, the future program will broadly take advantage of this aspect of beam handling. More details about the beam characteristics can be found in the CDR [1].

3. Research Programs

In most general terms, the scientific thrusts of the facility can be summarized by the following broad research goals:
The first goal is to achieve a comprehensive and quantitative understanding of all aspects of matter that are governed by the strong force, i.e., matter at the level of nuclei, nucleons, quarks and gluons. The research goal of the present facility thus encompasses all aspects of hadronic matter, including the investigation of fundamental symmetries and interactions that are relevant for this regime.

The second goal addresses many-body aspects of matter. The many-body aspects play an important and often decisive role at all levels of the hierarchical structure of matter. They govern the behavior of matter as it appears in our physical world.

These two broad science aspects, the structure and dynamics of hadronic matter and the complexity of the physical many-body system, transcend and determine the more specific research programs that will be pursued at the future facility:

i. Investigations with beams of short-lived radioactive nuclei, addressing important questions about nuclei far from stability, areas of astrophysics and nucleo-synthesis in supernovae and other stellar processes, and tests of fundamental symmetries;
ii. The study of hadronic matter at the sub-nuclear level with beams of antiprotons, including the two key aspects: confinement of quarks and the generation of the hadron masses. They are intimately related to the existence (and spontaneous breaking) of chiral symmetry, a fundamental property of strong interactions;
iii. The study of compressed, dense hadronic matter in nucleus-nucleus collisions at high energies;
iv. The study of bulk matter in the high-density plasma state, a state of matter of interest for inertial confinement fusion and astrophysical settings;
v. Studies of Quantum Electrodynamics (QED), of extremely strong (electro-magnetic) fields, and of ion-matter interactions and the study of very low energy anti-protons.

4. Parallel Operation and Synergy

An important consideration in the design of the facility was a high degree of truly parallel operation of the different research programs. With the proposed scheme of accelerator and storage rings, maximum integrated beam time, or integrated luminosity can be provided for each of the different programs operated in parallel. Figure 2 illustrates how parallel operation would be performed with a cooled and post-accelerated antiproton beam, in parallel to a fixed target experiment with radioactive beams and/or relativistic heavy-ion beams slowly extracted from the second synchrotron ring SIS300, and an additional beam pulse (every few minutes) for plasma physics. Thereby the facility operates for the different programs more or less like a dedicated facility. This high degree of synergy in facility performance will also lead to an increased scientific synergy between the different areas of research and to interdisciplinary intersections.

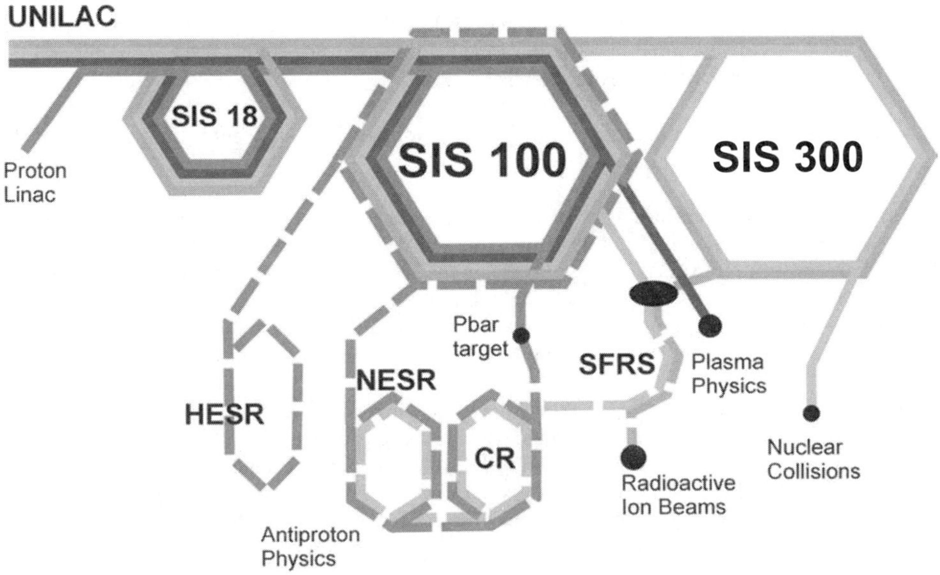

Figure 2: Schematic illustration of the highly efficient parallel operation at the new facility. In the example shown, all four different scientific programs are served in parallel: A proton beam (orange), accelerated in SIS100, produces antiprotons (orange dashed) in the antiproton target-station, which are then collected, accumulated and cooled in the CR/NESR storage-ring combination, injected and accelerated in SIS100, and then transferred to the HESR for in-ring experiments. In parallel, i.e. during the fraction of the SIS100 super-cycle not needed for the protons, a primary ion beam (blue) is accelerated in SIS100 and slowly extracted to the Super-FRS to produce radioactive secondary beams (blue dashed) for fixed target experiments. (Alternatively the radioactive beams could be sent to the CR and NESR instead of the antiprotons). In addition, every 10-100 seconds a high-energy heavy-ion beam (gray) is accelerated in SIS100/300 and slowly extracted for nuclear collision experiments; these experiments require lower beam intensities than the maximum possible from the accelerator. Moreover, intense beam pulses (green) are provided every few minutes for plasma physics experiments that require very low repetition rates.

5. Status of the Project

Since February 5th, 2003, when the proposal has been accepted for construction, important progress has been made despite large funding problems:

i) Within the superconducting magnet R&D program, model magnets for SIS 100 have been built and tested by an international collaboration at or above design specifications.

ii) The tedious process of getting construction permits has been started and a conditional permission has been obtained already to build and operate the facility as presented to the authorities

iii) Scientific collaborations and proto-collaborations, some with more than 300 collaborators, have been formed and started or continued the planning and the design of their experiments

A few collaborations are presented here in order to give an insight into the dynamics of the process, which defines the research program:

The PANDA collaboration [4] prepares a 4π detector for anti-proton reaction studies in the High Energy Storage ring. The following experiments, among others, are foreseen:
a) Charmonium (cc) spectroscopy: Precision measurements of mass, width, decay branches of all charmonium states, especially for extracting information on the quark confinement. The unequaled resolution in the pp_bar formation process and small systematic uncertainties give the unique opportunity to improve dramatically the knowledge, which cannot be achieved elsewhere.
b) Firm establishment of the QCD-predicted gluonic excitations (charmed hybrids, glueballs) in the charmonium mass range (3-5 GeV/c^2) using high statistics in combination with sophisticated spin parity analysis in fully exclusive measurements.
c) Search for modifications of meson properties in the nuclear medium, and their possible relationship to the partial restoration of chiral symmetry for light quarks. Particular emphasis is placed on mesons with open and hidden charm, which extends ongoing studies in the light quark sector to heavy quarks, and adds information on contributions of the gluon dynamics to hadron masses.
d) Precision gamma-ray spectroscopy of single and double hyper nuclei for extracting information on their structure and on the hyperon-nucleon and hyperon-hyperon interaction.

The CBM Collaboration [5] proposes to build a dedicated heavy-ion experiment to investigate the properties of highly compressed baryonic matter. The goal is to explore the QCD phase diagram in the region of moderate temperatures but very high baryon densities. The envisaged research program includes the study of key questions of QCD like confinement, chiral symmetry restoration and the nuclear equation of state at high densities. The most promising diagnostic probes are vector mesons decaying into dilepton pairs, strangeness and charm. The collaboration intends to perform comprehensive measurements of hadrons, electrons and photons created in collisions of heavy nuclei. CBM will be a fixed target experiment, which covers a large fraction of the populated phase space. The major experimental challenge is posed by the extremely high reaction rates of up to 10^7 events/second. These conditions require unprecedented detector performances concerning speed and radiation hardness. The detector layout comprises a high resolution Silicon Tracking System in a magnetic dipole field for particle momentum and vertex determination, Ring Imaging Cherenkov Detectors and Transition Radiation Detectors for the identification of electrons, an array of Resistive Plate Chambers for hadron identification via TOF measurements, and an electromagnetic calorimeter for the identification of electrons, photons and muons. The detector signals are processed by a high-speed data acquisition and trigger system.

The FLAIR community:

The low energy anti-proton community working at the Anti-Proton Decelerator Facility at CERN has put forward a proposal to move their program to Darmstadt and to join the Atomic Physics program. Within the FLAIR proposal [6], they point out, that the Future Accelerator Facility for Beams of Ions and Antiprotons at Darmstadt will produce the highest flux of antiprotons in the world. In the planned complex of storage rings, it is already possible to decelerate the anti-protons to about 30 MeV kinetic energy, opening up the possibility to create antiprotons at even lower energies. They propose to overcome the limitations of the CERN AD by using cooled beams at higher intensities and at a factor 10 or even lower energy. The new low energy facility should have the possibility of slow (*i. e.* continuous) extraction, which will allow nuclear/particle physics type experiments requiring coincidence measurements to be performed. A facility consisting of two storage rings, a magnetic (LSR) and an electrostatic (USR) one, and a universal trap facility (HITRAP) is proposed, recycling and upgrading existing storage rings. These components of the facility can provide fast and slow extracted cooled beams down to 300 keV (LSR), to 20 keV (HITRAP, USR), and cooled particles at rest (HITRAP). This will allow a large variety of new experiments to be performed. Among the unique experiments only possible at such a facility are nuclear physics studies, using antiprotons as a hadronic probe to investigate the structure of nuclei, including radioactive isotopes produced at the future facility, and many atomic collision type experiments with internal targets in both storage rings. An important synergetic aspect is that the whole structure will also be used to study highly charged ions, including trapping them in Penning traps like HITRAP and investigating them in the new Cave A, both are already part of the atomic physics program of the future facility as described in the Conceptual Design Report [1]

The NUSTAR Community

A group of scientists in nuclear structure and astro-physics research formed recently the NUSTAR community [7] in order to put forward the experimental program with radioactive beams behind the Superconducting Fragment Separator. There, slowly extracted radioactive beams allow in a series of experimental set-ups, e.g., studying Isospin phenomena in nuclear collisions up to 1.5 AGeV, or gamma spectroscopy on radioactive nuclei at low energies or even at rest. This community is also in charge of in-beam experiments in the Next Experimental Storage Ring (NESR). There, a e^--A mini-collider is foreseen to measure through elastic collisions of electrons with radioactive ions the electromagnetic radius of the new unstable nuclei. Quite a sophisticated apparatus is required to measure the electrons with precision and large solid angle. In order to identify potential neutron skin, these measurements are then compared with strong interaction radii measured through nuclear collisions on, e.g., a hydrogen gas target. Recently, it was suggested that the e-A mini collider should be complemented by functions of a p_bar-A mini collider [8].

There are new proposals ASSIA [9] and PAX [10], which focus on experiments with polarized beams or polarized targets. The scientific program proposed by the ASSIA collaboration could benefit in the future from a polarized proton beam of momentum up to 15 GeV/c, if the operation of the HESR as a polarized proton-anti-proton collider would provide reasonable luminosities due to the continuous electron cooling of the anti- proton beam.

Both the Plasma Physics program and the Atomic Physics program with highly charged high Z nuclei at this Facility is out of the scope of this conference, but well documented for the interested reader in the CDR [1]

6. Outlook

By end of 2004 a draft agreement is to be prepared to assure the construction of the International Facility as a truly international project. The letter of Intents, due in January and in April, are to be evaluated in Spring and Summer of 2004 to give green light to the collaborations for their presentation of a Technical Report end of 2004. The project will enable scientists already in a few years to work with an increased performance of the SIS18 complex. First use of higher beam energies of SIS100 is foreseen about 5 years after constructions starts. The downtime for Science due to the switching over from the existing SIS 18 facility to the new International Facility shall be kept to a minimum.

References

[1] An International Accelerator Facility for Beams of Ions and Antiprotons; Conceptual Design Report, November 2001 and References therein; http://www.gsi.de/GSI; Future/cdr/; this report is also available on CD.
[2] see: http://ikpp30.ikp.kfa-juelich.de/khuk/empfehlungen.shtml
[3] see: http://www.wissenschaftsrat.de/presse/pm_2002.htm
[4] PANDA Collaboration, http://www.gsi.de/zukunftsprojekt/experimente/hesr-panda/index.html
[5] CBM collaboration, http://www.gsi.de/zukunftsprojekt/experimente/CBM/index.html
[6] FLAIR community, http://www.gsi.de/zukunftsprojekt/experimente/flair/index.html
[7] NUSTAR community, http://www.gsi.de/forschung/kp/kp2/nustar.html
[8] P. Kienle, Nucl. Instr. Meth. B 214 (2004) 191
[9] ASSIA proposal
[10] PAX proposal
[11] DIRAC at Darmstadt proposal

STRANGENESS PRODUCTION AND PROSPECTS FOR LHC

Christian Kuhn for the ALICE collaboration
Institut de Recherches Subatomiques (IN2P3/CNRS - ULP), F-67037 Strasbourg

Abstract

Strangeness production represents one of the major probes for studying strong interacting matter in high energy density conditions and characterizing the Quark Gluon Plasma. In the light of some important results extracted from the past and present heavy ion experiments, at SPS and RHIC, we show that strange particles will also play an essential role at LHC and present new opportunities offered, both in the soft and the hard physics domains. Simulations of strangeness reconstruction within the ALICE experiment are presented, mainly focused on Λ measurements. Finally, we discuss the sensitivity of ALICE for the discovery of new forms of strange matter, namely metastable short-lived (weak decaying) strange dibaryons.

1. Introduction

Heavy ion collisions at (ultra-)relativistic energies are expected to reach sufficiently high energy densities to create a Quark Gluon Plasma (QGP) [1]. Over the large energy range going from SPS up to RHIC and the future LHC, strangeness production is one of the most sensitive probes of this new state of matter. Indeed, strange particles provide many relevant observables both in the soft physics regim, where they help to understand the global chemical and kinetic freeze-out conditions of the bulk matter, and in the hard physics regim, where it is essential to identify hadrons up to very high momenta. The interest for studying the bulk properties is that they may yield indirect information about the initial expected partonic medium. Particle abundances may reflect the temperature and the baryo-chemical potential at chemical freeze-out and probe the degree of chemical equilibration [2, 3]. Transverse momentum spectra and anisotropies tell us about the expansion dynamics, probing the degree of kinetic equilibration and the degree of collectivity (transverse and elliptic flow), hence indirectly the pressure built at the early time of the collision. With this respect, strange particles might represent a powefull tool to solve the question whether

W. Greiner et al. (eds.), Structure and Dynamics of Elementary Matter, 95–102.

the observed flow has originated in a partonic phase. Of course, all these studies conducted at SPS and RHIC will have to be also performed in priority at LHC, in order to see how the whole collision picture evolves over this important jump in energy (examples will be given in section 2). Beside the bulk of low p_T particles, if the collision energy is high enough, high p_T hadrons are produced, resulting from very energetic parton-parton collisions at the early stage of the reaction. At RHIC, a new era has opened up with the onset of these hard processes, allowing for the study of jet quenching effects [4]. This topic will be briefly examined in section 3. At LHC, the influence of these hard processes is expected to be much stronger, thus offering many new opportunities (see section 4). Section 5 will be devoted to a short description of strange particle reconstruction within the ALICE detector. Preliminary simulation results will be given, focused mainly on Λ measurements, with particular emphasis on the reconstruction performances for high p_T particles. In the last section, our interest for a more exotic aspect of strangeness is expressed. The proposal that strange dibaryons might exist as metastable states has stimulated a substantial activity both on the theoretical and experimental sides for many years. One may wonder about the ALICE capabilities for pursuing this research in the future. We have addressed this question by evaluating the sensitivity of the ALICE detector to the discovery of such exotic strange clusters, concentrating on two hypothetical bound states: the H^0, a six quark bag (uuddss) and the $\Xi^0 p$, made by coalescence of 2 baryons.

2. Chemical composition and expansion dynamics

Reproducing particle abundancies with statistical models, using the Hadron Resonance Gaz Hamiltonian, has met the same success with the SPS and the RHIC data [3], suggesting at first that a high degree of chemical equilibration was reached and that the appearance of the QGP in the initial state could be the driving force towards this equilibrium. Even if widely controversial, this approach has also allowed to build a unified statistical picture [5] from SIS up to RHIC, which rises the question: what about LHC ? If particle production at LHC is characterized by the same systematics as the one observed at SPS and RHIC, then the particle yields will follow statistical orders given by the values of the chemical freeze-out temperature (T) and the baryochemical potentiel (μ_B) , with T expected to be hardly different from the one obtained at RHIC. Deviations from this prediction, however, would be particularly interesting as they might evidence the importance of hard processes. Another crucial issue is the evolution of the Wroblevski factor (λ_s). According to the statistical model predictions, this quantity, defined as the ratio between the newly created strange quark-antiquark pairs and the newly created light quark-antiquark pairs, should also be almost constant between RHIC and LHC, because of the very low variations expected for T and μ_B at chemical freeze-out. But according to other predictions [6], implying a non-equilibrium scenario and the possibility to have a large oversaturation of strangeness, a very high temperature in the QGP phase should result in a large enhancement of $s\bar{s}$ pair creation, substantial enough to lead to an increase of λ_s. Therefore, it will be very interesting to measure this quantity at LHC.

Strange particles play also an essential role in the study of the expansion dynamics. Indeed, an "hydrodynamically motivated" fit (using the blast wave model [7]) of the m_T spectra obtained at RHIC indicates that the Ξ particles seem to depart from the common thermal freeze-out behaviour shown by the other particles [8]. The temperature associated to their kinetic freeze-out is found to be higher and close to the chemical freeze-out temperature. This might indicate that these particles are decoupled earlier from the system due to their low rescattering cross section with surrounding medium. But in the mean time, they develop a substantial transverse flow which suggests that this collective motion could have originated, at least partially, in the early partonic stage. Part of the answer could be given by measuring the elliptic flow of these multi-strange baryons. The magnitude of elliptic flow [9] has been found to increase substantially from AGS to SPS and RHIC where it reaches the values predicted by hydrodynamical models. This suggests that a very rapid and almost complete local kinetic equilibration is reached [10]. This conclusion is based on the intuitive argument that the hydrodynamic limit of infinitely fast rescattering, leading to instantaneous local thermal equilibrium produces the largest possible elliptic flow values [10] and that the data seem to saturate this limit. These proofs of high pressure and rapid thermalization are considered as strong indirect evidences for the initial state being a QGP. With this respect, the systematic study of elliptic flow for strange and especially multi-strange particles is really of great interest, because it might reveal directly the presence of partonic flow [11]. Indeed, if the elliptic flow of identified particles proves to be independent of their relative hadronic cross sections, it may evidence that v_2 is built during a partonic phase. New results from STAR show that Ξ particles are indeed submitted to an important elliptic flow, comparable to that of lighter particles (see J. Harris in this conference). At LHC, elliptic flow is expected to saturate in the QGP phase which means that this kind of effects should be more pronounced and thus much clearly interpretated there.

3. Onset of hard processes at RHIC

A clear supression of high p_T hadrons is observed in central Au + Au collisions at RHIC [12] while no suppression is observed in the hadronic reference d-Au [13, 14]. One may then conclude that this suppression is consistent with the one expected from partonic energy loss in a QGP, or more carefully that it is the result of final state effects in dense QCD matter and not that of initial state effects. This has been widely debated in this conference (see for example M. Gyulassy). What we want to stress here is that it is interesting to provide measurements of this suppression for different hadron species and this is possible for strange particles that can be identified by their decay topology up to high p_T. In [15], it has been shown that baryons (Λ) and mesons (K^0_s) clearly exhibit a different suppression pattern. Λ yields seem to scale with the number of binary collisions up to larger transverse momenta than K^0_s whose suppression occurs before. Besides, a strong correlation between elliptic flow saturation and high p_T suppression is observed. It has been shown that these behaviours are probably the result of a complex interplay between two competitive hadronisation processes, namely quark coalescence at low and intermediate p_T and parton fragmentation at high p_T [16, 17]. This new type of analysis illustrates well the fact that the systematic study of strangeness has entered a very sophisticated stage where the old observables combined with new ones allow for a much better understanding of the chemistry and the dynamics of heavy ion collisions. As a result, many new footprints of the QGP but also related new questions have raised which need a systematic knowledge of the hadronic reference and the study of all observables over the largest possible energy domain in order to cover at best the transition from soft to hard physics.

4. New opportunities at LHC

At LHC, there will be plenty of new opportunities due to the jump of energy. The higher particle multiplicity combined to a much larger influence of hard processes will offer novel kinds of observables and data analysis, within the frame of the dedicated ALICE experiment. The large number of newly produced particles will allow, for example, to extract the freeze-out conditions for single events and consequently to caracterize selected event classes: high T and low T event classes, in order to disentangle different dynamics evolutions in the T-μ_B plane. Using a selection of jet events, it will be possible by performing a chemical analysis of those particular events, to study the links between the initial and the freeze-out phases and determine the correlations between the soft and the hard processes. Concerning jet quenching effects, a part from high p_T hadron suppression and its dependence on nuclear geometry (impact parameter and orientation with respect to the reaction plane), several other promising observables will be available for the study of parton energy loss, up to very high transverse momenta. The measurement of particle ratios at high p_T is one of them. Indeed, hard gluons are expected to loose approximately twice more energy than hard quarks. Therefore, the ratio of some hadronic species should be modified, depending on the relative contribution of gluon fragmentation. It will also be possible to study the attenuation and p_T broadening of jets, including the measurement of the total energy radiated within the jet cone and the broadening of the transverse energy distribution in the jet. Jet tagging, by determining the hadronic activity in the direction opposite to a hard photon, will also represent a new possibility. For all these studies, hyperons play a major role because they are measured by their decay topology and thus can be identified up to very high momenta where traditional hadron identification (specific ionisation in gas detectors or time of flight measurements) fails.

5. Strangeness reconstruction in ALICE

In the central barrel of the ALICE detector, the two main tools for strange particle reconstrution are the Time Projection Chamber (TPC) and the Inner Tracking System (ITS) which compose the main tracking device of the experiment. Two different strategies have been applied so far for secondary vertex finding. The first one consists of restraining the reconstruction region to a very narrow fiducial zone inside the beam pipe ($R_{max} = 2.9cm$), in order to reduce as much as possible the background. In this case, only the tracks producing 6 clusters in the ITS (one per layer) are selected for the reconstruction. With this method, we get indeed a very pure signal ($S/B \sim 5$) but the counterpart is a very small acceptance. Assuming a production rate of 100 Λ's per central Pb+Pb event in the detector pseudorapidity range, this leads to a poor yield of only 1.3 reconstructed $\Lambda's$ per event.

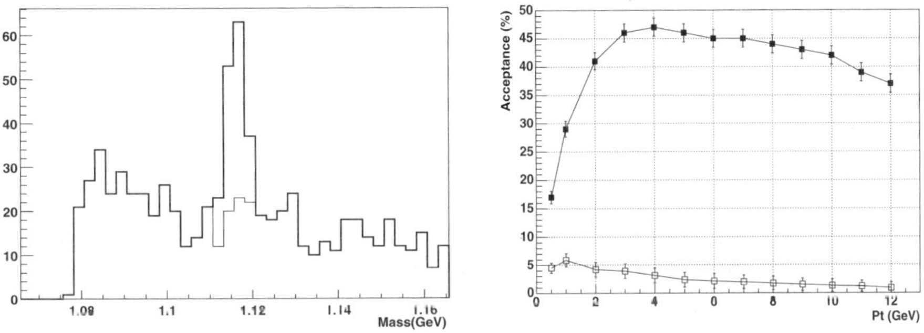

Figure 1. Left pannel: Λ invariant mass spectrum corresponding to the reconstruction of 20 Hijing events (with $dN_{ch}/dy = 4000$). Right pannel: Λ acceptances for two different fiducial zones: $R_{max} = 2.9cm$ (open squares) and $R_{max} = 100cm$ (full squares).

In order to increase the yields, especially at high p_T, a second strategy has been developped. It consists of performing the secondary vertex reconstruction with all the tracks found in the TPC, without any further selection based on the ITS clusters. In this case, the reconstruction zone radius can be larger than the radius of the last ITS strip layer. For this study, it has been set at 100 cm. The result is an important jump of acceptance. The price to pay is a strong increase of the background (see figure 1 (left)). The reconstruction cuts have been optimized in order to reduce it as much as possible ($S/B \sim 1.4$) without deteriorating too much the efficiency (16.5%) since the aim of this strategy is to approach the best possible reconstruction rate. As a result, it goes up to 5 reconstructed $\Lambda's$ per event. The improvement in acceptance is especially interesting at high p_T, as shown in figure 1 (right) where Λ acceptances are compared for $R_{max} = 2.9cm$ and $R_{max} = 100cm$. The gain for high p_T values represents almost a factor 20. The next step has consisted of improving also the efficiency for high p_T $\Lambda's$ by defining a set of p_T-dependent geometrical cuts which results in the efficiency profile shown in figure 2 (left).

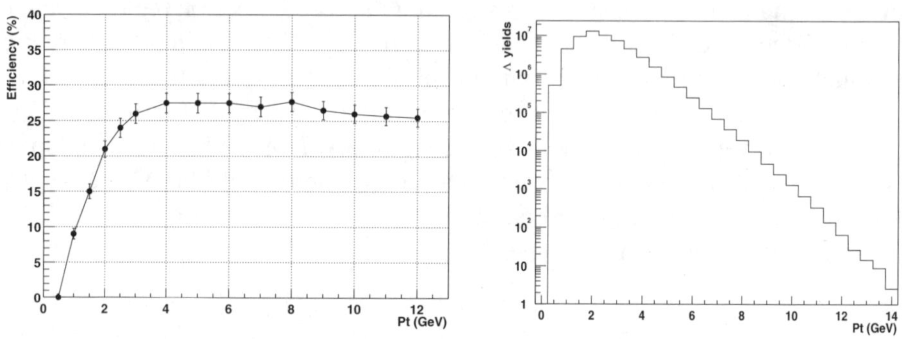

Figure 2. Λ reconstruction efficiency as a function of p_T (left). Transverse momentum distribution of reconstructed $\Lambda's$ for 10^7 central Pb+Pb events in ALICE (right)

The convolution of the Λ reconstruction rate (efficiency $\times$ acceptance) with the transverse momentum distribution expected for Λ production in central Pb+Pb collisions at LHC, supposed here to simply follow m_T scaling, can then provide the inclusive spectrum of reconstructed Λ's shown in figure 2 (right). We can conclude that, for a total number of 10^7 such events, a significant statistics of reconstructed Λ's can be obtained up to p_T= 13 GeV/c, which is of crucial interest for all the Physics aspects discussed before.

6. Search for exotic forms of strange matter in ALICE

Early predictions [18] suggest that a six quark bag bound state (uuddss) called H-dibaryon (H^0), may exist, because the color magnetic forces are attractive and thus allow its groundstate to be below the strong decay threshold ($M_{\Lambda\Lambda} = 2231$ MeV). Probably not stable against weak hadronic decay, its most probable mass range should be between 2055 and 2231 MeV. In this region, the H^0 can decay to Λ n or Σ N (nucleon) (if $m_{H^0} > 2134$ MeV) and $\Lambda N\pi$ (if $m_{H^0} > 2195$ MeV). The lifetimes associated with these various decay modes are predicted to be between 10^{-8} and 10^{-10}s. The hadronic counterpart to the H^0 is the so-called Dilambda $(\Lambda\Lambda)_b$, a bound state of two Λ's, with the same quantum numbers and the same decay channels as the H^0. Besides the Dilambda, according to the work of J. Schaffner-Bielich et al. [19], several other dibaryon bound states ($(\Xi^0 p)_b$, $(\Xi^0\Lambda)_b$,...) might exist. These authors have calculated the weak hadronic decays of various possible strange dibaryons and their lifetimes by using weak SU(3) symmetric contact interactions. They have also estimated their production rate by combining transport simulations (using RQMD) with wave function coalescence. At RHIC energy, the predicted yields per event are of the order of $10^{-2} - 10^{-3}$ per unit of rapidity. We hope to get soon also some estimates for the LHC energy. With decay lengths $(c\tau)$ predicted to be between 1 and 5 cm, all the proposed topologies are, in principle, findable in ALICE.

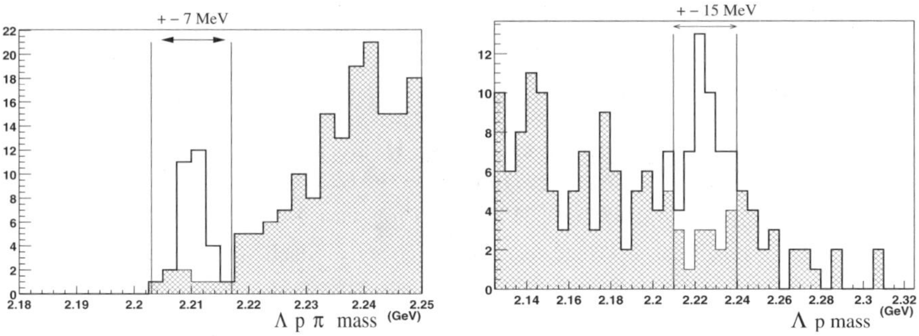

Figure 3. H^0 (left) and $(\Xi^0 p)_b$ (right) invariant mass spectra in ALICE (see text).

Among the proposed decay modes for the H^0, we have study so far the one where all final products are charged: $H^0 \longrightarrow \Lambda p\pi^- \longrightarrow p\pi^- p\pi^-$. As a first attempt, we have considered for the H^0 a mass of 2210 MeV, the same lifetime as for the Λ: $\tau = 2.6 \times 10^{-10} s$, a constant rapidity distribution limited to the midrapidity region ($-1 < y < +1$) and an exponential transverse momentum distribution. Figure 3 (left) shows the invariant mass spectrum obtained by the reconstruction of this decay chain in 135 000 Pb-Pb events, the H^0 multiplicity being of 1 per event. The acceptance is equal to about 0.5% and the efficiency to about 3.5%. This gives a rate of 1.7×10^{-4} reconstructed H^0 per event. By making a rough extrapolation of the background to the full expected statistics of 10^7 central events, one may deduce the detector sensitivity. We find that the observation of a H^0 signal, above the background at a 3σ confidence level, requires the reconstruction of about 80 H^0's. The sensitivity is defined by this number divided by the total number of analysed events and by the reconstruction rate. The sensitivity obtained is of about $5 \times 10^{-2} H^0$/event over the two units of rapidity covered, hence $2.5 \times 10^{-2} H^0$ per event and per unit of rapidity. For the reconstruction of the $(\Xi^0 p)_b$ (mass = 2225 MeV) in the $(\Xi^0 p)_b \longrightarrow \Lambda p$ channel, we find comparable results: a sensitivity of about 1.5×10^{-2} dibaryon per event and per unit of rapidity. Figure 3 (right) shows the invariant mass spectrum corresponding to the reconstruction of this decay chain in 60 000 events with a multiplicity of 1 $(\Xi^0 p)_b$ per event.

7. Conclusion

The study of high energy heavy ion collisions has reached a stage where serious evidences for a new phase of matter have been found, but work is still necessary to get a deep understanding of this complex matter. The excitation functions of all observables must be studied in the largest possible energy domain. After RHIC, we will benefit of a large gap in energy with the LHC program in which strangeness will keep on being one of the dominant probes.

References

[1] For a review, see: S.A. Bass, M. Gyulassy, H. Stoecker and W. Greiner, J. Phys. G **25** (1999) R1. See also H. Satz, Rep. Prog. Phys. **63** (2000) 1511

[2] R. Stock, Phys. Lett. B**456** (1999) 277 and Prog. Part. Nucl. Phys. **42** (1999) 295

[3] P. Braun-Munzinger, D. Magestro, K. Redlich and J. Stachel, Phys. Lett. B**518** (2001) 41

[4] X. N. Wang, M. Gyulassy and M. Plumer, Phys. Rev. D**51** (1995) 3436

[5] J. Cleymans and K. Redlich, Phys. Rev. Lett. **81** (1998) 5284 and references therein

[6] J. Rafelski and J. Letessier, hep-ph/0305284, submitted to J. Phys. G.

[7] E. Schnedermann, J. Sollfrank and U. Heinz, Phys. Rev. C**48** (1993) 2462.

[8] J. Adams et al., nucl-ex/0307024, submitted to Phys. Rev. Lett.

[9] J.Y. Ollitrault, Phys. Rev. D**46** (1992) 229.

[10] U. Heinz and P.F. Kolb, Nucl. Phys. A**702** (2002) 269

[11] N. Xu and Z. Xu, Nucl. Phys. A**715** (2003) 587c

[12] C. Adler et al., Phys. Rev. Lett **89** (2002) 202301 and S.S. Adler et al., nucl-ex/0308006

[13] J. Adams et al., Phys. Rev. Lett. **91** (2003) 072304

[14] S.S. Adler et al., Phys. Rev. Lett. **91** (2003) 072303

[15] J. Adams et al., nucl-ex/0306007, submitted to Phys. Rev. Lett.

[16] R. Fries, B. Mueller, C. Nonaka and S.A. Bass, nucl-th/0301087

[17] D. Molnar and S.A. Voloshin, nucl-th/0302014

[18] R.L. Jaffe, Phys. Rev. Lett. **38** (1977) 195

[19] J. Schaffner-Bielich, R. Mattiello and H. Sorge, Phys. Rev. Lett. **84** (2000) 4305

RESONANCE PRODUCTION IN RELATIVISTIC HEAVY ION COLLISIONS

C. Markert for the STAR Collaboration
Physics Department, Yale University, New Haven, CT 06520, USA

Abstract Resonances and their decay products may interact strongly with the hadrons in the fireball medium from the time of chemical freeze-out until the system breaks up at kinetic freeze-out. In order to understand the evolution and expansion of this hot and dense medium we compare resonance particle yields and spectra from elementary p+p and heavy ion collisions.

1. Introduction

Resonances have a lifetime of a few fm/c which is on the order of the expected lifetime of a fireball source. The measurements of resonances in heavy ion collisions may provide information about the early stage of the expanding source in terms of the influence of the medium on the resonance and decay particles. Comparisons between p+p and A+A interactions may show directly the influence of the medium in terms of yields, widths, mass and momentum distributions. Measurements from different short lifetime resonances may provide us with information about the time between chemical and kinetical freeze-out associated with the rescattering and regeneration of their decay daughters.

2. Resonances in Medium

Resonances are produced in a heavy ion collision. A fraction of them decay due to their short lifetimes inside the fireball medium. In addition elastic interactions can regenerate the resonances. Two freeze-out conditions of the fireball expansion can be characterized by the end of inelastic interactions (chemical freeze-out) and the end of elastic interactions (kinetic freeze-out).

Thermal models are successful in describing the production yield of stable particles with one temperature at chemical freeze-out. To make a statement about thermally produced resonances we have to take into account the elastic interactions that occur between the time of chemical freeze-out and kinetic

W. Greiner et al. (eds.), Structure and Dynamics of Elementary Matter, 103–110.

freeze-out. Some yield changes may take place due to rescattering of the decay daughters causing a loss of signal in the invariant mass reconstruction. The regeneration of resonances will have the opposite effect. For stable particles elastic interactions do not change their yields. If the time between chemical and thermal freeze-out is very short, the contribution of rescattering of decay daughters and regeneration of resonances is also small. Elastic interactions with the resonances can change their phase space distribution, for example the transverse momentum distribution. The time between chemical and thermal freeze out can be verified by comparing yields of resonances with different lifetimes. Transport model calculations may be able to describe medium effects at a microscopic level [1, 2]. These calculations have only been developed recently, since heavy ion experiments are capable of measuring resonances with short lifetimes. The initial motivation for the measurement of resonances was to determine mass shifts and width broadenings of the mass signal due to the influence of the medium (recent papers [13, 14]). The early signal can be changed during the expansion of the fireball source, when inelastic and elastic interactions may have an additional effect on the mass and width of the resonances.

3. Resonance Reconstruction

Experimentally resonances in heavy ion collisions have been measured at the NA49 experiment at $\sqrt{s_{\rm NN}} = 17$ GeV and at the STAR experiment at the STAR experiment at $\sqrt{s_{\rm NN}} = 200$ GeV. The resonances are reconstructed from their decay daughters (see table 1). Charged particles are identified via energy loss (*dE/dx*) and their measured momenta in the *Time Projection Chamber* (TPC). The Λs and Ξs are reconstructed via topological analysis [12]. The resonance signal is obtained by the invariant mass reconstruction of each daughter combination and subtraction of the combinatorial background calculated by the mixed event technique. The resonance ratios, spectra and yields are at mid-rapidity for the STAR data and 4π values for the NA49 data.

Particle	mass (MeV/ c^2)	width (MeV/ c^2)	lifetime (fm/ c)	decay channel
$K(892)^0$	896.1 ± 0.27	50.7 ± 0.6	3.89	$K + \pi$
$\Sigma(1385)^+$	1382.8 ± 0.4	35.8 ± 0.8	5.5	$\Lambda(\rightarrow p+\pi) + \pi^+$
$\Sigma(1385)^-$	1387.2 ± 0.5	39.4 ± 2.1	5.0	$\Lambda(\rightarrow p+\pi) + \pi^-$
$\Lambda(1520)$	1519.5 ± 1.0	15.6 ± 1.0	12.6	$p + K$
$\Xi(1530)^0$	1535.0 ± 0.6	9.9 ± 1.9	19.9	$\Xi + \pi$

Table 1. Resonances from PDG [9]

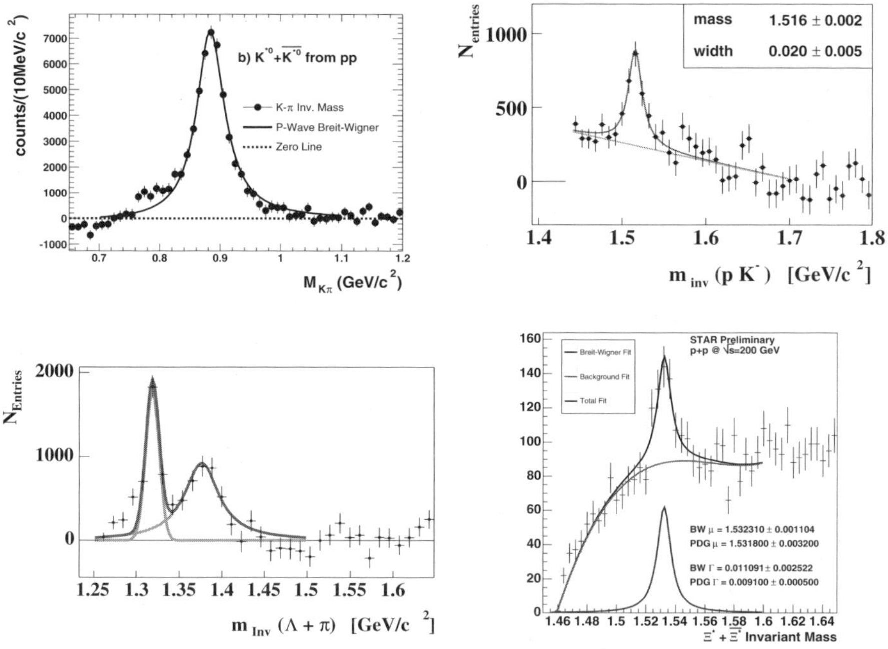

Figure 1. Invariant mass distributions of resonances in p+p collisions at $\sqrt{s_{\rm NN}}$ =200 GeV after mixed-event background subtraction. Top-Left: K(892) (π+K^- pairs), Top-Right: Λ(1520) (p+K^- pairs). Bottom-Left: Σ(1385)(Λ+π pairs), Right-Bottom: Ξ(1530) (Ξ +π pairs).

4. Strange Resonances in p+p Collisions at $\sqrt{s_{\rm NN}} = 200$ GeV

The masses and the experimental widths of the strange resonances K(892) [16], Σ(1385)[12], Λ(1520) [6, 8] and Ξ(1530) [15] of the integrated invariant mass spectrum are at mid-rapidity in p+p collisions. They are in agreement with the values from the PDG [9] and the expected width and mass from the simulation of the momentum resolution and the acceptance of the detector (figure 1).

Figure 2 shows the transverse mass distribution of K(892) (from p+p and Au+Au collisions) and Λ(1520) (from p+p collisions) where the inverse slope parameter T, the mean transverse momentum p_T and the yield for p+p collision systems are obtained with: $T_{K(892)}$ = 223 $\pm$ 9 MeV, $T_{\Lambda(1520)}$ = 326 $\pm$ 42 MeV and $\langle p_T\rangle_{K(892)}$ = 0.68 $\pm$ 0.03 $\pm$ 0.03, $\langle p_T\rangle_{\Lambda(1520)}$ = 1.04 $\pm$ 0.09$\pm$ 0.10.

The mean transverse momentum as a function of mass indicates that resonances with mass higher than 1 GeV/c^2 show a different trend than the stable particles (see figure 2). The black line in this figure is a fit to the ISR data

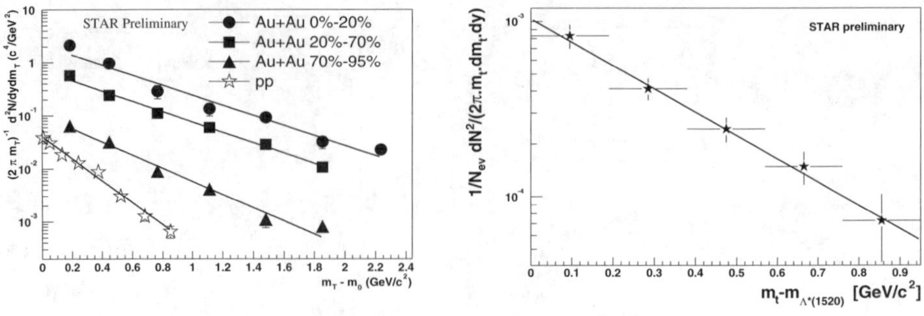

Figure 2. Transverse mass spectra for Left: K(892) in p+p and Au+Au collisions at $\sqrt{s_{NN}}$ =200 GeV and Right: Λ(1520)) in p+p collisions at $\sqrt{s_{NN}}$ =200 GeV.

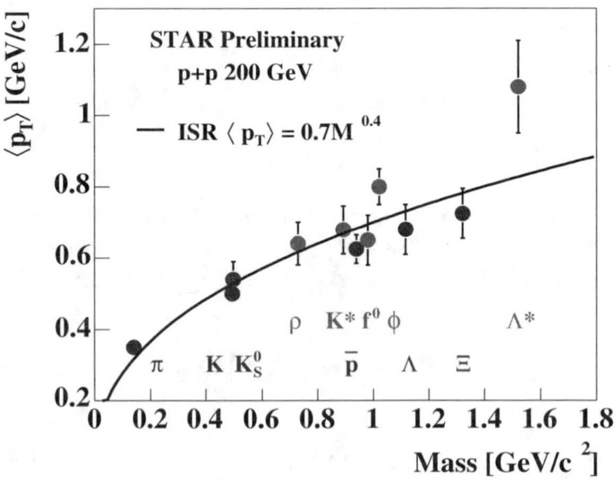

Figure 3. Mean transverse momentum as a function of particle mass in p+p collisions $\sqrt{s_{NN}}$ =200 GeV. Black line are the fit to measured ISR data (π, K and p) at collision energy of $\sqrt{s_{NN}}$ =26 GeV [4]

(π, K and p) for p+p collisions at $\sqrt{s_{NN}}$ =26 GeV [4]. The mean transverse momentum of the resonances are obtained via an exponential fit, while the Λ and the Ξ spectra are fitted with a power-law function. An exponential fit for this particles would increase their mean transverse momentum. The prediction derived from the ISR data for higher mass particles does not describe our resonances. This can be an indication that a fraction of resonances comes from jet fragmentation, which contributes to the higher momentum of the particles.

5. Strange Resonances in Au+Au Collisions at $\sqrt{s_{\mathrm{NN}}} =$ 200 GeV

The masses and the experimental widths of the strange resonances K(892), $\Sigma(1385)$ and $\Lambda(1520)$ [16, 12, 8] over the phase space integrated invariant mass spectrum in Au+Au collisions are compatible with the values from the PDG [9] and the width expected from the momentum resolution of the detector (see figure 4).

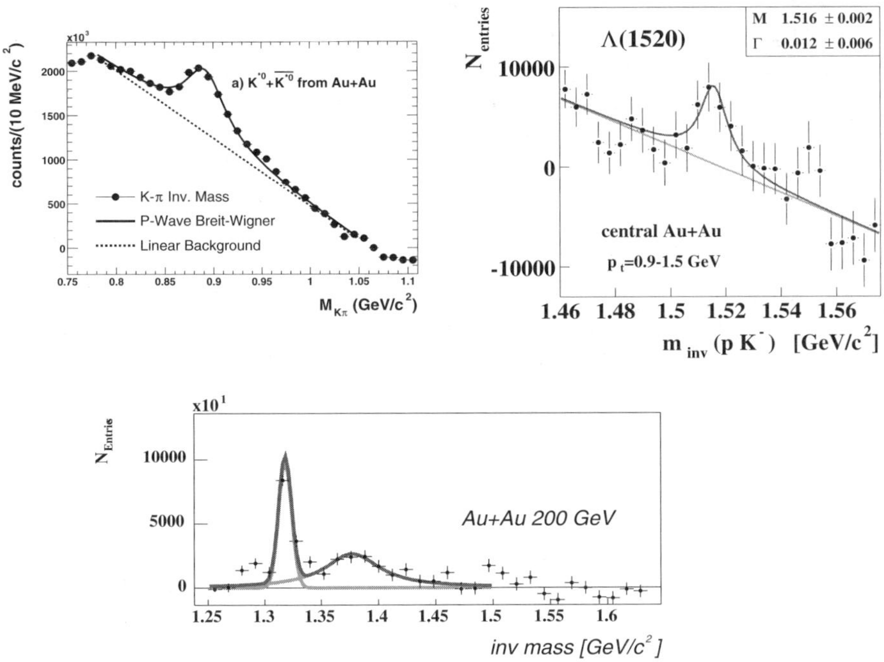

Figure 4. Invariant mass distributions of resonances in Au+Au collisions at $\sqrt{s_{\mathrm{NN}}}$ =200 GeV after mixed-event background subtraction. Up-Left: K(892) (π+K^- pairs), Up-Right: $\Lambda(1520)$ (p+K^- pairs). Bottom: $\Sigma(1385)$(Λ+π pairs).

For a comparison of resonance production in different collision systems, the resonance devided by non-resonance particle ratios eliminate the need for a volume and energy normalization. Figure 5 shows the K(892)/K and the $\Lambda(1520)/\Lambda$ ratios for p+p and Au+Au collisions as a function of the number of participants. The ratio decreases from p+p to Au+Au collision systems. This behavior shows that the resonance production in Au+Au is not a superposition of p+p interactions. This is an indication that the surrounding extended medium of a Au+Au collision may have an influence on the resonances and/or their decay particles. The question remains whether the initial resonance yield is changed and/or whether the decay particles taken to reconstruct the resonance have been affected by the medium.

The thermal model that fits all the ratios of stable particles very well predicts $\Lambda(1520)/\Lambda$ = 0.7 at the chemical freeze-out temperature of T=170 MeV. The measured value for central Au+Au collision at mid-rapidity is $\Lambda(1520)/\Lambda$ = 0.034 $\pm$ 0.011 $\pm$ 0.013, lower than the thermal model prediction. A thermal description of this ratio would lead to a 30-50 MeV smaller chemical freeze-out temperature. Medium effects like elastic interactions of the decay particles with particles of the medium (mostly pions) may result in a signal loss in the invariant mass reconstruction region of the resonances, if the contribution from rescattering is larger than the regeneration. This depends strongly on the time between chemical and thermal freeze-out. Resonances that decay into pions have a much larger contribution from regeneration due to all the surrounding pions in the medium compared to resonances like the $\Lambda(1520)$ that regenerate from a proton and kaon.

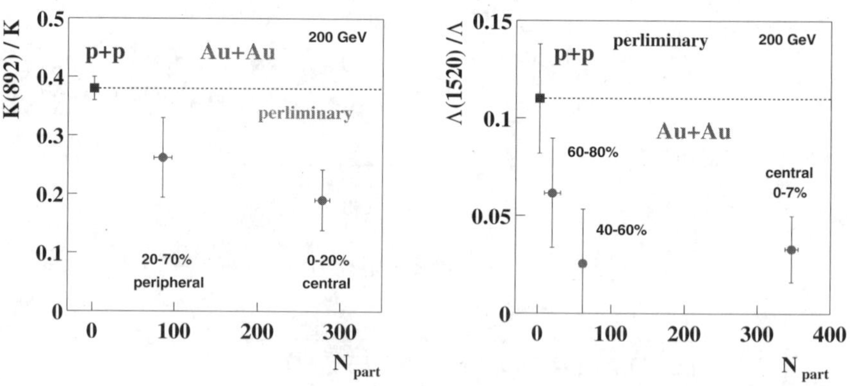

Figure 5. Resonance/non-resonance ratios (left: K*(892)/K [16], right: $\Lambda(1520)/\Lambda$) for p+p and Au+Au collisions at $\sqrt{s_{NN}}$ = 200 GeV, peripheral Au+Au data from [6]. Statistical and systematical errors are included.

Microscopic model calculations (UrQMD [1]) predict the ratio $\Lambda(1520)/\Lambda$ = 0.03 including a 30% signal loss from rescattering. The predicted value for K(892)/K in central Au+Au collisions are compatible with the data where K(892)/K = 0.19 $\pm$ 0.05. Furthermore, we would expect a more dominant signal loss in the low momentum region. This would lead to an increase in the inverse slope parameter in an exponential fit of the transverse momentum spectrum.

Figure 2 shows the transverse mass distribution of K(892) and $\Lambda(1520)$ where the inverse slope parameter T is obtained with $T_{K(892)} = 223 \pm 9$ MeV for p+p interactions and $T_{K(892)} = 350 \pm 23$ MeV for the 70-80% most peripheral

Au+Au collisions. This strong increase of 100 MeV in the inverse slope parameter is not obsereved for the stable particles.

6. Time Scale for A+A collisions

Model calculations that start with the particle yields from thermal production and includes a lifetime for the system after chemical freeze-out predict a signal loss for the measured resonances. The signal loss for depends on the length of the lifetime between chemical and thermal freeze-out where a cross section of rescattering for the decay daughters for the resonances is applied, but no regeneration. Using the measured values of K(892)/K and $\Lambda(1520)/\Lambda$ from RHIC energies, and a chemical freeze-out temperature of 160 MeV the lower limit for a lifetime interval between chemical and kinetic freeze-outs is estimated to be 5 fm/c [10, 11, 7]. More resonance measurements are necessary to put additional limiting conditions on the temperature and the lifetime between chemical and thermal freeze-out to verify this model. From thermal model predictions we expect a $\Sigma(1385)/\Lambda$ ratio of 0.5 which means that 50% of all the produced Λ's come from the $\Sigma(1385)$ decay.

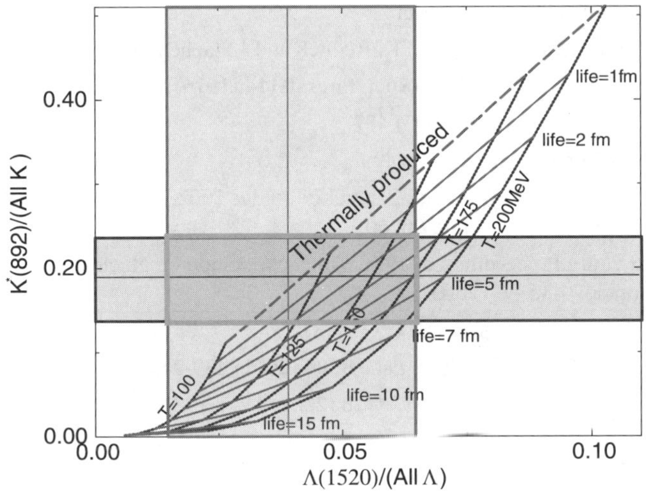

Figure 6. Dependence of lifetime and chemical freeze out temperature of a fireball source given by the two particle ratios $\Lambda(1520)/\Lambda$ and $K^*(892)/K$ including limits from the STAR data for cental Au+Au collisions at $\sqrt{s_{NN}} = 200$ GeV [10, 11].

7. Energy and System Size Dependence

The measured $\Lambda(1520)/\Lambda$ ratio in central Pb+Pb collisions at $\sqrt{s} = 17$ GeV is 0.028 ± 0.006 [5] which also cannot de described by the thermal model only.

A suppression factor of 2 is needed to reproduce the measured multiplicity measured by the NA49 Experiment. The $\Lambda(1520)/\Lambda$ ratios for p+p collisions are in agreement for the SPS and RHIC energies. The decrease of this ratio from p+p to heavy ion collisions is observed at both SPS and RHIC energies. The increase of the resonances over non-resonance ratio for K(892) and $\Lambda(1520)$ is observed in the very peripheral Au+Au collision at RHIC energies and stays constant up to the most central collisions (see figure 5). To test a very small medium we are currently analyzing the resonance production in d+Au collisions.

Acknowledgements

I would like to thank the organizers who invited me to such a well organized and pleasant summer school, with enough time for good discussion with experimentalist and theorist. I also would like to thank the STAR collaboration for support in presenting this data. And specially I would like to thank Sevil Salur, Ludovic Gaudichet, Haibin Zhang and Richard Witt for their contribution.

References

[1] M. Bleicher and J. Aichelin, *Phys. Lett.* **B530** (2002) 81, hep-ph/0201123.

[2] M. Bleicher private communication.

[3] P. Braun-Munzinger, D.Magestro, K. Redlich and J.Stachel, Phys. Lett. **B518** (2001) 41.

[4] M. Bourquin and J.-M. Gaillard, Nucl. Phys. **B114** (1976) 334.

[5] V. Friese, Nucl. Phys. **A698** (2002) 487.

[6] L. Gaudichet J.Phys. **G30** (2004) 549.

[7] C. Markert, G. Torrieri, J. Rafelski, Campos do Jordao 2002, New states of matter in hadronic interactions 533-552, hep-ph/0206260.

[8] C. Markert et al., Proceedings of 19th Winter Workshop on Nuclear Dynamics, Breckenridge, Colorado, 9-15 Feb 2003.

[9] Particle Data Group, Eur. Phys. J. **C3** (1998).

[10] G. Torrieri and J. Rafelski, *Phys. Lett.* **B509** (2001) 239-245.

[11] G. Torrieri and J. Rafelski, *J.Phys.* **G28** (2002) 1911-1920.

[12] S. Salur in this Proceedings.

[13] M.F.M. Lutz and C.L. Korpa, nucl-th/0105067.

[14] M.F.M. Lutz, *J.Phys.* **G28** (2002) 1729.

[15] R. Witt private communication.

[16] H. Zhang at al., J.Phys. **G30** (2004) 577.

CONSTRAINTS ON THE NUCLEAR EOS FROM HEAVY ION DATA IN THE SIS ENERGY RANGE

W. Reisdorf
Gesellschaft für Schwerionenforschung, D-64291, Darmstadt, Germany

Abstract We present the current status of extracting information on hot and compressed nuclear matter from nucleus-nucleus collisions with particular emphasis on the intermediate energy range, (0.1-2)A GeV. Our keywords are collective flow, subthreshold particle production, momentum dependence, degree of stopping, scaling and cluster formation.

1. Introduction

In the last two decades heavy ion collisions have become a laboratory tool to study the properties of hot and compressed nuclear matter. One of the aims is to obtain constraints on the relation between pressure, density and temperature, i.e. the Equation Of State (EOS). Besides being of fundamental interest, the Nuclear EOS is required for astophysical modelling of neutron stars, supernova dynamics *etc.* I introduce the two main observables from which information on the EOS has been inferred: collective flow (section 2) and subthreshold particle production (section 3). In section 4 I treat non-equilibrium effects. Data relevant to the liquid-vapour transition are presented in section 5.

2. Collective flow

One of the key observables in energetic nucleus-nucleus collisions is the phenomenon of collective flow [1]. To observe flow experimentally, large acceptance, high granularity detector systems are required that allow to study the full 3-dimensional momentum space distribution. In Fig. 1 we show such complete information for *He* fragments emitted in half-central reactions of Au on Au at 0.4A GeV, obtained with the FOPI [2] detector. Experience has shown that such data can be described in the center-of-mass (com) system by $dN/du_t dy d\phi = v_0[1 + 2v_1 cos(\phi) + 2v_2 cos(2\phi)]$, i.e. a Fourier series truncated after the second order term. Here y is the longitudinal rapidity (in the beam axis Z), u_t is the transverse component of the four-velocity $u = (\gamma, \vec{\beta}\gamma)$ and ϕ is the azimuthal angle with respect to the reaction plane, the orientation

W. Greiner et al. (eds.), *Structure and Dynamics of Elementary Matter*, 111–118.

of the latter being determined with a method outlined in [3]. The coefficient functions $v_i \equiv v_i(y, u_t)$ depend both on y and u_t. Fig. 1 shows, from top to bottom, the azimuthally averaged invariant distribution v_0/u_t, the 'sideflow' intensity field $v_0 \bullet 2v_1$ and the 'elliptic flow' intensity field $v_0 \bullet 2v_2$. Note that in the absence of 'flow' the two lower fields, that can have both positive and negative signs, would be empty and that we use scaled units: $u_t^{(0)} \equiv u_t/u_{1cm}$ and $y^{(0)} \equiv y/y_{1cm}$, the index $1cm$ relating to the incident projectile in the com.

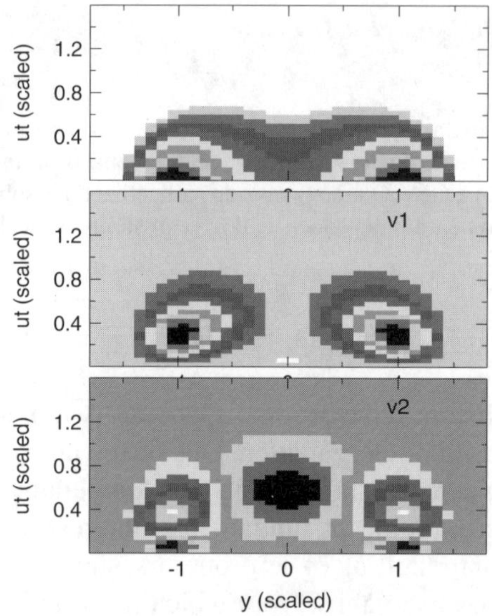

Figure 1. The three-dimensional momentum space distributions in half-central Au on Au collisions at 0.4A GeV. Shown are from the top the invariant distribution $dN/u_t du_t dy$ (scaled units), the sideflow intensity field and the elliptic flow intensity field. Contours differ by a factor 1.5, also see text.

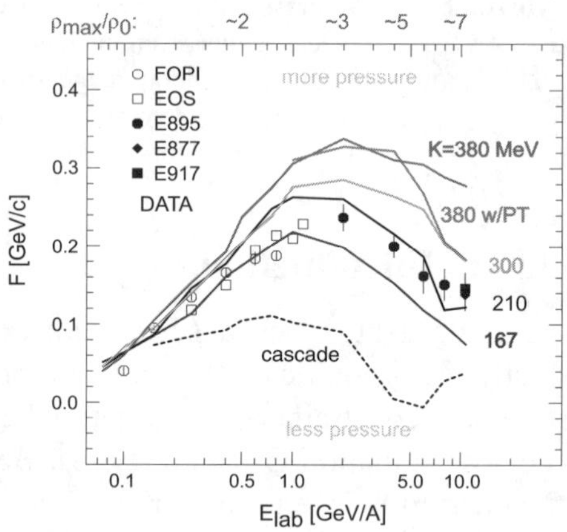

Figure 2. Excitation function for transverse flow. Transport theory predictions using different indicated compressibilities are confronted with the experimental data. The curve marked w/PT indicates the influence of a hypothetical phase transition. From ref. [4].

The participant-spectator structure in such half-overlap collisions is prominent in all three fields which for symmetric systems are symmetric, v_0, v_2, or antisymmetric, v_1 with respect to $y = 0$. In the v_0 field spectator matter is concentrated around beam ($y^{(0)} = 1$) and target ($y^{(0)} = -1$) rapidity, while participant matter concentrates around mid-rapidity, $y^{(0)} = 0$. Sideflow (middle panel) is created by the work of pressure gradients at the participant-spectator boundary; these gradients are larger if the EOS is 'stiff', a fact that is used to extract information on the incompressibility of nuclear matter. In principle the two islands differ by sign (which is ignored in the figure), the convention being that for repulsive action the forward spectator has positive 'flow'. The strong participant-spectator cross-talk is characteristic of this energy range ($0.1 < E/A < 2$ GeV) due to the fact that the expansion time of the initially hot and compressed participant (overlap-) matter is comparable to the 'passing-time', the time the spectator, that moves along with almost undegraded velocity, stays in contact with the participants and acts as some sort of recoil calorimeter.

>From the definition, the coefficients v_1 are given by $v_1(y, u_t) \equiv \langle u_x/u_t \rangle$ where u_x is the projection of $\vec{u}_t$ onto the reaction plane (XZ) and where the bracket indicates averaging over many events in a chosen centrality class. Often, the full information $v_1(y, u_t)$ is not used. Instead, the midrapidity ($y^{(0)} = 0$) slope parameter F is introduced, which is defined by $F \equiv d(m_n \langle u_x \rangle)/dy^{(0)}$, which implies scaled rapidities, but absolute momenta (in MeV/c) per nucleon, a somewhat inconsistent convention (m_n is the nucleon mass).

A confrontation of data for F with transport theoretical predictions [4] varying the EOS, characterized by its incompressibility K at saturation density ρ_0, is shown in Fig. 2. The following comments can be made: 1) while none of the calculations reproduce the data perfectly, it is clear that repulsive mean fields are needed, since the 'cascade' calculation (no mean field) fails to describe the data; 2) sensitivity to the incompressibility K requires $E_{beam}/A \geq 0.4$A GeV and around 2A GeV seems to exclude a very stiff EOS (K=380MeV); 3) a hypothetical phase transition (curve labelled '380 w/PT') would show up (for this observable) as a fast change in the slope dF/dE_{lab} which is however not readily visible in the shown data.

Coming back to the flow fields, Fig.1, the third (bottom) field represents, the 'elliptic' flow and is more complex in its structure; in the figure the background area separates islands of differing sign. One can show that $v_2(y, u_t) = \langle (u_x/u_t)^2 - (u_y/u_t)^2 \rangle_{y,u_t}$ where again the brackets indicate averaging over many events. Negative v_2 (mid-rapidity island in the Figure) indicate preferential out-of-plane emission (Y-direction). The amount of matter 'squeezed-out' in the reaction plane is influenced by the expansion velocity of the compressed matter, which in turn is connected with the sound velocity and compressibility of the strained matter. At higher energies ($E_{beam} >$ 4A GeV) v_2 changes sign [5] because the spectator matter has escaped before the expansion sets in and v_2 then reflects in momentum space the early shape of the participant overlap geometry. Again most analyses so far ignore the rich structure of the v_2 field and only a narrow interval around midrapidity is considered. An example will be shown in section 4.

3. Subthreshold particle production

Particle production in heavy ion reactions in the subthreshold regime requires multistep mechanisms which are favoured by high density. In an experiment that was performed by the KAOS Collaboration [6], the ratio of the K^+ production in Au+Au to that in C+C was measured as a function of beam energy, Fig. 3. The threshold energy for the reaction $NN \rightarrow K^+\Lambda N$ is 1.58A GeV. The strong rise with decreasing energy confirms the expectations: the higher density reached in the heavier system becomes increasingly important. In-medium effects on K^+ complicate the situation although it is argued that they rise only linearly with density, while the production rises quadratically or even steeper. Two independent transport theoretical calculations [7, 8] concur that a soft EOS (leading to relatively higher densities) reproduces the data best.

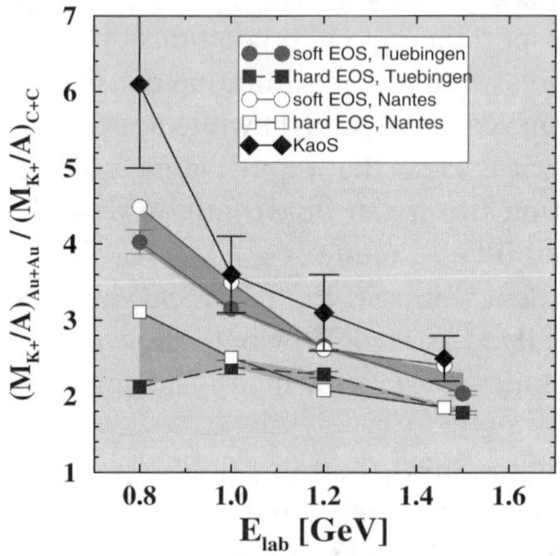

Figure 3. K^+ multiplicities per nucleon observed in Au+Au collisions divided by those measured in the C+C system versus beam energy. The data [6] are compared to theoretical simulations [7, 8] using soft (circles) and stiff EOS.

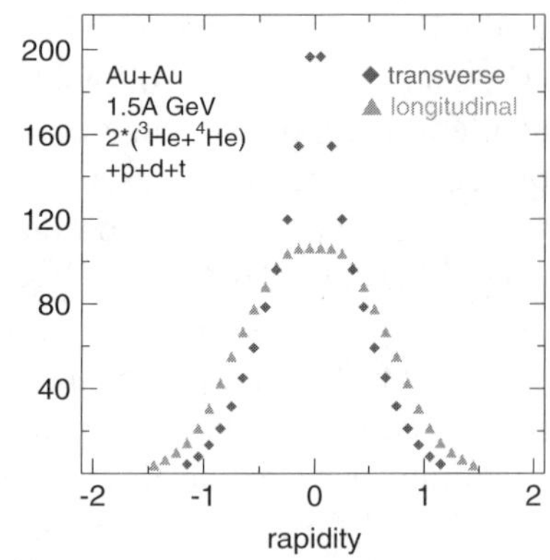

Figure 4. Scaled transverse and longitudinal rapidity distributions for Au+Au collisions at 1.5A GeV. All indicated particles, weighted with their charges are added up.

4. Coping with partial equilibrium

In principle transport models can handle, in contrast to ideal hydrodynamics, also the non-equilibrium situation, provided they correctly evaluate the collision integral, a highly non-trivial task because most microscopic processes, such as particle (π, K) production and properties and even elastic nucleon-nucleon scattering are expected to be influenced by the 'medium' in heavy-ion collisions, in contrast to the free scattering situation. This interesting side-product is of fundamental interest in its own right, but will not be discussed here.

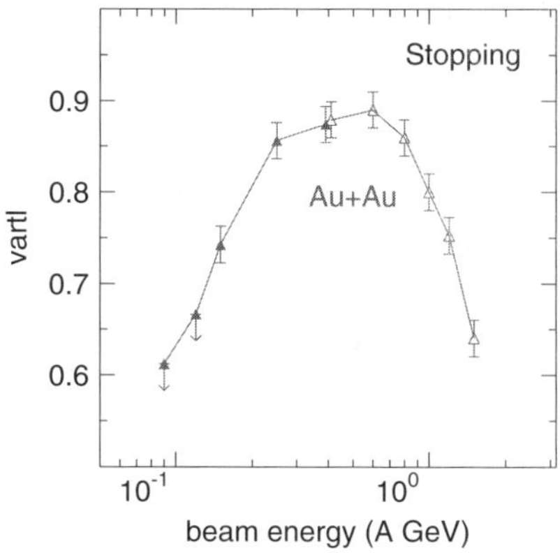

Figure 5. Excitation function for the degree of stopping in central Au on Au collisions

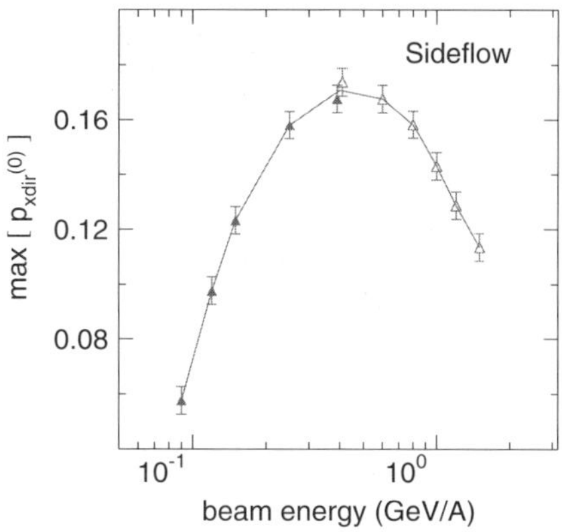

Figure 6. Excitation function for scaled global directed flow in Au on Au collisions

Experimental data that are sensitive to the degree of stopping are shown in Fig. 4 which shows rapidity ditributions measured at 1.5A GeV for central collisions of Au on Au in both the longitudinal and the transverse directions. All emitted charged particles, weighted with their nuclear charge Z, are added up. The ratio of variances $vartl$ of the transverse to the longitudinal rapidity distributions is highly sensitive to modifications in the collision integral. In the globally equilibrated thermal model $vartl = 1$. The excitation function for $vartl$, Fig. 5, shows a remarkable maximal plateau with a steep drop on both edges, a shape that is found to be highly correlated with global sideflow, $p_{xdir} = \sum sign(y) Z u_x / \sum Z$, Fig. 6. Note that consistent scaling is needed to evidence this important correlation. Clearly, such data need to be confronted with theoretical simulations before final conclusions on the EOS are drawn. While the steep drop below 0.2A GeV, is presumably due to Pauli blocking, stressing the Fermionic (degenerate) nature of nucleonic matter in this energy regime, the drop at the high energy end (beyond 0.8A GeV) is not yet understood.

Systematics of $vartl$ [9] suggest significant partial transparency implying that the *momentum* dependence of the mean fields will play an important role. Attempts to confine the momentum dependence phenomenologically by using elliptic flow in rather peripheral collisions can be found in ref. [10]. Fig. 7 shows theoretical efforts [11] to reproduce the measured dependence of the real part of the optical potential in proton-nucleus collisions and extrapolate it to higher density.

Non-equilibrium calls for a more involved effort to treat the dynamic field 'seen' by a nucleon, which is not only interacting with a moving, locally homogeneous environment (local density approximation) but with two fluids moving against each other (colliding nuclear matter, CNM). The difference is substantial, as is seen in Fig. 8. CNM fakes an effectively softer EOS [12].

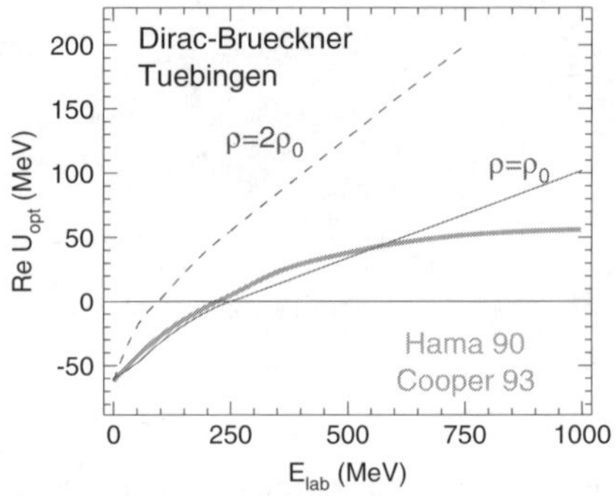

Figure 7. Energy dependence of the real part of the optical potential for proton-nucleus reactions (fat curve) and theoretical estimates for single and double saturation density. From ref. [11].

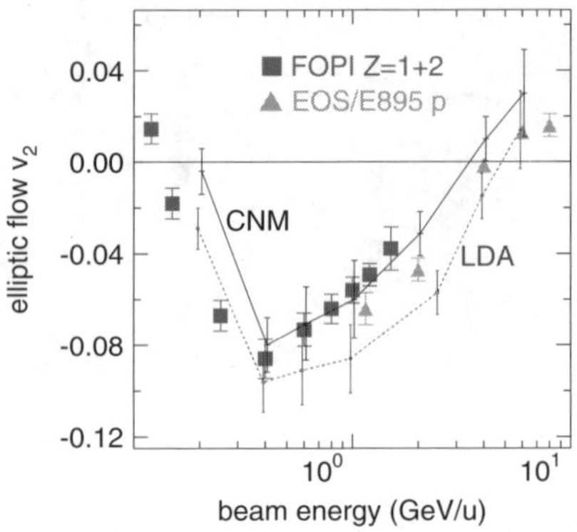

Figure 8. Excitation function for elliptic flow. Two calculations [12] using the local density approximation, LDA, resp. the colliding nuclear matter approximation, CNM, are compared to the data [13, 5].

Imperfect equilibrium also leads to a fractionation of flow onto different emitted paarticles types that *changes with beam energy* and goes beyond simple coalescence considerations. This introduces a 'chemical bias' into flow excitation functions that are retricted to one particle type, say protons, an effect that is not accounted for in simulations that ignore cluster formation. Due to space limitations I cannot disuss this in depth here, but wish to point out that the lower energy FOPI data ([13] with minor revisions) shown in Fig. 8, hold for the sum of Z=1 and 2 particles, while the EOS/E895 data [5] are for protons.

5. Liquid to vapour transition

To constrain the critical temperature T_c for the liquid-to-vapour transition using heavy ion collisions one needs to sample both the repulsive and the attractive parts of the nuclear forces. This can be done in semi-central, respectively central collisions at $0.4A$ GeV using system size dependences to follow both the development of repulsion, evidenced by sideflow, Fig. 9, and the clusterization, quantized by the charge distributions, Fig. 10.

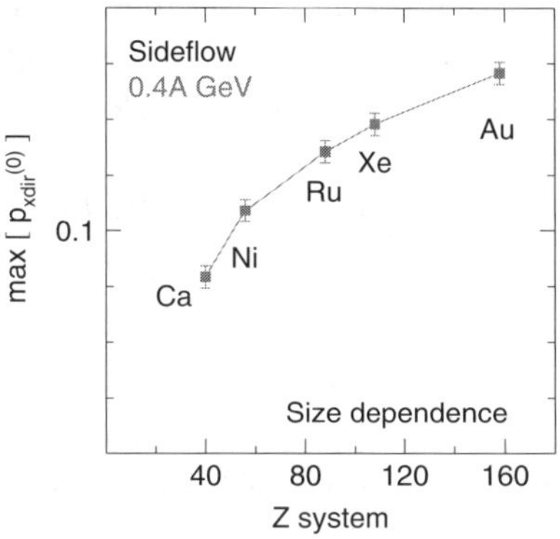

Figure 9. System size dependence of maximal scaled sideflow in symmetric collisions at 0.4A GeV.

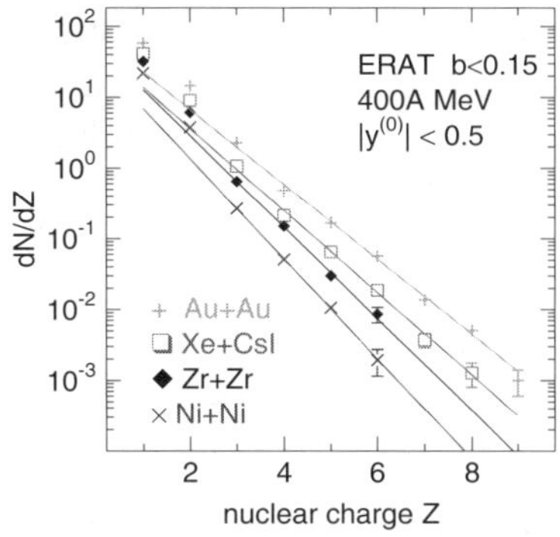

Figure 10. Charged particle multiplicity distributions measured in central collisions of Au+Au, Xe+CsI, Zr+Zr and Ni+Ni reactions at an incident energy of $0.4A$ GeV. The data are presented for c.m. rapidities : $|y^{(0)}| \leq 0.5$. The straight lines are exponential fits to the data in the interval $3 \leq Z \leq 6$.

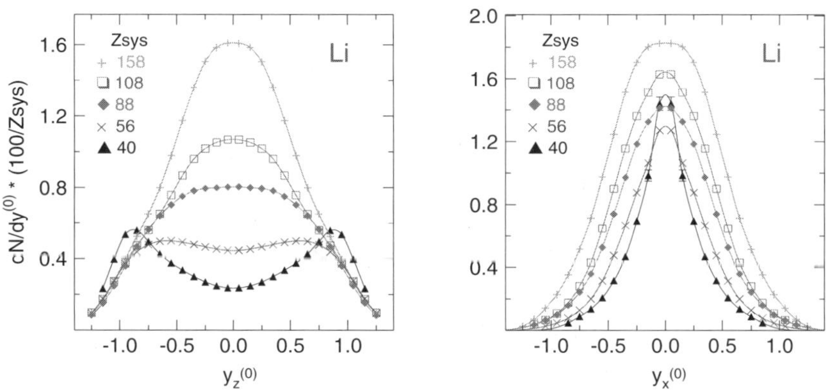

Figure 11. Scaled longitudinal (left) and transverse (right) rapidity distributions of clusters measured in central collisions for different symmetric reactions. The varying system charges Z_{sys} are indicated in the Figure. The ordinates are renormalized to a common reference system with charge $Z_{sys} = 100$.

Clusterization is favoured in larger systems. More details of the mechanism at work are seen in Fig. 11 which compares the evolution of the longitudinal and the transverse rapidity distributions with system size for the Li ($Z = 3$) 'droplets' which appear in increasing numbers at *mid-rapidity* as the system-size increases, accompanied by a broadening of the transeverse velocities. This broadening is associated with radial flow initiating the local cooling. We anticipate that future transport code simulations that reproduce all the features in Figs. 9-11 should help constraining the liquid-gas phase diagram.

6. Conclusion

Relatively soft EOS ($K \approx 200$) are currently favoured over stiff ones ($K \approx 400$), however before a firm conclusion is reached, more work is necessary to fully account for the degree of stopping and to better understand the related in-medium modified microscopic processes and particle properties, as well as the role of clusters in the balance of flow and in constraining the liquid-gas phase diagram.

I am grateful to to all my coworkers from the FOPI Collaboration for excellent team work and many discussions.

References

[1] W. Reisdorf, H. G. Ritter, Annu. Rev. Nucl. Part. Sci. 47 (1997) 663;
N. Herrmann, J. P. Wessels, T. Wienold, Annu. Rev. Nucl. Part. Sci. 49 (1999) 581.

[2] J. Ritman, Nucl. Phys. B 44 (1995) 708.

[3] P. Danielewicz, G. Odyniec, Phys. Lett. B (1985) 146.

[4] P. Danielewicz, R. Lacey, W. G. Lynch, Science 298 (2002) 1592, and priv. comm. .

[5] C. Pinkenburg et al., Phys. Rev. Lett. 83 (1999) 1274.

[6] C. Sturm et al., Phys. Rev. Lett. 86 (2001) 39.

[7] C. Fuchs et al., Phys. Rev. Lett. 86 (2001) 1974.

[8] C. Hartnack, J. Aichelin, J. Phys. G 28 (2002) 1649.

[9] W. Reisdorf et al. (FOPI Collaboration), submitted for publication.

[10] P. Danielewicz, Nucl. Phys. A 673 (2000) 375.

[11] T. Gaitanos et al., Eur. Phys. J. A 12 (2001) 421.

[12] C. Fuchs and T. Gaitanos, Nucl. Phys. A 714 (2003) 643.

[13] A. Andronic, Nucl. Phys. A 661 (1999) 333c.

BALANCE FUNCTIONS AT RHIC

Gary D. Westfall for the STAR Collaboration
National Superconducting Cyclotron Laboratory, Michigan State University, East Lansing, MIchigan 48823-1321 USA, E-mail: westfall@nscl.msu.edu

Abstract The balance function is based on the principle that charge is locally conserved when particles are pair produced. Balance functions have been measured for all charged pairs, identified pion pairs, and identified charged kaon pairs in Au+Au collisions at $\sqrt{s_{NN}}$ = 200 GeV and p+p collisions at $\sqrt{s}$ = 200 GeV at the Relativistic Heavy Ion Collider using STAR. Balance functions for all charged particles from Au+Au scale smoothly with centrality to the p+p value. Balance functions for charged particles and pions are narrower in central collisions than in peripheral collisions consistent with trends predicted by models incorporating the concept of late hadronization. Balance functions for kaon pairs represent a strangeness balance. Balance functions for kaons are narrower than those for pion pairs and may show less dependence on centrality.

1. Introduction

Collisions of Au nuclei at ultra-relativistic energies can produce large energy densities and high temperatures. Under these conditions, it is believed that the produced matter can be described by quark/gluon degrees of freedom. This phase would be followed by a transition to hadronic particles that are measured [1]. Numerous probes of the high density medium have been proposed [2], including observables related to fluctuations and correlations [3–7].

One such observable, the balance function [8], is sensitive to whether the transition to a hadronic phase was delayed, as expected if the quark-gluon phase were to persist for a substantial time. In ref. [8], the authors argue that as a result of local charge conservation, when particles and their anti-particles are produced, they are correlated initially in coordinate space. If hadronization occurs early, the members of a charge/anti-charge pair would be expected to separate in rapidity due to expansion and re-scattering in the strongly interacting medium. If, instead, the system exists in a deconfined phase for a substantial time, and the pairs are formed at hadronization, they will retain more of their correlation in rapidity. To measure this correlation,

W. Greiner et al. (eds.), Structure and Dynamics of Elementary Matter, 119–125.

the authors of ref. [8] proposed the balance function. This technique involves subtracting uncorrelated charge/anti-charge pairs on an event-by-event basis. The remaining charge/anti-charge particle pairs are examined to determine the correlation as a function of the relative rapidity, Δy, between the members of the pairs. The balance function is expected to be narrower for a scenario with delayed hadronization, and is therefore sensitive to the conjecture that a quark-gluon plasma may be produced.

The balance function is a new observable for heavy ion collisions and is similar in form to the observable used in [9] to study the associated charged density balance from both p+p and $e^+ + e^-$ collisions [10].

In this analysis the balance function is used to examine the pseudorapidity correlation of non-identified charged particles and the rapidity correlation of identified charged pions. The balance function is calculated as:

$$B = \frac{1}{2}\left\{\frac{\Delta_{+-} - \Delta_{++}}{N_+} + \frac{\Delta_{-+} - \Delta_{--}}{N_-}\right\} \tag{1}$$

where Δ_{+-} in the case of identified charged pion pairs denotes the number of identified charged pion pairs in a given rapidity range $\Delta y = |y(\pi^+) - y(\pi^-)|$, similarly for Δ_{++}, Δ_{--}, and Δ_{-+}. In the case of non-identified charged particle pairs, pseudorapidity (η) is used. The terms Δ_{+-}, Δ_{++}, Δ_{--}, and Δ_{-+} are calculated using pairs from a given event and the resulting distributions are summed over all events. Specifically, Δ_{+-} is calculated by taking in turn each positive particle in an event and incrementing a histogram of Δy ($\Delta\eta$) with respect to all the negative particles in that event. Δ_{+-} is then summed over all events. A similar procedure is followed for Δ_{++}, Δ_{--}, and Δ_{-+}. $N_{+(-)}$ is the number of positive(negative) pions or non-identified charged particles summed over all events.

The balance function is calculated for all charged pion, charged kaon, or non-identified charged particle pairs for a given centrality bin. The balance function, in addition to being proportional to opposite charge correlations, is also proportional to the acceptance for each Δy or $\Delta\eta$ bin. Thus the measured balance functions presented in this paper are narrower than model calculations assuming no detector acceptance effects.

The main detector used for this analysis was the Time Projection Chamber (TPC) [11] located in a solenoidal magnetic field of 0.5 T [12]. The TPC provided tracking information for charged particles having transverse momentum $p_t > 150$ MeV/c and $|\eta| < 1.1$ with complete azimuthal acceptance. Minimum bias events were defined by the coincidence of two Zero Degree Calorimeters (ZDCs) [13] located $\pm$ 18 m from the center of the interaction region. Central collisions were defined using both the ZDCs and the Central Trigger Barrel (CTB), an array of scintillator slats surrounding the outside of the TPC. A coincidence in the ZDCs accompanied by a signal in the CTB above a threshold empirically determined to correspond to approximately the 8% most central events was used to define the central collision trigger.

2. Balance Function Results

STAR previously has published balance functions for Au+Au reactions at $\sqrt{s_{NN}}$ = 130 GeV [14, 15]. Balance functions for all charged particle pairs and pion pairs were measured as a function of centrality. The balance function was observed to narrow in central collisions. The width of the balance function in peripheral collisions was consistent with HIJING calculations.

Here we present new preliminary balance function results for all charged particle pairs, charged pion pairs, and charged kaon pairs from Au+Au collisions at $\sqrt{s_{NN}}$ = 200 GeV and p+p collisions at $\sqrt{s}$ = 200 GeV. For this analysis 1.2 million minimum bias Au+Au events, 1.4 million centrally triggered Au+Au events, and 3 million p+p events were used. Only charged particle tracks having more than 15 space points along the reconstructed trajectory, which pointed to within 1 cm of the primary vertex, were used. The ratio of reconstructed space points to possible space points along the track was required to be greater than 0.52. Simulations of TPC performance indicate the tracking efficiency for tracks within this acceptance and momentum range is 85-90% [16].

Charged pion and kaon identification was accomplished by selecting particles within 2 standard deviations of the expected dE/dx for a given momentum, which provided pion and kaon identification for momenta less than 0.7 and 0.8 GeV/c respectively. Electrons were removed by requiring the measured dE/dx to be more than 1 standard deviation away from the expected value for an electron of the measured momentum.

Centrality was defined in terms of the multiplicity of charged particle in the TPC with $|\eta| < 0.5$. Nine bins in centrality were used for the balance functions for all charged particles and pions. The most central bin contained the 5% most central events and was designated as $0.95 < Centrality < 1.0$, while the most peripheral bin used the 70 - 80% most central events and was designated as $0.2 < Centrality < 0.3$. Five centrality bins were used for the kaon balance functions to improve the statistical accuracy of the kaon results. Fractions of the total observed cross section were related to the number of participating nucleons for Au+Au collisions using a Monte Carlo Glauber calculation.

In Fig. 1 the balance function for all charged particles is shown for nine centrality bins. In this figure two sets of mixed events are also shown. The first set is traditional mixed events in which tracks are chosen randomly from events having similar event vertex positions and multiplicities. These results are zero for all $\Delta\eta$ as expected. The balance function for mixed events integrates to zerobecause global charge conservation has been removed. The fact that the balance function is zero for all $\Delta\eta$ for all centralities demonstrates that STAR's

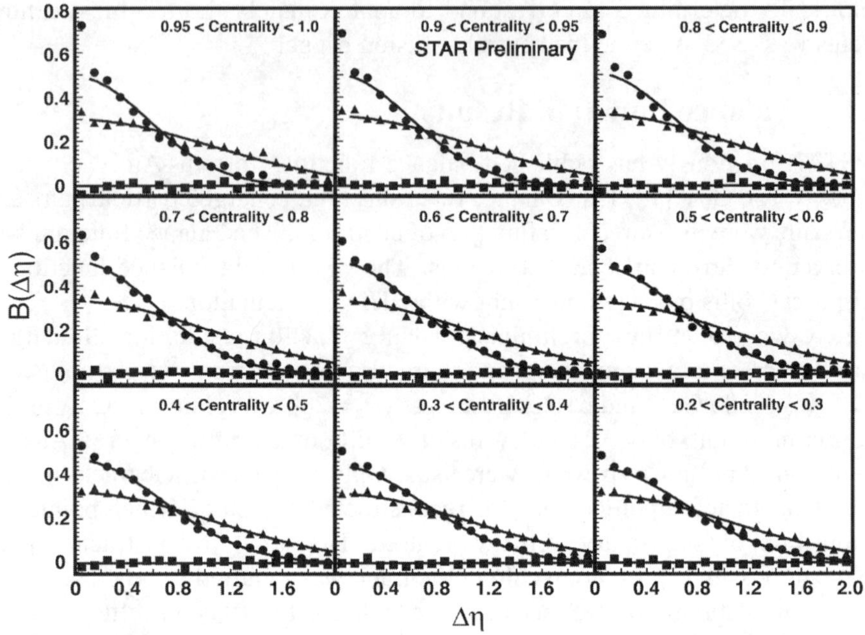

Figure 1. The balance function for all charged particle pairs from Au+Au at $\sqrt{s_{NN}} = 200$ GeV and p+p at $\sqrt{s} = 200$ GeV. The errors are statistical. Solid lines are drawn to guide the eye.

acceptance in $\Delta\eta$ is smooth. A second set of mixed events is also shown in which the pseudorapidities of the particles for each event are randomly shuffled. This set of events represents the widest balance function STAR can measure in its acceptance given no momentum correlations. The balance function for shuffled pseudorapidity events is always wider than the measured balance functions. One can see that the measured balance function narrows as the collisions become more central. For $\Delta\eta < 0.1$, the balance function is affected by HBT/Coulomb interactions and electron contamination so we exclude that bin when calculating the width of the balance function.

Balance functions for identified charged pion pairs are shown in Fig. 2 as a function of relative rapidity, Δy, for nine centrality bins. Balance functions for identified charged kaon pairs are shown in Fig. 3 as a function of relative rapidity, Δy, for five centrality bins.

To make a quantitative comparison of the centrality dependence of the balance function, we extract the width of the balance functions by taking the weighted average for the balance function for $0.1 \leq \Delta\eta \leq 2.0$ for charged particles and $0.2 \leq \Delta y \leq 2.0$ for pions and kaons.

We plot the width of balance function for charged particles from Au+Au and p+p collisions in the left panel of Fig. 4 as a function of the number of

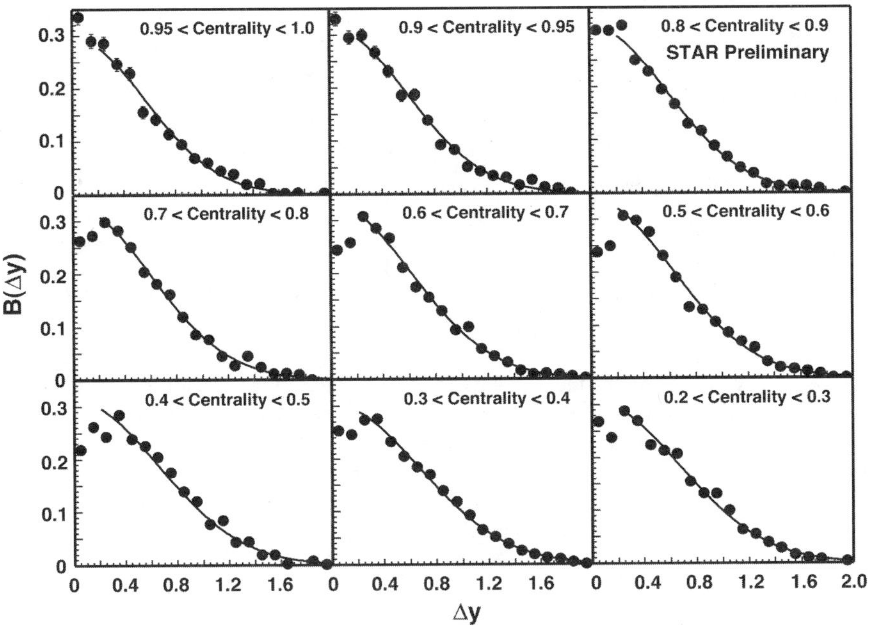

Figure 2. The balance function for identified pion pairs from Au+Au at $\sqrt{s_{NN}}$ = 200 GeV. The errors are statistical. Solid lines are drawn to guide the eye.

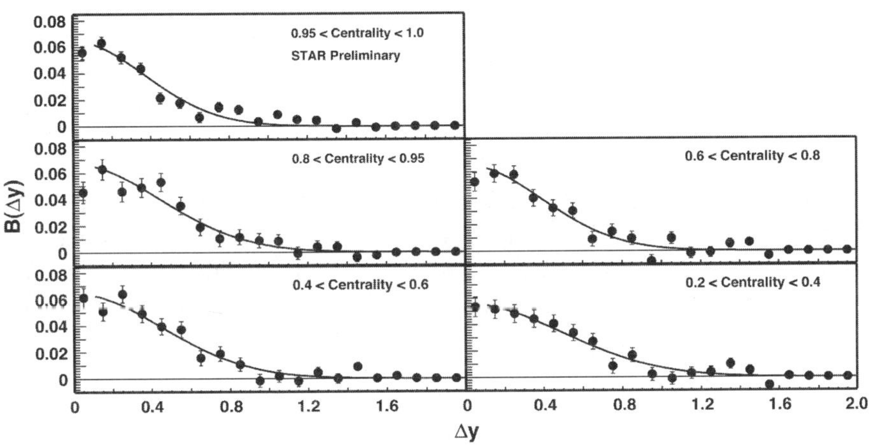

Figure 3. The balance function for identified kaon pairs from Au+Au at $\sqrt{s_{NN}}$ = 200. The errors are statistical. Solid lines are drawn to guide the eye.

participating nucleons, N_{part}. The width of the balance function is smaller in central collisions. The widths from Au+Au scale smoothly with N_{part} to the width from p+p collisions. Also plotted in the left panel of Fig. 4 are

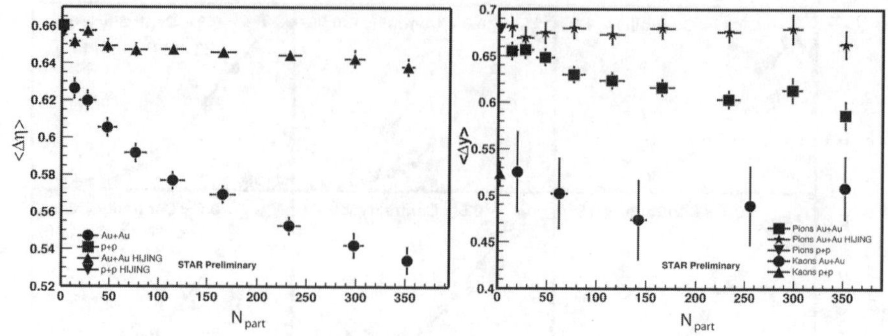

Figure 4. In the left panel the measured width of the balance function is shown for all charged pairs from Au+Au at $\sqrt{s_{NN}} = 200$ GeV and p+p at $\sqrt{s} = 200$ GeV compared with the predicted width for Au+Au collisions from HIJING. In the right panel the measured width of the balance function is shown for kaon and pion pairs from Au+Au at $\sqrt{s_{NN}} = 200$ GeV and p+p at $\sqrt{s}$ = 200 GeV compared with the predicted width for pions from Au+Au collisions from HIJING. The errors shown in both panels are statistical.

HIJING results for Au+Au at $\sqrt{s_{NN}} = 200$ GeV and p+p at $\sqrt{s} = 200$ GeV. The HIJING calculations were corrected using an acceptance filter incorporating the pseudorapidity and momentum acceptance of STAR as well as a realistic distribution of the position of the collision vertex inside STAR. The HIJING calculations for p+p agree with the measured value of the width of the balance function. The Au+Au predictions also are consistent with the p+p results but show little centrality dependence compared with the measured values.

In the right panel of Fig. 4 we show the widths of the balance function for pions and kaons as a function of the number of participating nucleons, N_{part} for Au+Au and p+p collisions. These experimental results are compared with filtered HIJING calculations for the pion balance function from Au+Au and p+p collisions. The width of the balance function for pions from Au+Au scales smoothly with N_{part} and scales smoothly to the p+p result. The balance function for pions is narrower in central Au+Au collisions than in peripheral Au+Au collisions as was observed for all charged particles. In contrast, the balance function predicted by HIJING for pions from Au+Au collisions shows little centrality dependence. The predicted balance function for p+p and Au+Au also have the same width. The balance function for kaons from Au+Au collisions is narrower than the balance function for pions at all centralities. However, there is little dependence of the balance function width on centrality given the statistical errors.

3. Summary

For all charged particles we observe that the balance function for Au+Au collisions gets narrower in central collisions. The widths of the balance function for Au+Au collisions scale smoothly with N_{part} to the p+p result. The balance function observed for identified pions shows behavior similar to that observed for all charged particles. However, the balance function observed for kaons from Au+Au does not show a strong dependence on centrality within the quoted errors. The balance function for kaons represents a strangeness balance as well as a charge balance. The observation that the kaon balance function may not narrow significantly with centrality may indicate that the hadronization process for kaons could be different than that for pions. Significant amounts of charge may be created at hadronization while little additional strangeness may be created leading to a lack of centrality dependence in the balance function for kaons. The lack of centrality dependence of the HIJING calculations and the observed narrowing of the balance function in central Au+Au collisions is consistent with trends predicted by models incorporating late hadronization. Interpretation of the observed narrowing requires more detailed study of its sensitivity to such other effects as flow, resonance production, and diffusion, in addition to late hadronization.

References

[1] J. W. Harris and B. Müller, Ann. Rev. Nucl. Part. Sci. **46**, 71 (1996).

[2] S. A. Bass, M. Gyulassy, H. Stöcker and W. Greiner, J. Phys. **G25**, R1 (1999).

[3] S. Jeon and V. Koch, Phys. Rev. Lett. **85**, 2076 (2000).

[4] M. Asakawa, U. Heinz and B. Müller, Phys. Rev. Lett. **85**, 2072 (2000).

[5] Zi-wei Lin and C. M. Ko, Phys. Rev. C**64**, 041901 (2001).

[6] H. Heiselberg and A. D. Jackson. Phys. Rev. C**63**, 064904 (2001).

[7] E. V. Shuryak and M. A. Stephanov. Phys. Rev. C**63**, 064903 (2001).

[8] S. A. Bass, P. Danielewicz and S. Pratt, Phys. Rev. Lett. **85**, 2689 (2000).

[9] D. Drijard *et al.* [ACCDHW Collaboration], Nucl. Phys. B **166**, 233 (1980).

[10] H. Aihara *et al.* [TPC/Two Gamma Collaboration], Phys. Rev. Lett. **53**, 2199 (1984).

[11] M. E. Beddo *et al.* [STAR Collaboration], BNL-PUB-5347 (1992).

[12] W. Betts *et al.*, IEEE Trans. Nucl. Sci. **44**, 592 (1997); H. Weiman *et al.*, IEEE Trans. Nucl. Sci. **44**, 671 (1997).

[13] C. Adler, A. Denisov, E. Garcia, M. Murray, H. Ströbele and S. White, Nucl. Instrum. Meth. A **461**, 337 (2001).

[14] M.B. Tonjes, Ph.D. Dissertation, Michigan State University, unpublished (2002).

[15] J. Adams **et al.** (STAR Collaboration), Phys. Rev. Lett. **90**, 172301 (2003).

[16] M. Calderon, Ph.D. Dissertation, Yale University, unpublished, (2001).

[17] X.N. Wang and M. Gyulassy, Phys. Rev. D**44**, 3501 (1991).

PHASE TRANSITIONS IN HIGH ENERGY HEAVY-ION COLLISIONS

L.P. Csernai[1,2], Andrea Anderlik[1], Cs. Anderlik[1], A. Keranen[3,4], V.K. Magas[5], J. Manninen[4], E. Molnár[1], Ágnes Nyíri[1], B.R. Schlei[6], D.D. Strottman[6] and K. Tamousiunas[1]

[1] *Section for Theoretical and Computational Physics, and Bergen Centre for Computational Science - Unifob, University of Bergen, Allégaten 55, 5007 Bergen, Norway*

[2] *Res. Inst. for Particle and Nuclear Physics, POB 49, 1525 Budapest, Hungary*

[3] *INFN, Sezione di Firenze, Via Sansone 1, 50019 Sesto F. - Florence, Italy*

[4] *Department of Physics, POBox 3000, 90014 Univeristy of Oulu, Finland*

[5] *Departamento de Fisica Teorica, Dr. Moliner 50, 46100 Burjassot, Valencia, Spain*

[6] *Los Alamos National Laboratory, Theory Division, Los Alamos, 87545 NM, USA*

Abstract Modelling Quark-Gluon Plasma formation and decay in high energy heavy ion reactions is presented in a framework of a multi-module setup. The collective features, governing the equlibrated fluid dynamical stages of the model are emphasized. Flow effects formed from the initial conditions are discussed. Particular attention is given to the improvement of the final hadronization and freeze-out part of the reaction which has strong effects on the observables.

1. Introduction

Phase Transitions imply that we have different phases of the matter we deal with, and it has an Equation of State (EoS). We can have an EoS if and only if the matter is in statistical equilibrium, or at least close to it. When this condition is satisfied we can apply a fluid- and thermodynamical description for the dynamics of the system. In high energy heavy ion reactions we observe many thousand particles produced in the reaction, so, we have all reasons to assume that in a good part of the reaction the conditions of the local equilibrium and continuum like behaviour are satisfied.

The initial and final stages are, on the other hand, obviously not in statistically equilibrated states, and must be described separately, in other theoretical approaches. The different approaches, as they describe different space-time (ST) domains and the corresponding approaches can be matched to each other across ST hyper-surfaces or across some transitional layers or fronts. The choice of realistic models at each stage of the collision, as well as the correct coupling

W. Greiner et al. (eds.), Structure and Dynamics of Elementary Matter, 127–142.

of the different stages or calculational modules are vital for a reliable reaction model. The most basic requirements are that conserved quantities should stay conserved when modules are interfaced to each other, and entropy should not decrease.

For the theoretical description and understanding of high energy heavy ion collisions it is important to mention that while our system is sufficiently large to assume statistical equilibrium, it is not a macroscopic system with Avogadro number of particles. We deal with a small, mesoscopic system, so discontinuities of the thermodynamical quantities, will not be observable, but at the same time the changes are sudden and rapid enough that they can be recognized.

The other basic feature of mesoscopic systems is also important to us: in such systems fluctuations are not negligible but dominant, these can be observed and these may help to identify the properties of the phase transition.

Dividing the reaction model into standardized modules and interfaces leads to additional advantages. Modules can be replaced easily, research groups can collaborate better and the GRIDS of High Performance Computers can increase the speed of complete reaction simulation by about an order of magnitude.

2. Initial Stage Module

Many fluid dynamical approaches use simple parametrizations for the initial state of the hydrodynamic stage with a few parameters. These can then be fitted or adjusted by comparing experimental data and model predictions.

A more physical approach is to start from a dynamical, pre-equilibrium model, e.g. Parton Cascade Model (PCM), or Collective (or Coherent) Yang-Mills model, or different versions of flux-tube models.

Our recent works are based on a latter type of approach [1], which was implemented in a Fire-Streak geometry streak by streak and upgraded to satisfy energy, momentum and baryon charge conservations exactly at given finite energies [2, 3]. The effective string-tension was different for each streak, stretching in the beam direction, so that central streaks with more color (and baryon) charges at their two ends had bigger string-tension and expanded less, than peripheral streaks. The expansion of the streaks was *assumed* to last until the expansion has stopped. Yo-yo motion, as known from the *Lund-model* was not assumed.

In the 1st version of the model at full stopping of string expansion, the baryon charge was distributed uniformly along each streak. This created an initial state for the hydrodynamic stage where each streak was stopped in its own CM frame, and had uniform baryon and energy density distribution. The sharp forward and backward end led, however, to a development of a large, Fwd/Bwd density peaks. [4]

Thus, we concluded that the assumption of uniform distribution and sharp cuts at the ends is overly simplified and has to be corrected. So, a Fwd/Bwd expansion was added to the model [3], which smoothed the Fwd/Bwd ends, and eliminated the artificial large, Fwd/Bwd density peaks.

Our calculations show that such a tilted initial state leads to the creation of of the third flow component [5], peaking at rapidities $|y| \approx 0.75$ [4]. Recent, STAR v_1 data [6] indicate that our *assumption* that the string expansion lasts until full stopping of each streak, may also be too simple and local equilibration may be achieved earlier, i.e. before the full uniform stopping of a streak. We did not explicitly calculate dissipative processes, some friction within and among the expanding streaks is certainly present and experiments seem to indicate that this friction is stronger.

3. Relativistic Fluid Dynamics

When parts of our system reach local statistical equilibrium in the ST, following a pre-equilibrium development, we can describe our system using Computational FLuid Dynamics (CFD) and an Equation of State. The equations of Fluid Dynamics (FD) can be most simply derived from the Boltzmann Transport Equation (BTE).

The relativistic BTE below, describes the time evolution of the single particle distribution function based on the assumptions that just two particle collisions are considered (binary collisions), the number of binary collisions at x is proportional to $f(x,p_1)\times\ f(x,p_2)$ and $f(x,p)$ is a smoothly varying function compared to mean free path. So, one can show that BTE has the form [see Chapter 3 of ref. [7]] :

$$p_k^\mu f_{k,\mu} = \sum_{l=1}^{N} C_{kl}(x,p_x)$$

where C is the collision integral, and k and l stand for different particle species or particle components. Then, using microscopic conservation laws in the collision integral, from the BTE we can derive the differential form of conservation laws [see section 3.6 of ref. [7]]:

$$N^\mu_{,\mu} = \sum_{k=1}^{N} N^\mu_{k,\mu} = 0 \quad : \text{conservation of particle number} \tag{1}$$

$$Q^\mu_{,\mu} = \sum_{k=1}^{N} Q^\mu_{k,\mu} = 0 \quad : \text{conservation of charge} \tag{2}$$

$$T^{\mu\nu}_{,\mu} = \sum_{k=1}^{N} T^{\mu\nu}_{k,\mu} = 0 \quad : \text{conservation of energy and momentum} \tag{3}$$

where $_{,\mu}$ denotes $\partial/\partial\mu$ and there is a summation for indexes occurring twice. These equations are valid for any distribution satisfying the BTE. These equations can be used if we know the solution and then we can evaluate the conserved quantities.

3.1 Perfect and Viscous Fluids

Equations (1-3) are also the equations of FD. But these equations must not be considered as consequences of BTE, rather these equations are **postulated**, as the differential forms of conservation laws. This can be done because such conservation laws can be derived from other theoretical approaches also.

These equations are not a closed set, because the energy momentum tensor and the particle four current should also be defined. In the Eulerian or perfect fluid dynamics we postulate the form of energy-momentum tensor, and an EoS, $P = P(e, n)$, must be given. This provides a closed set of partial differential equations that is solvable. Any EoS, which is consistent with thermodynamics, can be used. One does not have to assume dilute systems with binary collisions, or the assumption of molecular chaos. The only requirement is the existence of local equilibrium, otherwise we would not have an EoS.

In the case of Navier-Stokes fluid dynamics, we assume that the energy momentum tensor also contains the dissipative part $T^{\mu\nu(1)}$, which allows for small, first order deviations from local equilibrium. Thus, to have a solution we need not only the EoS, but also the transport coefficients that occur in the dissipative part of the energy-momentum tensor.

Boltzmann's H-theorem implies that if irreversible processes are present the entropy increases. In equilibrium the distribution is constant, the entropy is also constant, but it reached its maximum value.

It is important to know that this can be seen from fluid dynamics also. In perfect fluid dynamics the flow is adiabatic, no entropy is produced. This can be proven using the equations of perfect fluid dynamics and standard thermodynamical relations. Entropy production is strictly, and quantitatively connected to dissipative processes. Perfect flow is adiabatic even in the presence of phase transitions, if the two phases are in thermal, mechanical (P), chemical and phase equilibrium all the time! On the other hand if the phase transition deviates from equilibrium this leads to entropy production. See Assignment 9.4.a in ref. [7].

As shock waves, detonations, deflagrations, are idealizations of very sharp or very rapid processes as discontinuities, the above conclusions are valid for discontinuities also. Nevertheless, frequently, dissipative processes can be neglected for most of the flow except the sharp or rapid changes. Thus, as an end effect the dissipative processes are frequently localized in sharp fronts or in hypersurfaces of discontinuities. One should not, however, forget that dissipation is due to transport properties and characteristic constants of phase transition dynamics, even if it happens in "discontinuities".

3.2 Equations of Perfect Fluid Dynamics

We assume that our system is not homogeneous, but the gradients are small, so local distribution can be an equilibrium distribution, e.g. a Jüttner distribution. Now the local $n(x), e(x), s(x)$ and $P = P(n, e)$ are known, and we assume that in LR, $T^{\mu\nu}$ is diagonal. Now using the conservation laws with introducing the apparent density as:

$$N \equiv n\gamma \, ,$$

the continuity equation takes the familiar form

$$(\partial_t + \mathbf{v}\ \mathrm{grad})\ \mathrm{N} = -\mathrm{M}\ \mathrm{div}\ \mathbf{v} \, ,$$

similarly introducing

$$\mathbf{M} \equiv T^{0i} = w\,\gamma^2\mathbf{v} \tag{4}$$
$$\varepsilon \equiv T^{00} = (e + P\mathbf{v}^2)\gamma^2 \, , \tag{5}$$

the energy and momentum conservation will take the form

$$(\partial_t + \mathbf{v}\ \mathrm{grad})\mathbf{M} = -\mathbf{M}(\mathrm{div}\mathbf{v}) - \mathrm{grad\ P} \tag{6}$$
$$(\partial_t + \mathbf{v}\ \mathrm{grad})\varepsilon = -\varepsilon\ \mathrm{div}\mathbf{v} - \mathrm{div}(\mathrm{P}\mathbf{v}) \, . \tag{7}$$

Thus to solve the equations of relativistic fluid dynamics we have to solve the same partial differential equations as in the non-relativistic case. We cannot immediately apply the EoS to obtain the solution, as the EoS is given in terms of invariant scalars. So, we have to solve in addition the set of algebraic equations above, which connect N, $\mathbf{M}$, and ε to n, e, P and $\mathbf{v}$, for every fluid cell at every time step.

Important to note that the relativistic treatment is necessary not only at high velocities (i.e. if $v \sim c$) but also at low velocities when the pressure is not negligible compared to the energy density (i.e., $P \sim e$) as $w = e + P$ appears in the relativistic Euler equation! This is frequently the situation for ultra-relativistic gases, like the Stefan-Boltzmann or photon gas (or radiation pressure) where $P = e/3$. This is actually the indication of the fact that the constituents of the matter move with velocity c or close to it.

In principle the perfect fluid dynamics is absolutely unstable. Any spontaneous deviation from an exact solution, will grow exponentially, as their energy

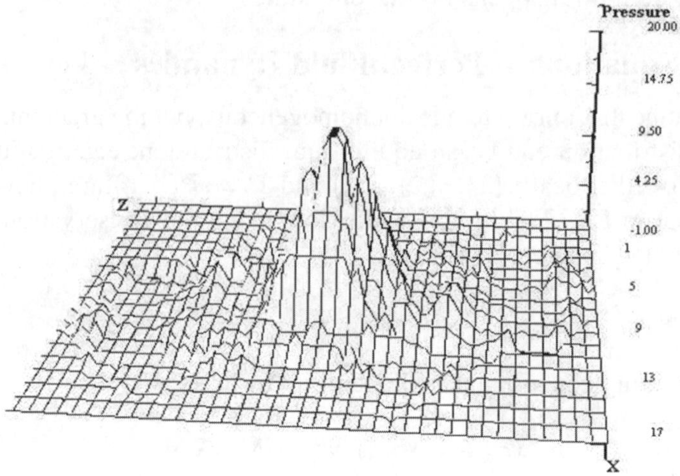

Figure 1. The pressure, in units of [GeV/fm^3], in the reaction plane at a late stage (13 fm/c) of a 65+65 Gev/nucl. Au+Au collision, at b=0.5 b-max, using the Bag Model EoS for QGP. The outer, supercooled regions have negative pressure, while in the very center it is still large. The directed flow, is visible as the outside, negative pressure regions in the upper-right and lower-left corners, close to the beam, $Z-$, direction. The third flow component [5] is shown by a smaller negative pressure peak at the lower side of the plot, slightly right of the middle. The section of the reaction plane, [x,z], shown is 12.1 $\times$ 9.6 fm.

cannot be dissipated away [8]. In other words the Reynolds number tends to infinity and the system tends to turbulence. Viscosity is needed to stabilize the flow.

There are several numerical methods to obtain a solution. In all cases the solution is discretized in some way and a coarse graining is introduced. This coarse graining automatically leads to dissipative and transport properties, and the parameters of these transport processes can be determined by numerical experiments. The result is that all CFD models are dissipative or viscous, which is an advantage as the these are usually stable. By decreasing the cell size, the dissipative effects can be decreased in CFD, to a limit when instabilities, or turbulence occurs. As a side effect, CFD models lead to entropy production, as part of the kinetic energy is dissipated away due to the coarse graining or smoothing.

4. Kinetic Theory of Particle Formation, Escape and Freeze-out

Most frequently relativistic kinetic theory describes the time, t, development of the single particle distribution function, $f(p;x) = f(p, \mathbf{r}, t)$, of conserved particles in the 6-dimensional Phase-Space (PS), $[\mathbf{p}, \mathbf{r}]$ and time. Then the total particle number does not change with time. We assume (usually) that the particles are on mass shell, so the 4-momentum has 3 independent variables. The phase space distribution is a density in the 6 dimensional PS and it is an invariant scalar, because a Lorentz boost contracts the system in the configuration space but elongates it in the momentum space.

If we want to describe a particle source (or drain) from a matter, the total number of newly crated particles can be given as

$$N_{new} = \int d^4p\, d^4x\; g(p,x),$$

where $g(p,x)$ is an invariant scalar density in the 4-dimensional PS and 4-dimensional space-time (ST). If the particles are on mass shell the momentum space integral can be reduced to

$$N_{new} = \int \frac{d^3p}{p^0}\, d^4x\; h(p,x), \tag{8}$$

where $h(p,x)$ is also an invariant scalar density, as $\frac{d^3p}{p^0}$ is also one.

The particle formation or source density $h(p,x)$ from a matter, can be characterized by 3 factors:
PS distribution, $f(p, \mathbf{r}, t)$,
PS emission probability, $P_{esc-PS}(p;x)$,
ST emission probability, $P_{esc-ST}(x;p)$.
Here f can be an arbitrary distribution, but we usually discuss matter in (or close to) statistical equilibrium, with known PS distribution. P_{esc-PS} is most frequently assumed to be unity, while P_{esc-ST} is the ST distribution if the source (e.g. a 4-dimensional Gaussian in the ST). The experimental analysis of two particle correlations from heavy ion reactions, is based on these very simple assumptions in most cases.

These are very strong assumptions. It is not supported by experiments that the emission probability is isotrope and homogeneous in the PS, or neither that the ST distribution of the source is Gaussian.

4.1 Non-isotropic particle sources

Not only in heavy ion reactions but in any dynamical process, particle creation, (or condensation) happens mostly in a directed way: the phenomenon propagates into some direction, i.e. it happens in some (space-like) layer or front (like detonations, deflagrations, shocks, condensation waves or freeze-out). The reason is that neighbouring regions in the front may interact, minimize the energy of the front by evening it out, provide energy to neigbouring regions to exceed the threshold conditions. Even in relativistic processes, that are time-like, (have time-like normal) and the neighbouring points of a front cannot be in causal connection, in the LR frame of the matter the dynamical process may and frequently has a direction. (See the example in ref. [9].)

These fronts or layers are not necessarily narrow, but they have a characteristic direction (or normal, $d^3\sigma^\mu$), thus the ST integral, d^4x, can be converted into

$$d^4x \longrightarrow ds^\mu \, d^3\sigma_\mu .$$

The front can actually be directed time-like, $d^3\sigma^\mu \, d^3\sigma_\mu = +1$, or space-like, $d^3\sigma^\mu \, d^3\sigma_\mu = -1$.

An example for emission distributed in the ST, but being time-directed can be described by the expression:

$$N_{new} = \int d^3p \int d\tau d^3\sigma_\mu \; f_{eq}(p,x) \; \frac{C_1 p^\mu}{p^0} \; \Theta(p^\nu \, d^3\sigma_\nu) \; G\left(\tau - \tau_{FO}(x)\right) \quad (9)$$

where

$$\frac{C_1 p^\mu}{p^0} \; \Theta(p^\nu \, d^3\sigma_\nu) \quad (10)$$

is the PS emission probability, where the step function eliminates the possibility of emission of particles, in the direction opposite to the normal of the front (for time-like normals it is always $+1$), and the term, $\frac{p^\mu}{p^0} = (1, \mathbf{v})$, together with the 6- dimensional PS density, f_{eq}, yield the generalized 4-dimensional flux of particles with momentum $\mathbf{p}$: $f_{eq}(p,x) \, \frac{p^\mu}{p^0}$.

Here G is a Gaussian time distribution or emission density for time-like directed fronts:

$$d\tau \, G(\tau - \tau_{FO}) = \frac{d\tau}{\sqrt{2\pi} \, \tau_{coll.}} \exp\left[-\frac{(\tau - \tau_{FO})^2}{2\tau_{coll.}^2} \right]$$

and for space-like fronts:

$$ds \, G(s - s_{FO}) = \frac{ds}{\sqrt{2\pi} \, \lambda_{m.f.p.}} \exp\left[-\frac{(s - s_{FO})^2}{2\lambda_{m.f.p.}^2} \right]$$

.

In more complex models the emission probabilities may take more realistic and more complicated forms. Furthermore, $f_{e.q.}$ can be space-time dependent, and can be determined *self-consistently* during the detonation, deflagration or freeze out process [10, 11, 12, 13] .

Let us rewrite the escape probability, (10), with the hyper-surface element, covering both the timelike and spacelike parts of the freeze-out surface,

$$P^*_{esc-PS} = \frac{p^\mu d^3\sigma_\mu}{(p^\nu u_\nu)} \, \Theta(p^\mu d^3\sigma_\mu). \tag{11}$$

Here $d^3\sigma_\mu(x)$ and $u_\mu(x)$ are ST dependent and this yields the secondary $x-$dependence of $P^*_{esc-PS}(p; x)$. If we take the four velocity equal to $u_\mu = (1, 0, 0, 0)$, in the Rest Frame of the Front (RFF), i.e. where $d^3\sigma_\mu = (1, 0, 0, 0)$, then P is unity. Otherwise, in the Rest Frame of the Gas (RFG), where $u_\mu = (1, 0, 0, 0)$, the escape probability P is $P^*_{esc-PS} = p^\mu d^3\sigma_\mu \, \Theta(p^\mu d^3\sigma_\mu)/p^0$.

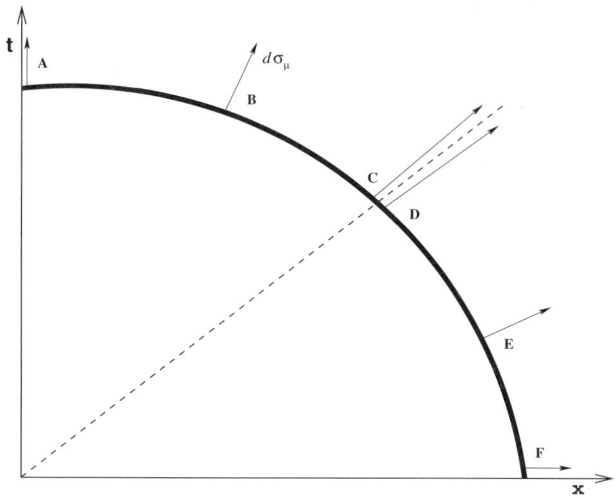

Figure 2. A simple FO-surface in the Rest Frame of the Gas (RFG), where $u^\mu = (1, 0, 0, 0)$, including time-like and space-like parts. Then the normal vector of the FO front, $d^3\sigma_\mu(x)$, is time-like at the time-like part and it is changing smoothly into a space-like 4-vector in the space-like part. On these two parts of the hyper-surface, in the Local Rest Frame of the Front (RFF), $d^3\sigma_\mu(x)$ points into the direction of the t' (x') axis respectively.

The nominator depends on $d^3\sigma_\mu$. In the following, we will take different typical values for $d^3\sigma_\mu$, for characteristic regions (A, B, C, D, E, F) of the freeze-out hyper-surface (see Figure 2):

- A $d^3\sigma_\mu = (1, 0, 0, 0)$, leads to $p^\mu d^3\sigma_\mu = p^0 \geq 0$, as we have seen above for FO along the time axis in RFG,
- B $d^3\sigma_\mu = \gamma(a, b, 0, 0)$,
- C $d^3\sigma_\mu = \gamma(1 + \epsilon, 1 - \epsilon, 0, 0)$, just above the light-cone,
- D $d^3\sigma_\mu = \gamma(1 - \epsilon, 1 + \epsilon, 0, 0)$, just below the light-cone,

- E $d^3\sigma_\mu = \gamma(c, d, 0, 0)$,
- F $d^3\sigma_\mu = (0, 1, 0, 0)$, leads to $p^\mu d^3\sigma_\mu = p^x$, for FO along the spatial axis x in RFG,

where the γ factors serve the unit normalization of the normal vectors (see Figure 2). The effect of this relativistically invariant FO factor leads to a smoothly changing behaviour as the direction of the normal vector changes in RFG (see Figure 2).

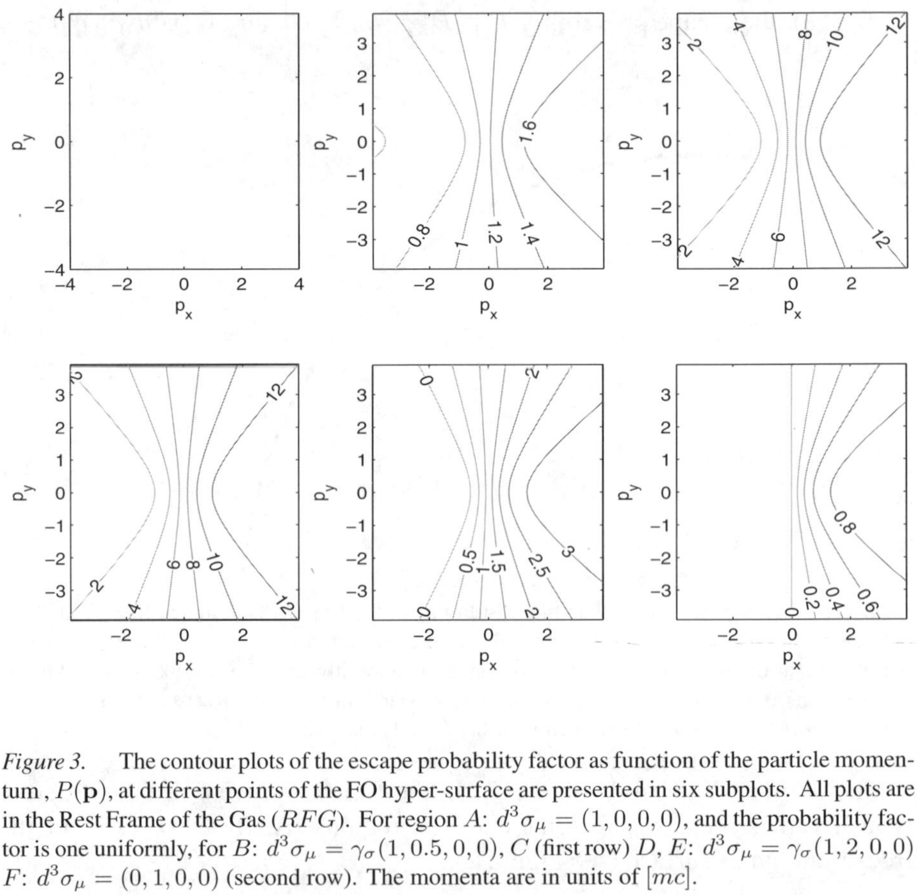

Figure 3. The contour plots of the escape probability factor as function of the particle momentum , $P(\mathbf{p})$, at different points of the FO hyper-surface are presented in six subplots. All plots are in the Rest Frame of the Gas (RFG). For region A: $d^3\sigma_\mu = (1, 0, 0, 0)$, and the probability factor is one uniformly, for B: $d^3\sigma_\mu = \gamma_\sigma(1, 0.5, 0, 0)$, C (first row) D, E: $d^3\sigma_\mu = \gamma_\sigma(1, 2, 0, 0)$ F: $d^3\sigma_\mu = (0, 1, 0, 0)$ (second row). The momenta are in units of $[mc]$.

To calculate the parameters of the normal vector $d^3\sigma_\mu$ for different cases listed above, we simply make use of the Lorentz transformation. The normal vector of the time-like part of the freeze-out hypersurface may be defined as the local t'-axis, while the normal vector for the space-like part may be defined as the local x'-axis. As the $d^3\sigma_\mu$ normal vector is normalized to unity its components may be interpreted in terms of γ_σ and v_σ, as $d^3\sigma_\mu = \gamma_\sigma(1, v_\sigma, 0, 0)$, where

$\gamma_\sigma = \frac{1}{\sqrt{1-v_\sigma^2}}$ for time-like normals and $\gamma_\sigma = \frac{1}{\sqrt{v_\sigma^2-1}}$ for space-like normals. This will lead to,

- A $d^3\sigma_\mu = (1,0,0,0)$, leads to $P = 1$,
- B $d^3\sigma_\mu = \gamma_\sigma(1, v_\sigma, 0, 0)$, leads to $P = \frac{\gamma_\sigma(p^0+v_\sigma p^x)}{p^0}$, and $\gamma_\sigma = \frac{1}{\sqrt{1-v_\sigma^2}}$,
- C $d^3\sigma_\mu = \gamma_\sigma(1+\epsilon, 1-\epsilon, 0, 0)$, leads to $P = \frac{4\gamma_\sigma^2(p^0+p^x)+\gamma_\sigma(p^0+p^x)}{4\gamma_\sigma p^0}$,
- D $d^3\sigma_\mu = \gamma_\sigma(1-\epsilon, 1+\epsilon, 0, 0)$, leads to $P = \frac{4\gamma_\sigma^2(p^0+p^x)-\gamma_\sigma(p^0-p^x)}{4\gamma_\sigma p^0} \times \Theta(p^\mu d^3\sigma_\mu)$,
- E $d^3\sigma_\mu = \gamma_\sigma(1, v_\sigma, 0, 0)$, leads to $P = \frac{\gamma_\sigma(p^0+v_\sigma p^x)}{p^0} \times \Theta(p^\mu d^3\sigma_\mu)$, and $\gamma_\sigma = \frac{1}{\sqrt{v_\sigma^2-1}}$,
- F $d^3\sigma_\mu = (0,1,0,0)$, leads to $P = \frac{p^x}{p^0} \times \Theta(p^\mu d^3\sigma_\mu)$, (see Figure 2).

The resulting Phase Space escape probabilities are shown in Figure 3 for the six cases described above.

In refs. [10, 11] the post FO distribution was evaluated for space-like gradual FO in a kinetic model. Initially we had an equilibrated, interacting, PS distribution, $f_{int}(p,x)$, and an escape probability, similar to eq. (11), but simplified so that it depended on the angle of the two vectors only. After some small fraction of particles were frozen out as the FO process progressed in the front, the interacting component were re-equilibrated, with smaller particle number, smaller energy and momentum, to account for the quantities carried away by the frozen out particles. This was then repeated many times in small steps along the FO front and the frozen out particles were accumulated in the post freeze out PS distribution, f_{free}. The resulting distribution was highly anisotropic and obviously non-equilibrated. The details of the post FO distribution depend on the details of the escape probability, and on the level of re-equilibration of the remaining, interacting component.

Bugaev assumed earlier [14], that the post FO distribution is a (sharply) "Cut-Juttner" distribution, but the above mentioned model shows that this can only be obtained if re-equilibration is not taking place. The kinetic model provided an asymmetric but smooth PS distribution [10], while the escape probability (11) yields a somewhat different, but also smooth PS distribution. These can be well approximated by the "Cancelling Juttner" distribution [15], Figure 4.

Up to now we analysed the FO probability in the momentum space only. At this point we should also mention, that if we discuss gradual FO across a finite layer, the distance from a *hypothetical* or *theoretical* FO surface must also be part of the FO probability, because the probability of a particle has no more

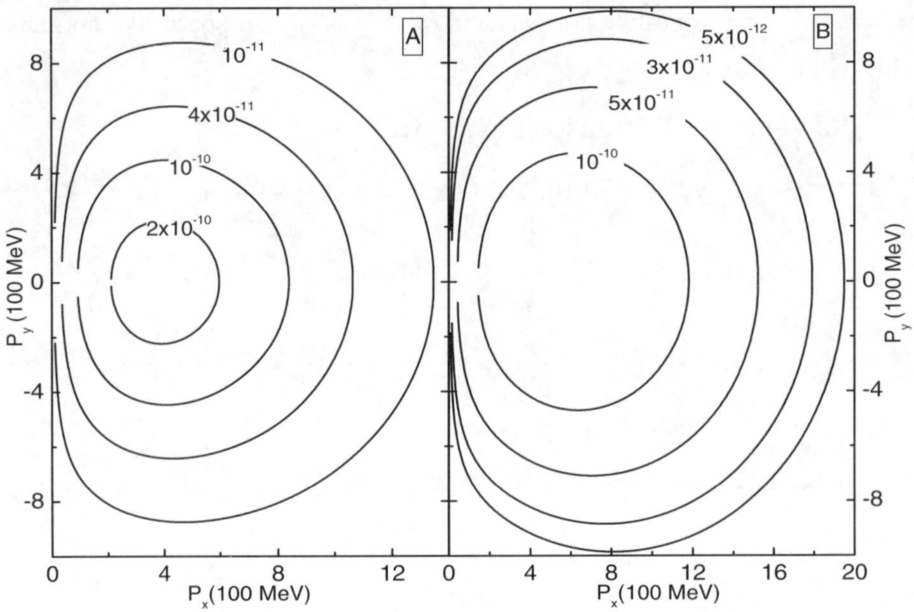

Figure 4. A simple FO-surface in the Rest Frame of the Gas (RFG) where $d^3\sigma_\mu = (1, 0, 0, 0)$, including timelike and spacelike parts. Then $d^3\sigma_\mu$ is $t' = d^3\sigma_\mu$ for the timelike part, changing smoothly into $x' = d^3\sigma_\mu$ for spacelike normal vectors.

collision on its way out depends on the local mean free path (m.f.p.) and on the distance from this surface. The m.f.p. is usually momentum dependent and this yields the secondary $p-$dependence of $P_{esc-ST}(x;p)$. Of course the position of such a *hypothetical* FO surface has to be determined self consistently or within a realistic kinetic model. However, if we assume simultaneous FO and hadronization [16, 17] from supercooled plasma, there are no realistic and reliable dynamical models up to know, which we could use, so, a now the first self consistent approximate method is our only choice.

4.2 Emission on an idealized hyper-surface

If the m.f.p. or the coll. time are negligibly small compared to the ST dimensions of the system we can idealize the source distribution further as the source is shrinking to a hyper-surface (HS) in the ST, i.e.:

$$\frac{d\tau}{\sqrt{2\pi}\;\tau_{coll.}} \exp\left[-\frac{(\tau - \tau_{FO})^2}{2\tau_{coll.}^2}\right] \longrightarrow \delta(\tau - \tau_{FO})\; d\tau : \text{time} - \text{like fronts},$$

$$\frac{ds}{\sqrt{2\pi}\;\lambda_{m.f.p.}} \exp\left[-\frac{(s - s_{FO})^2}{2\lambda_{m.f.p.}^2}\right] \longrightarrow \delta(s - s_{FO})\; ds : \text{space} - \text{like fronts},$$

and the integration in the normal direction to the front can be executed. Then

$$N_{new} = \int \frac{d^3p}{p^0} \int_{\Sigma} (p^\mu d^3\sigma_\mu)\, \Theta(p^\mu\, d^3\sigma_\mu)\, f_{FO}(p,x)\, C_1\,. \tag{12}$$

Here $_\Sigma$ denotes integration over a 3-dimensional HS. If all particles described by $f_{e.q.}$ are emitted during the process, $C_1 = 1$. If we do not perform the momentum integral we recover the modified "Cooper-Frye" freeze out formula [18, 14, 19]

$$p^0 \frac{dN}{d^3p} = \int_{\Sigma} p^\mu d^3\sigma_\mu\, \Theta(p^\mu\, d^3\sigma_\mu)\, f_{FO}(p,x) \tag{13}$$

Notice that we have changed the notation of the PS distribution from f_{eq} to f_{FO} denoting the post FO distribution integrated across the front.

Eq. (13) simplifies the process to a great extent, especially if we consider that f_{eq} changes during the detonation deflagration or freeze out process as we go across the front self-consistently. In a dynamical process across the front, even if f_{eq} is always an equilibrated distribution, its parameters change in a wide range [10, 11]. In the FO process the post FO distribution, after the process is completed is very far from any statistically equilibrated distribution. This should be taken into account if we apply the assumption of an idealized FO surface. The determination of the FO surface is an involved task, both physically and technically. In complex modeling tasks we perform the self-consistent determination of the surface in simplified models. In these models we can pinpoint the FO conditions and evaluate the post FO distributions.

Then based on this knowledge we determine the FO surface in the large scale, 3+1 dimensional system by analysing the complete ST history of the fluid dynamical stage calculated. This demanding task was completed recently by *Bernd R. Schlei* (of Los Alamos) [20], and this enables us to visualize the ST evolution of the surface where the emitted particles originate from.

5. Two particle correlations

In case of two particle correlations we have to return to eqs. (8) or (9), because the emission point in ST is crucial. Although, eqs. (12) or (13) also contain a ST dependence, this is reflecting the properties of the idealized hypersurface, where the process is taking place. This approximation is justified only if

- The ST domain of the emission is a negligibly narrow layer and we are sure that the internal processes within the front are not relevant.

- If we use a realistic estimate for the post FO distribution in these equations.

The PS density of created (or frozen out) particles, N_{cr}, is given in terms of a "Source function", $S(x,p)$, as [21] [1]

$$\frac{dN_{cr}}{d^3p} = \int S(x,p)\, d^4x \,, \tag{14}$$

which can be compared to eqs. (8,11), so that

$$p^0\, \frac{dN_{cr}}{d^3p} = \int d^4x\; f_{eq}(p,x)\;\; P_{esc-PH}(p;x)\;\; P_{esc-ST}(x;p) \;\; \text{or} \tag{15}$$

$$\frac{dN_{cr}}{d^3p} = \int d^4x\; \underbrace{f_{eq}(p,x)\;\; \overbrace{P_{esc-PH}(p;x)/p^0}^{P^*_{esc-PH}(p;x)}\;\; P_{esc-ST}(x;p)}_{S(p;x)} \,. \tag{16}$$

Thus, assuming emission or freeze-out from an interacting gas the source function is

$$S(x,p) = f_{eq}(p,x)\; P^*_{esc-PH}(p;x)\; P_{esc-ST}(x;p) \,, \tag{17}$$

here we assumed that the emission or escape probability is factored to a PS and a ST dependent part.

Two particle correlations provide information about the location of FO points in the ST, providing the possibility of deeper insight into the reaction mechanism. This obviously means that there is an essential difference between reaction models assuming (i) FO in an extended ST layer or (ii) FO accross an idealized ST hypersurface. The latter assumption is obviously not realistic.

Nevertheless, the idealized hypersurface assumption may not be so bad as one would think. Even in this case parts of this surface can represent early times while others late emissions. The spatial locations of different parts of the surface can be far from each other. If the ST dimensions of the idealized FO hypersurface are much larger than the thickness of the real FO layer, the hypersurface assumption may be satisfactory. In any case, two particle correlations are much more sensitive to this simplifying assumption than single particle data.

Note also that the thickness of the FO layer is dependent of the hadronic species measured. Thus, the hypersurface idealization is applicable for high multiplicity hadrons, protons and pions, while particles with low multiplicity, low cross-sections or low creation rates are not described well with this approximation.

6. Simultaneous Hadronization and Freeze-Out

If one assumes hadronization in thermal and fluid dynamical equilibrium via homogeneous nucleation [22] or similar processes, this leads to a lengthy, gradual hadronization and freeze-out. Such a long-lived particle source should be detected by pion interferometry, but recent data from RHIC ruled out this scenario. Therefore we do not assume a lengthy thermalization and chemical equilibration process at freeze-out either. As we described in ref. [23], we assume simultaneous FO and hadronization at the end of the heavy ion reaction.

We assume that pre-hadrons or quasi-hadrons are formed in the cooling and expanding plasma before we reach the critical temperature and density, and especially when the plasma supercooles. Supercooling is possible even if the phase transition is just a smooth, *but sharp* cross-over, which is always the case in small, finite systems. Then hadrons are formed out of thermal equilibrium, which are frozen out (i.e. never collide) after their formation. No detailed hadronization models exist for this kind of mechanism. The closest are coalescence or recombination models, which usually describe the hadron abundances well.

As discussed in [23] and [24], the similarity of the results arising from statistical (thermal) and coalescence models, is due to the fact that the Canonical Ensemble (CE) is generated if the average energy of the system is the same in the elements of the ensemble and the states included are generated with equal a priory probability. For most hadrons these conditions are satisfied, so the two models yield similar results and therefore we use the statistical model to estimate local hadron abundances at the end of the reaction [23].

The differences arise for hadrons where the assumption of equal a priory formation probability does not hold. An example is the $\Lambda(1520)$ resonance, an excited state where the radial wave function includes both s- and p- waves, and so the formation cross-section is much less than for other hadrons. Therefore the abundance of these hadrons is much smaller, and this is well described by coalescence type models, while statistical models fail to reproduce the smaller abundance (as the cross-section does not appear in these models).

For our purposes, however, the statistical model is adequate, because it includes all necessary statistical weights, (so it works well for the high multiplicity hadrons,) while low multiplicity resonances are not reproduced by fluid dynamical models anyway.

7. Summary

Recent experimental data indicate that detailed fluid dynamical data are becoming available at RHIC energies. This will provide us with the possibility to test the QGP Equation of State in heavy ion reactions. This will require further experimental efforts that provide us with the complete spectrum of collective flow data. To draw the right conclusion will also require a detailed and realistic theoretical reaction model which can simultaneously describe single particle and two-particle observables.

Notes

1. Note that in some cases the source function is defined differently [25], so that the p^0 factor is not included in the denominator: $p^0 \frac{dN_{cr}}{d^3p} = \int S'(x,p)d^4x$.

References

[1] M. Gyulassy, L.P. Csernai, *Nucl. Phys.* **A 460** (1986) 723.

[2] V.K. Magas, L.P. Csernai, D.D. Strottman, *Phys. Rev.* **C64** (2001), 014901.

[3] V.K. Magas, L.P. Csernai, D.D. Strottman, *Nucl. Phys.* **A 712** (2002) 167-204,

[4] V.K. Magas, L.P. Csernai, D.D. Strottman, Proceedings of the International Conference "New Trend in High-Energy Physics" (Crimea 2001), Yalta, Crimea, Ukraine, September 22-29, 2001, edited by P.N. Bogolyubov and L.L. Jenkovszky (Bogolyubov Institute for Theoretical Physics, Kiev, 2001), pp. 193-200; and hep-ph/0110347.

[5] L.P. Csernai and D. Röhrich, *Phys. Lett.* **B 458** (1999) 454.

[6] J. Adams *et al.* nucl-ex/0310029.

[7] L.P. Csernai: *Introduction to Relativistic Heavy Ion Collisions*, Willey (1994).

[8] L.D. Landau and E.M. Lifshitz: *Hydrodynamics* (Nauka, Moscow), Chapter 3, (1953).

[9] L.P. Csernai, *Sov. JETP* **65** (1987) 216; *Zh. Eksp. Theor. Fiz.* **92** (1987) 379.

[10] Cs. Anderlik, Z.I. Lázár, V.K. Magas, L.P. Csernai, H. Stöcker and W. Greiner, *Phys. Rev.* **C59** (1999) 388.

[11] V.K. Magas, Cs. Anderlik, L.P. Csernai, F. Grassi, W. Greiner Y. Hama, T. Kodama, Zs. Lázár and H. Stöcker, *Heavy Ion Phys.* **9** (1999) 193.

[12] V.K. Magas, A. Anderlik, Cs. Anderlik and L.P. Csernai, *Eur. Phys. J.* **C30** (2003) 255-261.

[13] E. Molnár et al. (2003) in preparation.

[14] K.A. Bugaev, *Nucl. Phys.* **A606** (1996) 559.

[15] K. Tamosiunas, and L.P. Csernai, *Eur. Phys. J.* **C** (2003) in press.

[16] T. Csörgő and L.P. Csernai, *Phys. Lett.* **B333** (1994) 494.

[17] L.P. Csernai and I.N. Mishustin, *Phys. Rev. Lett.* **74** (1995) 5005.

[18] F. Cooper and G. Frye, *Phys. Rev.* **D10** (1974) 186.

[19] Cs. Anderlik, L.P. Csernai, F. Grassi, W. Greiner, Y. Hama, T.Kodama, Zs. Lázár, V. Magas and H. Stöcker, *Phys. Rev.* **C59** (1999) 3309.

[20] B.R. Schlei: "Isosurfacing in Four Dimensions," in Theoretical Division quarterly - Winter 2003/04, Los Alamos, in print.

[21] B.R. Schlei, U. Ornik, M. Plümer, and R. Weiner, *Phys. Lett.* **B293** (1992) 275-281; and J. Bolz, U. Ornik, M. Plümer, B.R. Schlei, R.M. Weiner, *Phys. Rev.* **D47** (1993) 3860.

[22] L.P. Csernai and J.I. Kapusta, *Phys. Rev.* **D46** (1992) 1379; and L.P. Csernai and J.I. Kapusta, *Phys. Rev. Lett.* **69** (1992) 737.

[23] A. Keranen, L.P. Csernai, V. Magas and J. Manninen, *Phys. Rev.* **C 67** (2003) 034905,

[24] L.P. Csernai, *J. Phys.* **G 28** (2002) 1993.

[25] S. Chapman and U. Heinz, *Phys. Lett.* **B340** (1994) 250-253.

ASPECTS OF NON-EQUILBRIUM QUANTUM FIELD THEORY IN RELATIVISTIC HEAVY ION COLLISIONS

C. Greiner[1], S. Juchem[2] and Z. Xu[1,2]

[1]*Institut für Theoretische Physik der J. W. Goethe Universität, D-60054, Frankfurt am Main, Germany*

[2]*Institut für Theoretische Physik der Justus Liebig Universität, D-35392, Giessen, Germany*

Abstract In this lecture we review recent progress in various aspects of non-equilibrium QFT with respect to relativistic heavy ion collisions. As a first and rather general study we summarize our (numerical) investigations for a dissipative quantum time evolution of ϕ^4-field theory for a spatially homogeneous system in 2+1 space-time dimensions on the basis of the Kadanoff-Baym equations. The initial conditions can be chosen arbitrarily and far from equilibrium. The calculations demonstrate how thermalization and chemical equilibration is achieved ab initio from the underlying QFT. As a possible candidate for a non-equilibrium phase transition phenomena we then adress the stochastic formation of disoriented chiral condensates. Finally we elaborate on a new 3+1 dimensional Monte Carlo parton cascade solving kinetic Boltzmann processes including inelastic multiplication processes ($gg \leftrightarrow ggg$) in an unified manner, where the back reaction channel is treated for the first time fully consistently.

1. Introduction, overview and summary

The prime intention for present ultrarelativistic heavy ion collisions at CERN and at Brookhaven lies in the possible experimental identification of the quark gluon plasma (QGP). The evolving system in the reaction is at least initially far from any (quasi-)equilibrium configuration. Not only for this reason, nonequilibrium quantum field theory has become a major topic of research for describing transport processes in various areas of physics like in cosmological particle physics as well as condensed matter physics. The multidisciplinary aspect arises due to a common interest to understand the various real-time relaxation phenomena of quantum dissipative systems.

A seminal work in solving numerically for the first time the so called Kadanoff-Baym equations of evoltion equations for out-of-equilibrium one-particle Green's

W. Greiner et al. (eds.), Structure and Dynamics of Elementary Matter, 143–158.

function has been carried out by Danielewicz [1], who investigated a spatially homogeneous system with a deformed Fermi sphere in momentum space for the initial distribution of occupied momentum states in order to model the initial condition of a heavy-ion collision in the nonrelativistic domain. In comparison to a standard on-shell semi-classical Boltzmann equation the solutions showed quantitative differences like a larger collective relaxation time for achieving complete kinetic equilibration. Similar quantum modifications in the equilibration and momentum relaxation have been found for a relativistic situation in [2], where also a faster equilibration can occur at higher bombarding energies.

In section 2 we summarize our findings of a recent and very detailed study concerning the quantum time evolution of ϕ^4-field theory for a spatially homogeneous system in 2+1 space-time dimensions by including the mean-field (tadpole) and collisional (sunset) self-energies [3]. As an important and new step in the field, we point out the dynamics of the spectral ('off-shell') distributions of the excited quantum modes and the different phases in the approach to equilibrium described by Kubo-Martin-Schwinger (KMS) relations for full thermal equilibrium states, the latter being the only translation invariant solution and representing the stationary fixed points of the dissipative coupled equation of motions. Concentrating further on the importance of off-shell effects, far off-shell $1 \leftrightarrow 3$ decay processes are responsible for chemical equilibration. The physical process corresponds to a consistent treatment of Bremsstrahlung type processes in the medium by the interplay of production, propagation and subsequent decay of highly virtual mode excitations.

As an exotic candidate for a direct signature stemming from the QGP phase transition (and which can be adressed in RHIC experiments at Brookhaven) we summarize in section 3 ideas of stochastic formation of so called disoriented chiral condensates (DCC). The idea of DCC [4] first appeared in a work of Anselm but it was made widely known due to Bjorken, and Rajagopal and Wilczek. The spontaneous growth and subsequent decay of these configurations emerging after a rapid chiral phase transition from the QGP to the hadronic world would give rise to large collective fluctuations in the number of produced low momentum neutral pions compared to charged pions. Our work concentrates on the important question on the likelihood of an instability leading potentially to a large DCC yield of low momentum pions [5]. It reveals that an experimentally feasible DCC, if it does exist in nature, has to be a rare event with some finite probability following a nontrivial and nonpoissonian distribution on an event by event basis. DCCs could then (only) be revealed experimentally by inspecting higher order factorial cumulants θ_m $(m \geq 3)$ in the sampled distribution of low momentum pions.

In section 4 we finally detail on a new parton cascade scheme treating elastic and inelastic multiplication collisions in an unified manner [6]. For the first time the back reation channel ($ggg \rightarrow gg$) is treated fully consistently by re-

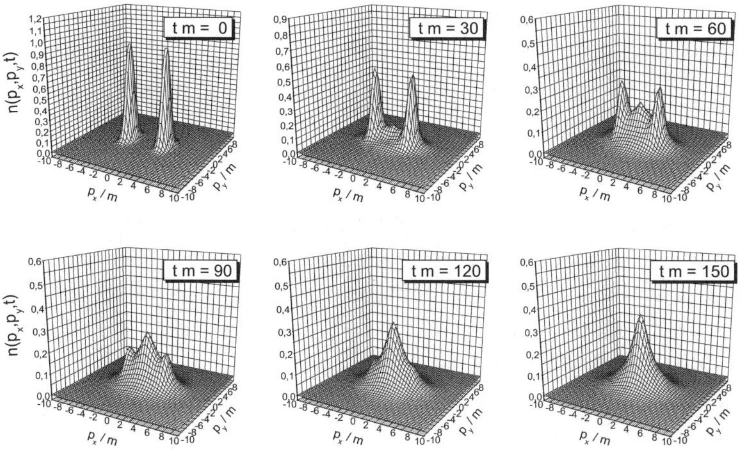

Figure 1. Characteristic evolution of the occupation number in momentum space within the full Kadanoff-Baym dynamics starting from an initially non-isotropic shape. The occupation number correspondes directly to the equal-time Green function via the relation 3 and is displayed for various times $\bar{t} \cdot m = 0, 30, 60, 90, 120, 150$.

specting detailed balance within the same algoritm. With this we can adress the important question of thermalization and early pressure build up on a partonic level for heavy ion collisions at RHIC. First (and preliminary) results show that indeed the gluon multiplication via Bremsstrahlung (and absorption) is of utmost importance, where thermalization is achievd on a timescale of about 1 fm/c.

2. Dissipative quantum dynamics for ϕ^4-theory

The starting point of our recent numerical investigation are the causal Dyson-Schwinger equations for certain real-time one-particle Green's functions, which are also called Kadanoff-Baym equations [3]:

$$
\begin{aligned}
-\left[\partial_\mu^x \partial_x^\mu + m^2\right] G^{>/<}(x,y) &= \Sigma^\delta(x)\, G^{>/<}(x,y) \qquad (1)\\
+ \int_{t_0}^{x_0} dz_0 \int d^dz &\qquad \left[\Sigma^>(x,z) - \Sigma^<(x,z)\right] G^{>/<}(z,y)
\end{aligned}
$$

$$- \int_{t_0}^{y_0} dz_0 \int d^d z \qquad \Sigma^{>/<}(x,z) \; [G^{>}(z,y) - G^{<}(z,y)]$$

Besides the Hartree term, to next order in the coupling constant the non-local sunset self-energy enters as

$$\Sigma^{>/<}(x,y) \;=\; -\frac{\lambda^2}{6} \left[G^{>/<}(x,y) \right]^3 . \tag{2}$$

With this the Kadanoff-Baym equations are in closed form and can thus be (numerically) integrated. In the following we summarize some major findings of such an analysis for a 2+1 dimensional and spatially homogenous system being prepared by some non-equilibrium initial configuration of excited modes in momentum space at $t = 0$ via

$$2\omega_{\mathbf{p}}\, i\, G^{<}_{\phi\phi}(\mathbf{p},t,t) \;=\; 2\,n(\mathbf{p},t) \,+\, 1. \tag{3}$$

A typical evolution for this Green's function is depicted in fig. 1. Starting from an initially far non-isotropic shape it develops towards a rotational symmetric distribution in momentum space. Initially there is no occupation along the p_y-momentum axis. As a function of time this area of momentum space is getting filled which takes about 100 time units. In fig. 2 the occupation number spectra along the transverse momentum axis is depicted. A quasi exponetial slope is experienced on the very onset of the ongoing evolution, where the system is still far from kinetic equilibrium, as most energy is still stored in excitations along the longitudinal axis. Hence, a 'thermal' spectra along the transverse direction as given by the present example does not necessarily imply that the (complex) system is close to kinetic and thermal equilibrium. (This could be also true for relativistic heavy ion collisions - see eg [7] - where the transverse spectra typically look 'thermal'.)

On the other hand, we observe that starting from very different initial conditions for systems containing the *same* energy that the single momentum modes converge to the same respective numbers for large times as characteristic for a system in equilibrium. The different momentum modes in Fig. 3 typically show a three-phase structure. For small times ($t \cdot m < 10$) one finds damped oscillations that can be identified with a typical switching on effect at $t = 0$, where correlations build up and dephasing of the various modes does occur. For 'intermediate' time scales ($10 < t \cdot m < 100$) one observes a strong change of all momentum modes in the direction of the final stationary state. We address this phase to 'kinetic' equilibration and point out, that the momentum modes can temporarily even exceed their respective equilibrium value (see the evolution of the distribution D1). We remark that this behaviour of 'overshooting' – as in the particular case of D1 – is *not* observed in a simulation with a kinetic

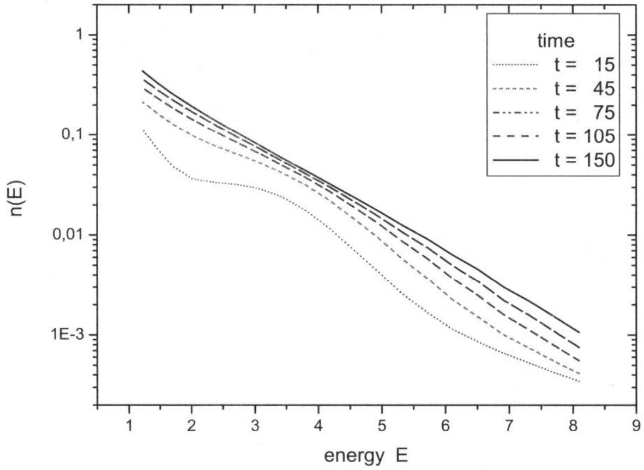

Figure 2. Occupation number spectra along the transverse momentum axis $(p_x = 0, p_y)$ plotted versus (transverse) energy. The time is taken in units of the inverse mass m, the energy is given in units of m.

Boltzmann equation. Hence this highly nonlinear effect must be attributed to quantal off-shell and memory effects not included in the standard Boltzmann limit. The third phase, i.e. the 'late' time evolution is characterized by a smooth approach of the single momentum modes to their respective equilibrium values. As we will now explain, this phase is adequately characterized by chemical processes changing the 'number' of (quasi-)particles or excitations. (It turns out that this rate for chemical equilibration is only slightly smaller than that of kinetic equilibration for the coupling parameter chosen [3].)

In principle one can now study the time evolution of the spectral function as a function of time. This is an important and lively topic, but we refer the interested reader to the paper [3]. In Fig. 4 (lower part) the spectral function $A(\mathbf{p}, p_0, \bar{t})$ for the initial distribution D2 at very late times for various momentum modes is depicted as a function of the energy p_0. The system has nearly equilibrated and the spectral function are rather broad, reflecting a strongly interacting system via the choice of a rather large coupling constant.

At perfect thermal equilibrium the famous KMS condition of the Green functions hold dictating that the equilibrium form of the complete and off-shell distribution function at temperature T is simply given via

$$N_{eq}(\mathbf{p}, p_0) = N_{eq}(p_0) = \frac{1}{\exp(p_0/T) - 1} = N_{bose}(p_0/T), \quad (4)$$

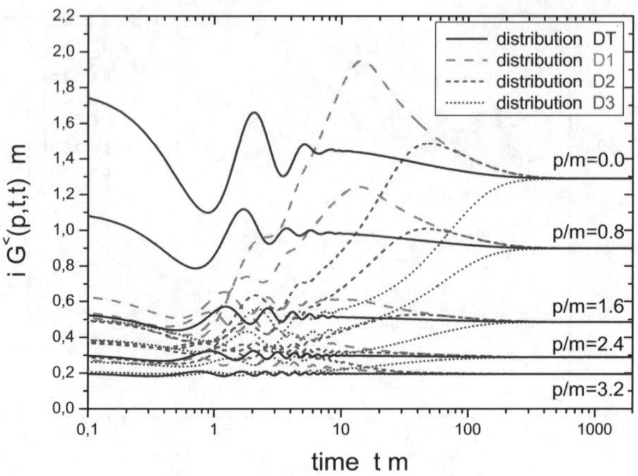

Figure 3. Time evolution of selected momentum modes of the equal-time Green function $|\,\mathbf{p}\,|/m = 0.0$, 0.8, 1.6, 2.4, 3.2 (from top to bottom) for four different initial configurations D1, D2, D3 and DT (characterized by the different line types) with the same energy density. All momentum modes assume the same respective equilibrium value for long times ($t \cdot m > 500$) independent of the initial state.

which is the well-known Bose distribution. The distribution function $N(p_0)$ as extracted from the calculation is displayed in Fig. 4 (upper part) for the same momentum modes as a function of the energy p_0. We find that $N(p_0)$ for all individual momentum modes can be fitted by a single Bose function with temperature $T/m = 1.835$. Thus the distribution function emerging from the Kadanoff-Baym time evolution for $t \to \infty$ approaches indeed a Bose function in the energy p_0 that is *independent* of the momentum as demanded by the equilibrium form. The KMS conditions are approached as a dynamical fixed point in the complex non-equilibrium evolution.

We note that the chemical potential μ – used as a second fit parameter – is indeed close to zero for these late times as expected for the correct equilibrium state of the neutral ϕ^4-theory which is characterized by a vanishing chemical potential μ in equilibrium. This is a consequence of the exact treatment. A kinetic Boltzmann equation in general would lead to a stationary state for $t \to \infty$ with a finite chemical potential.

In Fig. 5 the width Γ is displayed as a function of the energy p_0. The width shows two maxima for all momentum modes. The first is characterized by an increase towards a maximum around the on-shell energy and falling off beyond. This behaviour stems from the $2 \leftrightarrow 2$ processes in the self-energies.

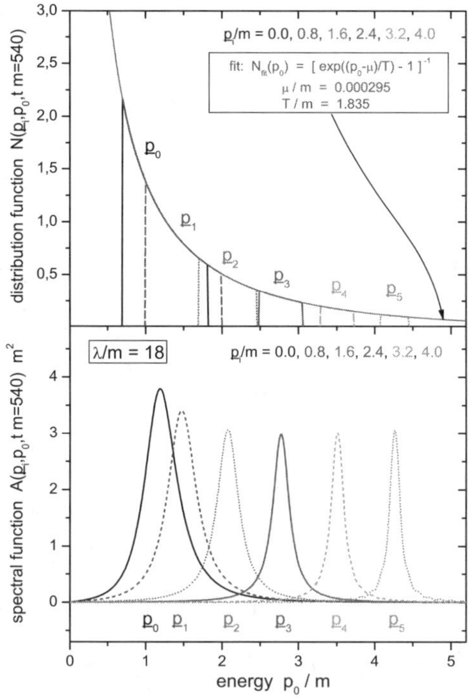

Figure 4. Spectral function A for various momentum modes $|\,\mathbf{p}\,|/m = 0.0$, 0.8, 1.6, 2.4, 3.2, 4.0 as a function of energy for late times $\bar{t} \cdot m = 540$ (lower part). Corresponding distribution function N at the same time for the same momentum modes (upper part). All momentum modes with respective off-shell equilibration can be fitted with a single Bose function of temperature $T_{eq}/m = 1.835$ and a chemical potential close to zero.

Particles can be scattered by other particles – present in the system at finite temperature – such that they achieve the shown (collisional) damping width. On the other hand, particles with sufficient energy can decay into three other particles. Above (approximately) the threshold of $p_{0,th}(\mathbf{p}) = \sqrt{\mathbf{p}^2 + (3\,m^*)^2}$ these $1 \leftrightarrow 3$ processes lead to an increase of the width (as marked for the highest momentum mode by the second thin line). The inclusion of this part of the particle width in the spectral function is responsible for the process of chemical equilibration. Within the Kadanoff-Baym scheme the off-shell $1 \leftrightarrow 3$ transitions lead to a violation of 'particle number'. The physical interpretation of the dynamics is a consistent treatment of Bremsstrahlungs processes in the medium via the interplay of production, propagation and subsequent decay of highly virtual mode excitations.

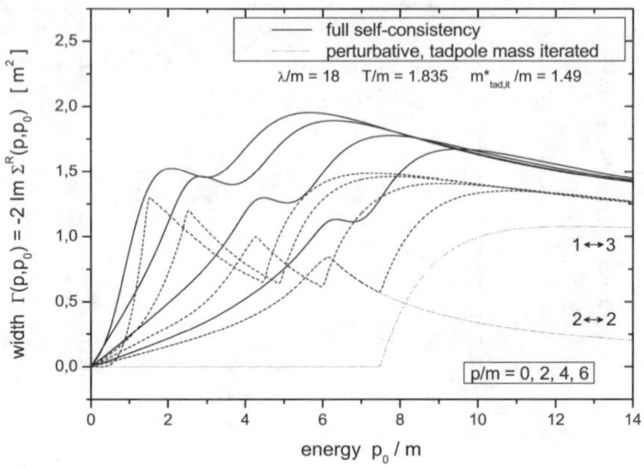

Figure 5. Self-consistent width (solid lines) and perturbative width (dashed lines) as a function of the energy p_0/m for various momentum modes $|\,\mathbf{p}\,|/m = 0, 2, 4, 6$ for a thermal system. For the highest momentum mode $|\,\mathbf{p}\,|/m = 6$ of the perturbative calculation, the collision contribution $(2 \leftrightarrow 2)$ and the decay contribution $(1 \leftrightarrow 3)$ to the width are explicitly displayed.

3. Stochastic disoriented chiral condensates

In this section we give a brief report on our findings on the stochastic nature of DCC formation and how to possibly identify their existence experimentally[5]. This work resulted from an investigation[8], where we adressed the potential likeliness of an instability leading to a sufficiently large DCC event during the evolution of a fireball undergoing a phase transition within the linear σ-model.

The main idea is that the final fluctuations depend critically on the initial conditions chosen for the evolving chiral order parameter, thus deciding to some extent whether the system enters temporarily the instable region during the 'roll-down' period of the order parameter[8, 5]. In fact, a semi-classical and dissipative dynamics of the order parameter and the pionic fields can be obtained by an effective and complex action, where the interaction with the thermal pions has been integrated out. To the end, we have utilized the following stochastic Langevin equations of motion for the order parameters $\Phi_a = \frac{1}{V}\int d^3x\, \phi_a(\mathbf{x},t)$ in a D-dimensional ('Hubble') expanding volume $V(\tau)$ to describe the evolution of collective pion and sigma fields[8, 5]:

$$
\begin{aligned}
\ddot{\Phi}_0 + \left(\frac{D}{\tau} + \eta\right)\dot{\Phi}_0 + m_T^2 \Phi_0 &= f_\pi m_\pi^2 + \xi_0, \\
\ddot{\Phi}_i + \left(\frac{D}{\tau} + \eta\right)\dot{\Phi}_i + m_T^2 \Phi_i &= \xi_i,
\end{aligned}
\tag{5}
$$

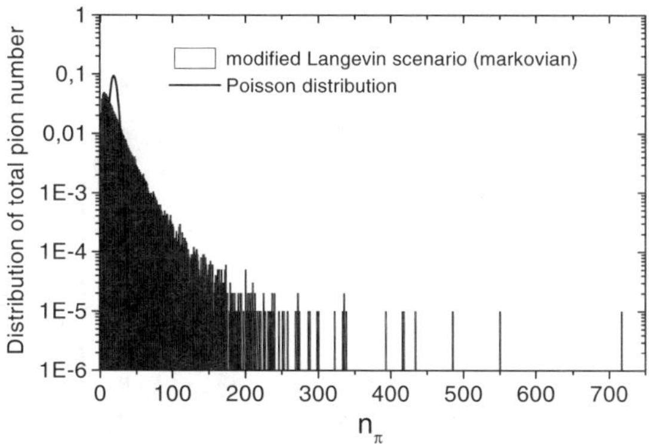

Figure 6. Statistical distribution $P(n_\pi)$ of the final yield n_π in low momentum pion number of a single DCC for a rapidly expanding situation (see ref.[5] for details) compared with a corresponding simple poissonian distribution.

with $\Phi_0 = \sigma$ and $\Phi_i = (\pi_1, \pi_2, \pi_3)$ being the chiral meson fields and $m_T^2 = \lambda\left(\Phi_0^2 + \sum_i \Phi_i^2 + \frac{1}{2}T^2 - f_\pi^2\right) + m_\pi^2$ denotes the effective transversal ('pionic') masses. These coupled Langevin equations resemble in its structure a phenomenological Ginzburg-Landau description of phase transition. Aside from a theoretical justification one can regard the Langevin equation as a practical tool to study the effect of thermalization on a subsystem, to sample a large set of possible trajectories in the evolution, and to address also the question of all thermodynamically possible initial configurations in a systematic manner.

In fig. 6 we show the statistical distribution in the number of produced long wavelength pions N_π out of the evolving chiral order fields within the DCC domain $V(\tau)$ for one particular set of parameters[5]. A rather rapid and ($D =$)3-dimensional expansion has been employed. (The results majorly depend on how fast the assumed cooling and expansion proceeds.) In general one finds that only for D=3 and sufficiently fast expansion individual unusual strong fluctuations of the order of 50 - 200 pions might occur, although the average number $\langle n_\pi \rangle$ of the emerging long wavelength pions only posesses a moderate and *undetectable* value of 5 -20.

In these interesting cases the final distribution does *not* follow a usual Poissonian distribution (comp. fig. 6), which represents a very important outcome. (Critical, dynamical) Fluctuations with a large number of produced pions are still likely with some small but finite probability! Unusual events out of sample contain a multiple in the number of pions compared to the average. One can interpret those particular events as semi-classical 'pion bursts' similar to the mystique Centauro candidates. If DCCs are being produced, an experimental

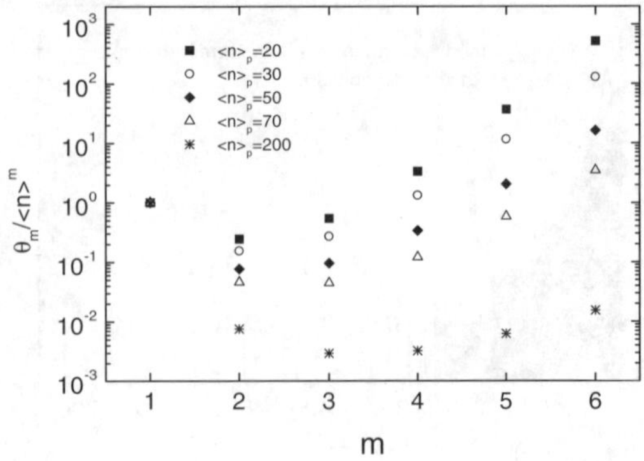

Figure 7. The reduced factorial cumulants for $m = 1$ to 6 for the pion number distribution of low momentum stemming from a single emerging DCC (of the previous figure) and an additional poissonian distributed background pion source with different mean values $\langle n \rangle_P = 20 - 200$.

finding will be a rare event following a strikingly, nontrivial and nonpoissonian distribution. A dedicated event-by-event analysis for the experimental programs (e.g. the STAR TPC at RHIC) is then unalterable.

The further analysis of this unusual distribution by means of the cumulant expansion shows that the reduced higher order factorial cumulants $\theta_m / < n_\pi >^m$ for $m \geq 3$ exhibit an abnormal, exponentially increasing tendency, as illustrated in fig. 7. There an additional incoherent Poissonian background of (low momentum) pions stemming from other possible sources has been added. We advocate that an analysis by means of the higher order cumulants serves as a powerful signature. In conclusion, the occurence of a rapid chiral phase transition (and thus DCCs) can be identified experimentally by inspecting higher order facorial cumulants θ_m $(m \geq 3)$ for taken distributions of low momentum pions.

4. Elastic and inelastic multi-particle collisions in a parton cascade

A major goal of the experiments at RHIC and LHC is the creation and observation of the quark-gluon-plasma. To describe the dynamics of ultra-relativistic heavy ion collisions (uRHICs), and to address the crucial question of thermalization and early pressure build up, we have developed a kinetic parton cascade algorithm [6] inspired by perturbative QCD including inelastic ('Bremsstrahlung') collisions $gg \leftrightarrow ggg$ besides elastic collisions. It is the aim to get a more detailed and quantitative understanding of the early dynamical stages of deconfined matter and to test various initial conditions for the liberated gluons, than envoking ad hoc varoius phenomenological, hydrodynamical ansaetze. An parton cascade analysis incorporating only binary $2 \leftrightarrow 2$ collisions described via direct pQCD scattering processes shows that thermalization and early quasi-hydrodynamic behaviour (for achieving sufficient elliptic flow) can not be buildt up or maintained, but only if a much higher cross section is being employed [9]. The possible importance of the inelastic reactions was indeed raised in the so called 'bottom up thermalization' picture [10]. It is intuitively clear that gluon multiplication should lead to a faster equilibration.

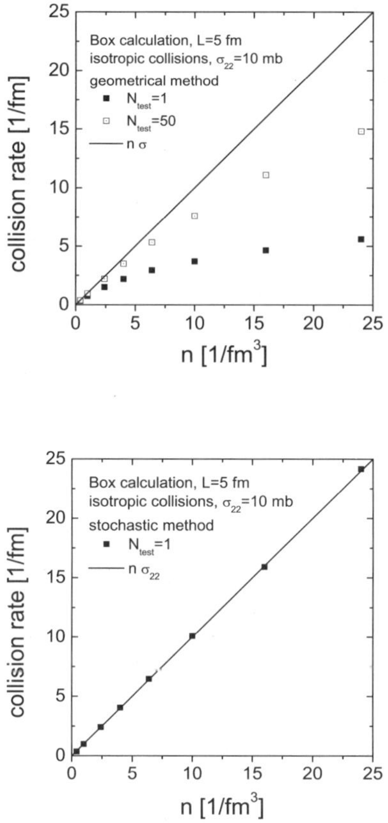

Figure 8. Collision rates in the simulation versus the ideal rates $1/(n\sigma)$. The total cross section is fixed to 10 mb. In the upper (lower) figure the geometrical (stochastic) method is employed.

In developing the algorithm special emphasis is put on obeying the principle of detailed balance among the gain and loss contributions. The standard incorporation of $2 \leftrightarrow 2$ scattering processes in a transport description is based on the geometric interpretation of the cross section [11]. For large particle densities, however, such an implementation leads to considerable problems to generate a causal collision sequence among the various partons, leading to a severe reduction of the collision rate compared to the one dictated by the Boltzmann equation, see fig. 8. In principle, this can be cured by the test particle method. However, for such a situation the stochastic method (see eg [12]) is better suited to solve the Boltzmann equation directly via transition rates in sufficiently small spatial cells, see fig. 8. Here the actual collision rate being realized in the simulation fully equals the ideal one of the Boltzmann equation which shall be simulated. One can go to arbitrary high density. Moreover, this method can, in principle, be generalized to (any) multi-particle scattering processes.

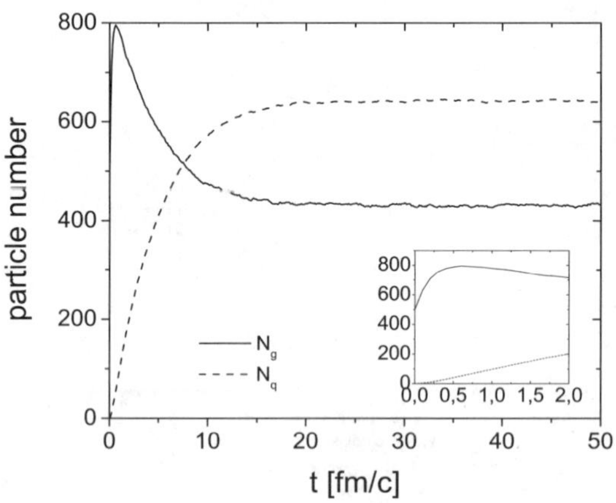

Figure 9. Time evolution of the gluon and quark numbers for the box calculation. The gluon number starts at 500.

To demonstrate the applicability of the method, we show in the following the formation of a quark gluon plasma within a fixed box and study the way of kinetic and chemical equilibration for different parton species. The initial conditions of the partons entering the cascade are given by the multiple minijets (partons with $p_t > 2$ GeV/c) production [13] from the binary nucleon-nucleon-scattering in a nucleus-nucleus-collision according to the differential jet cross section:

$$\frac{d\sigma_{jet}}{dp_T^2 dy_1 dy_2} = K \sum_{a,b} x_1 f_a(x_1, p_T^2) x_2 f_b(x_2, p_T^2) \frac{d\sigma_{ab}}{d\hat{t}}\,, \tag{6}$$

where p_T is the transverse momentum and y_1 and y_2 are the rapidities of the produced partons. The minijets are considered in the central rapidity interval $(-0.5 : 0.5)$ at RHIC energy of $\sqrt{s} = 200$ GeV. The initial partons are distributed uniformly in a cubic box of size 3 fm, which approximately corresponds to the central region of heavy ion collions at some early initial time.

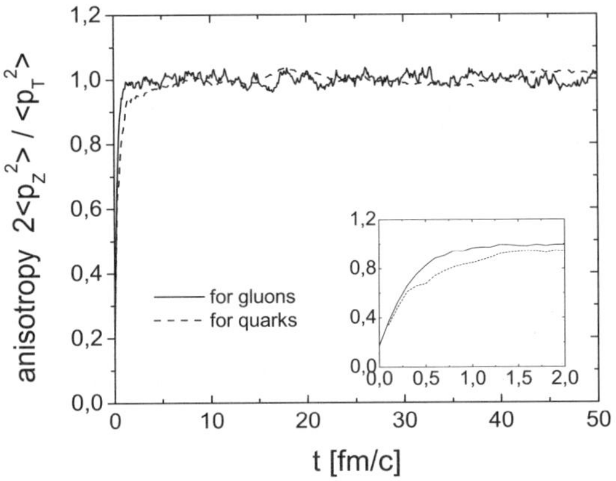

Figure 10. Time evolution of the momentum anisotropy for the QGP in a box.

We consider here gluon and quarks with two flavors as parton species. Collision processes are the two-body elementary parton-parton scatterings and three-body process $gg \leftrightarrow ggg$ in leading-order of pQCD, employing effective Landau-Pomeranchuk-Migdal suppression and standard screening masses for the infrared sector of the scattering amplitude.

Fig. 9 shows the time evolution of the gluon and quark number. Gluon equilibration undergoes two stages: at first the gluon number increases rapidly and then smoothly evolves to its final equilibrium value together with the quark number at a much lower rate. The early stage of gluon production also leads to an immediate kinetic equilibration of the momentum distribution (see fig. 10 and fig. 12) as well as to a rather abrupt lowering of the temperature by soft gluon emission until detailed balance among gain and loss contributions is reached (see fig. 11).

The later slower evolution is then governed by chemical equilibration of the quark degrees of freedom. The final temperature of fig. 11 is identical to the late slope parameter of the energy spectra depicted in fig. 12.

With the present cascade we are aiming to study in detail real HIC incorporating various initial conditions like minijets or the color glass condensate. In

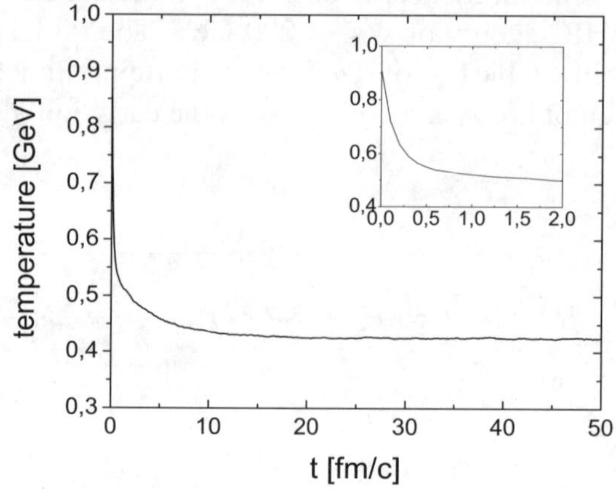

Figure 11. Time evolution of the effective temperature $T = \frac{E}{3N}$.

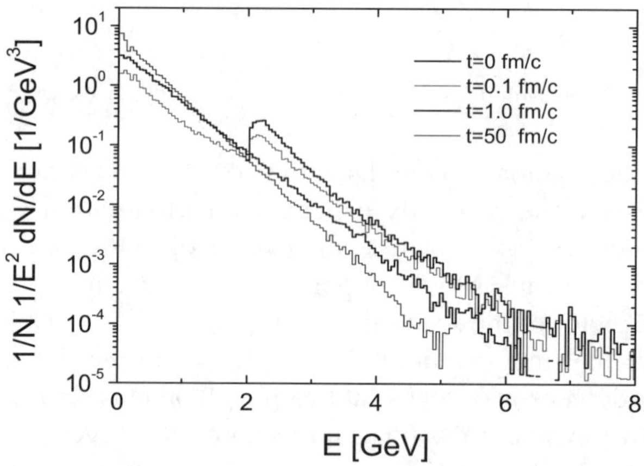

Figure 12. Occupation versus energy spectra at different times. within the box calculation. At $t = 0$ only minijets with $E > 2$ GeV are populated. Energy degradation to lower momenta proceeds rapidly by gluon emission.

fig. 13 we depict one such first (and still *preliminary*, as yet not fully tested) calculation with minijet initial conditions. Thermalization and chemical equilibration as propoesed in the bottom up picture [10] can then thoroghly be tested

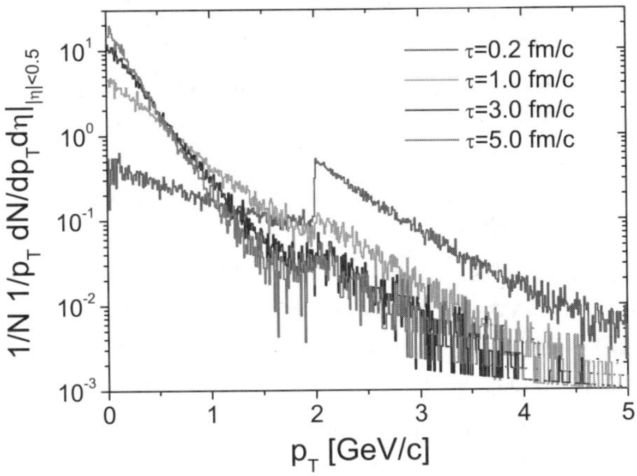

Figure 13. Transverse momentum spectrum at midrapidity at different times for a real, central fully 3-D (including transversal expansion) ultrarelativistic heavy ion collision. Only the partons residing in a cylinder of radius $R \leq 5$ fm are taken for the plot. The initial out-of-equilibrium conditions are given by minijets and are distributed via the corresponding overlap function in space-time. The calculation is *preliminary*. From $t = 0$ on first only minijets with $E > 2$ GeV are populated. Energy degradation to lower momenta proceeds rapidly by gluon emission within the first fm/c. Maintenance of (quasi-)kinetic and chemical equilibrium is given up to 4 fm/c, where longitudinal and transversal (quasi-hydrodynamical) work is done resulting in a continous lowering of the temperature. On the other hand the initial non-equilibrium high momentum tail following a standard power-law for mini-jet production is partly surviving.

within this approach. Furthermore we will study the impact parameter dependence of the transverse energy in order to understand elliptic and transverse flow at RHIC within this new scheme of a kinetic parton cascade including inelastic interactions by comparing to experimental data.

Acknowledgments

The work refered to in section 2 has been done together with W. Cassing and is gratefully acknowledged. The various work has been supported by BMBF, DFG and GSI Darmstadt.

References

[1] P. Danielewicz, *Ann. Phys.* (N.Y.) **152** (1984) 305.

[2] C. Greiner, K. Wagner, and P.-G. Reinhard, *Phys. Rev.* C **49**, 1693 (1994).

[3] S. Juchem, W. Cassing and C. Greiner, 'Quantum dynamics and thermalization for out of equilibrium ϕ^4 theory' , arXiv:hep-ph/037353, *Physical Review* D**69** (2004) in press.

[4] D. Anselm, *Phys. Lett.* B **217**, 169 (1989); J.D. Bjorken, *Int. J. Mod. Phys. A* **7**, 4819 (1992); K. Rajagopal and F. Wilczek, *Nucl. Phys.* B **404**, 577 (1993).

[5] Z. Xu and C. Greiner, *Phys. Rev.* D **62**, 036012 (2000).

[6] Z. Xu and C. Greiner, 'Elastic and inelastic multiplication collisions in a parton cascade treated in an unified manner', publication in preparation.

[7] B. Schenke and C. Greiner, 'Statistical description with anisotropic momentum distributions for hadron production in nucleus-nucleus collisions', *arXiv:nucl-th/0305008*

[8] T.S. Biró and C. Greiner, *Phys. Rev. Lett.* **79**, 3138 (1997).

[9] D. Molnar and M. Gyulassy, *Nucl. Phys.* A **697**, 495 (2002).

[10] R. Baier, A.H. Mueller, D. Schiff and D.T. Son, PLB **502**, 51 (2001).

[11] D. Molnar and M. Gyulassy, *Phys. Rev.* C **62**, 054907 (2000).

[12] A. Lang et al, *J.Comp.Phys.* **106** (1993), 391 (1993).

[13] K.J. Eskola et al., NPB **323**, 37 (1989); X.-N. Wang, PRep **280**, 287 (1997).

THE QGP DISCOVERED AT RHIC

M. Gyulassy
Physics Department, Columbia University, New York, USA
gyulassy@nt3.phys.columbia.edu

Abstract Three empirical lines of evidence, $(\mathbf{P_{QCD}}, \mathbf{pQCD}, \mathbf{dA})$, from RHIC have converged and point to the discovery of a strongly coupled Quark Gluon Plasma. The evidence includes (1) bulk collective elliptic flow and (2) jet quenching and mono-jet production, observed in Au+Au collisions at 200 AGeV, and (3) a critical control experiment using D+Au at 200 AGeV.

1. The Theoretical QGP

The Standard Model of strong interactions predicts the existence of a new phase of matter, called a Quark Gluon Plasma (QGP), in which the quark and gluon degrees of freedom normally confined within hadrons are mostly liberated. Lattice QCD calculations show that there is a rapid rise of the entropy density, $\sigma(T)$, of matter when the temperature reaches $T \approx T_c \sim 160$ MeV. Beyond T_c the effective number of degrees of freedom, $n(T)$, saturates near the number of quark and gluon helicity states $n_{QCD} = 8_c \times 2_s + \frac{7}{8} \times 3_c \times N_f \times 2_s \times 2_{q\bar{q}} \approx 37$. The entropy density $\sigma(T) = dP/dT = (\epsilon + P)/T \propto n(T)T^3$ approaches the Stefan Boltzmann limit $4P_{SB}(T)/3T$. The transition region is a smooth crossover when dynamical quarks are taken into account, but the width of the transition region remains relatively narrow, $\Delta T_c/T_c \sim 0.1$ [1]-[6].

The rapid rise of the entropy was predicted long before QCD by Hagedorn due to the observed exponential rise of the number of hadron resonances [5]. However, the saturation of the number of degrees of freedom near n_{QCD} is a unique feature of QCD. Even though the entropy density approaches the ideal, weakly interacting plasma limit, lattice calculations of correlators show that the QGP is far from ideal below $3T_c$. The nonideal nature of this strongly coupled QGP is also seen from the deviation of the pressure, $P(T)$, and energy density $\epsilon(T)$ from the Stefan Boltzmann limit as shown in Fig. 1 from [1].

The equation of state of the QGP, $P_{QCD}(T)$, is the bulk thermodynamic property that can be investigated experimentally via "barometric" observables.

W. Greiner et al. (eds.), Structure and Dynamics of Elementary Matter, 159–182.

A measure of its stiffness is given by the speed of sound squared, $c_s^2 = dP/d\epsilon = d\log T/d\log\sigma = (3 + d\log n/d\log T)^{-1}$ shown in Fig. 2. Note that c_s^2 drops rapidly below $1/3$ as the effective number of degrees of freedom drops when T approaches T_c. This softening of the QGP equation of state near T_c is a key feature can be looked for in the collective hydrodynamic flow patterns produced when the plasma expands.

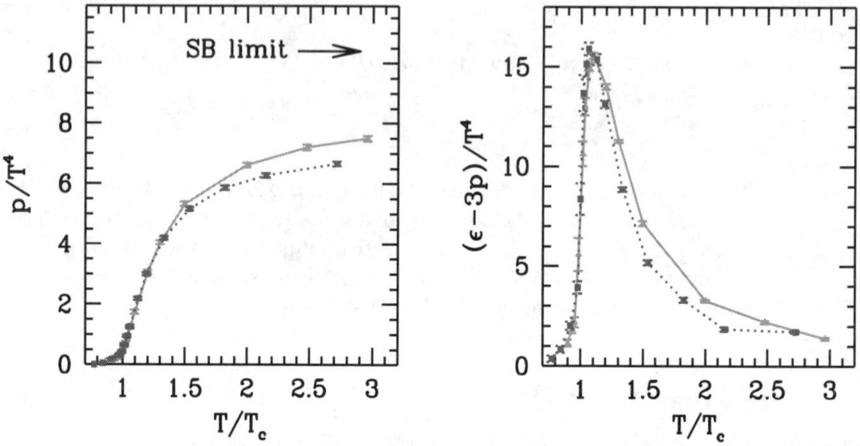

Figure 1. A recent Lattice QCD calculation [1] of the pressure, $P(T)/T^4$, and a measure of the deviation from the ideal Stefan-Boltzmann limit $(\epsilon(T) - 3P(T))/T^4$. Note that the scale on both graphs has not been corrected for finite lattice volume effects: see [2] for discussion.

Another distinctive feature of the QGP phase diagram is shown in the right panel in Fig. 2. Recent lattice QCD calculations [1] have begun to converge on numerical evidence that the QGP may have a second tricritical point [4, 6] at moderate baryon densities with $\mu_B = 3\mu_q \sim 360$ MeV and $T \sim T_c$.

QCD predictions of the QGP phase date back thirty years [7] and followed immediately after the discovery of asymptotic freedom of QCD. The experimental strategies to search for new forms of dense matter also date back thirty years when T.D. Lee proposed "vacuum engineering" [8]. It was then also realized by W. Greiner and collaborators [9–11] that extended regions of dense nuclear matter can be formed in high energy interactions of heavy nuclei, and that the measurement of collective flow patterns will provide the novel barometric probes of the equation of state of ultra-dense matter. The hunt for the QGP and other phases of nuclear matter has been underway since that time using several generations of higher energy accelerators, BEVALAC, AGS, SPS, and now RHIC, and covering an impressive energy range $\sqrt{s} - 2m_N = 0.2 - 200$ AGeV. In three years, LHC is expected to start vacuum engineering at 5500 AGeV.

The first conclusive evidence for (highly dissipative) collective nuclear flow was seen at the BEVALAC in 1984 [12], and at AGS and SPS there after.

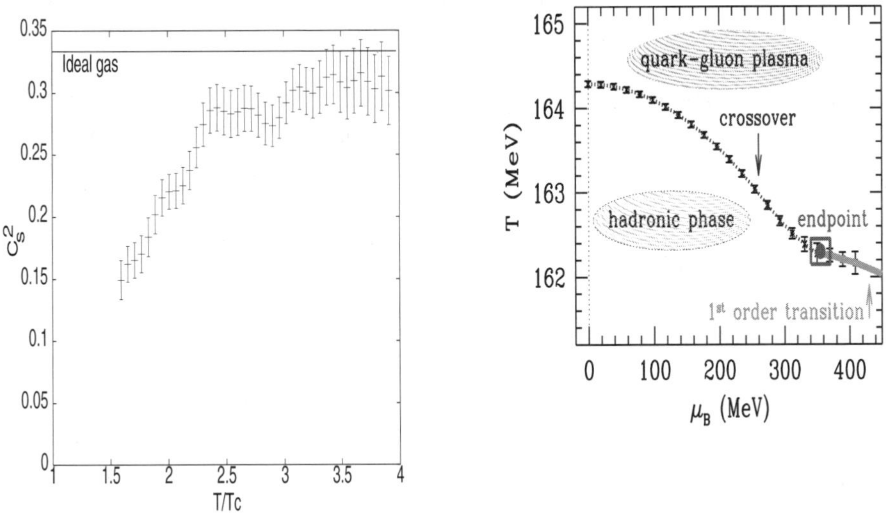

Figure 2. Important features of the QGP equation of state. The speed of sound [3] $c_s^2 = d\epsilon/dP$ drops below $1/3$ for $T < 2T_c \approx 300$ MeV. Right panel shows a current estimate of the location of the tricritical point at finite baryon density [1]

However, the first conclusive evidence for nearly dissipation free collective flow obeying P_{QCD} had to await RHIC. In these lectures, the discovery of the novel low dissipation elliptic flow pattern at RHIC is highlighted as the first of three lines of evidence for QGP production at RHIC. Together with two other convergent lines of evidence, jet quenching and the critical $D + Au$ null control, I conclude that the QGP has not only been discovered but that a few of its remarkable properties have already been established experimentally.

2. The Empirical QGP

The discovery of the gedanken QGP phase of matter in the laboratory requires an empirical definition of the minimal number of necessary and sufficient conditions in terms of experimentally accessible observables. My empirical definition is summarized by the following symbolic equation

$$\mathbf{QGP} = \mathbf{P_{QCD}} + \mathbf{pQCD} + \mathbf{dA} \quad . \tag{1}$$

Why are three independent lines of evidence needed? The first term, $\mathbf{P_{QCD}}$, stands for a class of observables that provide information about its bulk thermodynamic equation of state. The equation of state characterizes its *long wavelength* nonperturbative thermodynamic properties briefly reviewed in section 1.

The second term, **pQCD**, stands for class of observables that provide direct evidence about its *short-wavelength* dynamics predicted by perturbative QCD. The QCD plasma differs qualitative from familiar abelian QED plasmas due to its unique non-Abelian color field dynamics. The radiative energy loss of energetic short-wavelength partons was predicted to lead to striking quenching patterns [13]-[16] of moderate and high p_T hadrons. The high RHIC cm energy of 200 AGeV insures that $p_T \sim 10 - 20$ GeV jet production rates are large enough to measure via a wide array of inclusive and correlation observables. These hard partons serve as effective "external" tomographic probes of the the QGP and test its **pQCD** *chromo*-dynamics. Jets play the analogous role of neutrinos that probe the physics of stellar cores, while hadrons play the role of photons that probe the corona of the fireball.

Below RHIC energies, the **pQCD** line of evidence could not be fully developed because the jet rates decrease too rapidly with energy. However, even more importantly, at the lower $p_T < 4$ GeV available the effects of initial state nuclear dynamics and the final state hadronic dynamics could not be completely deconvoluted from the final spectra. This is the key point that I will repeatedly emphasize which differentiates the observables at the SPS and RHIC energies. The necessity to test for the same complications at RHIC is what gives rise to the third term in Eq.(1).

The third term, denoted by **dA**, stands for control experiments that can clearly differentiate between alternative nuclear dependences specific to *initial state* partonic wavefunctions as well as the production mechanisms. The control differential, **dA**, is critical at any energy because the QCD plasma must first be created from pure kinetic energy! There has been no hot QGP in the universe (except in cosmic ray collisions) since the last drop condensed into hadrons about 13 billion years ago. Cold crystalline quark matter may lurk in the cores of neutron stars, but the transient hot QGP must be "materialized" in the lab. The "matter" arises from decoherence of *virtual* quantum chromo fluctuations in the initial wavefuntions of high energy nuclei.

At ultra-relativistic energies, these virtual fluctuations are frozen out due to time dilation into what has been called a Color Glass Condensate (CGC) [17]. The CGC is the high density generalization of the Bjorken-Feynman dilute parton model. At high field strengths, the non-linear interactions of virtual quantum color fluctuations are predicted to limit the very small Bjorken $x_{BJ} \to 0$ Fourier components. The saturation property of CGC is related to unitarity constraints and determines the maximal entropy that can be produced in AA at a given $\sqrt{s}$ as also pointed out by EKRT [18].

The **dA** control is needed to characterize to what extent these nonlinear initial state physics effects can be differentiated from effects due to final state interactions in the QGP matter that forms from it. At RHIC, the best experimental handle on the **dA** term happens to be the study of $D + A$ reactions.

In such light-heavy ion reactions, the initial state CGC physics can be isolated because the produced QGP, if any, is too tenuous.

Why don't I add more terms in Eq.(1)? In fact, each term stands for many independent components, as I elaborate below. For the three required terms in Eq.(1) the published experimental evidence is now overwhelming and conclusive. Four independent experiments have converged to complementary very high quality data sets.

The three terms in Eq.(1) are necessary and sufficient for establishing that a discovery has been made of a uniquely different form of strongly interacting QGP. After discussing the three lines of evidence, I will elaborate on why I believe that direct photons, J/ψ, HBT, or other interesting observables do not need to be added to Eq.(1). Those observable provide valuable additional constraints on the *combined and convoluted* properties of the initial state, the QGP, *AND* the dense hadronic matter into which it condenses. However, the deconvolution of the initial and hadronic final effects has already proven to be very difficult at SPS energies and will continue to be at RHIC.

To avoid misunderstanding, the discovery of the QGP does not mean that its physical properties are now understood. In fact, it only signals the beginning of a long and well focused direction of research. The history of the neutron star discovery offers an instructive analogy. In 1934 Baade and Zwicky proposed the theoretical existence of neutron stars soon after Chadwick discovered free neutrons. Thirty years later in 1967 Hewish and Bell observed the first few pulsars when suitable radio telescopes could finally be constructed. An amusing anecdote is that they actually agonized for a time about whether LGM (little green men) were sending them encrypted messages from the cosmos. T. Gold in 1968 (as $D + Au$ did at RHIC in 2003) put the debate to rest. Gold proposed that radiative energy loss of a magnetized neutron star would cause a predictable spin down. Later precision measurements confirmed this. Seventy years after its proposal, neutron star research still remains a very active experimental and theoretical direction of physics. Current interest has focused on possible *color field* super-conductivity [19] recently predicted in the very high μ_B sector of the QCD phase diagram, beyond the boundaries [6] of Fig. 2b.

The critical $D + Au$ control experiments in 2003 could have found that the $Au + Au$ QGP observables were strongly distorted by the possible initial CGC state that created it. This would have certainly foiled Eq.(1). The search for the bulk QGP phase of matter would then have had to await higher energies and densities at LHC or for a better understanding of how to deconvolute that initial state physics. The large positive signatures in similar $p+Pb$ control experiments at SPS showed in fact initial effects strongly distort key observables. At SPS the physics of high p_T Cronin enhancement and $p + A \to J/\psi$ suppression remain the important open problems. In contrast, at RHIC energies, the absence of jet quenching at mid-rapidity and the "return of the jeti" correlations in $\mathbf{dA} = \mathbf{D} + \mathbf{Au}$ provided the check-mate completion of Eq.(1).

As emphasized by McLerran[17] in these proceedings, the $D + Au$ control at RHIC at high rapidities does in fact produce a positive signature for new initial state physics. In those kinematic ranges $x_{BJ} < 0.001$, $\mathbf{dA}$ fails as a null control for QGP, but may signal the onset of nonlinear CGC initial conditions. In this lecture, I concentrate on the mid-rapidity region, $x_{BJ} > 0.01$, where Eq.(1) was conclusively satisfied.

3. $\mathrm{P_{QCD}}$ and Bulk Collective Flow

The identification of a new form of "bulk matter" requires the observation of novel and uniquely different collective properties from ones seen before. This requirement is the first term in Eq.(1). In heavy ion collisions, the primary observables of bulk collectivity are the radial, azimuthal and longitudinal flow patterns of hundreds or now thousands of produced hadrons. Stöcker, Greiner, and collaborators were the first to predict [9–11, 20, 21] distinctive "side splash and squeeze-out" collective flow patterns in nuclear collisions. The different types of collective flows are conveniently quantified in terms of the first few azimuthal Fourier components [22], $v_n(y, p_T, N_p, h)$, of centrality selected triple differential inclusive distribution of hadrons, h. The centrality or impact parameter range is usually specified by a range of associated multiplicities, from which the average number of participating nucleons, N_p, can be deduced. The azimuthal angle of the hadrons are measured relative to a globally determined estimate for the collision reaction plane angle $\Phi(M)$. The "directed" v_1 and "elliptic" v_2 flow components [12, 22, 23]-[31] are readily identified from azimuthal dependence

$$\begin{aligned}\frac{dN_h(N_p)}{dy dp_T^2 d\phi} = \frac{dN_h(N_p)}{dy dp_T^2}\frac{1}{2\pi}(1 &+ 2v_1(y, p_T, N_p, h)\cos\phi \\ &+ 2v_2(y, p_T, N_p, h)\cos 2\phi + \cdots) \ . \quad (2)\end{aligned}$$

The "radial flow" component, "1" , is identified [32] from the hadron mass dependence of the blue shifted transverse momentum spectra

$$\frac{dN_h(N_p)}{dy dp_T^2} \sim \exp[-m_h \cosh(\rho_\perp - \beta(y))/T_f] \ , \quad (3)$$

where $m_h(\sinh(\rho_\perp), \cosh(\rho_\perp)) = (p_\perp, \sqrt{m_h^2 + p_\perp^2})$ and $\beta(y)$ is the mean collective transverse flow rapidity at y.

Figure (3) shows the striking bulk collectivity elliptic flow signature of QGP formation at RHIC. Unlike at SPS and lower energies, the observed large elliptic deformation $((1 + 2v_2)/(1 - 2v_2) \sim 1.5)$ of the final transverse momentum

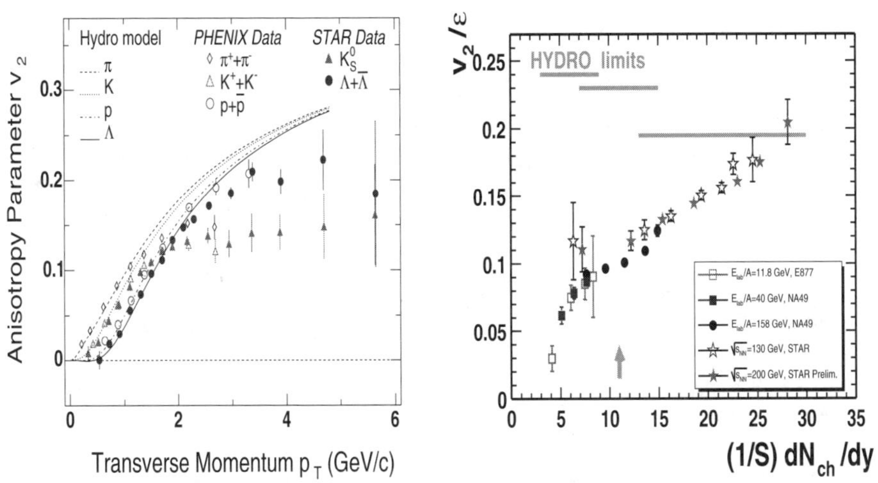

Figure 3. First line of evidence: Bulk collective flow is the barometric signature of QGP production. Left figure combines STAR [26]-[29] and PHENIX [30, 54] measurements of the azimuthal elliptic flow ($v_2(p_T)$) of π, K, p, Λ in Au+Au at 200 AGeV. The predicted hydrodynamic flow pattern from [33]-[37] agrees well with observations in the bulk $p_T < 1$ GeV domain. Right figure from [24] shows v_2 scaled to the initial elliptic spatial anisotropy, ϵ, as a function of the charge particle density per unit transverse area. The bulk hydrodynamic limit is only attained at RHIC.

distribution agrees for the first time with non-viscous hydrodynamic predictions [33]-[43] at least up to about $p_T \sim 1$ GeV/c. However, the right panel shows that when the local rapidity density per unit area [23, 24] drops below the values achieved at RHIC $\sim 30/\text{fm}^2$, then the elliptic flow (scaled by the initial spatial ellipticity, $\epsilon = \langle (y^2 - x^2)/(y^2 + x^2) \rangle$) falls below the perfect fluid hydrodynamic predictions. We will discuss in more detail the origin of the large discrepancy at SPS energies in the next section.

The most impressive feature in Fig. 3 is the agreement of the observed hadron mass dependence of the elliptic flow pattern for all hadron species, π, K, p, Λ, with the hydrodynamic predictions below 1 GeV/c. This is the QGP fingerprint that shows that there is a common bulk collective azimuthally asymmetric flow velocity field, $u^\mu(\tau, r, \phi)$.

The flow velocity and temperature fields of a perfect (non-viscous) fluid obeys the hydrodynamic equations:

$$\partial_\mu \left\{ [\epsilon_{QCD}(T(x)) + P_{QCD}(T(x))] u^\mu(x) u^\nu(x) - g^{\mu\nu} P_{QCD}(T(x)) \right\} = 0 \,, \tag{4}$$

where $T(x)$ is the local temperature field, $P_{QCD}(T)$ is the QGP equation of state, and $\epsilon_{QCD}(T) = (TdP/dT - P)_{QCD}$ is the local proper energy density.

The above equations apply in the rapidity window $|y| < 1$, where the baryon chemical potential can be neglected. Eq.(4) provides the barometric connection between the observed flow velocity and the sought after $\mathbf{P_{QCD}}$.

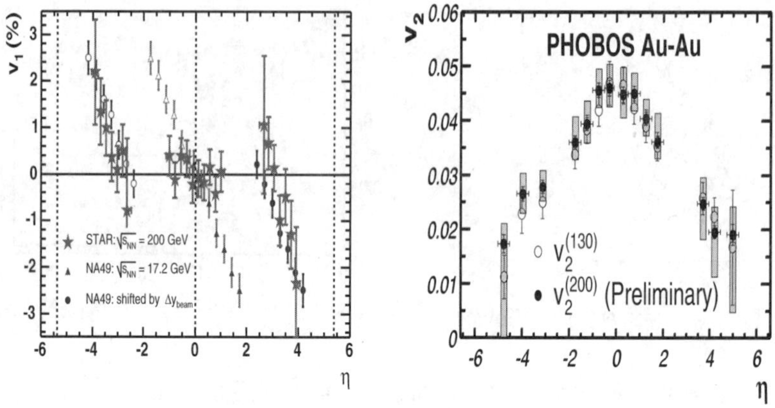

Figure 4. Left figure combines STAR and NA49 data [26] and shows that the directed sidewards flow, $v_1(y)$, is correlated over 8 units of rapidity at RHIC. At SPS collectivity is dominated by the overlapping fragmentation regions while at RHIC the nearly identical directed flow of in the fragmentation regions is shifted to $|y| > 2$. Right figure shows the pseudo rapidity dependence of elliptic from PHOBOS [31].

The long-range nature of collective flow has also been conclusively established by STAR and PHOBOS seen in Fig. 4. The sidewards flow is anti-correlated over 8 units of rapidity! In addition, its azimuthal orientation was shown to coincide with the azimuthal direction of the largest axis of elliptic deformation at $y = 0$. This provides an important test of the overall consistency of the hydrodynamic origin of flow. The rapidity dependence of the elliptic flow in Fig. 4 also shows the long range nature of bulk collectivity.

Why is v_2 more emphasized than v_1 or radial flow as a signature of QGP formation? The primary reason is that elliptic flow is generated mainly during the highest density phase of the evolution before the initial geometric spatial asymmetry of the plasma disappears. It comes from the azimuthal dependence of the pressure gradients, which can be studied by varying the centrality of the events [22]. Detailed parton transport [44] and hydrodynamic [39] calculations show that most of the v_2 at RHIC is produced before 3 fm/c and that elliptic flow is relatively insensitive to the late stage dissipative expansion of the hadronic phase. In contrast, radial flow has been observed at all energies [32] and has been shown to be very sensitive to late time "pion wind" radial pressure gradients [45], which continue to blow after the QGP condenses into hadronic resonances.

The observation of near ideal v_2 fluid collectivity as predicted with the P_{QCD} together with $v_1(y)$ and other consistency checks (**c.c.**) conclusively establish the first term in Eq.(1) :

$$\mathbf{P_{QCD}} = v_2(p_T; \pi, K, p, \Lambda) + v_1(y) + \textbf{ c.c. } . \tag{5}$$

Preliminary Quark Matter 2004 analyses of Ξ, Ω flow are consistent with the predicted $v_2(p_T, y = 0, M, h)$ and this information is lumped into the **c.c.** terms in Eq.(5). Other data which provide consistency checks of the hydrodynamic explanation of collective flow include the observed π, K, p radial flow data [32] for $p_T < 2$ GeV. In addition, predicted statistical thermodynamic distributions [46] of final hadron yields agree remarkably well with RHIC data. Had hadro-chemistry failed at RHIC, then a large question mark would have remained about bulk equilibration in the QGP phase.

4. QGP Precursors at SPS and Dissipative Collectivity

It is important to point out, that no detailed 3+1D hydrodynamic calculation [41]-[43] has yet been able to reproduce the rapid decrease of $v_2(|\eta| > 1)$ observed by PHOBOS in Fig. 4. This discrepancy is due, in my opinion, to the onset of *hadronic* dissipation effects as the comoving density decreases with increasing y. From the right panel of Fig. 3, we see that as a decrease of the local transverse density from mid-rapidity RHIC conditions leads to an increasing deviation from the perfect fluid limit. The initial density was also observed to decrease at RHIC as $|y|$ increases [47]. Therefore, from SPS data alone, we should have expect deviations from the perfect fluid limit away from the mid-rapidity region. It would be interesting to superpose the PHOBOS data on top of the NA49 systematics.

To elaborate on this point, Fig. 5 shows CERES data [48] on $v_2(p_T)$ at SPS energy $\sqrt{s} = 17$ AGeV. In agreement with the right panel of Fig. 3, the CERES data falls well below hydrodynamic predictions. At even lower energies, AGS and BEVALAC, the v_2 even becomes *negative* and this "squeeze out" of plane [11] is now well understood in terms of non-equilibrium BUU nuclear transport theory [25, 49]. In order to account for the smallness of v_2 at SPS, hydrodynamics has to be frozen out at unphysically high $T_f \approx T_c - 160$ MeV. However, the observed radial flow rules out this simple fix.

The discrepancy of v_2 and hydro at SPS energies can be traced to the important contribution of the dissipative final *hadronic* state interactions. The hadronic fluid is far from ideal. In approaches [45, 39] that combine perfect fluid QGP hydrodynamics with non-equilibrium hadronic transport dynamics, the importance of dissipative hadron dynamics at SPS was clearly demonstrated. The problem is that the QGP at lower initial densities condenses on a faster time scale cannot take advantage of the of the spatial asymmetry to generate large v_2. The subsequent dissipative hadronic fluid is very inefficient in exploiting spatial asymmetry. A factor of two reduction of the initial QGP density, therefore, leads to a significant systematic bias of the v_2 barometer, not only at SPS but

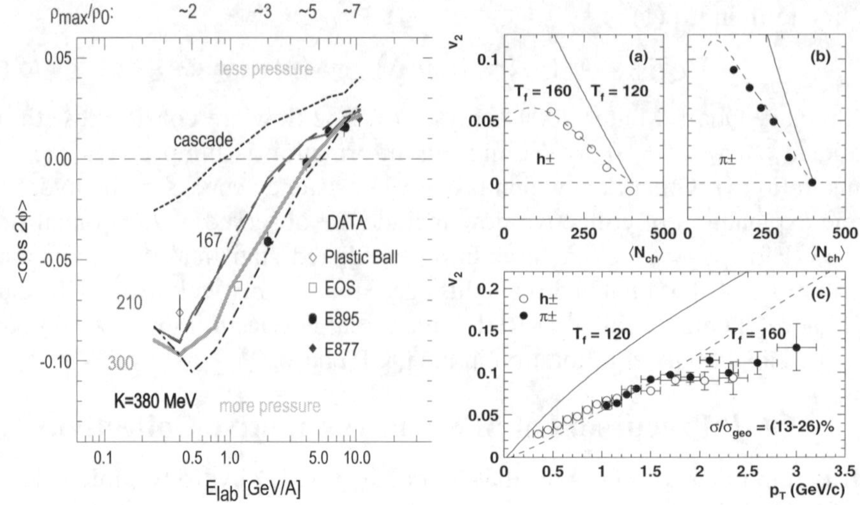

Figure 5. Evidence for dissipative collective flow below RHIC energies. Left: Non equilibrium BUU nuclear transport theory [49, 25] can explain the observed elliptic squeeze-out (negative v_2) collectivity below 4 AGeV. Right: CERES[48] data on elliptic flow at SPS is well below hydrodynamic predictions with freeze-out $T_f = 120$ MeV required to reproduce the single inclusive radial flow. Early freeze-out with $T_f = 160$ MeV, simulating effects of dissipation, is needed to reproduce the data.

also at high $|y|$ at RHIC. Current hadronic transport theory is not yet accurate enough to re-calibrate the barometer away from mid-rapdities at RHIC.

In light of the above discussion, the smallness of dissipative corrections in the central regions of RHIC is even more surprising. At mid-rapidities, the lack of substantial dissipation in the QGP phase is in itself remarkable. Calculations based on parton transport theory [44] predict large deviations from the ideal non-viscous hydrodynamic limit. Instead, the data show that the QGP is almost a perfect fluid. A Navier Stokes analysis [40] is consistent with [44] and shows that the viscosity of the QGP must be about ten times less than expected if the QGP were a weakly interactive Debye screened plasma. This unexpected feature of the QGP must be due to nonperturbative and hence strong coupling physics that persists to at least $3T_c$.

4.1 The Minimal Viscosity of the QGP

One intriguing theoretical possibility being explored in the literature [50] is that the shear viscosity, η, in the strongly coupled QGP may saturate at a universal super-string bound, $\eta/\sigma = 1/4\pi$. This conjectured duality between string theory and QCD may help to explain also the $\sim 20\%$ deviation of $P_{QCD}(T)$ from the ideal Stefan Boltzmann limit. The discovery of nearly perfect fluid flow of *long-wavelength* modes with $p_T < 1$ GeV at RHIC is certain to fuel more interest in this direction.

I propose that a simpler physical explanation of the lower bound on viscosity follows from the uncertainty principle, as derived in Eq.(3.3) of Ref. [51]. Standard kinetic theory derivation of shear viscosity leads to $\eta = (\rho\langle p\rangle\lambda)/3$ where ρ is the proper density, $\langle p\rangle$ is the average total momentum, and λ is the momentum degradation transport mean free path. The uncertainty principle implies that quanta with average momentum components $\langle p\rangle$ cannot be localized to better than $\Delta x \sim 1/\langle p\rangle$. Therefore the momentum degradation mean free path cannot be defined more accurately than $\lambda > 1/\langle p\rangle$. For an ultra-relativistic system, the entropy density is $\sigma \approx 4\rho$, therefore

$$\frac{\eta}{\sigma} > \frac{1}{12} \ , \tag{6}$$

which is within 5% of the string theory bound. It is the consequence of the universality of the Heisenberg uncertainty principle. Surprisingly, the QGP found at RHIC saturates this uncertainty bound and the data clearly rule out the order of magnitude larger predictions based on pQCD [51]. See again [44]. The long-wavelength modes in the QGP are as maximally coupled as $\hbar = 1$ allows.

It is "shear" good luck that the mid rapidity initial conditions at RHIC are dense enough to essentially eliminate the dilution of elliptic flow due to the imperfect hadron fluid formed after the spatial asymmetry vanishes. (Recall that $(\eta/\sigma)_H \sim (T_c/T)^{1/c_H{}^2} > 1$ for $T < T_c$ [51]). At lower energies or higher rapidities this good luck runs out and the mixture of near perfect QGP and imperfect hadronic fluid dynamics reduces the elliptic flow.

5. pQCD and Jet Quenching

In addition to the breakdown of perfect fluid collectivity at high rapidity seen in Fig. 4, Fig. 3 clearly shows that hydrodynamics also breaks down at very short wavelengths and high transverse momenta, $p_T > 2$ GeV. Instead of continuing to rise with p_T, the elliptic asymmetry stops growing and the difference between baryon vs meson v_2 reverses sign! Between $2 < p_T < 5$ GeV the baryon $v_2^B(p_T)$ exceeds the meson $v_2^M(p_T)$ by approximately 3/2. For such short-wavelength components of the QGP, local equilibrium simply cannot be maintained due the fundamental asymptotic freedom property of QCD. I return to the baryon dominated transition region $1 < p_T < 5$ GeV in a later section since this involves interesting but as yet uncertain non-equilibrium non-perturbative processes. In this section I concentrate on the $p_T > 2$ GeV meson observables that can be readily understood in terms QGP modified **pQCD** dynamics [13, 14].

The quantitative study of short-wavelength partonic **pQCD** dynamics focuses on the rare high p_T power law tails that extend far beyond the typical (long-wavelength) scales $p < 3T \sim 1$ GeV of the bulk QGP. The second major discovery at RHIC is that the non-equilibrium power law high p_T jet distributions remain power law like but are strongly quenched [52]-[60]. Furthermore, the quenching pattern has a distinct centrality, p_T, azimuthal angle, and hadron flavor dependence that can be used to test the underlying dynamics in many independent ways. Below RHIC energies, the initial state Cronin enhancement

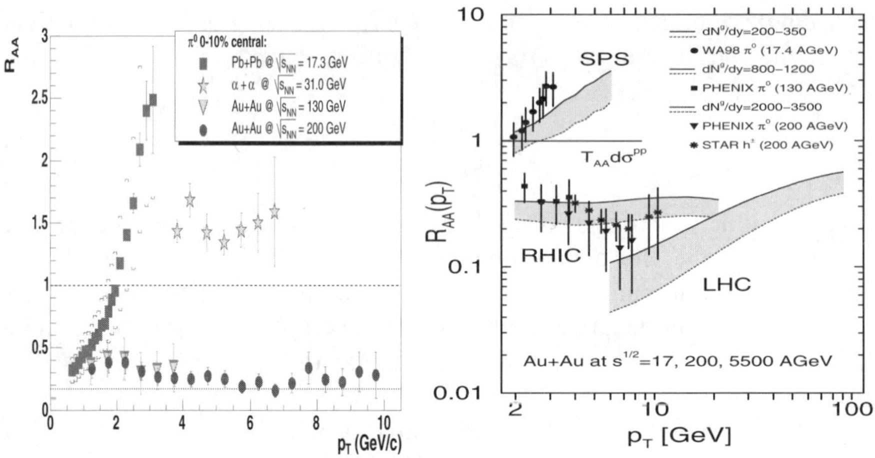

Figure 6. Jet Quenching at RHIC. Left [61] shows the jet quenching pattern of π^0 discovered by PHENIX [52, 53] at RHIC compared to previous observation of high p_T enhancement at ISR and SPS energies. The nuclear modification factor $R_{AA} = dN_{AA}/T_{AA}(b)d\sigma_{pp}$ measures the deviation of AA spectra from factorized pQCD. Right shows predictions [64] of the $\sqrt{s}$ and p_T dependence from SPS, RHIC, LHC based on the GLV theory [65] of radiative energy loss.

of moderately high p_T tails was observed in central $Pb + Pb$ reactions at the SPS. At the ISR a reduced Cronin enhancement in $\alpha + \alpha$ reactions was seen. In contrast, at RHIC a large suppression, by a factor of 4-5, was discovered in central $Au + Au$ that extends beyond 10 GeV for π^0.

Jet quenching in $A + A$ was proposed in [15, 16] as a way to study the dense matter produced at RHIC energies. As noted before, the pQCD jet production rates finally become large enough to measure yields up to high $p_T > 10$ GeV. Order of magnitude suppression effects were predicted based on simple estimates of induced gluon radiative energy loss. Ordinary, elastic energy loss [62] was known by that time to be too small to lead to significant attenuation.

As reviewed in [13, 14] refinements in the theory since then have opened the possibility of using the observed jet quenching pattern as a tomographic tool [63] to probe the parton densities in a QGP. The right panel shows a recent jet tomographic analysis [64] of the PHENIX π^0 data [52, 53] based on the

GLV opacity formalism [65]. Vitev and I concluded from Fig. 6b that the initial gluon rapidity density required to account for the observed jet quenching pattern must be $dN_g/dy \sim 1000 \pm 200$.

This jet tomographic measure of the initial dN_g/dy is in remarkable agreement with three other independent sources: (1) the initial entropy deduced via the Bjorken formula from the measured multiplicity, (2) the initial condition of the QGP required in hydrodynamics to produce the observed elliptic flow, and (3) the estimate of the maximum gluon rapidity density bound from the CGC gluon saturated initial condition [18].

These four independent measures makes it possible to estimate the maximal initial energy density in central collisions

$$\epsilon_0 = \epsilon(\tau \sim 1/p_0) \approx \frac{p_0^2}{\pi R^2}\frac{dN_g}{dy} \approx 20\frac{\text{GeV}}{\text{fm}^3} \sim 100 \times \epsilon_A \qquad (7)$$

where $p_0 \approx Q_{sat} \approx 1.0 - 1.4$ GeV is the mean transverse momentum of the initial produced gluons from the incident saturated virtual nuclear CGC fields [17, 18]. This scale controls the formation time $\hbar/p_0 \approx 0.2$ fm/c of the initially out-of-equilibrium (mostly gluonic) QGP. The success of the hydrodynamics requires that local equilibrium be achieved on a fast proper time scale $\tau_{eq} \approx (1-3)/p_0 < 0.6$ fm/c. The temperature at that time is $T(\tau_{eq}) \approx (\epsilon_0/(1-3) \times 12)^{1/4} \approx 2T_c$.

In HIJING model[66], the mini-jet cutoff $p_0 = 2-2.2$ GeV limits the number of mini-jets well below 1000. The inferred opacity of the QGP is much higher and consistent with the CGC and EKRT estimates.

5.1 $R_{AA}(p_T)$ and Single Jet Tomography

In order to illustrate the ideas behind jet tomography, I will simplify the discussion here to a schematic form. See [13] for details. The fractional radiative energy loss of a high energy parton in an expanding QGP is proportional to the position weighed line integral over color charge density $\rho(\vec{x}_\perp, \tau)$

$$\Delta E_{GLV}/E \approx C_2 \kappa(E) \int_0^{L(\phi)} d\tau \; \tau\rho(\vec{x}_\perp(\tau), \tau) \quad , \qquad (8)$$

where C_2 is the color Casimir of the jet parton and $\kappa(E)$ is a slowly varying function of the jet energy [13]. The azimuthal angle sensitive escape time, $L(\phi)$, depends on the initial production point and direction of propagation relative to the elliptic flow axis of the QGP [67]. For a longitudinal expanding QGP, isentropic perfect fluid flow implies that $\tau\rho(\tau) \approx (1/A_\perp) dN_g/dy$ is fixed by the initial gluon rapidity density. In this case

$$\frac{\Delta E(\phi)}{E} \propto C_2 \frac{L(\phi)}{A_\perp}\frac{dN_g}{dy} \propto C_2 N_{part}^{2/3} \frac{L(\phi)}{< L >} \quad . \qquad (9)$$

Therefore, gluon jets are expected to lose about $9/4$ more energy than quarks. In addition, since the produced gluon density scales as the number of wounded participating nucleons at a given impact parameter, (9) predicts a particular centrality and azimuthal dependence of the energy loss. Detailed numerical studies show that the actual GLV energy loss can account qualitatively for the saturation of $v_2(p_T > 2)$ [67], the unexpected p_T independence [64] of the quenching pattern, the centrality dependence [54] of the suppression factor [68, 69], and the rapidity dependence [74] of $R_{AA}(\eta, p_T)$ [42, 69] for pions and high $p_T > 6$ GeV inclusive charged hadrons at RHIC.

To further illustrate qualitatively how (9) influences the quench pattern, consider a simplified initial jet distribution rate, $d^2N/d^2p_0 = cp_0^{-n}$ where $n \sim 7$. In applications these are of course calculated numerically. After passing through the QGP, the final jet $p_T = p_f = p_0(1-\epsilon)$. The average over fluctuations constrains $\langle\epsilon\rangle = \Delta E(\phi)/E$. The quenched jet distribution is $d^2N/d^2p_T = (d^2p_0/d^2p_f)d^2N/d^2p_0 \approx (1-\epsilon(\phi))^{n-2}\ cp_f^{-n}$. The calculated hadron inclusive distribution is obtained by folding the quench jet distribution over the fragmentation function $D(z = p_h/p_f, Q^2 = p_f^2)$. However, in this illustrative example the fragmentation function dependence drops out, and the nuclear modification factor, $R_{AA}(p_h, \phi, N_{part}) = dN_h(\epsilon)/dN_h(0)$ reduces to

$$R_{AA} = \langle(1-\epsilon(\phi))^{n-2}\rangle \approx \left\langle\left(1 - \epsilon_c\frac{L(\phi)}{\langle L\rangle}\left(\frac{N_{part}}{2A}\right)^{2/3}\right)^{n-2}\right\rangle \tag{10}$$

The average over ϵ takes into account fluctuations of the radiative energy loss. The resulting R_{AA} is independent of p_T in this approximation. This was also found in the detailed numerical work in [64]. In central collisions, $N_{part} \approx 2A$ and $L(\phi) \approx \langle L\rangle$, and the magnitude of quenching is fixed by $\epsilon_c \propto dN_g/dy$. The centrality and azimuthal dependence for non-central collisions follows without additional calculations.

5.2 I_{AA} and Di-Jet Tomography

Measurements of near side and away side azimuthal angle correlations of di-jet fragments provide the opportunity to probe the evolution of the QGP color charge density in even more detail. Fig. 7 show the discovery [57–59] of mono-jet production [15] in central collisions at RHIC. In peripheral collisions, the distribution $dN/d\Delta\phi$ of the azimuthal distribution of $p_T \sim 2$ GeV hadrons relative to a tagged $p_T \sim 4$ GeV leading jet fragment shows the same near side and away side back-to-back jetty correlations as measured in $p + p$. This is a direct proof that the kinematic range studied tests the physics of pQCD binary parton collision processes. For central collisions, on the other hand, away side jet correlations are almost completely suppressed.

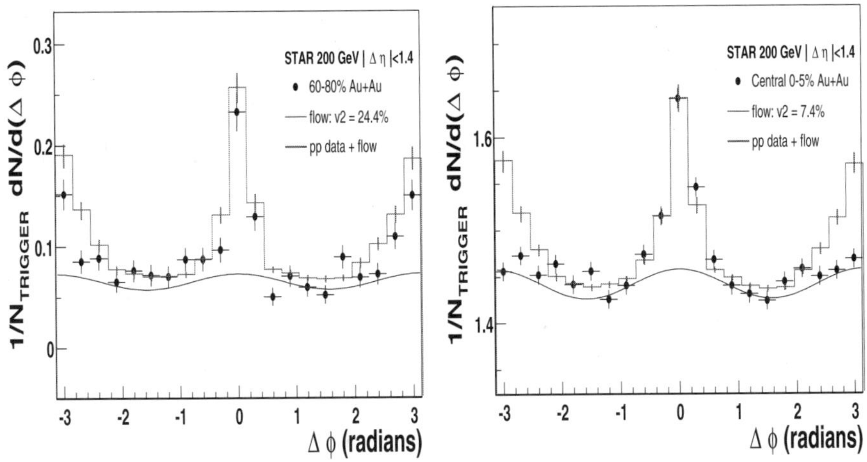

Figure 7. Monojets at RHIC from STAR [58, 59, 57]. Strongly correlated back-to-back di-jet production in $p + p$ and peripheral $Au + Au$ left side is compared to mono-jet production discovered in central $Au + Au$.

The quantitative measure of the nuclear modification of di-jet correlations in $A + B$ reactions at a given $\sqrt{s}$ is given by a formidable multi-variable function

$$C_{AB}(y_1, p_{T1}, \phi_1, y_2, p_{T2}, \phi_2; b, \Phi_b, h_1, h_2) \ , \tag{11}$$

where (y_1, p_{T1}, ϕ_1) is the trigger particle of flavor h_1 and the (y_2, p_{T2}, ϕ_2) is an associated particle of flavor h_2 for collisions at an impact parameter b with a collective flow axis, the reaction plane, fixing the azimuthal angle Φ_b. Obviously C_{AB} is a very powerful microscope to study the modification of short-wavelength correlations in the strongly interacting QGP.

The published data are as yet limited to $y_1 \approx y_2 \approx 0$, broad p_T cuts: $p_{T1} > 4$ GeV and $p_T \sim 2$ GeV, two bins of $\phi_1 - \phi_2$, and of course averaged over Φ_b. The measured modification of di-jet correlations is obtained by subtracting out the correlations due to bulk elliptic flow via the di-jet measure

$$\begin{aligned} I_{AA} &= \int_{\Delta_-}^{\Delta_+} d(\phi_1 - \phi_2) \, \{N(\phi_1 - \phi_2) \\ & \quad - N_B(1 + 2v_2(p_1)v_2(p_2)\cos(2(\phi_1 - \phi_2))\} \ , \end{aligned} \tag{12}$$

where the number of triggered pairs $N(\phi_1 - \phi_2)$ is normalized relative to the expected number based on $p + p$ measurements in the same $[\Delta_-, \Delta_+]$ relative azimuthal angle range. Wang [70, 71] has analyzed the centrality, N_{part}, dependence of I_{AA} as well as R_{AA} and showed that both can be understood from the same pQCD energy loss formalism. This provides another critical consistency check of jet tomography at RHIC.

Additional preliminary data from STAR presented by K. Filimonov at Quark Matter 2004 showed the first direct evidence that back-to-back di-jet quenching has a distinctive dependence on the azimuthal orientation of the jets relative to the reaction plane as expected from the obvious generalization of Eq.(10)

$$I_{AA}(\Delta\phi = \pi, \Phi_b) \approx \langle \{(1 - k_b L_1(\vec{r}_0, \hat{p}_T))(1 - k_b L_2(\vec{r}_0, -\hat{p}_T))\}^{n-2} \rangle \quad (13)$$

where $\vec{r}_0$ is the initial transverse production point of the approximately back-to-back di-jet moving in directions $\hat{p}_T$ relative to the reaction plane Φ_b. Here k_b the effective fractional energy loss per unit length in the QGP produced at impact parameter b. The high p_{T1} trigger naturally biases $\vec{r}_0$ to be near the surface and $\hat{p}_T$ is biased toward the outward normal direction. This means that on the average, $L_1 \ll L_2$, and the away side fragments should be strongly suppressed in the most central collisions, while the near side fragment correlations should be similar to that seen in pp. However, in non-central minimum biased events, Eq. (13) naturally predicts that the away side fragments are less quenched when the trigger hadron lies in the reaction plane than perpendicular to it, as observed by STAR.

5.3 The Empirical pQCD Line of Evidence

Single and di-jet data for pions above 3 GeV and protons above 6 GeV provide conclusive evidence that the QGP matter is partially opaque to short-wavelength probes with a quenching pattern as predicted by the **pQCD** radiative energy loss:

$$\mathbf{pQCD} = R_{AuAu}(p_T, \phi, N_{part}) + I_{AuAu}(\phi_1 - \phi_2; b, \Phi_b) \quad . \quad (14)$$

With the vastly increased statistics from the current RHIC RUN 4, the tests of consistency of the theory will be further extended to $p_T \sim 20$ GeV, and I_{AA} and C_{AA} will attain ever greater resolving power. In addition, heavy quark tomography [76–78] will provide new tests of the theory.

6. The dAu Control

Only one year ago [79] the interpretation of high p_T suppression was under intense debate because it was not yet clear how much of the quenching was due to initial state saturation (shadowing) of the gluon distributions and how much due to jet quenching discussed in the previous section. There was only one way to find out - eliminate the QGP final state interactions by substituting a Deuterium beam for one of the two heavy nuclei. In fact, it was long ago anticipated [16] that such a control test would be needed to isolate the unknown nuclear gluon shadowing contribution to the A+A quench pattern. In addition

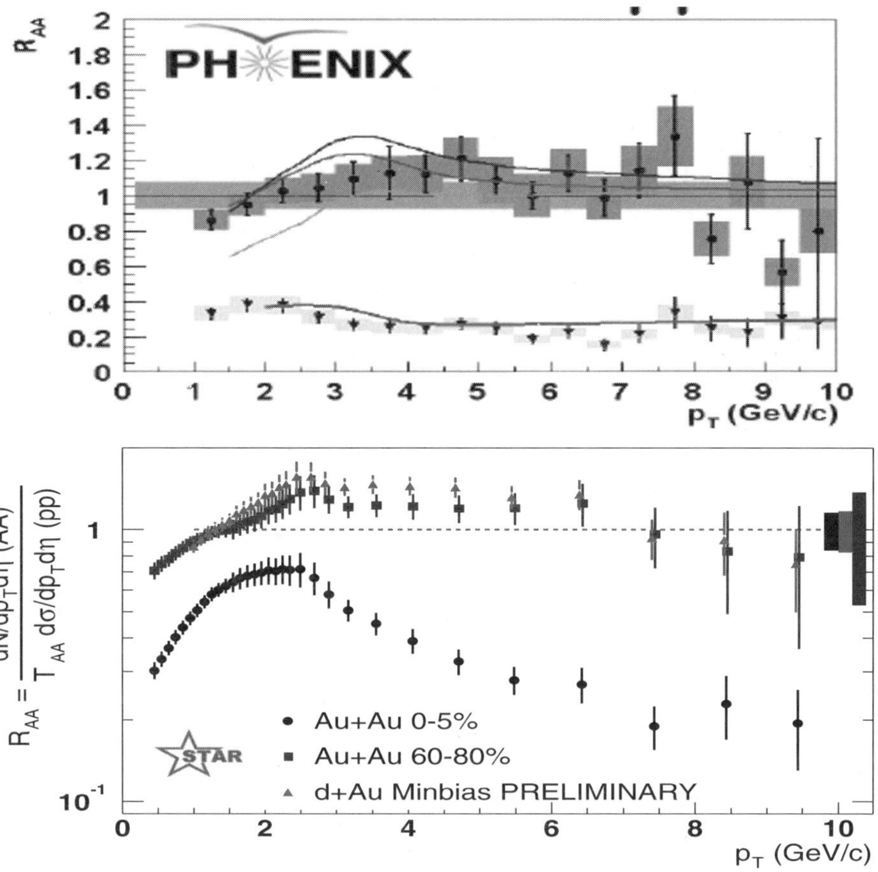

Figure 8. The **dA** control: PHENIX [72] π^0 and STAR [73] $h^{\pm}$ data compare R_{DAu} to R_{AuAu}. These and BRAHMS [74] and PHOBOS [75] data prove that jet quenching in $Au + Au$ must be due to final state interactions. Curves for π^0 show predictions from [64] for $Au + Au$ and from [82] $D + Au$. The curves for DAu show the interplay between different gluon shadow parameterizations (EKS, none, HIJING) and Cronin enhancement and are similar to predictions in [80–82]. In lower panel, the unquenching of charged hadrons is also seen in $D + Au$ relative to $Au + Au$ at high p_T.

$D + Au$ was required to test predictions of possible initial state Cronin multiple interactions [80–82, 84]. In contrast, one model of CGC [83] predicted a 30% suppression in central D+Au.

The data [72–75] conclusively rule out large initial shadowing as the cause of the $x_{BJ} > 0.01$ quenching in Au+Au and establish the empirical control analog of Eq.(14)

$$\mathbf{dA} = R_{DAu}(p_T, \phi, N_{part}) + I_{DAu}(\phi_1 - \phi_2, b) \quad . \tag{15}$$

6.1 The "Return of the Jeti"

The I_{DAu} measurement from STAR [73] is the check mate! The return

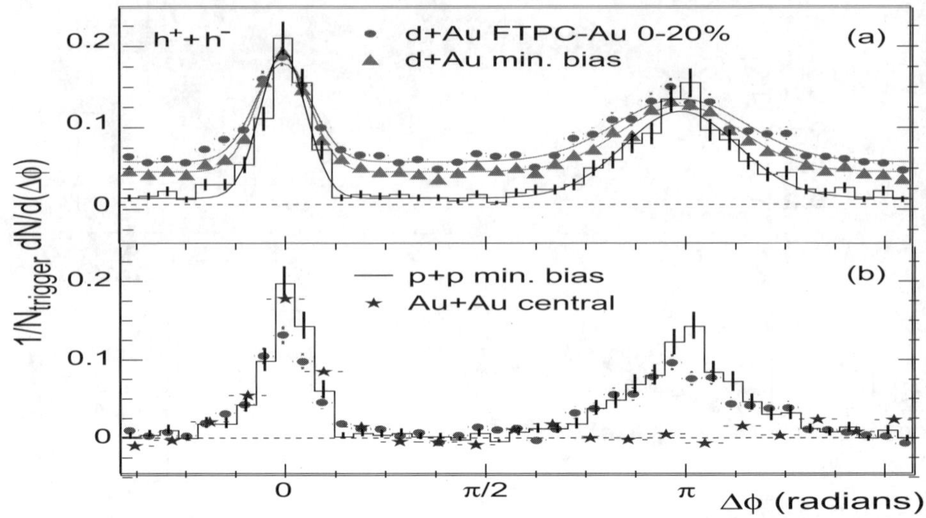

Figure 9. The **dA** "Return of the Jeti": Di-jet fragment azimuthal correlations from STAR [73] in DAu are unquenched relative to the mono jet correlation observed in central $Au + Au$.

of back-to-back jet correlation in $D + Au$ to the level observed in pp is seen in Fig. 9. The data rule out CGC gluon fusion models that predict mono-jets [86] correlations in the $x_{BJ} > 0.01$ region. These $D + Au$ data support the conclusion [70, 71] that the observed jet quenching in $Au + Au$ is due to parton energy loss.

Another independent check of the strikingly different nature of the nuclear matter created in $Au + Au$ versus $D + Au$ is provided by the width of the away side correlation function [87]. The transport properties of cold nuclear matter extracted from the Cronin enhancement effect [82] have only a small effect on the measured di-jet acoplanarity. Preliminary PHENIX data presented by J. Rak at Quark Matter 2004 confirm the qualitative similarity of $p + p$ and $D + Au$ for $p_T > 2$ GeV, but at the same time demonstrate a strong quantifiable increase in acoplanarity in central $Au + Au$ consistent with multiple semi-hard scattering [88] in a dense QGP.

7. Conclusions

The three lines of evidence have converged from the four RHIC experiments. The empirical QGP that has been found at RHIC via the combination of Eqs.(1,5,14,15). This QGP is, however, not the weakly interacting [7], color-dielectric "wQGP", that we have searched for. Because of its near perfect fluid long-wavelength properties, it must be very strongly coupled at least up to several times T_c. Symbolically, we should denote[1] the empirical QGP at RHIC by "**sQGP**" to emphasize its special properties. The **sQGP** is not only a near perfect fluid but it also retains part of its QCD asymptotic freedom character through its highly suppressed, but power-law, short-wavelength spectrum.

In summary, the **sQGP** found RHIC was seen through the following three convergent lines of evidence

$$\begin{aligned}\mathbf{sQGP} &= \mathbf{P_{QCD}} + \mathbf{pQCD} + \mathbf{dA} \\ &= \{v_2(p_T; \pi, K, p, \Lambda) + v_1(y) + \ \mathbf{c.c.}\} \\ &\quad + \{R_{AuAu}(p_T, \phi, N_{part}) + I_{AuAu}(\phi_1 - \phi_2; b, \Phi_b)\} \\ &\quad + \{R_{DAu}(p_T, \phi, N_{part}) + I_{DAu}(\phi_1 - \phi_2, b)\} \ . \qquad (16)\end{aligned}$$

Other surprising properties are already known. The anomalous $p_T = 2 - 5$ GeV baryon/meson ratios [27, 89], noted in connection with Fig. 3 and also seen indirectly in Fig. 8, already point to unexpected novel baryon number physics. Current speculations center around possible gluonic baryon junction dynamics [92], and possible multi-quark quark coalescence mechanisms [93]. The baryon number transport properties in the sQGP will certainly teach us new physics.

The experimental task of mapping out the novel properties of sQGP has only begun. It is important to concentrate, however, on those observables which are least distorted or "polluted" by uninteresting hadronic final interactions. For example, the severity of the HBT puzzle depends on which hadronic transport model is used. Pure hydrodynamics with late freeze-out times fails badly to reproduce final state soft pion correlations. Hybrid hydro+RQMD hadronic cascade does somewhat better [90], but there is at least one transport model [91] that reproduced the data. Non-equilibrium hadron resonance transport dynamics are unfortunately still not well enough understood at any energy to allow definitive conclusions to be drawn.

Thermal direct photons have yet to be measured, but it is already known that the pre-equilibrium [80] and hadronic final state contributions in the few GeV p_T range will produce large backgrounds on top of the thermal component. These must be deconvoluted if thermal photons are to serve as a sQGP thermometer. However, even the theoretical thermal photon rates are still under debate [94].

The J/ψ suppression discovery at SPS was originally attributed to Debye screening of $c\bar{c}$ in a wQGP paradigm. Recent lattice QCD results now indicate

[1] I thank T.D.Lee for discussions and suggesting the sQGP designation of the matter discovered at RHIC to distinguish it from weakly interacting, Debye screened plasmas, wQGP, which may only exist at temperatures $T \gg T_c$.

that heavy quark correlations persist perhaps up to $2T_c$. This is another indication for the strongly coupled nature of **sQCD**. The suppression of J/ψ in AA is also strongly influenced by initial state and final state hadronic (comover) effects. It will be no easier to deconvolute these competing effects at RHIC. These observables of course need to be measured at RHIC, but one should not expect an easy interpretation.

I believe that the most promising direction of future experiments at RHIC will be precision measurements of short-wavelength ($p_T > 2$ GeV) correlators C_{AB} illustrated in Eq.(11). These are very powerful six dimensional microscopes with four discrete (h_1, h_2, A, B) and two continuously adjustable geometric (b, Φ_b) experimental knobs in addition to the beam energy $\sqrt{s} = 20 - 200$ AGeV. One of the important correlators will be that of direct photon tagged jets [95]. Another will be open charm and possibly bottom quark tomographic probes. The available experimental knobs have hardly been varied yet. Much remains to be done to map out to clarify the properties of the **sQGP** found at RHIC.

Acknowledgments

This work was supported by the Director, Office of Energy Research, Office of High Energy and Nuclear Physics, Division of Nuclear Physics, of the U.S. Department of Energy under Contract DE-FG-02-93ER-40764. I also gratefully acknowledge partial support from an Alexander von Humboldt-Stiftung Foundation for continuation of collaborative work at the Institut für Theoretische Physik, Frankfurt.

References

[1] Z. Fodor and S. D. Katz, arXiv:hep-lat/0402006. F. Csikor et al arXiv:hep-lat/0401022.

[2] C. R. Allton et al, Phys. Rev. D **68**, 014507 (2003) [arXiv:hep-lat/0305007]. F. Karsch, E. Laermann and A. Peikert, Phys. Lett. B **478**, 447 (2000) [arXiv:hep-lat/0002003].

[3] S. Gupta, Pramana **61**, 877 (2003) [arXiv:hep-ph/0303072].

[4] M. A. Halasz et al, Phys. Rev. D **58**, 096007 (1998) [arXiv:hep-ph/9804290]. M. A. Stephanov, K. Rajagopal and E. V. Shuryak, Phys. Rev. Lett. **81**, 4816 (1998) [arXiv:hep-ph/9806219].

[5] F. Karsch, K. Redlich and A. Tawfik, Phys. Lett. B **571**, 67 (2003) [arXiv:hep-ph/0306208].

[6] D. H. Rischke, Prog. Part. Nucl. Phys. **52**, 197 (2004) [arXiv:nucl-th/0305030].

[7] J. C. Collins and M. J. Perry, Phys. Rev. Lett. **34**, 1353 (1975); B. A. Freedman and L. D. McLerran, Phys. Rev. D **16**, 1169 (1977); G. Chapline and M. Nauenberg, Phys. Rev. D **16**, 450 (1977).

[8] T. D. Lee and G. C. Wick, Phys. Rev. D **9**, 2291 (1974).

[9] J. Hofmann, H. Stö cker, W. Scheid and W. Greiner, On The Possibility Of Nuclear Shock Waves In Relativistic Heavy Ion Collisions, Bear Mountain Workshop, New York, Dec

1974. H. G. Baumgardt *et al.*, "Shock Waves And Mach Cones In Fast Nucleus-Nucleus Collisions," Z. Phys. A **273**, 359 (1975).

[10] H. Stö cker, J. A. Maruhn and W. Greiner, Z. Phys. A **293**, 173 (1979). L. Csernai and H. Stö cker, Phys. Rev. C **25**, 3208 (1981).

[11] H. Stö cker and W. Greiner, Phys. Rept. **137**, 277 (1986).

[12] W. Reisdorf and H. G. Ritter, Ann. Rev. Nucl. Part. Sci. **47**, 663 (1997).

[13] M. Gyulassy, I. Vitev, X. N. Wang and B. W. Zhang, arXiv:nucl-th/0302077. M. Gyulassy, Lect. Notes Phys. **583**, 37 (2002) [arXiv:nucl-th/0106072].

[14] R. Baier, D. Schiff and B. G. Zakharov, Ann. Rev. Nucl. Part. Sci. **50**, 37 (2000) [arXiv:hep-ph/0002198].

[15] M. Gyulassy and M. Plumer, Nucl. Phys. A **527**, 641 (1991). M. Gyulassy, M. Plumer, M. Thoma and X. N. Wang, Nucl. Phys. A **538**, 37C (1992)

[16] X. Wang and M. Gyulassy, Phys. Rev. Lett. **68**, 1480 (1992).

[17] L. McLerran, arXiv:hep-ph/0402137. E. Iancu and R. Venugopalan, arXiv:hep-ph/0303204. L. D. McLerran and R. Venugopalan, Phys. Rev. D **49**, 2233 (1994) [arXiv:hep-ph/9309289]. D. Kharzeev and E. Levin, Phys. Lett. B **523**, 79 (2001) [arXiv:nucl-th/0108006].

[18] K. J. Eskola, K. Kajantie, P. V. Ruuskanen and K. Tuominen, Phys. Lett. B **543**, 208 (2002) [arXiv:hep-ph/0204034]; Phys. Lett. B **532**, 222 (2002) [arXiv:hep-ph/0201256].

[19] M. G. Alford, J. Berges and K. Rajagopal, Nucl. Phys. B **571**, 269 (2000) [arXiv:hep-ph/9910254].

[20] H. Stö cker, J. A. Maruhn and W. Greiner, Phys. Rev. Lett. **44**, 725 (1980).

[21] H. Stö cker *et al.*, Phys. Rev. C **25**, 1873 (1982).

[22] J. Y. Ollitrault, Phys. Rev. D **46**, 229 (1992).

[23] S. A. Voloshin and A. M. Poskanzer, Phys. Lett. B **474**, 27 (2000) [arXiv:nucl-th/9906075].

[24] C. Alt *et al.* [NA49 Collaboration], Phys. Rev. C **68**, 034903 (2003) [arXiv:nucl-ex/0303001].

[25] G. Stoicea *et al.*, arXiv:nucl-ex/0401041.

[26] J. Adams *et al.* [STAR Collaboration], arXiv:nucl-ex/0310029, Phys. Rev. Lett. 92 (2004) 062301.

[27] P. R. Sorensen, arXiv:nucl-ex/0309003. Ph.D. thesis.

[28] J. Adams *et al.* [STAR Collaboration], arXiv:nucl-ex/0306007, Phys. Rev. Lett. 92 (2004) 052302

[29] C. Adler *et al.* [STAR Collaboration], Phys. Rev. C **66**, 034904 (2002) [arXiv:nucl-ex/0206001].

[30] S. S. Adler *et al.* [PHENIX Collaboration], Phys. Rev. Lett. **91**, 182301 (2003) [arXiv:nucl-ex/0305013].

[31] B. B. Back *et al.* [PHOBOS collaboration], Nucl. Phys. A **715**, 65 (2003) [arXiv:nucl-ex/0212009].

[32] Y. Cheng, F. Liu, Z. Liu, K. Schweda and N. Xu, Phys. Rev. C **68**, 034910 (2003). N. Xu *et al.* [NA44 Collaboration], Nucl. Phys. A **610**, 175C (1996).

[33] P. F. Kolb, P. Huovinen, U. W. Heinz and H. Heiselberg, Phys. Lett. B **500**, 232 (2001).

[34] P. Huovinen, P. F. Kolb, U. W. Heinz, P. V. Ruuskanen and S. A. Voloshin, Phys. Lett. B **503**, 58 (2001).

[35] P. F. Kolb, U. W. Heinz, P. Huovinen, K. J. Eskola and K. Tuominen, Nucl. Phys. A **696**, 197 (2001).

[36] P. Huovinen, arXiv:nucl-th/0305064.

[37] P. F. Kolb and U. Heinz, arXiv:nucl-th/0305084.

[38] D. Teaney, J. Lauret and E. V. Shuryak, nucl-th/0104041.

[39] D. Teaney, J. Lauret and E. V. Shuryak, arXiv:nucl-th/0110037.

[40] D. Teaney, Phys. Rev. C **68**, 034913 (2003). D. Teaney, arXiv:nucl-th/0301099.

[41] T. Hirano and Y. Nara, Phys. Rev. Lett. **91**, 082301 (2003) [arXiv:nucl-th/0301042].

[42] T. Hirano and Y. Nara, Phys. Rev. C **68**, 064902 (2003) [arXiv:nucl-th/0307087].

[43] T. Hirano and Y. Nara, arXiv:nucl-th/0307015.

[44] D. Molnar and M. Gyulassy, Nucl. Phys. A **697**, 495 (2002) [Erratum-ibid. A **703**, 893 (2002)] [arXiv:nucl-th/0104073]. B. Zhang, M. Gyulassy and C. M. Ko, Phys. Lett. B **455**, 45 (1999) [arXiv:nucl-th/9902016].

[45] S. A. Bass and A. Dumitru, Phys. Rev. C **61**, 064909 (2000) [arXiv:nucl-th/0001033].

[46] P. Braun-Munzinger, K. Redlich and J. Stachel, arXiv:nucl-th/0304013.

[47] I. G. Bearden *et al.* [BRAHMS Collaboration], Phys. Rev. Lett. **88**, 202301 (2002) [arXiv:nucl-ex/0112001].

[48] G. Agakichiev *et al.* [CERES/NA45 Collaboration], arXiv:nucl-ex/0303014.

[49] P. Danielewicz, R. Lacey and W. G. Lynch, Science **298**, 1592 (2002) [arXiv:nucl-th/0208016].

[50] G. Policastro, D. T. Son and A. O. Starinets, JHEP **0212**, 054 (2002) [arXiv:hep-th/0210220]. A. Buchel and J. T. Liu, arXiv:hep-th/0311175.

[51] P. Danielewicz and M. Gyulassy, Phys. Rev. D **31** (1985) 53.

[52] K. Adcox *et al.*, Phys. Rev. Lett. **88**, 022301 (2002); P. Levai *et al.*, Nucl. Phys. A **698**, 631 (2002).

[53] K. Adcox *et al.* [PHENIX Collaboration], Phys. Lett. B **561**, 82 (2003) [arXiv:nucl-ex/0207009].

[54] S. S. Adler *et al.* [PHENIX Collaboration], Phys. Rev. Lett. **91**, 072301 (2003) [arXiv:nucl-ex/0304022].

[55] J. Adams *et al.* [STAR Collaboration], arXiv:nucl-ex/0305015.

[56] C. Adler *et al.*, [STAR Collaboration] Phys. Rev. Lett. **89**, 202301 (2002) [arXiv:nucl-ex/0206011].

[57] P. Jacobs and J. Klay [STAR Collaboration], arXiv:nucl-ex/0308023.

[58] C. Adler *et al.* [STAR Collaboration], Phys. Rev. Lett. **90**, 082302 (2003) [arXiv:nucl-ex/0210033].

[59] D. Hardtke [The STAR Collaboration], Nucl. Phys. A **715**, 272 (2003) [arXiv:nucl-ex/0212004].

[60] C. Adler *et al.* [STAR Collaboration], Phys. Rev. Lett. **90**, 032301 (2003).

[61] D. d'Enterria [PHENIX Collaboration], arXiv:nucl-ex/0401001.

[62] J. D. Bjorken, FERMILAB-PUB-82-059-THY and erratum (unpublished); M. H. Thoma and M. Gyulassy, Nucl. Phys. B **351**, 491 (1991); E. Braaten and M. H. Thoma, Phys. Rev. D **44**, 2625 (1991); M. H. Thoma, J. Phys. G **26**, 1507 (2000) [arXiv:hep-ph/0003016].

[63] M. Gyulassy, P. Levai, and I. Vitev, Phys. Lett. B **538**, 282 (2002): E. Wang and X.-N. Wang, Phys. Rev. Lett. **89**, 162301 (2002); C. A. Salgado and U. A. Wiedemann, Phys. Rev. Lett. **89**, 092303 (2002);

[64] I. Vitev and M. Gyulassy, Phys. Rev. Lett. **89**, 252301 (2002) [arXiv:hep-ph/0209161].

[65] M. Gyulassy, P. Levai and I. Vitev, Nucl. Phys. B **594**, 371 (2001) [arXiv:nucl-th/0006010]. Phys. Rev. Lett. **85**, 5535 (2000) [arXiv:nucl-th/0005032]; Nucl. Phys. B **571**, 197 (2000) [arXiv:hep-ph/9907461].

[66] V. Topor Pop *et al.*, Phys. Rev. C **68**, 054902 (2003) [arXiv:nucl-th/0209089]. X. N. Wang and M. Gyulassy, Phys. Rev. D **44**, 3501 (1991).

[67] M. Gyulassy, I. Vitev and X. N. Wang, Phys. Rev. Lett. **86**, 2537 (2001) [arXiv:nucl-th/0012092]. M. Gyulassy, I. Vitev, X. N. Wang and P. Huovinen, Phys. Lett. B **526**, 301 (2002) [arXiv:nucl-th/0109063].

[68] M. Gyulassy, CIPANP Conference seminar, May 21, 2003, "http://www.phenix.bnl.gov/WWW/publish/nagle/CIPANP/"

[69] A. Adil, M. Gyulassy, I. Vitev, to be published.

[70] X. N. Wang, arXiv:nucl-th/0305010.

[71] X. N. Wang, Phys. Lett. B **579**, 299 (2004) [arXiv:nucl-th/0307036].

[72] S. S. Adler *et al.* [PHENIX Collaboration], Phys. Rev. Lett. **91**, 072303 (2003) [arXiv:nucl-ex/0306021].

[73] J. Adams *et al.* [STAR Collaboration], Phys. Rev. Lett. **91**, 072304 (2003) [arXiv:nucl-ex/0306024].

[74] I. Arsene *et al.* [BRAHMS Collaboration], Phys. Rev. Lett. **91**, 072305 (2003) [arXiv:nucl-ex/0307003].

[75] B. B. Back *et al.* [PHOBOS Collaboration], Phys. Rev. Lett. **91**, 072302 (2003) [arXiv:nucl-ex/0306025].

[76] Y. L. Dokshitzer and D. E. Kharzeev, Phys. Lett. B **519**, 199 (2001) [arXiv:hep-ph/0106202].

[77] M. Djordjevic and M. Gyulassy, Nucl. Phys. A **733**, 265 (2004) [arXiv:nucl-th/0310076].

[78] S. Batsouli, S. Kelly, M. Gyulassy and J. L. Nagle, Phys. Lett. B **557**, 26 (2003) [arXiv:nucl-th/0212068].

[79] Transverse Dynamics at RHIC, BNL March 6-8, 2003, "http://www.phenix.bnl.gov/phenix/WWW/publish/rak/workshop/int/program_TD.htm"

[80] X. N. Wang, Phys. Rev. C **61**, 064910 (2000) [arXiv:nucl-th/9812021].

[81] X. N. Wang, Phys. Rept. **280**, 287 (1997) [arXiv:hep-ph/9605214].

[82] I. Vitev, Phys. Lett. B **562**, 36 (2003) [arXiv:nucl-th/0302002]; A. Accardi and M. Gyulassy, arXiv:nucl-th/0308029; P. Levai, G. Papp, G. G. Barnafoldi and G. I. Fai, arXiv:nucl-th/0306019.

[83] D. Kharzeev, E. Levin and L. McLerran, Phys. Lett. B **561**, 93 (2003) [arXiv:hep-ph/0210332].

[84] J. W. Qiu and I. Vitev, arXiv:hep-ph/0309094; arXiv:hep-ph/0401062.

[85] S. A. Bass *et al.*, Prog. Part. Nucl. Phys. **41**, 225 (1998) [arXiv:nucl-th/9803035].

[86] V. R. Zoller, arXiv:hep-ph/0306038.

[87] J. W. Qiu and I. Vitev, Phys. Lett. B **570**, 161 (2003) [arXiv:nucl-th/0306039]; I. Vitev, arXiv:nucl-th/0308028.

[88] M. Gyulassy, P. Levai and I. Vitev, Phys. Rev. D **66**, 014005 (2002) [arXiv:nucl-th/0201078].

[89] S. S. Adler *et al.* [PHENIX Collaboration], arXiv:nucl-ex/0307022.

[90] S. Soff, S. A. Bass and A. Dumitru, Phys. Rev. Lett. **86**, 3981 (2001) [arXiv:nucl-th/0012085].

[91] Z. W. Lin, C. M. Ko and S. Pal, Phys. Rev. Lett. **89**, 152301 (2002) [arXiv:nucl-th/0204054].

[92] D. Kharzeev, Phys. Lett. B **378**, 238 (1996) [arXiv:nucl-th/9602027]. S. E. Vance et al, Phys. Lett. B **443**, 45 (1998) [arXiv:nucl-th/9806008]. I. Vitev and M. Gyulassy, Phys. Rev. C **65**, 041902 (2002) [arXiv:nucl-th/0104066].

[93] P. Csizmadia, et al J. Phys. G **25**, 321 (1999) [arXiv:hep-ph/9809456]. R. J. Fries, B. Muller, C. Nonaka and S. A. Bass, Phys. Rev. Lett. **90**, 202303 (2003) [arXiv:nucl-th/0301087]. D. Molnar and S. A. Voloshin, Phys. Rev. Lett. **91**, 092301 (2003) [arXiv:nucl-th/0302014]. V. Greco, C. M. Ko and P. Levai, Phys. Rev. C **68**, 034904 (2003) [arXiv:nucl-th/0305024]. Z. W. Lin and D. Molnar, Phys. Rev. C **68**, 044901 (2003) [arXiv:nucl-th/0304045].

[94] F. Gelis, Nucl. Phys. A **715**, 329 (2003) [arXiv:hep-ph/0209072].

[95] X. N. Wang, Z. Huang and I. Sarcevic, Phys. Rev. Lett. **77**, 231 (1996) [arXiv:hep-ph/9605213].

WHAT IS THE EVIDENCE FOR THE COLOR GLASS CONDENSATE?

Larry McLerran
Physics Department PO Box 5000, Brookhaven National Laboratory, Upton, NY 11973, USA

Abstract I introduce the concept of the Color Glass Condensate. I review data from HERA and RHIC which suggest that such a universal form of matter has been found.

1. What is the Color Glass Condensate?

The ideas for the Color Glass Condensate originate in the result for the HERA data on the gluon distribution function shown in Fig. 1(a) [1] The gluon density

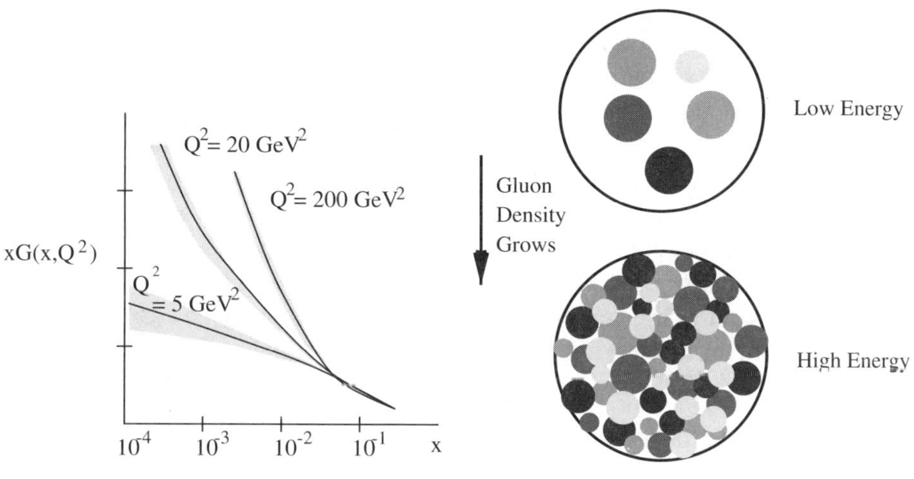

Figure 1. (a)The HERA data for the gluon distribution function as a function of x for various values of Q^2. (b) A physical picture of the low x gluon density inside a hadron as a function of energy

is rising rapidly as a function of decreasing x. This was expected in a variety of theoretical works,[2]-[4] and has the implication that the real physical transverse density of gluons must increase.[2]-[3],[5]. This follows because total cross

W. Greiner et al. (eds.), Structure and Dynamics of Elementary Matter, 183–205.

sections rise slowly at high energies but the number of gluons is rising rapidly. This is shown in Fig. 1(b). This led to the conjecture that the density of gluons should become limited, that is, there is gluon saturation. [2]-[3], [5]

The low x gluons therefore are closely packed together. The strong interaction strength must become weak, $\alpha_S \ll 1$. Weakly coupled systems should be possible to understand from first principles in QCD.[5]

This weakly coupled system is called a Color Glass Condensate for reasons we now enumerate:[6]

- **Color** The gluons which make up this matter are colored.
- **Glass** The gluons at small x are generated from gluons at larger values of x. In the infinite momentum frame, these larger momentum gluons travel very fast and their natural time scales are Lorentz time dilated. This time dilated scale is transferred to the low x degrees of freedom which therefore evolve very slowly compared to natural time scales. This is the property of a glass.
- **Condensate** The phase space density

$$\rho = \frac{1}{\pi R^2} \frac{dN}{dyd^2p_T} \tag{1}$$

  is generated by a trade off between a negative mass-squared term linear in the density which generates the instability, $-\rho$ and an interaction term $\alpha_S\rho^2$ which stabilizes the system at a phase space density $\rho \sim 1/\alpha_S$. Because $\alpha_S << 1$, this means that the quantum mechanical states of the system associated with the condensate are multiply occupied. They are highly coherent, and share some properties of Bose condensates. The gluon occupation factor is very high, of order $1/\alpha_S$, but it is only slowly (logarithmically) increasing when further increasing the energy, or decreasing the transverse momentum. This provides saturation and cures the infrared problem of the traditional BFKL approach.[7]

Implicit in this definition is a concept of fast gluons which act as sources for the colored fields at small x. These degrees of freedom are treated differently than the fast gluons which are taken to be sources. The slow ones are fields. There is an arbitrary X_0 which separates these degrees of freedom. This arbitrariness is cured by a renormalization group equation which requires that physics be independent of X_0. In fact this equation determines much of the structure of the resulting theory as its solution flows to a universal fixed point. [6]-[9]

There is evidence which supports this picture. One piece is the observation of limiting fragmentation. This phenomena is that if particles collide at some fixed center of mass energy and the distribution of particles are measured as a function of their longitudinal momentum from the longitudinal momentum of one of the colliding particles, then these distributions do not change as one goes to higher energy, except for the new degrees of freedom that appear. This is true near zero longitudinal momentum in the center of mass frame because new degrees of freedom appear as the center of mass energy is increased. In the analogy with the CGC, the degrees of freedom, save the new ones added in at low longitudinal momentum, are the sources. The fields correspond to the new degrees of freedom. The sources are fixed in accord with limiting fragmentation. One generates an effective theory for the low longitudinal momentum degrees of freedom as fixed sources above some cutoff, and the fields generated by these sources below the cutoff. A recent measurement of limiting fragmentation comes from the Phobos experiment at RHIC shown in Fig. 2 [10]

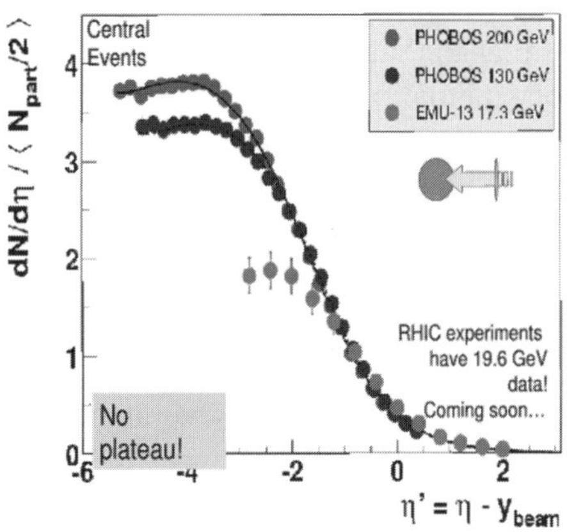

Figure 2. Limiting fragmentation and the RHIC data.

Of course the perfect scaling of the limiting fragmentation curves is only an approximation. As shown by Jalilian-Marian, the limiting fragmentation curves are given by the total quark, antiquark and gluon distribution functions of the fast particle measured at a momentum scale Q_{sat}^2 appropriate for the particle that it collides with.[11] The saturation momentum Q_{sat} will play a crucial role in our later discussion. It is a momentum scale which is determined by the density of gluons in the CGC

$$\frac{1}{\pi R^2}\frac{dN}{dy} \sim \frac{1}{\alpha_S} Q_{sat}^2 \tag{2}$$

The saturation momenta turns out to depend on the total beam energy because the longitudinal momentum scale of the target particle at fixed x of the projectile will depend upon the beam energy. It is nevertheless remarkable how small these violations appear to be.

The CGC may be defined mathematically by a path integral:

$$Z = \int_{X_0} [dA][dj] exp\,(iS[A,j] - \chi[j]) \tag{3}$$

What this means is that there is an effective theory defined below some cutoff in x at X_0, and that this effective theory is a gluon field in the presence of an external source j. This source arises from the quarks and gluons with $x \geq X_0$, and is a variable of integration. The fluctuations in j are controlled by the weight function $\chi[j]$. It is $\chi[j]$ which satisfies renormalization group equations which make the theory independent of X_0.[8]-[14],[6]. The equation for χ is called the JIMWLK equation. This equation reduces in appropriate limits to the BFKL and DGLAP evolution equations.[4], [15] The theory above is mathematically very similar to that of spin glasses.

There are a variety of kinematic regions where one can find solutions of the renormalization group equations which have different properties. There is a region where the gluon density is very high, and the physics is controlled by the CGC. This is when typical momenta are less than a saturation momenta which depends on x,

$$Q^2 \leq Q^2_{sat}(x) \tag{4}$$

The dependence of x has been evaluated by several authors, [2],[16]-[18], and in the energy range appropriate for current experiments has been determined by Triantafyllopoulos to be

$$Q^2_{sat} \sim (x_0/x)^\lambda \; GeV^2 \tag{5}$$

where with about 30% uncertainty $\lambda = 0.3$. The value of x_0 is not determined from the renormalization group equations and must be found from experiment.

There is also a region of very high Q^2 at fixed x, where the density of gluons is small and perturbative QCD is reliable. It turns out there is a third region intermediate between high density and low where there are universal solutions to the renormalization group equations and scaling in terms of Q^2_{sat}.[17] In this region and in the region of the CGC, distribution functions are universal functions of only $Q^2/Q^2_{sat}(x)$. The extended scaling region is when

$$Q^2_{sat} \leq Q^2 \leq Q^4_{sat}/\Lambda^2_{QCD} \tag{6}$$

These various regions are shown in Fig. 3 in the plane of $ln(1/x)$ and $ln(Q^2)$

2. Why is the CGC Important?

The Color Glass Condensate is a new universal type of matter:

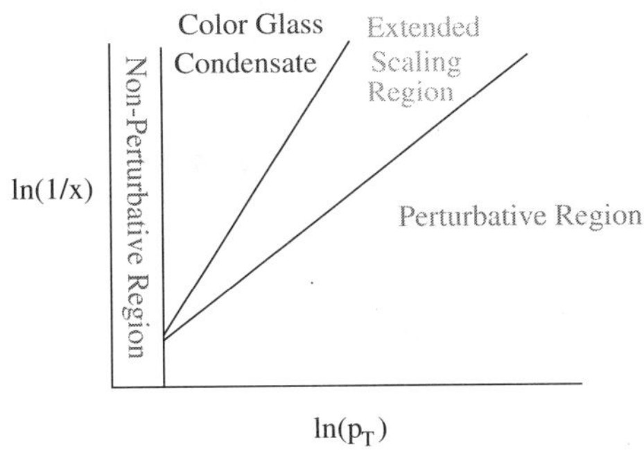

Figure 3. The various regions where there are a CGC, extended scaling, and low density of glue.

- **Matter:** The separation between the gluons in the CGC is small compared to the size of the system. Due to Lorentz time dilation, the lifetime of this matter is long compared to natural time scales.
- **New:** This matter can only be probed at high energy, and it may be produced in ultrarelativistic nuclear collisions.
- **Universal:** The CGC is universal independent of the type of hadron which generated it. Universality of this matter implies it is of fundamental interest.

The Color Glass Condensate is a theory of:

- The origin of glue and sea quarks in hadrons.
- The origin and nature of cross sections and particle production.
- The distribution of valence quantum numbers at small x.
- The initial conditions for the matter which evolves into the Quark Gluon Plasma at RHIC.

3. What does a high energy hadron look like?

In Fig. 4, a hadron is shown in its rest frame and as viewed from the infinite momentum frame. The picture does not look Lorentz invariant. This is because the hadron is made of many different Fock space components, some containing valence quarks and a few gluons and sea quarks and some containing valence quarks plus many sea quarks and gluons. At low energies, one is sensitive to matrix elements which involve the valence quarks and a few gluons and sea quarks. For high energy collisions, typical matrix elements involve the valence

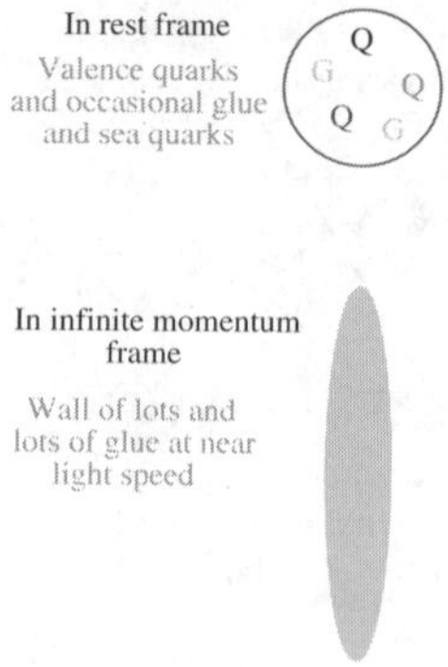

Figure 4. A hadron is shown at rest and in the infinite momentum frame.

quarks plus many sea quarks and gluons. At high energies, the hadron appears as if it is a gluon wall.

Of course the Color Glass Condensate description is Lorentz invariant. In fact, there is a very subtle duality of description. If one views the hadron in the infinite momentum frame, one is scattering from high density of gluonic matter. In the rest frame of the hadron, where the probe has very high energy, the Color Glass Condensate appears through coherent multiple scattering of the probe valence quarks. One can prove these descriptions are mathematically equivalent.[19]-[20]

The form of the fields associated with the CGC can also be easily seen. In the infinite momentum frame, the entire hadron is Lorentz contracted into a small region of $x^- = t - z$. The glassy nature of the fields makes them independent of the light cone time $x^+ = t + z$. Therefore the only large component of $F^{\mu\nu}$ is $F^{i+} \sim \partial/\partial x^- A^i$. A little algebra shows that

$$\vec{E} \perp \vec{B} \perp \vec{z} \tag{7}$$

In Fig. 5, these fields are shown. They correspond to the Lienard-Wiechart potential for a boosted electron, except that they are colored and they have a random polarization and color.

The density of gluons per unit area defines a momentum scale, the saturation momenta,

$$\frac{1}{\pi R^2} \frac{dN}{dy} \sim \frac{1}{\alpha_S} Q^2_{sat}(x) \tag{8}$$

We insert the extra factor of $1/\alpha_S$ because the fields are classical, and therefore their density should scale in this way. This form is guaranteed because the system is almost scale invariant. Small violations of scaling arise because α_S is measured at Q_{sat}.

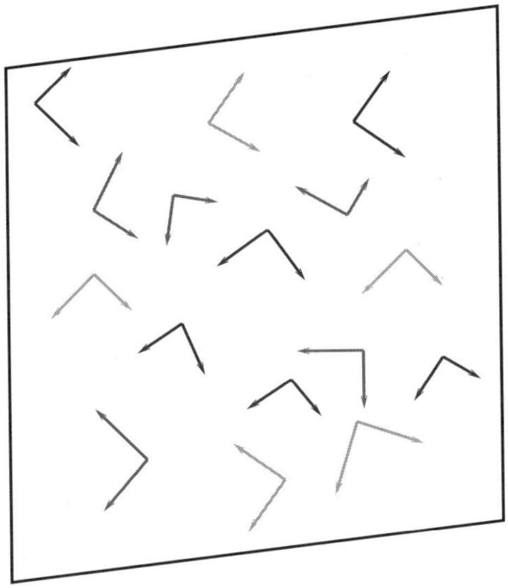

Figure 5. The distribution of colored fields within the CGC.

4. Experimental Evidence in Support of CGC

In this section, I discuss the accumulated evidence from HERA and RHIC, and elsewhere, in support of the hypothesis of a Color Glass Condensate.

4.1 Geometrical Scaling

Geometrical scaling is the observation[21]-[22] that the deep inelastic cross section for virtual photon scattering as a function of Q^2 and x is really only a function of

$$\sigma^{\gamma^* p} \sim F(Q^2/Q^2_{sat}) \tag{9}$$

where the saturation momentum is taken to be

$$Q^2_{sat} \sim (x_0/x)^\lambda \, 1GeV^2 \tag{10}$$

and $\lambda \sim 0.3$ and $x_0 \sim 10^{-4}$. This scaling works for $x \leq 10^{-2}$ and for the available data in Q^2. The data is shown in Fig. 6

It is straightforward to understand why this scaling works for the small $Q^2 \leq Q^2_{sat}$. This is the region of the CGC, and there is only one dimensionful scale which characterizes the system: the saturation momentum.[16] The surprise is that there is an extended scaling window for $Q^2_{sat} \leq Q^2 \leq Q^4_{sat}/\Lambda^2_{QCD}$.[17] This can be proven analytically. As well, one now has reliable computation of the dependence on x of the saturation momentum, that is, one knows the

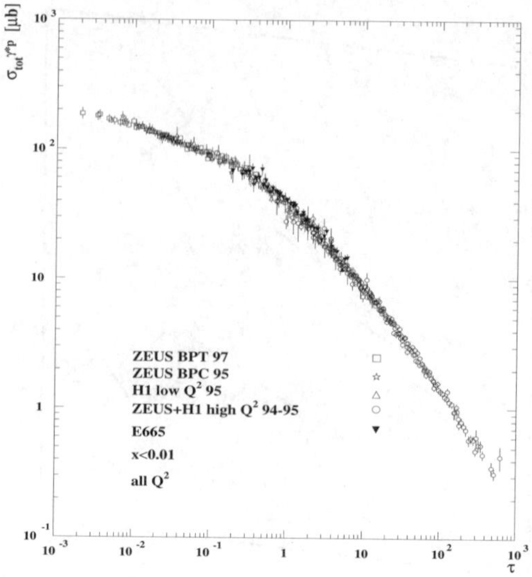

Figure 6. The cross section $\sigma^{\gamma^* p}$ as a function of the scaling variable $\tau = Q^2/Q^2_{sat}$.

exponent λ to about 15% accuracy, and it agrees with what is seen from the geometrical scaling curve.[18] What is not determined from the theory of the CGC is the scale x_0, and this must be found by experiment. This comes from the boundary conditions for the renormalization group equations.

4.2 The Structure Function F_2

Using the dipole description of the virtual photon wavefunction, the structure function F_2 can be related to the gluon distribution function which arises from the CGC. One can compute and compare to data. There are 3 unknown parameters in this description: the hadron size, the scale x_0 and the quark mass. In addition, the parameter λ which controls the energy dependence of the saturation momentum is determined by experiment to better accuracy than it is currently known theoretically.[16]-[24] The results for the description of the data are remarkably good for $x \leq 10^{-2}$ and $Q^2 \leq 45\ GeV^2$, as shown below in Figs. 7-8[24]

One should note that this description includes both the high and low Q^2 data. Descriptions based on DGLAP evolution can describe the large Q^2 points. The CGC description is very economical in the number of parameters which are used.

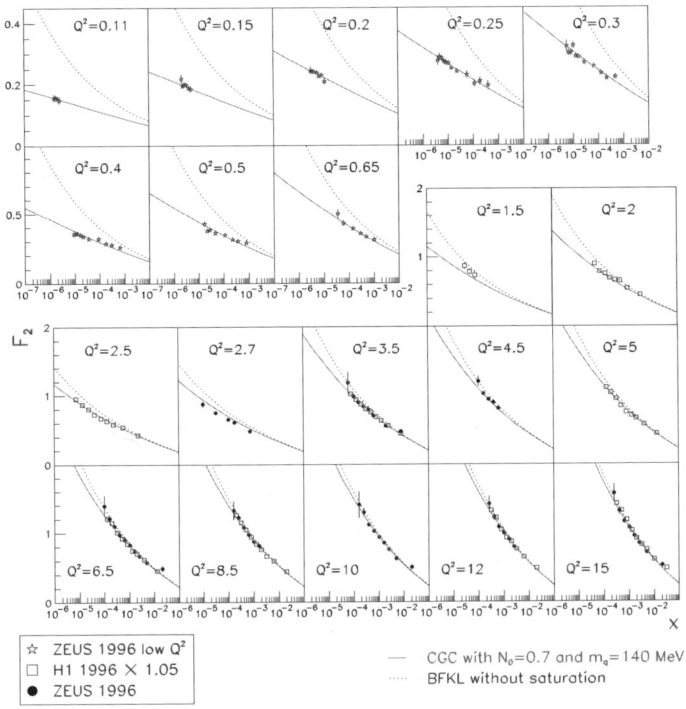

Figure 7. The CGC description of F_2.

4.3 Diffraction and Quasi-Elastic Processes

The CGC provides a description of the underlying structure of gluonic matter inside a hadron. As such, it should be sensitive to probes of the transverse extent of this matter, which can be experimentally studied in diffraction and related quasi elastic particle production.[25]

For diffraction by a virtual photon, a good theoretical first principles computation is only available for small mass states which are produced by the virtual photon.[26]-[33]. In Fig. 9, a computation of the ratio of the diffractive to total deep inelastic cross section for various Q^2 and produced masses is shown. The agreement is good for small masses, and even reasonable for large masses.

There are additional computations of quasi-elastic ρ meson production and J/Ψ production[32]-[33]. These are shown in Figs. 10-11. They agree well up to an overall normalization uncertainty for the ρ meson associated with knowledge of the ρ meson wavefunction. The various curves correspond in the figures correspond to different models for these wavefunctions.

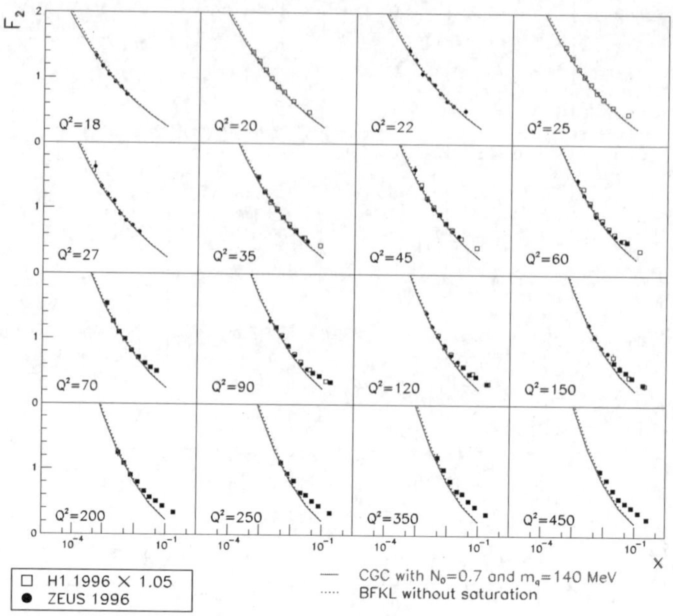

Figure 8. More of the CGC description of F_2.

4.4 Qualitative Understanding of Total Hadronic Cross Sections

The elastic and total cross section of *pp* scattering as a function of energy is shown in Fig. 12. The cross section is slowly varying as a function of energy and is believed to grow as $ln^2(E)$. To understand this behaviour, imagine that we take some probe and try to penetrate the hadron. The cross section is defined by the impact parameter at which the hadron become opaque. We expect that at fixed x, the impact parameter distribution of matter inside a hadron fall off at large b like $e^{-2m_\pi b}$. At fixed impact parameter, we expect that the number of gluons grows as $(x_0/x)^\lambda$. Setting

$$(x_0/x_{min})^\lambda \, exp\,(-2m_\pi b) \sim 1 \tag{11}$$

gives

$$\sigma \sim b^2 \sim ln^2(1/x_{min}) \sim ln^2(E) \tag{12}$$

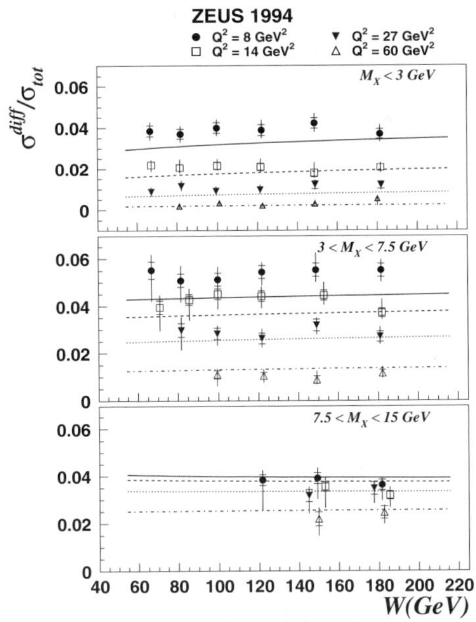

Figure 9. The ratio of diffractive to total cross sections for deep inelastic scattering.

Here x_{min} is the minimal value of x accessible for some energy E and goes as $x_{min} \sim \Lambda_{QCD}/E$.

This simple physically motivated picture give the expected $ln^2(E)$ growth of the total cross section. It has proven difficult to make these simple arguments rigorous.[34]-[35].

5. Heavy Ion Collisions

The collision of two ultrarelativistic heavy ions can be visualized as the scattering of two sheets of colored glass, as shown in Fig. 13. [36]-[39] At very early times after the collision the matter is at very high energy density and in the form of a CGC. As time goes on, the matter expands. As it expands the density of gluons decreases, and gluons begin to propagate with little interaction. At later times, the interaction strength increases and there is sufficient time for the matter to thermalize and form a Quark Gluon Plasma. This scenario is shown in Fig. 14, with realistic estimates for energy density and time scales appropriate for the RHIC heavy ion accelerator.

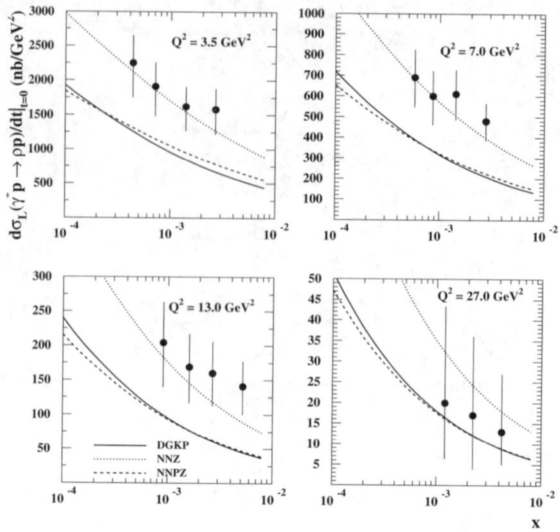

Figure 10. The CGC description of quasi-elastic ρ meson production.

5.1 The Multiplicity

The CGC allows for a direct computation of the particle multiplicity in hadronic collisions. If one naively tries to compute jet production, the total multiplicity is infrared divergent. This follows because of the $1/p_T^4$ nature of the perturbative formula for gluon production

$$\frac{1}{\pi R^2} \frac{dN}{dyd^2p_T} \sim \frac{1}{\alpha_S} \frac{Q_{sat}^4}{p_T^4} \tag{13}$$

In the CGC, when $p_T \leq Q_{sat}$, this formula is cutoff. This means that the total gluon multiplicity goes as

$$\frac{1}{\pi R^2} \frac{dN}{dy} \sim \frac{1}{\alpha_S} Q_{sat}^2 \tag{14}$$

One can compute the proportionality constant and before the RHIC data appeared, predictions were made for the gluon multiplicity. In Fig. 15, the predictions for the first RHIC run are presented. The CGC was one of the few models which got the multiplicity correct.

Also, the dependence of the multiplicity on the number of participants can also be computed, realizing that the saturation momentum should be (for not

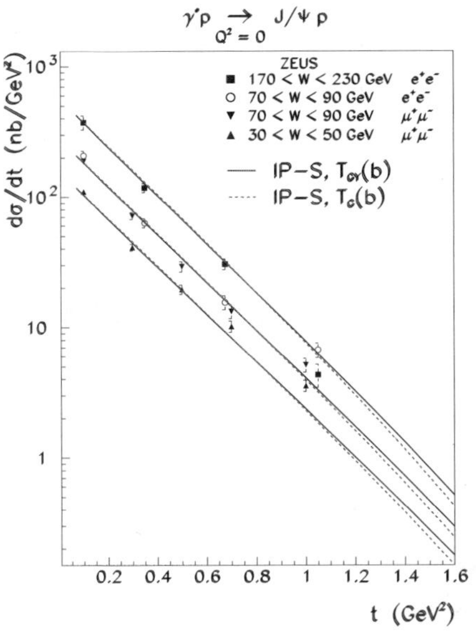

Figure 11. Quasi-elastic J/Ψ production.

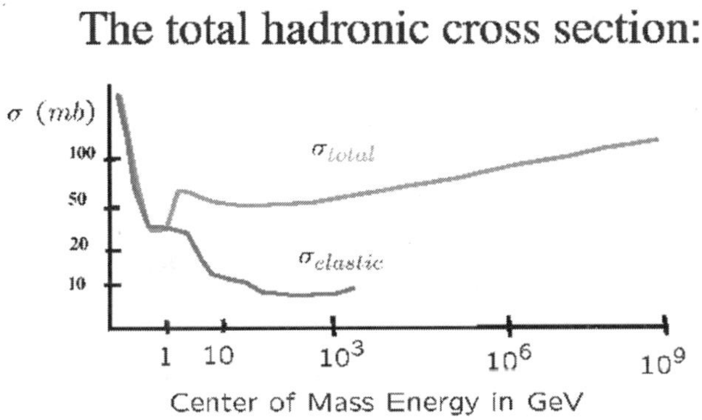

Figure 12. The total *pp* cross section.

too small x) proportional $N_{part}^{1/3}$. This leads to

$$\frac{dN}{dy} \sim \frac{1}{\alpha_S} \tag{15}$$

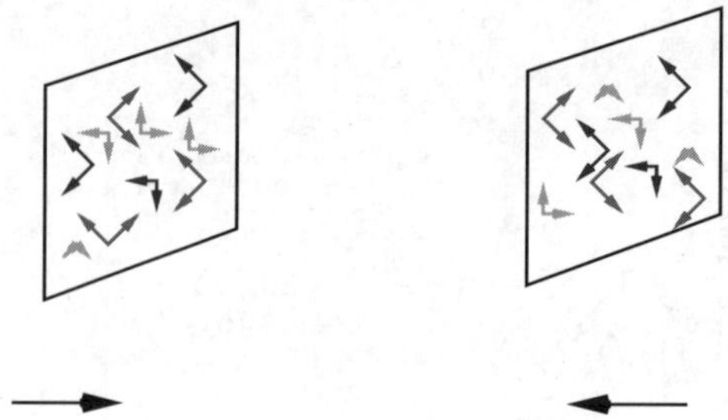

Figure 13. A collision of two ultra-relativistic nuclei.

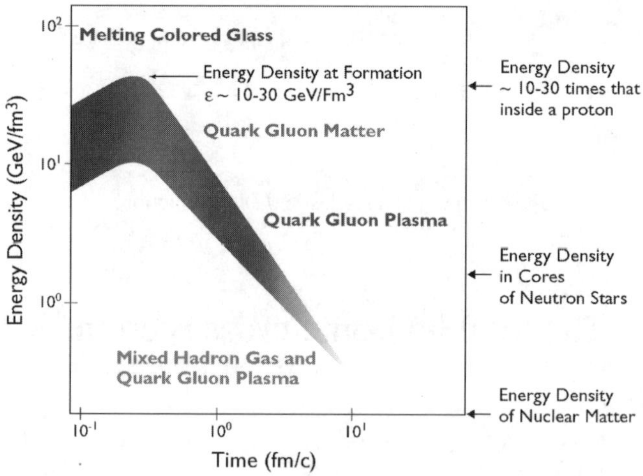

Figure 14. The evolution of a heavy ion collision at RHIC energy.

so that we have a very slow logarithmic dependence on the number of participants. This was a prediction of the CGC and it agreed with experiment, as shown in Fig. 16.[40]-[43]

One can go even further and compute the dependence of the multiplicity on rapidity and centrality, and the transverse momentum distribution of produced hadrons by using CGC initial conditions matched together with a hydrodynamic calculation which evolves the matter thought the Quark Gluon Plasma.[44] The results describe the data remarkably.[45]-[46] This hydrodynamic simulation of Nara and Hirano does remarkably well.

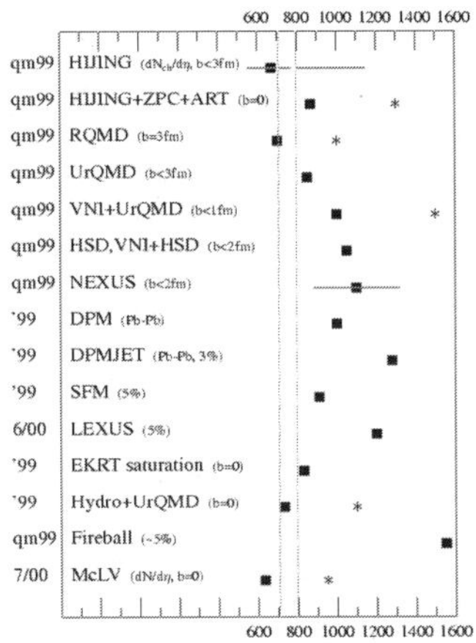

Figure 15. Predictions for the total multiplicity as measured at RHIC. The band is the experimentally measured region. The CGC prediction is marked McLV.

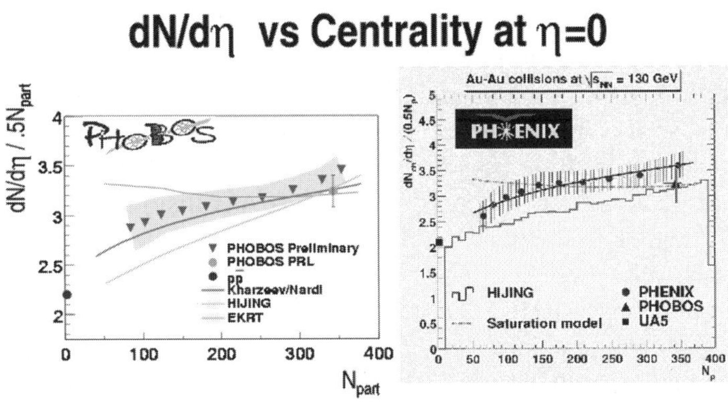

Figure 16. The total multiplicity as a function of the number of participants as measured by Phobos and Phenix.

5.2 High p_T Particles

The early results from RHIC on gold-gold collisions revealed that the high p_T production cross sections were almost an order of magnitude below that expected for jet production arising from incoherent parton-parton scattering.[47] This could be either due to initial state shadowing of the gluon distribution inside the nuclei,[48] or to final state jet quenching.[49] For centrally produced jets, the x of the parton which produces a 5-10 GeV particle is of order 10^{-1},

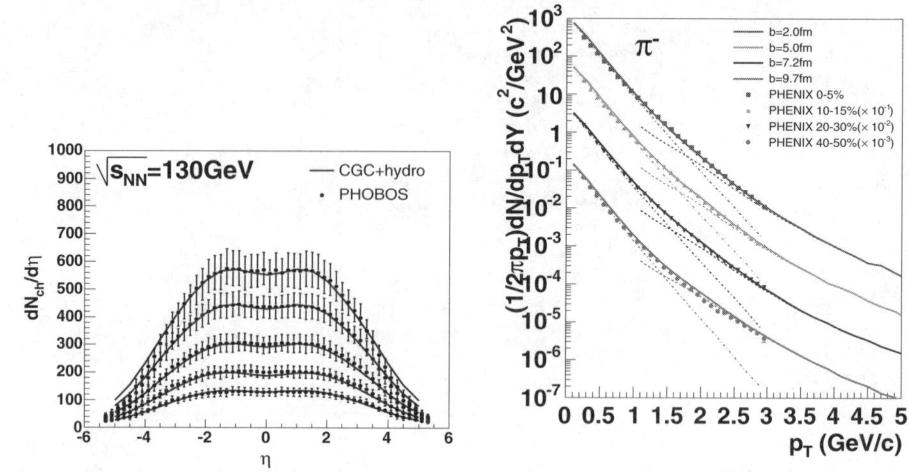

Figure 17. Results of hydrodynamic simulation with a CGC initial condition for the multiplicity as a function of y and centrality, and for the transverse momentum distribution at zero rapidity as a function of transverse momentum and centrality.

and this is outside the region where on the basis of the HERA data one expects the effects of the CGC to be important. Nevertheless, nuclei might be different than protons, so it is not a priori impossible.

The crucial test of these two different mechanisms is the comparison of dA scattering to pp. If there is suppression of jet in dA collisions, then it is an initial state effect. The experiments were performed, and all there is little initial state effect for centrally produced jets.[50] The suppression of centrally produced jets in AA collisions at RHIC is indeed due to final state interactions, that is jet quenching.

This is not in contradiction with the existence of a CGC. The particles which control the multiplicity distribution in the central region are relatively soft, and arise from $x \sim 10^{-2}$. To probe such small x degrees of freedom at high transverse momentum at RHIC requires that one go to the forward region.[51]-[52]

If one uses naive Glauber theory to compute the effects of shadowing by multiple scattering, one expects that if one goes into the forward region of the deuteron, the probe propagates through more matter in the nucleus. This is because we probe all of the gluons with x greater than the minimum x of the nucleus which can be seen by the deuteron. Going more forward makes this minimum x smaller. Now multiple scattering will produce more particles at high at some intermediate value of p_T. (At very high p_T, the effects of multiple scattering will dissappear.) This is the source of the Cronin peak and it is expected to occur at p_T of $2 - 4\ GeV$. Clearly the height of this peak should increase as one goes more forward on the side of the deuteron, and should increase with the centrality of the collision.[53] A result of such a computation is shown in Fig. 18.

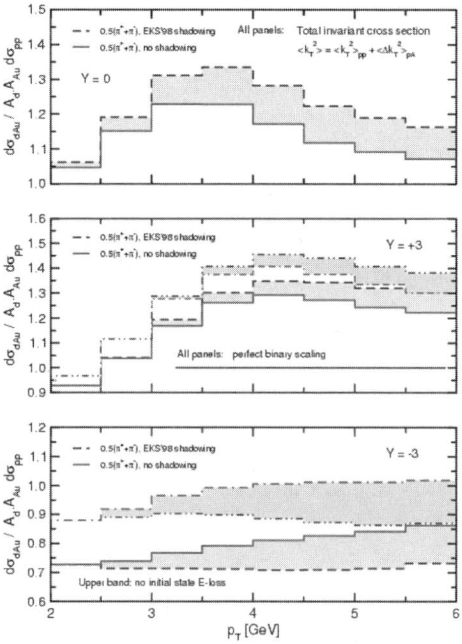

Figure 18. The expectations of classical multiple scattering for the p_T distribution in dA collisions.

Classical rescattering effects are included in the computation of the properties of the CGC. There is another effect however and that is quantum evolution generated by the renormalization group equations. It was a surprise that when one computed the evolution of the gluon distribution function including both effects, the quantum evolution dominated. This means that the height of the Cronin peak, and the overall magnitude of the gluon distribution decreased as one went from backwards to forward angles.[54]- [56] The results of one such computation are shown in Fig. 19 It was also a surprise how rapid the effect set in.

The Brahms experiment at RHIC recently presented data on the ratio of central to peripheral transverse momentum distributions.[57] The ratio R_{CP} is defined in such a way that if the processes were due to incoherent production of jets, then $R_{CP} = 1$. A value less than one indicates suppression, and a value larger than one indicates a Cronin type enhancement. The results for a variety of forward angles for R_{CP} as a function of p_T is shown in Fig. 20 a. There is clearly a decrease in R_{CP} as one goes to forward angles, in distinction

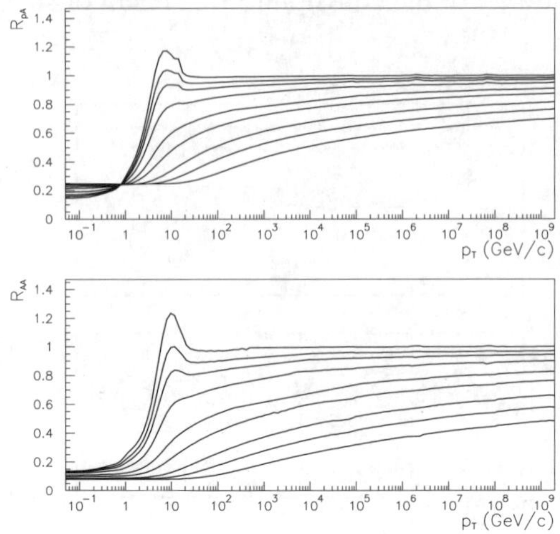

Figure 19. The gluon intrinsic gluon distribution function as a function of p_T for different pseudo-rapidities.

from the predictions of classical multiple scattering. The effect is very rapid in rapidity, as was expected from computations of the CGC. In Fig. 20 b, the ratio

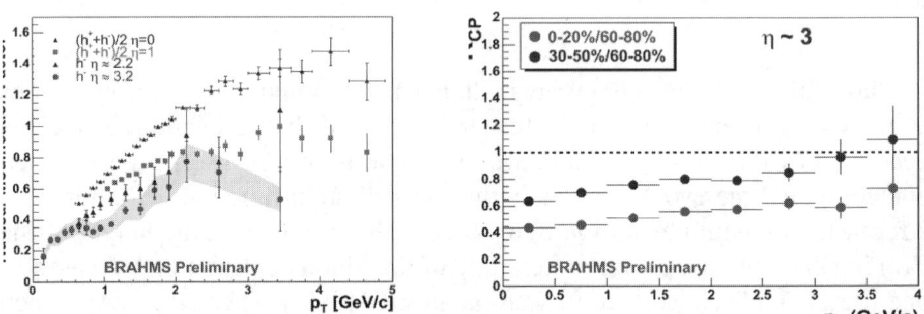

Figure 20. (a) The ratio R_{CP} as a function of p_T for various forward pseudorapidities. (b) The ratio R_{CP} at a fixed forward pseudorapidity as a function of p_T for less central and more central dA collisions

R_{CP} is shown as a function of p_T for the forward pseudorapidity $\eta \sim 3$ for less central and more central events. The ratio decreases for more central collisions, against the expectation of classical multiple scattering and consistent with the CGC hypothesis.

Similar results have been seen in the Star and Phobos experiment as shown in Fig.21 [58]-[59] Phenix has also shown very dramatically the dependence

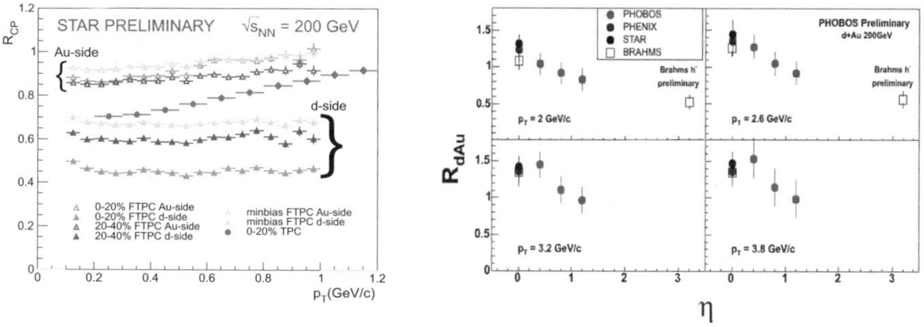

Figure 21. (a) The p_T distributions as measured in Star for dA collisions at various pseudo-rapidities. (b) The ratio R_{dAu} as a function of pseudorapidity at fixed p_T values and different centralities

upon centrality of R_{dAu} and pseudorapidity for hadron with $1\ GeV \le p_T \le 3\ GeV$,[60] as seen in Fig. 22 This data beautifully illustrates the Cronin

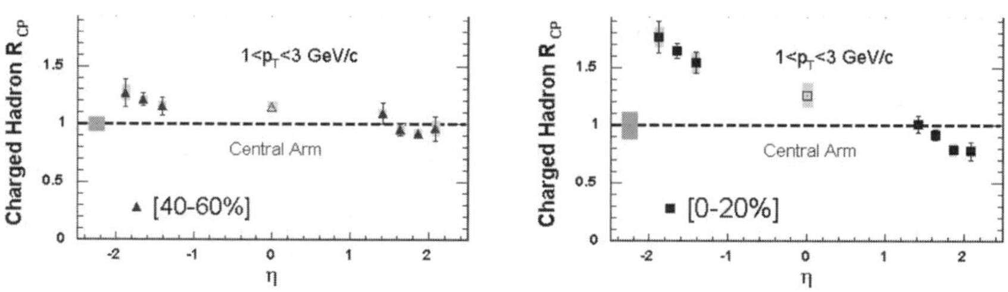

Figure 22. The pseudorapidity distribution of the ratio R_{CP} for stopped particles with $1\ GeV \le p_T \le 3\ GeV$ as a function at different centralities as measured in Phenix.

enhancement on the gold side and the depletion on the deuteron side, and as well the dependence on centrality. It appears that classical multiple scattering dominates on the gold side, and quantum evolution on the deuteron side. By a happy coincidence, these effects nearly cancel in the mid-rapidity region, making RHIC a very good machine for studying QGP effects at midrapidity for hard probes. The J/Ψ has also been measured in the Phenix experiment and shows similar dependence on centrality and pseudorapidity as does the hadron spectra[60].

5.3 Alternative Explanations

Although classical multiple scattering cannot explain the effects seen at RHIC in the forward dA experiments, is it possible that some other theory of shadowing can do it? It has not been the expectation from a number of computations which build in shadowing.[61] When any of these models are corrected to include the effects of classical multiple scattering, they fail to describe the dA data.[62]-[63] Nevertheless, someone will surely find some parameterization of shadowing based on somewhat shadowy assumptions and modeling which when combined with classical multiple scattering will describe the data. The question which must be asked is how robust is the underlying model which predicts these distributions. The CGC provides a robust theoretical framework based on first principles from QCD in which such shadowing and a variety of other effects are predicted. The CGC also explains a variety of other phenomena, as we have seen.

One of the test of the CGC hypothesis will be the forward backward correlations for jets in dA collisions, where one of the jets is at forward pseudorapidity. If there is a CGC present, the p_T of the jet will be broadened on a scale of order of the saturation momentum. At present, this measurement has not been reported[64], nor have there been reliable theoretical computations.

6. Summary

My colleague Dima Kharzeev was quoted in the press as saying about the CGC and the recent RHIC results

- This is nothing short of a major discovery.
- it's going to trigger a real revolution in nuclear physics

By now, it should be clear why I support his position. The measurements at RHIC which support the idea that there is ultradense matter, and that at early times this matter has the properties predicted for a Color Glass Condensate is without doubt a major discovery. I think however in order to make the case for the Color Glass Condensate compelling, one needs to supplement the RHIC data with electron scattering data from HERA. Additional tests will come from hard processes measured in LHC and potentially definitive tests from experiments at eRHIC.

A revolution involves at least the following:

- Revolutions involve major realignments of traditional relationships between large groups of people.
- Friends try to kill one another. Sometimes successfully.
- There are bad consequences if you try to make a revolution and fail.

The kind of debate over the CGC hypothesis is less whether the CGC can describe phenomena observed in HERA and RHIC, but whether or not there are alternative explanations. These alternative explanations often involve model computations, and do not try to unify the wide range of phenomena described in the talk. It is always very difficult to falsify a model because there is always some arbitrariness in any model. Also, models are a bit like a hydra: If you rule out one model two new ones will appear. On the other hand a theory, such as that of the CGC, must pass much more stringent tests, and if it is wrong, it must be discarded. Historically, such debates are never resolved on the basis of whether a theory or a model with free parameters can best describe data. The conclusion will almost always be that some model can. The reason that theories become accepted is because they have a simple and unifying intellectual framework, that the arguments which motivate the theory are compelling, and that it describes, within the accuracy of various approximations, a wide variety of diverse phenomena.

7. Acknowledgements

I gratefully acknowledge conversations with Edmond Iancu, Kazu Itakura, Miklos Gyulassy, Dima Kharzeev, Genya Levin, and Raju Venugopalan on the subject of this talk

This manuscript has been authorized under Contract No. DE-AC02-98H10886 with the U. S. Department of Energy.

References

[1] J. Breitweg et. al. *Eur. Phys. J.* **67**, 609 (1999).

[2] L. V. Gribov, E. M. Levin and M. G. Ryskin, *Phys. Rept.* **100**, 1 (1983).

[3] A. H. Mueller and Jian-wei Qiu, *Nucl. Phys.* **B268**, 427 (1986); J.-P. Blaizot and A. H. Mueller, *Nucl. Phys.* **B289**, 847 (1987).

[4] L.N. Lipatov, *Sov. J. Nucl. Phys.* **23** (1976), 338;
E.A. Kuraev, L.N. Lipatov and V.S. Fadin, *Sov. Phys. JETP* **45** (1977), 199;
Ya.Ya. Balitsky and L.N. Lipatov, *Sov. J. Nucl. Phys.* **28** (1978), 822.

[5] L. D. McLerran and R. Venugopalan, *Phys. Rev.* **D49**, 2233(1994); 3352 (1994); **D50**, 2225 (1994).

[6] E. Iancu, A. Leonidov and L. D. McLerran, *Nucl. Phys.* **A692**, 583 (2001); E. Ferreiro E. Iancu, A. Leonidov and L. D. McLerran, *Nucl. Phys.* bf A710,373 (2002).

[7] E. Iancu and L. McLerran, *Phys.Lett.* **B510**, 145 (2001).

[8] J. Jalilian-Marian, A. Kovner, L. McLerran and H. Weigert, *Phys. Rev.* **D55** (1997), 5414.

[9] J. Jalilian-Marian, A. Kovner, A. Leonidov and H. Weigert, *Nucl. Phys.* **B504** (1997), 415; *Phys. Rev.* **D59** (1999), 014014.

[10] B. Back et. al. *Phys. Rev. Lett.* **91**, 052303 (2003).

[11] J. Jalilian-Marian, nucl-th/0212018

[12] I. Balitsky, *Nucl. Phys.* **B463** (1996), 99.

[13] Yu. V. Kovchegov, *Phys. Rev.* **D60** (1999), 034008; *ibid.* **D61** (2000), 074018.

[14] A. H. Mueller, *Phys.Lett.* **B523**, 243 (2001)

[15] V.N. Gribov and L.N. Lipatov, *Sov. Journ. Nucl. Phys.* **15** (1972), 438; G. Altarelli and G. Parisi, *Nucl. Phys.* **B126** (1977), 298; Yu. L. Dokshitzer, *Sov. Phys. JETP* **46** (1977), 641.

[16] E. Levin and K. Tuchin, *Nucl. Phys.* **A691**, 779 (2001)

[17] E. Iancu, K. Itakura and L. McLerran, Nucl. Phys. A **708**, 327 (2002).

[18] A. H. Mueller and V. N. Triantafyllopoulos, *Nucl.Phys.* **B640**, 331 (2002). D. N. Triantafyllopoulos, Nucl. Phys. B **648**, 293 (2003).

[19] A. H. Mueller, *Nucl. Phys.* **B415** (1994), 373; *ibid.* **B437** (1995), 107; A. H. Mueller and B. Patel, Nucl. Phys. B **425**, 471 (1994).

[20] Y. Kovchegov and A. H. Mueller, *Nucl. Phys.* **B259**, 451 (1998).

[21] D. Schildknecht and Bernd Surrow, *Phys. Lett.* **B499**, 116 (2001)

[22] A. M. Stasto, K. Golec-Biernat and J. Kwiecinski, *Phys. Rev. Lett.*, **86**, 596 (2001).

[23] E. Gotsman, E. Levin, M. Lublinsky and U. Maor, *Eur. Phys. J.* **C27**, 411 (2003).

[24] E. Iancu, K. Itakura, and S. Munier, hep-ph/0310338

[25] L. Frankfurt, V. Guzey, M. McDermott, *Phys.Rev.Lett.* **87**, 192301 (2001) L. Frankfurt, M. Strikman and M. Zhalov, *Phys.Lett.* **B537**, 51 (2002); T. Rogers, V. Guzey, M. Strikman and Z. Xu, hep-ph/0309099.

[26] K. Golec-Biernat and M. Wüsthoff, *Phys. Rev.* **D59** (1999), 014017; *ibid.* **D60** (1999), 114023; *Eur. Phys. J.* **C20** (2001) 313.

[27] W. Buchmuller and A. Hebecker, *Nucl.Phys.* **B476**, 203 (1996). W. Buchmuller, T. Gehrmann and A. Hebecker, *Nucl.Phys.* **B537**, 477 (1999)

[28] Y. Kovchegov and L. McLerran, *Phys.Rev.* **D60**, 054025 (1999).

[29] J. Bartels, K. Golec-Biernat and H. Kowalski, *Phys.Rev.* **D66**, 014001 (2002);

[30] J. Bartels, E. Gotsman, E. Levin, M. Lublinsky and U. Maor, *Phys. Lett.* **B556**, 114 (2003)

[31] E. Levin and M. Lublinsky, *Eur.Phys.* **J.C22**, 647 (2002); *Nucl. Phys.* **A712**, 95 (2002); *Phys. Lett.* **B521** 233 (2001)

[32] A. H. Mueller, S. Munier and A. Stasto *Nucl.Phys.* **B603**, 427 (2001)

[33] H. Kowalski and D. Teaney, *Phys. Rev.* **D68**,114005 (2003).

[34] E. Ferreiro, E. Iancu, K. Itakura, and L. McLerran, *Nucl. Phys.* **A710**, 373 (2002).

[35] A. Kovner and U. Wiedemann, *Phys. Lett.* B551 331, (2003)

[36] A. Kovner, L. McLerran and H. Weigert, Phys. Rev **D52** 3809 (1995); **D52** 6231 (1995).

[37] A. Krasnitz and R. Venugopalan, Nucl. Phys. **B557** 237 (1999); *Phys. Rev. Lett.* **84** (2000), 4309; Phys. Rev. Lett. **86** (2001).

[38] T. Lappi, *Phys. Rev.* **C67**, 054903 (2003).

[39] A. Krasnitz, Y. Nara, and R. Venugopalan, *Phys. Rev. Lett.* **87** (2001) 192302.

[40] K. Eskola, K. Kajantie and K. Tuominen, *Phys. Lett.* **B497**, 39 (2001).

[41] D. Kharzeev and M. Nardi, *Phys. Lett.* **B507**, 121 (2001); D. Kharzeev, E. Levin, *Phys. Lett.* **B523**, 79 (2001).

[42] Phys. Rev. C65, 31901R (2002).

[43] K. Adcox et. al. *Phys. Rev. Lett.* **86**, 3500 (2001)

[44] T. Hirano and Y. Nara, nucl-th/0307015.

[45] I. Beardon et. al. *Phys. Rev. Lett.* **88**, 202301(2002); *Phys. Lett.* **B523**, 227 (2001); B. Back et. al. *Phys. Rev. Lett.* **87**, 102303 (2001).

[46] K. Adcox et. al. *Phys. Rev. Lett.* **88**

[47] B. Back. et. al. *Phys. Lett.* **B578**, 297 (2004); C. Adler et. al. *Phys. Rev. Lett.* **89** 202301, (2002).

[48] D. Kharzeev, E. Levin and L. McLerran, *Phys. Lett.* **B561**, 93 (2003)

[49] J. D. Bjorken, Fermilab-Pub-82-059-THY; D. Appell, *Phys. Rev.* **D33**, 717 (1986);M. Gyulassy and M. Plumer, *Phys. Lett.* **B243**, 432 (1990); M. Gyulassy, M. Plumer, M. Thoma and X. N. Wang, *Nucl. Phys.* **A538**, 37C (1992);

[50] I. Arsene et. al. *Phys. Rev. Lett.* **91**, 072305 (2003); B. Back et. al. *Phys. Rev. Lett.* **91**, 072302 (2003); S.S. Adler et al., *Phys. Rev. Lett.* **91** , 072303 (2003); J. Adams, *Phys. Rev. Lett.* **91**, 072304 (2003); M. Gyulassy, I. Vitev and X. N. Wang, *Phys. Rev. Lett.* **86**, 2537 (2001)

[51] A. Dumitru and J. Jalilian-Marian, *Phys.Lett.* **B547**, 15 (2002); *Phys.Rev.Lett.***89**, 022301(2002)

[52] F. Gelis and J. Jalilian-Marian, *Phys.Rev.* **D67**, 074019 (2003).

[53] I. Vitev and M. Gyulassy, *Phys. Rev. Lett.* **89**, 252 (2002); M. Gyulassy, P. Levai and I. Vitev, *Phys. Rev.* **D66**, 014005 (2002); *Nucl. Phys.* **B594**, 371 (2001); *Nucl. Phys.* **B571**, 197 (2000) I. Vitev, *Phys. Lett.* **B562**, 36 (2003).

[54] D. Kharzeev, Y. Kovchegov and K. Tuchin, *Phys.Rev.* **D68**, 094013 (2003).

[55] R. Baier, A. Kovner and U. Wiedemann, *Phys.Rev.* **D68**, 054009 (2003).

[56] J. Albacete, N. Armesto, A. Kovner, C. Salgado and U. Wiedemann, hep-ph/0307179

[57] I. Aresene et. al. nucl-ex/0401025;. Talks presented at Quark Matter 2004: Rachid Noucier for the Phobos Collaboration;

[58] Zhangbu Xu for the Star Collaboration at Quark Matter 2004.

[59] Rachid Noucier for the Phobos Collaboration;

[60] Talks presented at Quark Matter 2004: R. G de Cassagnac for the Phenix Collaboration at Quark Matter 2004, M. X. Liu for the Phenix Collaboration;

[61] M. Gyulassy, Plenary session talk at Quark Matter 2004, Oakland, Ca. Jan. 2004;

[62] A. Accardi, Parallel session talk at Quark Matter 2004, Oakland, Ca. Jan. 2004;

[63] G. Bernafoldy, Parallel session talk at Quark Matter 2004, Oakland, Ca. Jan. 2004;

[64] L. Bland, private communication.

CLUSTERS OF MATTER AND ANTIMATTER

I.N. Mishustin
Kurchatov Institute, Russian Research Center, Moscow 123182, Russia
Niels Bohr Institute, University of Copenhagen, DK-2100 Copenhagen, Denmark
Institut für Theoretische Physik, Goethe Universität, D–60054 Frankfurt am Main, Germany
mish@mbslab.kiae.ru

Abstract In the Introduction I present the Dirac picture of fermionic vacuum and show how it is modified in nuclear environment. The main part of the lecture is devoted to the possibility of producing new types of nuclear systems by implanting an antibaryon into ordinary nuclei. The structure of nuclei containing one antiproton or antilambda is investigated within the framework of a relativistic mean-field model. Self-consistent calculations predict an enhanced binding and considerable compression in such systems as compared with normal nuclei. Life times of such nuclei with respect to the antibaryon annihilation are estimated on the basis of a kinetic approach. It is demonstrated that due to significant reduction of the reaction Q-value, in-medium annihilation rates should be strongly suppressed. This may lead to the life times of bound antibaryon-nuclear systems which are long enough for their observation. Possible experimental signatures of such unusual states are discussed.

1. Introduction

At present significant ampunts of antimatter can be produced in laboratory by colliding heavy nuclei with highest energy. In this way not only elementary antiparticles but also light antinuclei like $\overline{\mathrm{d}}$, $\overline{\mathrm{t}}$, $\overline{\mathrm{He}}$ can be formed. One may speculate that more exotic clusters of baryons and antibaryons, like $\overline{B}+B$, $\overline{B}+d$, $\overline{B}+\alpha$ or even $\overline{d}+d$, can also be formed in such reactions. On the other hand, more heavy matter-antimatter clusters like $\overline{B}+A$ where A stands for normal nuclei (^{4}He, ^{16}O, ^{40}Ca,...) can be produced by using high-energy antiproton beams. If their life times are long enough, they will show up themselves by the delayed annihilation. Below I discuss theoretical predictions on properties of such clusters and their possible signatures. But first I present general framework for unified description of baryons and antibaryons.

Now it is generally accepted that physical vacuum has nontrivial structure. This conclusion was first made by Paul Dirac on the basis of his famous equation for a fermion field which described simultaneously particles and antiparticles.

W. Greiner et al. (eds.), Structure and Dynamics of Elementary Matter, 207–224.

The Dirac equation in the vacuum has a simple form

$$(i\gamma^\mu \partial_\mu - m)\Psi(x) = 0\,, \tag{1}$$

where $\gamma^\mu = (\gamma^0, \boldsymbol{\gamma})$ are Dirac matrices, m is the fermion mass and $\Psi(x)$ is a 4-component spinor field. For a plane wave solution $\Psi(x) = e^{-ipx}u_p$ this equation is written as

$$(\widehat{p} - m)u_p = 0\,, \tag{2}$$

where $\widehat{p} = \gamma^0 E - \boldsymbol{\gamma}\mathbf{p}$. Multiplying by $(\widehat{p} + m)$ and requiring that $u_p \neq 0$ one obtains the equation $E^2 - \mathbf{p}^2 - m^2 = 0$ which has two solutions

$$E^\pm(\mathbf{p}) = \pm\sqrt{\mathbf{p}^2 + m^2}\,. \tag{3}$$

Here the $+$ sign corresponds to particles with positive energy $E_N(\mathbf{p}) = E^+(\mathbf{p})$, while the $-$ sign corresponds to solutions with negative energy. To ensure stability of the physical vacuum Dirac has assumed that these negative-energy states are occupied, forming what is called now the Dirac sea. Then the second solution of Eq. (3) receives natural interpretation: it describes holes in the Dirac sea. These holes are identified with antiparticles. Their energies are obviously given by $E_{\overline{N}}(\mathbf{p}) = -E^-(-\mathbf{p}) = \sqrt{\mathbf{p}^2 + m^2}$. Unfortunately, the Dirac sea brings divergent contributions to physical quantities such as energy density, and one should introduce a proper regularization scheme to rid off these divergences. This picture has received numerous confirmations in quantum electrodynamics and other fields.

2. Relativistic description of nuclear matter

It has been noticed already many years ago (see e.g. ref. [1]) that nuclear physics may provide a unique laboratory for investigating the Dirac picture of vacuum. The basis for this is given by relativistic mean-field models which are widely used now for describing nuclear matter and finite nuclei. Within this approach nucleons are described by the Dirac equation coupled to scalar and vector meson fields. Scalar S and vector V potentials generated by these fields modify plane-wave solutions of the Dirac equation as follows

$$E^\pm(\mathbf{p}) = V \pm \sqrt{\mathbf{p}^2 + (m - S)^2}\,. \tag{4}$$

Again, the $+$ sign corresponds to nucleons with positive energy $E_N(\mathbf{p}) = E^+(\mathbf{p}) = V + \sqrt{\mathbf{p}^2 + (m - S)^2}$, and the $-$ sign corresponds to antinucleons with energy $E_{\overline{N}}(\mathbf{p}) = -E^-(-\mathbf{p}) = -V + \sqrt{\mathbf{p}^2 + (m - S)^2}$. It is remarkable that changing sign of the vector potential for antinucleons is exactly what is expected from the G-parity transformation of the nucleon potential. As follows from Eq. (4), the spectrum of single-particle states of the Dirac equation in nuclear environment is modified in two ways. First, the mass gap between

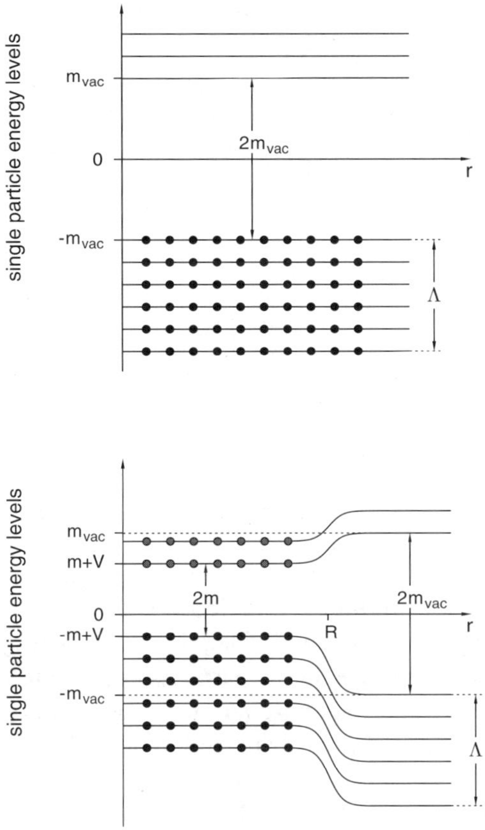

Figure 1. Schematic spectrum of Dirac equation in vacuum (upper panel) and in a nucleus of radius R (lower panel). A divergent contribution of negative-energy states is often regularized by introducing a cut-off momentum Λ

positive- and negative-energy states, $2(m - S)$, is reduced due to the scalar potential and second, all states are shifted upwards due to the vector potential. These changes are illustrated in Fig. 1.

It is well known from nuclear phenomenology that good description of nuclear ground state is achieved with $S \simeq 350$ MeV and $V \simeq 300$ MeV so that the net potential for nucleons is $V - S \simeq -50$ MeV. Using the same values one obtains for antinucleons a very deep potential, $-V - S \simeq -650$ MeV. Such a potential would produce many strongly bound states in the Dirac sea. However, if these states are occupied they are hidden from the direct observation. Only creating a hole in this sea, i.e. inserting a real antibaryon into the nucleus, would produce an observable effect. If this picture is correct one should expect the existence of strongly bound states of antinucleons, or antibaryons in general, with nuclei.

The structure of such systems can be investigated within the framework of the Relativistic Mean Field (RMF) model [2, 3]. The general Lagrangian of the RMF model is written as

$$\begin{aligned}\mathcal{L} = &\sum_{j=N,\overline{B}} \overline{\psi}_j \left(i\gamma^\mu\partial_\mu - m_j\right)\psi_j \\ &+ \frac{1}{2}\partial^\mu\sigma\partial_\mu\sigma - \frac{1}{2}m_\sigma^2\sigma^2 - \frac{b}{3}\sigma^3 - \frac{c}{4}\sigma^4 \\ &- \frac{1}{4}\omega^{\mu\nu}\omega_{\mu\nu} + \frac{1}{2}m_\omega^2\omega^\mu\omega_\mu + \frac{d}{4}\left(\omega^\mu\omega_\mu\right)^2 \\ &- \frac{1}{4}\vec{\rho}^{\mu\nu}\vec{\rho}_{\mu\nu} + \frac{1}{2}m_\rho^2\vec{\rho}^\mu\vec{\rho}_\mu \\ &+ \sum_{j=N,\overline{B}} \overline{\psi}_j \left(g_{\sigma j}\sigma + g_{\omega j}\omega^\mu\gamma_\mu + g_{\rho j}\vec{\rho}^\mu\gamma_\mu\vec{\tau}_j\right)\psi_j \end{aligned} \tag{5}$$

plus Coulomb part. Here summation includes valence nucleons ($j = N$), i. e. the nucleons forming a nucleus, and valence antibaryons ($j = \overline{B}$) inserted in the nucleus. They are treated as Dirac particles coupled to the scalar-isoscalar (σ), vector-isoscalar (ω) and vector-isovector ($\vec{\rho}$) meson fields. The calculations are carried out within the mean-field approximation where the meson fields are replaced by their expectation values. Also a "no-sea" approximation is used. This implies that all occupied states of the Dirac sea are "integrated out" so that they do not appear explicitly. It is assumed that their effect is taken into account by nonlinear terms in the meson Lagrangian.

Below we study static finite systems consisting of A nucleons and one or more antibaryons $\overline{B} = \overline{N}, \overline{\Lambda}$,... Corresponding single particle energies ϵ_j^α and wave functions ψ_j^α are found from the Dirac equation

$$\epsilon_j^\alpha\psi_j^\alpha = \left(-i\gamma\cdot\nabla + m_j - S_j + \gamma^0 V_j\right)\psi_j^\alpha\,. \tag{6}$$

The scalar and vector potentials acting on (anti)baryons are defined as

$$S_j = g_{\sigma j}\sigma\,,\ V_j = g_{\omega j}\omega^0 + g_{\rho j}\rho_3^0\tau_3 + e_j A^0\frac{1+\tau_3}{2}\,. \tag{7}$$

Equations of motion for the meson fields read

$$(-\Delta + m_\sigma^2 + b\sigma + c\sigma^2)\sigma \ = \ \sum_j g_{\sigma j}\rho_{Sj}\,, \tag{8}$$

$$(-\Delta + m_\omega^2 + d\omega_0^2)\omega_0 \ = \ \sum_j g_{\omega j}\rho_j\,, \tag{9}$$

$$(-\Delta + m_\rho^2)\,\rho_3^0 \;=\; \sum_j g_{\rho j}\rho_{Ij}\,, \tag{10}$$

$$\Delta A_0 \;=\; \sum_j e_j \rho_{Qj} \tag{11}$$

The scalar, vector, isovector and charge densities are defined as

$$\rho_{Sj} = \langle \overline{\psi}_j \psi_j \rangle\,, \tag{12}$$

$$\rho_j = \langle \psi_j^\dagger \psi_j \rangle\,, \tag{13}$$

$$\rho_{Ij} = \langle \psi_j^\dagger \tau_3 \psi_j \rangle\,, \tag{14}$$

$$\rho_{Qj} = \frac{1}{2}\langle \psi_j^\dagger (1+\tau_3)\psi_j \rangle\,, \tag{15}$$

where angular brackets mean averaging over the ground state wave function.

To investigate sensitivity of the results to the model parameters, we have considered several versions of the RMF model, namely the NL3 [4], NLZ2 [5] and TM1 [6] models. The corresponding parameter sets were found by fitting binding energies and radii of spherical nuclei from ^{16}O (not included in the TM1 fit) to Pb isotopes. Most calculations are done with antibaryon coupling constants which are given by the G-parity transformation ($g_{\sigma\overline{N}} = g_{\sigma N}$, $g_{\omega\overline{N}} = -g_{\omega N}$) and $SU(3)$ flavor symmetry ($g_{\sigma\overline{\Lambda}} = \frac{2}{3}g_{\sigma\overline{N}}$, $g_{\omega\overline{\Lambda}} = \frac{2}{3}g_{\omega\overline{N}}$).

3. Nuclear systems containing antibaryons

Now I report on our recent study of antibaryon-doped nuclear systems [7]. Unlike some previous works, we take into account the rearrangement of nuclear structure due to the presence of a real antibaryon. Following the procedure suggested in Ref. [8] and assuming the axial symmetry of the nuclear system, we solve effective Schrödinger equations for nucleons and an antibaryon together with differential equations for mean meson and Coulomb fields. We explicitly take into account the antibaryon contributions to the source terms for these fields in the r.h.s. of Eqs. (8)-(11). It is important that antibaryons give a negative contribution to the vector density, while a positive contribution to the scalar density. This leads to increased attraction and decreased repulsion for surrounding nucleons. To maximize attraction, nucleons move to the center of the nucleus, where the antibaryon has its largest occupation probability. This gives rise to a strong local compression of the nucleus and leads to a dramatic rearrangement of its structure.

These features are illustrated by several examples presented in Figs. 2-3. Figure 2 shows single-particle states in upper and lower wells for ^{4}He and ^{16}O nuclei with and without an antiproton. One can clearly see the rearrangement of nuclear structure due to the presence of a hole in the Dirac sea. It is interesting

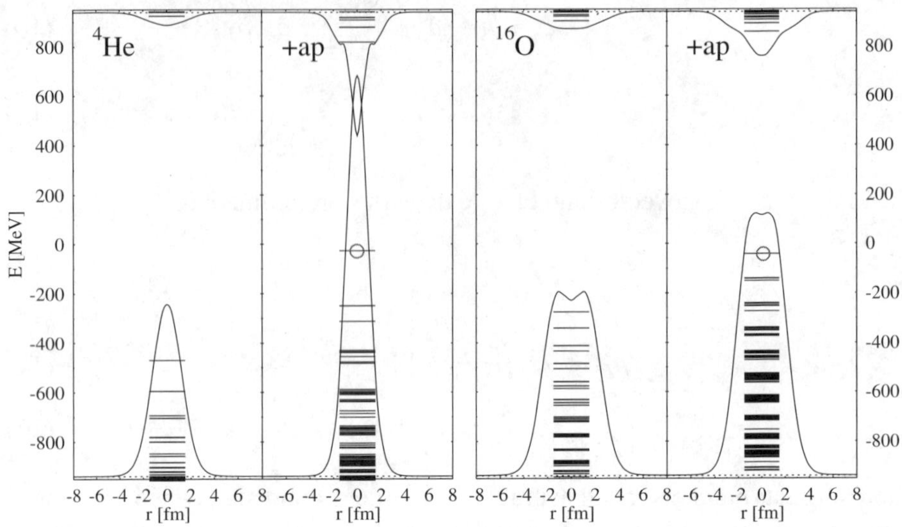

Figure 2. Single-particle levels in the upper and lower wells for ^{4}He and ^{16}O nuclei calculated with the parametrization NL-Z2. Left figures show ordinary nuclei with all levels in the lower well occupied. Right figures show the same nuclei but containing one antiproton (labelled as "+ap") i. e. one hole in the lower well.

to note that the highest hole level (lowest antiproton level) moves to about zero energy. This means that the whole antiproton rest mass is eaten up by the attractive potential.

Figure 3 shows 3d plots of nucleon density distributions for ^{16}O nucleus with and without an antibaryon ($\overline{p}$, $\overline{\Lambda}$) in a lowest energy state. The calculations show that inserting an antiproton into the ^{16}O nucleus leads to the increase of central nucleon density by a factor 2–4 depending on the RMF parametrization. Due to a very deep antiproton potential the binding energy of the whole system is increased significantly as compared with 130 MeV for normal ^{16}O. The calculated binding energies of the $\overline{p}$+^{16}O system are 830, 1050 and 1160 MeV for the NL–Z2, NL3 and TM1, respectively. Due to this anomalous binding we call such systems super bound nuclei (SBN). In the case of antilambdas we rescale the coupling constants with a factor 2/3 that leads to the binding energy of 560÷700 MeV for the $\overline{\Lambda}$+^{16}O system.

As second example, we investigate the effect of a single antiproton inserted into the ^{8}Be nucleus. The normal ^{8}Be nucleus is not spherical, exhibiting a clearly visible 2α structure with the ground state deformation $\beta_2 \simeq 1.20$. As seen in Fig. 3, inserting an antiproton in ^{8}Be results in a much less elongated

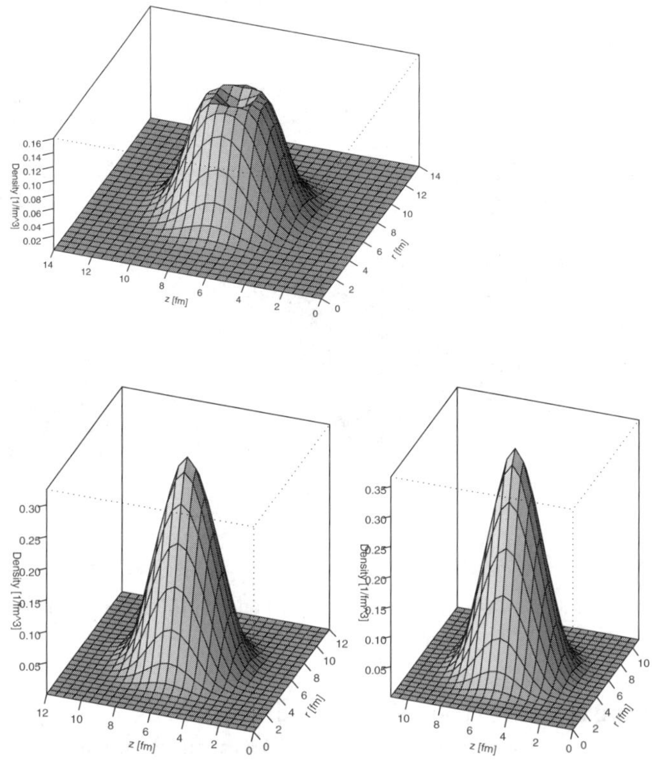

Figure 3. Sum of neutron and proton densities for ^{16}O (top), ^{16}O with $\overline{p}$ (bottom right) and ^{16}O with $\overline{\Lambda}$ (bottom left) calculated with the parametrization NL-Z2.

shape ($\beta_2 \simeq 0.23$) and disappearance of its cluster structure. The binding energy increases from 53 MeV to about 700 MeV. Similar, but weaker effects have been predicted [9] for the K^- bound state in the ^{8}Be nucleus.

The calculations have been performed also with reduced antinucleon coupling constants as compared to the G-parity prescription. We have found that the main conclusions about enhanced binding and considerable compression of $\overline{p}$-doped nuclei remain valid even when these coupling constants are reduced by factor 3 or so.

4. Antibaryon annihilation in nuclear medium

The crucial question concerning possible observation of the bound antibaryon-nuclear systems is their life time. The main decay channel for such states is the annihilation of antibaryons on surrounding nucleons. We have performed detailed calculations within the framework of kinetic theory. Let us first consider annihilation of slow antibaryons in homogeneous nucleonic matter with a

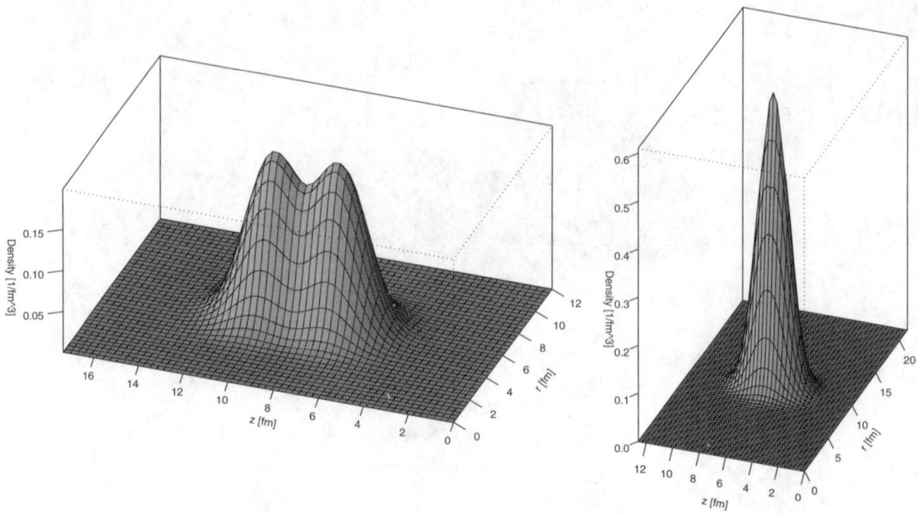

Figure 4. Nucleon densities for ^{8}Be without (left) and with (right) an antiproton calculated with the parametrization NL3.

small admixture of antibaryons. Using again the mean–field approach we treat the $\overline{B}+N$ matter as a mixture of Fermi gases of nucleons and antibaryons interacting with mean meson fields. Assuming that annihilation on multinucleon clusters is relatively small, we consider here only the binary $\overline{B}N$ annihilation. Then the local rate of $\overline{B}$ annihilation per unit volume can be written as a sum over all $\overline{B}N$ annihilation channels with production of $n \geq 2$ mesons $M_1, ...M_n$ in the final state:

$$\frac{dN_{\text{ann}}}{dtdV} = \sum_n \frac{dN_{\overline{B}N \to M_1 \dots M_n}}{dtdV} \,. \tag{16}$$

In the following we disregard in–medium modifications of these mesons, assuming that their 4–momenta k_l satisfy the vacuum mass–shell constraints, $k_l^2 = m_l^2$, with vacuum masses m_l.

In the quasiclassical approximation one can describe the phase–space density of antibaryons ($i = \overline{B}$) and nucleons ($i = N$) by distribution functions $f_i(x, \boldsymbol{p})$, where $x = (t, \boldsymbol{r})$ is the space–time coordinate and $\boldsymbol{p}$ is the 3–momentum of the i–th particle. Their vector and scalar densities are determined by integrals over 3–momenta:

$$\rho_i(x) = \int d^3p \, f_i(x, \boldsymbol{p}) \,, \tag{17}$$

$$\rho_{Si}(x) = \int d^3p \, \frac{m_i^*}{E_i} f_i(x, \boldsymbol{p}) \,, \tag{18}$$

where $E_i = \sqrt{m_i^{*2} + \boldsymbol{p}^2}$. At zero temperature we use Fermi distributions $f_i = \frac{\nu_i}{(2\pi)^3}\Theta(p_{Fi} - |\boldsymbol{p}|)$, where ν_i is the spin–isospin degeneracy factor of the i–th particles, $p_{Fi} = (6\pi^2\rho_i/\nu_i)^{1/3}$ is their Fermi momentum and $\Theta(x) \equiv (1 + \mathrm{sgn}\,x)/2$.

Within the kinetic approach (see e.g. Ref. [10]), the rate of the reaction $\overline{B}N \to c\,(c = M_1 \ldots M_n)$ can be written as

$$\frac{dN_{\overline{B}N\to c}}{dtdV} = \int \frac{d^3p_{\overline{B}}}{E_{\overline{B}}} f_{\overline{B}}(x, \boldsymbol{p}_{\overline{B}}) \int \frac{d^3p_N}{E_N} f_N(x, \boldsymbol{p}_N) \times$$
$$\int \prod_{l=1}^{n} \frac{d^3k_l}{E_l} W_c(P = p_{\overline{B}} + p_N | k_1, \ldots, k_n)\, \delta^{(4)}(P - k_1 - \ldots - k_n)\,. \quad (19)$$

Here W_c is the $\overline{B}N \to c$ transition probability, $p_i = (E_i, \boldsymbol{p}_i)$ and $k_l = (E_i, \boldsymbol{k}_l)$ are the 4–momenta of the initial particles and the final mesons.

Following standard arguments of a statistical approach we assume that the transition probabilities do not depend sensitively on the particles' momenta so that W_c can be replaced by constants,

$$W_c(P|k_1, \ldots, k_n) \simeq W_c = \mathrm{const}\,. \quad (20)$$

Then the contributions of different reaction channels are determined mainly by their invariant phase space volumes

$$R_c(s) = \int \prod_{l=1}^{n} \frac{d^3k_l}{E_l}\, \delta^{(4)}(P - \sum_{l=1}^{n} k_l)\,, \quad (21)$$

where $s = P^2 = (p_{\overline{B}} + p_N)^2$ is the c.m. energy squared available for the reaction.

Now Eq. (19) can be written as

$$\frac{dN_{\overline{B}N\to c}}{dtdV} = \rho_{\overline{B}}\rho_N W_c \left\langle \frac{R_c(s)}{E_{\overline{B}}E_N} \right\rangle \equiv \rho_{\overline{B}}\Gamma_c\,, \quad (22)$$

where Γ_c is the partial annihilation width for channel c. The angular brackets denote averaging over the momentum distribution of incoming particles,

$$\langle \mathcal{O} \rangle \equiv \int \prod_{i=\overline{B},N} \left(d^3p_i \frac{f_i(x, \boldsymbol{p}_i)}{\rho_i(x)} \right) \mathcal{O}\,. \quad (23)$$

where $\mathcal{O}$ is an arbitrary function of 3–momenta $\boldsymbol{p}_{\overline{B}}, \boldsymbol{p}_N$.

Let us consider first the limit of dilute matter, i.e. $\rho_{\overline{B}}, \rho_N \to 0$. In this case, $p_{Fi} \to 0$, $m_i^* \to m_i$ $(i = \overline{B}, N)$ and the distribution functions $f_i(x, \boldsymbol{p})$ can be formally replaced by $\rho_i\delta^{(3)}(\boldsymbol{p})$. As a consequence, at $\rho_{\overline{B}}, \rho_N \to 0$ we get the following relation:

$$\frac{dN_{\overline{B}N\to c}}{dtdV} \simeq \rho_{\overline{B}}\rho_N W_c \frac{R_c(s_0)}{m_{\overline{B}}m_N}\,. \tag{24}$$

Here $s_0 = (m_{\overline{B}} + m_N)^2$ corresponds to the energy available for $\overline{B}N$ annihilation at rest in vacuum. On the other hand, in this limit the reaction rate can be expressed [10] through the vacuum cross section of the $\overline{B}N$ annihilation at $v_{\rm rel} \to 0$:

$$\frac{dN_{\overline{B}N\to c}}{dtdV} \simeq \rho_{\overline{B}}\rho_N\,(\sigma_{\overline{B}N\to c}v_{\rm rel})_0\,, \tag{25}$$

where the subscript "0" indicates that the quantity in brackets is taken at $\boldsymbol{p}_{\overline{B},N} \to 0$. By comparing Eqs. (24) and (25) one can express the transition probabilities W_c by experimental cross sections of the $\overline{B}N$ annihilation:

$$W_c = \frac{m_{\overline{B}}m_N}{R_c(s_0)}\,(\sigma_{\overline{B}N\to c}v_{\rm rel})_0\,. \tag{26}$$

In a general case of finite densities, the partial annihilation width, defined in Eq. (22), can be rewritten as

$$\Gamma_c = \Gamma_{0c}\left\langle \lambda_c(s)\,\frac{m_{\overline{B}}m_N}{E_{\overline{B}}E_N}\right\rangle\,, \tag{27}$$

where

$$\Gamma_{0c} = \rho_N\,(\sigma_{\overline{B}N\to c}v_{\rm rel})_0\,, \tag{28}$$

is the auxiliary width, corresponding to the vacuum cross section, and

$$\lambda_c(s) = \frac{R_c(s)}{R_c(s_0)} \tag{29}$$

is the phase–space suppression factor, which plays a central role in our estimates.

In nuclear medium the single particle energies of antibaryons and nucleons are modified by the scalar and vector potentials. The minimum energy of a $\overline{B}N$ pair available for annihilation, i.e. the reaction Q value, is

$$Q = \sqrt{s_*} = m^*_{\overline{B}} + V_{\overline{B}} + m^*_N + V_N\,, \tag{30}$$

where m^*_i and V_i are effective masses and vector potentials of particles ($i = \overline{B}, N$). In the case of antinucleons with the G–parity transformed potentials $m^*_{\overline{N}} = m^*_N$, $V_{\overline{N}} = -V_N$ and hence $Q = 2m^*_N$. This value might be significantly reduced as compared to the minimal energy in vacuum, $\sqrt{s_0} = 2m_N$,

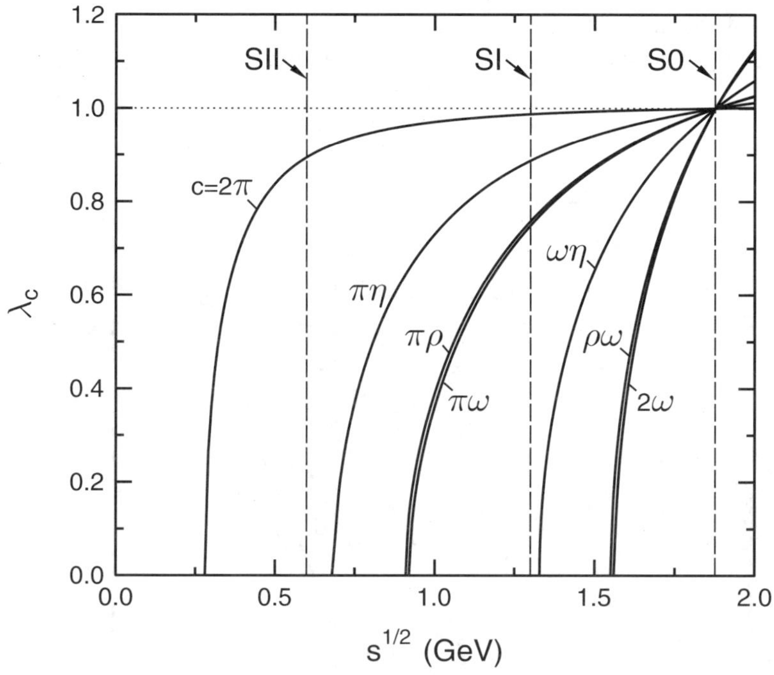

Figure 5. Phase space suppression factors for $\overline{N}N$ annihilation into two mesons as functions of c.m. energy in the subthreshold region. The vertical dashed lines marked as SI ands SII indicate $\sqrt{s_*}$ values adopted in the calculations for infinite nuclear matter with densities ρ_0 and 2ρ, respectively.

leading to strong suppression of the available phase space for annihilation products. Actual c.m. energies of $\overline{B}N$ pairs have a certain spread due to the Fermi motion of nucleons. It is easy to check that even at $\rho_N \sim 2\rho_0$ the variation of the Q values does not exceed 10%. Taking into account that uncertainties in scalar and vector potentials are of the same order, we simplify the general expression (27) by replacing $\lambda_c(s)$ with $\lambda_c(s_*)$. Then the partial annihilation width in the medium can be represented in a simple form

$$\Gamma_c \simeq \Gamma_{0c}\lambda_c(s_*)\,J_{\overline{B}N}\,. \tag{31}$$

Here we have introduced an important geometric quantity

$$J_{\overline{B}N} = \left(\frac{m_{\overline{B}}}{m^*_{\overline{B}}}\frac{\rho_{S\overline{B}}}{\rho_{\overline{B}}}\right)\cdot\left(\frac{m_N}{m^*_N}\frac{\rho_{SN}}{\rho_N}\right)\,. \tag{32}$$

which, besides λ_c, determines the annihilation rate in medium. Returning to the initial expression (22) we see that the annihilation rate is in fact proportional to the product of the scalar densities of annihilating particles. This is demanded by the Lorentz invariance of Eq. (22), because factors W_c and λ_c are Lorentz invariants.

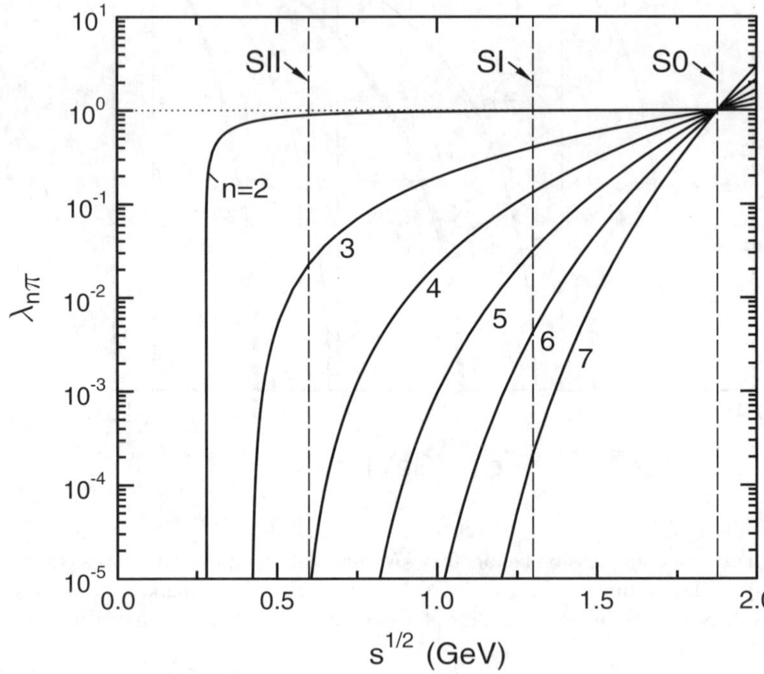

Figure 6. Same as Fig. 5, but for $\overline{N}N$ annihilation into n direct pions. Note the change to a log scale on the vertical axis.

Let us now consider in more detail the phase space suppression factors λ_c introduced by Eq. (29). In the particular case of the two–body annihilation channels λ_c can be calculated analytically. Direct calculation using Eq. (21) gives

$$R_{\overline{B}N\to M_1M_2}(s) = 2\pi\sqrt{\left[1-\frac{(m_1+m_2)^2}{s}\right]\left[1-\frac{(m_1-m_2)^2}{s}\right]} \times \Theta(\sqrt{s}-m_1-m_2)\,, \qquad (33)$$

where m_1 and m_2 are meson masses. Figure 5 presents the results for the two–body channels involving pions and the heavy mesons η, ρ, ω. One can see a strong decrease of $\lambda_c(s)$ in the subthreshold region, $s < s_0$, which is getting more prominent for heavier final states.

For channels with more than two secondary mesons, the multidimensional integrals in Eq. (21) are evaluated numerically using the Monte Carlo method. Figure 6 shows the energy dependence of factors λ_c for nonresonant channels $\overline{N}N \to (n\pi)_{\rm dir}$ with different pion multiplicities n. One can see that at given s the reduction of phase space becomes more and more important with increasing n. In the case of the lightest annihilation channel, $c = 2\pi$, the phase space is only slightly reduced in the subthreshold energy region. The noticeable deviation of $\lambda_{2\pi}$ from unity takes place only in the vicinity of the 2π threshold, i.e. at $\sqrt{s} \to 2m_\pi$. On the other hand, for $n > 2$ factors $\lambda_{n\pi}(s)$ decrease very rapidly with decreasing s. For example, at $\sqrt{s} = 1$ GeV, the phase-space suppression factors $\lambda_{n\pi}$ are $0.97,\ 0.19, 0.014$ for $n = 2, 3, 4$, respectively.

It is interesting to note that all strong annihilation channels would be closed when $m_N^*(\rho_N) < m_\pi$. In this case only electromagnetic ($c = 2\gamma$) or multinucleon annihilation channels would limit the life time of an antinucleon in nuclear medium.

To calculate the Γ_{0c} factors which enter into the in–medium annihilation widths (31) we use vacuum branching ratios, $B_c = \sigma_{\overline{B}N\to c}/\sigma_{\overline{B}N}^{\rm ann}$, which are extracted from the experimental data. In the case of antinucleons ($\overline{B} = \overline{N}$) one has

$$\Gamma_{0c} = B_c\rho_N(\sigma_{\overline{N}N}^{\rm ann}v_{\rm rel})_0 = \Gamma_0\, B_c\frac{\rho_N}{\rho_0}\ \text{MeV}\,. \tag{34}$$

The numerical value $\Gamma_0 = 104$ MeV is obtained when the parametrization of Ref. [11],

$$\sigma_{\overline{p}p}^{\rm ann} = C + \frac{D}{v_{\rm rel}}\,,\ C = 38\ \text{mb}\,,\ D = 35\text{mb}\cdot\text{c}\,, \tag{35}$$

is used for the total $\overline{N}N$ annihilation cross section.

In the end of this section we estimate the life times of bound antibaryon–nucleus systems with respect to annihilation. The minimal energy available for annihilation is now given by $\sqrt{s_*} = Q = \min(E_{\overline{B}}+E_N) = m_{\overline{B}}+m_N-B_{\overline{B}}-B_N$, where $B_{\overline{B}}$ and B_N are the binding energies of the annihilating partners. Since the antibaryon is localized in a central core of the nucleus, our former assumption of a homogeneous antibaryon density is not valid. The annihilation rates can be estimated by averaging the local expression (22) over the volume where the antibaryon wave function $\psi_{\overline{B}}(x)$ is essentially nonzero. Taking into account that the antibaryon density distribution $\rho_{\overline{B}}$ is normalized to unity, after a simple calculation we get

$$\Gamma = \sum_c \Gamma_c, \quad \Gamma_c = \Gamma_0 B_c\lambda_c(s_*)\widetilde{J}_{\overline{B}N}\,. \tag{36}$$

Here we have defined the overlap integral appropriate for localized density distributions,

$$\widetilde{J}_{\overline{B}N} = \frac{m_{\overline{B}} m_N}{\rho_0} \int dV \frac{\rho_{S\overline{B}}}{m^*_{\overline{B}}} \frac{\rho_{SN}}{m^*_N} . \tag{37}$$

In particular, for the ${}^{16}_{\bar{p}}$O system discussed above, numerical calculations give for $\widetilde{J}_{\overline{B}N}$ the values 9.5 and 43.4 for NLZ2 and NL3 parameter sets, respectively. Large variations in the numerical values of $\widetilde{J}_{\overline{N}N}$ are caused by its high sensitivity to the in–medium effective masses. From numerical calculations one can see that m^*_N becomes rather small in the central region, especially for the NL3 parameter set. This gives rise to large value of $\widetilde{J}_{\overline{N}N}$ in this case.

The estimated total annihilation widths for the ${}^{16}_{\bar{p}}$O system are in the range 7–22 MeV, depending on the RMF parametrization. The corresponding life times are 10 – 30 fm/c. Such large values seem absolutely counter–intuitive, if in–medium effects are ignored. These predictions should be considered as rough estimates based on extrapolations of nuclear matter properties into the high density region. The large margin in the life times is mainly caused by uncertainties in the overlap integral between antinucleon and nucleon scalar densities. Longer life times may be expected for nuclei containing antihyperons. The reason is that instead of pions at least one kaon must be produced in this case. We have also analyzed multi-nucleon annihilation channels (Pontecorvo-like reactions) and have found their contribution to be less than 40% of the single-nucleon annihilation.

5. Observable signatures

We believe that such exotic antibaryon-nuclear systems can be produced by using antiproton beams of multi-GeV energy, e. g. at the future GSI facility [12]. It is well known that low-energy antiprotons annihilate on the nuclear periphery (at about 5% of the normal density). Since the annihilation cross section drops significantly with energy, a high-energy antiproton can penetrate deeper into the nuclear interior. Then it can be stopped there in an inelastic collision with a nucleon, e. g. via the reaction $A(\bar{p}, N\pi)_{\bar{p}}A'$, leading to the formation of a $\bar{p}$-doped nucleus. Reactions like $A(\bar{p}, \Lambda)_{\overline{\Lambda}}A'$ can be used to produce a $\overline{\Lambda}$-doped nuclei. Fast nucleons or lambdas can be used for triggering such events. In order to be captured by a target-nucleus final antibaryons must be slow in the lab frame. Rough estimates of the formation probability in a central $\bar{p}A$ collision give the values $10^{-5} - 10^{-6}$. With the $\bar{p}$ beam luminosity of $2{\cdot}10^{32}$ cm^{-2}s^{-1} planned at GSI this will correspond to the reaction rates from tens to hundreds of desired events per second.

If bound antibaryon–nuclear systems exist and live long enough they can manifest themselves in several ways which are discussed below.

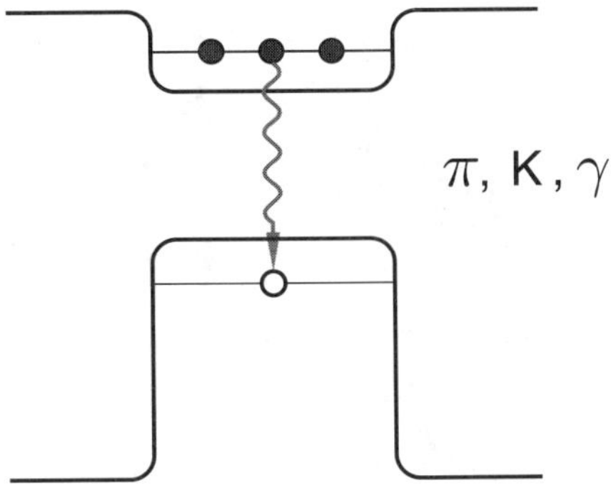

$$\Delta E = 2m_N - B_N - B_{\bar{N}} \simeq 1 \text{ GeV}$$

$$\Gamma \simeq 10\text{-}20 \text{ MeV}$$

Figure 7. One-body annihilation leading to sharp lines in the energy spectrum of emitted particles.

Transitions from atomic to nuclear states

If a slow antiproton is first captured on a Coulomb orbit forming an antiproton atom, it can later on make transition on a deep nuclear bound state. This process was first proposed in Ref. [13] where antiproton–nuclear bound states were studied within the nonrelativistic optical model. Such a transition will be accompanied by the emission of a monoenergetic photon, pion or kaon, depending on the antibaryon type. According to our calculations, many discrete states may exist in the relativistic antiproton potential. Their binding energies and thus the transition energies may vary from a few tens to several hundred MeV. The probability of such transitions is determined by the matrix element of the corresponding transition operator between the Coulomb and nuclear antiproton wave functions. As shown in Ref. [13], the partial width associated with such transitions could be as small as $10^{-4} - 10^{-5}$ of the total width of the atomic level. The total width is obviously determined by the annihilation on the nuclear surface. Therefore, a special effort should be made to find this signal in the huge background of direct annihilation events.

Super–transitions from upper to lower well

As pointed out earlier, high–energy antiprotons have a better chance to penetrate deep into the nucleus. Under certain conditions discussed earlier they can be trapped into a deep bound state. We have demonstrated above that the annihilation can be delayed in this case due to the reduction of energy released in this process. An antibaryon sitting in the nucleus can be viewed as a hole in otherwise filled Dirac sea. This state represents a strong excitation of the

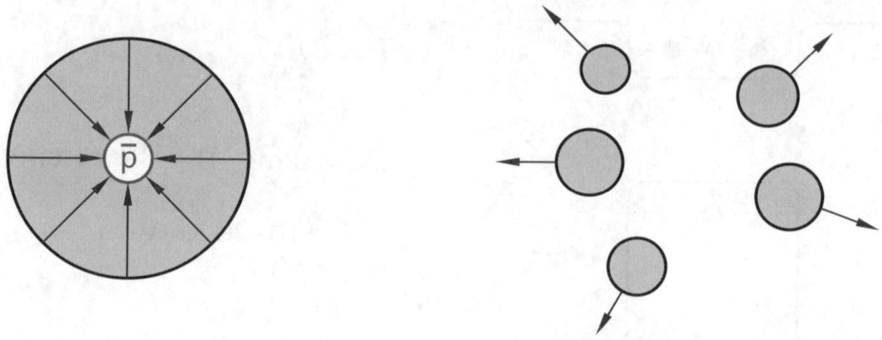

Figure 8. Left: compression of a nucleus induced by an antibaryon. Right: collective flow of nuclear fragments resulting from the expansion and disintegration of the nucleus after antibaryon annihilation.

nucleus which soon or later will decay. Among other decay modes most interesting is a *super-transition* when a nucleon from a discrete level of the Fermi sea jumps into the hole state in the Dirac sea. This can be achieved by emitting a single photon, pion or kaon with energy of about $Q = E_N + E_{\overline{B}}$ and isotropic angular distribution in the nucleus rest frame. Such process is forbidden in vacuum by the energy–momentum conservation laws. But in the considered case the recoil momentum can be carried away by the residual nucleus. This process is illustrated in Fig. 7. The appearance of relatively narrow lines with energies of about $0.5 - 1$ GeV and width $\Delta E \simeq 20 \div 50$ MeV in the spectrum of secondary photons or mesons would be a direct signature of the deeply–bound antibaryon–nuclear states.

Explosive multifragmentation

Another signal may come from explosive disintegration of the compressed nucleus after the antibaryon annihilation. When the driving force for the compression disappears, the nucleus will expand as a compressed spring. As a result of the collective expansion the nucleus will be torn apart into fragments. This process can be observed by measuring collective velocities of these fragments. This process is illustrated in Fig. 8. Multifragmentation of nuclei induced by high–energy antiprotons has been studied earlier [14], but with limited statistics. So far only minimum bias events have been analyzed. In this case the distribution of fragments follows closely predictions of the statistical multifragmentation model [15] where no collective effects are included. We emphasize again that a dedicated study with proper triggering is needed to find explosive events.

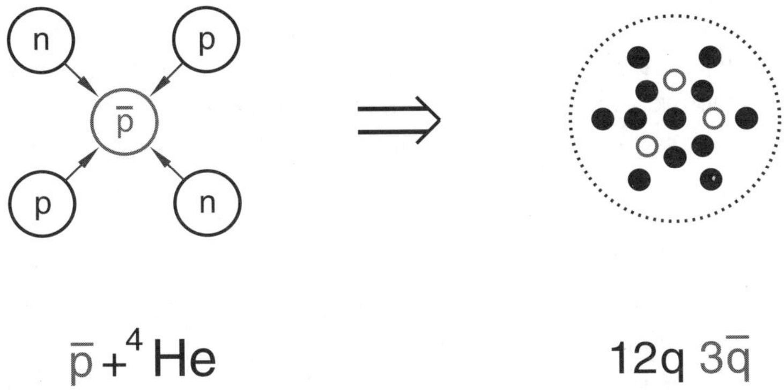

Figure 9. Formation of a multi-quark-antiquark cluster at an intermediate stage of the antibaryon annihilation in a ^{4}He nucleus.

Multi–quark–antiquark clusters

It is interesting to look at the antibaryon–nuclear systems from a somewhat different point of view. An antibaryon implanted into a normal nucleus acts as a strong attracting center for nearby nucleons. Due to uncompensated attractive force these nucleons acquire acceleration towards the center. As the result of this inward collective motion the nucleons will pile up and produce a compression zone around the antibaryon. If such a process were completely elastic, it would look like a monopole–type oscillations around the equilibrium configuration found by solving static equations (see Sect. 3). In this dynamical process even stronger compression can be reached as compared with the one predicted by this static configuration. The maximum compression is achieved when the initial potential energy generated by the antibaryon is transformed into the compressional energy. Simple estimates show that local baryon densities up to $5\rho_0$ may be obtained in this way. It is most likely that the deconfinment transition will occur at this stage and a high–density cloud containing an antibaryon and a few nucleons will appear in the form of a multi–quark–antiquark cluster. One may speculate that the whole ^{4}He or even ^{16}O nucleus can be transformed into the quark phase by this mechanism. This process is illustrated in Fig. 9.

As shown in Ref. [16], an admixture of antiquarks to cold quark matter is energetically favorable. The problem of annihilation is now transferred to the quark level. But the argument concerning the reduction of available phase space due to the entrance-channel nuclear effects should work in this case too. One may hope to produce relatively cold droplets of the quark phase by the inertial compression of nuclear matter initiated by an antibaryon.

Thus, the annihilation process in nuclear environment may serve as a breeder for creating new multi–quark–antiquark structures like the baryonium ($3q\,3\overline{q}$), deuteronium ($6q\,3\overline{q}$), tritonium ($9q\,3\overline{q}$) and even helionium ($12q\,3\overline{q}$) (see Fig. 9). Discovery of such states would significantly extend our knowledge about the quark–gluon structure of matter. Such a study will be complimentary to the active search for exotic multi–quark systems stimulated by the discovery of the pentaquark.

6. Conclusions

My most general conclusion is that the antibaryon–nuclear physics is a broad field of research which is not yet explored sufficiently well. It may bring about new interesting phenomena ranging from unusual annihilation channels to exotic antibaryon–nuclear bound states. This is a nonperturbative domain of strong interactions and therefore the progress in this field can be achieved only by close cooperation of theory and experiment.

I am grateful to T. Bürvenich, W. Greiner and L.M. Satarov for fruitful discussions and help in the preparation of this lecture. This work was supported in part by the GSI (Germany), the RFBR grant 03-02-04007 and the NSH grant 1885.2003.2 (Russia).

References

[1] N. Auerbach, A.S. Goldhaber, M.B. Johnson, L.D. Miller, and A. Picklesimer, Phys. Lett. **B182** (1986) 221.

[2] B.D. Serot and J.D. Walecka, Adv. Nucl. Phys. **16**, 1 (1985).

[3] P.G. Reinhardt,, Rep. Prog. Phys. **52**,439 (1989).

[4] G. Lalazissis, J. König, and P. Ring, Phys. Rev. **C55**, 540 (1997).

[5] M. Bender, K. Rutz, P.–G. Reinhard, J.A. Maruhn, and W. Greiner, Phys. Rev. **C60** (1999) 34304.

[6] Y. Sugahara and H. Toki, Nucl. Phys. **A579** (1994) 557.

[7] T. Bürvenich, I.N. Mishustin, L.M. Satarov, J.A. Maruhn, H. Stöcker, and W. Greiner, Phys. Lett. **B542** (2002) 261.

[8] G. Mao, H. Stöcker, and W. Greiner, Int. J. Mod. Phys. **E8** (1999) 389.

[9] Y. Akaishi and T. Yamazaki, Phys. Rev. **C65** (2002) 044005.

[10] W. Cassing, Nucl. Phys. **A700**, 618 (2002).

[11] C.B. Dover, T. Gutsche, M. Maryama, A. Faessler, Prog. Part. Nucl. Phys. **29**, 87 (1992).

[12] GSI Future Project, see http://www-new.gsi.de/zukunftsprojekt/index_e.html.

[13] C.Y. Wong, A.K. Kerman, G.R. Sachler, and A.D. Mackellar, Phys. Rev. C **29**, 574 (1984).

[14] L. Beaulieu et al., Phys. Lett. **B463**, 159 (1999).

[15] J.P. Bondorf, A.S. Botvina, A.S. Iljinov, I.N. Mishustin, K. Sneppen, Phys. Rep. **257**,133 (1995).

[16] I.N. Mishustin, L.M. Satarov, H. Stöecker, and W. Greiner, Phys. Rev. **C 59** (1999) 3343.

HADRONIZATION AT RHIC: INTERPLAY OF RECOMBINATION AND FRAGMENTATION

S. A. Bass[1,2], R.J. Fries[3] and C. Nonaka[1]

[1] *Department of Physics, Duke University, Durham, NC 27708, USA*

[2] *RIKEN BNL Research Center, Brookhaven National Laboratory, Upton, NY 11973, USA*

[3] *Department of Physics, University of Minnesota, Minneapolis, MN 55455, USA*

Abstract We discuss hadron production in heavy ion collisions at RHIC. We argue that hadrons at transverse momenta $P_T < 5$ GeV are formed by recombination of partons from the dense parton phase created in central collisions at RHIC. We provide a theoretical description of the recombination process for $P_T > 2$ GeV. Below $P_T = 2$ GeV our results smoothly match a purely statistical description. At high transverse momentum hadron production is well described in the language of perturbative QCD by the fragmentation of partons. We give numerical results for a variety of hadron spectra, ratios and nuclear suppression factors. We also discuss the anisotropic flow v_2 and give results based on a flow in the parton phase. Our results are consistent with the existence of a parton phase at RHIC hadronizing at a temperature of 175 MeV and a radial flow velocity of $0.55c$.

1. Introduction

Recent data from the Relativistic Heavy Ion Collider (RHIC) have shown a strong nuclear suppression of the pion yield at transverse momenta larger than 2 GeV/c in central Au + Au collisions, compared to $p + p$ interactions [1]. This is widely seen as the experimental confirmation of jet quenching, the phenomenon that high energy partons lose energy when they travel through the hot medium created in a heavy ion collision [2–4], entailing a suppression of intermediate and high P_T hadrons.

However, the experiments at RHIC have provided new puzzles. The amount of suppression seems to depend on the hadron species. In fact, in the production of protons and antiprotons between 2 and 4 GeV/c the suppression seems to be completely absent. Generally, pions and kaons appear to suffer from a strong energy loss while baryons and antibaryons do not. Two stunning experimental facts exemplify this [5–8]. First, the ratio of protons over positively charged

W. Greiner et al. (eds.), Structure and Dynamics of Elementary Matter, 225–237.

pions is equal or above one for $P_T > 1.5\text{GeV}/c$ and is approximately constant up to 4 GeV/c. Second, the nuclear suppression factor R_{AA} below 4 GeV/c is close to one for protons and lambdas, while it is about 0.3 for pions.

There have been recent attempts to describe the different behavior of baryons and mesons through the existence of gluon junctions [9] or alternatively through recombination as the dominant mechanism of hadronization [10, 11]. The recombination picture has attracted additional attention due to the observation that the elliptic flow pattern of different hadron species can be explained by a simple recombination mechanism [12–15]. The anisotropies v_2 for the different hadrons are compatible with a universal value of v_2 in the parton phase, related to the hadronic flow by factors of two and three depending on the number of valence quarks [16].

The competition between recombination and fragmentation delays the onset of the perturbative/fragmentation regime to relatively high transverse momentum of 4–6 GeV/c, depending on the hadron species. This is the explanation for several key observations at RHIC:

- the two component form of hadron transverse momentum spectra, including an exponential part and a power law tail with a transition between 4 and 6 GeV/c.
- the very different behavior of the nuclear suppression factors R_{AA} of mesons and baryons.
- the particle dependence of the elliptic flow.
- the unusually large baryon/meson ratios.

2. Formalism

In the fragmentation process a parton with transverse momentum p_T is leaving the interaction zone while still being connected with other partons by a color string. The breaking of the string creates quark antiquark pairs which finally turn into hadrons. The distribution of one of these hadrons, which is bound to have less transverse momentum $P_T = zp_T$, is then described by a fragmentation function. The average value $\langle z \rangle$ is about 0.5 for pions in $p + p$ collisions. In other words the production of a, say, 5 GeV/c pion has to start with a 10 GeV/c parton in average, which are rare to find due to the steeply falling parton spectrum. Jet quenching even enhances the lack of high p_T partons. On the other hand, the 5 GeV/c pion could be produced by the recombination of a quark and an antiquark with about 2.5 GeV/c each in average. 2.5 GeV/c and 10 GeV/c are separated by orders of magnitude in the parton spectrum. The price to pay is of course that two of these partons have to be found close to each other in phase space. However we do have a densely populated phase space in central heavy ion collisions at RHIC where we even expect the existence of a thermalized quark gluon plasma.

Recombination of quarks has been considered before in hadron collisions [17] and was also applied to heavy ion collisions [18]. In QCD the leading particle effect in the forward region of a hadron collision is well known. This is the phenomenon that the production of hadrons that share valence partons with the beam hadrons are favored in forward direction. It has been realized that this can only be explained by recombination. This has fueled a series of theoretical work, see e.g. [19] and references therein. Recently the recombination idea for heavy ion collisions, stimulated by the RHIC results, has been revived for elliptic flow [13] and hadron spectra and ratios [10, 11].

The formalism of recombination has already been developed in a covariant setup for the process of baryons coalescing into light nuclei and clusters in nuclear collisions [20, 21]. We give a brief derivation for the case of mesons. By introducing the density matrix $\hat{\rho}$ for the system of partons, the number of quark-antiquark states that will be measured as mesons is given by

$$N_M = \sum_{ab} \int \frac{d^3P}{(2\pi)^3} \langle M; \mathbf{P} | \, \hat{\rho}_{ab} \, | M; \mathbf{P} \rangle \tag{1}$$

Here $|M; \mathbf{P}\rangle$ is a meson state with momentum $\mathbf{P}$ and the sum is over all combinations of quantum numbers — flavor, helicity and color — of valence partons that contribute to the given meson M. This can be cast in covariant form using a hypersurface Σ for hadronization [22]

$$E\frac{dN_M}{d^3P} = C_M \int_{\Sigma} \frac{d\sigma \, P \cdot u(\sigma)}{(2\pi)^3} \int \frac{d^3q}{(2\pi)^3} w_a\Big(\sigma; \frac{\mathbf{P}}{2} - \mathbf{q}\Big) \Phi_M^W(\mathbf{q}) w_b\Big(\sigma; \frac{\mathbf{P}}{2} + \mathbf{q}\Big). \tag{2}$$

Here E is the energy of the four vector P, $d\sigma$ a measure on Σ and $u(\sigma)$ is the future oriented unit vector orthogonal to the hypersurface Σ. w_a and w_b are the phase space densities for the two partons a and b, C_M is a degeneracy factor and

$$\Phi_M^W(\mathbf{q}) = \int d^3r \, \Phi_M^W(\mathbf{r}, \mathbf{q}) \tag{3}$$

where $\Phi_M^W(\mathbf{r}, \mathbf{q})$ is the Wigner function of the meson [21].

Since the hadron structure is best known in a light cone frame, we write the integral over $\mathbf{q}$ in terms of light cone coordinates in a frame where the hadron has no transverse momentum but a large light cone component P^+. This can be achieved by a simple rotation from the lab frame. Introducing the momentum $k = P/2 - q$ of parton a in this frame we have $k^+ = xP^+$ with $0 < x < 1$. We make an ansatz for the Wigner function of the meson in terms of light cone wave functions $\phi_M(x)$. The final result can be written as [10]

$$E\frac{N_M}{d^3P} = C_M \int_\Sigma d\sigma \frac{P \cdot u(\sigma)}{(2\pi)^3} \int_0^1 dx\, w_a(\sigma; xP^+)\, |\phi_M(x)|^2\, w_b(\sigma; (1-x)P^+). \tag{4}$$

For a baryon with valence partons a, b and c we obtain

$$\begin{aligned} E\frac{N_B}{d^3P} &= C_B \int_\Sigma d\sigma \tfrac{P \cdot u(\sigma)}{(2\pi)^3} \int_0^1 dx_1\, dx_2\, dx_3 \delta(x_1 + x_2 + x_3 - 1) \\ &\quad \times w_a(\sigma; x_1P^+) w_b(\sigma; x_2P^+) w_c(\sigma; x_3P^+)\, |\phi_B(x_1, x_2, x_3)|^2 . \end{aligned} \tag{5}$$

$\phi_B(x_1, x_2, x_3)$ is the effective wave function of the baryon in light cone coordinates.

It turns out that for a purely exponential spectrum the shape of the wave function does not matter. In that case the dependence on x drops out of the product of phase space densities

$$w_a(\sigma; xP^+) w_b(\sigma; (1-x)P^+) \sim e^{-xP^+/T} e^{-(1-x)P^+/T} = e^{-P^+/T}. \tag{6}$$

One can show that a narrow width approximation, using δ peaked wave functions that distribute the momentum of the hadron equally among the valence quarks — 1/2 in the case of a meson and 1/3 in the case of a baryon — deviates by less then 20% from a calculation using light cone functions. That there is a small deviation can be attributed to the violation of energy conservation. Since recombination is a $2 \to 1$ or $3 \to 1$ process, energy will generally not be conserved if we enforce momentum conservation and a mass shell condition for all particles. However, one can show that violations of energy conservation are of the order of the effective quark masses or Λ_{QCD} and can therefore be neglected for transverse hadron momenta larger than 2 GeV/c.

Fragmentation of partons is given by [23]

$$E\frac{d\sigma_h}{d^3P} = \sum_a \int_0^1 \frac{dz}{z^2} D_{a\to h}(z) E_a \frac{d\sigma_a}{d^3P_a}. \tag{7}$$

The sum runs over all parton species a and σ_a is the cross section for the production of parton a with momentum $P_a = P/z$. We use a leading order (LO) pQCD calculation of σ_a [24] together with LO KKP fragmentation functions [25]. Energy loss of the partons is taken into account by a shift of the parton spectrum by

$$\Delta p_T = \sqrt{\lambda p_T}. \tag{8}$$

3. Implementation

Assuming longitudinal boost invariance, we fix the hypersurface Σ by choosing $\tau = \sqrt{t^2 - z^2} = \text{const.}$ for

$$R^\mu = (t, x, y, z) = (\tau \cosh \eta, \rho \cos \phi, \rho \sin \phi, \tau \sinh \eta). \tag{9}$$

It is convenient to introduce the space time rapidity η and the radial coordinate ρ, since the measure for the hypersurface Σ then takes the simple form $d\sigma_R = \tau\, d\eta\, \rho\, d\rho\, d\phi$ and the normal vector is given by

$$u^\mu(R) = (\cosh \eta, 0, 0, \sinh \eta). \tag{10}$$

Using a similar parametrization of the parton momentum

$$p^\mu = (m_T \cosh y, p_T \cos \Phi, p_T \sin \Phi, m_T \sinh y) \tag{11}$$

with rapidity y and transverse mass $m_T = \sqrt{m^2 + p_T^2}$, we obtain $p \cdot u(R) = m_T \cosh(\eta - y)$. We remind the reader that we will use capitalized variables (P, P_T, M, M_T, etc.) for hadrons.

Now we have to specify the spectrum of partons. As already discussed above we assume that the parton spectrum consists of two domains. At large p_T, the distribution of partons is given by perturbative QCD and follows a power law. For the transverse momentum distribution at midrapidity for central collisions (b=0) we use the parametrization [26, 27]

$$\left.\frac{dN_a^{\text{pert}}}{d^2p_T dy}\right|_{y=0} = K \frac{C}{(1 + p_T/B)^\beta}. \tag{12}$$

The parameters C, B, and β are taken from a leading order (LO) pQCD calculation and can be found in [27] for the three light quark flavors and gluons. A constant K factor of 1.5 is included to roughly account for higher order corrections in α_s [28]. The calculation includes nuclear shadowing of the parton distributions, but no higher twist initial state effects. Higher twist effects, like the Cronin effect, will fade like $A^{1/3}\Lambda_{\text{QCD}}^2/P_T^2$ for high transverse momentum [29]. Since we will show that fragmentation and pQCD are only dominant for transverse momenta above 5 GeV/c in the hadron spectrum for RHIC, it is safe to omit the Cronin effect.

Energy loss of partons, resulting in a shift of the transverse momentum spectrum [3, 4], is taken into account and parameterized as

$$\Delta p_T(b, p_T) = \epsilon(b) \sqrt{p_T}\, \frac{\langle L \rangle}{R_A}. \tag{13}$$

For central collisions we take $\langle L \rangle = R_A$ and therefore $\Delta p_T(0, p_T) = \epsilon(0)\sqrt{p_T}$. The choice of $\langle L \rangle = R_A$ neglects the fact that for the strong quenching observed at RHIC energies, jet emission becomes a surface effect [4, 30]. However, we note that only the product $\epsilon\langle L\rangle$ as a whole is a parameter. We have no ambition here to make a connection to the microscopic parameters of jet quenching, therefore we will not disentangle ϵ and $\langle L\rangle$. This would require a more sophisticated model of the emission geometry. We also do not use a radial profile for the emission and the density of the medium. This can be found discussed elsewhere in the literature [4, 31, 32]. Nevertheless, our "minimal" description of energy loss is quite successful to describe the available data on high-P_T spectra ($P_T > 5$ GeV/c) for π^0, K_s^0 and charged hadrons $(h^+ + h^-)/2$ in central Au+Au collisions. From a fit to these data we find $\epsilon_0 = \epsilon(0) = 0.82\mathrm{GeV}^{1/2}$ for central Au+Au collisions. This value corresponds to an average energy loss of 3 GeV for a 10 GeV parton in a central Au+Au collision. For fragmentation of pions, kaons, protons and antiprotons we use LO KKP fragmentation functions with the scale set to the hadron transverse momentum P_T [25]. Λ fragmentation is calculated with the LO fragmentation functions of de Florian, Stratmann and Vogelsang [33].

Besides the perturbative tail of the parton spectrum that will turn into hadrons via fragmentation, we assume the existence of a spectrum of thermalized partons that are recombining at hadronization and dominate at low and intermediate values of p_T. In this phase we assume the effective degrees of freedom to be constituent quarks without dynamical gluons. We take the spectrum to be exponential with a given temperature T

$$w_a(R;p) = \gamma_a e^{-p\cdot v(R)/T} e^{-\eta^2/2\Delta^2} f(\rho,\phi). \tag{14}$$

γ_a is a fugacity factor for each parton species a. We also include longitudinal and radial flow through the velocity vector

$$v^\mu(R) = (\cosh\eta_L \cosh\eta_T, \sinh\eta_T\cos\phi, \sinh\eta_T\sin\phi, \sinh\eta_L\cosh\eta_T). \tag{15}$$

$\eta_L(R)$ and $\eta_T(R)$ are the rapidities of the longitudinal and radial flow which still could depend on the space time point $R \in \Sigma$. For the longitudinal expansion we choose a Bjorken scenario where the longitudinal rapidity is simply fixed by the space time rapidity

$$\eta_L(R) = \eta. \tag{16}$$

The transverse flow is given by a velocity $v_T(R)$ with $v_T = \tanh\eta_T$. For practical purposes we will not work with a radial profile but assume v_T to be independent of ρ and ϕ. However, for collisions with finite impact parameter b we will later allow a dependence of v_T on the azimuthal angle ϕ in order to describe the measured elliptic asymmetry in the spectra. The space-time structure of the parton source in (14) is given by a transverse distribution $f(\rho,\phi)$ and a wide Gaussian rapidity distribution with a width Δ.

We assume that hadronization occurs at $\tau = 5$ fm at a temperature $T = 175$ MeV in the parton phase. This is consistent with predictions of the phase transition temperature at vanishing baryon chemical potential from lattice QCD [34]. For the spread of the parton distribution in longitudinal direction we choose $\Delta = 2$. The constituent quark masses are taken to be 260 MeV for u and d quarks and 460 MeV for s quarks.

In the region where contributions from recombination and fragmentation are of the same size we expect other mechanisms to play a role, which interpolate between the two pictures. This could include partial recombination and higher twist fragmentation. In the absence of a consistent description of these mechanisms we simply add both contributions to the hadron spectrum – recombination from the exponential part and the fragmentation from the pQCD part – for $P_T > 2$ GeV/c.

4. Spectra and Nuclear Modification Factors

Figure 1 shows the inclusive charged hadron spectrum using recombination and fragmentation for pions, kaons, protons and antiprotons. The recombination plus fragmentation approach nicely reproduces the two component form of the hadron transverse momentum spectra. We note that the crossover between the recombination dominated and fragmentation dominated regimes is around 5 GeV/c. A detailed particle species decomposition shows that the crossover is in fact earlier for pions ($\sim$4 GeV/c) than for protons ($\sim$6 GeV/c).

The nuclear modification factor is defined as the ratio of the hadron yield in Au+Au collisions to the one in $p+p$ scaled with the number of collisions

$$R_{AA} = \frac{d^2 N_{\mathrm{Au+Au}}(b)/dP_T^2}{N_{\mathrm{coll}}(b)\, d^2 N_{\mathrm{p+p}}/dP_T^2}. \tag{17}$$

Similarly one can consider scaled ratios of different centrality bins like central to peripheral

$$R_{CP} = \frac{N_{\mathrm{coll}}(b)\, d^2 N_{\mathrm{Au+Au}}(0)/dP_T^2}{N_{\mathrm{coll}}(0)\, d^2 N_{\mathrm{Au+Au}}(b)/dP_T^2}. \tag{18}$$

The upper frame of fig. 2 displays the scaled ratio R_{CP} for neutral pions and protons. The ratio of impact parameters 0 and 12 fm is used and compared to data from the PHENIX collaboration [35]. The data for protons shows R_{CP} to be between 0.8 and 1 below 4 GeV/c, which is quite surprising, considering the strong suppression suffered by the pions in that momentum domain. As already noticed above, recombination is more effective for protons than for pions. Therefore our calculation for the protons yields a similarly value of 0.8 as observed by the experiment. This implies that protons from recombination make up for the loss suffered from jet quenching at intermediate transverse momenta. Our calculations predicts sharp drops in R_{CP} and R_{AA} for protons

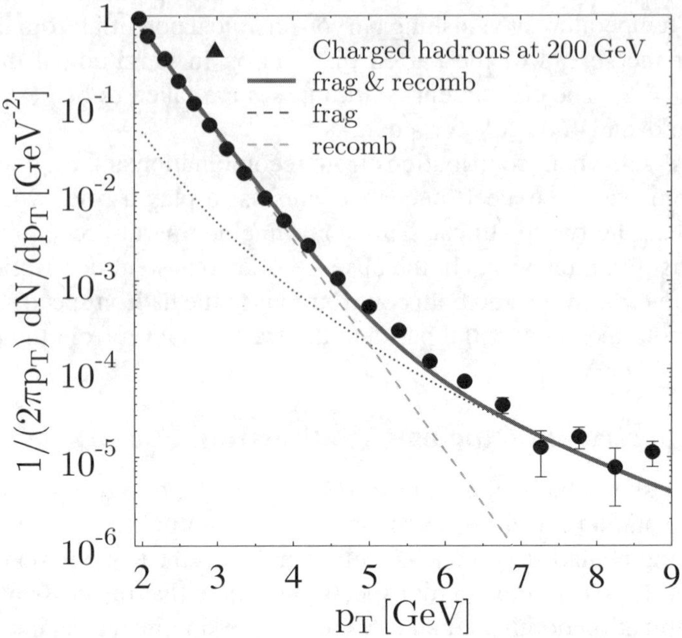

Figure 1. Charged hadron spectrum in Au+Au collisions at $\sqrt{S} = 200$ GeV as a function of P_T. Contributions from fragmentation (dashed), recombination (dotted) and the sum of both (solid line) are shown. Data are from the PHENIX collaboration.

and antiprotons beyond 4 GeV/c where fragmentation with jet quenching start to dominate.

In the lower frame of fig. 2 we compiled $R_{\rm CP}$ for K_s^0 and $\Lambda + \bar{\Lambda}$ together with data from STAR [36]. The different behavior of mesons and baryons is again impressively confirmed. For the first time experimental data indicate that a steep decrease in $R_{\rm CP}$ for baryons will occur beyond 4 GeV/c. The data on $\Lambda + \bar{\Lambda}$ suggest a drop to the perturbative value even sharper than what our results show. This could be due to too less Λ baryons from fragmentation using this particular set of fragmentation functions [33].

5. Hadron Ratios and Elliptic Flow

One of the main motivations at the onset of our investigation was to find an explanation for the surprisingly large proton over pion ratios that are of order one above 1.5 GeV/c. We have already shown in a previous publication [10] that recombination naturally provides a p/π^0 of order one for hadron transverse momenta up to 4 GeV. In addition we predict that a sharp drop beyond 4 GeV/c should be seen when the fragmentation process takes over. The value predicted

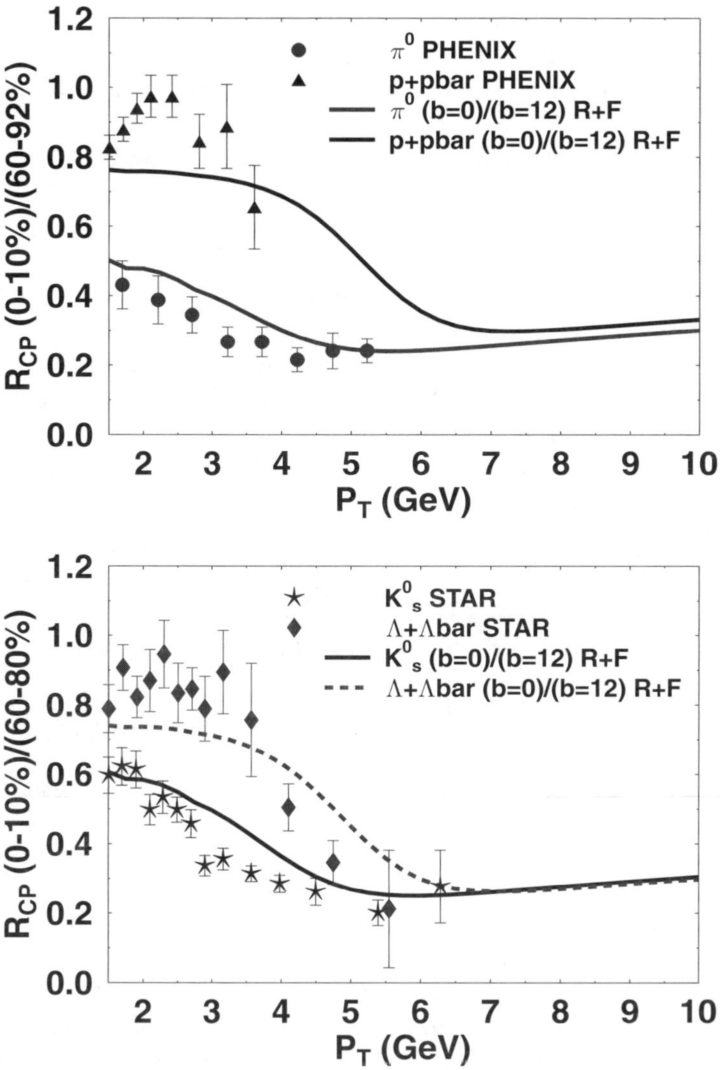

Figure 2. Upper frame: $R_{\rm CP}$ for neutral pions (bottom) and protons (top) given by the ratio of particle yields at impact parameters 0 and 12 fm compared to data from PHENIX. Lower frame: $R_{\rm CP}$ for K_s^0 (bottom) and $\Lambda + \bar{\Lambda}$ (top) given by the ratio of particle yields at impact parameters 0 and 12 fm compared to data from STAR.

for the ratio in the fragmentation domain is about 0.1. At small transverse momenta, the statistical model describes the data well but continues to rise beyond 4 GeV/c.

The left panel of fig. 3 shows hadron ratios. The top frame displays the ratio of charged hadrons to neutral pions reported by PHENIX [37] in comparison with our results. Our calculation agrees very well with the PHENIX data in the recombination region but slightly underestimates the ratio in the fragmentation region.

Preliminary results for the ratios $(\Lambda + \bar{\Lambda})/4K_s^0$ and were presented by the STAR collaboration [36]. Recombination predicts a large peak in the Λ/kaon ratio and a sharp decrease beyond 4 GeV/c similar to the p/π^0 ratio. First indications of such a sharp transition can be seen in the STAR data. This observation supports the recombination picture including a transition to the fragmentation regime beyond 4 GeV/c.

The predicted decrease in the proton/pion ratio, if confirmed, and first observations of a similar drop in the Λ/kaon ratio will be strong arguments in favor of the recombination+fragmentation picture.

Anisotropic flow v_2 is defined as the second Fourier coefficient of the azimuthal hadron spectrum

$$v_2(P_T) = \langle \cos 2\Phi \rangle = \frac{\int d\Phi \cos 2\Phi d^2N/P_T dP_T d\Phi}{\int d\Phi d^2N/P_T dP_T d\Phi}, \tag{19}$$

and measures the elliptic asymmetry in the spectrum (sometimes called elliptic flow)

$$\frac{dN}{P_T dP_T d\Phi} = \frac{dN}{2\pi P_T dP_T}[1 + 2v_2(P_T)\cos 2\Phi + \ldots], \tag{20}$$

Such an asymmetry arises in heavy ion collisions with non-vanishing impact parameter $b > 0$.

Our results on the particle dependence of elliptic flow are summarized in fig. 3. One can see the different behavior of mesons and baryons by comparing protons with pions and kaons with Λs. v_2 for baryons saturates at a higher value than for mesons in the recombination domain. At higher P_T, when fragmentation takes over, the results rapidly approach each other. In our calculation, where we do not take into account the binding energies, we cannot resolve the splitting between protons and mesons coming from the mass difference. Nevertheless the agreement with data from PHENIX [38] and preliminary data from STAR [39] is good.

It can be shown that in the region where recombination of partons dominates the hadronization process (i.e. for $P_T < 4(6)$ GeV for mesons (baryons)) and mass effects are suppressed (i.e. $P_T > 1$ GeV), v_2 obeys a simple scaling law

$$v_2(P_T) \approx n v_2^q(P_T/n) \tag{21}$$

where n is the number of valence quarks of the hadron. Here it is assumed that strange quarks and light quarks have the same elliptic flow v_2^q. At higher transverse momentum, $P_T > 6$ GeV, where fragmentation of partons dominates, the v_2 of all hadrons lies on a universal curve since the effect of fragmentation functions largely cancels in Eq. (19). We refer the reader to Refs. [40, 41] for a more detailed discussion.

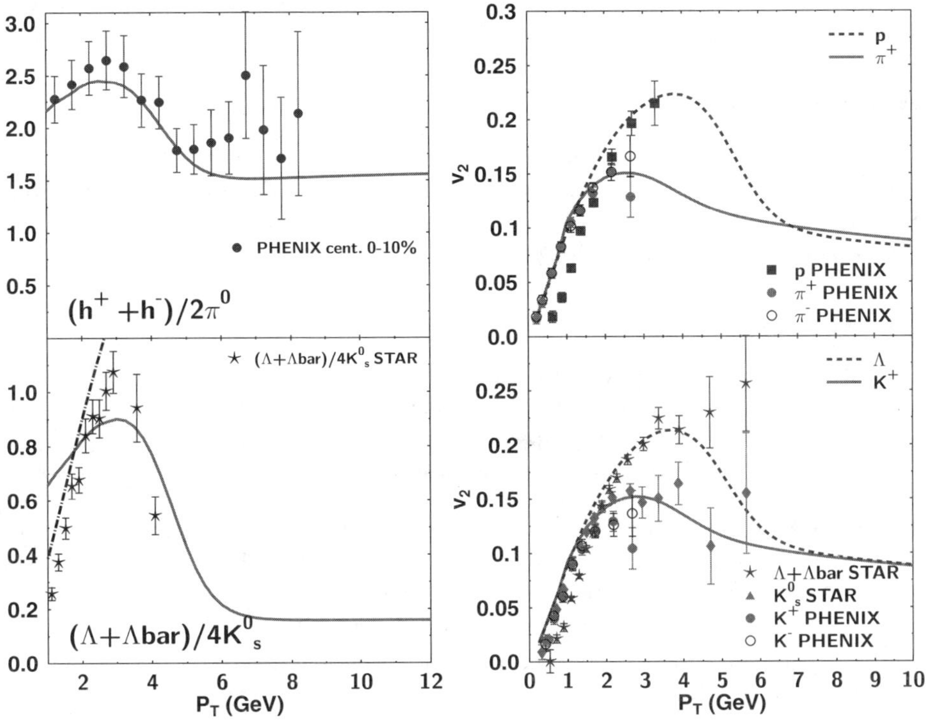

Figure 3. Left frame: $(h^+ + h^-)/2\pi^0$ and $(\Lambda + \bar{\Lambda})/4K_s^0$ and $2(\Xi^- + \Xi^+)/(\Lambda + \bar{\Lambda})$ ratios as functions of transverse momentum P_T. We show data from STAR and PHENIX as well as results from the statistical model (dash dotted lines). Right frame: anisotropic flow for p and π^+ compared to PHENIX data (top) and for K^+ and $\Lambda + \bar{\Lambda}$ compared to STAR and PHENIX data (bottom).

6. Conclusion

In summary, we propose a two component behaviour of hadronic observables in heavy ion collisions at RHIC. These components include fragmentation from a high-p_T perturbative parton phase and recombination from a thermal parton phase. We have found that the competition between recombination and fragmentation of partons as hadronization mechanisms can simultaneously explain several of the surprising features of the published data. In particular, the proton excess at intermediate P_T, the different nuclear suppression observed in pion and proton spectra, and the different saturation thresholds in the elliptic flow, can be consistently explained. We predict that all baryon spectra will exhibit a rapid transition around 5 GeV/c to a region dominated by parton fragmentation. Finally, we emphasize that our scenario requires the assumption of a thermalized partonic phase characterized by an exponential momentum spectrum and without strong multi-parton correlations. Such a phase may be appropriately called a quark-gluon plasma.

Acknowledgments

This work was in part supported in part by RIKEN, Brookhaven National Laboratory and DOE grants DE-FG02-96ER40945 and DE-AC02-98CH10886. S.A.B. acknowledges support from an Outstanding Junior Investigator Award (DOE grant DE-FG02-03ER41239).

References

[1] K. Adcox *et al.* [PHENIX Collaboration], Phys. Rev. Lett. **88** (2002), 022301; C. Adler *et al.* [STAR Collaboration], Phys. Rev. Lett. **90** (2003), 082302.

[2] J. D. Bjorken, FERMILAB-PUB-82-059-THY (1982); M. H. Thoma and M. Gyulassy, Nucl. Phys. B **351** (1991), 491; X. N. Wang and M. Gyulassy, Phys. Rev. Lett. **68** (1992), 1480; M. Gyulassy and X. Wang, Nucl. Phys. B **420** (1994), 583; R. Baier, Y. L. Dokshitzer, A. H. Mueller, S. Peigne and D. Schiff, Nucl. Phys. B **483** (1997), 291; B. G. Zakharov, JETP Lett. **65** (1997), 615; M. Gyulassy, P. Levai and I. Vitev, Phys. Rev. Lett. **85** (2000), 5535; U. A. Wiedemann, Nucl. Phys. B **588** (2000), 303.

[3] R. Baier, Y. L. Dokshitzer, A. H. Mueller and D. Schiff, JHEP **0109** (2001), 033;

[4] B. Muller, Phys. Rev. C **67** (2003), 061901 [arXiv:nucl-th/0208038].

[5] K. Adcox *et al.* [PHENIX Collaboration], Phys. Rev. Lett. **88** (2002), 242301.

[6] C. Adler *et al.* [STAR Collaboration], Phys. Rev. Lett. **86** (2001), 4778.

[7] C. Adler *et al.* [STAR Collaboration], Phys. Rev. Lett. **89** (2002), 092301.

[8] K. Adcox *et al.* [PHENIX Collaboration], Phys. Rev. Lett. **89** (2002), 092302.

[9] I. Vitev and M. Gyulassy, Phys. Rev. C **65** (2002), 041902.

[10] R. J. Fries, B. Müller, C. Nonaka and S. A. Bass, Phys. Rev. Lett. **90** (2003), 202303.

[11] V. Greco, C. M. Ko and P. Levai, Phys. Rev. Lett. **90** (2003), 202302.

[12] Z. W. Lin and C. M. Ko, Phys. Rev. Lett. **89** (2002), 202302.

[13] S. A. Voloshin, Nucl. Phys. A **715** (2003), 379c.

[14] D. Molnár and S. A. Voloshin, nucl-th/0302014;

[15] Z. w. Lin and D. Molnar, Phys. Rev. C **68** (2003), 044901 [arXiv:nucl-th/0304045].

[16] P. Sorensen for the STAR Collaboration, J. Phys. G, *to appear* (2003), nucl-ex/0305008.

[17] K. P. Das and R. C. Hwa, Phys. Lett. B **68** (1977), 459; Erratum *ibid.* **73** (1978), 504; R. G. Roberts, R. C. Hwa and S. Matsuda, J. Phys. G **5** (1979), 1043.

[18] C. Gupt *et al.*, Nuovo Cim. A **75** (1983), 408; T. S. Biro, P. Levai and J. Zimanyi, Phys. Lett. B **347** (1995), 6; T. S. Biro, P. Levai and J. Zimanyi, J. Phys. G **28** (2002), 1561.

[19] J. C. Anjos, J. Magnin and G. Herrera, Phys. Lett. B **523** (2001), 29; E. Braaten, Y. Jia and T. Mehen, Phys. Rev. Lett. **89** (2002), 122002.

[20] C. B. Dover *et al.*, Phys. Rev. C **44** (1991), 1636.

[21] R. Scheibl and U. W. Heinz, Phys. Rev. C **59** (1999), 1585.

[22] F. Cooper and G. Frye, Phys. Rev. D **10** (1974), 186.

[23] J. F. Owens, Rev. Mod. Phys. **59** (1987), 465.

[24] R. J. Fries, B. Müller, and D. K. Srivastava, Phys. Rev. Lett. **90** (2003), 132301; D. K. Srivastava, C. Gale and R. J. Fries, Phys. Rev. C **67** (2003), 034903.

[25] B. A. Kniehl, G. Kramer and B. Pötter, Nucl. Phys. B **582** (2000), 514.

[26] R. J. Fries, B. Müller and D. K. Srivastava, Phys. Rev. Lett. **90** (2003), 132301.

[27] D. K. Srivastava, C. Gale and R. J. Fries, Phys. Rev. C **67** (2003), 034903.

[28] K. J. Eskola and K. Tuominen, Phys. Rev. D **63** (2001), 114006; G. G. Barnafoldi, G. Fai, P. Levai, G. Papp and Y. Zhang, J. Phys. G **27** (2001), 1767.

[29] M. Luo, J. Qiu and G. Sterman, Phys. Lett. B **279** (1992), 377; J. Qiu and G. Sterman, Int. J. Mod. Phys. E **12** (2003), 149.

[30] E. V. Shuryak, Phys. Rev. C **66** (2002), 027902.

[31] E. Wang and X. N. Wang, Phys. Rev. Lett. **89** (2002), 162301.

[32] X. N. Wang, nucl-th/0305010.

[33] D. de Florian, M. Stratmann and W. Vogelsang, Phys. Rev. D **57** (1998), 5811.

[34] F. Karsch, Nucl. Phys. A **698** (2002), 199.

[35] J. Velkovska for the PHENIX Collaboration, Talk at SQM 2003, J. Phys. G *to be published.*

[36] H. Long for the STAR Collaboration, Talk at SQM 2003, J. Phys. G *to be published.*

[37] S. S. Adler *et al.* [PHENIX Collaboration], Phys. Rev. Lett. **91** (2003), 172301 [arXiv:nucl-ex/0305036].

[38] . S. Adler *et al.* [PHENIX Collaboration], Phys. Rev. Lett. **91** (2003), 182301 [arXiv:nucl-ex/0305013].

[39] R. Snellings for the STAR Collaboration, J. Phys. G *to be published*, nucl-ex/0305001.

[40] R. J. Fries, B. Muller, C. Nonaka and S. A. Bass, Phys. Rev. C **68** (2003), 044902 [arXiv:nucl-th/0306027].

[41] C. Nonaka, R. J. Fries, S. A. Bass, nucl-th/0308051.

REVIEW OF QGP SIGNATURES - IDEAS VERSUS OBSERVABLES *

E. L. Bratkovskaya, M. Bleicher, A. Dumitru, K. Paech, M. Reiter, S. Soff, H. Stöcker, H. Weber
Institut für Theoretische Physik, Universität Frankfurt, 60054 Frankfurt, Germany

M. van Leeuwen
NIKHEF, Amsterdam, Netherlands

W. Cassing
Institut für Theoretische Physik, Universität Giessen, 35392 Giessen, Germany

Abstract We investigate hadron production and transverse hadron spectra in nucleus-nucleus collisions from 2 A·GeV to 21.3 A·TeV within two independent transport approaches (UrQMD and HSD) based on quark, diquark, string and hadronic degrees of freedom. The enhancement of pion production in central Au+Au (Pb+Pb) collisions relative to scaled pp collisions (the 'kink') is described well by both approaches without involving a phase transition. However, the maximum in the K^+/π^+ ratio at 20 to 30 A·GeV (the 'horn') is missed by $\sim 40\%$. Also, at energies above $\sim$ 5 A·GeV, the measured $K^\pm$ m_T-spectra have a larger inverse slope than expected from the models. Thus the pressure generated by hadronic interactions in the transport models at high energies is too low. This finding suggests that the additional pressure - as expected from lattice QCD at finite quark chemical potential and temperature - might be generated by strong interactions in the early pre-hadronic/partonic phase of central heavy ion collisions. Finally, we discuss the emergence of density perturbations in a first-order phase transition and why they might affect relative hadron multiplicities, collective flow, and hadron mean-free paths at decoupling. A minimum in the collective flow v_2 excitation function was discovered experimentally at 40 A·GeV - such a behavior has been predicted long ago as signature for a first order phase transition.

*Supported by DFG, BMBF, GSI

W. Greiner et al. (eds.), Structure and Dynamics of Elementary Matter, 239–254.

1. Introduction

The phase transition from partonic degrees of freedom (quarks and gluons) to interacting hadrons is a central topic of modern high-energy physics. In order to understand the dynamics and relevant scales of this transition laboratory experiments under controlled conditions are presently performed with ultra-relativistic nucleus-nucleus collisions. Hadronic spectra and relative hadron abundancies from these experiments reflect important aspects of the dynamics in the hot and dense zone formed in the early phase of the reaction. Furthermore, as has been proposed early by Rafelski and Müller [1] the strangeness degree of freedom might play an important role in distinguishing hadronic and partonic dynamics.

In fact, estimates based on the Bjorken formula [2] for the energy density achieved in central Au+Au collisions suggest that the critical energy density for the formation of a quark-gluon plasma (QGP) is by far exceeded during a few fm/c in the initial phase of the collision at Relativistic Heavy Ion Collider (RHIC) energies [3], but sufficient energy densities ($\sim$ 0.7-1 GeV/fm^3 [4]) might already be achieved at Alternating Gradient Synchrotron (AGS) energies of $\sim$ 10 $A{\cdot}$GeV [5, 6]. More recently, lattice QCD calculations at finite temperature and quark chemical potential μ_q [7] show a rapid increase of the thermodynamic pressure P with temperature above the critical temperature T_c for a phase transition to the QGP. The crucial question is, however, at what bombarding energies the conditions for the phase transition (or cross-over) might be fulfilled.

Presently, transverse mass (or momentum) spectra of hadrons are in the center of interest. It is experimentally observed that the transverse mass spectra of kaons at AGS and SPS energies show a substantial *flattening* or *hardening* in central Au+Au collisions relative to pp interactions (cf. Refs. [8, 9]). In order to quantify this effect, the spectra are often parametrised as:

$$\frac{1}{m_T}\frac{dN}{dm_T} \sim \exp(-\frac{m_T}{T}) \qquad (1)$$

where $m_T = \sqrt{m^2 + p_T^2}$ is the transverse mass and T is the inverse slope parameter. This hardening of the spectra is commonly attributed to strong collective flow, which is absent in the pp or pA data.

The authors of Refs. [10] have proposed to interpret the approximately constant $K^{\pm}$ slopes above $\sim$ 30 A·GeV – the 'step' – as an indication for a phase transition following an early suggestion by Van Hove [11]. This interpretation is also based on a rather sharp maximum in the K^+/π^+ ratio at $\sim$ 20 to 30 A·GeV in central Au+Au (Pb+Pb) collisions (the 'horn' [10]). However, it is presently not clear, if the statistical model assumptions invoked in Refs. [10] hold to be reliable.

We will demonstrate in this contribution that the pressure needed to generate a large collective flow – to explain the hard slopes of the $K^{\pm}$ spectra as well as the 'horn' in the K^+/π^+ ratio – is not produced in the present models by the interactions of hadrons in the expansion phase of the hadronic fireball. In our studies we use two independent transport models that employ hadronic and string degrees of freedom, i.e., UrQMD (v. 1.3) [12, 13] and HSD [14, 15]. They take into account the formation and multiple rescattering of hadrons and thus dynamically describe the generation of pressure in the hadronic expansion phase. This involves also interactions of 'leading' pre-hadrons that contain a valence quark (antiquark) from a 'hard' collision (cf. Refs. [14, 16]).

The UrQMD transport approach [12, 13] includes all baryonic resonances up to masses of 2 GeV as well as mesonic resonances up to 1.9 GeV as tabulated by the Particle Data Group [17]. For hadronic continuum excitations a string model is used with hadron formation times in the order of 1-2 fm/c depending on the momentum and energy of the created hadron. In the HSD approach nucleons, Δ's, N*(1440), N*(1535), Λ, Σ and Σ^* hyperons, Ξ's, Ξ^*'s and Ω's as well as their antiparticles are included on the baryonic side whereas the 0^- and 1^- octet states are included in the mesonic sector. High energy inelastic hadron-hadron collisions in HSD are described by the FRITIOF string model [18] whereas low energy hadron-hadron collisions are modeled based on experimental cross sections. Both transport approaches reproduce the nucleon-nucleon, meson-nucleon and meson-meson cross section data in a wide kinematic range. We point out, that no explicit parton-parton scattering processes (beyond the interactions of 'leading' quarks/diquarks) are included in the studies below contrary to the multi-phase transport model (AMPT) [19], which is currently employed from upper SPS to RHIC energies.

2. Hadron excitation functions and ratios

2.1 pp versus central AA reactions – the 'kink'

In order to explore the main physics from central AA reactions it is instructive to have a look at the various particle multiplicities relative to scaled pp collisions as a function of bombarding energy. For this aim we show in Fig. 1 the total multiplicities of π^+, K^+ and K^- (i.e., the 4π yields) from central Au+Au (at AGS) or Pb+Pb (at SPS) collisions (from UrQMD and HSD) in comparison to the scaled total multiplicities from pp collisions versus the kinetic energy per particle $E_{\rm lab}$.

The general trend from both transport approaches is quite similar: we observe a slight absorption of pions at lower bombarding energy and a relative enhancement of pion production by rescattering in heavy-ion collisions above $\sim$10 A$\cdot$GeV. Kaons and antikaons from AA collisions are always enhanced in central reactions relative to scaled pp multiplicities, which is a consequence of

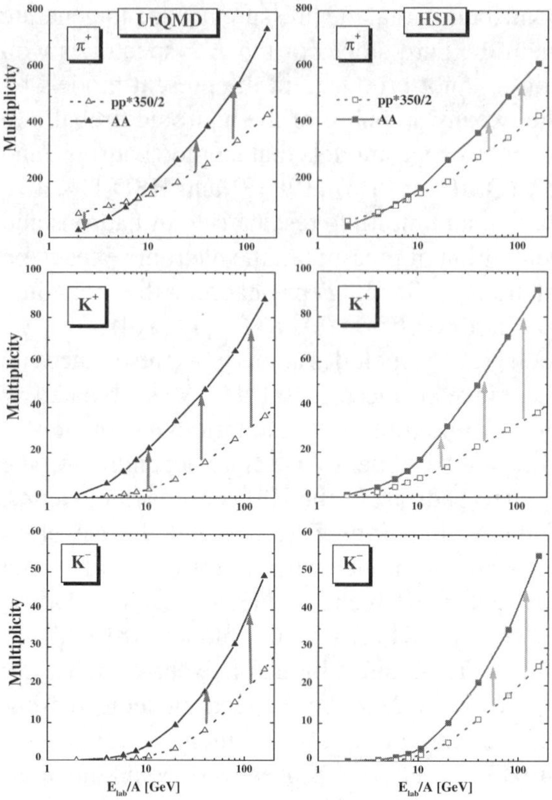

Figure 1. Total multiplicities of π^+, K^+ and K^- (i.e., 4π yields) from central Au+Au (at AGS) or Pb+Pb (at SPS) collisions in comparison to the total multiplicities from *pp* collisions (scaled by a factor 350/2) versus kinetic energy $E_{\rm lab}$. The solid lines with full triangles and squares show the UrQMD (l.h.s.) and HSD results (r.h.s.) for AA collisions, respectively. The dotted lines with open triangles and squares correspond to the *pp* multiplicities calculated within UrQMD (l.h.s.) and HSD (r.h.s.). The figure is taken from Ref. [16].

strong final state interactions. Thus, the 'kink' in the pion ratio as well as the $K^\pm$ enhancement might result from conventional hadronic final state interactions.

2.2 Particle yields in central collisions of heavy nuclei

Fig. 2 shows the excitation function of π^+, π^-, K^+, K^- and $\Lambda + \Sigma^0$ yields (midrapidity (l.h.s.) and rapidity integrated (r.h.s)) from central Au+Au (Pb+Pb) collisions in comparison to the experimental data. Note that all data from the NA49 Collaboration at 30 A·GeV have to be considered as 'preliminary'. As can be seen from Fig. 2 the differences between the independent transport models are less than 20%. The maximum deviations between the models and the experimental data are less than $\sim 30\%$. In addition, a systematic analysis of the results from both models and experimental data for central nucleus-nucleus collisions from 2 to 160 A·GeV in Ref. [16] has shown that also the 'longitudinal' rapidity distributions of protons, pions, kaons, antikaons and hyperons are quite similar in both models and in reasonable agreement with available data. The exception are the pion rapidity spectra at the highest

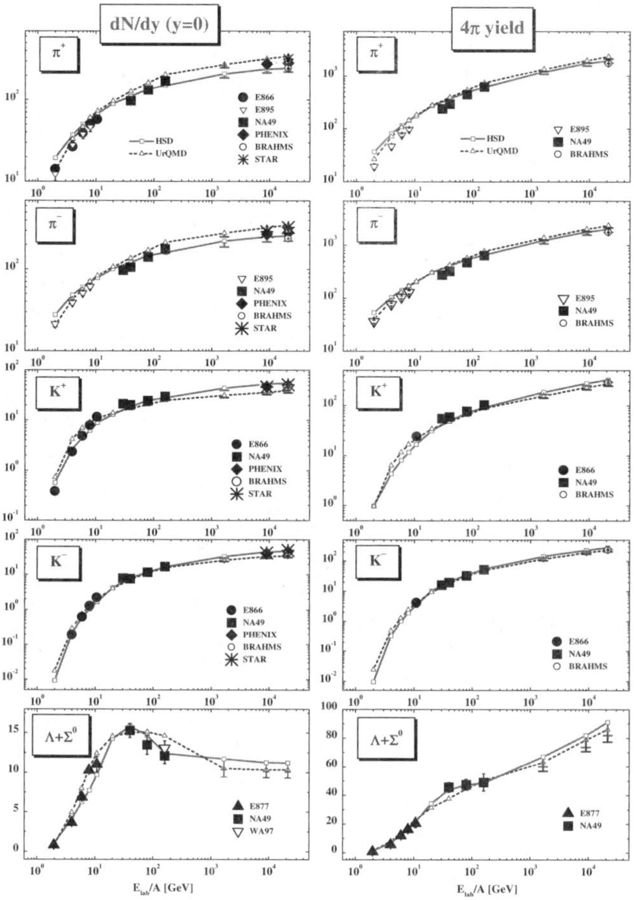

Figure 2. The excitation function of π^+, π^-, K^+, K^- and $\Lambda + \Sigma^0$ yields from central Au+Au or Pb+Pb collisions in comparison to the experimental data from Refs. [20–22] (AGS), [23–25] (SPS) and [26–28] (RHIC) for midrapidity (left column) and rapidity integrated yields (right column). The solid lines with open squares show the results from HSD whereas the dashed lines with open triangles indicate the UrQMD calculations. The lower theoretical errorbars at RHIC energies correspond to the yields for 10% central events. The figure is taken from Ref. [29].

AGS energy and lower SPS energies, which are overestimated by both models [16]. For a more detailed comparison of HSD and UrQMD calculations with experimental data at RHIC energies we refer the reader to Refs. [30, 31].

2.3 Particle ratios – the 'horn'

In Fig. 3 we present the excitation function of the particle ratios K^+/π^+, K^-/π^- and $(\Lambda+\Sigma^0)/\pi$ from central Au+Au (Pb+Pb) collisions in comparison to experimental data. The deviations between the transport models and the data are most pronounced for the midrapidity ratios (left column) since the ratios are very sensitive to actual rapidity spectra. The K^+/π^+ ratio in UrQMD shows a maximum at ~ 8 A·GeV and then drops to a constant ratio of 0.11 at top SPS and RHIC energies. In the case of HSD a continuously rising ratio with bombarding energy is found for the midrapidity ratios which partly is due to a dip in the pion pseudo-rapidity distribution at RHIC energies (cf. Fig. 1 in Ref. [30]). The 4π ratio in HSD is roughly constant from top SPS to RHIC energies, however, larger than the ratio from UrQMD due to the lower amount of pion production in HSD

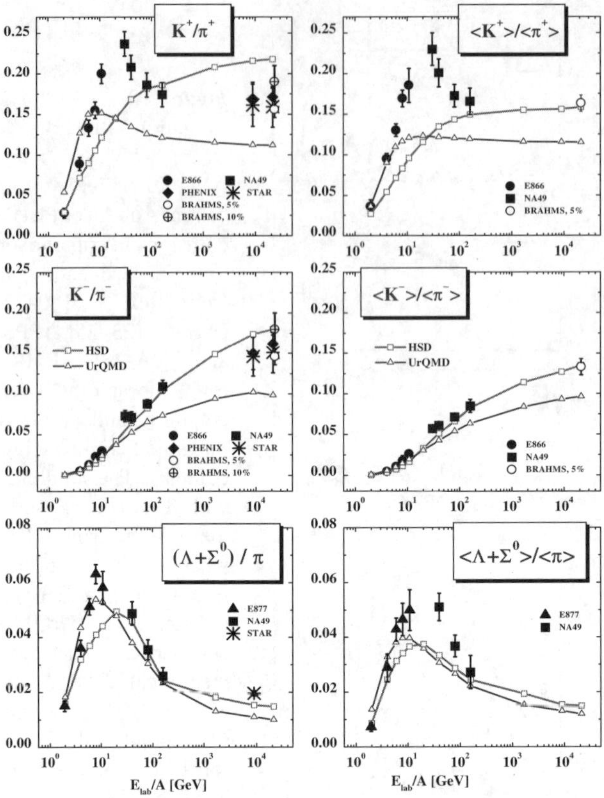

Figure 3. The excitation function of $K^+/\pi^+, K^-/\pi^-$ and $(\Lambda + \Sigma^0)/\pi$ ratios from 5% central (AGS energies, SPS at 160 A·GeV and at RHIC energies), 7% central (20, 30, 40 and 80 A·GeV), 10% central for $\Lambda + \Sigma^0$ at 160 A·GeV Au+Au (AGS and RHIC) or Pb+Pb (SPS) collisions in comparison to the experimental data from Refs. [20, 22] (AGS), [23–25] (SPS) and [26–28] (RHIC) for midrapidity (left column) and rapidity integrated yields (right column). The solid lines with open squares show the results from HSD whereas the dashed lines with open triangles indicate the UrQMD calculations. The figure is taken from Ref. [29].

essentially due to an energy-density cut of 1 GeV/fm^3, which does not allow to form hadrons above this critical energy density [16] and a slightly higher K^+ yield (cf. Fig. 2). Nevertheless, the experimental maximum in the K^+/π^+ ratio is missed, which we address dominantly to the excess of pions in the transport codes rather than to missing strangeness production. Qualitatively, the same arguments - due to strangeness conservation - also hold for the $(\Lambda + \Sigma^0)/\pi$ ratio, where the pronounced experimental maxima are underestimated due to the excess of pions in the transport models at top AGS energies (for HSD) and above $\sim$ 5 A·GeV (for UrQMD). Since the K^- yields are well reproduced by both approaches (cf. Fig. 2) the deviations in the K^-/π^- ratios at SPS and RHIC energies in UrQMD can be traced back to the excess of pions (see discussion above).

We stress that the maximum in the $(\Lambda + \Sigma^0)/\pi$ ratio is essentially due to a change from baryon to meson dominated dynamics with increasing bombarding energy. Similar arguments hold for the experimentally observed maxima in the ratio Ξ/π (cf. Ref. [32]). However, the 'horn' in the K^+/π^+ ratio at $\sim$30 A·GeV is not described by neither of our transport models.

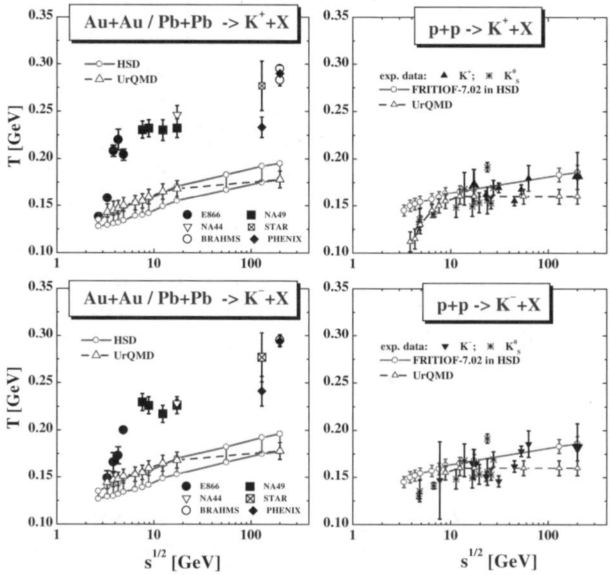

Figure 4. Comparison of the inverse slope parameters T for K^+ and K^- mesons from central Au+Au (Pb+Pb) collisions (l.h.s.) and pp reactions (r.h.s.) as a function of the invariant energy $\sqrt{s}$ from HSD (upper and lower solid lines) and UrQMD (open triangles) with data from Refs. [20, 8, 40, 28, 26, 27] for AA and [39, 41, 28] for pp collisions. The upper and lower solid lines result from different limits of the HSD calculations as discussed in the text. The figure is taken from Ref. [29].

3. Transverse mass spectra – the 'step'

We now focus on transverse mass spectra of pions and kaons/antikaons from central Au+Au (Pb+Pb) collisions from 2 A·GeV to 21.3 A·TeV and compare to recent data (cf. Ref. [33]). Without explicit representation we mention that the agreement between the transport calculations and the data for pp and for central C+C and Si+Si is quite satisfactory [33]; no obvious traces of 'new' physics are visible. The situation, however, changes for central Au+Au (or Pb+Pb) collisions. Whereas at the lowest energy of 4 A·GeV the agreement between the transport approaches and the data is still acceptable, severe deviations are visible in the $K^{\pm}$ spectra at SPS energies of 30 and 160 A·GeV [33]. We note that the $\pi^{\pm}$ spectra are reasonably described at all energies while the inverse slope T of the $K^{\pm}$ transverse mass spectra in Eq. (1) is underestimated severely by about the same amount in both transport approaches (within statistics). The increase of the inverse $K^{\pm}$ slopes in heavy-ion collisions with respect to pp collisions, which is generated by rescatterings of produced hadrons in the transport models, is small because the elastic meson-baryon scattering is strongly forward peaked and therefore gives little additional transverse momentum at midrapidity.

The question remains whether the underestimation of the $K^{\pm}$ slopes in the transverse mass spectra [33] might be due to conventional hadronic medium effects. In fact, the m_T slopes of kaons and antikaons at SIS energies (1.5 to 2 A·GeV) were found to differ significantly [34]. As argued in [15] the different slopes could be traced back to repulsive kaon-nucleon potentials, which lead to a hardening of the K^+ spectra, and attractive antikaon-nucleon potentials, which lead to a softening of the K^- spectra. However, the effect of such potentials was calculated within HSD and found to be of minor importance at AGS and SPS energies [15] since the meson densities are comparable to or even larger than the baryon densities at AGS energies and above. Additional self energy contributions stem from $K^{\pm}$ interactions with mesons; however, s-wave kaon-pion interactions are weak due to chiral symmetry arguments and p-wave interactions such as $\pi + K \leftrightarrow K^*$ transitions are suppressed substantially by the approximately 'thermal' pion spectrum [35].

Furthermore, we have pursued the idea of Refs. [36, 37] that the $K^{\pm}$ spectra could be hardened by string-string interactions, which increase the effective string tension σ and thus the probability to produce mesons at high m_T [19, 37]. In order to estimate the largest possible effect of string-string interactions we have assumed that for two overlapping strings the string tension σ is increased by a factor of two, for three overlapping strings by a factor of three etc. Here the overlap of strings is defined geometrically assuming a transverse string radius R_s, which according to the studies in Ref. [38] should be $R_s \leq 0.25$ fm. Based on these assumptions (and R_s=0.25 fm), we find only a small increase of the inverse slope parameters at AGS energies, where the string densities are low. At 160 $A{\cdot}$GeV the model gives a hardening of the spectra by about 15%, which, however, is still significantly less than the effect observed in the data.

Our findings are summarized in Fig. 4, where the dependence of the inverse slope parameter T (see Eq. (1)) on $\sqrt{s}$ is shown and compared to the experimental data [8, 39] for central Au+Au (Pb+Pb) collisions (l.h.s.) and pp reactions (r.h.s.). The upper and lower solid lines (with open circles) on the l.h.s. in Fig. 4 correspond to results from HSD calculations, where the upper and lower limits are due to fitting the slope T itself, an uncertainty in the repulsive $K^{\pm}$-pion potential or the possible effect of string overlaps. The slope parameters from pp collisions (r.h.s. in Fig. 4) are seen to increase smoothly with energy both in the experiment (full squares) and in the HSD calculations (full lines with open circles). The UrQMD results for pp collisions are shown as open triangles connected by the solid line and systematically lower than the slopes from HSD at all energies.

We mention that the RQMD model [36] gives higher inverse slope parameters for kaons at AGS and SPS energies than HSD and UrQMD, which essentially might be traced back to the implementation of effective resonances with masses above 2 GeV as well as 'color ropes' which decay isotropically in their rest frame [42]. A more detailed discussion of this issue is presented in Ref.[29].

4. Proton elliptic flow - the minimum

The NA49 Collaboration [44] has recently observed a vanishing elliptic flow of protons in Pb+Pb collisions at 40 A·GeV at midrapidity for all centralities

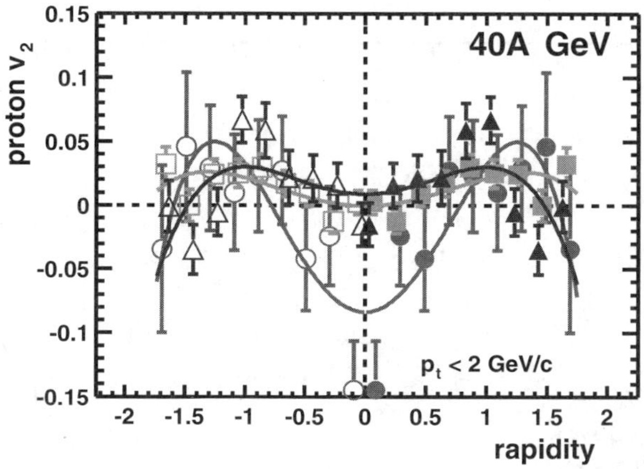

Figure 5. Elliptic flow v_2 of protons versus rapidity from 40 A·GeV Pb+Pb collisions [44] measured for three centrality bins: central (dots), mid-central (squares) and peripheral (triangles). The solid lines are polynomial fits to the data [44].

(Fig. 5). This observation of the apparent collapse of the collective flow v_2 is remarkable because the proton elliptic flow v_2 at top AGS (11 A·GeV) [43] and top SPS energies (160 A·GeV) [44] is non-zero for mid-central collisions and large for peripheral collisions. This experimental observation of a minimum of the collective flow excitation function has been predicted as a signature for a first order phase transition [45].

5. Thermodynamics in the $T - \mu_B$ plane

This still leaves us with the question of the origin of the rapid increase of the $K^{\pm}$ slopes with invariant energy for central Au+Au collisions at AGS energies and the constant slope at SPS energies (the 'step'), which is missed in both transport approaches. We recall that higher transverse particle momenta either arise from repulsive self energies – in mean-field dynamics – or from collisions, which reduce longitudinal momenta in favor of transverse momenta [5, 46]. As shown above in Fig. 4 conventional hadron self-energy effects and hadronic binary collisions are insufficient to describe the dramatic increase of the $K^{\pm}$ slopes as a function of $\sqrt{s}$. This indicates additional mechanisms for the generation of the pressure that is observed experimentally.

Here we propose that additional pre-hadronic/partonic degrees of freedom might be responsible for this effect already at ~ 5 A·GeV. Our arguments are based on a comparison of the thermodynamic parameters T and μ_B extracted from the transport models in the central overlap regime of Au+Au collisions [48] with the experimental systematics on chemical freeze-out configurations [47] in the T, μ_B plane. The solid line in Fig. 6 characterizes the universal chemical freeze-out line from Cleymans et al. [47] whereas the full dots with errorbars denote the 'experimental' chemical freeze-out parameters - determined from the fits to the experimental yields - taken from Ref. [47]. The various symbols (in vertical sequence) stand for temperatures T and chemical potentials μ_B

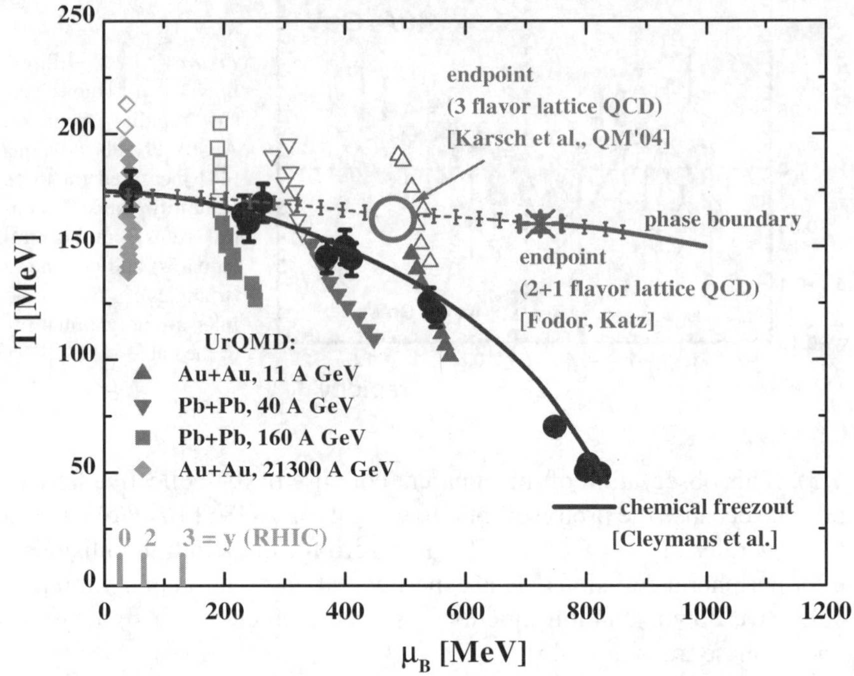

Figure 6. Schematic phase diagram in the $T - \mu_B$ plane. The solid line characterizes the universal chemical freeze-out line from Cleymans et al. [47] whereas the full dots (with errorbars) denote the 'experimental' chemical freeze-out parameters from Ref. [47]. The various symbols stand for temperatures T and chemical potentials μ_B extracted from UrQMD transport calculations in central Au+Au (Pb+Pb) collisions at 21.3 A·TeV, 160, 40 and 11 A·GeV [48] (see text). The stars indicate the tri-critical endpoints from lattice QCD calculations by Karsch et al. [49] (large open circle) and Fodor and Katz [7] (star). The 'horizontal' line with errorbars is the phase boundary (from the endpoint) given in Ref. [7]. The 'vertical' lines indicate μ_B for different rapidity intervals at RHIC energies from Ref. [50].

extracted from UrQMD transport calculations in central Au+Au (Pb+Pb) collisions at 21.3 A·TeV, 160, 40 and 11 A·GeV [48] as a function of the reaction time (from top to bottom). The open symbols denote nonequilibrium configurations and correspond to T parameters extracted from the transverse momentum distributions, whereas the full symbols denote configurations in approximate pressure equilibrium in longitudinal and transverse direction.

During the nonequilibrium phase (open symbols) the transport calculations show much higher temperatures (or energy densities) than the 'experimental' chemical freeze-out configurations at all bombarding energies ($\geq$ 11 A·GeV). These numbers are also higher than the tri-critical endpoints extracted from lattice QCD calculations by Karsch et al. [49] (large open circle) and Fodor and Katz [7] (star). Though the QCD lattice calculations differ substantially in the value of μ_B for the critical endpoint, the critical temperature T_c is in the range of 160 MeV in both calculations, while the energy density is in the order of 1 GeV/fm^3 or even below. Nevertheless, this diagram shows that at RHIC energies one encounters more likely a cross-over between the different phases when stepping down in temperature during the expansion phase of the 'hot fireball'. This situation changes at lower SPS or AGS (as well as new GSI SIS-300) energies, where for sufficiently large chemical potentials μ_B the cross over should change to a first order transition [51], i.e., beyond the tri-critical point in the (T, μ_B) plane. Nevertheless, Fig. 6 demonstrates that the transport calculations show temperatures (energy densities) well above the phase boundary (horizontal line with errorbars) in the very early phase of the collisions, where hadronic interactions practically yield no pressure, but pre-hadronic degrees of freedom should do. This argument is in line with the studies on elliptic flow at RHIC energies, that is underestimated by 30% at midrapidity in the HSD approach for all centralities [30]. Only strong early stage pre-hadronic interactions might cure this problem.

In Fig. 6 we also show the baryon chemical potential μ_B for different rapidity intervals at RHIC energies as obtained from a statistical model analysis by the BRAHMS Collaboration based on measured the antihadron to hadron yield ratios [50]. For midrapidity, $\mu_B \simeq 0$, whereas for forward rapidities μ_B increases up to $\mu_B \simeq 130$ MeV at $y = 3$. Thus, the forward rapidity measurement allows to probe large μ_B at the same bombarding energies. Hence, at RHIC only a rather limited chemical potential range is accessible experimentally. To reach the probable first order phase transition region, the International Facility at GSI seems to be the right place to go.

6. Density perturbations from dynamical symmetry breaking

It is of great interest, of course, to investigate whether the above-mentioned observations could be due to a phase transition of strongly interacting matter. The natural effective theory for exploring the effects from phase transitions on the production and phase-space distribution of hadrons is hydrodynamics: the equation of state enters directly by closing the system of continuity equations for energy, momentum and charge conservation. Typically, first-order phase transitions are modelled by matching the pressure of the low-density massive (symmetry broken) phase to that of the high-density massless (symmetric) phase along a 'phase boundary' in the $T - \mu_B$ plane (cf. Fig. 6). On the phase transition line, the system is in a mixed state where both phases coexist and where their relative fractions are determined from Gibbs's conditions of phase equilibrium. This construction assumes that the phase transition is a quasi-static, reversible process (entropy is conserved) near equilibrium. Entropy is produced only in the initial compression stage which ends with the formation of a locally equilibrated fireball of hot and dense matter which subsequently expands and cools.

The crossover from suppressed to increased pion production in central nuclear collisions relative to scaled *pp* collisions reflects the excess entropy produced at higher energies [52], as also seen in Fig. 1. Somewhat surprisingly though, the excitation function of entropy production turns out to be rather smooth, without exhibiting 'discontinuities' from crossing the phase boundary at some energy. Aside from some dynamical effects, the main reason for this behavior is that in *baryon-rich* matter the specific entropy is a smooth function of temperature without a pronounced 'jump' [52].

Since the entropy produced right at impact (on a time scale of order $2R_A/\gamma$ in the CM frame) increases smoothly with energy, all hadron abundance ratios will behave correspondingly and the sharp 'horn' in K^+/π^+ seen in the data can not be reproduced. This holds for typical hydro models with a first-order phase transition [52] as well as for hadronic transport models (see discussion above).

If early-stage entropy poduction can not account for the sharp peak of K^+/π^+ ratio then perhaps the phase transition *back* to the broken phase (which occurs later on after some cooling) can ? This might be possible indeed if one abandons the equilibrium phase transition based on the macroscopic Gibbs construction and, in turn, introduces a dynamical microscopic treatment of phase transitions into hydrodynamics. It is well-known that first-order phase transitions lead to inhomogeneities such as high-density 'nuggets', surrounded by low-density 'voids'. Analogous effects are frequently discussed within the context of the QCD transition in the early universe, where inhomogeneities of the entropy (or baryon to photon ratio) might affect BBN. The usual mixed-phase construction applies on scales much larger than the size and separation of inhomogeneities, and on such scales the matter and entropy distributions appear smooth. On small scales however, for example in heavy-ion collisions, inhomogeneous density distributions have significant effects on observables which are non-linear functions of the density: take a large homogeneous system, split it in half, and move all baryons into one half, then let each half equilibrate. Suppose you can not measure the hadron multiplicities in each half separately, just the total. The obvious measurement, namely of baryon number, doesn't reflect the presence of the high-density nugget because the total baryon number is the same as for the homogeneous distribution. However, the total yield of K^+ over the total yield of π^+ will be larger than for the homogeneous system ! This is because the ratio is enhanced in the high-density nugget by a much bigger factor than it is suppressed in the low-density half of the system. (Other hadrons like (multi-)strange baryons of course compensate the strangeness and are also more abundant than for the homogeneous system.) The effect diminishes rapidly when

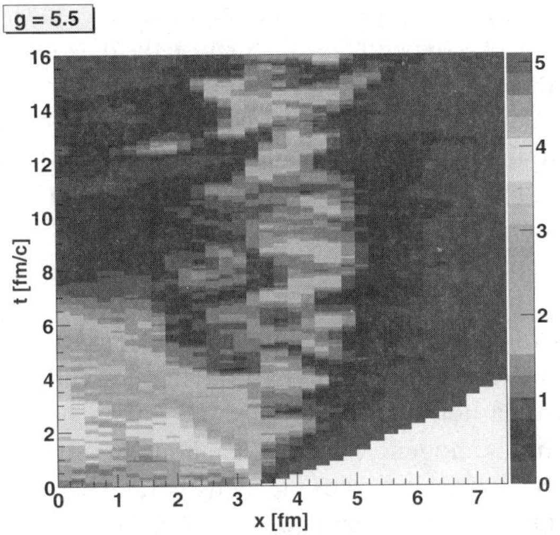

Figure 7. Space-time evolution of the fluid energy density along the x-axis at $y = z = 0$. The scale on the right specifies the energy density in units of nuclear matter density $\epsilon_0 \simeq 150$ MeV/fm^3.

the entropy per (net) baryon becomes large, that is, in the meson-dominated high-energy regime. To investigate the formation of inhomogeneities during the phase transition we solve for the coupled evolution of an order parameter field such as the chiral condensate ϕ and the thermalized matter fields [53]:

$$\Box\phi + \partial V_{\text{eff}}/\partial\phi = 0, \quad \partial_\mu\left(T^{\mu\nu}_{\text{fl}} + T^{\mu\nu}_{\phi}\right) = 0. \tag{2}$$

Here, $T^{\mu\nu}_{\text{fl}}$ is the energy-momentum tensor of the fluid, $T^{\mu\nu}_{\phi}$ that of the classical modes of the chiral condensate, and V_{eff} is the effective potential obtained by integrating out the thermalized degrees of freedom. We focus first on energy-density inhomogeneities and present solutions of these coupled equations for vanishing baryon density [53]. As initial condition we chose a homogeneous energy density above the critical energy density for the transition to the broken phase. However, the condensate ϕ exhibits 'primordial' Gaussian fluctuations on length scales $\sim$ 1 fm on top of a smoothly varying mean field. These fluctuations are then propagated through a first-order chiral phase transition and leave a rather inhomogeneous (energy-) density distribution in the wake of the transition, as seen in Fig. 7. Evidently, the scale for such fluctuations is not tiny and so it would not be appropriate to assume a homogeneous density distribution. On the other hand, they are too small to be resolved in rapidity space because the scale factor is large at times long after the initial impact. To resolve individual hot/dense spots would require a resolution better than one unit of rapidity, which is roughly equal to the thermal width of the local particle momentum distributions.

However, additional hints for the existence of large density inhomogeneities created in the course of the transition to the broken phase remain to be explored. (Inhomogeneities from fluctuations of particle production in the primary nucleon-nucleon collisions should be largely washed out until decoupling by hydrodynamic transport of matter due to pressure gradients, see e.g. [54].) Clearly, the yields of other hadron species depend non-linearly on the density as well, and their behavior has to be tested for consistency. Moreover, coordinate-space fluctuations of the energy-momentum tensor of matter produced by a phase transition are uncorrelated to the reaction plane and therefore should act to reduce out-of-plane collective flow ($v_2/\langle p_t\rangle$) as compared to equilibrium hydrodynamics, cf. the discussion in [53]. Finally, Hanbury-Brown–Twiss correlations could provide valuable coordinate-space information on the regions from which particles are emitted. In this regard, note the stunning result of CERES [55] according to which pions decouple when their mean-free path is ~ 1 fm. This is inconceivable in standard equilibrium hydrodynamics without density perturbations because there particles decouple only when their mean-free path exceeds the scale of spatial homogeneity, which is about an order of magnitude larger [56]. The CERES analysis indicates that density (and perhaps velocity) gradients in coordinate space are 1/1 fm rather than 1/10 fm.

7. Conclusions

Summarizing this contribution, we point out that baryon stopping [57] and hadron production in central Au+Au (or Pb+Pb) collisions is quite well described in the independent transport approaches HSD and UrQMD. Also the 'longitudinal' rapidity distributions of protons, pions, kaons, antikaons and hyperons are similar in both models and in reasonable agreement with available data. The exception are the pion rapidity spectra at the highest AGS energy and lower SPS energies, which are overestimated by both models [16]. As a consequence the HSD and UrQMD transport approaches underestimate the experimental maximum of the K^+/π^+ ratio ('horn') at $\sim$ 20 to 30 A·GeV. However, we point out that the maxima in the K^+/π^+ and $(\Lambda + \Sigma^0)/\pi$ ratios partly reflect a change from baryon to meson dominated dynamics with increasing bombarding energy.

We have found that the inverse slope parameters T for $K^\pm$ mesons from the HSD and UrQMD transport models are practically independent of system size from pp up to central Pb+Pb collisions and show only a slight increase with collision energy, but no 'step' in the $K^\pm$ transverse momentum slopes. The rapid increase of the inverse slope parameters of kaons for collisions of heavy nuclei (Au+Au) found experimentally in the AGS energy range, however, is not reproduced by neither model (see Fig. 4). Since the pion transverse mass spectra – which are hardly effected by collective flow – are described sufficiently well at all bombarding energies [29], the failure has to be attributed to a lack of pressure. This additional pressure should be generated in the early phase of the collision, where the 'transverse' energy densities in the transport approaches are higher than the critical energy densities for a phase transition (or cross-over) to the QGP. The interesting finding of our analysis is, that pre-hadronic degrees of freedom might already play a substantial role in central Au+Au collisions at AGS energies above ~ 5 A·GeV. The more astonishing is the experimentally observed collapse of the collective flow v_2 at 40 A·GeV - such a behavior has been predicted as signature for a first order phase transition long ago.

References

[1] J. Rafelski and B. Müller, *Phys. Rev. Lett.* **48** (1982) 1066.

[2] J.D. Bjorken, *Phys. Rev.* D **27** 1983 140.

[3] *Quark Matter 2002*, *Nucl. Phys.* A **715** (2003) 1.

[4] F. Karsch *et al.*, *Nucl. Phys.* B **502** (2001) 321.

[5] H. Stöcker and W. Greiner, *Phys. Rep.* **137** (1986) 277.

[6] W. Cassing, E.L. Bratkovskaya, and S. Juchem, *Nucl. Phys.* A **674** (2000) 249.

[7] Z. Fodor and S. D. Katz, JHEP **0203**, 014 (2002); Z. Fodor, S. D. Katz, and K. K. Szabo, Phys. Lett. B **568**, 73 (2003).

[8] V. Friese *et al.*, NA49 Collaboration, *J. Phys.* G **30** (2004) 119.

[9] M. I. Gorenstein, M. Gaździcki, and K. Bugaev, *Phys. Lett.* B **567** (2003) 175.

[10] M. Gaździcki and M. I. Gorenstein, *Acta Phys. Polon.* B **30** (1999) 2705.

[11] L. Van Hove, *Phys. Lett.* B **118** (1982) 138.

[12] S.A. Bass *et al.*, *Prog. Part. Nucl. Phys.* **42** (1998) 255.

[13] M. Bleicher *et al.*, *J. Phys.* G **25** (1999) 1859.

[14] J. Geiss, W. Cassing, and C. Greiner, *Nucl. Phys.* A **644** (1998) 107.

[15] W. Cassing and E. L. Bratkovskaya, *Phys. Rep.* **308** (1999) 65.

[16] H. Weber, E.L. Bratkovskaya, W. Cassing, and H. Stöcker, *Phys. Rev.* C **67** (2003) 014904.

[17] K. Hagiwara *et al.*, (Review of Particle Properties), *Phys. Rev.* D **66** (2002) 010001.

[18] B. Andersson *et al.*, *Z. Phys. C* **57** (1993) 485.

[19] Z. W. Lin *et al.*, *Nucl. Phys.* A **698** (2002) 375.

[20] L. Ahle *et al.*, E866 and E917 Collaboration, *Phys. Lett.* B **476** (2000) 1; *ibid.* **490** (2000) 53.

[21] J. L. Klay *et al.*, E895 Collaboration, *Phys. Rev.* C **68** (2003) 054905.

[22] S. Ahmad *et al.*, E891 Collaboration, *Phys. Lett.* B **382** (1996) 35; C. Pinkenburg *et al.*, E866 Collaboration, *Nucl. Phys.* A **698** (2002) 495c.

[23] S. V. Afanasiev *et al.*, NA49 Collaboration, *Phys. Rev.* C **66** (2002) 054902.

[24] A. Mischke *et al.*, NA49 Collaboration, *J. Phys.* G. **28** (2002) 1761; *Nucl. Phys.* A **715** (2993) 453.

[25] F. Antinori *et al.*, WA97 Collaboration, *Nucl. Phys.* A **661** (1999) 130c.

[26] D. Ouerdane *et al.*, BRAHMS Collaboration, *Nucl. Phys.* A **715** (2003) 478; J. H. Lee *et al.*, *J. Phys.* G **30** (2004) S85.

[27] S. S. Adler *et al.*, PHENIX Collaboration, preprint nucl-ex/0307010; preprint nucl-ex/0307022.

[28] C. Adler *et al.*, STAR Collaboration, preprint nucl-ex/0206008; O. Barannikova *et al.*, *Nucl. Phys.* A **715** (2003) 458; K. Filimonov *et al.*, preprint hep-ex/0306056.

[29] E. L. Bratkovskaya *et al.*, preprint nucl-th/0402026.

[30] E. L. Bratkovskaya, W. Cassing and H. Stöcker, *Phys. Rev.* C **67** (2003) 054905.

[31] S. Soff *et al.*, *Phys. Lett.* B **551** (2003) 115.

[32] K. Redlich, J. Cleymans, H. Oeschler, and A. Tounsi, *Acta Phys. Polonica* B **33** (2002) 1609.

[33] E. L. Bratkovskaya, S. Soff, H. Stöcker, M. van Leeuwen, and W. Cassing, *Phys. Rev. Lett.* **92** (2004) 032302.

[34] A. Förster *et al.*, KaoS Collaboration, *J. Phys.* G **28** (2002) 2011.

[35] B. V. Martemyanov *et al.*, nucl-th/0212064.

[36] H. Sorge, *Phys. Rev.* C **52** (1995) 3291.

[37] S. Soff *et al.*, *Phys. Lett.* B **471** (1999) 89.

[38] J. Geiss *et al.*, *Phys. Lett.* B **447** (1999) 31.

[39] I. Kraus *et al.*, NA49 Collaboration, *J. Phys.* G **30** (2004) 5583.

[40] I.G. Bearden *et al.*, NA44 Collaboration, preprint nucl-ex/0202019.

[41] M. Kliemant, B. Lungwitz, and M. Gaździcki, preprint hep-ex/0308002.

[42] H. van Hecke *et al.*, *Phys. Rev. Lett.* **81** (1998) 5764.

[43] H. Liu *et al.*, E895 Collaboration, *Phys. Rev. Lett.* **84** (2000) 5488.

[44] C. Alt *et al.*, NA49 Collaboration, preprint nucl ex/0303001, *Phys. Rev.* C, in press.

[45] J. Hofmann, H. Stöcker, U. Heinz, W. Scheid, and W. Greiner, *Phys. Rev. Lett.* **36** (1976) 88.

[46] W. Cassing and U. Mosel, *Prog. Part. Nucl. Phys.* **25**, 235 (1990).

[47] J. Cleymans and K. Redlich, *Phys. Rev.* C **60** (1999) 054908.

[48] L. V. Bravina *et al.*, *Phys. Rev.* C **60** (1999) 024904. *Nucl. Phys.* A **698** (2002) 383.

[49] F. Karsch, talk given in *Quark Matter 2004*, Oakland, January 11-17, 2004; http://qm2004.lbl.gov

[50] I. G. Bearden *et al.*, BRAHMS Collaboration, *Phys. Rev. Lett.* **90** (2003) 102301.

[51] E. V. Shuryak, *Nucl. Phys.* A **661** (1999) 119c.

[52] M. Reiter *et al.*, *Nucl. Phys.* A **643** (1998) 99.

[53] K. Paech, H. Stöcker and A. Dumitru, *Phys. Rev.* C **68** (2003) 044907; K. Paech, preprint nucl-th/0308049.

[54] M. Bleicher *et al.*, *Nucl. Phys.* A **638** (1998) 391.

[55] D. Adamova *et al.*, CERES Collaboration, *Phys. Rev. Lett.* **90** (2003) 022301.

[56] S. Soff, S. A. Bass and A. Dumitru, *Phys. Rev. Lett.* **86** (2001) 3981.

[57] H. Weber, E.L. Bratkovskaya and H. Stöcker, *Phys. Lett.* B **545** (2002) 285.

TOWARDS A COMPREHENSIVE DESCRIPTION OF HEAVY ION REACTIONS

F. Cerutti[1], F. Ballarini[2], G. Battistoni[1], M. Cavinato[1], A. Empl[3], E. Fabrici[1], A. Fassò[4], A. Ferrari[5], E. Gadioli[1], E. Gadioli Erba[1], M. V. Garzelli[1], A. Ottolenghi[2], V. Parini[1], L. S. Pinsky[3], J. Ranft[6], P. R. Sala[7]
[1]*Milan University and INFN, Italy,* [2]*Pavia University and INFN, Italy,* [3]*Houston University, Texas, USA,* [4]*SLAC, Stanford, USA,* [5]*CERN, Switzerland (on leave from INFN, Milan),* [6]*Siegen University, Germany,* [7]*ETH, Zurich, Switzerland (on leave from INFN, Milan)*

Abstract Comprehensive analysis of heavy ion reactions at low and intermediate energies is made using the Boltzmann Master Equation theory and a Relativistic Quantum Molecular Dynamics model, respectively. As an example of the large variety of applications of such a study, a result concerning space radiation protection is presented.

1. Introduction

A considerable experimental information on energy, spin and parity of nuclear levels and their decay properties is available [1], and is of paramount importance for the study of nuclear structure. However, a comparable systematic information in nuclear dynamics does not exist. It should basically consist, for each pair of interacting nuclei, of the knowledge of elastic scattering and reaction cross sections, of the residue excitation functions, of emitted particle spectra, correlation and polarization. Researches aiming at this goal have been done to different degrees of systematicity for nucleon induced reactions, especially in the case of neutrons, and provided the data on which our understanding of nuclear reaction mechanisms is based (see for instance Refs. [2, 3] and references therein). Much less is known in the case of heavy ion reactions where even the measurements of the most basic quantity, i.e., the reaction cross section, are rather scarce. This lack of information reflects on our present understanding of the reaction mechanisms for heavy ions which in most cases is only qualitative and does not allow precise cross section predictions.

On the other hand, accurate knowledge of reactions which may occur when two heavy ions interact is of crucial importance in many trans-disciplinary fields,

W. Greiner et al. (eds.), Structure and Dynamics of Elementary Matter, 255–263.

particularly in medicine and space physics (see a collection of recent papers on the subject in Ref. [4]). In these cases one needs to know not just the results of particular processes, as investigated in many experiments, but what happens in a natural process to which all possible reaction mechanisms contribute. The collection of this information requires a considerable experimental and theoretical effort. In fact, a theoretical calculation to be really usable should be able to reproduce large sets of data in wide energy and mass ranges. Such calculations are actually performed for nucleon induced reactions, and examples of them and their relevance in interdisciplinary fields may be found in the specialized literature [5–9], while heavy ion interactions are much less studied.

In this paper a short account is given of some results which we have recently obtained in a series of investigations aimed to contribute to fill this gap in our knowledge, considering heavy ion interactions from the Coulomb barrier up to relativistic energies and adopting different approaches. Reactions induced by light projectiles at few tens of MeV/n were studied by very simplified mean field calculations coupled with the Boltzmann Master Equation (BME) theory (Section 2). Reactions between medium and heavy nuclei at few hundreds of MeV/n were analysed interfacing FLUKA[1] with a modified version of a Relativistic Quantum Molecular Dynamics (RQMD) model developed by the Frankfurt group for energies higher than 100 MeV/n [13] (Section 3).

In the last Section we discuss possible applications of these models in the field of radiation biological effects, and we present preliminary calculations concerning space radiation protection.

2. Mean field and BME calculations

When two heavy ions collide, in the lowest energy range there is a continuous interplay between one-body (mean field) and two-body (nucleon–nucleon) interactions. However, to make tractable this process we assume the mean field interaction to be dominant until the two ions come in contact or, in the case of a light projectile, one of its fragments comes in contact with the target. Afterwards we assume that a composite nucleus is created which thermalizes by means of a cascade of nucleon–nucleon interactions, a process which we simulate with the BME's by evaluating the time variation of the momentum distribution of the nucleons belonging to the composite nucleus as a result of their mutual interactions and their emission into the continuum as separate entities or as a part of a cluster.

In the case of reactions induced by light projectiles, it seems a fair approximation to consider a rather small number of reaction mechanisms. Namely

[1]FLUKA is a transport and interaction MonteCarlo code capable of describing in any complex geometry the propagation of showers induced by hadronic and electromagnetic particles ranging from thermal neutrons up to cosmic rays with several TeV incident energy [10–12].

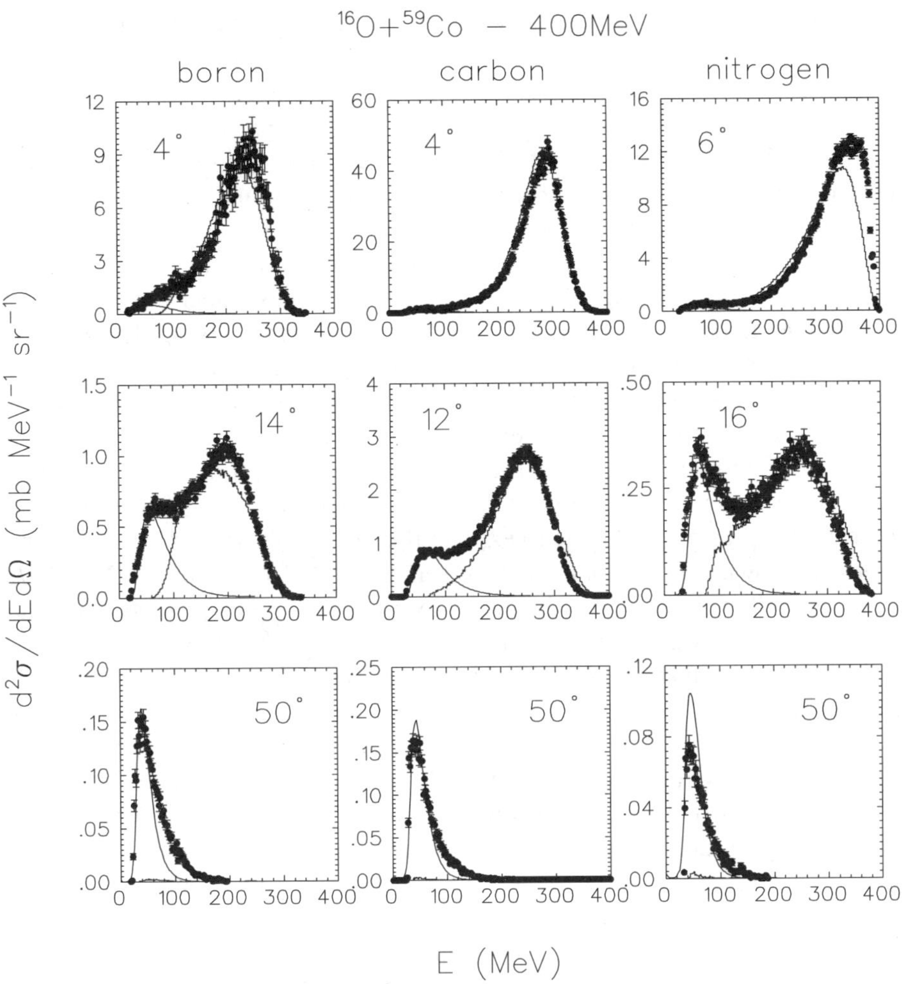

Figure 1. Spectra of boron, carbon and nitrogen fragments emitted at different angles in the interaction of 400 MeV ^{16}O with ^{59}Co [17, 18]. The experimental data are given by the black circles, the calculated binary fragmentation and coalescence contributions are given by the histograms and the full lines, respectively.

complete fusion, binary fragmentation of the projectile followed in most cases by the fusion of one fragment with the target (break-up-fusion reactions), transfer of a single nucleon from the projectile to the target, and inelastic scattering [14–18]. Once a composite nucleus is excited by one of these mechanisms, the BME theory [19–21] allows one to calculate the time evolution of the occupation probability $n(\epsilon, \theta, t)$ of the nucleon states by integrating a set of coupled differential equations, which express the variation of $n(\epsilon, \theta, t)$ in the time interval between t and $t + dt$ as a function of the occupation probabilities at time t and the decay rates for nucleon–nucleon scatterings and emissions into the continuum.

To do away with the intrinsic limitation of averaging over many different reaction paths, the pre-equilibrium spectra obtained by solving the BME's may be used as an input for a MonteCarlo calculation assuming that the probability of emitting a particle i with energy between ϵ and $\epsilon + d\epsilon$ at a polar angle between θ and $\theta + d\theta$ in the time interval $(t,\ t+dt)$ be equal to its differential multiplicity $d^3M_i(\epsilon, \theta, t)$ [22, 23]. The probability of any possible sequence of events may be evaluated as a joint probability using these *elementary* probabilities.

The analysis of the data indicates that one has to take into account the full complexity of the interaction even to account for observables which might be thought to be related to a well defined reaction mechanism. This is the case of intermediate mass fragment (IMF) production which one could ascribe to projectile fragmentation. This mechanism is responsible for their production at forward angles but is unable to explain their quite sizeable emission at large angles, as shown in Fig. 1 for the IMF's from the ^{16}O+^{59}Co system at 25 MeV/n. At those angles the major role is played by a more complex process which has been interpreted as the nucleon coalescence in the course of the thermalization of the composite nuclei created in the complete fusion and break-up-fusion reactions [17, 18].

Fig. 2 shows how these calculations may also reproduce quite demanding observables, like excitation functions of individual residues. In the interaction of ^{12}C with ^{103}Rh, the heaviest residues such as ^{110}Sn and ^{111g}In are produced by complete fusion and the high energy tail in their excitation functions is indicative of pre-equilibrium emissions prior to evaporation of equilibrated nuclei. The production of $^{108m+g}$In and ^{105g}Ag is mainly due to break-up-fusion reactions, while the production of near target nuclei (which at the highest considered energies turn out to have the largest cross sections) is due to mechanisms in which the projectile transfers very little mass, energy and momentum to the target, such as i) break-up of the projectile into α-type fragments (^{8}Be and α) followed by incomplete fusion and subsequent re-emission of absorbed α's [24], ii) single nucleon transfer, and iii) inelastic scattering. To the production of residues with mass considerably smaller than the target mass, all possible reaction mechanisms contribute to a various degree.

It has to be stressed that the calculations shown in Fig. 2 are the same reproducing the IMF spectra and in spite of their complexity require a small number of free parameters displaying a weak dependence on the target mass, thus suggesting that they may be used with a fair confidence to predict the cross sections of unmeasured reactions.

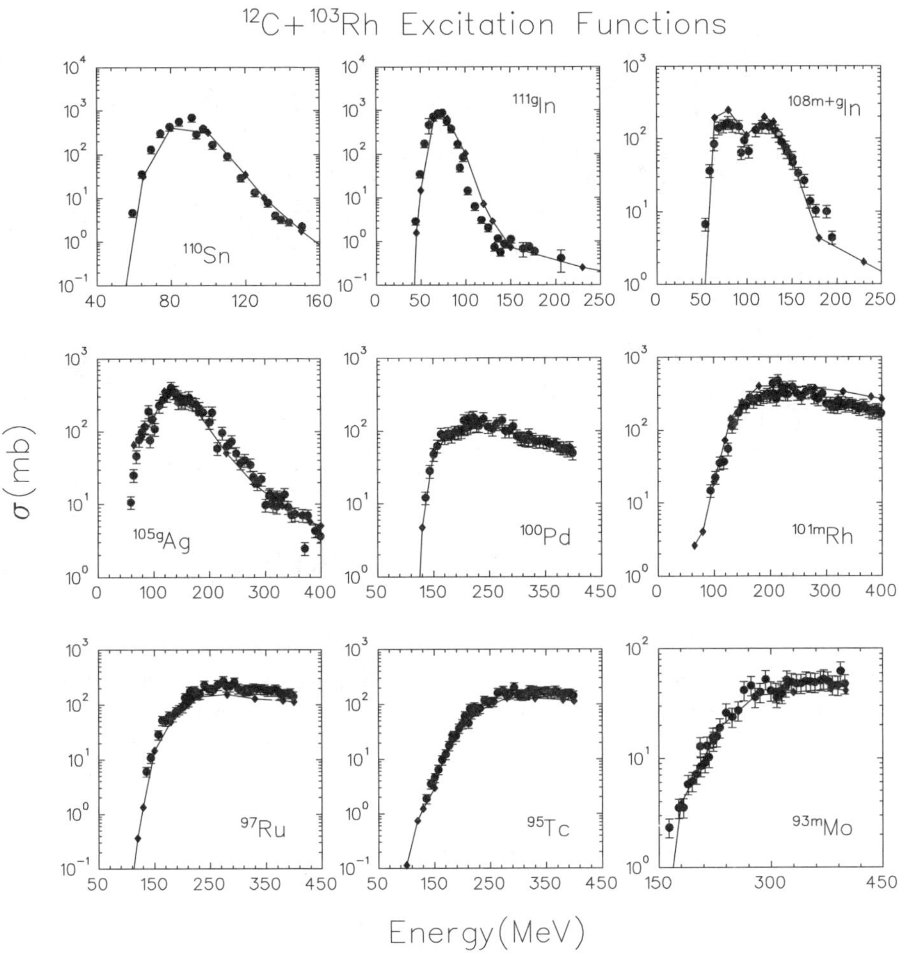

Figure 2. Residue excitation functions in the interaction of ^{12}C with ^{103}Rh up to 400 MeV bombarding energy. The experimental data [14, 15] are represented by black circles, the theoretical predictions by black diamonds connected by straight lines.

3. RQMD calculations

At higher incident energies different approaches seem to be more appropriate. The QMD model represents a viable solution thanks to its dynamic description of the force field among nucleons during the reaction. Its implementation in the code RQMD-2.4 by H. Sorge turned out to be successful in simulating relativistic nucleus–nucleus particle production over a wide energy range [25].

Interfacing of RQMD-2.4 with FLUKA, allowing the latter to treat heavy ion nuclear interactions between 100 MeV/n and 5 GeV/n, required some further

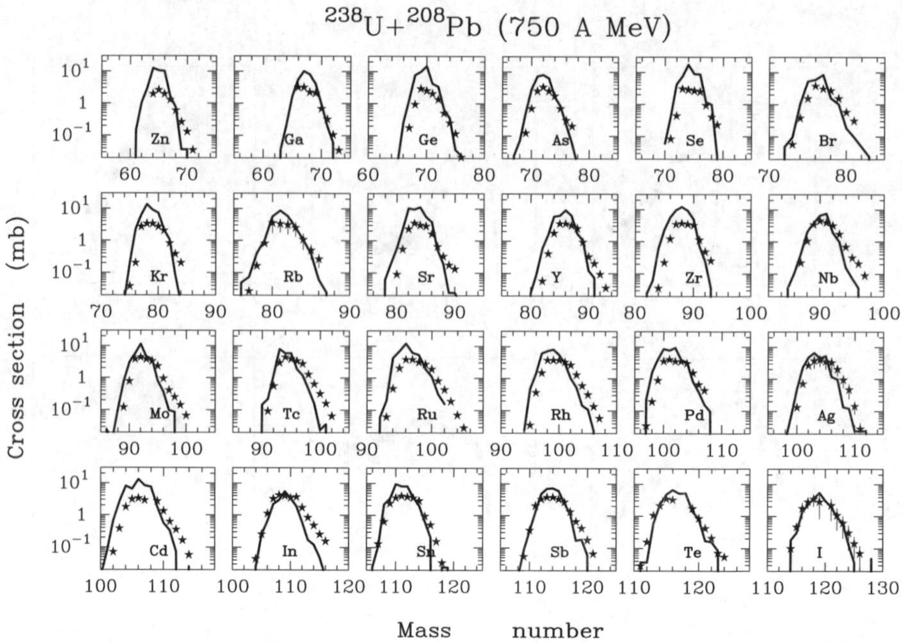

Figure 3. Isotopic distributions of fragmentation products for the $^{238}U+^{208}Pb$ system at 750 MeV/n. Data [26] are represented by stars. The fission products have been excluded in the experimental analysis and so in our calculation (full lines) too.

developments. Originally, RQMD-2.4 was not giving any information on the residues and fragments produced at the end of the fast stage of the reaction. We assume that projectile- and target-like nuclei may be identified by gathering spectator nucleons, i.e. those not involved in the primary interactions. Their excitation energy is that corresponding to the holes left by the hit nucleons. Exact energy-momentum conservation is assured event-by-event taking into account experimental binding energies. Finally, the last stage of the deexcitation process of projectile- and target-like nuclei is calculated using the FLUKA evaporation/fission/fragmentation module.

Fig. 3 shows a recent result obtained by coupling this modified version of RQMD-2.4 with FLUKA. Other examples, regarding double differential neutron yields too, can be found in Ref. [27]. In spite of the achieved good reproduction of the experimental data, it might be more proper to assign to the two residues also the nucleons whose kinetic energy is below a given threshold in the rest nucleus frame and to describe the subsequent interactions of these nucleons within projectile- and target-like nuclei by a suitable pre-equilibrium module. The implementation of this development is under way.

4. Heavy ions and biological targets

During the last decade, the interaction of protons and heavier ions in biological targets has raised a renewed interest, also due to progress in the field of hadrontherapy. In fact ions, due to the localization of energy deposition in the Bragg peak region, allow for a better conformation of the dose. This makes them particularly suitable for the treatment of localized tumors. Ion therapy, initially performed with proton beams of energies of some hundreds MeV, is now made also with Carbon ions, which provide a higher biological effectiveness due to their higher Linear Energy Transfer (LET). Heavy ion interactions are of concern also for radiation protection in space, where astronauts are continuously exposed (1 mSv/day) to Galactic Cosmic Rays. The GCR flux consists of 87% protons, 12% Helium ions and 1% heavier ions ("HZE particles"), all with energies peaked around 1 GeV/n; Carbon and Iron are among the most abundant elements in the HZE particle spectrum. In spite of the small contribution of HZE particles to the GCR flux, these particles provide a large contribution to the absorbed dose, up to 50%.

The original radiation field is significantly modified by interactions with the beam line constituents (in the case of hadrontherapy), with the spacecraft walls and shielding structures (in the case of space radiation) and with the human body (in both cases). A physical and biophysical characterization of the radiation fields of interest is therefore necessary. Radiation transport codes and models simulating radiation-induced biological damage are a powerful tool in this field, where experimental data are currently scarce. In this framework, the FLUKA code has been used to calculate distributions of absorbed and "biological" organ doses in two anthropomorphic phantoms (a mathematical model and a "voxel" model) . "Biological dose" refers to the average number of radiation-induced "Complex Lesions" (CL) per cell in the organ of interest. CL are a clustered DNA damage type that plays a fundamental role both in cell killing (of interest in radiation therapy) and in chromosome aberration induction, as well as in cell conversion to malignancy (of interest in radiation protection). Yields of CL Gy^{-1} $cell^{-1}$ induced by radiation of different qualities have been previously calculated with "event-by-event" radiation track-structure simulations [28]. These yields have subsequently been integrated in FLUKA, that allowed us to characterize therapeutic proton beams [29] and to calculate organ doses due to the proton component of Solar Particle Events, which are episodic but intense injections of charged particles - mainly protons - from the Sun [30]. Following the recent implementation in FLUKA of nucleus–nucleus interactions below 5 GeV/n, simulations aimed to calculate organ doses due to the complex GCR spectra are under way. We present herein results of preliminary tests relative to irradiation with monochromatic 1 GeV/n ion beams, including Helium, Carbon and Iron. The calculations were repeated for different

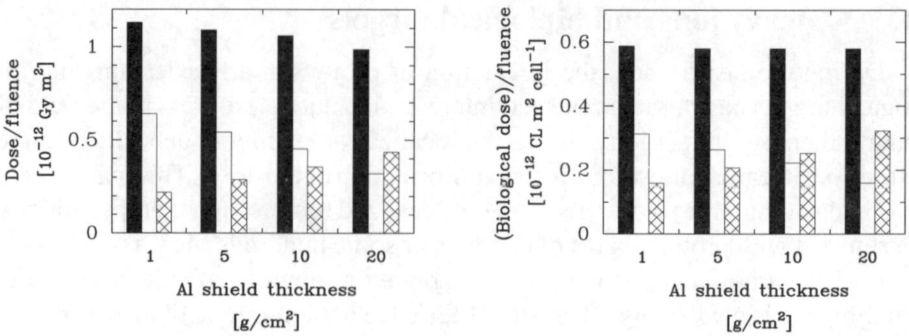

Figure 4. Skin absorbed (on the left) and biological (on the right) dose per unit fluence as a function of the Al shield thickness calculated by simulating a mathematical phantom irradiated with a 1 GeV/n Carbon ion beam. White bars and shaded bars represent the contributions of primary ions and secondary hadrons, respectively, to the total dose (black bars), which includes also the electromagnetic contribution.

values of Al shield thickness, i.e. 1, 5 and 10 or 20 g cm^{-2} (nominal spacesuit, spacecraft and storm shelter, respectively).

Fig. 4 shows on the left the absorbed dose to skin per unit fluence as a function of the Al shield thickness following irradiation of the mathematical phantom with a 1 GeV/n Carbon-ion beam. The contributions of primary ions and secondary particles produced in nuclear reactions (white bars and shaded bars, respectively) to the total dose (black bars) are shown separately. By increasing the shield thickness, the contribution of primary ions decreases, whereas that of nuclear reaction products increases. The role of nuclear reactions is even more important when looking at the biological dose (on the right). In fact, at 5 and 10 g cm^{-2} shielding the contribution of secondaries is close to that of primary ions, whereas at 20 g cm^{-2} it is about 2-fold with respect to primary ions, thus leading to a constant total dose with increasing shielding.

Acknowledgments

This work was partially supported under NASA Grants NAG8-1658 and 01-OBPR-05, ASI contract 1/R/320/02, DoE contract DE-AC03-76SF00515 and EC contract FIGH-CT1999-00005.

References

[1] R. B. Firestone, *Table of Isotopes*, Eight Edition, John Wiley & Sons Inc., 1999.

[2] G. R. Satchler, *Direct Nuclear Reactions*, Clarendon Press, Oxford, Oxford University Press, New York, 1983.

[3] E. Gadioli and P. E. Hodgson, *Pre-Equilibrium Nuclear Reactions*, Oxford Science Publications, Clarendon Press, Oxford, 1992.

[4] E. Gadioli ed., *Proceedings of the 10th International Conference on Nuclear Reaction Mechanisms*, Varenna, June 9-13, 2003, Ricerca Scientifica ed Educazione Permanente, Suppl. 122, 2003.

[5] IAEA-TECDOC-1034, *Handbook for calculation of nuclear reaction data, Reference input parameter library*, 1998.

[6] M. B. Chadwick *et al.*, Nucl. Sci. Eng. **131**, 293 (1999).

[7] A. Gandini and G. Reffo eds., *Nuclear Reaction Data and Nuclear Reactors - Physics, Design, and Safety*, World Scientific, Singapore, 1999.

[8] ICRU Report 63, *Nuclear Data for Neutron and Proton Radiotherapy and for Radiation Protection*, 2000.

[9] INDC(NDS)-431, *Nuclear model parameter testing for nuclear data evaluation*, 2002.

[10] A. Fasso, A. Ferrari, J. Ranft, and P.R. Sala, Proceedings of SARE-3, KEK-Tsukuba, H. Hirayama ed., May 7–9 1997, KEK report Proceedings 97-5, 32 (1997).

[11] A. Fasso, A. Ferrari, J. Ranft, and P.R. Sala, Proceedings of the MonteCarlo 2000 Conference, Lisbon, October 23–26 2000, A. Kling, F. Barao, M. Nakagawa, L. Tavora, P. Vaz eds., Springer-Verlag Berlin, 955 (2001).

[12] A. Fasso, A. Ferrari and P.R. Sala, Proceedings of the MonteCarlo 2000 Conference [11], 159 (2001).

[13] H. Sorge, H. Støcker and W. Greiner, Nucl. Phys. **A498**, 567c (1989); Ann. Phys. **192**, 266 (1989).

[14] E. Gadioli *et al.*, Phys. Lett. **B394**, 29 (1997).

[15] E. Gadioli *et al.*, Nucl. Phys. **A641**, 271 (1998).

[16] E. Gadioli *et al.*, Eur. Phys. J. **A11**, 161 (2001).

[17] E. Gadioli *et al.*, Nucl. Phys. **A708**, 391 (2002).

[18] E. Gadioli *et al.*, Eur. Phys. J. **A17**, 195 (2001).

[19] C. Brusati *et al.*, Z. Phys. **A353**, 57 (1995).

[20] M. Cavinato *et al.*, Z. Phys. **A347**, 237 (1994).

[21] M. Cavinato *et al.*, Nucl. Phys. **A643**, 15 (1998).

[22] M. Cavinato *et al.*, Phys. Lett. **B382**, 1 (1996).

[23] M. Cavinato *et al.*, Nucl. Phys. **A679**, 753 (2001).

[24] E. Gadioli *et al.*, Nucl. Phys. **A654**, 523 (1999).

[25] H. Sorge, Phys. Rev. **C52**, 3291 (1995).

[26] J. Benlliure *et al.*, Eur. Phys. J. **A2**, 193 (1998).

[27] V. Andersen *et al.*, *The* FLUKA *code for space applications: recent developments*, Adv. Space Res., in press.

[28] A. Ottolenghi *et al.*, Radiat. Environ. Biophys. **34**, 239 (1995).

[29] M. Biaggi *et al.*, Nucl. Instr. Meth. Phys. Res. **B159**, 89 (1999).

[30] F. Ballarini *et al.*, *Role of shielding in modulating the effects of Solar Particle Events: Monte Carlo calculation of physical and "biological" dose in different organs*, Adv. Space Res., in press.

SIGNALS OF DECONFINEMENT TRANSITION IN NUCLEUS-NUCLEUS COLLISIONS

Mark Gorenstein
Bogolyubov Institute for Theoretical Physics, Kiev, Ukraine
mark@mgor.kiev.ua

Abstract We discuss the energy dependence of hadron production in relativistic nucleus-nucleus collisions. Several 'anomalies' in the energy dependence have been predicted as signals of the deconfinement phase transition and three these signals are observed at the CERN SPS indicating that the onset of deconfinement in Pb+Pb collisions is located at about 30 A·GeV.

Keywords: Nucelus-nucleus collisions, energy scan programme, quark-gluon plasma.

1. Introduction

The data on heavy nucleus-nucleus (A+A) collisions suggested that there is a significant change in the energy dependence of the pion and strangeness yields which is located between the top AGS (11 A·GeV) and SPS (158 A·GeV) energies. Based on the statistical approach it was speculated that the change is related to the onset of deconfinement at the early stage of A+A collisions, and a simplified quantitative model was developed, the Statistical Model of the Early Stage (SMES) [1]. The SMES predicted a sharp maximum in the multiplicity ratio of strange hadrons to pions at the begining of the transition region. This prediction triggered a new experimental programme at the SPS – the energy scan programme – in which the NA49 experiment recorded central Pb+Pb collisions at several energies: the results from the run at 40, 80 and 158 A·GeV are published [2], the results from the 30 A·GeV run were shown for the first time at the conference SQM03 [3], the data at 20 A·GeV are still being analysed. The energy scan program at the CERN SPS results an observation of several 'anomalies' in the energy dependence of hadron production in the same domain of the collision energy. These 'anomalies' are interpreted as signals of the deconfinement phase transition in A+A collisions. In this report we review the physical arguments which lead us to the proposed signals as well as discuss their experimental status.

W. Greiner et al. (eds.), Structure and Dynamics of Elementary Matter, 265–276.

2. Signals of Deconfinement: Model Predictions

Originally, two signals of the deconfinement phase transition were proposed within SMES [1]: the energy dependence of the mean pion and strangeness multiplicities. Recently, two new signals were suggested within SMES: the energy dependence of the inverse slope of the transverse mass spectrum of kaons [4] and the energy dependence of properly filtered multiplicity fluctuations [5, 6].

An exact nature of the deconfinement phase transition is still debated. On the other hand, it is rather well established in the lattice QCD at zero baryonic chemical potential that very strong changes of the energy density ε take place in a narrow temperature interval $\Delta T = 5 \div 10$ MeV. Within this temperature interval the energy density changes by about an order of magnitude, whereas the pressure remains approximately unchanged. One may refer to this temperature interval as a 'generalised mixed phase'.

The Pion Multiplicity. The majority of all particles produced in high energy interactions are pions. Thus, pions carry basic information on the entropy created in the collisions. On the other hand, the entropy production should depend on the form of matter present at the early stage of collisions. A deconfined matter is expected to lead to the final state with higher entropy than that created by a confined matter. Consequently, it is natural to expect that the onset of creation of a deconfined matter should be signalled by an enhancement of the pion production. Clearly, a trivial dependence of the pion multiplicity on the size of colliding nuclei should be removed and thus a relevant observable is the ratio of the mean pion multiplicity $\langle\pi\rangle$ to the mean number of wounded nucleons $\langle N_W\rangle$. The simple intuitive argumentation can be further quantified within SMES assuming the generalised Fermi–Landau initial conditions: the initial volume is Lorenz-contracted, $V \propto (\sqrt{s})^{-1}$ ($\sqrt{s}$ is the c.m.s. energy of the nucleon pair), the initial energy density is given by $\varepsilon \propto gT^4 \propto (\sqrt{s} - 2m_N) \cdot \sqrt{s}$ (T is the initial temperature and g is an effective number of internal degrees of freedom at the early stage, m_N is the nucleon mass). The pion multiplicity is proportional to the initial entropy, and the $\langle\pi\rangle/\langle N_W\rangle$ ratio can be thus calculated outside the transition region as

$$\langle\pi\rangle/\langle N_W\rangle \propto V\, g\, T^3 \propto g^{1/4}(\sqrt{s} - m_N)^{3/4}(\sqrt{s})^{-1/4} \equiv g^{1/4}\, F\,. \quad (1)$$

Therefore, the $\langle\pi\rangle/\langle N_W\rangle$ ratio increases linearly with F outside the transition region, and the slope parameter is proportional to $g^{1/4}$ [7]. In the transition region, a steepening of the pion energy dependence is predicted, because of activation of partonic degrees of freedom, i.e. an effective number of internal degrees of freedom in the quark gluon plasma (QGP) is larger than in the hadron gas (HG): $g_{QGP} > g_{HG}$.

The Strangeness to Pion Ratio. The energy dependence of the strangeness to entropy ratio is a crucial signal of the deconfinement due to its weak dependence on the assumed initial conditions. Within SMES at low collision energies, when the confined matter is produced, the strangeness to entropy ratio steeply increases with collision energy. Due to a low temperature at the early stage ($T < T_C \cong 170$ MeV) and the high mass of the carriers of strangeness ($m_S \cong 500$ MeV, the kaon mass) the total strangeness is $\propto \exp(-m_S/T)$. On the other hand, the total entropy is approximately $\propto T^3$. Therefore, the strangeness to pion ratio is $\propto \exp(-m_S/T) \cdot T^{-3}$ in the HG and strongly increases with the collision energy. When the transition to a deconfined matter is crossed ($T > T_C$), the mass of the strangeness carriers is significantly reduced ($m_s = 130 \div 170$ MeV, the strange quark mass). Due to the low mass ($m_s < T$) the strangeness yield becomes (approximately) proportional to the entropy (both are proportional to T^3), and the strangeness to entropy (or pion) ratio becomes independent of energy in the QGP. This leads to a "jump" in the energy dependence from the larger value for confined matter at T_C to the value for deconfined matter. Thus, within the SMES, the non-monotonic energy dependence of the strangeness to entropy ratio is followed by a saturation at the deconfined value which is a direct consequence of the onset of deconfinement taking place at about 30 AGeV [1].

The Inverse Slope of Transverse Mass Spectra. We discuss another well known observable, which may be sensitive to the onset of deconfinement, the transverse momentum, p_T, spectra of produced hadrons. It was suggested by Van Hove [8] more than 20 years ago to identify the deconfinement phase transition in high energy proton-antiproton interactions with a plateau-like structure of the average transverse momentum as a function of the hadron multiplicity[1]. According to the general concepts of the hydrodynamical approach the hadron multiplicity reflects the entropy, whereas the transverse hadron activity reflects the combined effects of temperature and collective transverse expansion. The entropy is assumed to be created at the early stage of the collision and is approximately constant during the hydrodynamic expansion. The multiplicity is proportional to the entropy, $S = s \cdot V$, where s is the entropy density and V is the effective volume occupied by particles. During the hydrodynamic expansion, s decreases and V increases with $s \cdot V$ being approximately constant. Large multiplicity at high energies means a large entropy density at the beginning of the expansion (and consequently a larger volume at the end). Large value of s at the early stage of the collisions means normally high temperature T at this stage. This, in turn, leads to an increase of the transverse hadron activity, a

[1] In the original suggestion [8] the correlation between average transverse momentum and hadron multiplicity was discussed for $p + \bar{p}$ at fixed collision energy. Today we have an advantage to use central A+A collisions at different energies [4].

flattening of the transverse momentum spectrum. Therefore, with increasing collision energy one expects to observe an increase of both the hadron multiplicity and the average transverse momentum per hadron. However, presence of the deconfinement phase transition would change this correlation. In the phase transition region, the initial entropy density (and hence the final hadron multiplicity) increases with the collision energy, but temperature $T = T_C$ and pressure $p = p_C$ remain constant. The equation of state (EoS) presented in a form $p(\varepsilon)/\varepsilon$ versus ε shows a minimum (the 'softest point' [9]) at the boundary of the *generalised mixed phase* and the QGP. Consequently the shape of the p_T spectrum is approximately independent of the multiplicity or the collision energy. The transverse expansion effect may even decrease when crossing the transition region [8]. Thus one expects an 'anomaly' in the energy dependence of the transverse hadron activity: the average transverse momentum increases with the collision energy when the early stage matter is either in a pure confined or in a pure deconfined phase, and it remains approximately constant when the matter is in a mixed phase [4, 8]. A simplified picture with $T = T_C$ inside the mixed phase is changed if the created early stage matter has a non-zero baryonic density. It was however demonstrated [10] that the main qualitative features ($T \cong const$, $p \cong const$, and a minimum of the function $p(\varepsilon)/\varepsilon$ versus ε) are present also in this case. In the SMES model [1], which correctly predicted energy dependence of the pion and strangeness yields, the modification of the EoS due to the deconfinement phase transition is located between 30 and about 160 $A{\cdot}$GeV. Thus an anomaly in the energy dependence of the transverse hadron activity may be expected in this energy range.

The energy density at the early stage increases with increasing collision energy. At low and high energies, when a pure confined or deconfined phase is produced, this leads to an increase of the initial temperature and pressure. This, in turn, results in an increase of the transverse expansion of a matter and consequently a flattening of the transverse mass spectra of final state hadrons. The experimental data on the transverse mass spectra ($m_T = (m^2 + p_T^2)^{1/2}$, m is a particle mass) are usually parametrized by a simple exponential dependence:

$$\frac{dN}{m_T dm_T} = C \exp\left(-\frac{m_T}{T^*}\right) , \tag{2}$$

where the inverse slope parameter T^* is sensitive to both the thermal and collective motion in the transverse direction. In the parameterisation (2), the shape of the m_T spectrum is fully determined by a single parameter, the inverse slope T^*. In particular, the average transverse mass $\langle m_T \rangle$ can be expressed as:

$$\langle m_T \rangle = T^* + m + \frac{(T^*)^2}{m + T^*} . \tag{3}$$

Hydrodynamical transverse flow with collective velocity v_T modifies the Boltzmann m_T-spectrum of hadrons. At low transverse momenta, it leads to the result

(T_{kin} is a kinetic freeze-out temperature):

$$T^*_{low-p_T} = T_{kin} + \frac{1}{2} m\, v_T^2 . \tag{4}$$

A linear mass dependence (4) of T^* is supported by the data for hadron spectra at small p_T. However, for $p_T >> m$ the hydrodynamical transverse flow leads to the mass-independent blue-shifted 'temperature':

$$T^*_{high-p_T} = T_{kin} \cdot \sqrt{\frac{1+v_T}{1-v_T}} . \tag{5}$$

A simple one parameter exponential fit (2) is quite accurate up to $m_T - m \cong 1$ GeV for K^+ and K^- mesons in A+A collisions at all energies. This means that $T^*_{low-p_T} \approx T^*_{high-p_T}$ for kaons and the energy dependence of the average transverse mass $\langle m_T \rangle$ (3) and the average transverse momentum $\langle p_T \rangle$ for kaons is qualitatively the same as that for the parameter T^*. Note that a simple exponential fit (2) neither works for light π-mesons, $T^*_{low-p_T} < T^*_{high-p_T}$, nor for heavy (anti)protons and (anti)lambdas, $T^*_{low-p_T} > T^*_{high-p_T}$. This means that the average transverse masses, $\langle m_T \rangle$, and their energy dependence for these hadrons are not connected to the behavior of the slope parameters in the simple way described by Eq. (3): one should separately consider both $T^*_{low-p_T}$ and $T^*_{high-p_T}$ slopes (see Refs. [11, 12] for details).

The Dynamical Event-by-Event Fluctuations. In thermodynamics, the energy E, volume V and entropy S are related to each other through the EoS. Thus, various values of the energy of the initial equilibrium state lead to different, but uniquely determined, initial entropies. When the collision energy is fixed, the energy, which is used to hadron production still fluctuates. These fluctuations of of the inelastic energy are caused by the fluctuations in the dynamical process which leads to the particle production. They are called the dynamical energy fluctuations [5]. Clearly the dynamical energy fluctuations lead to the dynamical fluctuations of any macroscopic parameter of the matter, X, like its entropy and strangeness content. The relation between the dynamical energy fluctuations and dynamical fluctuations of the macroscopic parameter X is given by the EoS. Consequently, simultaneous measurements of the event-by-event fluctuations of both the energy and the parameter X yield information on the EoS. Since EoS shows an anomalous behavior in the phase transition region, the anomaly should be visible in the ratio of entropy to energy fluctuations [5].

According to the first and the second principles of thermodynamics the entropy change δS is given as $T\delta S = \delta E + p\delta V$. If we fix the collision geometry, choosing e.g. only a sample of central A+A collisions, we can expect $\delta V \cong 0$. Within SMES the ratio of entropy to energy fluctuations can be then easily calculated and presented as a simple function of the p/ε ratio [5]:

$$R_e \equiv \frac{(\delta S)^2/S^2}{(\delta E)^2/E^2} = \left(1 + \frac{p}{\varepsilon}\right)^{-2} . \qquad (6)$$

Thus it is easy to predict a qualitative dependence of the R_e ratio on the collision energy. Within the model, the confined matter, which is modelled as an ideal gas, is created at the collision early stage below the energy of 30 A·GeV. In this domain, the ratio p/ε, and consequently the R_e ratio, are approximately independent of the collision energy and equal about 1/3 and 0.56, respectively. The model assumes that the deconfinement phase–transition is of the first order. Thus, there is the mixed phase region, corresponding to the energy interval 30÷60 A·GeV. At the end of the mixed phase the p/ε ratio reaches minimum (the "softest point" of EoS [9]). Thus in the transition energy range the R_e ratio increases and reaches its maximum, $R_e \approx 0.8$, at the end of the transition domain. Further on, in the pure deconfined phase, which is represented by an ideal quark-gluon gas under bag pressure, the p/ε ratio increases and approaches its asymptotic value 1/3 at the highest SPS energy 160 A·GeV. An estimate of entropy fluctuations can be obtained form the analysis of multiplicity fluctuations as proposed in [5].

At the stage of particle freeze-out, the system's entropy is related to the mean particle multiplicity. We assume that the multiplicity of negatively charged hadrons is proportional to the system entropy, $S \propto \overline{N}_-$. Thus the initial entropy fluctuations are transformed into the the fluctuations of the *mean* multiplicity. It is important to distinguish these dynamical fluctuations of $\overline{N}_-$ formed at the initial stage of A+A reaction, from the statistical fluctuations of N_- around $\overline{N}_-$ at the freeze-out (see Ref. [5] for details).

In Ref. [6] we study within the SMES the energy dependence of the dynamical strangeness fluctuations caused by the dynamical energy fluctuations. We define $\overline{N}_s$ as a total number of strange quark-antiquark pairs created in A+A collision, and consider the fluctuation ratio:

$$R_s = \frac{(\delta \overline{N}_s)^2/\overline{N}_s^2}{(\delta E)^2/E^2} . \qquad (7)$$

When $T \rightarrow \infty$ the system is in the QGP phase. Strange (anti)quarks can be considered as massless and the bag constant can be neglected. Then $\varepsilon \propto T^4$ and $n_s \propto T^3$ and consequently $d\varepsilon/\varepsilon = 4 \cdot dT/T$ and $dn_s/n_s = 3 \cdot dT/T$, which result in $R_s = (3/4)^2 \cong 0.56$. In the confined phase, $T < T_C$, the energy density is still approximately given by $\propto T^4$ due to the dominant contributions of non-strange hadron constituents. However, the deppendence of the strangeness density on T is dominated in this case by the exponential factor, $n_s \propto \exp(-m_S/T)$, as $T << m_S \cong 500$ MeV. Therefore, at small T one finds $d\varepsilon/\varepsilon = 4 \cdot dT/T$ and $dn_s/n_s = m_S \cdot dT/T^2$, so that the ratio

$R_s = m_S/(4T)$ decreases with T. The strangeness density n_s is small and goes to zero at $T \to 0$, but the fluctuation ratio R_s (7) is large and goes to infinity at zero temperature limit.

3. Signals of Deconfinement: Experimental Results

The Pion Kink. A recent compilation of the data on the pion multiplicity in central Pb+Pb (Au+Au) collisions and p+p interactions is shown in Fig. 1.

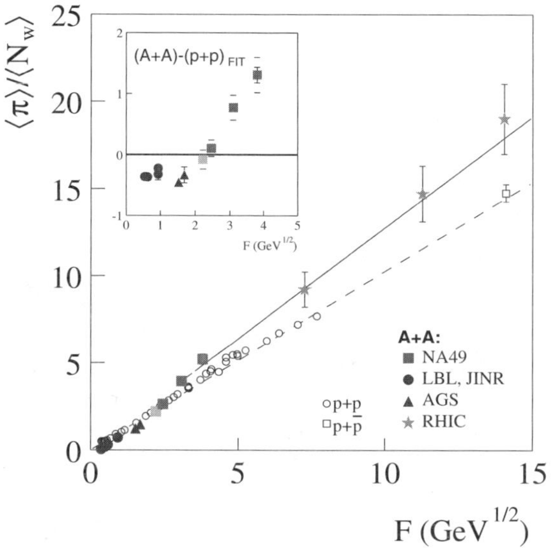

Figure 1. The dependence of the total pion multiplicity per wounded nucleon, $\langle\pi\rangle/\langle N_W\rangle$, on the Fermi's energy measure $F \equiv (\sqrt{s} - 2m_N)^{3/4} \cdot (\sqrt{s})^{-1/4}$ for central A+A collisions (closed symbols) and inelastic p+$p(\overline{p})$ interactions (open symbols). The lines are calculated with SMES [1]. The dashed line indicates a fit of the form $a \cdot F$ to the $p + p(\overline{p})$ data ($a = 1.01 \pm 0.04$ GeV$^{-1/2}$). The solid line corresponds to the fit of A+A data for initial energy above 40 A·GeV ($a = 1.36 \pm 0.03$ GeV$^{-1/2}$).

One observes that the mean pion multiplicity per wounded nucleon in $p + p(\overline{p})$ interactions is approximately proportional to F. For central A+A collisions, the dependence is more complicated and cannot be fitted by a single linear function. Below 40 A·GeV the ratio $\langle\pi\rangle/\langle N_W\rangle$ in A+A collisions is lower than in $p + p$ interactions (pion suppression), while at higher energies $\langle\pi\rangle/\langle N_W\rangle$ is larger in A+A collisions than in $p + p(\overline{p})$ interactions (pion enhancement). In the region between the AGS and the low SPS energy, the slope changes from $a \cong 1.01$ GeV$^{-1/2}$ (the dashed line in Fig. 1) to $a \cong 1.36$ GeV$^{-1/2}$ (the full line in Fig. 1).

The measured increase of the slope for A+A collisions, by a factor of about 1.3, is interpreted within the SMES as due to an increase of the effective number of the internal degrees of freedom by a factor of $(1.3)^4 \cong 3$ and is caused by the creation of a transient state of deconfined matter at energies higher than 30 A·GeV. A transition from the pion suppression [13] to pion enhancement is demonstrated more clearly in the insert of Fig. 1, where the difference between $\langle\pi\rangle/\langle N_W\rangle$ for A+A collisions and the straight line parametrisation of the $p+p$ data is plotted as a function of F up to the highest SPS energy.

The Strange Horn. One can argue that the strangeness to entropy ratio is closely proportional to the two ratios directly measured in experiments: the $\langle K^+\rangle/\langle\pi^+\rangle$ ratio and the $E_S = (\langle\Lambda\rangle + \langle K + \overline{K}\rangle)/\langle\pi\rangle$ ratio. The energy dependence of both ratios is plotted in Fig. 2 for central Pb+Pb (Au+Au) collisions and p+p interactions.

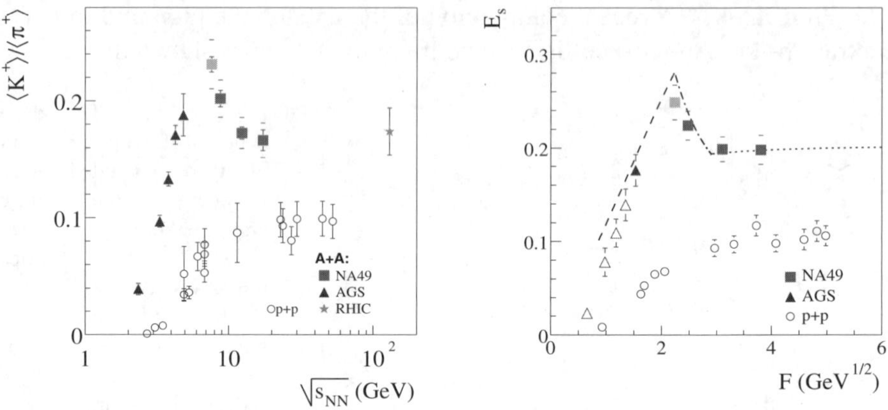

Figure 2. The dependence of the $\langle K^+\rangle/\langle\pi^+\rangle$ (left) and E_S (right) ratios on the collision energy for central A+A collisions (closed symbols) and inelastic $p + p$ interactions (open symbols). The predictions of SMES [1] for the E_S ratio are shown by a line. Different line styles indicate predictions in the energy domains in which the confined matter (dashed line), the mixed phase (dashed–dotted line) and the deconfined matter (dotted line) are created at the early stage of the A+A collision.

For p+p interactions both ratios show monotonic increase with energy. However, very different behavior is observed for central Pb+Pb (Au+Au) collisions. The steep threshold rise of the ratio characteristic for confined matter then settles into saturations at the level expected for deconfined matter. In the transition region (at low SPS energies) a sharp maximum is observed caused by a higher strangeness to entropy ratio in the confiend matter than in the deconfiend matter. As seen in Fig. 2 the measured dependence is consistent with that expected within the SMES.

The Step in Slopes. The energy dependence of the inverse slope parameter fitted to the K^+ and K^- transverse mass spectra for central Pb+Pb (Au+Au) collisions is shown in Fig. 3 [4]. The striking features of the data can be summarised and interpreted as follows. The T^* parameter increases strongly with collision energy up to the lowest (30 A·GeV) SPS energy point. This is an energy region where the creation of confined matter at the early stage of the collisions is expected. Increasing collision energy leads to an increase of the early stage temperature and pressure. Consequently the transverse activity of produced hadrons, measured by the inverse slope parameter, increases with increasing energy. The T^* parameter is approximately independent of the collision energy in the SPS energy range. In this energy region the transition between confined and deconfined matter is expected to be located. The resulting modification of the equation of state "suppresses" the hydrodynamical transverse expansion and leads to the observed plateau structure in the energy dependence of the T^* parameter. At higher energies (RHIC data), T^* again increases with the collision energy. The equation of state at the early stage becomes again stiff, the early stage temperature and pressure increase with collision energy, and this results in increase of T^* too.

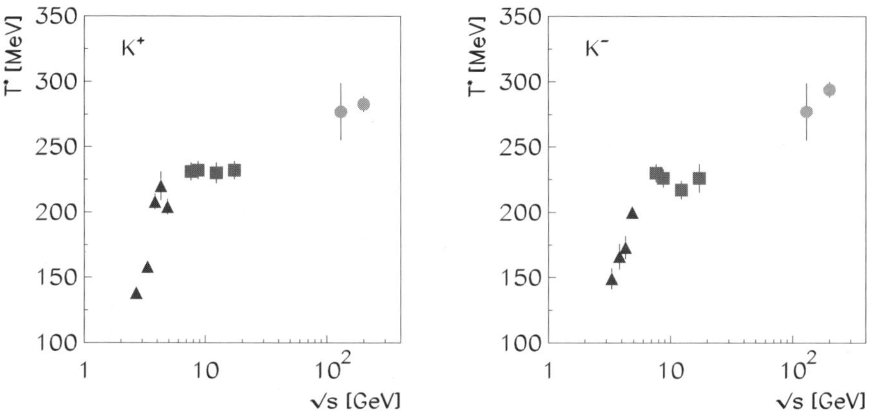

Figure 3. The energy dependence of the inverse slope parameter T^* for K^+ (left) and K^- (right) mesons (see Ref. [4]) produced at mid-rapidity in central Pb+Pb (Au+Au) collisions at AGS (triangles), SPS (squares) and RHIC (circles) energies.

Kaons are the best and unique particles among measured hadron species for observing the effect of the modification of the equation of state due to the onset of the deconfinement in hadron transverse momentum spectra. The arguments are the following.

1). The kaons m_T–spectra are only weakly affected by the hadron re-scattering and resonance decays during the post-hydrodynamic hadron cascade at the SPS and RHIC energies [11].

2). A simple one parameter exponential fit (3) is quite accurate for kaons in central A+A collisions at all energies. This simplifies strongly an analysis of the experimental data.

3). The high quality data on m_T-spectra of K^+ and K^- mesons in central Pb+Pb (Au+Au) collisions are available in the full range of relevant energies.

The Dynamical Event-by-Event Fluctuations. The ratios of entropy to energy (6) and strangeness to energy (7) dynamical fluctuations calculated within SMES are presented in Fig. 4. We find a non-monotonic energy dependence of R_e with a maximum at the boundary between the mixed phase and the QGP [5]. A pronounced minimum-structure is expected in the dependence of R_s on the collision energy [6]. It is located at $30 \div 60$ A·GeV, where the mixed phase is created at the early stage of A+A collision.

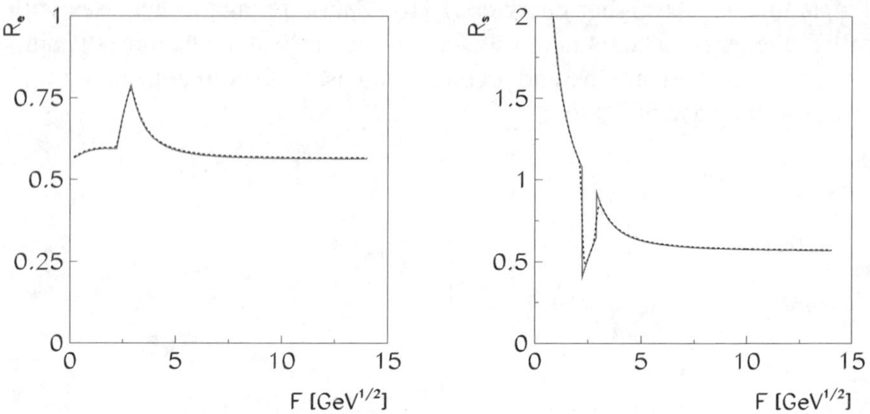

Figure 4. The collision energy dependence of the relative entropy to energy fluctuations R_e (6), left (see Ref. [5]), and strangeness to energy fluctuations R_s (7), right (see Ref. [6]), calculated within SMES.

The experimental data on the energy dependnece of R_e and R_s ratio are not yet available. Both entropy and strangeness fluctuation measures, R_e and R_s, show anomalous behavior in the transition region: the maximum is expected for R_e and the minimum for R_s. Consequetly, even a stronger anomaly is predicted for the ratio:

$$R_{s/e} \equiv \frac{R_s}{R_e} = \frac{(\delta \overline{N}_s)^2 / \overline{N}_s^2}{(\delta \overline{N}_-)^2 / \overline{N}_-^2} . \tag{8}$$

Experimental measurements of $R_{s/e}$ may be easier than the measurements of R_e and R_s because the ratio $R_{s/e}$ requires measurements of particle multiplicities only, whereas both R_e and R_s involve also measurements of particle energies.

4. Conclusions

The energy scan program at the CERN SPS together with the measurements at lower (LBL, JINR, SIS, BNL AGS) and higher (BNL RHIC) energies yielded systematic data on energy dependence of hadron production in central Pb+Pb (Au+Au) collisions. Predicted signals of the deconfinement phase transition, namely anomalies in the energy dependence of hadron production – the *pion kink* [1, 7], *strange horn* [1] and the *step in slope* of K-mesons [4] – are simultaneously observed in the same domain of the low SPS energies and presented in the left panel of Fig. 5.

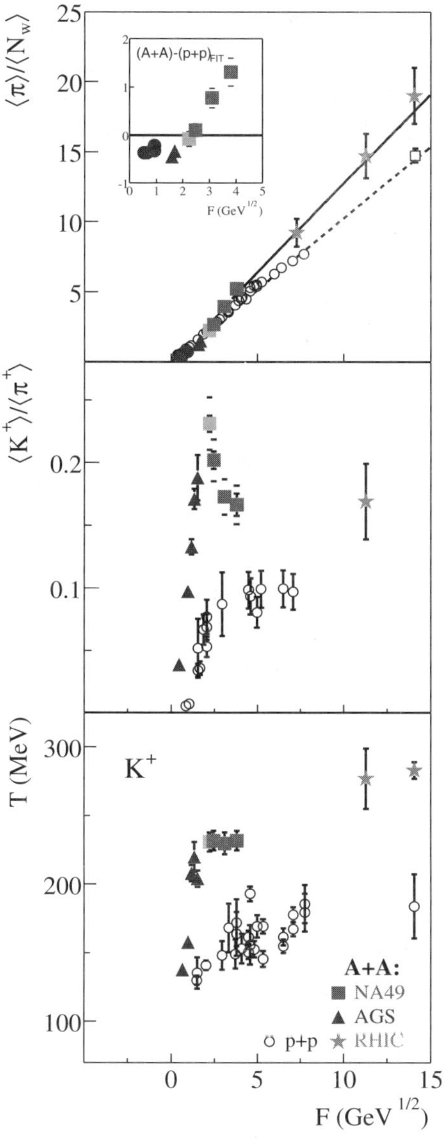

Figure 5 . The collision energy dependence of $\langle\pi\rangle/\langle N_W\rangle$, $\langle K^+\rangle/\langle\pi^+\rangle$, and $T^*(K^+)$. The predicted signature of the deocnfinement transition – the pion kink [1, 7], the strange horn [1] and the step in m_T-slopes [4] are observed simultaneously in the same domain of collision energies $F = 2 \div 3$ GeV$^{1/2}$. The open symbols correspond to the data in $p+p$ collisions. The right panel shows the prediction of SMES for $R_{s/e}$ (8) constructed from fluctuations of strange and non-strange hadron multiplicities [6]. A pronounced minimum of $R_{s/e}$ is predicted in the collision energy domain in which deconfinement transition is located.

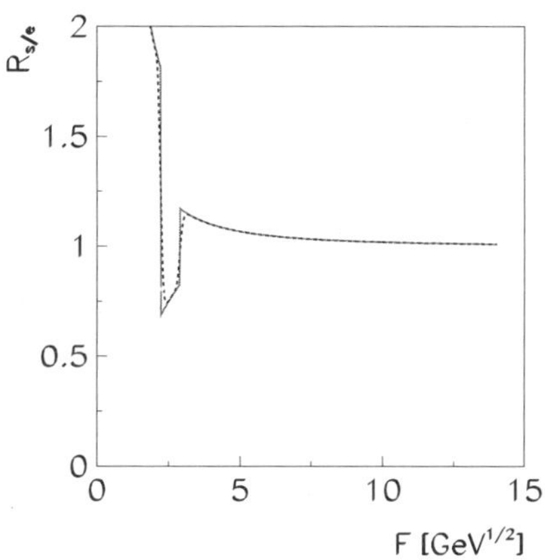

The anomalies in the energy dependence of the hadron production are seen in central A+A collisions and absent in the data of $p+p$ reactions. They indicate that the onset of the deconfinement in central Pb+Pb collisions is located at about 30 A·GeV. The theoretical picture is however far from being complete. It seems that the Landau-type initial conditions are not appropriated to describe both the multiplicity and transverse momentum data at the same time. The initial conditions with spatially spread energy and longitudinal velocity distributions are needed to reproduce all main features of the data simultaneously [14]. The analysis of the data from the energy scan program is also still in progress. In particular, first results at 20 A·GeV are soon expected. We hope that the properly analysed event–by-event fluctuations (see Ref. [5]) may also be sensitive to the onset of the deconfinement. Especially promising looks a new measure $R_{s/e}$ (8) constructed from the fluctuations of strange and non-strange hadron multiplicities [6] and shown in the right panel of Fig. 5.

Acknowledgements. The results presented in this report were obtained thanks to my close collaboration with Marek Gaździcki. I would like to thank him for a fruitful period of joint work. I am also thankful to Walter Greiner for his permanent support and encouraging of these studies.

References

[1] M. Gaździcki and M. I. Gorenstein, Acta Phys. Polon. **B 30**, 2705 (1999) [arXiv:hep-ph/9803462].

[2] S,V. Afanasiev et al (NA49 Collaboration), Phys. Rev. **C 66**, 054902 (2002).

[3] V. Friese et al (NA49 Collaboration), nucl-ex/0305017.

[4] M. I. Gorenstein, M. Gaździcki and K. Bugaev, Phys. Lett. **B 567**, 175 (2003) [arXiv:hep-ph/0303041].

[5] M. Gaździcki, M. I. Gorenstein and St. Mrówczyński, hep–ph/0304052.

[6] M.I. Gorenstein, M. Gaździcki and O.S. Zozulya, hep-ph/0309142.

[7] M. Gaździcki, Z. Phys. **C 66**, 659 (1995) and J. Phys. **G 23**, 1881 (1997).

[8] L. Van Hove, Phys. Lett. **B 118**, 138 (1982).

[9] C.M. Hung and E. Shuryak, Phys. Rev. Lett. **75**, 4003 (1995).

[10] C.M. Hung and E. Shuryak, Phys. Rev. **57**, 1891 (1997).

[11] D. Teaney, J. Lauret and E.V. Shuryak, Phys. Rev. Lett. **86**, 4783 (2001) and nucl–th/0110037.

[12] M. I. Gorenstein, K. Bugaev, M. Gaździcki, Phys. Rev. Lett. **88**, 132301 (2002), Phys. Lett. **B 544**, 127 (2002), Phys. Rev. **C 68**, 01790 (2003).

[13] M. Gaździcki, M. I. Gorenstein and St. Mrówczyński, Eur. Phys. J. **C 5**, 129 (1998).

[14] M. Gaździcki, M.I. Gorenstein, F. Grassi, Y. Hama, T. Kodama, O. Socolowski Jr., hep-ph/0309192.

3D CORE-COLLAPSE SUPERNOVA CALCULATIONS

Tobias Bollenbach
Max-Planck-Institut für Physik komplexer Systeme, Nöthnitzer Straße 38, 01187 Dresden, Germany

Terrance Strother, Wolfgang Bauer
National Superconducting Cyclotron Laboratory and Department of Physics and Astronomy, Michigan State University East Lansing, MI 48824 - 1116, USA

Abstract Recent observations of the polarization of the light emitted by supernova explosions indicate that there are large deviations from spherical symmetry in the very heart of these explosions. Asymmetries may well play a key role in the explosion mechanism. So far there is no convincing theoretical explanation for these observations. In this work the impact of angular momentum on the core collapse which is possibly the origin of large asymmetries is studied. We introduce a new approach to the supernova problem: a three dimensional test particle based simulation. The infall phase of the collapse of a typical iron core is investigated using numerical calculations. Our main focus is the impact of angular momentum. Significant deviations from spherical symmetry are found for rapidly rotating supernova cores.

1. Introduction

The delayed shock mechanism suggested by [1] is the commonly accepted paradigm for core collapse supernovae. The success of this explosion mechanism crucially depends on the cross sections for neutrino capture on nucleons, the neutrino production rates, the details of the neutrino transport, convection behind the stalled shock, and other factors many of which are not known with certainty. Depending on these ingredients simulations of the delayed shock mechanism yielded successful explosions [2, 3] while they failed in other recent cases [4–7].

Several two dimensional simulations have been carried out in the last ten years in which especially the impact of convection between the neutrino sphere and the stalled shock has been studied [2, 8, 3, 9]. It is widely believed today

W. Greiner et al. (eds.), Structure and Dynamics of Elementary Matter, 277–287.

that convection in this region helps the success of the explosion by making a larger fraction of the energy converted to neutrinos available. [10] recently extended their simulations to three dimensions and found that their results remain qualitatively the same as in two dimensions.

The reason for the inability of many of the most realistic recent simulations to produce an explosion may lie in a partly incorrect treatment of the underlying microphysics [11]. However, it is also conceivable that phenomena that can only occur in three dimensions play a key role in the explosion. Moreover, the impact of angular momentum on the core collapse and the explosion mechanism has been studied relatively little. Some studies came to the result that rotation does not have dramatic effects on the explosion mechanism [3, 9, 12]. But well known phenomena from earth such as vortices in flowing liquids or tornadoes in the atmosphere could hardly occur in less than three dimensions. In fact, rotation can become extremely rapid in the late stages of core collapse, because the inner core contracts progressively, enforcing higher angular velocities due to global angular momentum conservation.

Recent observations of the polarization of the light emitted by supernovae were made by [13, 14], see also [15]. Using a method in which they observe the polarization and the wavelength of the light simultaneously ("spectropolarimetry"), they found that the light emitted from the center of core collapse supernovae is significantly polarized. They conclude that there must be large asymmetries in the explosion to cause this polarization. Rotation is a very good candidate to explain such deviations from spherical symmetry.

2. Three Dimensional Simulations Using the Test Particle Method

Here we introduce a new technique for modeling core collapse supernovae – the test particle method [16]. Using this method, three dimensional simulations of the infall phase are performed with emphasis on the influence of angular momentum.

Our simulation uses the test particle (or pseudoparticle) method that has been utilized extensively in nuclear physics in the last two decades [17–24]. The underlying microphysics is simplified using input from one and two dimensional simulations. This is necessary because of the vast computational reqirements of three dimensional simulations. While in nuclear physics usually up to one thousand test particles represent one nucleon, the scales are opposite in our model for the collapse of an iron core with a mass around $1.3M_{\odot}$: assuming that we use 10^6 test particles, one test particle represents a mass which is a little less than that of the entire earth.

In our model all N_{tp} particles have a mass $M_{\mathrm{IC}}/N_{\mathrm{tp}} =: m_{\mathrm{tp}}$, where M_{IC} denotes the mass of the whole iron core. For each particle, position $\mathbf{r}_{\mathrm{j}}$ and momentum $\mathbf{p}_\mathrm{j}$ are tracked. The test particles obey relativistic Newtonian dynamics:

$$\dot{\mathbf{r}}_\mathrm{j} = \frac{d\mathbf{r}_\mathrm{j}}{dt} = \frac{\mathbf{p}_\mathrm{j}}{\sqrt{m_\mathrm{tp}^2 + (\frac{\mathbf{p}_\mathrm{j}}{c})^2}} \tag{1}$$

$$\dot{\mathbf{p}}_\mathrm{j} = \frac{d\mathbf{p}_\mathrm{j}}{dt} = \mathbf{F}_\mathrm{G,j}(\mathbf{r_1}, \ldots, \mathbf{r}_{\mathbf{N}_\mathrm{tp}}) + \mathbf{F}_\mathrm{EOS,j}(\mathbf{r_j}) \tag{2}$$

$$j = 1, \ldots, N_\mathrm{tp}.$$

Here, $\mathbf{F}_\mathrm{G,j}$ denotes the force on particle j due to gravity and $\mathbf{F}_\mathrm{EOS,j}$ the force due to the equation of state. Gravity is modeled using the Newtonian monopole approximation. The calculation of $\mathbf{F}_\mathrm{EOS,j}$ will be explained below.

Thermodynamic quantities such as the local mass density are defined on a spherical coordinate grid in our simulation. The boundaries between the cells of this grid are defined by the surfaces of constant r, constant ϕ, and constant θ in spherical coordinates (r, ϕ, θ) using the standard notation. For the θ coordinate the boundaries are chosen so that the difference between $\cos\theta$ of two arbitrary neighboring boundaries is constant, i.e.

$$|\cos\theta_\mathrm{j} - \cos\theta_\mathrm{i}| = \mathrm{const.}$$

if $\theta_\mathrm{j}, \theta_\mathrm{i}$ are neighboring θ-boundaries. This way the volume of a grid cell is independent of θ. During core collapse an enormous change in the length scale occurs: the inner iron core contracts roughly by a factor of 10^2 to 10^3. Hence we scale the grid down simultaneously with the core.

In principle, the local mass density can be calculated for each cell of the grid by dividing the total mass of the test particles in that cell by its volume. In order to minimize errors due to fluctuations and to smooth the density distribution, we used a more sophisticated method in which each test particles' mass is smeared over the cell it is in and the neighboring cells which are located nearest to the test particle.

To calculate $\mathbf{F}_\mathrm{EOS,j}$ in equation (2), the calculation of spatial derivatives of the internal energy per beryon u_int on the grid is needed. To numerically approximate the gradient $\nabla u_\mathrm{int}(r, \phi, \theta)$, we use a modification of a standard technique for calculating numerical derivatives [25] and the well known expression for the gradient in spherical coordinates:

$$\nabla u_\mathrm{int}(r, \phi, \theta) = \left.\frac{\partial u_\mathrm{int}}{\partial r}\right|_{(r,\phi,\theta)} \vec{e}_r + \frac{1}{r\sin\theta} \left.\frac{\partial u_\mathrm{int}}{\partial \phi}\right|_{(r,\phi,\theta)} \vec{e}_\phi - \frac{\sin\theta}{r} \left.\frac{\partial u_\mathrm{int}}{\partial(\cos\theta)}\right|_{(r,\phi,\theta)} \vec{e}_\theta. \tag{3}$$

Here $\vec{e}_r, \vec{e}_\phi, \vec{e}_\theta$ denote the orthonormal basis vectors for spherical coordinates. What needs to be done is the numerical definition of

$$\left.\frac{\partial u_{\rm int}}{\partial r}\right|_{(r,\phi,\theta)}, \quad \left.\frac{\partial u_{\rm int}}{\partial \phi}\right|_{(r,\phi,\theta)}, \text{ and } \left.\frac{\partial u_{\rm int}}{\partial (\cos\theta)}\right|_{(r,\phi,\theta)}.$$

We will exemplarily describe our technique for $\frac{\partial u_{\rm int}}{\partial r}$ (in principle, $\frac{\partial u_{\rm int}}{\partial \phi}$ and $\frac{\partial u_{\rm int}}{\partial \cos\theta}$ are obtained the same way). We used the following linear interpolation:

$$\begin{aligned} \left.\frac{\partial u_{\rm int}}{\partial r}\right|_{(r,\phi,\theta)} &= \frac{r - r_{\rm S}}{r_{\rm B} - r_{\rm S}} \times \left(\left.\frac{\partial u_{\rm int}}{\partial r}\right|_{(r,\phi,\theta)}\right)_{\rm right} \\ &+ \frac{r_{\rm B} - r}{r_{\rm B} - r_{\rm S}} \times \left(\left.\frac{\partial u_{\rm int}}{\partial r}\right|_{(r,\phi,\theta)}\right)_{\rm left}, \end{aligned} \tag{4}$$

where $r_{\rm S}$ and $r_{\rm B}$ denote the r-coordinates of the cell boundaries of the grid cell in which (r, ϕ, θ) is located, $r_{\rm S} \leq r \leq r_{\rm B}$, and $(\partial u_{\rm int}/\partial r|_{(r,\phi,\theta)})_{\rm left/right}$ are numerically calculated finite-difference approximations at the two r-boundaries of this cell, see [26] for details.

As a realistic EOS for core collapse conditions we use a combination of the nuclear EOS by [27, 28] and the Helmholtz EOS for the electron/positron gas by [29], see also [30]. The former is used for $\rho \geq 10^{11}\text{g/cm}^3$, the latter for $\rho < 10^{11}\text{g/cm}^3$ where the nuclear contribution to the pressure is negligible. The input for the LS EOS is the mass denisty ρ, temperature T, and electron fraction Y_e. The Helmholtz EOS needs the composition of the matter given by the average charge and mass numbers of the present nuclei instead of Y_e. We calculate these quantities by mapping data from the one dimensional simulations performed by [31] (see also [32]) on our model. Their tabulated data enables us to define the temperature $T(\rho)$ and electron fraction $Y_e(\rho)$ as a function of density. This should be a good approximation for the state of the matter until core bounce.

We use the EOS to obtain $\mathbf{F}_{\rm EOS,j}$ as introduced in equation (2). For this the internal energy per baryon $u_{\rm int}$ is calculated for every grid cell. This is converted into an internal energy per test particle by multiplying with $m_{\rm tp}/m_{\rm B}$, the number of baryons per test particle. We then calculate $\mathbf{F}_{\rm EOS,j}$ as the gradient of this quantity at the location of test particle j using the method described above:

$$\mathbf{F}_{\rm EOS,j} = -\frac{m_{\rm tp}}{m_{\rm B}} \left.\nabla u_{\rm int}\right|_{(r,\phi,\theta)_j} \tag{5}$$

where $(r, \phi, \theta)_j$ denote the spherical coordinates of test particle j's position.

This technique is justified by energy conservation: during the infall phase energy losses due to neutrinos radiating away from the core are negligible. Thus the sum of the core's kinetic energy $E_{\rm kin}$, gravitational energy $E_{\rm G}$, and internal energy $E_{\rm int}$

$$E_{\rm tot} = E_{\rm kin} + E_{\rm G} + E_{\rm int} \tag{6}$$

is approximately conserved during collapse (this is no longer valid when neutrino losses become large after bounce). Note that our method of calculating $\mathbf{F}_{\mathrm{EOS,j}}$ implies the (approximate) conservation of E_{tot}. In our simulations, energy (and angular momentum) is kept track of to assure that the method just described works properly.

Equatorial and cylindrical symmetry are assumed on the grid (but not for the test particles). Mimicking the presence of the star's outer layers, a minimum density ρ_{min} is enforced throughout the grid. The time development of the collapsing core is obtained by numerically integrating equations (1) and (2) using a fourth-order Runge-Kutta algorithm.

This method enables us to simulate the collapse of a rotating iron core in three dimensions until core bounce. The code conserves angular momentum. Energy conservation is given in good approximation. Work on the model described here is continuously carried on. At present, we have not yet implemented test particle collisions that lead to heating, stopping, and neutrino production. Thus our simulation is currently limited to the infall phase.

It is desirable to use a realistic density profile of a presupernova core like that calculated with the stellar evolution code of [33], see also [34]. Due to problems caused by the enormous scale change that occurs during core collapse and because the effects of rotation should mainly play a role during the late stages of collapse (when angular momentum conservation leads to the highest angular velocities), the simulation is started when collapse is already going on and the (inner) core has already contracted by a factor of 5. Therefore we used a contracted variant of the mass distribution of the progenitor calculated by [33] as our initial condition. The fact that the matter is already falling inward is mimicked by imprinting an initial velocity profile $v(r) = k_{\mathrm{v}} r$ on the test particles, where a realistic value $k_{\mathrm{v}} = 89.1\ \mathrm{s}^{-1}$ estimated using the results of previous simulations [35] is chosen. Since this is a good model only for the inner core, M_{IC} is chosen to be $0.7 M_{\odot}$ – a typical value for the inner core mass.

Stellar evolution calculations for a rotating progenitor done by [36] indicate that it is a very good approximation to assume that the inner core (initially) rotates like a rigid body. Therefore we used a constant initial angular velocity ω_0. $N_{\mathrm{tp}} = 10^6$ test particles were used in all simulations presented here. Table 1 shows the remaining data ($|E_{\mathrm{rot}}/E_{\mathrm{G}}|_{\mathrm{init}}$ is the ratio of the initial rotational to the initial gravitational energy).

3. Results

As the data obtained from the simulation runs of this series in table 1 show, the time of core bounce t_{bounce} and the maximum density at bounce ρ_{max} depend on the initial angular momentum $|\mathbf{L_0}|$ in a strictly monotonous way as shown in Fig. 1. More rapid rotation leads to a later core bounce and a

Name of Run	ω_0 [$\frac{\text{rad}}{\text{s}}$]	$\lvert L_0 \rvert$ [10^{41}Js]	$\lvert\frac{E_{\text{rot}}}{E_{\text{G}}}\rvert_{\text{init}}$	t_{bounce} [ms]	$\rho_{\max}$ [ρ_0]
HLSEOS01	10	0.466	0.027%	3.44	2.99
HLSEOS07	70	3.26	1.3%	3.61	2.35
HLSEOS10	100	4.67	2.7%	3.76	1.63
HLSEOS13	130	6.06	4.5%	3.89	1.03
HLSEOS16	160	7.46	6.8%	4.03	0.56
HLSEOS19	190	8.86	9.6%	4.17	0.34
HLSEOS22	220	10.3	13%	4.31	0.21

Table 1. Simulation runs using the combination of the Helmholtz and the LS EOS. ρ_0 is the nuclear matter density.

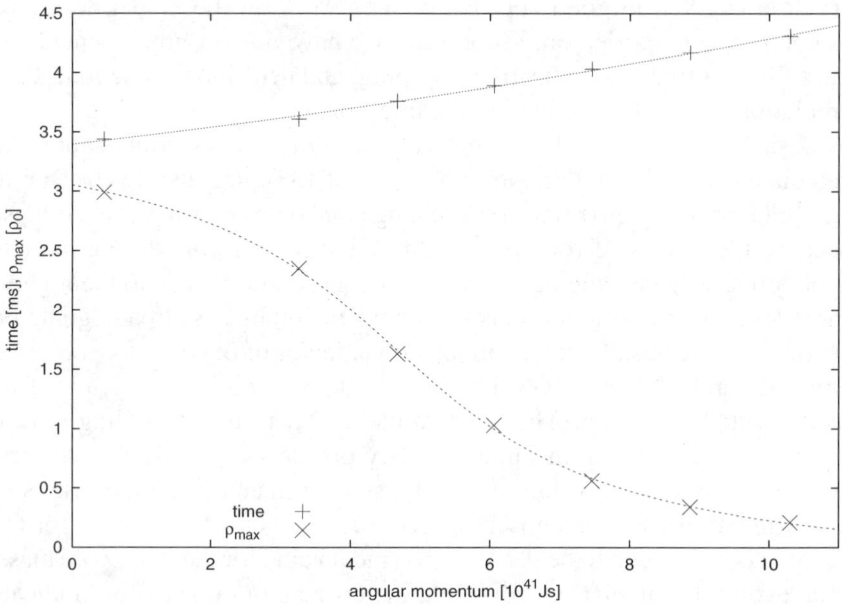

Figure 1. Maximum density reached at core bounce (×-symbols) and time of core bounce (+-symbols) as a function of the angular momentun of the core. The lines that connect the data points are only there to guide the eye.

reduced maximum compression of the core. Fig. 2 shows the development of the mass density distribution during the infall phase till shortly after bounce for models HLSEOS07 (slowly rotating), HLSEOS16 (moderately rotating), and HLSEOS22 (rapidly rotating). The same scale is used in all plots in this figure.

The abrupt change between the light yellow and the white regions is located at the radial edge of the grid. An oblate shape (due to centrifugal forces) of the more rapidly rotating models is obvious in (b) and (c). It can also be seen that the mass distributions at core bounce get more diffuse with increasing rotation (d).

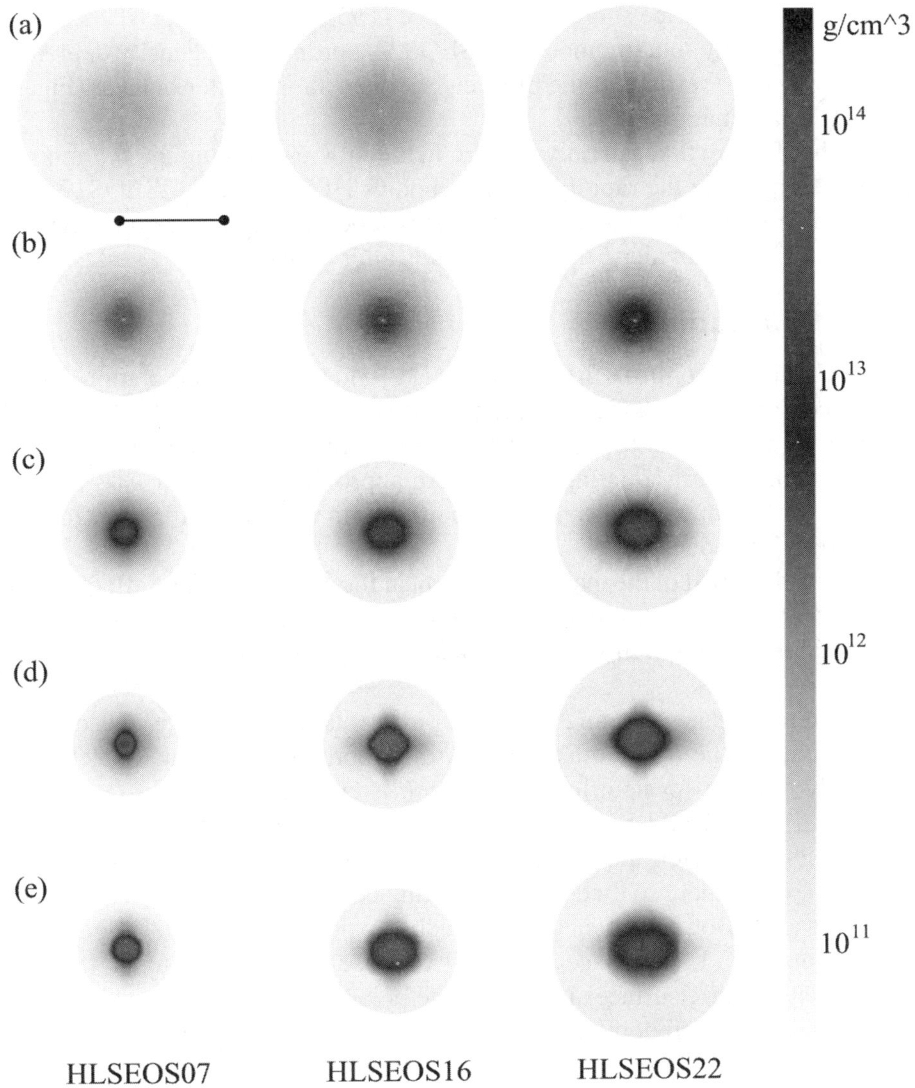

Figure 2. Mass density in a slice in the x-z-plane at five different times for models HLSEOS07 (left), HLSEOS16 (center), and HLSEOS22 (right). (a) onset of simulation, (b) after 2 ms, (c) presence of centrifugal forces becomes apparent (after 3 ms), (d) core bounce, (e) shortly after core bounce. All plots have the same radius (123 km) indicated by the black line in the top left. The density scale is only approximately valid: the highest density (indicated by black) decreases from left to right. The density scale extends from $5.13 \times 10^{10}\,\mathrm{g\,cm^{-3}}$ to $6.15 \times 10^{14}\,\mathrm{g\,cm^{-3}}$ for HLSEOS07, from $5.30 \times 10^{10}\,\mathrm{g\,cm^{-3}}$ to $1.48 \times 10^{14}\,\mathrm{g\,cm^{-3}}$ for HLSEOS16, and from $5.30 \times 10^{10}\,\mathrm{g\,cm^{-3}}$ to $5.93 \times 10^{13}\,\mathrm{g\,cm^{-3}}$ for HLSEOS22.

A slightly prolate shape shortly after bounce is apparent in model HLSEOS07. In the very rapidly rotating model HLSEOS22 a pronounced density depletion along the z-axis (vertical in the plot) can be seen shortly after bounce (Fig. 3 shows this in a magnified way).

Another feature that can only be seen in the movies showing the full motion of core collapse for the rapidly rotating models HLSEOS22, HLSEOS19, and HLSEOS16 is that significantly more mass rebounds from the core near the equatorial plane than along the rotation axis after bounce.

4. Implications

Our simulation confirms the findings of [37] and [9] that faster rotation delays core bouce and reduces the maximum density at bounce. Like [37] and [12] we find "centrifugal bounces" which occur before nuclear matter density is reached in our rapidly rotating models. An approximate value of $\rho_{\max} \approx 3\rho_0$ [35, 38] for the maximum density at bounce can also be confirmed (for the slowly rotating models).

The fact that most of the mass rebounds near the equatorial plane after bounce implies that the created shockwave might be more vigorous there. Thus the shock might not stall in this plane and lead to a "prompt shock explosion" that starts there.

An interesting new feature of our rapidly rotating models is the density depletion along the rotation axis shortly after core bounce. This depletion is extremely relevant for neutrino transport: it is clear that neutrinos trapped in the central core would be able to escape much easier by diffusing through the low density region along the rotation axis. Hence, a large neutrino flux might be created there so that neutrino heating could lead to a jet-like explosion along this axis. Both phenomena are possible explanations for the polarization observations by [13, 14], see also [15] because the light scattered from or through asymmetric surfaces is likely to be polarized.

If neutrinos are predominantly emitted along the rotation axis, the effects of parity violation in the neutrino emission might lead to an asymmetry in the number of neutrinos emitted from the "north pole" relative to those emitted from the "south pole". This would lead to a recoil of a neutron star resulting from the explosion. Such a macroscopic "kick" has indeed been observed by [39], and interpreted in terms of parity violation in neutrino emission [40, 41]. However, these calculations were performed for spherically symmetric systems, and the effects noted by us will amplify this effect.

The results shown here can be argued to be mainly of academic interest, due to the rather unrealistic mean field approximation employed. Work is currently in progess to extend the results presented here. Of particular importance is the implementation of collisions between individual test particles in order to

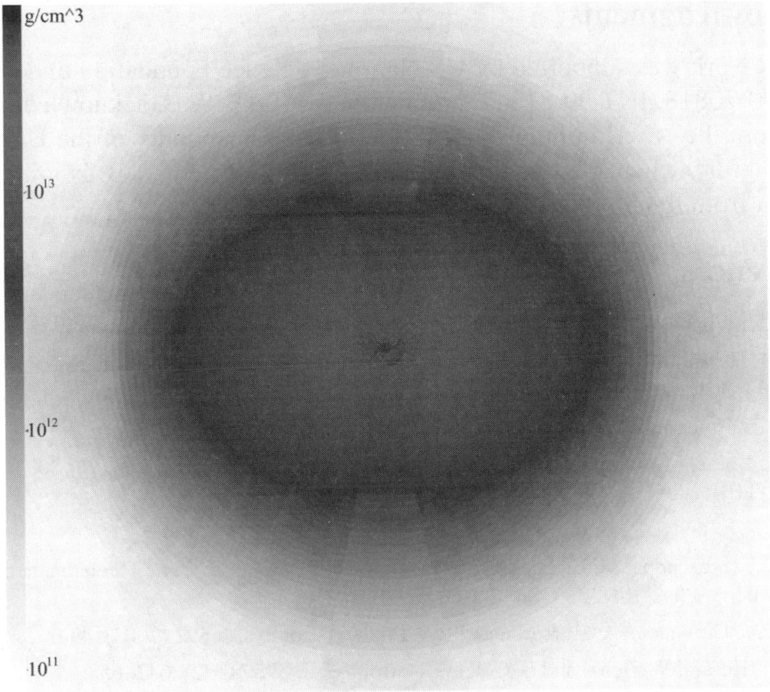

Figure 3. Density depletion along the z-axis after bounce in model HLSEOS22 at $t = 5.21$ms. The radius in this plot (measured from the center horizontally to the right) is 51km.

understand the heating, entropy production, and density pile-up in the central region, as well as to follow the core's development after bounce. Neutrino and radiation transport will also be included in our model by introducing a different species of test particles. The resulting coupled kinetic equations can be developed in complete analogy to [42], see also our presentation at the first Argonne - INT - MSU - JINA RIA Workshop [43]. The method we are developing relies in direct simulation Monte Carlo techniques [44]. This will allow us to drawer firmer conclusions about the effects of neutrinos in the presence of rotation.

Additionally, a more sophisticated way of mapping data obtained from lower dimensional simulations on our model is needed after bounce. Eventually, we also need to address the effects of the huge magnetic fields on particle trajectories and ways to generate these magnetic fields self-consistently in our simulations.

The methods and their extensions sketched here have the advantage that they are computationally less involved than a full three-dimensional hydrodynamics simulation, coupled to the radiation and neutrino transport problem. In addition, we have developed algorithms that lend themselves to a relatively straightforward utilization of parallel processors. So there is cause for optimism.

Acknowledgments

This work was supported by US National Science Foundation under grants PHY-00-70818, INT-99-81342, and PHY-02-45009. W.B. acknowledges support from the A.v.Humboldt Foundation and the hospitality of the University of Frankfurt, where part of this research was conducted. T.B. is grateful for support from the Studienstiftung des Deutschen Volkes.

References

[1] J.R. Wilson, *Numerical astrophysics*, p. 422, Jones and Bartlett, Boston, 1985.

[2] M. Herant, W. Benz, W.R. Hix, C.L. Fryer, and S.A. Colgate, Astrophys. J. **435** (1994), 339–361.

[3] C.L. Fryer and A. Heger, Astrophys. J. **541** (2000), 1033–1050.

[4] R. Buras, M. Rampp, H.-Th. Janka, and K. Kifonidis, Phys. Rev. Lett. **90** (2003), no. 24, 241101.

[5] H.-Th. Janka, R. Buras, and M. Rampp, Nucl. Phys. A **718** (2003), 269c.

[6] A. Mezzacappa, M. Liebendörfer, O.E.B. Messer, W.R. Hix, F.-K. Thielemann, and S.W. Bruenn, Phys. Rev. Lett. **86** (2001), no. 10, 1935.

[7] T.A. Thompson, A. Burrows, and P.A. Pinto, Astrophys. J. **592** (2003), 434.

[8] M. Herant, W. Benz, and S. Colgate, Astrophys. J. **395** (1992), 642–653.

[9] S. Yamada and K. Sato, Astrophys. J. **434** (1994), 268–276.

[10] C.L. Fryer and M.S. Warren, Astrophys. J. **574** (2002), L65.

[11] K. Langanke, G. Martínez-Pinedo, J.M. Sampaio, D.J. Dean, W.R. Hix, O.E.B. Messer, A. Mezzacappa, M. Liebendörfer, H.-Th. Janka, and M. Rampp, Phys. Rev. Lett. **90** (2003), no. 24, 241102.

[12] K. Kotake, S. Yamada, K. Sato, and T.M. Shimizu, Nucl. Phys. A **718** (2003), 629c.

[13] L. Wang, D.A. Howell, P. Höflich, and J.C. Wheeler, Astrophys. J. **550** (2001), 1030–1035.

[14] L. Wang, J.C. Wheeler, Z. Li, and A. Clocchiatti, Astophys. J. **467** (1996), 435–445.

[15] L. Wang and J.C. Wheeler, Sky and Telescope (2002).

[16] T. Bollenbach and W. Bauer, in "Exotic Clustering", edited by S. Costa, A. Insolia, and C. Tùve, American Institute of Physics Conference Proceedings, Volume 644 (Melville, New York, 2002), p. 219-232.

[17] C.-Y. Wong, Phys. Rev. C **25** (1982), no. 3, 1460–1475.

[18] G.F. Bertsch et al., Phys. Rev. C **29** (1984) 673.

[19] H. Kruse et al., Phys. Rev. Lett. **54** (1985) 289.

[20] W. Bauer, G.F. Bertsch, W. Gassing, and U. Mosel, Phys. Rev. C **34** (1986), 2127.

[21] H. Stöcker, and W. Greiner, Phys. Rep. **137** (1986) 277.

[22] G.F. Bertsch and S. Das Gupta, Phys. Rep. **160** (1988) 189.

[23] P. Schuck et al., Prog. Part. Nucl. Phys. **22** (1989) 181.

[24] W.G. Gong, W. Bauer, C.K. Gelbke, and S. Pratt, Phys. Rev. C **43** (1991), 781.

[25] W.H. Press, S.A. Teukolsky, W.T. Vetterling, and B.P. Flannery, *Numerical recipes in c*, second ed., Press Syndicate of the University of Cambridge, Cambridge, UK, 1988.

[26] Tobias Bollenbach, *Numerical study of rotating core collapse supernovae*, Master's thesis, Michigan State University, East Lansing, MI, 2002.

[27] J.M. Lattimer and F.D. Swesty, http://www.ess.sunysb.edu/dswesty/lseos.html, 1993.

[28] J.M. Lattimer and F.D. Swesty, Nucl. Phys. A **535** (1991), 331–367.

[29] F.X. Timmes, http://flash.uchicago.edu/ fxt/code_pages/eos.shtml, 2000.

[30] F.X. Timmes and F.D. Swesty, Astrophys. J. Suppl. S. **126** (2000), 501–516.

[31] J. Cooperstein and J. Wambach, Nucl. Phys. A **420** (1984), 591–620.

[32] J. Cooperstein, Nucl. Phys. A **438** (1985), 722–739.

[33] S.E. Woosley and T.A. Weaver, Ann. Rev. Astron. Astrophys. **24** (1986), 205–253.

[34] T.A. Weaver, G.B. Zimmermann, and S.E. Woosley, Astrophys. J. **225** (1978), 1021–1029.

[35] H.A. Bethe, Rev. Mod. Phys. **62** (1990), no. 4, 801–866.

[36] A. Heger, N. Langer, and S.E. Woosley, Astrophys. J. **528** (2000), 368–396.

[37] T. Zwerger and E. Müller, Astron. Astrophys. **320** (1997), 209–227.

[38] E. Baron, J. Cooperstein, and S. Kahana, Nucl. Phys. A **440** (1985), 744–754.

[39] A.G. Lyne and D.R. Lorimer, Nature **369** (1994), 127.

[40] C.J. Horowitz and Gang Li, Phys. Rev. Lett. **80** (1998), 3694–3697.

[41] C.J. Horowitz and J. Piekarewicz, Nucl. Phys. A **640** (1998), 281–290.

[42] S.J. Wang, B.-A. Li, W. Bauer, and J. Randrup, Annals of Physics **209**, 251 (1991).

[43] W. Bauer, http://mocha.phys.washington.edu/∼int_talk/WorkShops/RIA/People/Bauer_W/Bauer.-RIA.pdf (2004).

[44] G. Kortemeyer, F. Daffin, and W. Bauer, Phys. Lett. **B374**, 25 (1996).

DARK MATTER, MASSIVE NEUTRINOS AND SUSY PARTICLES

K. Pretzl
Laboratory of High Energy Physics,
University of Bern,
Sidlerstrasse 5,
3012 Bern, Switzerland
E-mail: pretzl@lhep.unibe.ch

Abstract Recent experimental observations show that the universe is flat and consists of 27% matter and 73% dark energy. Most of the matter is invisible and of unknown nature. Weakly interacting massive particles (so-called WIMPs) like massive neutrinos and/or particles predicted by Super Symmetry (SUSY) are possible candidates for the dark matter in the universe. The latest developments in this field of research and the possible direct search for WIMPs with earthbound detectors is discussed in this paper.

1. Introduction

The story of the dark matter started in 1933, when the Swiss-American astronomer Fritz Zwicky published astonishing results [1] from his studies of galactic clusters. He measured the velocity dispersion of individual galaxies in a cluster and conjectured that large amounts of invisible matter are necessary to keep the galaxies gravitationally bound to the cluster. His astonishing findings were received with scepticism by many astronomers at that time. However, they were confirmed more than 60 years later by gravitational lensing observations, a technique which was also proposed by Zwicky for this purpose [2]. In the 1970s dark matter in the halos of spiral galaxies was discovered by observing constant orbital velocities of stars and hydrogen clouds (21 cm radiation) independent from their radial distance from the galactic center [3][4]. Measurements performed on hundreds of spiral galaxies strongly indicated that all galaxies are embedded in a massive halo of dark matter. The real dimensions of the halo, however, could not be determined and remained unknown until very recently when the analysis of galaxy-galaxy gravitational lensing images seemed to indicate that the dark halo of galaxies may exceed the visible diameter by more

W. Greiner et al. (eds.), Structure and Dynamics of Elementary Matter, 289–302.

than an order of magnitude [5]. Further investigations are necessary to tell how large the dark halos really are and whether they have dimensions close to the mean distance between galaxies in a cluster.

Recent results from gravitational lensing measurements of galaxy clusters [6], from supernovae type 1a surveys [7][8] and cosmic microwave background (CMB) anisotropy measurements [9] provide us with a rather precise knowledge of the total matter budget in our universe (fig.1). According to these, the matter density of $\Omega_m = 0.27 \pm 0.04$ and the vacuum energy density of $\Omega_\Lambda = 0.73 \pm 0.04$ add up to a total density of $\Omega_{total} = \Omega_m + \Omega_\Lambda = 1.02 \pm 0.02$, which is consistent with a flat inflationary universe. The density parameter for the vacuum energy Ω_Λ, which is also called dark energy, is associated with the cosmological term Λ in the Einstein equation, giving rise to an accelerated expansion of the universe, as suggested by the recent supernova type 1a distance measurements. From the visible matter density $\Omega_{vis} = 0.01$ and the baryonic matter density $\Omega_b = 0.044 \pm 0.04$ we conjecture that most of the matter in the universe is dark and that there is 6 times more non-baryonic than baryonic matter. The question was raised whether a considerable fraction of the invisible baryonic matter could hide in the galactic halos. Gravitational micro-lensing observations on the Large and Small Magellanic Clouds by the EROS, MACHO and OGLE collaborations have, however, not found a significant abundance of baryonic dark matter objects in the mass range of 10^{-7} to 10 solar masses [10].

2. Dark Matter Candidates

Cosmological models describing structure formation in the universe seem to favor a cocktail consisting of 70% cold and 30% hot dark matter. Primordial particles which are relativistic at freeze-out are candidates for hot dark matter, while non-relativistic particles, generally known as WIMPs (weakly interacting massive particles), are candidates for cold dark matter.

Because of their high number density, primordial neutrinos with a mass of several eV would be good candidates for hot dark matter. Their contribution to the matter density would be

$$\Omega_\nu \cdot h^2 = \sum_{\nu=1}^{3} \frac{m_\nu}{92eV}$$

with the Hubble parameter h=0.7. Unfortunately only upper limits from direct neutrino mass measurements are available so far:

$$m_{\nu_e} \leq 2.2eV, m_{\nu_\mu} \leq 170keV, m_{\nu_\tau} < 18MeV$$

Neutrino oscillations first observed with atmospheric neutrinos by the Super-Kamiokande experiment [11] and recently also with solar neutrinos by the SNO

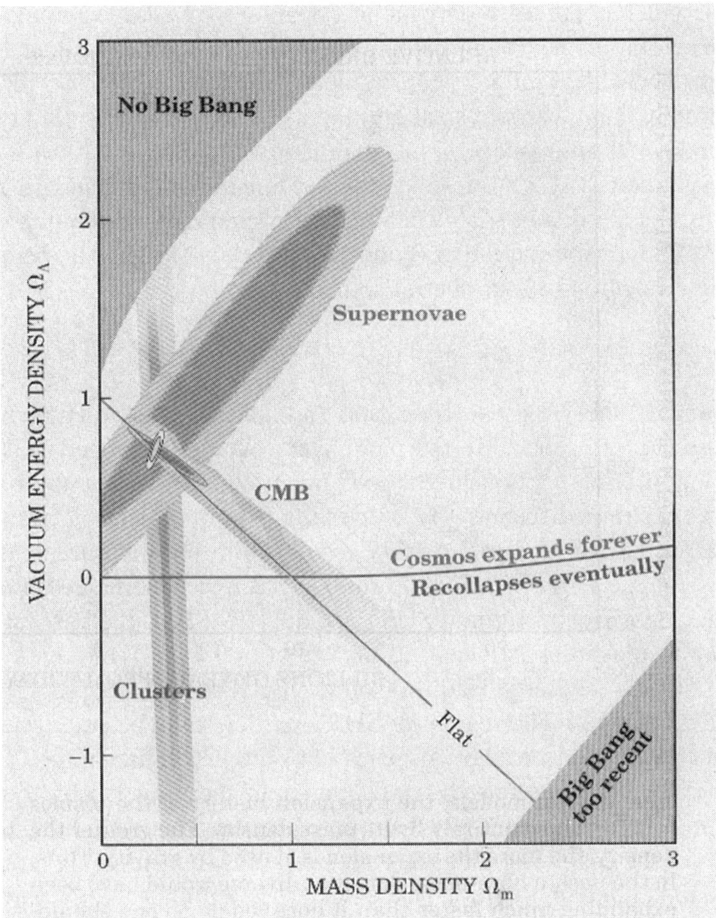

Figure 1. Cosmological parameter space of the normalized mass and vacuum energy densities Ω_m and Ω_Λ.

experiment [12] as well as with reactor neutrinos with the KamLAND experiment [13] provide only information about the neutrino mass differences, but not absolute mass values. Nevertheless, if one uses the values obtained by Super-Kamiokande of $\Delta^2 m \approx 2 \cdot 10^{-3}(eV)^2$, assuming $\nu_\mu \leftrightarrow \nu_\tau$ oscillations, and by SNO and KamLAND of $\Delta \mathrm{m}^2 \approx 7 \cdot 10^{-5}(eV)^2$, assuming $\nu_e \leftrightarrow \nu_\mu$ oscillations, and if one makes some assumptions about the neutrino mass hierarchy on the basis of Grand Unified Theories (GUT), the heaviest neutrino would have a mass of the order of $10^{-2} eV$. This corresponds to $\Omega_\nu \approx 10^{-4}$, which seems to be too small to be relevant for the missing mass in the universe.

Axions, which are predicted by models dealing with charge conjugation and parity (CP) violation in strong interactions, would be potential cold dark matter candidates. Their mass range is constrained to the interval of 10^{-6} to 10^{-5} eV by cosmological and astrophysical arguments. The axion detection technique is based on Primakoff conversion in a magnetic field using microwave cavities.

Super Symmetry (SUSY), a well studied extension of the Standard Model of particle physics, predicts a viable candidate for cold dark matter, the so-called neutralino. In the Minimal Super Symmetric Model (MSSM) the neutralino is a linear superposition of four neutral spin 1/2 gauginos:

$$\chi = n_{11}\tilde{W}^3 + n_{12}\tilde{B} + n_{13}\tilde{H}_1 + n_{14}\tilde{H}_2$$

with $\tilde{W}^3$ and $\tilde{B}$, the Wino and Bino, and $\tilde{H}_1$ and $\tilde{H}_2$, the two Higgsinos. Because of its complex structure the parameter space is large and makes mass and interaction cross section predictions uncertain. However, the parameter space was constrained recently by the results from the electron positron collider (LEP) at CERN and preliminary results from the measurements of the anomalous magnetic moment of the muon (g-2 experiment) at Brookhaven. Within the Constrained Minimal SUSY Model (CMSSM) the best estimate of the neutralino mass lies between 50 GeV and 600 GeV. This mass range is well accessible to the Large Hadron Collider (LHC), which is presently under construction at CERN. The search for SUSY particles will be amongst the most prominent subjects for the LHC. If neutralinos are first found at the LHC, they still need to be confirmed as the dark matter in the universe by their direct detection in WIMP experiments. Compared to previous estimates, the latest predictions of the cosmic abundance of neutralinos is considerably reduced due to neutralino-stau coannihilation processes [14]. Another SUSY particle, the sneutrino, does not qualify for a dark matter candidate, since the relic density of light sneutrinos would be much too small to be of cosmological significance and heavy sneutrinos are already excluded by direct detection experiments [15].

3. WIMP Elastic Scattering

Dark matter WIMPs can interact with nuclei in a detector via spin-independent coherent scatterings with Higgs and squark exchange as well as spin-dependent scatterings with Z and squark exchange. The corresponding cross sections are [16]:

(a) cross section for spin-independent coherent scattering:

$$\sigma_0^{scalar} = \frac{4m_{red}^2}{\pi}[Zf_p + (A-Z)f_n]^2$$

with m_{red}, the reduced mass of the projectile and the target particles, Z and A the charge and mass of the target nucleus, f_p and f_n, the effective WIMP proton and WIMP neutron couplings. Since $f_p \approx f_n$, the coherent cross section becomes $\sigma_0^{scalar} \sim m_{red}^2 A^2$, which allows to relate the WIMP nucleus to the WIMP nucleon cross section:

$$\sigma_{nucleon} = \left[\frac{1 + m_x/m_N}{1 + m_x/m_p}\right]^2 \frac{\sigma_0^{scalar}}{A^2}$$

with m_x, the WIMP mass, m_p the proton and m_N the nucleus mass. Since this expression of the cross section does not depend on the target material composition, it allows for a direct comparison of the results of various WIMP detectors.

(b) cross section for spin-dependent scattering:

$$\sigma_0^{spin} = 4G_F^2 m_{red}^2 C_A$$

with the enhancement factor $C_A = \frac{8}{\pi}(a_p < S_p > +a_n < S_n >)^2 \frac{(J+1)}{J}$, where G_F is the Fermi coupling, $a_{p,n}$ are the effective WIMP-nucleon couplings, $< S_{p,n} >$ are the expectation values of the proton and neutron spins within the nucleus and J is the spin of the nucleus. An extraction of a spin-independent WIMP-nucleon cross section in a model-independent way is not possible, since the nuclear and the supersymmetric degrees of freedom do not decouple from each other. Nevertheless, when using the “odd" group model [17], which assumes that all the nuclear spin is carried by the “odd” group, either the protons or the neutrons (whichever is unpaired), WIMP-nucleon cross sections can be deduced from neutralino-nucleus interactions, as described in reference [18]. WIMP detectors with the capability of using several target materials will be able to distinguish between the spin-dependent and the spin-independent nature of the interaction.

4. Direct Detection of Dark Matter

The direct detection of WIMPs is based on the measurement of nuclear recoils in elastic WIMP scattering processes. In order to obtain high detection efficiencies, devices with high sensitivity to low nuclear recoil energies (possibly below 1 keV) are needed. WIMP detectors can be categorized in conventional and cryogenic devices. Most of the conventional WIMP detectors use NaI, Ge crystals or liquid Xenon (LXe). Conventional devices have the advantage that large detector masses ($\sim$ 100 kg) can be employed, which makes them sensitive to annual modulations of the WIMP signal owing to the movement of the earth with respect to the dark halo rest frame. Annual modulation, if observed, would provide strong evidence for a WIMP signal, assuming it is not faked by spurious modulated background signals. However, due to quenching of the ionization signals, conventional detectors have lower nuclear recoil detection efficiencies than cryogenic devices.

Table 1. Cryogenic WIMP Detectors in Operation

Collaboration	Location (m.w.e.)	Eff. Target Mass (kg)	Exposure (kg d)
CDMS	Stanford (17)	Ge (1.0) Si (0.2)	28.3
CRESST	Gran Sasso (3800)	Al_2O_3 (0.262)	1.5
EDELWEISS	Frejus (4800)	Ge (1.0)	30.5
MIBETA	Gran Sasso (3800)	TeO_2 (0.68)	60
ORPHEUS	Bern (70)	Sn (0.206)	0.68
ROSEBUD	Canfranc (2450)	$CaWO_4$ (0.054) BGO (0.046) Ge (0.067) Al_2O_3 (0.050)	0.034 0.034
Tokyo group	Kamioka Observatory (2700)	LiF (0.168)	4.1

Table 2. Conventional WIMP Detectors in Operation

Collaboration	Location (m.w.e.)	Eff. Target Mass (kg)	Exposure (kg d)
DAMA	Gran Sasso (3800)	NaI (100)	107731
LSM Saclay	Frejus (4800)	NaI (10)	805
UKDMC	Boulby Mine (2530)	NaI (6)	709
ELEGANT	Kamioka (2700)	NaI (662)	241630
Heidelberg-Moscow	Gran Sasso (3800)	Ge (2.7)	252
IGEX	Canfranc (2450)	Ge (2.2)	60
ZEPLIN I	Boulby Mine (2530)	Liquid Xe (3.1)	230
SIMPLE	Rustrel (63)	Superheated Droplets ($15\ 10^{-3}$)	0.19
PICASSO	SNO (4755)	Superheated Droplets ($21.7\ 10^{-3}$)	0.35

Cryogenic detectors are able to measure small recoil energies with high efficiency because they measure the total deposited energy in form of ionization and of phonons. A small energy loss $\Delta \mathrm{E}$ can lead to an appreciable temperature increase $\Delta \mathrm{T} = \frac{\Delta \mathrm{E}}{C}$, provided the detector is operated at low enough temperatures (typically mK), where the heat capacity C is small. Cryogenic detectors for dark matter searches have been developed in the last 15 years. At a first low temperature detector workshop (LTD1), held in 1987 at the Ringberg Castle in Tegernsee in the South Bavarian Alps, several innovative detector ideas were presented [19]. Some of these ideas were materialized and are now used in direct WIMP detection experiments [20].

Cryogenic detectors are made of many different materials, like Ge, Si, TeO_2, sapphire (Al_2O_3), LiF, $CaWO_4$ and BGO, including superconductors, like Sn, Zn, Al, etc. This turns out to be an advantage for the WIMP search, since the resulting recoil spectra are characteristically different for detectors with different materials, a feature which helps to effectively discriminate a WIMP signal against background. In comparison to conventional detectors, however, cryogenic detectors are rather limited in target mass. Cryogenic WIMP experiments now in operation use detectors with a mass of about 1 kg. Many plan to increase the target mass to 10 kg (CDMS, CRESST, EDELWEISS) and beyond (CUORE) in the near future. Tables 1 and 2 contain a list of actually running WIMP detectors. Also listed in these tables is their location, their depth in meters of water equivalent (m.w.e.), their detector material, their target mass in kg and their exposure in kg days.

The most frequently used cryogenic devices are bolometers, which consist of an absorber and a sensitive thermometer attached to it. Essentially two different kinds of thermometers have proven to be suitable: a) Neutron Transmuted Doped Germanium (NTDG) thermistors, which are heavily doped semiconductors slightly below the metal insulator transition. These thermometers have the advantage that they are easy to handle and commercially available. WIMPs interacting in the absorber increase the temperature and thus change the resistance in the thermistor, which can be measured (CDMS, EDELWEISS, ROSEBUD, MIBETA, Tokyo). b) Superconducting Transition Edge Sensors (TES), which are made out of thin tungsten strips attached to the surface of the absorber and are operated close to their superconducting to normal phase transition temperature of about 15 mK (CDMS, CRESST). Another technique uses superheated superconducting granules (SSG). An SSG WIMP detector (ORPHEUS) consists of billions of spherical Sn granules with diameters of about 30 μm. The detector is operated in a magnetic field. WIMPs interacting in a granule can cause a phase transition from the superconducting to the normal state. This phase transition of individual granules can be detected by pickup loops which measure the flux change due to the disappearance of the Meissner-Ochsenfeld effect. The energy threshold of the detector is adjustable by setting the external magnetic field just below the phase transition boundary.

WIMP detectors are located in underground laboratories to be protected from cosmic ray background. In addition, they employ passive shielding against radioactivity from the surrounding walls and materials. The shielding and the detector itself have to be made from radio-poor materials. However, background suppression using passive shielding alone is limited in its effectiveness and is expensive. Nevertheless, large factors in WIMP detection sensitivity can be gained by using in addition active background recognition methods. CDMS and EDELWEISS are employing cryogenic Ge and Si absorbers with two different circuits for reading out phonon and charge signals from the detector. The simultaneous readout of both signals allows to discriminate genuine nuclear recoils from electron-induced background events. For the same deposited recoil energy, the ionisation signal from nuclear recoils is highly quenched compared to the signals from electrons, a feature which allows to separate genuine nuclear recoils from electron background. However, it was recognized that background electrons interacting in a thin surface layer of the detector show a reduced ionization yield due to incomplete charge collection and therefore were misidentified as nuclear recoils.

By coating the surface of the absorber crystal with a thin ($\sim 60\mu$m) layer of amorphous silicon, the CDMS group was able to considerably improve the charge collection. This method was also successfully applied by EDELWEISS, which achieved so far the best sensitivity. With this technique, CDMS and EDELWEISS reached an excellent γ background discrimination, in the case of CDMS 99.99% above 5 keV. Surface event rejection can be further improved by ballistic phonon detection with NbSi thin layers (EDELWEISS) attached to opposite surfaces of the Ge bolometer, which will allow to discriminate events with full charge collection from those with a charge deficit. Applying a rise-time cut to the nonthermal phonon signals in their ZIP (Z-dependent Ionization and Phonon) detector, the CDMS group was able to obtain a further improvement on γ-induced background and surface event rejection. Due to the fact that the signal rise-time of electrons is faster than that of nuclear recoils, a background rejection of better than 95% could be achieved - at the expense, however, of a 45% signal loss of nuclear recoils.

In order to eliminate the surface event problem while maintaining a powerful active background discrimination, the ROSEBUD and CRESST collaborations are using scintillating crystals ($CaW0_4$, BGO in the case of ROSEBUD and $CaWO_4$ in the case of CRESST) which allow to collect the light and phonon signals on an event to event basis. This technique, which was first suggested in [21], has shown encouraging results with a 99.9% background rejection for energies E >20 keV, and it will be used in the future CRESST Phase II detector.

Active background rejection will also be practiced with the ORPHEUS SSG detector, since minimum ionizing particles and gammas cause many granules to flip, while WIMPs cause only one granule to flip (flip meaning a transition from the superconducting to the normal state). Although the background rejection power of SSG is still under study, preliminary results of the ORPHEUS SSG detector are shown in fig.3. Signals from single granule flips could be clearly distinguished from multiple flips [22].

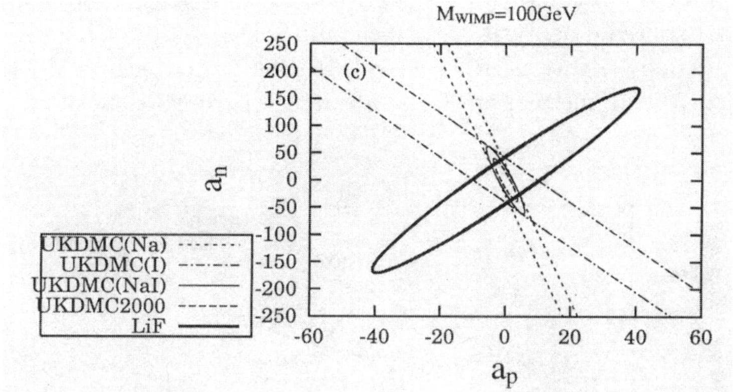

Figure 2. The limits in the $a_p - a_n$ plane for a WIMP mass of 100 GeV.

The Tokyo group reported first results with 8 LiF bolometers with a total mass of 168 g [23]. Using the method of ref. [18], they were able to derive limits in the $a_p - a_n$ WIMP-nucleon coupling plane, as shown in fig.2. When including the data of UKDMC [24], the limits in the $a_p - a_n$ plane can be considerably reduced due to the sign of the spin factors $< Sp(N) > / < Sn(N) >$ of ^{19}F, which differs from ^{7}Li, ^{23}Na and ^{127}I. In the future, they plan to use NaF crystals in order to eliminate the contamination of tritium, which is produced by neutron capture in ^{6}Li during production and handling of the crystals at ground level.

5. Results

The experimental results are usually presented as exclusion plots, which show the WIMP-nucleon cross section versus the WIMP mass.

They are derived from the expected nuclear recoil rates for a given set of parameters:

$$\frac{dR}{dE_{recoil}} = \frac{\rho_{halo} N_A}{m_x A} \int_{v_{min}(E_{recoil})}^{v_{max}} \frac{d\sigma}{dE_{recoil}} \mid \mathrm{F}(E_{recoil}) \mid^2 \mathrm{f}(v) v \mathrm{d}v$$

with m_x, the mass of the WIMP, N_A, Avogadro's number, ρ_{halo} = 0.3 GeV cm^{-3}, the dark matter halo density in the solar neighborhood, F(E_{recoil}), the nuclear form factor, $f(v)$, an assumed isothermal Maxwell-Boltzmann velocity distribution of the WIMPs in the halo, $v_{min} = \sqrt{E_{recoil} m_N / 2m_{red}^2}$, and σ, the elastic cross section for spin-independent WIMP scattering. The expected nuclear recoil spectrum for interactions with WIMPs of a given mass will then be folded with the detector response, which was obtained experimentally from calibration measurements with neutron sources or in neutron beams. From a maximum likelihood analysis, an upper limit cross section value (90% C.L.) can be extracted for several different WIMP masses. Exclusion plots for spin independent WIMP interactions from different experiments are shown in fig.3.

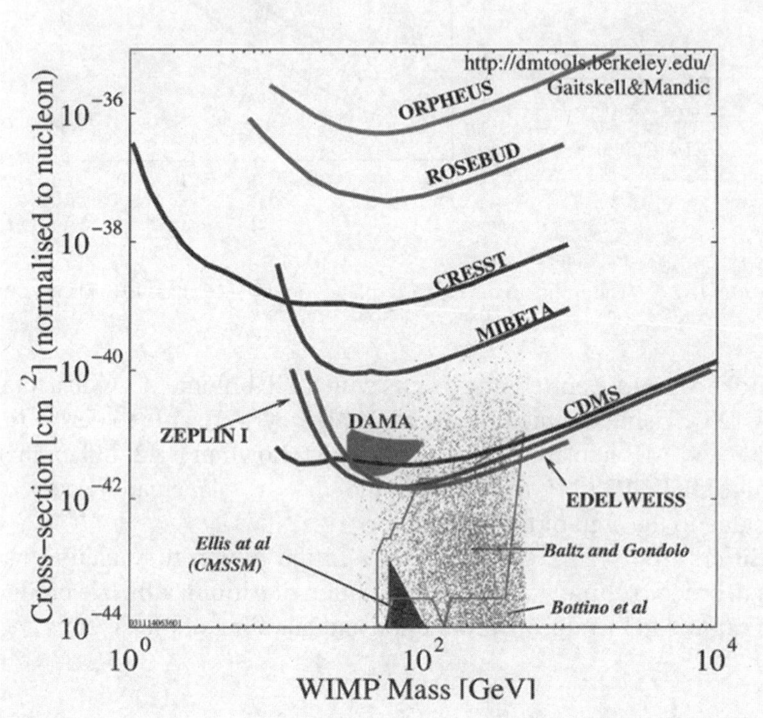

Figure 3. Exclusion plot for spin-independent WIMP interactions.

A first positive signal has been claimed by the DAMA group [25]. They observed an annual modulation consistent with a signal expected from a WIMP with a mass of $m_x = (52^{+10}_{-8})$ GeV and a WIMP nucleon cross section of $\xi\sigma = (7.2^{+0.4}_{-0.9})10^{-6}$ pb with $\xi = \rho_{WIMP}/\rho_{halo}$. The signal did not disappear on altering the standard halo model assumptions and giving the halo an orbital twist.

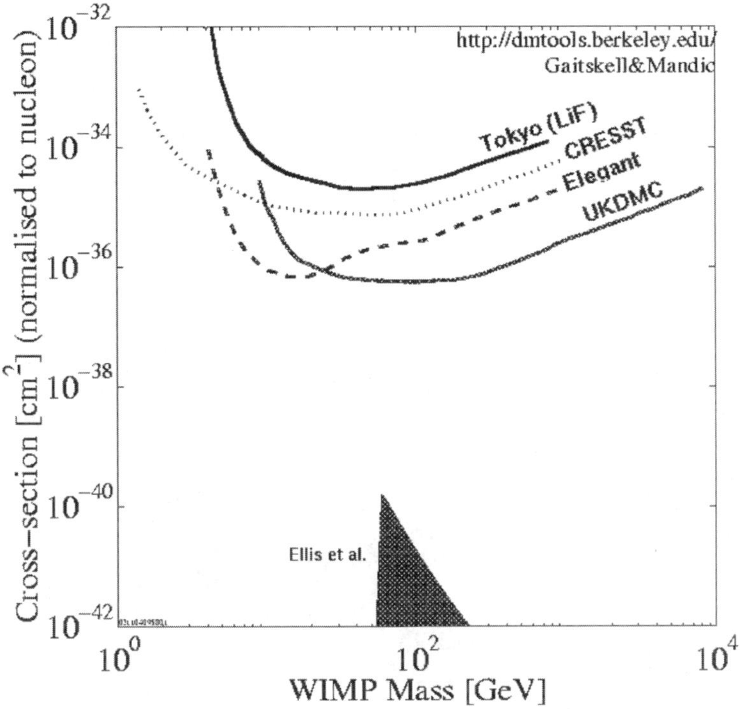

Figure 4. Exclusion plot for spin-dependent WIMP interactions.

First the CDMS group published results [26] which seem to be incompatible with the DAMA results at 99.8% C.L. CDMS does not observe any excess events above background in the corresponding region of the recoil spectrum. Recently the EDELWEISS group published results [27] obtained with a 320 g Ge cryogenic detector and an effective exposure of 30.5 kg days. Their result is incompatible at more than 99.8% C.L. with a WIMP of mass 44 GeV and a WIMP nucleon cross section of $5.4 \cdot 10^{-6}$ pb. Also shown in fig.3 are the results of the LXe dark matter detector ZEPLIN, which seem to be in disagreement with the DAMA observations. In view of these results, the question remains: is there seasonally changing background which could fake a modulated signal as seen by the DAMA experiment with the right phase and amplitude? Seasonal modulation with a similar phase and amplitude has been measured with cosmic ray muons by the MACRO collaboration [28]. These muons could give rise to secondary neutrons which in turn cause recoil events in a WIMP detector. The DAMA experiment has investigated this possibility and sees no background modulation for recoils above 90 keV [29].

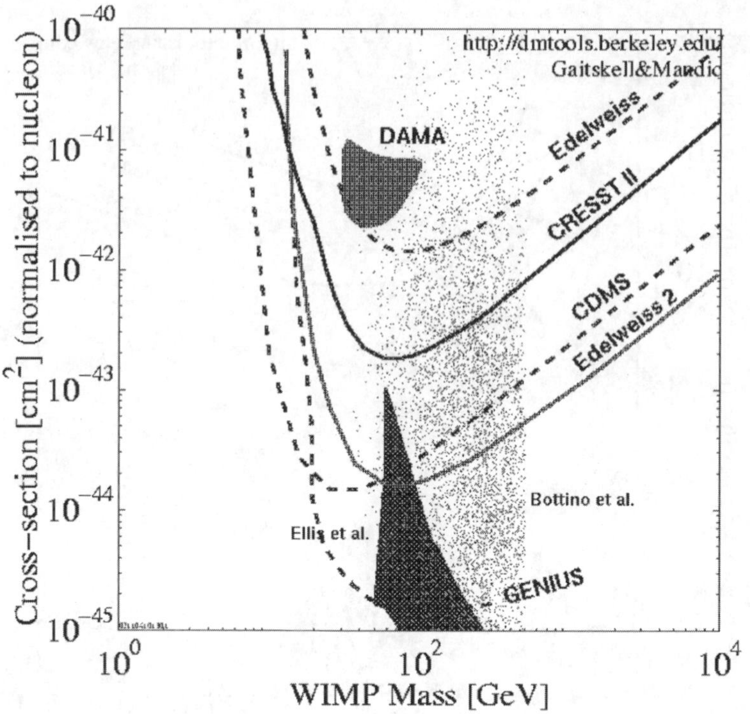

Figure 5. Exclusion plot for spin-independent WIMP interactions as aimed for by future experiments.

In fig.3 some model predictions for neutralinos are also shown [30] [31] [32]. They depend strongly on the assumed SUSY parameters and differ considerably from model to model. The CMSSM model predictions turned out to be more than two orders of magnitude below the presently reached experimental sensitivities, demanding a new generation of detectors with much higher sensitivities for the future. Exclusion plots for spin-dependent WIMP interactions, derived from the cryogenic detector experiments Tokyo (LiF) [23] and CRESST [33] as well as ELEGANT [34] and UKDMC [24] using NaI crystals, are shown in fig.4.

6. Outlook

The CDMS experiment is taking data with 5 kg Ge and 2 kg Si absorbers in the Soudan Mine (2000 m.w.e.). EDELWEISS will increase its detector mass to about 7 kg in 2004 and to 40 kg later on. CRESST will replace its sapphire absorbers with $CaWO_4$ crystals (10 kg), which will enable them to measure simultaneously scintillation light and heat for a better background recognition [35]. The CUORICINO detector of the MIBETA collaboration with 40 kg (TeO_2) is presently under construction. A big step in sensitivity is expected from the proposed GENIUS (100 kg Ge) detector [36]. The projected sensitivities of some of the future direct WIMP detectors are shown in fig.5.

7. Conclusions

Small mass (<1kg) cryogenic detectors have reached respectable WIMP detection sensitivities. The active background rejection capability, by detecting ionization (or light) and phonons simultaneously, has proven to be essential. The sensitivities of the cryogenic detectors CDMS and EDELWEISS as well as the LXe detector ZEPLIN have exceeded those of DAMA, showing incompatible results. Constrained Super Symmetric models predict WIMP masses between 50 and 600 GeV and interaction cross sections which are several orders of magnitude below present experimental limits. WIMP detectors with larger target masses are planned for the near future. With the increased sensitivity of the new generation detectors, WIMPs may be discovered soon. If not, serious constraints will be imposed on supersymmetric models. Although we know more about the total amount of dark matter in the universe, its true nature still remains a mystery. I will end with a citation of W. Shakespeare: "There is no darkness, only ignorance."

Acknowledgments

I would like to thank very warmly Walter Greiner for his kind invitation to participate in this conference. My thanks also go to Mikhail Itkis, Cem Güclü and Joachim Reinhardt, who greatly helped to organize this conference.

References

[1] F. Zwicky, Helv. Phys. Act. **6**, 110 (1933).

[2] F. Zwicky, Phys. Rev. **51**, 67 (1937).

[3] S.M. Faber and J.S. Gallagher, Ann. Rev. Astron. Astrophys. **17**, 135 (1979)

[4] V. Rubin, Comm. Astrophys. **8**, 79 (1979).

[5] Ph. Fischer et al, astro-ph/9912119 (1999).

[6] Y. Mellier, Space Science Reviews **100**, 73-87 (2002).

[7] S. Perlmutter et al, Astrophys. J. **483**, 565 (1997).

[8] A.G. Riess et al, Astron. J. **116**, 1009 (1998).

[9] D.N. Spergel et al, ApJs **148**, 175 (2003).

[10] A. Milsztajn, Space Science Reviews **100**, 103-114 (2002).

[11] Y. Fukuda et al, Phys. Rev. Lett. **Vol. 81/8**, 1562 (1998).

[12] Q.R. Anmad et al, Phys. Rev. Lett. **Vol. 89**, 011301 (2002).

[13] K. Eguchi et al, Phys. Rev. Lett. **Vol. 90**, 021802-1 (2003).

[14] J. Ellis, T. Falk, K.A. Olive and M. Srednicki, Astropart. Phys. **13**, 181 (2000).

[15] T. Falk, K.A. Olive and M. Srednicki, Phys. Lett. **B339**, 248 (1994).

[16] G. Jungmann, M. Kamionkowski, K. Griest, Physics Reports **Vol. 267**, Nos. 5/6 (March 1996).

[17] A. de Shalit and H. Feschbach, Theoretical Nuclear Physics **Vol. I**, 754 (Wiley, New York, 1975).

[18] D.R. Tovey et al, Phys. Lett. **B488**, 17-26 (2000).

[19] K. Pretzl, N. Schmitz, L. Stodolsky (eds.), Low Temperature Detectors for Neutrinos and Dark Matter (Springer, Berlin, 1987).

[20] K. Pretzl, NIM **A454**, 114-127 (2000).

[21] L. Gonzales-Mestres and D. Perret-Gallix, NIM **A279**, 382-387 (1989).

[22] K. Borer et al, Proceedings of 4th Int. Workshop on the Identification of Dark Matter (IDM2002), 2-6 September 2002 in York UK, eds. Neil J.C. Spooner and Vitaly Kudryavtsev, World Scientific, 332-337 (2002).

[23] K. Miuchi et al, astro-ph/0204411.

[24] P.F. Smith et al, Phys. Lett. **B379**, 299 (1996).

[25] R. Bernabei et al, Phys. Lett. **B480**, 23 (2000).

[26] R. Abusaidi et al, Phys. Rev. Lett. **84**, 5699 (2000).

[27] A. Benoit et al, astro-ph/0206271.

[28] M. Ambrosio et al, Astropart. Phys. **7**, 109 (1997).

[29] R. Bernabei et al, Eur. Phys. J. **C18**, 283 (2000).

[30] A. Bottino et al, hep-ph/0001309.

[31] E. Baltz and P. Gondolo, hep-ph/0102147.

[32] J. Ellis, A. Ferstl and K. Olive, Phys. Lett. **B481**, 304 (2000).

[33] G. Angloher et al, Astropart. Phys. **18**, 43-55 (2002).

[34] S. Yoshida et al, Nucl. Phys. **B (Proc.Suppl.) 87**, 58 (2000).

[35] M. Bravin et al, Astropart. Phys. **12**, 107 (1999).

[36] L. Baudis et al, Phys. Rep. **307**, 301 (1998).

STRUCTURE OF STARS AND NUCLEI

Jürgen Schaffner–Bielich
Institut für Theoretische Physik, J. W. Goethe Universität, D-60054 Frankfurt am Main, Germany

Abstract In these lectures, the properties of dense hadronic and quark matter and its relation to compact stars will be discussed. In a bottom–up approach we start with nuclear and hypernuclear physics at low density and extrapolate hadronic matter to large densities. The matching to the quark matter phase is performed in a top–down approach starting at asymptotically large densities. Implications for the mass–radius relation of compact stars and the existence of a new family of solutions will be outlined.

1. Introduction: Elementary Matter and Neutron Stars

Neutron stars are created by supernovae type II explosions and are the final endpoint of evolution of massive stars. The compact remnants of the core collapse supernovae have masses in the range of 1–2 solar masses and radii of the order of 10 km. The interior of neutron stars consists of matter under extreme densities, several times the density of normal nuclear matter, $n_0 = 3 \cdot 10^{14}$ g/cm^3. The study of neutron stars has considerably advanced during the last years. With new telescopes, ground-based and in satellites, one measures spectra of supernova remnants like the crab nebula not only in the optical but also in the x-ray (Chandra, XMM–Newton), in radio as well as in the infrared band. The Hubble Space Telescope and the Chandra Telescope has even published a movie of the crab nebula on–line! These movies demonstrate how the rotating neutron star, the crab pulsar, pushes out energetic wisps into the crab nebula in the equatorial plane as well as jets of matter along the polar axis.

More than 1000 pulsars are known today. The masses of pulsars were measured most precisely from a few binary neutron star systems [1], especially from the Hulse-Taylor pulsar with $M = (1.4411 \pm 0.00035) M_\odot$. The shortest known rotation period so far measured is 1.557 ms for the pulsar PSR 1937+21.

Our knowledge about compact stars got a new twist by the discovery of isolated, non-pulsating neutron stars. The first one seen and the closest one known is RX J1856 [2] being radio-quiet and with no pulsations. The thermal spectrum indicates a temperature of 49 eV in the optical band. The x-ray spectrum, however, shows a nice Planck curve with a temperature of 60 eV

W. Greiner et al. (eds.), Structure and Dynamics of Elementary Matter, 303–320.

as measured by the Chandra satellite [3]. Most surprisingly, no spectral lines from elements in the atmosphere of the neutron star have been found in the spectra! These puzzles of the spectra of the isolated neutron star are still not fully resolved (see [4] for a possible explanations). The revised and improved parallax measurement of RX J1856 gives a distance of $D = 117 \pm 12$ parsec [5]. An ideal black-body emitter would have a very small radius of $R_\infty = 4 - 8$ km at that distance which is much smaller than the canonical value for a neutron star of 10 km [3]. Corrections from an atmosphere resulted in an apparent radius of $R_\infty = 15 \pm 3$ km which would be compatible with most of the modern neutron star models on the market [5, 6].

The structure of neutron stars encompasses several distinctly different zones when going from the surface to the centre (for a review see e.g. [7]). First, there is a thin atmosphere up to about 10^4 g/cm^3, mainly iron but could be also hydrogen or helium by accretion. Then the outer crusts or the envelope begins which consists of free electrons and nuclei forming a Coulomb lattice. The sequence of nuclei starts with iron and then continues to neutron-rich nuclei stopping at the neutron drip-line at about 10^{11} g/cm^3 [8]. The inner crusts consists of free neutrons in addition to nuclei and electrons where the neutrons are in a superfluid state. The nuclei can now form the pasta in the crust, various inhomogeneous phase structures as bubbles (meat-balls), rods (spaghetti), and plates (lasagna) immersed in the neutron and electron fluid. The sequence can then reverse, so that the neutron fluid forms the geometrical structures immersed in a background of extremely neutron-rich and superheavy nuclei [9].

At about half times normal nuclear matter density, the matter distribution will be uniform, consisting of mainly neutrons with a small admixture of protons and electrons, where the protons can now form a superconductor. Still, the end of the crust is located far away from the centre, as the crust is only a few hundred meters thick for massive neutron stars!

The core of the neutron star consists of matter under extreme densities. New forms of matter have been proposed to exist under these condition in the very core of compact stars: Bose condensation of pions or kaons, phase transitions to hyperon matter (hyperon star) and quark matter (hybrid star). If strange quark matter is absolutely stable, then the corresponding compact star is dubbed a strange star and the quark phase extends all the way from the centre to the neutron-drip line, as all free neutrons are swallowed by the true ground state of matter, while nuclei are saved by virtue of the Coulomb barrier.

Now let us start with a simple consideration: let us assume that the high-density equation of state can be described by free (known) particles. The stable baryons known in vacuum are the nucleons (n,p) and hyperons (the Λ and $\Sigma^{-,0,+}$ with one strange quark, the $\Xi^{-,0}$ with two strange quarks, and the Ω^- with three strange quarks). The masses of the hyperons increases with the number of strange quarks. Besides those, there are spinless mesons which are stable against strong interactions (charged pions, and kaons with one anti-strange or one strange quark). They can form a Bose condensate. A calculation of neutron star matter for a free gas of particles shows, that first the negatively charged Σ^- appears at about $4n_0$ and then the Λ at about $8n_0$ [10]. The Σ^- appears before the Λ despite the fact that it is slightly heavier, because it is negatively charged. The presence of the Σ^- can then take over the rôle of the electron in making the overall matter charge neutral, thereby lowering the Fermi energy of electrons and the total energy of the system. One can compute that there will be no other particles in the composition of a free gas of hadrons up to $20n_0$ which is well beyond the maximum density in the interior of a compact star (besides that our approach will be not applicable anymore, as hadrons are composite particles). The corresponding equation of state will be even slightly softer than that of a a free neutron gas due to the appearance of additional particles. Hence, the maximum mass is the one given by Oppenheimer and Volkoff for a free gas of neutrons which is about $0.7M_\odot$ [11]. That maximum mass is more than a factor two smaller than required by the measurement of the mass of the Hulse-Taylor pulsar of $1.44M_\odot$! Neutron stars can *not* be described by a approximately free gas of particles, contrary to white dwarfs. Interactions between hadrons are crucial for explaining such a large neutron star mass.

2. Nuclei with baryons: hypernuclei and strange hadronic matter

As hyperons seem to appear in dense hadronic matter relevant for compact stars, we study in the following the interactions between baryons of the baryon octet (nucleons, Λ, Σ's, and Ξ's) in dense matter. The nucleon-nucleon interaction is very well known from nuclear properties (the general problem is, of course, how to extrapolate the interaction to densities beyond normal nuclear matter density). For hyperons the situation is far less clear but a lot of information can be gathered from hypernuclear data and hyperonic atoms (for reviews see e.g. [12, 13]). Most of the hyperon-nucleon and hyperon-hyperon interaction known experimentally indicates an attractive potential.

The measured single particle levels of Λ hypernuclei from mass numbers of $A = 3$ to 209 result in a potential of $U_\Lambda = -30$ MeV at n_0 (see e.g. [14, 15]). It is well known, that the ΛN interaction gets repulsive at densities of about $2n_0$ and above [14, 16], so this nonlinear behaviour with density has to be taken into account for the modelling of neutron star matter.

For Σ's, there is only one bound Σ hypernucleus known, ${}^4_\Sigma$He, which is, however, bound by isospin forces. Σ^- atomic data indicates a repulsive potential at n_0 [17] assuming again a nonlinear density dependence of the optical potential. This finding is consistent with the absence of narrow Σ hypernuclear states (even in the continuum) for the reaction ^{9}Be(K^-,π^-) as measured at BNL [18].

There are seven Ξ hypernuclear events reported in the literature, which are summarised in [19]. Correcting the Ξ vacuum masses for the older events, one arrives at a relativistic potential of about -28 MeV at n_0 which is comparable to that of the Λ. However, more recent indirect estimates of the ΞN interaction by final state interactions in the reaction (K^-,K^+) on ^{12}C point towards a reduced attraction of about -14 MeV [20].

Double Λ hypernuclear states have been also seen experimentally (see [13] for a summary). Besides the three older candidates, there are two new measurements reported in 2001. Experiment E906 at BNL finds indications of about 400 produced ${}^{4}_{\Lambda\Lambda}$H on a ^{9}Be target by the correlated emission of pions of the sequential weak decays. Experiment E373 at KEK reconstructs the production of a ${}^{6}_{\Lambda\Lambda}$He and its subsequent weak mesonic decays. There are difficulties to reconcile the old and the new data in detail, but the overall consensus is that the $\Lambda\Lambda$ interaction must be attractive (see the discussion in [13]).

Note, that all other hyperon-hyperon interactions, as $\Lambda\Sigma$, $\Sigma\Sigma$, $\Lambda\Xi$, $\Sigma\Xi$, and $\Xi\Xi$ are essentially unknown experimentally!

The hyperon-hyperon interaction strength can be also probed by two particle correlations of e.g. two Λs in relativistic heavy-ion collisions as it was measured recently by the NA49 collaboration [21]. Unfortunately, the statistics does not allow for a definite conclusion about the size of the s-wave scattering length of two Λs, although the data indicates a small scattering length. As a lot of hyperons are produced in a single central heavy-ion collision at relativistic bombarding energies, multiply strange nuclear systems can be produced in the laboratory [22–25]. Indeed, the hypernuclei ${}^{3}_{\Lambda}$H and ${}^{4}_{\Lambda}$H can be seen in the invariant mass spectra of the weak decay products of Helium and a π^- [26].

Building up more and more hyperons in a "nucleus", the total binding energy of the system increases, as new degrees of freedom are filled up with a lower Fermi energy than nucleons. However, in the multi-hyperonic medium new reaction channels are possible, as $\Xi^- + p \to 2\Lambda$ and $\Xi^0 + n \to 2\Lambda$. These reactions are Pauli-blocked, as the possible lowest lying Λ hyperon states are filled up by Λ's. Hence, the whole multi-hyperonic nucleus can be stabilised by virtue of the Pauli-blocking in the hyperon world! Relativistic shell model calculations find that the binding energy can be -13 MeV/A or even -21 MeV/A, depending on the strength of the hyperon-hyperon interactions [23, 24, 27]. Even purely hyperonic "nuclei", consisting of only Λ, Ξ^0, and Ξ^- hyperons can exist with up to -8 MeV/A binding energy. More importantly, the charge of these systems is negative, while the baryon number is positive — indeed, a quite unusual nuclear property.

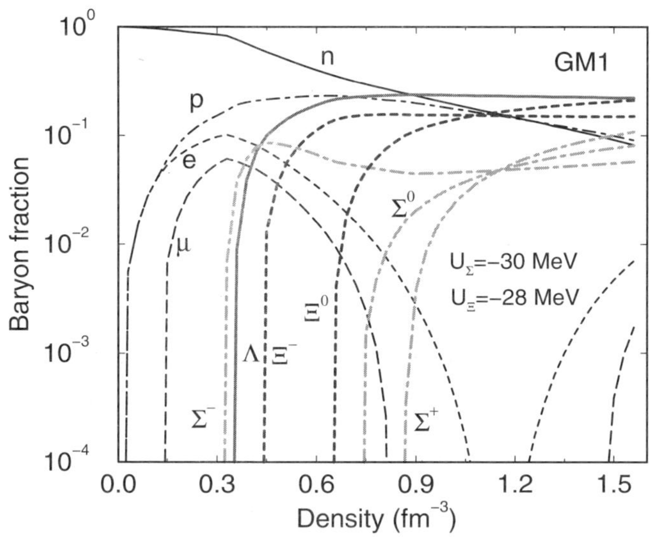

Figure 1. The composition of neutron star matter with hyperons within the relativistic mean-field model using parameter set 1 of [37] and an attractive potential for the Σ hyperons at n_0.

3. Neutron Stars with baryons

As outlined in the introduction, hyperons will appear within a free gas of hadrons but interactions have to be taken into account to be compatible with the observed neutron star masses. Our more or less profound knowledge about the nucleon-nucleon, nucleon-hyperon, and hyperon-hyperon, the latter being the least well known of all, can be utilised to derive models for neutron star matter. In many different approaches emerges that hyperons, either the Λ or the Σ^-, appear in neutron star matter at about $2n_0$: in relativistic mean-field models [28–30], in a nonrelativistic potential model [31], in the quark-meson coupling model [32], in relativistic Hartree–Fock models [33], in Brueckner–Hartree–Fock calculations [34, 35], and within chiral effective Lagrangians [36]. Nevertheless, as we will show in the following, the details of the hyperon composition of neutron star matter is rather sensitive to the chosen hyperon potentials.

Fig. 1 shows the composition of neutron star matter as a function of density. At low densities, neutrons are dominating and there is a rapid rise of the proton and electron fraction. At about $2n_0$, first the Σ^- appears closely followed by the Λ. The electron fraction starts to drop at that density. At $3n_0$, the Ξ^- is present in neutron star matter before the lighter hyperons Σ^0 and the Σ^+ appear by virtue of its negative charge. The hyperon fraction considerably rises with increasing density. At $6n_0$ the Λ fraction even exceeds that of the neutron, being then the most populated baryon in matter in β equilibrium (which may more

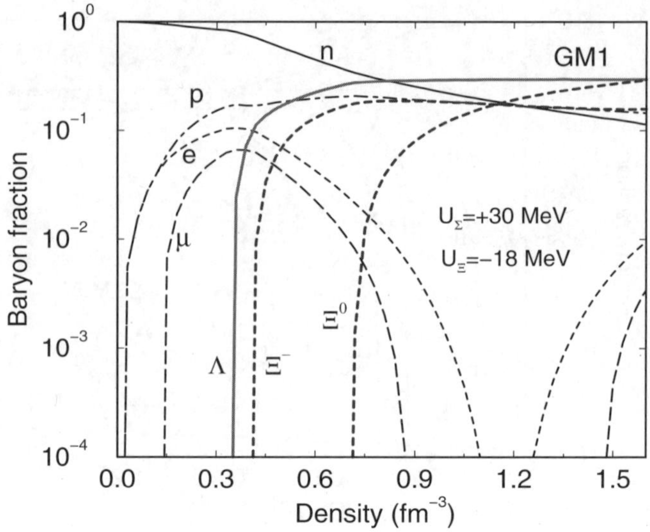

Figure 2. The composition of neutron star matter with hyperons, but now with a repulsive potential for the Σ hyperons at n_0.

aptly be called hyperon star matter). Neutron stars are giant multi-hypernuclei [28]! Note, that at about $7n_0$, electrons disappear and the isospin partners have equal fractions, as the isospin potential is negligibly small at these densities. At that point, the negatively charged Ξ^- balances the positive charge of the protons! At even larger densities, the matter distribution approaches about equal fractions for all baryons and positrons can be present. The positrons can not annihilate as there are no more electrons in the system left. The overall effect of the presence of additional degrees of freedom in the composition of a neutron star, be it hyperons or any other particle, is to lower the pressure and therefore to soften the equation of state. The maximum allowed mass for a neutron star will then decrease, typically from say $2M_\odot$ for a purely nucleonic star to about $1.5M_\odot$ for a hyperonic star depending on the parameter set used for the nucleon–nucleon interaction.

Now what happens if one just change the hyperon potentials, in particular switch the sign of the potential of the Σ hyperons from attraction to repulsion. Note that one adjusts the coupling constant of the Σ hyperons to the scalar meson slightly so as to get a repulsive potential at n_0. All hyperon potentials are repulsive above, say $2n_0$, as pointed out above, due to the nonlinear density dependence of the baryon potentials. Furthermore, it is interesting to know, that the baryon potentials are dominated by its coupling to the vector mesons at large densities which are unaltered as their coupling strengths are fixed solely by symmetry constraints (SU(6) symmetry). Fig. 2 depicts the matter composition for a repulsive potential of the Σ (at n_0) as indicated by Σ^- atomic data and a reduced attraction for the Ξ as extracted from final state interactions of Ξs with ^{12}C. Here, a relativistic potential depth of $U_\Xi = -18$ MeV is taken, which is slightly larger than the non-relativistic one. The Σ hyperons are not present at all up to $10n_0$, demonstrating how a small change in the parameters can drastically shift its critical density. The only hyperons appearing in matter are the Λ, Ξ^-, and the Ξ^0, in that order. Nevertheless, the onset of the Ξ^- population is shifted slightly towards lower densities to $2.5n_0$ compared to Fig. 1. The Ξ^- takes over the rôle of the Σ^- to diminish the fraction of electrons and its presence is even more favoured, if the Σ^- hyperons are absent. The Ξ hyperons constitute now a substantial fraction of the whole composition of hyperon star matter. Note, that the electron fraction starts to drop even before the critical density for the Ξ^- is reached. Once the neutral Λ fraction rises, the system lowers its energy by replacing protons with Λs, thereby reducing again the Fermi energy of electrons which are needed to balance the charge of the protons. One might think, that a repulsive potential for the Ξ would also remove them completely from the matter distribution as the Σs. However, this is not the case, the Ξ^- still appears, now at $3n_0$. The underlying reason is, that the repulsive vector potential for the Ξs is a factor of two smaller than for the other hyperons and a factor three smaller than for nucleons as demanded by SU(6) symmetry for the vector coupling constants. The repulsive contribution to the overall potential, the vector potential, has to dominate at large densities to avoid a collapse of matter. Hence, the Ξ is more favoured than every other baryon at large densities due to its decreased repulsive potential and less sensitive to the changes of the hyperon potential at n_0 than the Σ.

So far we have not discussed the interactions between hyperons. As indicated by the scarce double hypernuclear data, the $\Lambda\Lambda$ interaction is attractive, maybe even more than the Λ–nucleon interaction. There is definitely an additional attractive contribution between two Λs in nuclei, as the binding energy of two Λs is more than just twice the binding energy of a single Λ. So additional potentials between hyperons have to be considered. Here we follow the discussion in [24, 30, 38]. It is natural to assume that there is scalar and a vector potential only between hyperons, as for nucleons and the nucleon–hyperon interaction. Utilising SU(3) symmetry arguments for the meson exchanges, the additional potentials appear to be just mediated by the missing members of the scalar and vector meson nonet: a hidden strange scalar meson, the $f_0(980)$, which we denote as σ^* in the following, and the hidden strange vector meson $\phi(1020)$. The vector coupling constant to the ϕ–meson can be fixed by SU(6) symmetry, so that all vector coupling constants are given in terms of the nucleon one. In SU(6) symmetry, there is ideal mixing between the singlet and octet states and the vector coupling constants are proportional to the number of light or strange quarks of the baryon, respectively. The conserved nonstrange vector current is

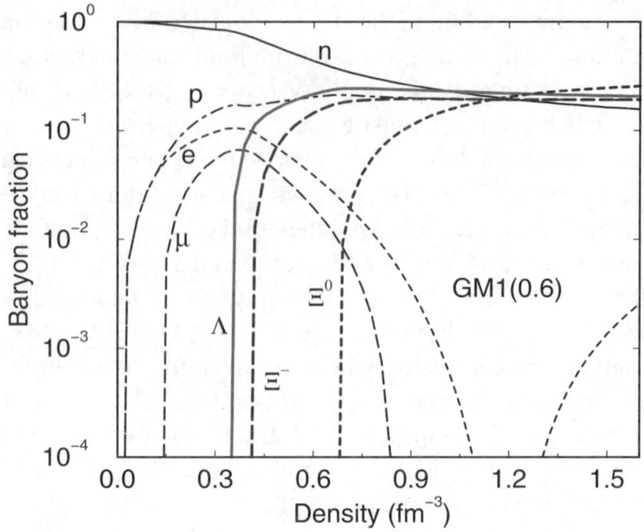

Figure 3. The composition of neutron star matter with hyperons including effects from an explicit hyperon–hyperon interaction ($g_{\sigma*\Lambda}/g_{\sigma N} = 0.6$).

then proportional to the light quark number current, while the hidden strange vector current is proportional to the strange quark number current. The scalar meson coupling constant to the σ^* will be adjusted to the hyperon–hyperon interaction strength. We assume that the σ^* meson coupling constants scale with the number of strange quarks, like for the ϕ meson coupling constants. As a consequence of these arguments, the hyperon–hyperon interaction between Ξs should then be twice as strong as between Λs and Σs. The overall strength of the hyperon–hyperon interaction is then controlled by one parameter, the hidden strange scalar coupling constant of the Λ to σ^*.

Fig. 3 shows the composition, if the additional potentials for the hyperon–hyperon interaction are taken into account with $g_{\sigma*\Lambda}/g_{\sigma N} = 0.6$. As one can see, there is not a substantial change in the composition of neutron star matter compared to the case without explicit hyperon–hyperon interactions. The appearance of hyperons is hardly modified. One notices, however, that the hyperon fraction at large densities is reduced, so that the hyperon fraction does not exceed the neutron fraction anymore. The composition at large densities is now uniform, there are about equal amounts of all baryons present (except for the Σ hyperons). The reduced fraction of hyperons at large densities is due to the additional vector potential between hyperons which dominates over the attractive hidden strange scalar potential at large hyperon densities. The equal abundance of all baryons, except for the Σs, is due to the fact that the sum of vector potentials of each baryon scales with the total number of quarks, not

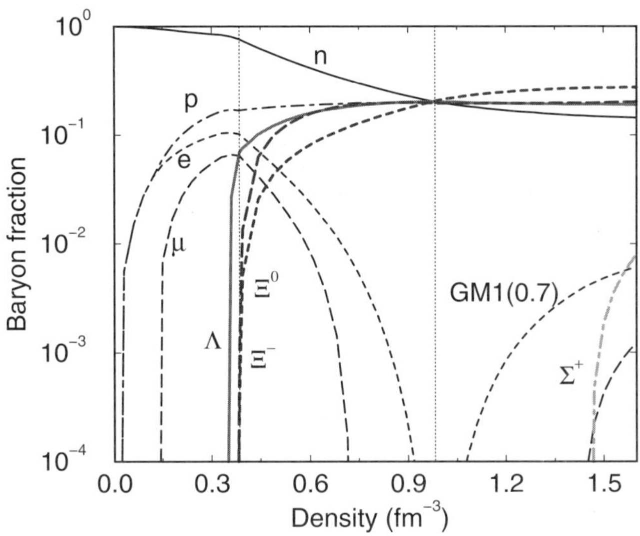

Figure 4. The composition of neutron star matter with hyperons, now for a slightly increased hyperon–hyperon interaction ($g_{\sigma * \Lambda}/g_{\sigma N} = 0.7$). A first order phase transition appears, the boundaries of the mixed phase are marked by the vertical dotted lines.

with the number of light quarks. The strange twist by including the attractive hyperon–hyperon interaction is that the hyperon fractions are reduced. The hyperon–hyperon interaction is attractive at normal nuclear density and for a small hyperon population, but it turns repulsive for a large hyperon density i.e. in the core of a neutron star. The additional repulsions helps to stabilise the system at large densities in the sense that the equation of state gets stiffer at large densities.

Now let us increase the strength of the hyperon–hyperon interaction slightly from $g_{\sigma * \Lambda}/g_{\sigma N} = 0.6$ to $g_{\sigma * \Lambda}/g_{\sigma N} = 0.7$. Fig. 4 depicts the composition of neutron star matter. Surprisingly, the composition has drastically changed. The Λ population sets in slightly before the appearance of the Ξ hyperons, which appear together at about $2.5 n_0$. At $6.5 n_0$, the baryon fractions are equal to each other and the electrons, as well as the muons, have disappeared in the medium. A first order phase transition from nucleon dominated to hyperon dominated hadronic matter appeared in the neutron star matter at $2.5 n_0$! Using the general Gibbs criterion to model the phase transition (see e.g. [39]), the electron fraction is continuous through the boundaries of the mixed phase. Negatively charged bubbles of hyperon dominated matter form which are immersed in slightly positively charged nucleon dominated matter. At the end of the mixed phase at $6.5 n_0$, the situation is reversed, now positively charged bubbles of nucleon dominated matter is present in the background of slightly negatively charged

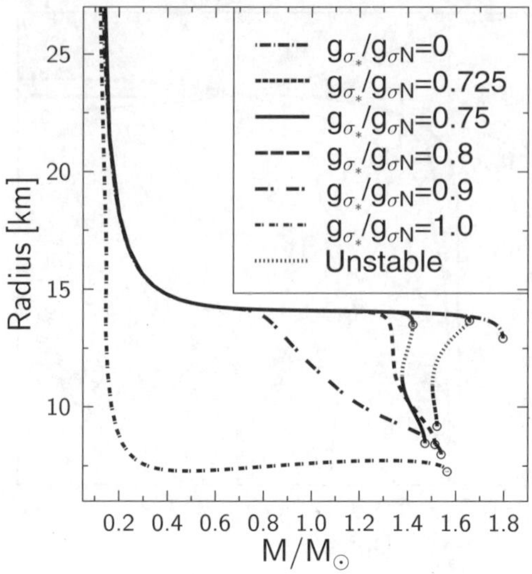

Figure 5. The mass–radius relation of compact stars for different interaction strengths of the hyperon–hyperon potential. For increasing hyperon–hyperon attraction, the radius of the maximum mass compact star decreases down to only $R = 7 - 8$ km. In some intermediate cases, an additional branch of stable solutions appears (taken from [38]).

hyperon dominated matter. At each stage, the criterion of total global (not local) charge neutrality is fulfilled. The boundaries of the mixed phase are marked by the vertical dotted lines in the figure. Note that the mixed phase extends over a quite large region of density, thereby constituting a substantial fraction of the total matter inside a neutron star. Geometrical phases appear in the mixed phase, bubbles, rods and slabs, like in the liquid–gas phase transition of nuclear matter in the crust. The charged structures will form a Coulomb lattice, thereby changing the liquid to a solid and modifying the transport properties of the inner parts of a neutron star.

The presence of a first order phase transition in neutron star matter has also an important impact on the global properties of compact stars [38]. The mass–radius relation for compact stars is shown in Fig. 5 for different hyperon–hyperon interaction strengths. For a moderate hyperon–hyperon interaction strength, $g_{\sigma*\Lambda}/g_{\sigma N} = 0.725$ and 0.75, the maximum mass is reduced while the corresponding radius is unaltered in the ordinary neutron star branch. In addition, there appears a new branch of solutions at smaller radii but similar (maximum) masses. The dotted lines denote the unstable regions of the solution, where the compact star is unstable with respect to radial oscillations. The dashed and full lines then indicate that the solution is stable against radial oscillations, as the mass increases with decreasing radii. The maximum mass of this new branch of solution amounts to about $1.5M_\odot$ with a radius of about only $7-8$ km. The compact stars representing this new family of solutions are hypercompact stars. For comparison, the maximum mass of the ordinary neutron star branch lies at similar values while the radius is much larger, around 13 km. Note, that hypercompact stars can have larger or lower masses than the maximum mass of the ordinary neutron star branch.

Increasing the hyperon–hyperon interaction, the new family of solution disappears and there is a continuous stable line in the mass–radius diagram. The maximum mass compact star is, however, now always located close to the hypercompact star branch, i.e. at about $1.5M_\odot$ with a radius of about $7-8$ km. A new effect happens for the case $g_{\sigma*\Lambda}/g_{\sigma N} = 1.0$. For asymptotically large radii, that means for low densities, the mass and radius is like in the other cases. The masses of the compact stars stay, however, below $0.2M_\odot$ down to radii of $7-8$ km and increase with a rather constant radius up to a maximum mass of about $1.6M_\odot$. The underlying equation of state for this case is one which exhibits a vanishing pressure at a finite value of the energy density, a feature characteristic for absolutely stable matter, like absolutely stable strange quark matter [40, 41]. The corresponding compact stars are so called selfbound, as they are stabilised by strongly attractive interactions, contrary to ordinary compact stars which are hold together by the gravitational force. The outermost layer of material of a selfbound compact star consists of the same material as for the outer crust of an ordinary neutron star: a lattice of nuclei in a sea of electrons. The maximum density of that crust must be below the neutron drip density, as free neutrons will be eaten up by the absolutely stable matter which is more deeply bound than ordinary matter. Nuclei survive due to the Coulomb barrier between them and the phase of absolutely stable matter. There is a strong jump in baryon density, accompanied by a similar jump in the electron density which constitutes the Coulomb barrier, of three orders of magnitude at the phase boundary of nuclear matter and absolutely stable matter. As the crust material is the same for selfbound stars and neutron stars, the mass-radius curves approach each other for low central energy densities and large radii, where most of the material is residing in the crust. The properties of selfbound strange hadronic matter are similar to the ones of selfbound strange quark matter [42, 43].

Let us come back to the issue of the new branch of stable solutions. Fig. 6 demonstrates the general feature of the mass–radius relation for compact stars. For white dwarfs, the mass increases for decreasing radii up to the Chandrasekhar mass, the maximum mass of white dwarfs. Beyond that limiting mass, the solutions are unstable against radial oscillations, until the mass increases again for decreasing radii. The family of neutron stars has been reached now, which is again stable up to some maximum mass which depends sensitively on the hadronic interactions. Beyond that maximum mass for neutron stars, the mass–radius relation curves either in a (unstable) spiral down to some

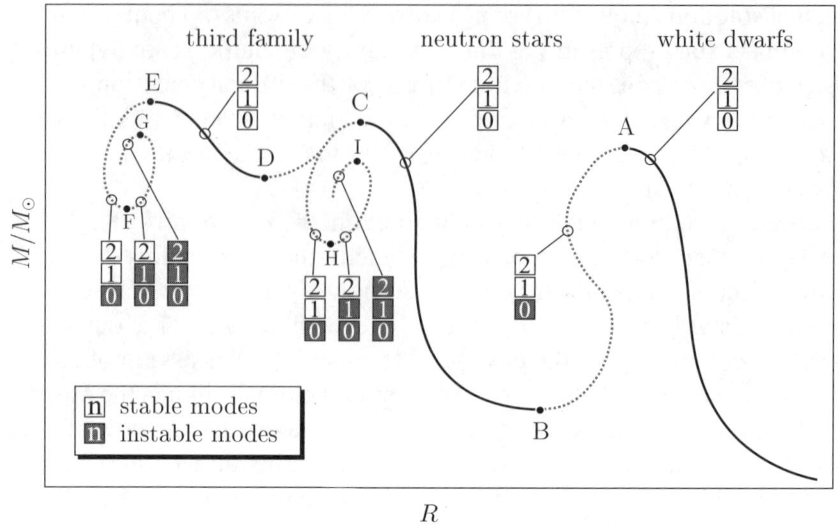

Figure 6. A sketch of the mass–radius relation of compact stars. The curves with an increasing mass while lowering the radius are stable solutions. A third solution can appear for radii below the ones for ordinary neutron stars (taken from [44]).

fixed point or continues again to increase the mass with lower radii. The latter case is stable against radial oscillations and marks the onset of a new family of stable solution of compact stars. The masses of these new branch can have similar values as those of ordinary neutron stars, so that they can be dubbed neutron star twins, but the radii are considerably smaller. After the maximum mass of the third family is reached, the unstable final spiral appears which marks the end of Tolman–Oppenheimer–Volkoff (TOV) solutions for hadronic matter.

The existence of a new family of compact stars besides white dwarfs and ordinary neutron stars stems from the presence of a strong first order phase transition in the equation of state. The input for the TOV equations is just the macroscopic relation between energy density and pressure. The microscopical details are not relevant for the global properties of compact stars. Hence, the first order phase transition can be to anything, like to strange quark matter in hybrid stars [45, 44, 46], to a pion condensed phase [47], to a kaon condensed phase [48], or, as discussed above, to strange hadronic matter [38]. The criterion for the existence of a new branch in the mass–radius relation of compact stars has already been discussed by Gerlach in 1968 [49]. The new branch appears to be stable, if the equation of state changes rapidly from a rather soft one to a much harder one. Exactly this feature is present at the end of the mixed phase of a strong first order phase transition, no matter what the underlying reason for that strong phase transition is.

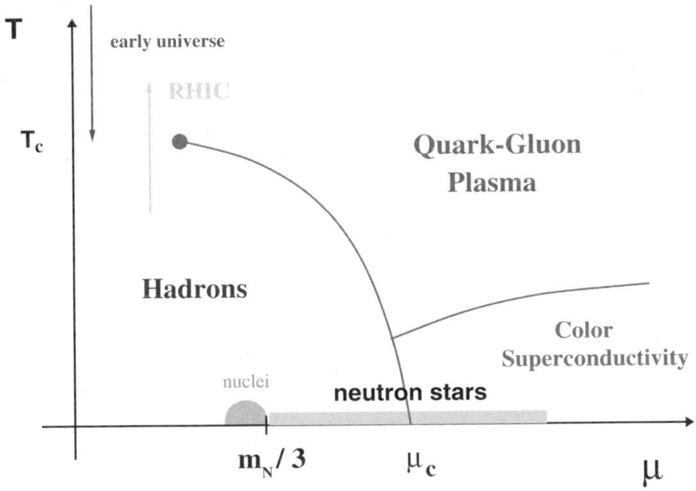

Figure 7. The phase diagram of QCD for finite temperature and quark chemical potentials. There is a line of a first order phase transition starting at zero temperature and a critical quark chemical potential μ_c and stopping in a critical endpoint where the phase transition is of second order (taken from [51]).

4. Neutron stars with quarks

In the previous section, we started with some interaction fixed to properties of matter at low densities, that means at n_0, which is a bottom–up approach for the description of dense, cold matter. In this section, we want to discuss the issue of the strong first order phase transition in terms of a top–down approach. At sufficiently large densities, the more appropriate description of dense matter will be in terms of quarks, which are, according to our present understanding of QCD, asymptotically free.

For large temperatures and vanishing net baryon density, lattice gauge simulations indicate, that the phase transition is first order for pure gluon theory and likely to be a rapid crossover at about $T_c = 170$ MeV in full QCD including dynamical quarks [50]. On the lattice, it was seen that when the quark condensate drops at T_c the Polyakov loop increases. The quark condensate is an order parameter for the chiral phase transition, where the quarks are getting massless for vanishing current quark masses for $T \geq T_\chi$. The Polyakov loop is an order parameter for pure gluon theory and has a nonvanishing value in the deconfined phase for $T \geq T_d$. Hence, the lattice data demonstrates that $T_c = T_\chi = T_d$ at zero quark-chemical potential. Why these two entirely different phase transitions happen at the same temperature is presently not fully understood. In principle, the chiral and the deconfinement phase transitions can happen at entirely different scales in QCD.

Neutron stars probe a very different region of the phase diagram of QCD (see Fig. 7). Neutron star matter is cold on nuclear scales and has an extremely large baryon density. At some critical value of the quark chemical potential μ_c, there is a first order phase transition to colour superconducting quark matter [52–55] which is based on theoretical arguments (see [56–59] for reviews about the many phases of QCD). Matter with colour superconducting quarks occupies the phase diagram at large chemical potentials and moderate temperatures. For large chemical potentials and temperatures, there will be a quark–gluon plasma. The line of first order phase transition which started at $T = 0$ and $\mu = \mu_c$ stops in a critical point. At that point the phase transition is of second order, beyond the point at larger temperatures the phase transition is a rapid crossover. The precise location of that critical endpoint is not known at present.

Recently, lattice simulations started to explore the phase diagram of QCD at finite quark chemical potential attempting to find the critical endpoint. The extension to finite μ is performed by reweighting techniques or a Taylor expansion which will fail, of course, if μ/T gets large. So far, first lattice calculations find values of about $\mu_{\text{end}} = 400 - 700$ MeV [60, 50]. In any case, it is quite likely that there is a first order phase transition at finite density and low temperature which is right in the region of interest for neutron star matter.

We now study the following picture of cold and dense matter: at low densities there are hadrons and at intermediate densities the description of dense matter will be more appropriate in terms of massive quarks. The transition from hadrons to massive quarks is assumed to be smooth. Then there is a phase transition from massive quarks to massless quarks at the chiral phase transition $\mu = \mu_\chi$. For $\mu > \mu_\chi$, perturbative QCD is used as a model of the equation of state of QCD and for $\mu < \mu_\chi$ a matching to the low density equation of state has to be performed [46, 59]. The thermodynamic potential can be computed up to order α_s^2 for massless quarks from which all thermodynamic observables are derived self-consistently (pressure, energy density, and baryon number density). The renormalisation subtraction point needs to be fixed in the perturbative treatment. It should be proportional to the overall scale of the system and is chosen to be $\bar{\Lambda}/\mu = 2, 3$. One finds then, that the resulting pressure starts at some finite value of the chemical potential and approaches very slowly the Stefan–Boltzmann limit of a free gas. Even at rather large densities, say of $n = 30n_0$, there are still sizable corrections to a simple free gas of quarks.

The matching to the low density equation of state can happen basically in two ways: either the matching appears to be smoothly from the low to the high density part or the matching incorporates a drastic change of the slope of the pressure as a function of chemical potential. Fig. 8 depicts the situation schematically. If the pressure in the low density equation of state increases rapidly with density, the matching will be smooth and the phase transition is weakly first order. On the other hand, if the pressure rises slowly with density

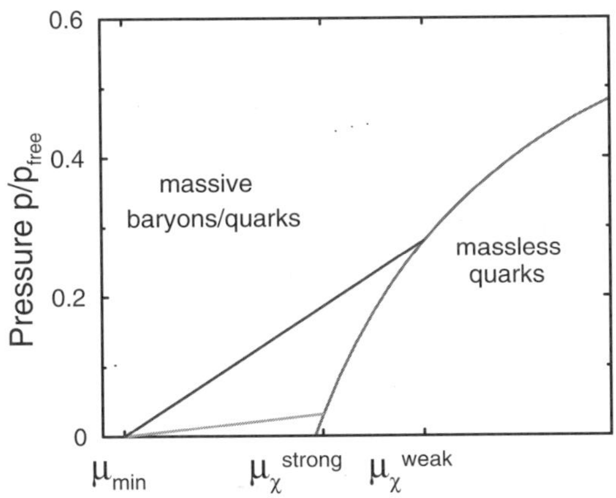

Figure 8. The matching of the low density equation of state to the high density matter of massless quarks (taken from [51]). Depending on the rise of the pressure in the low density phase, the matching results in a weak phase transition at $\mu_{\rm weak}$ (upper line) or a strong phase transition at $\mu_{\rm strong}$ (lower line).

the phase transition will be strongly first order. Now the matching of the two equation of states results in a significant change of the slope of the curves in Fig. 8. As the slope reflects just the number density, the baryon density as well as the energy density will jump at the matching point causing a strong first order phase transition. Which situation is realized in nature, is not clear. Nonrelativistic model calculations of asymmetric matter [61] hint at that the pressure in the low density regime stays small up to about $2n_0$, being only a few percent of that of a free gas of quarks, so that a strong first order phase transition is not excluded a priori.

Fig. 9 shows the mass–radius relation for various equation of states. The solid lines stand for pure quarks stars calculated in perturbative QCD, the long–dashed ones are calculated within the MIT bag model, a free gas of massless quarks only modified by a constant bag pressure. The curves for quark stars start at zero mass and radius as the pressure vanishes at a finite energy density — no hadronic mantle has been taken into account which would change the mass–radius relation at small masses to large radii (see e.g. the discussion of Fig. 5). The mass–radius relation for a purely hadronic equation of state is depicted on the right side of the plot. At some chosen critical density, the phase transition to quark matter occurs. The corresponding hybrid stars have smaller radii. For $n_c = 2.5n_0$, a new family of stable solution appears where the short–dashed line turns into a solid line. The matching is done for the quark matter equation of state producing quark stars with a rather low mass of only $1M_\odot$ and a corresponding radius of only 6 km. Depending on the matching of the hadronic to the quark matter equation of state, the new family of solution lies

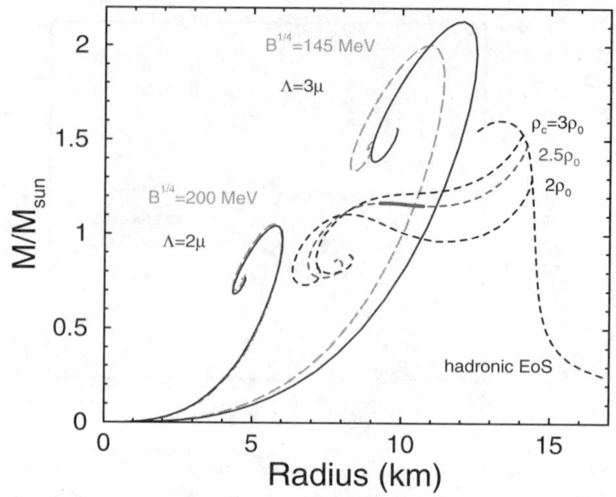

Figure 9. The mass radius relation of pure quark stars are shown by the solid lines (perturbative QCD) and long–dashed lines (MIT bag model). The mass–radius curves for a hadronic equation of state (EoS) with a phase transition to quark matter are plotted with short–dashed lines for different critical densities. A new stable solution appears where the curve turns from a short–dashed to a solid line.

somewhere between the curves for pure hadronic and pure quark stars, i.e. radii down to 6 km are in principle possible for the new family of compact stars. We note in passing, that there is a recent flurry of activity studying effects from color superconducting quark matter on the mass-radius relation of compact stars [62–68].

It is clear from the previous discussion, that the studies of compact stars is far from being completed. Further investigations of the nonperturbative treatments of the quark phase and how it matches to the low density equation of state are needed to elucidate the possible existence of a new family of compact stars. The recent data from present x-ray satellites, Chandra and XMM Newton, and the Hubble Space telescope have considerably advanced our knowledge of the properties of compact stars with more surprising news to come. Future telescopes, like the x-ray satellite XEUS and the Next Generation Space Telescope, are poised to finally pin down the mass–radius relation of compact stars and to determine the equation of state of cold and dense, strongly interacting matter. Excitingly, this means also that the possible existence of hypercompact stars might be confirmed within the coming years!

Acknowledgments

It is a pleasure to thank Eduardo Fraga, Norman Glendenning, Matthias Hanauske, Rob Pisarski, and Igor Mishustin with whom I collaborated on the topics addressed in these lectures and Mark Alford for comments. I am indebted to Walter Greiner and Horst Stöcker for their continuous support which made this work possible. I especially thank Walter Greiner for inviting me to give these lectures.

References

[1] S. E. Thorsett and D. Chakrabarty, Astrophys. J. **512**, 288 (1999).

[2] F. M. Walter, Astrophys. J. **549**, 433 (2001).

[3] J. J. Drake *et al.*, Astrophys. J. **572**, 996 (2002).

[4] V. Burwitz *et al.*, Astron. Astrophys. **399**, 1109 (2003).

[5] F. M. Walter and J. Lattimer, Astrophys. J. **576**, L145 (2002).

[6] J. A. Pons *et al.*, Astrophys. J. **564**, 981 (2002).

[7] P. Haensel, EAS Publications Series, volume 7, Final Stages of Stellar Evolution, Proceedings of the conference held 16-21 September, 2001 in Aussois, France, edited by C. Motch and J.-M. Hameury, p. 249 (2003).

[8] G. Baym, C. Pethick, and P. Sutherland, Astrophys. J. **170**, 299 (1971).

[9] J. W. Negele and D. Vautherin, Nucl. Phys. **A207**, 298 (1973).

[10] V. A. Ambartsumyan and G. S. Saakyan, Sov. Astron. **4**, 187 (1960).

[11] J. R. Oppenheimer and G. M. Volkoff, Phys. Rev. **55**, 374 (1939).

[12] C. J. Batty, E. Friedman, and A. Gal, Phys. Rep. **287**, 385 (1997).

[13] A. Gal, Nucl. Phys. **A721**, 945 (2003).

[14] D. J. Millener, C. B. Dover, and A. Gal, Phys. Rev. C **38**, 2700 (1988).

[15] M. Rufa *et al.*, Phys. Rev. C **42**, 2469 (1990).

[16] J. Schaffner-Bielich, I. N. Mishustin, and J. Bondorf, Nucl. Phys. **A625**, 325 (1997).

[17] J. Mares, E. Friedman, A. Gal, and B. K. Jennings, Nucl. Phys. **A594**, 311 (1995).

[18] S. Bart *et al.*, Phys. Rev. Lett. **83**, 5238 (1999).

[19] C. B. Dover and A. Gal, Ann. Phys. (N.Y.) **146**, 309 (1983).

[20] P. Khaustov *et al.*, Phys. Rev. C **61**, 054603 (2000).

[21] C. Blume *et al.*, Nucl. Phys. **A715**, 55 (2003).

[22] J. Schaffner, C. Greiner, and H. Stöcker, Phys. Rev. C **46**, 322 (1992).

[23] J. Schaffner *et al.*, Phys. Rev. Lett. **71**, 1328 (1993).

[24] J. Schaffner *et al.*, Ann. Phys. (N.Y.) **235**, 35 (1994).

[25] J. Schaffner-Bielich, R. Mattiello, and H. Sorge, Phys. Rev. Lett. **84**, 4305 (2000).

[26] L. E. Finch et al. (E864 collaboration), Nucl. Phys. **A661**, 395c (1999).

[27] J. Schaffner-Bielich and A. Gal, Phys. Rev. C **62**, 034311 (2000).

[28] N. K. Glendenning, Astrophys. J. **293**, 470 (1985).

[29] R. Knorren, M. Prakash, and P. J. Ellis, Phys. Rev. C **52**, 3470 (1995).

[30] J. Schaffner and I. N. Mishustin, Phys. Rev. C **53**, 1416 (1996).

[31] S. Balberg and A. Gal, Nucl. Phys. **A625**, 435 (1997).

[32] S. Pal *et al.*, Phys. Rev. C **60**, 015802 (1999).

[33] H. Huber, F. Weber, M. K. Weigel, and C. Schaab, Int. J. Mod. Phys. **E7**, 301 (1998).

[34] M. Baldo, G. F. Burgio, and H. J. Schulze, Phys. Rev. C **61**, 055801 (2000).

[35] I. Vidana *et al.*, Phys. Rev. C **62**, 035801 (2000).

[36] M. Hanauske *et al.*, Astrophys. J. **537**, 958 (2000).

[37] N. K. Glendenning and S. A. Moszkowski, Phys. Rev. Lett. **67**, 2414 (1991).

[38] J. Schaffner-Bielich, M. Hanauske, H. Stöcker, and W. Greiner, Phys. Rev. Lett. **89**, 171101 (2002).

[39] N. K. Glendenning, Phys. Rev. D **46**, 1274 (1992).

[40] E. Witten, Phys. Rev. D **30**, 272 (1984).

[41] E. Farhi and R. L. Jaffe, Phys. Rev. D **30**, 2379 (1984).

[42] P. Haensel, J. L. Zdunik, and R. Schaeffer, Astron. Astrophys. **160**, 121 (1986).

[43] C. Alcock, E. Farhi, and A. Olinto, Astrophys. J. **310**, 261 (1986).

[44] K. Schertler, C. Greiner, J. Schaffner-Bielich, and M. H. Thoma, Nucl. Phys. **A677**, 463 (2000).

[45] N. K. Glendenning and C. Kettner, Astron. Astrophys. **353**, L9 (2000).

[46] E. S. Fraga, R. D. Pisarski, and J. Schaffner-Bielich, Phys. Rev. D **63**, 121702(R) (2001).

[47] B. Kämpfer, J. Phys. A **14**, L471 (1981).

[48] S. Banik and D. Bandyopadhyay, Phys. Rev. C **64**, 055805 (2001).

[49] U. H. Gerlach, Phys. Rev. **172**, 1325 (1968).

[50] F. Karsch, hep-lat/0401031 (2004).

[51] E. S. Fraga, Y. Hatta, R. D. Pisarski, and J. Schaffner-Bielich, nucl-th/0301062 (2003).

[52] B. C. Barrois, Nucl. Phys. B **129**, 390 (1977).

[53] D. Bailin and A. Love, Phys. Rep. **107**, 325 (1984).

[54] M. Alford, K. Rajagopal, and F. Wilczek, Phys. Lett. B **422**, 247 (1998).

[55] R. Rapp, T. Schäfer, E. V. Shuryak, and M. Velkovsky, Phys. Rev. Lett. **81**, 53 (1998).

[56] K. Rajagopal and F. Wilczek, hep-ph/0011333 (2000).

[57] T. Schäfer, hep-ph/0304281 (2003).

[58] D. H. Rischke, Prog. Part. Nucl. Phys. **52**, 197 (2004).

[59] M. Alford, nucl-th/0312007 (2003).

[60] Z. Fodor and S. D. Katz, JHEP **03**, 014 (2002).

[61] A. Akmal, V. R. Pandharipande, and D. G. Ravenhall, Phys. Rev. C **58**, 1804 (1998).

[62] M. Alford and S. Reddy, Phys. Rev. D **67**, 074024 (2003).

[63] M. Baldo *et al.*, Phys. Lett. B **562**, 153 (2003).

[64] S. Banik and D. Bandyopadhyay, Phys. Rev. D **67**, 123003 (2003).

[65] D. Blaschke *et al.*, AIP Conf. Proc. **660**, 209 (2003).

[66] I. Shovkovy, M. Hanauske, and M. Huang, Phys. Rev. D **67**, 103004 (2003).

[67] S. B. Ruster and D. H. Rischke, Phys. Rev. D **69**, 045011 (2004).

[68] M. Buballa, F. Neumann, M. Oertel, and I. Shovkovy, nucl-th/0312078 (2003).

SPIN-ONE COLOR SUPERCONDUCTIVITY IN NEUTRON STARS

Andreas Schmitt
Institut für Theoretische Physik der J. W. Goethe Universität, D-60054, Frankfurt am Main, Germany
aschmitt@th.physik.uni-frankfurt.de

Abstract We study color superconductivity in quark matter where quarks of the same flavor form Cooper pairs of total spin one. We discuss two phases, the polar and the color-spin locked phase. Both phases exhibit an electromagnetic Meissner effect, contrary to spin-zero color superconductors.

1. Introduction

Quantum chromodynamics (QCD) is an asymptotically free theory [1] and, thus, quark matter at large quark chemical potential μ is a weakly coupled system. In this case, the dominant interaction between two quarks is single-gluon exchange. Single-gluon exchange is attractive in the color-antitriplet channel. Consequently, at sufficiently low temperatures, the quark Fermi surface is unstable with respect to the formation of Cooper pairs. Since this is analogous to what happens in ordinary superconductors [2], this phenomenon was termed *color* superconductivity [3].

Color superconductivity was studied from first principles in the framework of QCD at weak coupling [4–6] as well as in more phenomenological Nambu–Jona-Lasinio-type models [7]. The former approach is rigorously valid at asymptotically large densities, while the latter is valid at moderate densities of the order of a few times the nuclear matter ground state density. Both approaches indicate that the color-superconducting state is the true ground state of quark matter at any density beyond the quark-hadron phase transition and at sufficiently low temperature. They also agree in the magnitude of the color-superconducting gap parameter, ϕ, which they predict to be of the order of tens of MeV (for quark Cooper pairs with total spin zero) at densities of the order of ten times the nuclear matter ground state density. Gap parameters of this order of magni-

W. Greiner et al. (eds.), Structure and Dynamics of Elementary Matter, 321–328.

tude may have enormous phenomenological implications, since the transition temperature to the normal conducting phase is typically of the order of ϕ [6, 8, 9]. For instance, during the evolution of a neutron star, the temperature ranges from a few tens of MeV down to a few keV [10]. If its core is sufficiently dense to consist of quark matter, this quark matter core is then very likely a color superconductor.

Color superconductivity is more complex than ordinary superconductivity, because quarks do not only carry electric, but also color and flavor charge. Two different quark flavors may form Cooper pairs in the color-antitriplet, flavor-singlet, total spin-zero channel (the so-called 2SC phase) [3]. For three different quark flavors, the favored state is the so-called color-flavor locked (CFL) phase [11], with spin-zero Cooper pairs in the color-antitriplet, flavor-antitriplet representation.

A necessary condition for pairing of quarks of different flavor is that their respective Fermi surfaces are equal. In physical systems, however, this condition may be hard to achieve. For instance, compact stellar objects are neutral with respect to electric and color charge. This requires the introduction of separate chemical potentials for quarks which differ in color and flavor [12]. Imposing the constraints of color and electric charge neutrality leads to different values for these chemical potentials. This effect and the mass difference between the light up and down quarks and the heavier strange quark [13] may then lead to different Fermi surfaces for each quark species. If this difference is of the order of, or larger than, the color-superconducting gap parameter ϕ, a color-superconducting state where quarks with different flavor form spin-zero pairs with zero momentum is no longer favored. Then, besides a transition to the normal conducting state there is the possibility that quarks of the same flavor form Cooper pairs with total spin one [3, 6, 14, 15]. Quarks with the same flavor have the same mass and the same electric charge, and thus pairing is neither affected by a mass difference nor a nonzero electric chemical potential which may be required to fulfill the constraint of electric neutrality. Moreover, a potentially nonzero color chemical potential does not destroy the Cooper pairs, because we expect it to be parametrically much smaller than the gap, $\mu_{\text{color}} \sim \phi^2/(g\mu) \ll \phi$, where g is the strong coupling constant [16].

2. Spin-one color superconductivity

Contrary to Cooper pairs of quarks with different flavors, single flavor pairs cannot be formed in the spin-zero channel. This fact can be seen from a simple symmetry argument. Since the attractive color channel $[\bar{\mathbf{3}}]_c$ is antisymmetric, and the two-fermion wave function of a Cooper pair has to be antisymmetric in total, the symmetric spin-triplet channel $[\mathbf{3}]_J$ is exclusively preferred. Note that in the case of the 2SC or CFL phases there is an additional flavor structure which allows for the pairing in the spin-singlet channel. In these cases, this spin-zero channel is preferred compared to the spin-one channel.

A well-known example of spin-one Cooper pairs in condensed matter physics is superfluid ^{3}He [17]. Unlike superfluid ^{4}He, where the bosonic atoms form a Bose-Einstein condensate, ^{3}He atoms are fermions, and thus superfluidity is theoretically explained by a BCS-like mechanism. Whereas in ordinary superconductivity electrons form Cooper pairs in an $S = 0$, $L = 0$ state (where S and L denote spin and orbital momentum, respectively), atoms in superfluid ^{3}He form pairs in an $S = 1$, $L = 1$ state. Therefore, the order parameter corresponding to the superfluid phase spontaneously breaks a total symmetry group G that includes the spin and orbital momentum symmetries, $G_{He} = SO(3)_S \times SO(3)_L \times U(1)$. Here, $U(1)$ corresponds to particle number conservation. Due to the triplets in orbital momentum as well as in spin space, the order parameter is a complex 3×3 matrix. There are three experimentally observed superfluid phases in ^{3}He, namely the B and A phases, and, including systems with an external magnetic field, the A_1 phase [17].

With the necessary modifications, many of the ideas in the theory of superfluid ^{3}He can be used for spin-one color superconductivity. The main modifications are related to the relativistic nature of cold and dense quark matter, the local gauge symmetries (i.e. color, electromagnetism), and the different group structure of the color gauge group $SU(3)_c$ compared to $SO(3)_S$ and $SO(3)_L$. In spin-one color superconductors, the original symmetry group, spontaneously broken by the order parameter, is given by

$$G_1 = SU(3)_c \times SO(3)_J \times U(1)_{em} \times U(1)_B \,, \tag{1}$$

where $SO(3)_J$ is the symmetry related to the total spin $J = L+S$, $U(1)_{em}$ is the electromagnetic gauge group, and $U(1)_B$ is a global symmetry corresponding to baryon number conservation. The general form of the gap matrix $\Phi^+(K)$ (defined in [9]) for pairing in the even-parity channel can be written as

$$\Phi^+(K) = \sum \phi^e(K) \mathcal{M}_{\mathbf{k}} \Lambda^e_{\mathbf{k}} \,, \tag{2}$$

where

$$\mathcal{M}_{\mathbf{k}} \equiv \sum_{ij} \psi_{ij} J_i \left[\alpha \, \hat{k}^j + \beta \, \gamma^j_\perp(\hat{\mathbf{k}}) \right] \tag{3}$$

Here, the gap function $\phi^e(K)$ depends on the quark four-momentum $K = (k_0, \mathbf{k})$, ψ_{ij} is the complex 3×3 order parameter, $(J_i)_{jk} = -i\epsilon_{ijk}$ the color-triplet basis, and $\Lambda^e_{\mathbf{k}} \equiv (1 + e\gamma_0 \boldsymbol{\gamma} \cdot \hat{\mathbf{k}})/2$, $e = \pm$ are projectors onto positive and negative energy states. The terms in the angular brackets in Eq. (3) account for a general linear combination (with real coefficients α, β) of pairing of quarks of the same chirality (first term) and of opposite chirality (second term), where we use the notation $\boldsymbol{\gamma}_\perp(\hat{\mathbf{k}}) \equiv \boldsymbol{\gamma} - \hat{\mathbf{k}} \, \boldsymbol{\gamma} \cdot \hat{\mathbf{k}}$. In the following we refer to these two pairing patterns as "longitudinal" and "transverse" gaps, respectively (cf. Ref. [9]). Note that the appearance of the transverse gaps is a consequence of the relativistic treatment. In ^{3}He, which is a nonrelativistic system, there are only longitudinal gaps. Which linear combination actually leads to a stable state, must be investigated using the thermodynamic potential [14, 18]. Here, we only discuss the properties of various possible states.

2.1 The polar phase

The polar phase is determined by the order parameter

$$\psi_{ij}^{polar} = \delta_{i3}\delta_{j3} \ . \tag{4}$$

Denoting the residual group by H, we have

$$H_{polar} = SU(2)_c \times SO(2)_J \times \tilde{U}(1)_{em} \times \tilde{U}(1)_B \ . \tag{5}$$

Since the order parameter picks a special direction in color space, $\psi_{ij}^{polar} \sim \delta_{i3}$, the color gauge group $SU(3)_c$ is broken to $SU(2)_c$. Physically, this means that all quarks of one color, say blue, remain unpaired. Consequently, a Cooper pair carries the color charge anti-blue. The discussion of the electric charge of a Cooper pair is more subtle. The residual $\tilde{U}(1)_{em}$ is generated by

$$\tilde{Q} = Q - 2\sqrt{3}\, q\, T_8 \ , \tag{6}$$

where Q and T_8 are generators of $U(1)_{em}$ and $SU(3)_c$, respectively, and q is the electric charge of the quarks. This unbroken group, which is a combination of the color and electromagnetic gauge groups, indicates that there is a new charge with respect to which the Cooper pairs are neutral. This phenomenon is similar to the color superconducting phases where quarks of different flavors form Cooper pairs [19]. For the polar phase, it was investigated in Ref. [20]. It leads to the conclusion that there is no electromagnetic Meissner effect in this phase (for a one-flavor system; below, we comment on a many-flavor system in the polar phase). This conclusion can be confirmed by a calculation of the Meissner masses for the gluons and the photon. A nonvanishing Meissner mass gives rise to a finite penetration depth for magnetic gauge bosons (=expulsion of magnetic fields). In Table 1 we collect the Meissner masses for the polar phase and the color-spin locked phase (see next section). Note that all results correspond to an equal mixing of longitudinal and transverse gaps, i.e., $\alpha = \beta = 1$ in Eq. (3). The Meissner masses are computed from the zero-energy, low-momentum limit of the spatial components of the polarization tensors,

$$m_{ab}^2 \equiv \lim_{p\to 0} \Pi_{ab}^{ii}(0,p) \ , \quad a,b = 1,\ldots,8,9 \ , \tag{7}$$

The masses $\tilde{m}_{88}$ and $\tilde{m}_{\gamma\gamma} \equiv \tilde{m}_{99}$ arise from a mixing between the eighth gluon field and the photon field [20]. The absence of the electromagnetic Meissner

	m_{aa}^2							$\tilde{m}_{88}^2$	$\tilde{m}_{\gamma\gamma}^2$
a	1	2	3	4	5	6	7	8	9
polar	0			$\frac{1}{2}g^2$				$\frac{1}{3}g^2 + 4\,q^2e^2$	0
CSL	ζg^2	ηg^2	ζg^2	ζg^2	ηg^2	ζg^2	ηg^2	ζg^2	$6\,q^2e^2$

Table 1. Zero-temperature Meissner masses (squared) for gluons, photons and mixed gauge bosons in the polar and CSL phases. g and e are the strong and electromagnetic coupling constants, respectively. All masses are given in units of $\mu^2/(6\pi^2)$. We use the abbreviations $\eta \equiv (3 + 4\ln 2)/27$, and $\zeta \equiv (6 - 4\ln 2)/9$.

effect in the polar phase due to a new charge is only true for a single flavor system where only one electric quark charge, q, is present. In a system with more than one quark flavor, $N_f > 1$, where quarks of the same flavor separately form Cooper pairs, there is also a mixing between gluons and photons but the Meissner mass for the new (rotated) photon is nonzero (if the quarks carry different electric charges).

Let us now turn to the structure and the magnitude of the gap in the polar phase. As one can see from Eq. (4), the condensate picks one direction in real space, aligned with the 3– (or z–) axis. This gives rise to an anisotropy of the gap, expressed by an angular dependent gap function.

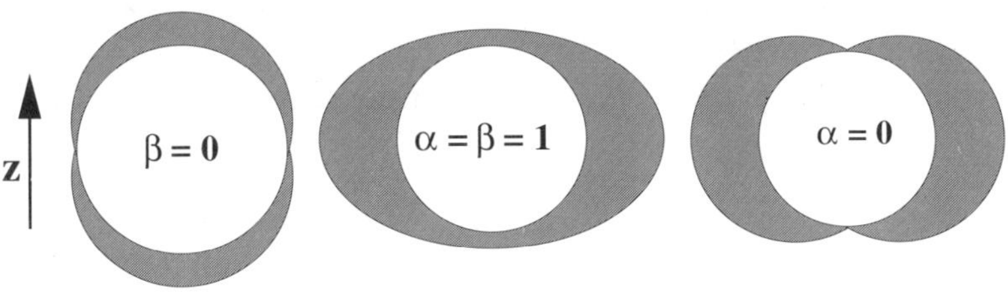

Figure 1. Angular dependence (schematically) of the gap function in the polar phase for a longitudinal gap (left), a transverse gap (right), and an equal admixture of longitudinal and transverse gaps (middle). The gap functions are rotationally symmetric with respect to the z-axis.

In Fig. 1 we show this angular dependence schematically. In the case of a longitudinal gap, there is a line (= the equator) on the Fermi sphere where the gap vanishes. This structure is known from the polar phase in ^{3}He [17]. It originates from the angular-dependent eigenvalues λ_r ($r = 1, 2$) of the matrix $\gamma_0 \mathcal{M}_{\mathbf{k}}^\dagger \mathcal{M}_{\mathbf{k}} \gamma_0$. Each of these eigenvalues defines one quasiparticle excitation branch ($k \equiv |\mathbf{k}|$),

$$\epsilon_{r,\mathbf{k}} = \sqrt{(k-\mu)^2 + \lambda_r\,\phi^2}\ . \tag{8}$$

For all combinations of longitudinal and transverse gaps, there is one gapped and one ungapped branch in the polar phase. In Fig. 1 we show the functions

	polar			CSL		
	$\beta = 0$	$\alpha = 0$	$\alpha = \beta = 1$	$\beta = 0$	$\alpha = 0$	$\alpha = \beta = 1$
λ_1	$\cos^2\theta$	$\sin^2\theta$	1	1	1	4
λ_2	0	0	0	0	0	1
ϕ/ϕ^{2SC}	e^{-6}	$e^{-4.5}$	$e^{-d(\theta)}$	e^{-6}	$e^{-4.5}$	$2^{-2/3}e^{-5}$

Table 2. Eigenvalues λ_r and the ratio ϕ/ϕ^{2SC} for the polar and CSL phases. We abbreviate $d(\theta) \equiv 3(3 + \cos^2\theta)/2$.

$\lambda_1(\theta)\phi(\theta)$ (where θ is the angle between $\hat{\mathbf{k}}$ and the z-axis) corresponding to the gapped excitation. In the case of a transverse gap, which has no analogue in the polar phase of ^{3}He, there are two nodal points of the gap function at the north and south pole of the Fermi sphere. For a gap ansatz consisting of an equal admixture of longitudinal and transverse gaps, there are no nodal points. Nevertheless, there is still an anisotropy which originates from an angular-dependent factor occuring in the gap function. In Table 2 we show this factor for the three combinations of longitudinal and transverse gaps discussed in this talk. The value of the gap ϕ, shown in this table, was computed from the QCD gap equation for $T = 0$ at the Fermi surface [9, 14]. It is normalized to the gap in the 2SC phase [4, 8]. In all phases considered here the gap in spin-one color superconductors is smaller than the one in the 2SC phase by a factor varying from $e^{-6} \sim 2 \cdot 10^{-3}$ to $e^{-4.5} \sim 10^{-2}$.

The transition temperature T_c in the polar phase for $\alpha = \beta = 1$ is given by [9]

$$T_c^{polar} = 0.57\,\phi^{polar}\ . \tag{9}$$

This relation between the transition temperature and the gap is well-known from BCS theory.

2.2 The color-spin locked phase

The color-spin locked (CSL) phase is defined by the order parameter

$$\psi_{ij} = \delta_{ij}\ . \tag{10}$$

In this phase, each spatial direction is assigned a direction in color space. Therefore, the residual symmetry group is the group of joint color and spatial rotations,

$$H_{CSL} = SO(3)_{c+J}\ . \tag{11}$$

This is similar to the CFL phase, where color is locked to flavor. Contrary to the polar phase there is neither a residual $\tilde{U}(1)_B$ nor a $\tilde{U}(1)_{em}$. Physically, the absence of a $\tilde{U}(1)_B$ corresponds to the superfluidity of the state. The absence of a $\tilde{U}(1)_{em}$ means that Cooper pairs are electrically charged and thus there is an electromagnetic Meissner effect in the CSL phase. In Table 1 we show the non-zero Meissner mass of the photon. Also all eight gluon Meissner masses are nonvanishing since the color symmetry is completely broken. Note that the residual H_{CSL} is a global symmetry. Another difference to the polar phase is that there is no mixing between magnetic gluons and photons. The reason for this fact is that the polarization tensor in Eq. (7) is diagonal in the indices a, b. Therefore, in the CSL phase, the masses in Table 1 are the masses of the original gluon and photon fields, $\tilde{m}_{88}^2 = m_{88}^2$ and $\tilde{m}_{\gamma\gamma}^2 = m_{\gamma\gamma}^2$.

The gap function in the CSL phase is isotropic for all three combinations of α and β shown in Table 2. Neither the eigenvalues λ_r nor the exponential factors that reduce the gap values compared to the 2SC phase depend on $\hat{\mathbf{k}}$. The CSL phase with $\alpha = \beta = 1$ has a particular gap structure since both eigenvalues λ_1, λ_2 are nonzero. Therefore, there are two gapped quasiparticle excitation energies. This two-gap structure gives rise to an additional factor $2^{-2/3}$ (cf. last column of Table 2). Moreover, it is the origin of a modified relation between the transition temperature and the zero-temperature gap. We find the following deviation from the BCS relation in the CSL phase with $\alpha = \beta = 1$ [9]:

$$T_c^{CSL} = 2^{2/3} \cdot 0.57\, \phi^{CSL} \,. \tag{12}$$

3. Conclusions

From the results presented in the previous sections we conclude that the gap in spin-one color superconductors is of the order of 20 - 400 keV (assuming that the gap in the 2SC phase is of the order of 10 - 100 MeV). The critical temperature is therefore of the order of 10 - 400 keV [9]. Consequently, when the core of a neutron star cools below this temperature, it could very well consist of quark matter in a spin-one color-superconducting state. A natural question is how such a state influences the properties of the star. For instance, the presence of the electromagnetic Meissner effect in spin-one color superconductors might have observable consequences for the magnetic field of a neutron star. Ordinary neutron star matter is commonly believed to be an electromagnetic superconductor of type II, while a spin-one color superconductor is of type I [20]. The question whether neutron star matter is a type-I or type-II superconductor has recently stirred a lot of attention [21], because it was shown that type-II superconducting matter is incompatible with the observation of pulsars with precession periods of order 1 year [22]. The presence of spin-one color-superconducting quark matter in the pulsars' core could explain this observation.

Acknowledgments

I would like to thank the organizers of *"Structure and Dynamics of Elementary Matter"* for an interesting and fruitful conference. I thank Qun Wang and Dirk H. Rischke who substantially contributed to the work I presented in this talk.

References

[1] D.J. Gross and F. Wilczek, Phys. Rev. Lett. **30**, 1343 (1973); H.D. Politzer, *ibid.* 1346.

[2] J. Bardeen, L.N. Cooper, and J.R. Schrieffer, Phys. Rev. **108**, 1175 (1957).

[3] D. Bailin and A. Love, Phys. Rep. **107**, 325 (1984).

[4] D.T. Son, Phys. Rev. D **59**, 094019 (1999);

[5] T. Schäfer and F. Wilczek, Phys. Rev. D **60**, 114003 (1999); D.K. Hong *et al.*, Phys. Rev. D **61**, 056001 (2000); Erratum *ibid.* D **62**, 059903 (2000).

[6] R.D. Pisarski and D.H. Rischke, Phys. Rev. D **61**, 051501 (2000); *ibid.*, 074017.

[7] M. Alford, K. Rajagopal, and F. Wilczek, Phys. Lett. B **422**, 247 (1998); R. Rapp *et al.*, Phys. Rev. Lett. **81**, 53 (1998).

[8] W. Brown, J.T. Liu, and H.-c. Ren, Phys. Rev. D **61**, 114012 (2000), *ibid.* **62**, 054013, 054016 (2000); Q. Wang and D.H. Rischke, Phys. Rev. D **65**, 054005 (2002).

[9] A. Schmitt, Q. Wang, and D.H. Rischke, Phys. Rev. D **66**, 114010 (2002).

[10] J.A. Pons *et al.*, Astrophys. J. **513**, 780 (1999).

[11] M. Alford, K. Rajagopal, and F. Wilczek, Nucl. Phys. **B537**, 443 (1999).

[12] K. Rajagopal and F. Wilczek, Phys. Rev. Lett. **86**, 3492 (2001); M. Alford and K. Rajagopal, JHEP **0206**, 031 (2002); A.W. Steiner, S. Reddy, and M. Prakash, Phys. Rev. D **66**, 094007 (2002); F. Neumann, M. Buballa, and M. Oertel, Nucl. Phys. **A714**, 481 (2003).

[13] M.G. Alford, J. Berges, and K. Rajagopal, Nucl. Phys. **B558**, 219 (1999).

[14] T. Schäfer, Phys. Rev. D **62**, 094007 (2000).

[15] M. Iwasaki and T. Iwado, Phys. Lett. B **350**, 163 (1995); M.G. Alford *et al.*, Phys. Rev. D **67**, 054018 (2003); M. Buballa, J. Hosek, and M. Oertel, Phys. Rev. Lett. **90**, 182002 (2003).

[16] A. Gerhold and A. Rebhan, Phys. Rev. D **68**, 011502 (2003); A. Kryjevski, Phys. Rev. D **68**, 074008 (2003); D.D. Dietrich and D.H. Rischke, (in preparation).

[17] A.J. Leggett, Rev. Mod. Phys. **47**, 331 (1975); D. Vollhardt and P. Wölfle, *The Superfluid Phases of Helium 3* (Taylor & Francis, London, 1990).

[18] A. Schmitt, Q. Wang, and D.H. Rischke, (in preparation).

[19] M.G. Alford, J. Berges, and K. Rajagopal, Nucl. Phys. **B571**, 269 (2000); E.V. Gorbar, Phys. Rev. D **62**, 014007 (2000); D.F. Litim and C. Manuel, Phys. Rev. D **64**, 094013 (2001).

[20] A. Schmitt, Q. Wang, and D.H. Rischke, nucl-th/0301090 (to appear in Phys. Rev. Lett.); A. Schmitt, Q. Wang, and D.H. Rischke, nucl-th/0311006.

[21] K.B.W. Buckley, M.A. Metlitski, and A.R. Zhitnitsky, astro-ph/0308148.

[22] B. Link, Phys. Rev. Lett. **91**, 101101 (2003).

THEORY OF GAPLESS SUPERCONDUCTIVITY IN QUARK MATTER

Igor Shovkovy* and Mei Huang
Institut für Theoretische Physik, J. W. Goethe Universität, D-60054, Frankfurt/Main, Germany

Abstract The unusual properties of the recently proposed gapless color superconducting phase of dense quark matter are briefly reviewed.

Keywords: Dense quark matter, Gapless color superconductivity.

1. Introduction

Sufficiently cold and dense quark matter is a color superconductor [1]. The corresponding ground state is characterized by a condensate of Cooper pairs made of quarks. Since quarks carry color charge, the $SU(3)_c$ color gauge group of strong interactions is partially or completely broken through the Anderson-Higgs mechanism. At asymptotic densities, this phenomenon was studied from first principles in Refs. [2–5]. These studies suggest that dense quark matter has a rich phase structure, consisting of many different phases (see Ref. [6] for reviews).

It is natural to expect that some color superconducting phases may exist in the interior of compact stars [7]. The estimated central densities of such stars might be as large as $10\rho_0$ (where $\rho_0 \approx 0.15$ fm^{-3} is the saturation density), while their temperatures are in the range of tens of keV. This could provide ideal conditions for the diquark Cooper pairing that leads to color superconductivity.

We start our discussion by pointing that matter in the bulk of a compact star should be neutral with respect to electrical as well as color charges. Also, such matter should remain in β-equilibrium. Satisfying these requirements impose nontrivial relations between the chemical potentials of different quarks [8, 9]. In turn, such relations influence the pairing dynamics between quarks, for instance, by suppressing conventional 2SC phase and favoring the so-called gapless 2SC (g2SC) phase [10].

*The speaker. On leave from Bogolyubov Institute for Theoretical Physics, 03143, Kiev, Ukraine.
E-mail: shovkovy@th.physik.uni-frankfurt.de

W. Greiner et al. (eds.), Structure and Dynamics of Elementary Matter, 329–336.

2. Model

We use the simplest SU(2) Nambu-Jona-Lasinio model of Ref. [11] to describe dense quark matter,

$$\begin{aligned}\mathcal{L} &= \bar{q}(i\gamma^\mu\partial_\mu - m_0)q + G_S\left[(\bar{q}q)^2 + (\bar{q}i\gamma_5\vec{\tau}q)^2\right] \\ &+ G_D\left[(i\bar{q}^C\varepsilon\epsilon^b\gamma_5 q)(i\bar{q}\varepsilon\epsilon^b\gamma_5 q^C)\right], \end{aligned} \tag{1}$$

where $q^C = C\bar{q}^T$ is the charge-conjugate spinor and $C = i\gamma^2\gamma^0$ is the charge conjugation matrix. The quark field is a 4-component Dirac spinor that carries flavor ($i = 1, 2$) and color ($\alpha = 1, 2, 3$) indices. $\vec{\tau} = (\tau^1, \tau^2, \tau^3)$ are Pauli matrices in the flavor space, and $(\varepsilon)^{ik} \equiv \varepsilon^{ik}$, $(\epsilon^b)^{\alpha\beta} \equiv \epsilon^{\alpha\beta b}$ are antisymmetric tensors in flavor and color, respectively. We introduce two independent coupling constants in the quark-antiquark and diquark channels, G_S and G_D. Also we restrict ourselves to the chiral limit ($m_0 = 0$) only.

The values of the parameters in the NJL model are the same as in Ref. [12]: $G_S = 5.016$ GeV^{-2} and the cut-off $\Lambda = 653$ MeV. The strength of the diquark coupling G_D is taken to be proportional to the quark-antiquark coupling constant, i.e., $G_D = \eta G_S$ with η being around 0.75 [10].

In β-equilibrium, the diagonal matrix of quark chemical potentials is given in terms of baryon ($\mu_B \equiv 3\mu$), electrical (μ_e) and color (μ_8) chemical potentials as follows:

$$\mu_{ij,\alpha\beta} = (\mu\delta_{ij} - \mu_e Q_{ij})\delta_{\alpha\beta} + \frac{2}{\sqrt{3}}\mu_8\delta_{ij}(T_8)_{\alpha\beta}, \tag{2}$$

where Q and T_8 are the generators of U(1)$_{\text{em}}$ of electromagnetism and the U(1)$_8$ subgroup of SU(3)$_c$ group.

In the mean field approximation, the finite temperature effective potential for quark matter in β-equilibrium with electrons takes the form:

$$\begin{aligned}\Omega &= \Omega_0 - \frac{1}{12\pi^2}\left(\mu_e^4 + 2\pi^2T^2\mu_e^2 + \frac{7\pi^4}{15}T^4\right) + \frac{m^2}{4G_S} \\ &+ \frac{\Delta^2}{4G_D} - \sum_a \int \frac{d^3p}{(2\pi)^3}\left[E_a + 2T\ln\left(1 + e^{-E_a/T}\right)\right], \end{aligned} \tag{3}$$

where Ω_0 is a constant added to make the pressure of the vacuum zero. The sum in the second line of Eq. (3) runs over all (6 quark and 6 antiquark) quasiparticles. The dispersion relations and the degeneracy factors of the quasiparticles read

$$E_{ub}^{\pm} = E(p) \pm \mu_{ub}, \qquad [\times 1] \tag{4}$$

$$E_{db}^{\pm} = E(p) \pm \mu_{db}, \qquad [\times 1] \tag{5}$$

$$E_{\Delta^\pm}^{\pm} = E_\Delta^{\pm}(p) \pm \delta\mu. \qquad [\times 2] \tag{6}$$

Here we introduced the following shorthand notation:

$$E(p) \equiv \sqrt{\mathbf{p}^2 + m^2}, \tag{7}$$

$$E_{\Delta}^{\pm}(p) = \sqrt{[E(p) \pm \bar{\mu}]^2 + \Delta^2}, \tag{8}$$

$$\bar{\mu} \equiv \frac{\mu_{ur} + \mu_{dg}}{2} = \frac{\mu_{ug} + \mu_{dr}}{2} = \mu - \frac{\mu_e}{6} + \frac{\mu_8}{3}, \tag{9}$$

$$\delta\mu \equiv \frac{\mu_{dg} - \mu_{ur}}{2} = \frac{\mu_{dr} - \mu_{ug}}{2} = \frac{\mu_e}{2}. \tag{10}$$

The thermodynamic potential that determines the pressure of quark matter, $\Omega_{\text{phys}} = -P$, is obtained from Ω in Eq. (3) after substituting μ_8, μ_e, m and Δ that solve the color and electrical charge neutrality conditions, i.e.,

$$n_8 \equiv -\frac{\partial\Omega}{\partial\mu_8} = 0, \quad \text{and} \quad n_Q \equiv -\frac{\partial\Omega}{\partial\mu_e} = 0, \tag{11}$$

as well as the gap equations, i.e.,

$$\frac{\partial\Omega}{\partial m} = 0, \quad \text{and} \quad \frac{\partial\Omega}{\partial\Delta} = 0. \tag{12}$$

3. Charge neutrality condition

Here we discuss the charge neutrality condition that plays a paramount role in producing a stable gapless phase of quark matter.

To start with, let us note that creating a macroscopic chunk of matter with a nonzero electrical charge density is very costly energetically. This is because of a large Coulomb potential energy associated with charged matter. Indeed, the corresponding *energy density* grows with increasing the volume of the system (it is proportional to $n_Q^2 V^{2/3}$ where n_Q is the charge density and V is the volume of the system). As a result, it is almost impossible for matter inside stars to remain charged over macroscopic distances.

The neutrality in a macroscopic system could be imposed in two different ways: (i) *globally* (on average), or (ii) *locally*. For example, matter is globally neutral inside a mixed phase that is made of alternating layers of different coexisting components, each of which is charged, but the charges cancel on average. In such a mixed phase, the Coulomb energy density does not grow with the volume of the system. It gives only a finite contribution to the total energy density. In addition, an extra price should be paid to compensate the formation of boundaries between the components. Below, however, we do not study such phases of dense quark matter in detail.

In a homogeneous (one component) phase of matter, the neutrality condition is enforced locally. In this case, there is no extra energy cost to compensate the

Coulomb energy or the surface tension. By imposing the local neutrality in the model at hand, we find that the gapless 2SC phase is the ground state of dense quark matter provided the diquark coupling constant is neither too strong nor too weak.

By omitting the details of the derivation [10], we note that the approximate solutions to the electrical and color charge neutrality conditions in Eq. (11) are given by $\mu_8 \approx 0$ and

$$\mu_e \approx \frac{3}{5}\mu, \quad \text{for} \quad \Delta > \delta\mu, \tag{13}$$

$$\Delta \approx \sqrt{\frac{\mu_e^2}{4} - \left(\frac{(\mu - 2\mu_e/3)^3 - \mu_e^3}{6(\mu - \mu_e/6)^2}\right)^2}, \quad \text{for} \quad \Delta \leq \delta\mu. \tag{14}$$

It is remarkable that the solutions in the regions of parameters $\Delta > \delta\mu$ and $\Delta \leq \delta\mu$ are very different. One could check that this reflects a qualitatively different nature of the quasiparticle spectra in the two regions. While all the modes described by the dispersion relations in Eq. (6) are gapped when $\Delta > \delta\mu$, there appear two gapless modes when $\Delta \leq \delta\mu$.

Now, the effective potential in Eq. (3) in the case of neutral matter should be considered as a function of the gap parameter Δ only (here we take into account that, in the chiral limit, the dynamical Dirac mass m is vanishing at sufficiently large densities). Indeed, after the neutrality is imposed, the chemical potentials μ_e and μ_8 are not free parameters, but functions of Δ. To avoid a confusion, we add that the corresponding functions $\mu_e = \mu_e(\Delta)$ and $\mu_8 = \mu_8(\Delta)$ may not necessarily be given by explicit analytical expressions [see, for example, the relation in Eq. (14)]. The effective potential as a function of Δ for neutral matter is shown graphically in Fig. 1 (solid line).

To emphasize the role of the neutrality condition, in Fig. 1, we also show the formal result for the effective potential of non-neutral quark matter at a fixed value of the electrical chemical potential $\mu_e \approx 148$ MeV (dashed line). The chemical potential μ_e is chosen so that the electrical charge density of matter vanishes when the value of the gap corresponds to the maximum of the potential. This effective potential (labeled "$\mu_e = 148$ MeV") describes negatively charged matter to the left from the maximum, and positively charged matter to the right from the maximum. On both sides, it is unphysical because a large (infinite in an infinite volume) contribution of the Coulomb energy was not included there.

From the location of the minimum of the effective potential for neutral quark matter (solid line in Fig. 1), one determines the value of the gap parameter in the ground state, $\Delta \approx 68$ MeV. It appears that $\Delta < \delta\mu \equiv \mu_e/2 \approx 74$ MeV in such a ground state. This is a very interesting solution because it corresponds to the so-called gapless superconductivity [10]. We discuss some unusual properties of this phase of matter in the next section.

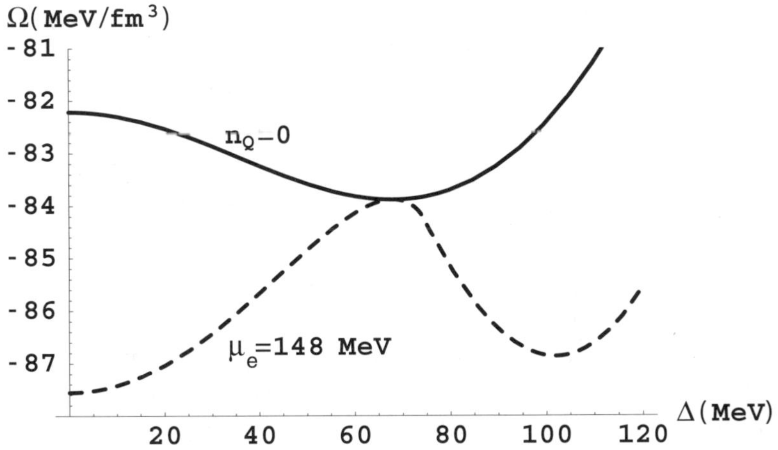

Figure 1. The effective potential as a function of the diquark gap Δ calculated at a fixed value of the electrical chemical potential $\mu_e \approx 148$ MeV (dashed line), and the effective potential defined along the neutrality line (solid line). The results are plotted for $\mu = 400$ MeV and $G_D = \eta G_S$ with $\eta = 0.75$.

Before concluding this section, we would like to comment that the other existing examples of stable gapless phases in the literature [13–15] make use of a restriction that is similar in spirit to the charge neutrality condition used here. In particular, it is the ratio of the fermion densities $n_1/n_2 = \lambda \neq 1$ that is kept fixed in Refs. [13–15]. The last condition, written in the form $n_1 - \lambda n_2 = 0$, resembles closely the electrical charge neutrality condition, $\frac{2}{3}n_u - \frac{1}{3}n_d - n_e = 0$, that we employ. It is only a nonzero electron contribution n_e that makes it different. This difference is not so important, however, because the typical conditions in neutral quark matter in β-equilibrium are such that $n_e \ll n_{u,d}$.

4. Gapless superconductivity

Here we review the main properties of the gapless 2SC phase of quark matter. By definition, this is a ground state that corresponds to a solution with $\Delta < \delta\mu$. The low energy quasiparticle spectrum in this phase contains *four* gapless and only two gapped modes, see Eqs. (4)–(6). This is in contrast to the ordinary 2SC phase (that appears in the strong coupling regime) whose spectrum contains only *two* gapless and four gapped modes.

The most remarkable property of the quasiparticle spectrum in the g2SC phase is that the low energy excitations ($E \ll \delta\mu - \Delta$) are very similar to those in the normal phase. The only difference is that the values of the chemical potentials of the free up and down quarks are replaced by $\mu^{\pm} \equiv \bar{\mu} \pm \sqrt{(\delta\mu)^2 - \Delta^2}$. This simple observation suggests that the low energy (large distance scale) properties of the g2SC phase should look similar to those in the normal phase [10]. For example, there is probably no Meissner effect in the g2SC phase, and the Debye screening mass of the gluons of the $SU(2)_c$ subgroup is nonzero. The latter, in fact, should be proportional to the density of the gapless modes, i.e.,

$$m_{1,D}^2 \simeq \frac{2\alpha_s}{\pi}\left(\frac{(\mu^-)^2\delta\mu}{\sqrt{(\delta\mu)^2-\Delta^2}}+\frac{(\mu^+)^2\delta\mu}{\sqrt{(\delta\mu)^2-\Delta^2}}\right), \quad \text{(g2SC)}. \qquad (15)$$

We remind that the corresponding value for the Debye screening mass in the ordinary 2SC phase is vanishing [16].

The fact that the low energy spectrum of quasiparticles in the g2SC phase looks nearly the same as in the normal phase has yet another consequence. Namely, at sufficiently low temperatures, there should appear two additional spin-1 condensates on top of the g2SC ground state. This conclusion is a result of a simple observation. Around the effective Fermi momenta $p = \mu^{\pm}$, the gapless quasiparticles are double degenerate. This is the reflection of the degeneracy with respect to the red and green colors of the unbroken $SU(2)_c$ subgroup. By taking into account the high degeneracy of states at $p = \mu^{\pm}$ and the attraction in the color antisymmetric channel, we conclude that the Cooper instability should develop in the g2SC ground state. This instability is removed spontaneously by the appearance of two secondary condensates of Cooper pairs. In view of the Pauli principle, these are spin-1 condensates of the so-called polar phase.

The finite temperature properties of the g2SC phase are very unusual too. The most striking are the results for the temperature dependence of the gap parameter for various values of the diquark coupling strengths. This is demonstrated by the plot in Fig. 2. The most amazing are the results for weak coupling. It appears that the gap function could take finite values at nonzero temperature even if it is exactly zero at zero temperature. This possibility comes about only because of the strong influence of the neutrality condition on the choice of the ground state in quark matter.

By looking at the results in Fig. 2, we immediately realize that the value of the ratio of the critical temperature T_c to the value of the gap at zero temperature Δ_0 is not a universal number in the g2SC phase. In contrast to the Bardeen-Copper-Schrieffer (BCS) type theories, the ratio T_c/Δ_0 in the g2SC phase depends on the diquark coupling constant. Moreover, it can be arbitrarily large, and even remain strictly infinite for a range of coupling strengths.

5. Summary

In this contribution, we reviewed some unusual zero and finite temperature properties of the g2SC phase of neutral two-flavor quark matter in β-equilibrium [10]. This new state of dense quark matter combines the properties of color

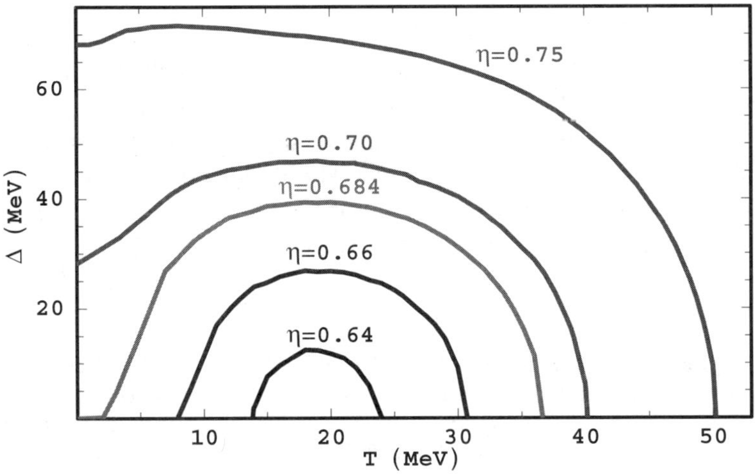

Figure 2. The temperature dependence of the diquark gap in neutral quark matter calculated for several values of the diquark coupling strength $\eta = G_D/G_S$.

superconducting and normal phases in a subtle way. As in a 2SC superconductor, the original color symmetry $SU(3)_c$ is broken down to $SU(2)_c$ through the Higgs mechanism. At the same time, the low energy spectrum of quasiparticles looks very similar to the spectrum in the normal phase. As a reflection of this, the Debye screening masses of the gluons of the $SU(2)_c$ subgroup are nonzero. In addition, there is probably no Meissner effect in the g2SC phase.

We find that the g2SC phase has rather unusual finite temperature properties. For example, the ratio of the critical temperature to the gap at zero temperature is not a universal number. It is determined by the value of the coupling constant and, in general, it is larger than the ratio $r_{BCS} \approx 0.567$ in the BCS theory. Note that the corresponding ratio in high-T_c superconductors is also larger than r_{BCS}. Whether this is a coincidence, we cannot answer yet.

At sufficiently low temperature, the situation gets even more complicated when the Cooper instability develops around the effective Fermi surfaces, and two additional spin-1 condensates develop on top of the gapless ground state. In this case, in accordance with the symmetry breaking pattern $SU(3)_c \rightarrow SU(2)_c$, five gluons are subject to the Meissner effect. The spin-1 condensates would also affect the transport properties of quark matter.

In nature, the gapless phase of quark matter may exist in cores of compact stars. This possibility was the main motivation for our study in Ref. [10]. It should be mentioned, however, that some gapless phases may also appear in nonrelativistic systems, such as asymmetric nuclear matter [13] and cold fermionic atoms [14].

Acknowledgments

I.S. would like to thank the organizers of the 2003 NATO Advanced Study Institute "Structure and Dynamics of Elementary Matter" for their invitation to attend the meeting, as well as for a financial support. M.H. acknowledges the financial support from the Alexander von Humboldt-Foundation. The work of I.S. was supported in part by Gesellschaft für Schwerionenforschung (GSI) and by Bundesministerium für Bildung und Forschung (BMBF).

References

[1] B. C. Barrois, Nucl. Phys. B **129** (1977) 390; S. C. Frautschi, in "Hadronic matter at extreme energy density", edited by N. Cabibbo and L. Sertorio (Plenum Press, 1980); D. Bailin and A. Love, Phys. Rep. **107** (1984) 325; M. G. Alford, K. Rajagopal and F. Wilczek, Phys. Lett. B **422** (1998) 247; R. Rapp, T. Schafer, E. V. Shuryak and M. Velkovsky, Phys. Rev. Lett. **81** (1998) 53.

[2] D.T. Son, Phys. Rev. D **59** (1999) 094019; T. Schäfer and F. Wilczek, Phys. Rev. D **60** (1999) 114033; D.K. Hong, V.A. Miransky, I.A. Shovkovy, and L.C.R. Wijewardhana, Phys. Rev. D **61** (2000) 056001; R.D. Pisarski and D.H. Rischke, Phys. Rev. D **61** (2000) 051501; S.D.H. Hsu and M. Schwetz, Nucl. Phys. **B572** (2000) 211; W.E. Brown, J.T. Liu, and H.-C. Ren, Phys. Rev. D **62** (2000) 054016; V. A. Miransky, I. A. Shovkovy and L. C. Wijewardhana, Phys. Rev. D **62** (2000) 085025.

[3] I.A. Shovkovy and L.C.R. Wijewardhana, Phys. Lett. B **470** (1999) 189; T. Schäfer, Nucl. Phys. **B575** (2000) 269; V. A. Miransky, I. A. Shovkovy and L. C. Wijewardhana, Phys. Rev. D **63** (2001) 056005.

[4] Q. Wang and D. H. Rischke, Phys. Rev. D **65** (2002) 054005.

[5] D. K. Hong, Nucl. Phys. B **582** (2000) 451; T. Schaefer, hep-ph/0307074.

[6] K. Rajagopal and F. Wilczek, hep-ph/0011333; M. G. Alford, Ann. Rev. Nucl. Part. Sci. **51** (2001) 131; T. Schafer, hep-ph/0304281; D. H. Rischke, nucl-th/0305030.

[7] M. Alford and S. Reddy, Phys. Rev. D **67** (2003) 074024; S. Banik and D. Bandyopadhyay, Phys. Rev. D **67** (2003) 123003; G. Lugones and J. E. Horvath, Astron. Astrophys. **403** (2003) 173; M. Baldo, M. Buballa, F. Burgio, F. Neumann, M. Oertel and H. J. Schulze, Phys. Lett. B **562** (2003) 153; D. Blaschke, S. Fredriksson, H. Grigorian and A. M. Oztas, nucl-th/0301002; I. Shovkovy, M. Hanauske and M. Huang, Phys. Rev. D **67** (2003) 103004; hep-ph/0310286; S. B. Ruster and D. H. Rischke, nucl-th/0309022.

[8] M. Alford and K. Rajagopal, JHEP **0206** (2002) 031.

[9] A. W. Steiner, S. Reddy and M. Prakash, Phys. Rev. D **66** (2002) 094007.

[10] I. Shovkovy and M. Huang, Phys. Lett. B **564** (2003) 205; Nucl. Phys. **A729** (2004) 835, hep-ph/0307273.

[11] M. Huang, P. F. Zhuang and W. Q. Chao, Phys. Rev. D **67** (2003) 065015.

[12] T. M. Schwarz, S. P. Klevansky and G. Papp, Phys. Rev. C **60** (1999) 055205.

[13] A. Sedrakian and U. Lombardo, Phys. Rev. Lett. **84** (2000) 602.

[14] W. V. Liu and F. Wilczek, Phys. Rev. Lett. **90** (2003) 047002.

[15] J. F. Liao and P. F. Zhuang, cond-mat/0307516.

[16] D. H. Rischke, Phys. Rev. D **62** (2000) 034007.

EXOTIC NUCLEI – THE EXPERIMENTAL QUEST

S. Hofmann
Gesellschaft für Schwerionenforschung (GSI), D–64220 Darmstadt and
Physikalisches Institut, J.W. Goethe-Universität, D–60054 Frankfurt, Germany
S.Hofmann@gsi.de

Abstract The nuclear shell model predicts that the next doubly magic shell-closure beyond ^{208}Pb is at a proton number Z = 114, 120, or 126 and at a neutron number N = 172 or 184. The outstanding aim of experimental investigations is the exploration of this region of spherical 'Super-Heavy Elements' (SHEs). The measured decay data reveal that for the heaviest elements, the dominant decay mode is α emission, not fission. Decay properties as well as reaction cross-sections are compared with results of theoretical investigations. Finally, plans are presented for the further development of the experimental set-up and the application of new techniques. At a higher sensitivity, the exploration of the region of spherical SHEs now becomes feasible, almost forty years after its prediction.

1. Introduction

In recent years the exploration of superheavy elements (SHE) received increasing interest both from theoretical as well as experimental investigation and both from chemical as well as physical studies. The reasons for the activity awakened again are based mainly on technical developments in the field of computer power, accelerator techniques and detection sensitivity.

Using faster and more powerful computers the *properties* of heavy nuclei are studied from multi-dimensional macroscopic-microscopic calculations [1–17], self consistent Skyrme-Hartree-Fock and relativistic mean field models [18–21]. The results reveal a rather complex structure of shell effects which determine the stability of nuclei in the superheavy region.

The most difficult problem, however, which is awaiting a theoretical solution, is the understanding of the *synthesis* of superheavy nuclei. The calculation of the involved dynamical processes on a microscopic level is presently the most challenging and work intensive task [22–35].

Successful methods for the laboratory synthesis of heavy elements have been fusion-evaporation reactions using heavy-element targets; recoil-separation techniques [36]; and the identification of the nuclei by generic ties to known

W. Greiner et al. (eds.), Structure and Dynamics of Elementary Matter, 337–352.

daughter decays after implantation into position-sensitive detectors [37, 38]. Experiments at low cross-sections necessitate projectile beams of high intensity and stability. Although the intensity limits have not presently been reached, considerable improvements of accelerator techniques have been made in recent years.

In the following section a description is given of studies of elements 110 to 112 performed at GSI Darmstadt. In subsequent sections the measured data on the decay properties of heavy and superheavy nuclei are compared with theoretical descriptions. Nuclear reactions are discussed for the synthesis of SHEs using stable and radioactive beams. Finally, a summary and outlook are given.

2. Experimental results

In this section, recent results are presented dealing with the confirmation of elements 110 to 112. Detailed presentations of the properties of elements 107 to 109 and of earlier results on elements 110 to 112 were given in previous reviews [38–40].

Element 110, now officially named darmstadtium [41], Ds, was discovered in 1994 using the reaction ^{62}Ni + ^{208}Pb $\rightarrow$ ^{270}Ds* [42]. A total of three decay chains were measured (see also remarks at the end of this section). The main experiment was preceded by a thorough study of the excitation functions for the synthesis of ^{257}Rf and ^{265}Hs in order to determine the optimum beam energy for the production of darmstadtium. The data revealed that the maximum cross-section for the synthesis of hassium was shifted to a lower excitation energy, different from the predictions of reaction theories.

The heavier isotope ^{271}Ds was synthesized with a beam of the more neutron-rich isotope ^{64}Ni [40]. A summary of the measured decay data and a comparison with literature data is given in Fig. 1. The data were fully confirmed by the results of recent experiments performed at RIKEN, Saitama [46] and LBNL, Berkeley [47]. The important result for the further production of elements beyond meitnerium was that the cross-section was enhanced from 2.6 pb to 15 pb by increasing the neutron number of the projectile by two, which gave hope that the cross-sections for production of heavier elements could decrease less steeply with more neutron-rich projectiles. However, this expectation was not proven in the case of element 112.

The even-even nucleus ^{270}Ds was synthesized using the reaction ^{64}Ni + ^{207}Pb [48]. A total of eight α-decay chains was measured during an irradiation time of seven days. Decay data were obtained for the ground-state and a high spin K isomer, for which calculations predict spin and parity 9^-, 10^- or 8^+ [50]. The relevant single particle Nilsson levels are $\nu[613]_{7/2+}$ and $\nu[615]_{9/2+}$ below the Fermi level and $\nu[725]_{11/2-}$ above the Fermi level. Configuration

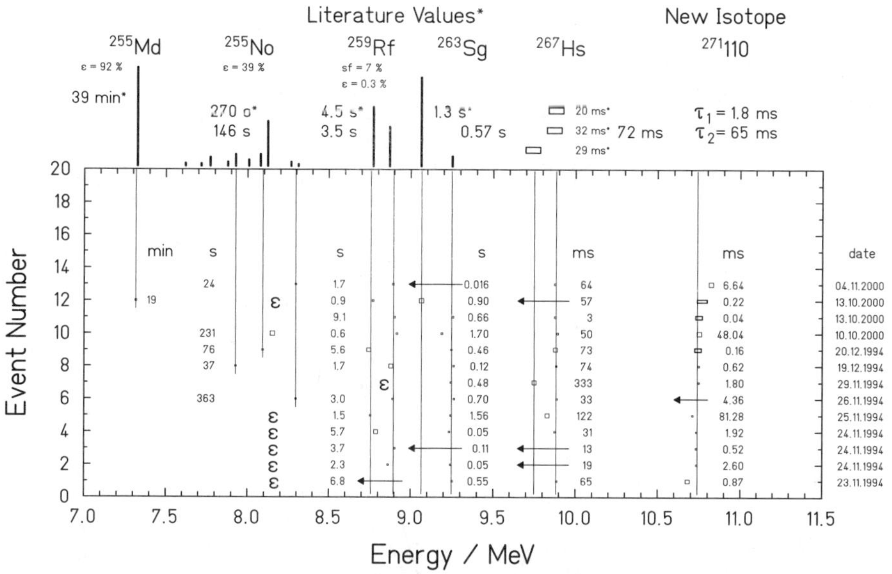

Figure 1. Energies and lifetimes of the thirteen α-decay chains resulting from the reaction ^{64}Ni + ^{208}Pb $\rightarrow$ ^{271}Ds + 1n. The isotope ^{271}Ds was identified by comparison of the decay properties of the daughter products with literature data marked by an asterisk (^{267}Hs [43], ^{263}Sg [44], ^{259}Rf [45], ^{255}No [45], ^{255}Md [45]). New decay data of the isotope ^{263}Sg could be deduced. The decay chains are arranged chronologically with the date of production given at the righthand ordinate. The size of the symbols reflects the detector resolution. Arrows mark escaping α particles for which only an energy loss signal, but still time and position information, was obtained.

and calculated energy of the excited states are $\{\nu[613]_{7/2+} \otimes \nu[725]_{11/2-}\}9^-$ at 1.31 MeV, $\{\nu[615]_{9/2+} \otimes \nu[725]_{11/2-}\}10^-$ at 1.34 MeV, and $\{\nu[613]_{7/2+} \otimes \nu[615]_{9/2+}\}8^+$ at 1.58 MeV.

The new nuclei ^{266}Hs and ^{262}Sg were identified as daughter products after α decay. Spontaneous fission of ^{262}Sg terminates the decay chain. A proposed partial decay scheme of ^{270}Ds is shown in Fig. 2.

Element 111 was synthesized in 1994 using the reaction ^{64}Ni + ^{209}Bi $\rightarrow$ ^{273}111*. A total of three α chains of the isotope ^{272}111 were observed [51]. Another three decay chains were measured in a confirmation experiment in October 2000 [52]. Also these results were confirmed in recent experiments by Morita et al. [53].

Element 112 was investigated at SHIP using the reaction ^{70}Zn + ^{208}Pb $\rightarrow$ ^{278}112* [54]. The irradiation was performed in January-February 1996. Over a period of 24 days, a total of 3.4×10^{18} projectiles were collected. One α-decay chain, shown in the left side of Fig. 3, was observed resulting in a cross-section of 0.5 pb. The chain was assigned to the one neutron-emission channel.

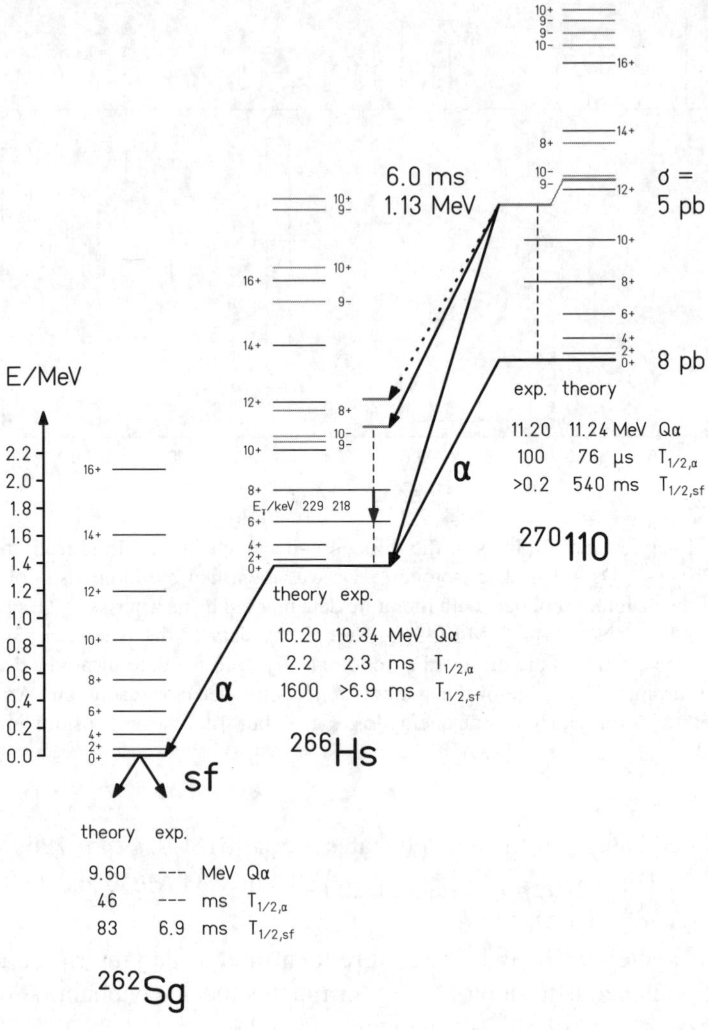

Figure 2. Tentative assignment of measured α and γ decay and spontaneous fission data (bold arrows) observed in the reaction ^{64}Ni + ^{207}Pb $\rightarrow$ ^{271}Ds*. The data were assigned to the ground-state decays of the new isotopes ^{270}Ds, ^{266}Hs, and ^{262}Sg and to a high spin K isomer in ^{270}Ds. The proposed partial level schemes are taken from theoretical studies of Muntian et al. [49] for the rotational levels, of Cwiok et al. [50] for the K isomers and of Smolanczuk [28] and Smolanczuk et al. [14] for the α energies and spontaneous fission half-lives, respectively. For a detailed discussion see Ref. [48].

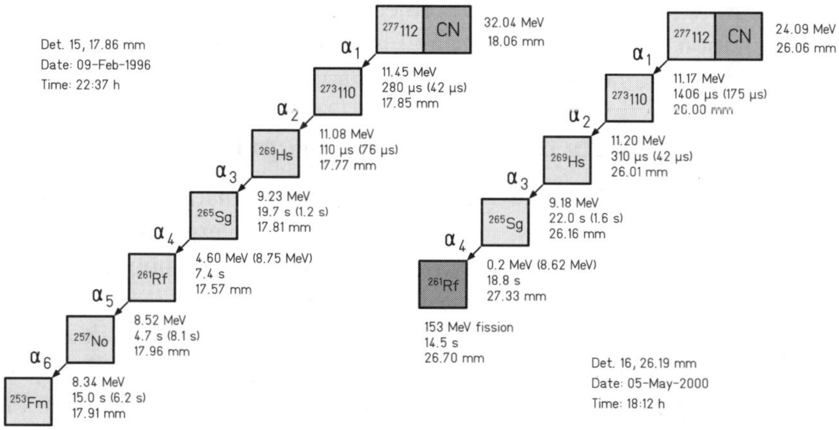

Figure 3. Two decay chains measured in experiments at SHIP in the cold fusion reaction ^{70}Zn + ^{208}Pb $\rightarrow$ $^{278}112^*$. The chains were assigned to the isotope ^{277}112 produced by evaporation of one neutron from the compound nucleus. The lifetimes given in brackets were calculated using the measured α energies. In the case of escaped α particles the α energies were determined using the measured lifetimes.

The experiment was repeated in May 2000 aiming to confirm the synthesis of ^{277}112 [52]. During a similar long measuring time, but using slightly higher beam energy, one more decay chain was observed, also shown in Fig. 3.

The measured decay pattern of the first four α decays is in agreement with the one observed in the first experiment. A new result was the occurrence of fission which ended the second decay chain at ^{261}Rf. A spontaneous-fission branch of this nucleus was not yet known, however, it was expected from theoretical calculations. The new results on ^{261}Rf were proven in a recent chemistry experiment [55, 56], in which this isotope was measured as granddaughter in the decay chain of ^{269}Hs.

A reanalysis of all decay chains measured at SHIP since 1994, a total of 34 decay chains was analyzed, revealed that the previously published first decay chain of ^{277}112 [54] (not shown in Fig. 3) and the second of the originally published four chains of ^{269}Ds [42] were spuriously created. Details of the results of the reanalysis are given in Ref. [52].

The excitation function of the reaction ^{54}Cr + ^{208}Pb was studied recently (June 2003, see Sect. 4.2). At high beam currents of up to 1 pμA the target was continuously monitored using the scattering of 20 keV electrons at the target material [57]. Successfully tested was a PbS target (melting point 1118° C) produced by depositing the target material on a carbon backing which was heated to several 100° C [58]. By heating the backing during evaporation, the formation of a crystalline needle structure of PbS was avoided, which would result in uncontrolled energy loss of the projectiles. Using the 'heated' PbS target, a 1n-excitation function was measured, which was identical to the previously measured one obtained with a metallic Pb target.

3. Nuclear structure and decay properties

The basic step which is necessary for the determination of the stability of SHEs is the calculation of the ground-state binding energy. As a signature for shell effects, we can extract the shell-correction energy by subtracting a smooth macroscopic part (derived from the liquid-drop model) from the measured or calculated total binding energy. In macroscopic-microscopic models the shell-correction energy is of course the essential input value which is calculated directly from the shell model. The shell-correction energy is plotted in Fig. 4a using the data from Ref. [59]. Two equally deep minima are obtained, one at Z = 108 and N = 162 for deformed nuclei with deformation parameters $\beta_2 \approx 0.22$, $\beta_4 \approx -0.07$ and the other at Z = 114 and N = 184 for spherical SHEs. Different results are obtained from self-consistent Hartree-Fock-Bogoliubov calculations and relativistic mean-field models [18–21]. They predict for the spherical nuclei shells at Z = 114, 120 or 126 (indicated as dashed lines in Fig. 4) and N = 184 or 172, with shell strengths being also a function of the amount of nucleons of the other type.

For the calculation of partial spontaneous fission half-lives the knowledge of ground-state binding energies is not sufficient. It is necessary to determine the fission barrier over a wide range of deformation. The most accurate data were obtained for even-even nuclei using the macroscopic-microscopic model [14, 59]. Partial spontaneous fission half-lives are plotted in Fig. 4b.

Partial α half-lives decrease almost monotonically from 10^{12} s down to 10^{-9} s near Z = 126 (Fig. 4c). The valley of β-stable nuclei (marked by black squares in Fig. 4d) passes through Z = 114, N = 184 [16]. At a distance from the bottom of the valley, the β half-lives decrease gradually down to values of one second.

The interesting question arises, if and how the uncertainty related with the location of the proton and neutron shell closures will change the half-lives of SHEs. Partial α and β half-lives are only insignificantly modified by shell effects, because the decay process occurs between neighboring nuclei. This is different for fission half-lives which are primarily determined by shell effects. However, the uncertainty related with the location of nuclei with the strongest shell-effects and thus longest partial fission half-life at Z = 114, 120 or 126 and N = 172 or 184, is inconsequential concerning the longest 'total' half-life of SHEs. The regions for SHEs in question are dominated by α decay. And α decay will be modified by only a factor of up to approximately 100, if the double shell closure will not be located at Z = 114 and N = 184.

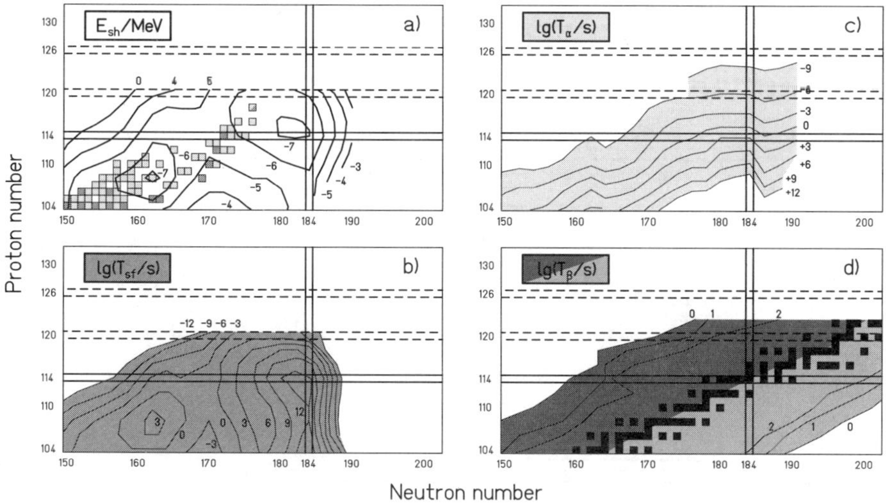

Figure 4. Shell-correction energy (a) and partial spontaneous fission, α and β half-lives (b)–(d). The calculated values in (a)–(c) are taken from Ref. [59, 14] and in (d) from Ref. [16]. The squares in (a) mark the nuclei presently known or under investigation, the filled squares in (d) mark the β stable nuclei.

The line of reasoning is, however, different concerning the production cross-section. The survival probability of the compound nucleus (CN) is determined among other factors significantly by the fission-barrier. Therefore all present calculations of cross-sections suffer from the uncertainty related with the location and strength of closed shells.

4. Nuclear reactions

4.1 Projectile-target combinations

Compound nuclei that could be produced concerning the availability of beams and targets are plotted in Fig. 5. The graphs also demonstrate the extension in the region of SHEs which will become possible with radioactive beams. The nuclei presently known or under investigation are marked by squares.

In Fig. 5a the curve marked with dots (•) shows the most neutron rich CN that can be produced with ^{208}Pb or ^{209}Bi targets and beams of the most neutron rich stable isotopes of the elements from Ti to Sn, given in the first column at the right ordinate. The double magic SHEs could be reached only if located at Z = 126 and N = 184.

On the average the accessible region is extended by 4 to 5 neutrons to the right using radioactive isotopes of the elements from Ti to Sn (symbol $\otimes$ in Fig. 5a). As possible most neutron rich radioactive projectiles those isotopes

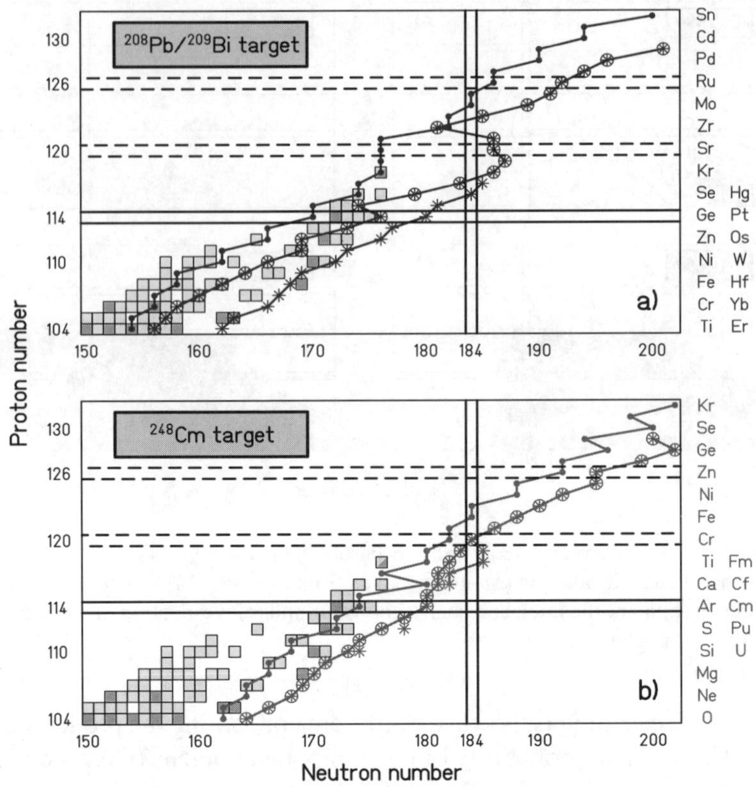

Figure 5. Most neutron rich compound nuclei produced in reactions using stable and radioactive beams and targets (see text for an explanation of the symbols).

were taken into account that could be produced with intensities of at least 10^9 /s according to the data presented in the RIA proposal [60].

Striking is the wide extension of possible CN to the neutron rich side at Z = 120 using beams of Kr, Rb and Sr. The reason is that these nuclei are available as fission fragments and can be extracted from ion sources with high yield, too. SHEs both at Z = 120 and 126 are well covered using reactions with Pb and Bi targets and radioactive beams.

More neutron rich nuclei of elements below Z = 118 can be produced using the radioactive beams of ^{96}Kr or ^{98}Rb and targets of stable neutron rich isotopes of elements below Pb from Hg down to Er (second column on the right ordinate, curve marked with symbol $*$). In these reactions evaporation residues could be produced in a region, where the new chains from element 114 and 116 are ending in the unknown. In this region α decay and spontaneous fission is expected with half-lives from seconds to hours (see Fig. 4).

In the lower plot, Fig. 5b, the equivalent combinations are given for reactions using a ^{248}Cm target and stable and radioactive beams. Concerning the location of CN, no significant extension into the direction of neutron rich SHEs results from using radioactive beams compared with the reactions given in Fig. 5a. Apparent from the graph (Fig. 5b) is the extraordinary use of ^{48}Ca for the synthesis of neutron rich SHEs. The surplus of neutrons beyond ^{48}Ca can be only slightly increased using radioactive beams.

Radioactive beams of ^{47}K and ^{46}Ar are likely produced with high yield. Using these beams and in principle feasible actinide targets from U to Fm (the elements are given in the second column at the right ordinate) the surplus of neutrons can be further increased (CN marked by symbol $*$). However, selecting a reaction using actinide targets, one has to consider the availability of the material and the tremendous increase of safety problems in the handling of targets from U to Cf, Es or Fm.

Finally, we have to notice that SHEs at $Z = 114$ and $N = 184$ can be reached also not in reactions with radioactive actinide targets and radioactive beams. Concerning the CN on the right from $N = 184$ one has to keep in mind that the fission barrier vanishes rapidly with neutron number, and therefore these nuclei cannot be synthesized.

4.2 Cross-sections, fusion valleys and excitation energy

The main features which determine the fusion process of heavy ions are (1) the fusion barrier and related beam energy and excitation energy, (2) the ratio of surface tension versus Coulomb repulsion, which determines the fusion probability and which strongly depends from the degree of asymmetry of the reaction partners (the product Z_1Z_2 at fixed $Z_1 + Z_2$), (3) the impact parameter and related angular momentum, and (4) the ratio of neutron evaporation versus fission probability of the CN. In fusion of SHEs the product Z_1Z_2 reaches extremely large and the fission barrier extremely small values. In addition, the fission barrier is fragile at increasing excitation energy and angular momentum, because it is solely built up from shell effects. For these reasons the fusion of SHEs is drastically hampered, whereas the fusion of lighter elements is advanced through the contracting effect of surface tension.

The effect of Coulomb repulsion on the cross-section starts to act severely for fusion of elements beyond Fm. From there on a continuous decrease of cross-section was measured from microbarns for the synthesis of nobelium down to picobarns for the synthesis of element 112. The data obtained in reactions with ^{208}Pb and ^{209}Bi for the 1n evaporation channel at low excitation energies of about 10–15 MeV (therefore named *cold fusion*) and in reactions with actinide targets for the 4n channel at excitation energies of 35–45 MeV (*hot fusion*) are plotted in Fig. 6. Interesting for further investigation of SHEs are the

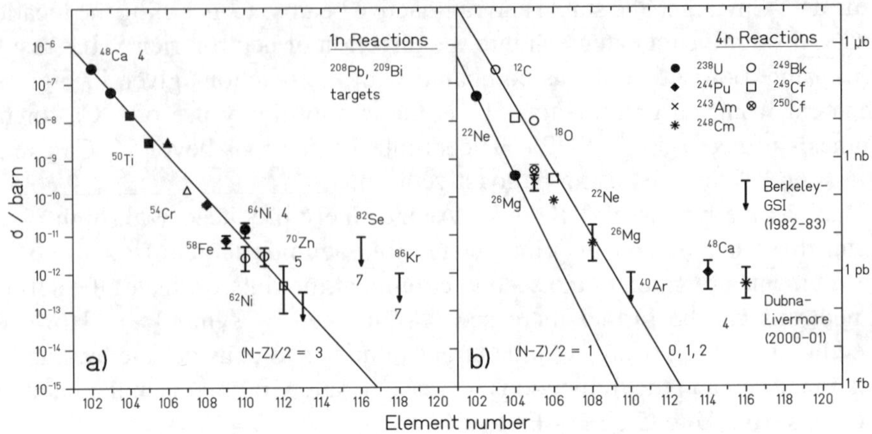

Figure 6. Measured cross-sections and cross-section limits for reactions using ^{208}Pb and ^{209}Bi targets and one neutron evaporation (a) and for reactions using actinide targets and four neutron evaporation (b). A new cross-section limit of 4.3 pb (at 7.4×10^{18} projectiles) was measured recently (11.9. - 17.10. 2003) at SHIP for the synthesis of ^{278}113 in the reaction ^{70}Zn + ^{209}Bi at $E^* = 12.2$ MeV. A mean upper limit of 0.33 pb (shown in Fig. 6a; 1 event would be 0.18 pb) was obtained from this experiment and a previous run (April-May 1998, $E^* = 11.57$ MeV) at a mean excitation energy of $E^* = 12.02$ MeV. The obtained limit is still in agreement with the expectations.

relatively high cross-sections measured for the synthesis of elements 114 and 116 (4n channel) [61–63]. In both cases the obtained values of about 0.5 pb are considerably larger than expected from the trend set by fusion of the lighter elements. An explanation could be a relatively high and wide fission barrier of the CN, which is created by strong shell effects in the region of spherical SHEs. Note in this context that the experimental sensitivity increased by three orders of magnitude since the 1982-83 search experiments for element 116 using a hot fusion reaction [64].

A number of excitation functions was measured for the synthesis of elements from rutherfordium to darmstadtium using Pb and Bi targets [38]. For the even elements these data are shown in Fig. 7. The figure includes the recently (June 2003) measured excitation function of the reaction ^{54}Cr + ^{208}Pb and an update of the previously obtained data of ^{50}Ti + ^{208}Pb [42, 65]. Due to a more accurate beam current measurement using a transformer technique instead of Faraday cups the old data of the ^{50}Ti + ^{208}Pb reaction given in [42] have to be multiplied by a factor of 1.5. All other data measured since 1994 were obtained with the new transformer technique.

The maximum evaporation residue cross-section (1n channel) was measured at beam energies well below a fusion barrier calculated in one dimension [22].

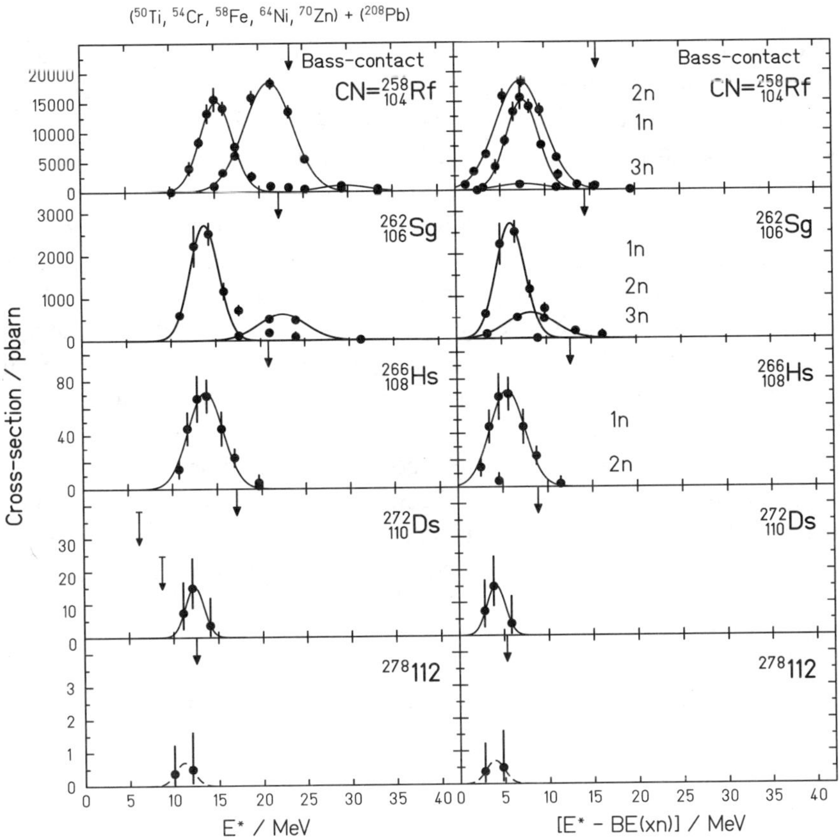

Figure 7. Measured even element excitation functions based on ^{208}Pb targets.

At the optimum beam energy projectile and target nuclei are just reaching the contact configuration in a central collision. The relatively simple fusion barrier based on the Bass model [22] is too high and a tunnelling process through this barrier cannot explain the measured cross-section.

Various processes are possible and are discussed in the literature which result in a lowering of the fusion barrier. Among these transfer of nucleons and excitation of vibrational degrees of freedom are the most important [25–34]. The theoretical studies are also aimed at reproducing the known cross-section data and further extrapolating the calculations into the region of spherical superheavy nuclei. The measured cross-sections for the formation of ^{257}Rf up to ^{277}112 are reproduced within about a factor of 2 by the various models. However, there are significant differences in the predicted cross-section values for the synthesis of spherical SHEs at and beyond Z = 114.

In the case of actinide targets, the target nucleus is strongly deformed and the height of the Coulomb barrier is a function of the orientation of the deformation axes. Reactions of a ^{48}Ca beam with various actinide targets were studied in Dubna [61, 62], and evidence for the 3n and 4n channels was obtained at beam energies resulting in excitation energies of 30 – 45 MeV.

It was pointed out in the literature [23] that closed shell nuclei as projectile and target are favorable for fusion of SHEs. The reason is not only a low reaction Q value and thus low excitation energy, but also that fusion of such systems is connected with a minimum of energy dissipation. The fusion path propagates along cold fusion valleys on the potential energy surface, where the reaction partners keep kinetic energy up to the closest possible distance. In this view the difference between *'cold'* and *'hot'* fusion is not only a result from different values of the excitation energy, but there exists also a qualitative difference. This is on the one side based on a well ordered fusion process along paths of minimum dissipation of energy (cold fusion), and on the other side on a process governed by the formation of a more or less energy equilibrated CN (hot fusion). This qualitative explanation is well in agreement with the results from experimental studies of quasi-fission and compound-nucleus fission [66].

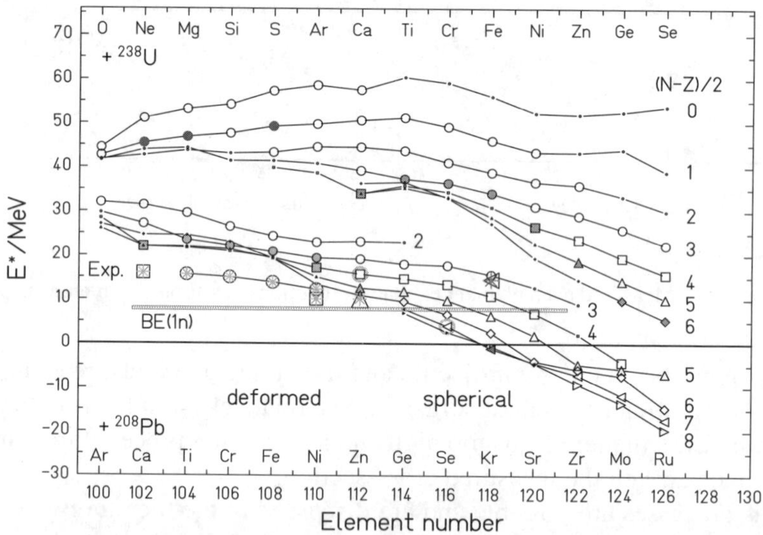

Figure 8. Excitation energy of compound nuclei for fusion from a contact configuration.

Although a realistic fusion barrier is not yet available and details of the fusion process resulting in SHE's are not yet understood, we already can investigate various reactions concerning the Q value, including radioactive projectiles. For the interesting cases the projectiles are not extremely neutron rich and their binding energy is known. The binding energy of SHE's from various models is in good agreement. Therefore the uncertainty of the Q value is estimated to be a few MeV only.

In a first step we consider the excitation energy of various reactions using targets of ^{208}Pb and ^{238}U and preferably stable projectiles. The results are plotted in Fig. 8. The lines connect projectiles with the same isospin. Larger symbols denote stable, the smaller dots radioactive projectiles. The excitation energy was calculated for beam energies so that a contact configuration is reached according to the Bass model [22]. Data for which maximum cross-sections were measured (marked by 'Exp.') are even below that energy, halfway to the limiting (for the 1n channel) one neutron binding energy.

Striking is the decrease of the excitation energy below the 1n-binding energy at about element 114. For that reason we studied the reaction ^{82}Se + ^{208}Pb $\rightarrow$ $^{290}116^*$ at excitation energies from 0 to 10 MeV in order to search for a possible radiative capture channel. The result was negative, cross-section limits of 5 pb were reached at four different beam energies [40].

In the case of element 118 synthesis the reaction ^{86}Kr + ^{208}Pb even gets endothermic. A surplus of kinetic energy is needed beyond the minimum necessary for fusion in order to evaporate one neutron. This feature was interpreted as 'unshielded' fusion reaction [67].

The extraordinary low excitation energy for the reaction using ^{48}Ca beam and ^{238}U target is obvious from the upper bunch of curves. Also with uranium target the reaction gets colder with increasing element number. The excitation energy reaches values close to the 1n-binding energy near element 124 using a ^{76}Ge beam. The reactions using actinide targets which are often described as 'hot' fusion according to the excitation energy of about 40 MeV for synthesis of elements up to about 110, change into 'cold' fusion for the heavier elements. This could be a promising aspect for the synthesis of SHE's near $Z = 126$ (which is one of the predicted closed proton shells) using actinide targets.

In the case of ^{208}Pb as a target a surplus of 30 MeV of kinetic energy is necessary in order to obtain an excitation energy high enough for emission of one neutron. However, such high kinetic energies will probably lead to dissipation of energy already in early stages of the fusion process. An increase of deep inelastic processes will result.

Using more neutron rich radioactive projectiles (higher values of isospin as shown in Fig. 8) does not, in general, result in lower excitation energies at the fusion barrier. In contrary, the excitation energy rises again (see Fig. 7 in Ref. [68]). An open question is how the stronger binding energy of the protons and, vice versa, the low binding energy of the neutrons will influence the fusion process, especially the first stages which are governed by transfer of nucleons.

Summary and outlook

The experimental work of the last two decades has shown that cross-sections for the synthesis of the heaviest elements decrease almost continuously. However, recent data on the synthesis of element 114 and 116 in Dubna using hot fusion seem to break this trend when the region of spherical superheavy elements is reached.

The progress towards the exploration of the island of spherical SHEs is difficult to predict. Despite the exciting new results, many questions of more general character are still awaiting an answer. New experimental developments will not only make it possible to perform experiments aimed at synthesizing new elements in reasonable measuring times, but will also allow for a number of various other investigations covering reaction physics and spectroscopy.

One can hope that, during the coming years, more data will be measured in order to promote a better understanding of the stability of the heaviest elements and the processes that lead to fusion. A microscopic description of the fusion process will be needed for an effective explanation of all measured phenomena in the case of low dissipative energies. Then, the relationships between fusion probability and stability of the fusion products may also become apparent.

An opportunity for the continuation of experiments in the region of SHEs at decreasing cross-sections afford, among others, further accelerator developments. High current beams and radioactive beams are options for the future. At increased beam currents, values of tens of particle μA's may become accessible, the cross-section level for the performance of experiments can be shifted down into the region of tens of femtobarns, and excitation functions can be measured on the level of tenths of picobarns. High currents, in turn, call for the development of new targets and separator improvements. Radioactive ion beams, not as intense as the ones with stable isotopes, will allow for approaching the closed neutron shell $N = 184$ and for studying the transition from deformed heavy nuclei to spherical SHEs already at about darmstadtium. The study of fusion reactions with regard to the initiating transfer processes using radioactive neutron rich beams is of high interest.

The half-lives of spherical SHEs are expected to be relatively long. Based on nuclear models, which are effective predictors of half-lives in the region of the heaviest elements, values from microseconds to years have been calculated for various isotopes. This wide range of half-lives encourages the application of a wide variety of experimental methods in the investigation of SHEs, from the safe identification of short lived isotopes by recoil-separation techniques to atomic physics experiments on trapped ions, and to the investigation of chemical properties of SHEs using long-lived isotopes.

5. Acknowledgments

The recent experiments at SHIP were performed in collaboration with D. Ackermann, F.P. Hessberger, B. Kindler, J. Kojouharova, B. Lommel, R. Mann, G. Münzenberg, H.J. Schött (GSI Darmstadt); O.N. Malyshev, A. Popeko, A. Yeremin (JINR Dubna); S. Antalic, P. Cagarda, S. Saro, B. Streicher (University Bratislava); P. Kuusiniemi, M. Leino, J. Uusitalo (University Jyväskylä).

References

[1] W.D. Myers and W.J. Swiatecki, Nucl. Phys. **81** (1966), 1.

[2] A. Sobiczewski *et al*, Phys. Lett. **22** (1966), 500.

[3] H. Meldner, Ark. Fys. **36** (1967), 593.

[4] S.G. Nilsson *et al*, Nucl. Phys. A **115** (1968), 545.

[5] U. Mosel and W. Greiner, Z. Phys. A **222** (1969), 261.

[6] E.O. Fiset and J.R. Nix, Nucl. Phys. A **193** (1972), 647.

[7] A. Sobiczewski, Physica Scripta **10A** (1974), 47.

[8] J. Randrup *et al*, PC **13** (1976), 229.

[9] P. Möller and J.R. Nix, Atomic Data and Nucl. Data Tables **26** (1981), 165.

[10] A. Sobiczewski *et al*, Phys. Lett. B **224** (1989), 1.

[11] A. Sobiczewski, Phys. Part. Nucl. **25** (1994), 295.

[12] A. Sobiczewski *et al*, J. Alloys Comp. **213/214** (1994), 38.

[13] P. Möller and J.R. Nix, J. Phys. G: Nucl. Part. Phys. **20** (1994), 1681.

[14] R. Smolanczuk *et al*, Phys. Rev. C **52** (1995), 1871.

[15] P. Möller *et al*, Atomic Data and Nucl. Data Tables **59** (1995), 185.

[16] P. Möller *et al*, Atomic Data and Nucl. Data Tables **66** (1997), 131.

[17] I. Muntian *et al*, Phys. Lett. B **500** (2001), 241.

[18] S. Cwiok *et al*, Nucl. Phys. A **611** (1996), 211.

[19] G.A. Lalazissis *et al*, Nucl. Phys. A **608** (1996), 202.

[20] K. Rutz *et al*, Phys. Rev. C **56** (1997), 238.

[21] A.T. Kruppa *et al*, Phys. Rev. C **61** (2000), 034313.

[22] R. Bass, Nucl. Phys. A **231** (1974), 45.

[23] R.K. Gupta *et al*, Z. Phys. A **283** (1977), 217.

[24] von Oertzen, W.: Z. Phys. A **342** (1992), 177.

[25] Y. Aritomo *et al*, Phys. Rev. C **59** (1999), 796.

[26] E.A. Cherepanov, Pramana J. Phys. **53** (1999), 619.

[27] R. Smolanczuk, Phys. Rev. C **59** (1999), 2634.

[28] R. Smolanczuk, Phys. Rev. Lett. **83** (1999), 4705.

[29] G. Giardina *et al*, Eur. Phys. J. A **8** (2000), 205.

[30] V.Yu. Denisov and S. Hofmann, Phys. Rev. C **61** (2000), 034606.

[31] G.G. Adamian *et al*, Nucl. Phys. A **678** (2000), 24.

[32] G.G. Adamian *et al*, Phys. Rev. C **62** (2000), 064303.

[33] R. Smolanczuk, Phys. Rev. C **63** (2001), 044607.
[34] V.I. Zagrebaev, Phys. Rev. C **64** (2001), 034606.
[35] V.Yu. Denisov and W. Nörenberg, Eur. Phys. J. A **15** (2002), 375.
[36] G. Münzenberg *et al*, Nucl. Instr. Meth. **161** (1979), 65.
[37] S. Hofmann, Z. Phys. A **291** (1979), 53.
[38] S. Hofmann and G. Münzenberg, Rev. Mod. Phys. **72** (2000), 733.
[39] G. Münzenberg, Rep. Prog. Phys. A **51** (1988), 57.
[40] S. Hofmann, Rep. Prog. Phys. A **61** (1998), 639.
[41] IUPAC 42nd General Assembly, Ottawa, Canada, 9-17 August 2003.
[42] S. Hofmann et al., Z. Phys. A **350** (1995), 277.
[43] Yu.A. Lazarev et al., Phys. Rev. Lett. **75** (1995), 1903.
[44] A. Ghiorso et al., Phys. Rev. Lett. **33** (1974), 1490.
[45] M.R. Schmorak, Nuclear Data Sheets **59** (1990), 507.
[46] K. Morita et al., RIKEN Accel. Prog. Rep. **35** (2003), 90.
[47] T.N. Ginter et al., Phys. Rev. C **67** (2003), 064609.
[48] S. Hofmann et al., Eur. Phys. J. A **10** (2001), 5.
[49] I. Muntian et al., Phys. Rev. C **60** (1999), 041302.
[50] S. Cwiok et al., Phys. Rev. Lett. **83**, (1999), 1108 and private communication.
[51] S. Hofmann et al., Z. Phys. A **350** (1995), 281.
[52] S. Hofmann et al., Eur. Phys. J. A **14** (2002), 147.
[53] Morita, K., Proc. Tours Symp. on *Nuclear Physics V*, Tours, France, 2003, AIP Conf. Proc., to published.
[54] S. Hofmann et al., Z. Phys. A **354** (1996), 229.
[55] Ch.E. Düllmann et al., Nature **418** (2002), 859.
[56] A. Türler et al., Eur. Phys. J. A **17** (2003), 505.
[57] R. Mann et al., to be published.
[58] B. Kindler et al., to be published.
[59] R. Smolanczuk and A. Sobiczewski, Proc. XV. Nucl. Phys. Divisional Conf. on *Low Energy Nuclear Dynamics*, St.Petersburg, Russia, 1995, p.313, World Scientific, Singapore, 1995.
[60] RIA, *The Rare Isotope Accelerator Project,* www.phy.anl.gov/ria (2000).
[61] Yu.Ts. Oganessian et al., Phys. Rev. C **63** (2000), 011301 and contribution to this Conference.
[62] Yu.Ts. Oganessian et al., Phys. Atomic Nuclei **64**, 1349 (2001).
[63] Yu.Ts. Oganessian et al., Nature **400** (1999), 242.
[64] P. Armbruster et al., Phys. Rev. Lett. **54** (1985), 406.
[65] S. Hofmann et al., Proc. of the VIII Int. Conference on *Nucleus-Nucleus Collisions*, Moscow, Russia, 2003, to be published.
[66] M.G. Itkis et al., in Proc. Int. Workshop on *Fusion Dynamics at the Extremes*, Dubna, Russia, 2000, p. 93, World Scientific, 2001.
[67] W.D. Myers and W.J. Światecki, Phys. Rev. C **62** (2000), 044610.
[68] S. Hofmann, Prog. Part. Nucl. Phys. **46** (2001), 293.

EXPERIMENTAL SEARCH FOR SUPERHEAVY NUCLEI

Yu. Ts. Oganessian
Flerov Laboratory of Nuclear Reactions
Joint Institute for Nuclear Research
141980 Dubna, Moscow region
E-mail: oganessian@flnr.jinr.ru

1. Introduction

The problem of experimental verification of the existence of the hypothetical domain of superheavy nuclides is one of the fundamental outcomes of the nuclear shell model and it has been extensively discussed in recent years.

The properties of heavy nuclides, produced in cold-fusion reactions, are in qualitative (in some cases also quantitative) agreement with the theoretical predictions what concerns the significant influence of structure on their stability when approaching the closed shells with Z=108 and N=162. The status of this region of nuclei is given in detail and, in my opinion, in a very interesting way, in the presentation of Prof. S. Hofmann at this conference [1]. The magic nucleus $^{270}_{108}$Hs and the daughter nucleus ^{266}Sg, resulting in its α-decay, has been synthesized also in the hot fusion reactions ^{248}Cm(^{26}Mg,4n) [2] and ^{248}Cm(^{22}Ne,4n) [3]; these two nuclides undergo α-decay with T_α ~2-7 s and 21 s, respectively. It should be noted that without the shell effects, and consequently, in the absence of the fission barrier, these nuclei would undergo spontaneous fission (SF) with $T_{SF} \sim 10^{-19}$ s.

Another problem is the existence of even stronger spherical shells beyond ^{208}Pb, in the domain of heavier neutron-rich nuclei with Z=114 (or possibly 120) and N=184. Here the nuclei are expected to demonstrate highly enhanced stability with respect to all the radioactive decay modes. They should form the so-called "stability island" of superheavy elements. Since 1998 we make attempts to verify this non-trivial theoretical hypothesis.

The problem lies in the necessity of synthesizing nuclides with $Z \geq 114$ and high neutron excess that are hard to be reached in fusion reactions with stable partner nuclei. In order to approach the N=184 shell, one needs maximum neutron excess in both the target and projectile nuclei. Therefore, instead of ^{208}Pb, which is successfully used as a target in cold-fusion reactions, we use

W. Greiner et al. (eds.), Structure and Dynamics of Elementary Matter, 353–376.

targets of enriched heavy isotopes of actinides like ^{244}Pu, ^{243}Am, ^{248}Cm, ^{249}Cf, while the rare and expensive isotope ^{48}Ca is used as the projectile.

The compound nuclei ^{292}114 (N=178), ^{296}116 (N=180) or ^{297}118 (N=179) formed in complete fusion have excitation energy $E^* \sim$27-33 MeV near the reaction threshold. Therefore, their transition to the ground state may take place via emission of three or four neutrons and γ-rays. This finally brings us to the region of neutron numbers N=174-177, yet we gain up to 6 to 8 neutrons in the evaporation residues (ER) as compared to combinations like ^{208}Pb+HI. According to theoretical models such a neutron excess should result in a considerable enhancement of nuclear stability, provided the predicted position and effect of the spherical neutron shell N=184 are correct.

This is a simple and perhaps the only motivation for our program aimed at the synthesis of new elements.

2. Experimental technique

The technique of our experiments on the synthesis of superheavy elements with the ^{48}Ca ion beam has been described elsewhere. Here we give only a few details considering the experimental setup and methods.

The beam of ^{48}Ca with an intensity of about 1.5 pμA hits a rotating target that consists of a thin (0.35 mg/cm^2) layer of Pu, Cm or Cf oxides on a 1.5-μm thick titanium backing. In the present experiments the compound nuclei recoiling from the target were separated in flight from ^{48}Ca beam ions, scattered particles and transfer-reaction products by the Dubna Gas-filled Recoil Separator (Fig.1). The transmission efficiency of the separator for Z=114 and 116 nuclei was estimated to be about 35-40% [4]. The separator is operated with hydrogen filling at a pressure of about 1 Torr.

Evaporation recoils pass through a time-of-flight system (TOF) and are implanted in a 4$\times$12-cm^2 semiconductor detector array with 12 vertical position-sensitive strips, located in the separator's focal plane, 4 meters downstream the target. This detector is surrounded by eight 4$\times$4-cm^2 side detectors without position sensitivity, forming a box of detectors open from the front (beam) side. The position-averaged detection efficiency for α-decays of implanted nuclei is 87% of 4π. Single spontaneous fission (SF) fragments are detected with 100% efficiency, and coincident pairs of fission fragments are detected with about 45% efficiency. The energy resolution for α-particles absorbed in the focal-plane detector is 60-90 keV. Alpha-particles escaping the focal-plane detector at different angles and registered by the side detectors have energy resolution of the summed signals of 140-200 keV. The FWHM position resolution of the signals of correlated decays of nuclei implanted in the detectors are 0.8-1.3 mm for ER-α signals and 0.5-0.8 mm for ER-SF signals.

In the position-sensitive strips detector signals from implanted recoils should be accompanied by TOF signals; signals from the decay of implanted nuclei

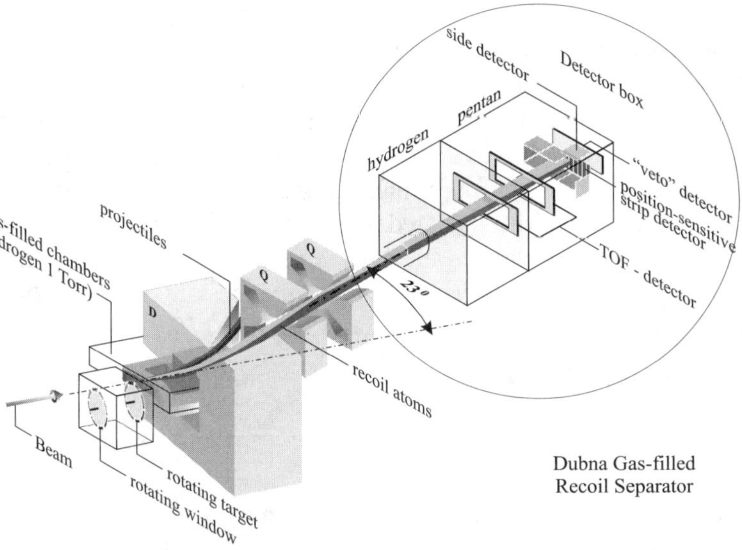

Figure 1. A schematic view of the Dubna Gas-filled Recoil Separator.

have no TOF mark. Sequences of genetically linked decays (e.g., ER$-\alpha-\alpha\ldots$SF) are characterized by coordinates (strip number and position on strip) that coincide within the detector position resolution. In identical decay sequences, the α-particle energies measured for the corresponding members of the decay chains should also coincide within the detector energy resolution and the distribution of decay times should correspond to the exponential radioactive decay law.

In many experiments we used a special detection mode for registering two or more sequential decays. The beam was switched off after a recoil signal was detected with parameters of implantation energy and TOF expected for the evaporation residues, followed by an α-like signal within a preset energy and time interval, in the same strip, within a position window of 1.4-1.9 mm. In this way all the expected sequential decays of the daughter nuclides should be observed in the absence of beam-associated background. Such a technique appeared to be especially efficient in detecting long-lived multi-step decay chains like those we obtained in our experiments on the synthesis of elements 115 and 113.

3. Synthesis of isotopes of elements 114 and 116 in the reactions ^{244}Pu,^{248}Cm+^{48}Ca

The first experiments of 1998-99 that aimed at the synthesis of the isotopes of element 114 in the reaction ^{244}Pu+^{48}Ca were performed with the beam energy of 236 MeV, which corresponds to the excitation energy of the compound nucleus

$^{292}114$, $E^* \sim 35$ MeV [5]. Here and on we use the calculated masses by W.D. Myers and W.J. Swiatecki [6]. Accordingly, one could expect $3n$- and $4n$ evaporation to be the most probable reaction channels, which lead to the isotopes $^{288,289}114$. The two events shown in Fig. 2a illustrate the observations. In both ER-$\alpha_1 - \alpha_2 -$SF decay chains detected in strips 2 and 8, all the four decays have positions that differ by not more than 0.5 mm. The measured decay energies and times meet all the criteria stated above. In further experiments three more events of the same type were detected. The decay pattern, as well as decay energies and times in the observed decay sequences are close to what could be expected for isotopes of element 114 with N=174-175.

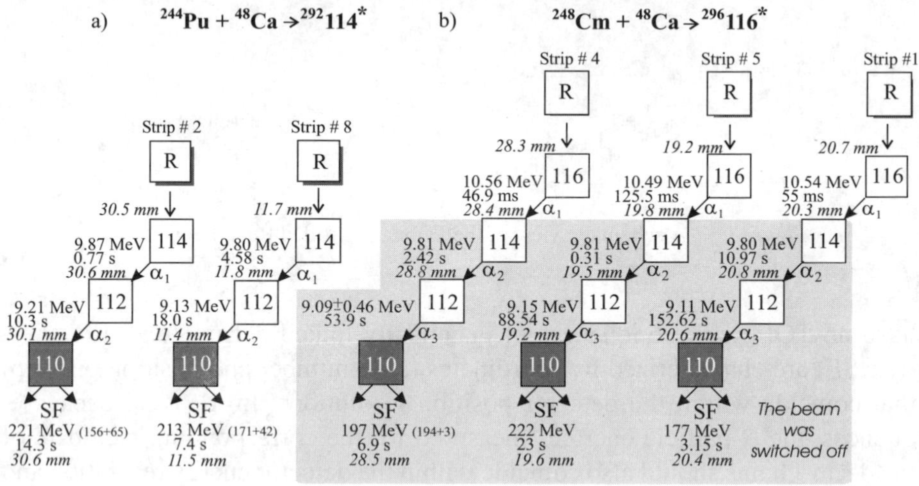

Figure 2. Decay chains observed in the reactions a) ^{244}Pu+^{48}Ca; b) ^{248}Cm+^{48}Ca.

In order to verify this assignment we performed another experiment with a ^{248}Cm target [7]. In this case, at $E^* \sim$31-35 MeV the reaction leads to the production of the isotopes of element 116 in the $3n$ and $4n$ evaporation channels.

Following the α-decay of the mother nucleus, its daughter products, i.e., the isotopes of elements 114, 112 and 110, should reproduce the decay pattern observed in the reaction ^{244}Pu+^{48}Ca. In other words, the mother nucleus produced in the ^{244}Pu+^{48}Ca reaction becomes the α-decay daughter of the heavier nucleus with Z=116 that is produced in the reaction with the ^{248}Cm target.

The decay chains presented in Fig. 2b show the complete coincidence of the decay characteristics of the nuclides with Z=114, 112 and 110 observed in both reactions. From the three observed events we obtain that the isotope of element 116 undergoes α-decay with E_α=10.53 MeV and T_α=53 ms. Three decay events of its daughter with Z=114 agree with the five events detected in the reaction ^{244}Pu+^{48}Ca, E_α=9.82 MeV, T_α=2.2 s.

All the eight decays referring to Z=112 show E_α=9.16 MeV and T_α=39 s. Finally, in eight cases also the final nucleus with Z=110 decays by spontaneous fission with T_{SF}=9.6 s and is characterized by a high value of the fission fragments energy released in the detectors, E_{tot} ~210 MeV. If one takes into account the loss of the kinetic energy of fission fragments due to the pulse-height defect of the detectors and the energy loss in the detectors' entrance windows, dead layers, and the pentane gas filling the detection system (ΔE_{tot} ~20 MeV as it was for the SF of ^{252}No), the measured value corresponds to a total kinetic energy TKE ~230 MeV. The latter value exceeds the TKE of 180 MeV typical for SF of ^{252}Cf by some 50 MeV and also gives evidence for the fission of a heavy nuclide (see below).

The α-particle energy spectrum presented in Fig. 3a shows the well-defined energy of all the three α-decays in the decay chain Z=116-114-112-110. These points to the decay via unhindered α-transitions that is typical of the α-decay of even-even nuclides.

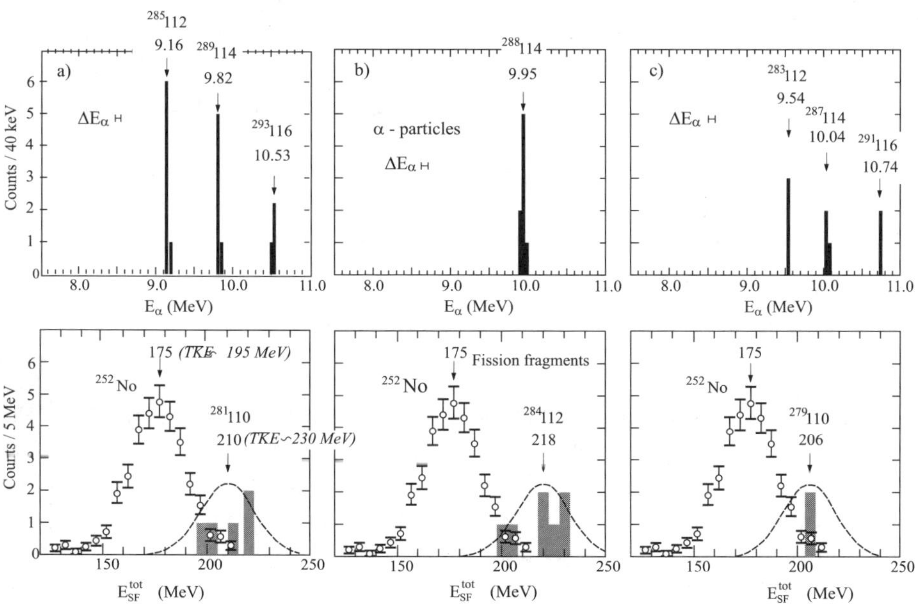

Figure 3. Energy spectra of the α-particles and of the sum energy of SF fragment pairs measured in various decay chains: a) $\alpha-\alpha$-SF chain in the reaction ^{244}Pu(^{48}Ca, $3n$)^{289}114 and $\alpha-\alpha-\alpha-$SF chain in the reaction ^{248}Cm(^{48}Ca, $3n$)^{293}116; b) α-SF chain in the reaction ^{244}Pu(^{48}Ca, $4n$)^{288}114; c) $\alpha-\alpha-$SF chain in the reaction ^{244}Pu(^{48}Ca, $5n$)^{287}114 and $\alpha-\alpha-\alpha-$SF chain in the reaction ^{245}Cm(^{48}Ca, $2n$)^{291}116. Lower plots show the sum energies of the SF fragment pairs measured by the detector array. Dots show the spectrum of the fragments of the SF of ^{252}No produced in the reaction ^{206}Pb(^{48}Ca, $2n$).

In the absence of hindrance, the α-decay energy Q_α and the decay probability (or the half-life T_α) are known to follow the Geiger-Nuttall relation (Log $T_\alpha \sim Z \times Q_\alpha^{-1/2}$). This relation, in form of the Viola-Seaborg formula [8], is fulfilled quite well for the 65 known even-even nuclides with Z>82 for which the α-decay energies and half-lives have been measured.

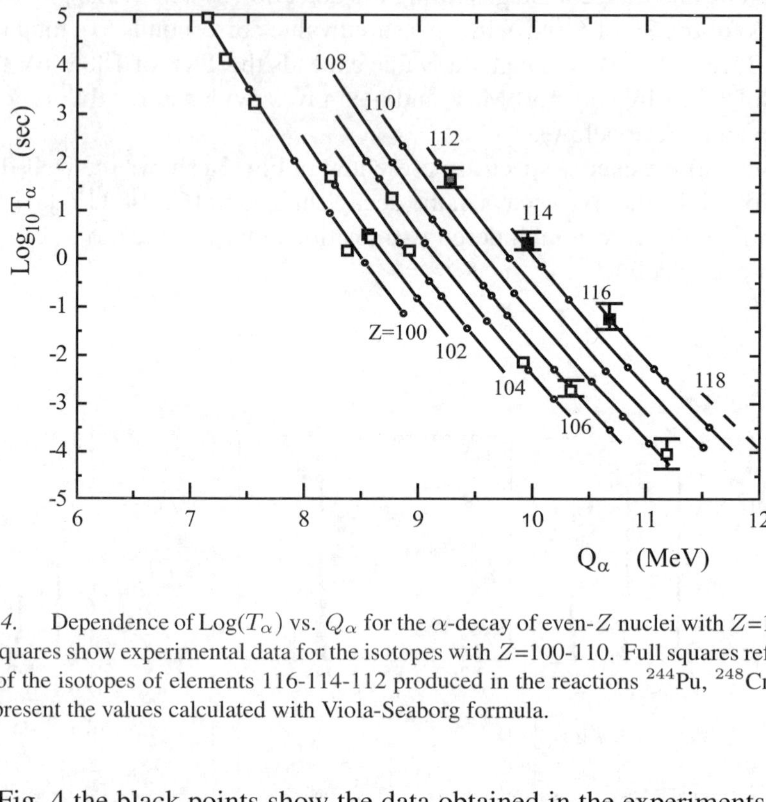

Figure 4. Dependence of Log(T_α) vs. Q_α for the α-decay of even-Z nuclei with Z=100-118. Open squares show experimental data for the isotopes with Z=100-110. Full squares refer to the decay of the isotopes of elements 116-114-112 produced in the reactions ^{244}Pu, ^{248}Cm+^{48}Ca. Lines present the values calculated with Viola-Seaborg formula.

In Fig. 4 the black points show the data obtained in the experiments ^{244}Pu, ^{248}Cm+ ^{48}Ca. Using the relationship LogT_α *vs.* Q_α, the Z value for every decay in the observed $\alpha - \alpha - \alpha-$SF sequence can be calculated. Taking into account the genetic correlation of the three decays, one can estimate the probability that the observed sequential decays refer to the nuclei with Z=116-114-112. This value exceeds 99.2%. Thus the probability of missing the first α-particle and that the decays refer to the chain with Z=114-112-110 is less than 0.8%.

Unfortunately, the definite mass-identification of the new nuclide is problematic. In the domain of neutron-rich nuclei under consideration the decay chains start from and end in nuclides with unknown decay properties. Thus, the widely used method of genetic correlations to known nuclei for the identification of the parent nuclide has a limited application.

In our case, for the mass-identification one needs to determine the number of neutrons emitted by the excited compound nucleus. The energy dependence of the cross section of the *xn*-evaporation reactions $\sigma_{xn}(E)$ for the given x number, is known to have a bell-like shape with a typical width of about 6-8 MeV (for ^{48}Ca beam energies about 240 MeV) and with a maximum at the corresponding excitation energy.

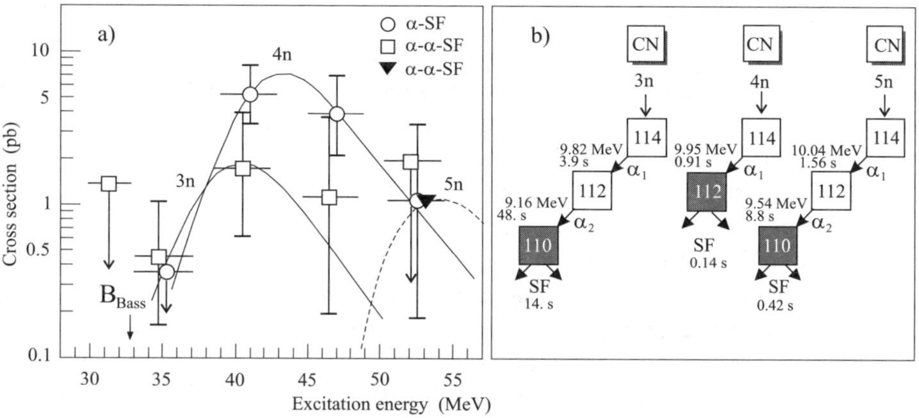

Figure 5. Dependence of the production cross sections of the isotopes of element 114 on the excitation energy of the compound nucleus ^{292}114 (a). Lines are drawn through the experimental points to guide the eye. The right drawing (b) shows the decay chains of the nuclei observed in the reaction ^{244}Pu+^{48}Ca.

Fig. 5a shows the excitation functions for the nuclides produced in the reaction ^{244}Pu+^{48}Ca. For the above described isotope of element 114 (open squares), the increase in excitation energy from E^*=35 MeV to 41 MeV results in an increase of the production cross section from 0.5 pb to 1.7 pb. The further energy increase leads to the decrease of cross section to ~1 pb at E^*=47 MeV. Finally, at E^*=53 MeV no decays of this type were detected, with the accumulated beam dose of $3{\cdot}10^{18}$.

From these data, one could arrive to a conclusion that the decays of the $\alpha_1 - \alpha_2$−SF type with typical time duration of one minute observed in the ^{244}Pu+^{48}Ca reaction refer to the production of the even-even isotope ^{288}114, produced via $4n$-evaporation. Indeed, in our first works, performed at $E^* \approx$ 35 MeV we put forward this identification as the most probable one [5, 7].

However, at E^* ~41 MeV and at higher energies we detected also other nuclides characterized by shorter-lived decays. At E^*=41 MeV their yield corresponded to a higher cross section of about 5 pb. On the basis of the 12 events detected in the excitation energy interval 41-53 MeV, it follows that this nuclide emits an α-particle (E_α=9.95 MeV, T_α=0.63 s) that is followed by spontaneous fission of the daughter nuclide with T_{SF}=0.1 s. Surprisingly enough, the corresponding excitation function (shown in Fig. 5a by open circles) is similar to that observed in the previous case. The dependence $\sigma(E^*)$ for the new emitter is also similar to what could be expected for the $4n$-evaporation channel of the reaction ^{244}Pu+^{48}Ca.

At the same time, both candidates for ^{288}114 that undergo $\alpha - \alpha-$SF and $\alpha-$SF decays differ substantially in their decay properties. Despite the small difference in the α-decay energy and half-life of the mother nuclides with Z=114, the difference in half-lives T_α and T_{SF} of their daughters (Z=112) is definitely beyond the experimental uncertainties. Therefore, the decay chains of the two types originate from two different neighboring isotopes, which result from the evaporation of a different number of neutrons.

Finally, at the maximum used energy of the ^{48}Ca ions of 257 MeV (E^*= 53 MeV), we observed yet another decay chain of the $\alpha - \alpha-$SF type that differs from both that were previously described. The decay chains of all the three isotopes observed in the ^{244}Pu+^{48}Ca reaction are presented in Fig. 5b.

It is reasonable to consider that the short $\alpha-$SF decay chains refer to the decay of the even-even nuclide ^{288}114, the product of $4n$-evaporation. Then the two observed $\alpha - \alpha-$SF decay sequences originate from the neighboring ^{289}114 and ^{287}114, they are produced in $3n$- and $5n$-evaporation channels with cross sections of about 1.7 pb and 1 pb, respectively.

One should note that here all the observed evaporation residues, viz. the isotopes of element 114 are produced at energies exceeding the Coulomb barrier (E^*>E^*_{min}=33 MeV). On the contrary, in the reaction ^{208}Pb(^{48}Ca,*xn*)$^{256-x}$No, which is used as a model for calculations of the cross sections of ER, the maximum ER yield is observed at the sub-barrier and near-barrier energy values [9], as well as in the other studied cold-fusion reactions with heavier projectiles (see [10] and Refs. therein).

Such an effect can be associated with the orientation of the deformed target nucleus. If the major contribution to formation of the compound nucleus corresponds to the compact system in the entrance channel (the minimum distance between the centers of the interacting nuclei), then as shown in [11], the fusion threshold should increase accordingly. Therefore, the channels with evaporation of smaller number of neutrons ($x \leq$3) should be suppressed and the cross-section maximum of $3n$-evaporation should shift to higher energies. The shift of excitation functions of the $2n$ and $3n$ channels can be different depending on the reaction Q value.

What concerns the mass-number identification, we should note that this effect does not introduce uncertainty if one observes in the experiment the production of several isotopes, rather than one. Moreover, a more reliable mass identification is possible if the mass number of the target nucleus is varied (e.g., by using ^{242}Pu and ^{244}Pu targets).

4. Experiments on the synthesis of element 118

The properties of the isotopes of elements 114 and 116, produced in the reactions ^{244}Pu,^{248}Cm+^{48}Ca, are close to those calculated in the framework of the macroscopic-microscopic model. These calculations are known to predict closed nuclear shells at Z=114 and N=184 [12, 13]. At the same time, in other models, such as Hartree-Fock-Bogoliubov (HFB) [14, 15] or relativistic mean-field calculations (RMF) [16], the maximum shell effect for N=184 is expected at Z=120, 122. Therefore, the attempt to synthesize heavier nuclei, in particular element 118, is the next logical step in studying superheavy nuclei.

Table I. Predicted decay properties of the ^{294}118.

Model	Q_α (MeV)	T_α (ms)
WS+YPE [13]	12.11	0.9
RMF [16]	11.03	40
HFB [14]	11.3	7
Extrapolation from exp. data	11.8	0.5

Aiming at element 118 and the choice of the synthesis reaction are also motivated by that we still can use ^{48}Ca projectiles, while highly enriched ^{249}Cf, a product of the β-decay of ^{249}Bk ($T_{1/2}$=320 days), is available as a target material. The excitation energy of the ^{297}118 compound nucleus at the Coulomb barrier of the reaction is 6.3 MeV lower than in the reaction ^{244}Pu+^{48}Ca. Here the maximum yield of the ERs is expected in the $3n$-evaporation channel resulting in the formation of the even-even nuclide ^{294}118. The properties of ^{294}118 calculated in various models (see Table I) differ considerably. Nevertheless, this does not prevent the performing of an experiment on the synthesis of ^{294}118, since the employed gas-filled separator and its detection system are characterized by a short dead time (t_{min} ~6 μs) and narrow energy resolution (~0.06 MeV) when detecting α-particles in a wide energy range. Because of the limited amount of enriched ^{249}Cf (97.3%) we used a 0.23-mg/cm^2 thick rotating target instead of the usual 0.35 mg/cm^2. The experiment was run from February through June 2002. In the course of the 2300-hour irradiation we accumulated a total beam dose of 2.5×10^{19} of ^{48}Ca ions [17].

Since the decay chains of the Z=118 nuclides are expected to terminate by SF, we first searched through the data for spontaneous fission events. A total of 18 SF events were detected in our experiment, see Fig. 6 for the corresponding measured summed energies. The 16 events in the lower-energy part of the spectrum (E_{tot}=125-175 MeV) originate from the decay of long-lived SF- nuclides (t_{SF}>0.5 hour). Most probably these events can be ascribed to the SF of long-lived nuclides around Cf-Fm that are produced in transfer reactions. The two

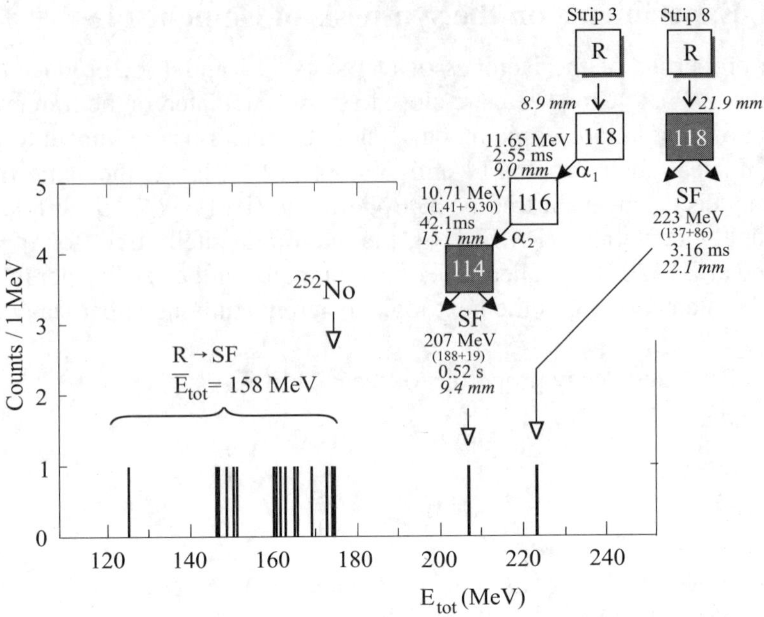

Figure 6. Kinetic energies of the SF fragments registered in the reaction ^{249}Cf+^{48}Ca. The sum energy is given if the two fragments were detected. The decay chains are shown for the two SF events with E_{tot}>200 MeV.

SF events with E_{tot}>200 MeV belong to relatively short-lived decay sequences: they are preceded by recoil signals, detected in appropriate position windows with time distances of 0.56 s and 3.16 ms. In the first case the recoil is followed by the emission of two alphas: the first with $E_{\alpha 1}$=11.65 MeV, t_α=2.55 ms (in the front detector) and the second with $E_{\alpha 2}$=10.71 MeV, $t_{\alpha 2}$=42.1 ms, this one detected by both the front and side detectors. The SF with E_{tot}=207 MeV, detected as two coincident fragments, followed in 0.52 s. The probability that the 4 signals ER$-\alpha_1 - \alpha_2-$SF in this sequence are of random origin is 1.5×10^{-6}, even not taking into account that they are position-correlated.

For the second SF event no alphas were detected in a 3.16-ms ER-SF interval. Here we also detected two coincident SF fragments with E_{tot}=223 MeV.

Both events were detected at the beam energy corresponding to E^* ~30 MeV. Here $2n$ and $3n$-evaporation channels are the most probable, leading to the formation of the isotopes of element 118 with mass numbers of 295 and 294, respectively.

One cannot exclude that both decay sequences ER$-\alpha_1 - \alpha_2-$SF and ER-SF represent two different decay modes of the same nuclide with $T_{1/2}$ ~2 ms. These two events correspond to a production cross section of about 0.7 pb. The ER$-\alpha_1$ $-$ α_2-SF chain is of special interest. If it originates from the odd nuclide $^{295}118$ ($2n$-evaporation), the two sequential α-decays should lead to the isotope $^{287}114$, which has been observed as a single event in the reaction ^{244}Pu(^{48}Ca,$5n$) at E^*=53 MeV. This nuclide, as one can see from Fig. 5b is an α-emitter, while the granddaughter nucleus shown in Fig. 6 undergoes SF.

However, this consideration is based only on a single event, whereas obtaining higher statistics is a difficult task because of the low production cross sections of the discussed reactions.

In these conditions it would be useful to go a step down in Z, to the α-decay daughters of Z=118 nuclides, i.e., to the isotopes of element 116 with mass numbers 290 and 291. In principle, they could be synthesized with a higher yield in the reactions ^{245,246}Cm+^{48}Ca.

5. Study of the reaction ^{245}Cm+^{48}Ca. New isotopes of element 116.

In the experiments aimed at synthesizing $^{290,291}116$ in the reaction ^{245}Cm+^{48}Ca we used a ^{245}Cm target material with isotopic enrichment of 97.8%. The beam energy in the middle of the target was set to 243 MeV, which corresponds to E^*=31-35 MeV. According to calculations, this energy range corresponds to a maximum yield of the $2n$- and $3n$-evaporation channels. In the present experiment, with the total ^{48}Ca dose of 1.2×10^{19} ions, we detected five decay chains of two types.

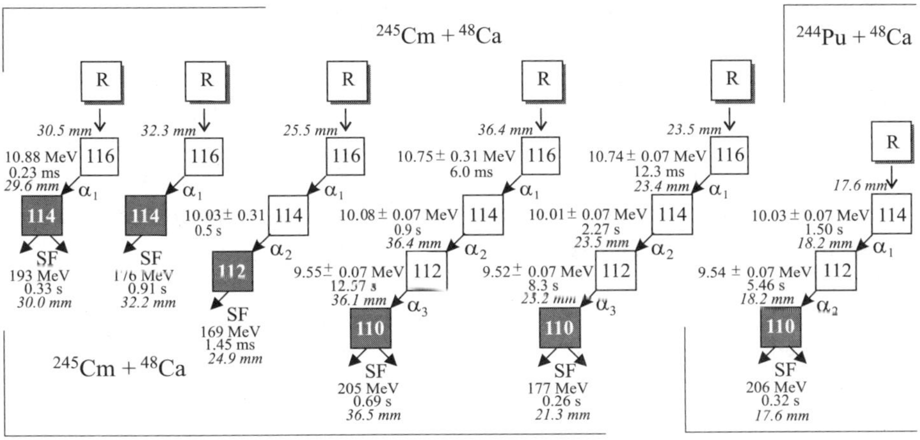

Figure 7. Decay chains observed in the reactions ^{245}Cm+^{48}Ca at the excitation energy of the compound nucleus $^{293}116$ E^*=33 MeV and ^{244}Pu+^{48}Ca at E^*=53 MeV.

Two events of the first type represent two-step decay sequences (ER$-\alpha-$SF) with a duration of 0.3-0.9 s, while the third event points to a ER$-\alpha$ $-$ $\alpha-$SF decay pattern, see Fig. 7. The decay pattern of the other two events differs from this type. Here we observed the sequential emission of three α-particles that is also terminated by SF. The ER$-\alpha-\alpha-\alpha-$SF sequences last for about 10 s, their longer time duration being due to the last α-decay.

New decay chains detected in the reaction ^{245}Cm+^{48}Ca differ from ER$-\alpha-\alpha-\alpha-$SF decays observed in the ^{248}Cm+^{48}Ca reaction. Indeed, in the experiment with ^{245}Cm, the energy of alphas emitted by the mother nucleus (Z=116) is 0.2-0.3 MeV higher and its lifetime is less that it was observed for the more neutron-rich isotope of element 116. For the daughter (Z=114) and especially the granddaughter (Z=112) nucleus this difference becomes even more evident.

The difference in decay properties of the daughter nuclei (Z=114) can be seen also when comparing the data from the ^{245}Cm+^{48}Ca experiment with the short ER$-\alpha-$SF decay chain that we ascribed to the decay of the even-even nuclide ^{288}114 produced in the reaction ^{244}Pu(^{48}Ca,4n)^{288}114.

Therefore, the decay chains observed in the reaction ^{245}Cm+^{48}Ca cannot originate from the isotopes ^{293}116 or ^{292}116, the products of 0n- or 1n -evaporation. However when comparing the daughter (Z=114) nuclei in the ER$-\alpha-\alpha-\alpha-$SF chain from the reaction ^{245}Cm+^{48}Ca with the properties of the single event, detected in the ^{244}Pu+^{48}Ca reaction at E^*=53 MeV and assigned to the decay of ^{287}114, we see a good agreement, Fig. 7. Then the longer-lived chain ER$-\alpha-\alpha-\alpha-$SF from the ^{245}Cm+^{48}Ca experiment should belong to ^{291}116, the 2n -evaporation product. Accordingly, the shorter-lived chain ER-$\alpha-\alpha$-SF (or ER$-\alpha-$SF) should be assigned to the decay of the even-even nuclide ^{290}116 produced via 3n-evaporation in the reaction ^{245}Cm+^{48}Ca.

Now, which of these decays was observed in the experiment on the synthesis of element 118 in the reaction ^{249}Cf+^{48}Ca?

Most probably, the α-decay of ^{294}118 leads to ^{290}116. However, decay of the primary Z=118 nuclide ends in the Z=114 granddaughter that undergoes SF with T_{SF}=0.3 s. However, the α-energies of the two neighbouring isotopes of element 116 with mass numbers 290 and 291 differ by just 0.1 MeV, and the α-particle energies of their descendants with Z=114 are rather close, too. One should also add here that the daughter nuclide ^{286}114 can undergo both α-decay and SF. The real difference shows up only at Z=112: T_α=6 s and T_{SF}=1 ms.

Therefore, for the final identification of element 118 it is necessary to continue experiments making another step to lower Z value, i.e. to isotopes with Z=114 that can be produced in the reaction ^{242}Pu+^{48}Ca.

6. Study of the reaction ^{242}Pu+^{48}Ca

This experiment has just been started. For this reason, I can here report only the first very preliminary results.

The target of enriched ^{242}Pu(99.9%) was irradiated with a ^{48}Ca beam of E=244 MeV. At this energy the compound nucleus ^{290}114, taking into account the target thickness, has excitation energy E^*=40.2$\pm$2.2 MeV, for which the evaporation products of the $3n$- and $4n$ evaporation channels will have a maximum yield. The total beam dose was $4.7 \cdot 10^{18}$ ions.

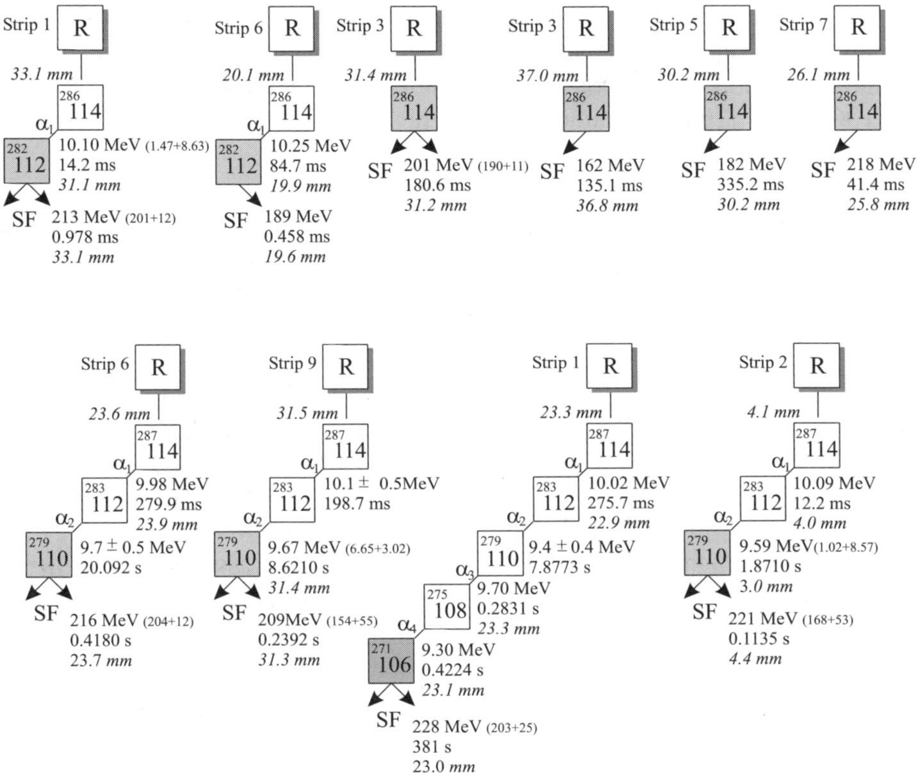

Figure 8. Decay chains observed in the reaction ^{242}Pu+^{48}Ca at the excitation energy of the compound nucleus $^{290}114$ E^{*}=40 MeV

In the given experiment 10 decay chains were observed. Depending on their properties they can be subdivided into two groups (Fig. 8).

The first group includes the short decays: ER-SF (4 events) and ER$-\alpha-$SF (2 events); the total decay time in this case amounts to $\sim$0.2 s. The spontaneous fission in the $\alpha-$SF chain takes place within a very short time: T_{SF} $\sim$0.5 ms. It is quite probable that these chains originate from the decay of one and the same isotope $^{286}114$, which can undergo both α-decay and SF. Its properties are in agreement with the above mentioned decay chains of the isotope $^{290}116$, and the new events define more precisely the decay properties of the $^{286}114$ nucleus.

The second group includes 4 other, longer decay chains with a long decay time. The energy of the first 2 α-transitions $E_{\alpha 1}$=10.03 MeV and $E_{\alpha 2}$=9.54 MeV are in agreement with the reported above data on the decay of the $^{287}114$ isotope. As in the previous cases, here 3 out of the 4 events consist of 2 α-decays terminating by SF with T_{SF}=0.3 s.

In one case, spontaneous fission competes with α-decay. This leads to a longer decay chain ER$-\alpha_1 - \alpha_2 - \alpha_3 - \alpha_4-$SF. The final nucleus in this chain, as we suppose it to be ^{271}Sg, fissions spontaneously for t_{SF}=381 s. It should be noted that spontaneous fission of the neutron-rich nucleus with Z=106 and N=165 are characterized by high fission fragment kinetic energy: E_{tot}=228 MeV.

This is a single, but quite an interesting case.

Thus, two isotopes of element 114 with mass number 287 and 286, formed in the $3n$- and $4n$-evaporation channels, have been observed in the ^{242}Pu+^{48}Ca reaction at E^*=40 MeV. For an additional confirmation of this conclusion we plan to measure their excitation functions in the ^{242}Pu+^{48}Ca reaction at different beam energies.

However, it is possible on the basis of the already obtained data create a picture of the decay pattern of the isotopes ^{294}118 and ^{295}118.

In the decay of the even-even isotope ^{294}118 ($3n$-channel) a decay chain ER$-\alpha_1 - \alpha_2-$SF is expected. As mentioned above, this chain was observed in the first experiment ^{249}Cf+^{48}Ca. A chain of the type ER$-\alpha_1 - \alpha_2 - \alpha_3-$SF can correspond to about 30% of the cases of decay of the isotope ^{294}118. In the decay of the neighboring even-odd nuclide ^{295}118 ($2n$-channel) a longer decay chain: ER$-\alpha_1 - \alpha_2 - \alpha_3 - \alpha_4-$SF is expected; with probability of $\leq$10% this decay may lead to the SF nuclide - ^{271}Sg.

In conclusion, we can say that for the daughter nuclei – the isotopes ^{260,261}116 – both the scenario of their decay and the basic decay characteristics have been determined: energies, α-transition times and SF-lifetimes.

The next experiment aimed at the synthesis of element 118 is planned for the middle of 2004.

7. Experiments on the synthesis of odd-Z nuclei in the ^{243}Am(^{48}Ca,xn)$^{291-x}$115 reaction

Our previous experiments were aimed at the synthesis of even-Z superheavy elements (114-118) in ^{48}Ca-induced reactions. For the neighboring odd-Z elements, especially their odd-odd isotopes, the probability of α-decay with respect to SF should increase due to hindrance for SF. For such odd-Z nuclei one might expect longer consecutive α-decay chains terminated by SF of relatively light descendant nuclides ($Z \leq$105).

The decay pattern of these superheavy nuclei is of interest for nuclear theory. In the course of α-decays, the increased stability of nuclei, caused by the predicted spherical neutron shell N=184 (or perhaps sub shell N=172), should gradually become weaker for descendant isotopes. However, the stability of these nuclei at the end of the decay chains should increase again due to the influence of the deformed shell at N=162.

The observation of nuclei passing from spherical to deformed shapes in the course of their consecutive α-decay could provide valuable information about the influence of nuclear structure changes on the decay properties of nuclei. For these investigations, we chose the fusion-evaporation reaction ^{243}Am+^{48}Ca, leading to isotopes of element 115.

According to calculations based on the results of experiments on the synthesis of even-Z nuclei [5, 7], the $3n$- and $4n$-evaporation channels leading to the isotopes ^{288}115 (N=173) and ^{287}115 (N=172) should be observed with the highest yields.

The experiments were performed between July 14 and August 10, 2003. Over this period, equal beam doses of 4.3×10^{18} ^{48}Ca projectiles were delivered to the target at two bombarding energies for the ^{48}Ca ions of 248 MeV and 253 MeV in the middle of the target. With the beam energy resolution and energy losses in the target ($\sim$3.3 MeV), we expected the resulting compound nuclei ^{291}115 to have excitation energies of 38.0-42.3 MeV and 42.4-46.5 MeV, respectively [18].

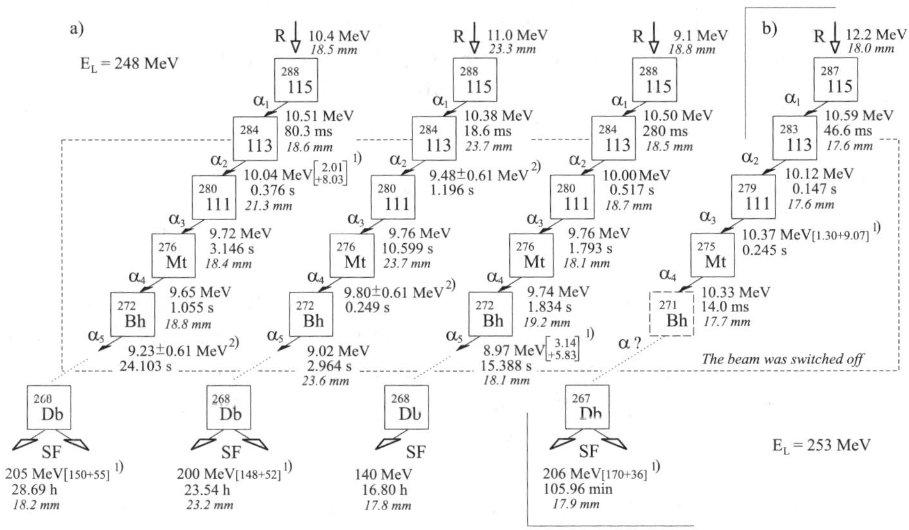

Figure 9. Time sequences in the decay chains observed at two ^{48}Ca energies: a) E_L=248 MeV and b) E_L=253 MeV. Measured energies, time intervals and positions of the observed decay events are shown. 1) Energies of events detected by both the focal-plane and side detectors. 2) Energies of events detected by side detectors only. We have inserted an unobserved nuclide in the fourth decay chain.

The three similar decay chains observed at 248 MeV are shown in Fig. 9a. The implantations of recoils in the focal-plane detector were followed by α-

particles with E_α=10.46±0.06 MeV. These sequences switched the ion beam off, and four more α-decays were detected in time intervals of 29 s, 15 s, and 20 s, respectively, in the absence of a beam-associated background. During the remainder of the 2.5 h beam-off period following the last position-correlated α-particle, no α-particles with E_α>7.6 MeV were registered by the focal-plane detectors.

The spontaneous fission of the final nuclei in these chains was detected 28.7 h, 23.5 h and 16.8 h, respectively, after the last α-decay. A search was performed to identify α-decays correlated closely in time (<60 s) and position to each of the three SF-events in Fig. 9a. No correlations were found.

In the first two decay chains, the focal-plane and side detectors simultaneously detected fission fragments with sum energies of 205 MeV and 200 MeV. In the third chain only the focal-plane detector registered a fission fragment. All three SF-events were registered in the corresponding strips and positions where the three ER$-\alpha_1-$...$-\alpha_5$ decay chains were observed. Thus, these SF-events should be assigned to the spontaneous fission of the descendant nuclei in these observed chains.

At 253 MeV, the aforementioned ER$-\alpha_1$-...-α_5-SF decay chains were not observed. However, a different decay chain, consisting of four α-decays and spontaneous fission, was registered (see Fig. 9b).

The beam was switched off after the detection of an ER signal followed in 46.6 ms by an α-particle with E_α=10.59 MeV in the same position of strip 7. Three other α decays were detected in a time interval of about 0.4 s in the absence of beam-associated background. After 106 minutes, the terminal SF-event was detected in beam with sum energy of 206 MeV in the same position of the strip 7. A more detailed description of the data analysis is given in our paper [18].

The radioactive properties of nuclei in this decay chain differ from those of the nuclei observed at the lower bombarding energy. The total decay time of this chain is about 10 times shorter and the α-decays are distinguished by higher α-particle energies and shorter lifetimes. Its production also required increasing the beam energy by about 5 MeV, so this decay chain must originate from another parent nucleus.

It is most probable that the different decay chains originate from neighboring parent isotopes of element 115, produced in the complete fusion reaction ^{243}Am+^{48}Ca followed by evaporation of three and four neutrons from the compound nucleus ^{291}115.

Indeed, at the excitation energy E^*=40 MeV, close to the expected maximum for the $3n$-evaporation channel, we observed longer decay chains of the odd-odd isotope ^{288}115. Increasing the beam energy by 5 MeV results in reducing the ^{288}115-isotope yield and, at the same time, increasing the yield of the $4n$-evaporation channel leading to the even-odd isotope ^{287}115. Corresponding cross sections for the $3n$- and $4n$-evaporation channels at the two projectile energies were measured to be $\sigma_{3n} = 2.7^{+4.8}_{-1.6}$ pb and $\sigma_{4n} = 0.9^{+3.2}_{-0.8}$ pb.

Table II. Decay properties of nuclei.

Isotope	Decay mode	Half-life	Q_α (MeV)	Isotope	Decay mode	Half-life	Q_α (MeV)
$^{288}115$	α	87^{+105}_{-30}ms	10.61±0.06	$^{287}115$	α	32^{+155}_{-14}ms	10.74±0.09
$^{284}113$	α	$0.48^{+0.58}_{-0.17}$s	10.15±0.06	$^{283}113$	α	100^{+490}_{-45}ms	10.26±0.09
$^{280}111$	α	$3.6^{+4.3}_{-1.3}$s	9.87±0.06	$^{279}111$	α	170^{+810}_{-80}ms	10.52±0.16
^{276}Mt	α	$0.72^{+0.87}_{-0.25}$s	9.85±0.06	^{275}Mt	α	9.7ms	10.48±0.09
^{272}Bh	α	9.8s	9.15±0.06	^{271}Bh	α	-	-
^{268}Db	SF (α/EC)	16h		^{267}Db	SF	73min	

The decay properties of the product nuclei are presented in Table II. In the decay chains shown in Fig. 9a, we assigned SF-events to the isotope ^{268}Db following five consecutive α-decays. This isotope also has the possibility of undergoing α-decay or electron capture (EC). In the case of the α-decay of ^{268}Db with T_α>2.5 h (after the beam-off period), one would more reasonably expect SF or EC of ^{264}Lr, because the α-decay of ^{264}Lr seems improbable due to the expected low α-decay energy (Q_α=6.84 MeV [13], T_α>100 d).

The electron capture of ^{268}Db or ^{264}Lr leads to the even-even isotopes ^{268}Rf or ^{264}No, for which rapid spontaneous fission can be expected (e.g., T_{SF}=1.4 s is predicted for ^{268}Rf [12]). Since both fission fragments of these final nuclei can have the Z=50 and N=82 closed shell configurations, one would expect their SF-decays to result in narrow symmetric mass distributions with rather high total kinetic energies, TKE≈230-240 MeV (see, e.g., Ref. [19]).

Indeed, the sum fission fragment energies observed for the final nuclei are close to this value. It is noteworthy that the likelihood that EC-decay occurs earlier in the observed decay chains is small because of the short T_α for the observed α-decays of elements 115, 113, 111, Mt, and Bh.

In the decay chain originating from $^{287}115$ (see Fig. 9b. and Table II), we propose that we must have missed the α-decay of ^{271}Bh. The expected T_α value for ^{271}Bh should be ~10 s (Q_α=9.07 MeV [13], T_α ≈5 s for allowed transition), which is much shorter than the interval between the last observed α-particle and the terminating SF-event, but is much longer than the intervals between the observed correlated α-particles. A search was performed to identify α-particles with E_α>7.0 MeV correlated in position to the observed α-decays. No correlated α-particles were observed an hour after the last α-decay.

In these experiments we registered 19 α-particles by the detector with 87% efficiency and the loss of one α-particle seems rather probable. Such a decay scheme of $^{287}115$ is also supported by the agreement of the observed decay properties of the other nuclides in the decay chains with the expectations of theory. This means that SF occurs directly in the decay of ^{267}Db since the calculated α-decay and EC-energies for this isotope are rather low (Q_α=7.41 MeV, Q_{EC}=1 MeV) and their expected partial half-lives significantly exceed the observed time interval of 106 min.

For the measured α-decay energies of the newly-produced isotopes, one can estimate expected half-lives for allowed transitions and compare them with the experimental values. The ratios between experimental T_{exp} and calculated T_{calc} half-lives define the hindrance factors caused by the odd numbers of protons and/or neutrons in the newly synthesized nuclei. The measured T_{exp} values closely reproduce the calculated ones for the first two nuclei of these chains; thus the element 115 and element 113 isotopes have rather low hindrance factors, if any, for α-decay. For the isotopes of elements 111, Mt and Bh, the difference between measured and calculated T_α values results in hindrance factors of 3-10. These match the hindrance factors that can be extracted for the deformed odd-odd nuclei $^{272}111$ and descendants produced in the experiments in GSI [10] and RIKEN [20].

One can suppose that in this region of nuclei, a noticeable transition from spherical to deformed shapes occurs at Z=109-111, resulting in complication of the level structure of these nuclei and in increased probability of α-transitions going through excited states. This assumption is in agreement with the macroscopic-microscopic calculations [13]. The deformation parameter β_2 was calculated to be 0.072 and 0.138 for $^{288}115$ and $^{284}113$. As the decay chain recedes from the shell closure at Z=114, the deformation parameter β_2 increases to 0.200, 0.211 and 0.247 for $^{280}111$, ^{276}Mt and ^{272}Bh, respectively.

8. Short discussion and conclusions

From the recent experimental results given in the present paper together with the already available data we can get an idea of the current state of the problem of the synthesis of superheavy elements and outline the future plans.

1. In the ^{242,244}Pu+^{48}Ca reactions the formation of 4 isotopes of element 114 with A=286-289 has been observed. The isotopes with Z=114 and A=287-289 undergo α-decay; the α-decay branching ratio for the even-even nucleus $^{286}114$ amounts to $\sim$30%. The daughter nuclei – the isotopes of element 112 with N=171 and 173 also undergo α-decay, which is terminated by spontaneous fission of the isotopes of element 110 with N=169 and 171. The even-even isotopes of element 112 with N=170 and 172 undergo spontaneous fission with T_{SF}=0.5 ms and 100 ms, respectively.

Alpha-transitions in the decay chains of the Z=114 isotopes have well-defined energies pointing to the absence of hindrance in the observed decays that is characteristic of α-decay of spherical nuclei. The measured values Q_α and T_α support the Z-assignment of the decay chains. In order to define the isotope mass numbers, we measured the energy dependence of their production cross section in the reaction ^{244}Pu+^{48}Ca in the range of excitation energy E^*=31-53 MeV of the compound nucleus ^{292}114.

2. The maximum cross section of the ER formation in the reaction ^{244}Pu+^{48}Ca corresponds to E^* ~40-42 MeV. A major contribution to it is from the channels with evaporation of 3 and 4 neutrons, their cross sections being about 1.7 pb and 5.3 pb, respectively.

The experimentally observed shift of the $3n$-evaporation maximum to higher energy can be interpreted as being due to the occurrence of selected states that are associated with the orientation of the deformed target nucleus in the reaction entrance channel. Such an effect should result in hindering of the $2n$ channel and partial overlap of the excitation functions of $3n$ and $4n$ channels.

At the same time, in the reaction ^{245}Cm+^{48}Ca, which is characterized by about 4-MeV lower excitation of the compound nucleus ^{293}116 compared to that of ^{292}114 in the ^{244}Pu+^{48}Ca case, new isotopes of element 116 are produced in $2n$- and $3n$-evaporation channels with cross sections of about 0.9 pb and 1.3 pb, respectively. It is evident that the problem of fusion of deformed actinide nuclei with massive projectiles like ^{48}Ca is of special interest by itself.

3. In view of the new data we should return to the analysis of the results of out first experiments carried out at lower excitation energy E^* $\leq$35 MeV. In our present measurements of the excitation functions of the xn-evaporation channels in the reaction ^{244}Pu+^{48}Ca at E^*=41-53 MeV, with integral beam dose of 1.0×10^{19}, as well as in our previous measurement at E^*=35 MeV (beam dose of 1.0×10^{19}) we did not observe a long-lived decay chain ER$-\alpha-\alpha-\alpha-$SF that we considered a candidate for the decay of ^{289}114 [21]. The cross section of producing this nucleus at E^* ~35 MeV can be estimated as 0.2 pb.

From the measured excitation functions (Fig. 5a) one can consider this chain to originate from the decay of the neighboring even-even isotope ^{290}114, the $2n$-evaporation product. The low production cross section at low energy supports such a hypothesis. The observed decay properties are also close to what is expected for ^{290}114. On the other hand, the HFB calculations of the structure of ^{289}114 [14] point that such a chain could represent a different, probably a rare decay mode of the even-odd ^{289}114 isotope.

In our experiment with the ^{242,244}Pu+^{48}Ca reaction we did not observe the 3-minute $\alpha-$SF decay chain that was earlier reported from the experiments performed with the separator VASSILISSA in the ^{242}Pu+^{48}Ca reaction, also at low excitation energy of the compound nucleus ^{290}114 (E^*=32.6 MeV) and ascribed to ^{287}114 [22]. It cannot be excluded that this is also another decay mode of ^{287}114. Yet this discussion is based on the few events observed in the first experiments. The planned measurements of the excitation functions in the reaction ^{242}Pu+^{48}Ca with higher statistics at lower beam energy will be able to give an answer to this question. We also think that these measurements will give a possibility to make an additional and independent assignment of masses of the isotopes observed in the reaction ^{244}Pu+^{48}Ca.

4. We considered using the reaction ^{249}Cf+^{48}Ca with the aim of synthesizing element 118 to be a more realistic approach with respect to the earlier attempt undertaken in the ^{208}Pb+^{86}Kr experiment [10, 23]. Unfortunately, despite the high accumulated ^{48}Ca beam dose because of the limited amounts of the target material we detected only two events, the ER-SF and ER-$\alpha-\alpha-$SF chains, and both of them can be ascribed to the decay of an isotope with Z=118. Before resuming this experiment with higher sensitivity we turned to investigating the decay properties of the daughter nuclides, the isotopes of element 116 produced in the reactions ^{245}Cm(^{48}Ca,xn)$^{293-x}$116. Here we synthesized two new isotopes of element 116 with mass numbers 291 and 290 formed in the $2n$ and $3n$-evaporation channels.

5. The even-even isotope ^{290}116 decays giving origin to ($\alpha-$SF) or ($\alpha-\alpha-$SF) chains, in accord with the one observed $\alpha-\alpha-$SF decay event of an isotope of element 118. The even-odd nuclide ^{291}116 has a longer-lived decay chain. As mentioned earlier, since in the reaction ^{249}Cf+^{48}Ca the larger Q value results in 6.3-MeV lower minimum excitation energy of the compound nucleus ^{297}118, as compared with ^{292}114 formed in the reaction ^{244}Pu+^{48}Ca, the maximum yield is expected for $2n$- and $3n$-evaporation products, in particular. Now the scenario and the properties of their α-decay daughters are established.

6. The α-decay energies Q_α and half-lives T_α of the Z-even nuclides with $Z \geq$108 are presented in Fig. 10 together with calculations made in the macroscopic -microscopic model. The available experimental data is in general agreement with the theoretically predicted considerable enhancement of nuclear stability in the vicinity of the deformed shells Z=108 and N=162.

For the isotopes with Z=112-116 and $N \geq$170 the expected increase of nuclear stability due to the effect of the spherical N=184 shell is also clearly observed. The experimental Q_α values of the isotopes with Z=110-116 are 0.1-0.5 MeV lower than the calculated values. Correspondingly, their half-lives are 2 to 10 times higher than theoretically predicted. In fact, such an insufficient deviation supports the theoretical concept of the decisive effect of spherical nuclear shells on the stability of superheavy nuclides, not only qualitatively but in some sense quantitatively.

7. We should note that the properties of the superheavy nuclei in the ground states observed in our experiments are also in agreement with calculations performed within microscopic models like HFB or RMF. At present the properties of the synthesized nuclides with Z=110-116 and N=170-177 do not allow to define the exact position of the closed spherical proton shell. Further investiga-

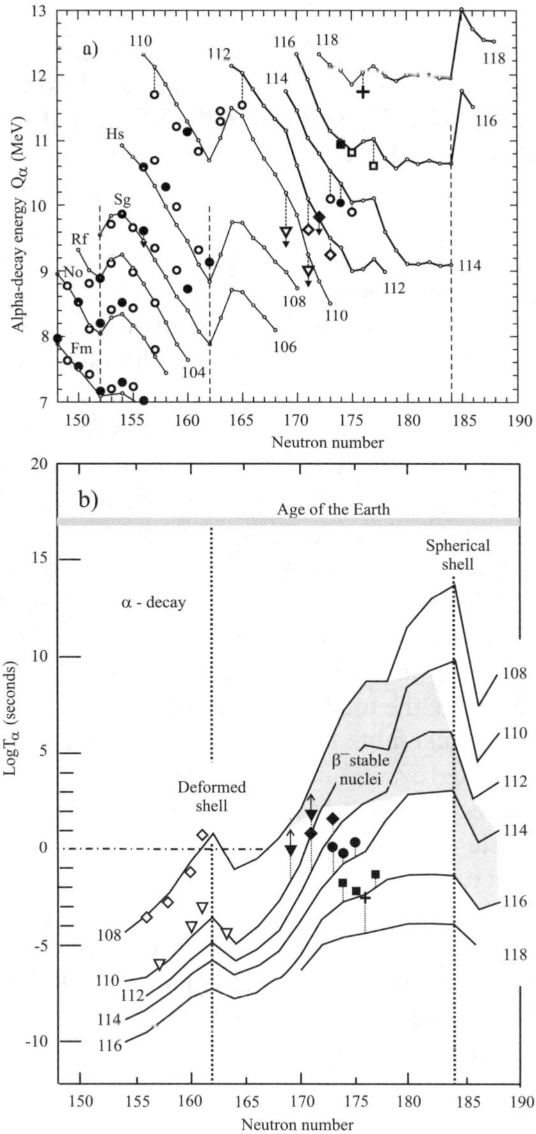

Figure 10. a) Decay energy Q_α of the even-Z isotopes with Z=100-118 vs. neutron number. Full symbols show data for even-even nuclei; open – for even-odd ones. Data for Z=110-118 and $N \geq 170$ are from experiments with ^{48}Ca projectiles. b) half-lives $\mathrm{Log}(T_\alpha)$ for the nuclei with Z=108-118 vs. neutron number. Solid lines present the calculations in the macroscopic-microscopic model [12,13]. Dashed lines indicate the closed neutron shells N=152, 162 and 184.

tion of Z>116 nuclei, in particular the isotopes of element 118, could probably make more definite conclusions possible.

8. In the Z-even nuclei we have up till now a single case of a 4-fold α-decay of the isotope $^{287}114$ down to $^{271}106$, whereas in the case of the Z-odd nuclei such a picture appears to be a typical one. In the ^{243}Am+^{48}Ca reaction 11 new nuclides with Z=105-115 and N=162-173, situated between the closed deformed and spherical shells, were obtained. The formation cross sections at the level of $\sim$ 1-3 pb make it possible to hope that in the future we shall be able to produce them in greater quantities.

The investigation of the properties of these nuclides can be a direct check of theoretical models on the influence of nuclear structure on the decay properties of nuclei, as soon as these models allow calculations not only of the ground state properties, but also of the excited states of the superheavy elements.

On the other hand, the long lifetimes of nuclei with Z=103-105, found close to the shell with N=162, allow detailed investigations not only of the radioactive, but also of the chemical properties of elements at and beyond the actinides. Other reactions with targets of ^{237}Np or ^{231}Pa can lead to the formation of new isotopes in this, to our opinion, very interesting region of nuclei.

9. From the above consideration it follows that both the even-even and even-odd new nuclides with $Z \geq 114$ undergo α-decay as the most probable decay mode. In their daughter α-decay products with Z=110-112 we observe a competition of α-decay and SF. From the events with both fission fragments detected we can estimate the total kinetic energy of fission fragments in SF of the new nuclides. Not claiming high accuracy of these figures because of the small number of detected events and the necessity of introducing corrections for losses in the measured fragment energies, we show them, however, in Fig. 11 with the general systematic of TKE *vs.* $Z^2/A^{1/3}$.

Our experimental data point to asymmetric mass-division in the spontaneous fission of superheavy nuclei that is in agreement with the data on mass distributions in fission of excited nuclei (E^* $\sim$30-40 MeV) [24]. On the contrary, the data shown in Fig. 11 for the neutron-rich isotopes ^{268}Db(Rf) and ^{271}Sg point out their symmetric fission, due to the shell effects at Z=50 and N=82 in the spontaneous fission of these nuclei.

Acknowledgements

Presenting the results of our joint work, I would like to acknowledge the enormous contribution of my colleagues and co-authors in obtaining the reported data.

In performing the experiments on the synthesis of superheavy nuclides we for the first time used targets made of ^{245}Cm, obtained at the reactor of the RIAR (Dimitrovgrad) and consequently enriched after isotopic separation at RFNC-VNIIEF (Sarov). We are grateful to Prof. R.I. Ilkaev, Drs. S.P. Vesnovskii and Yu.G. Toporov for doing this work in the shortest possible time.

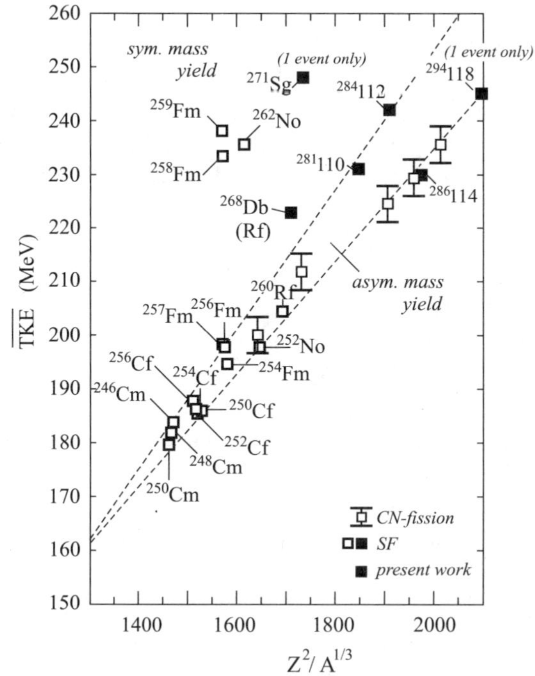

Figure 11. Dependence of <TKE> vs. the parameter $Z^2/A^{1/3}$ for the spontaneous fission of the nuclides with Z=96-118. Open squares present literature data; full squares – the SF detected in the decay chains of the superheavy nuclides (see text).

We are indebted to the Chairman of this session Prof. W. Greiner, and also to Profs. S. Hofmann, G. Muenzenberg and A. Sobiczewski for the interesting discussions and valuable remarks.

The experiments were carried out at the heavy-ion cyclotron U400 of the Flerov Laboratory of Nuclear Reactions in the frame of the collaboration FLNR (Dubna) - LLNL (Livermore), according to the plan for scientific international collaboration of JINR for the years 2000-2003, as well as in the frame of the decisions of the coordination committee RF-USA in the field of fundamental investigations of matter. This work was performed with the considerable support of the Ministry of Atomic Energy of the Russian Federation, the Russian Foundation for Basic Research and INTAS. Our colleagues from LLNL were also supported by the USA Department of Energy.

References

[1] S. Hofmann, This proceedings.

[2] Ch.E. Duellmann *et al.*, Nature **418**, 859 (2002); A. Tuerler *et al.*, GSI Scientific Report 2001, GSI 2002-1, p.1.

[3] Yu.A. Lazarev *et al.*, Phys. Rev. Lett. **73**, 624 (1994); A. Tuerler *et al.*, Phys. Rev. C **57**, 1648 (1998); A. Türler, Eur. Phys. J A **15**, 271 (2002).

[4] Yu.Ts. Oganessian *et al.*, *Proceedings of the Fourth International Conference on Dynamical Aspects of Nuclear Fission*, Casta-Papiernicka, Slovak Republic, 1998 (World Scientific, Singapore, 2000), p.334; K. Subotic *et al.*, Nucl. Instr. Meth. Phys. Res. A **481**, 71 (2002).

[5] Yu.Ts. Oganessian *et al.*, Phys. Rev. C **62**, 041604(R) (2000); Phys. At. Nucl. **63**, 1679 (2000).

[6] W.D. Myers and W.J. Swiatecki, Nucl. Phys. A **601**, 141 (1996).

[7] Yu.Ts. Oganessian *et al.*, Phys. Rev. C **63**, 011301(R) (2001); Phys. At. Nucl. **64**, 1349 (2001); Eur. Phys. J. A **15**, 201 (2002).

[8] V.E. Viola Jr., G.T. Seaborg, J. Inorg. Nucl. Chem. **28**, 741 (1966).

[9] Yu.Ts. Oganessian *et al.*, Phys. Rev. C **64**, 054606 (2001); A.V. Yeremin *et al.*, JINR Rapid Communications No. 6[92]-98, JINR, Dubna, 21 (1998).

[10] S. Hofmann and G. Muenzenberg, Rev. Mod. Phys. **72**, 733 (2000); S. Hofmann *et al.*, Eur. Phys. J. A **14**, 147 (2002).

[11] P. Moeller and J.R. Nix, *Proceedings of the Tours Symposium on Nuclear Physics III*, Tours, France, 1997, edited by M. Arnould, M. Lewitowicz, Yu.Ts. Oganessian, M. Ohta, H. Utsunomiya, and I. Wada, AIP Conf. Proc. No. 425 (AIP, New York, 1998), p.75; K. Nishio *et al.*, J. Nucl. Sci. Tech., Supplement 3, *Proceedings of the Actinides 2001 International Conference*, Hayama, Japan, 2001 (2002 Atomic Energy Society of Japan), p.26.

[12] R. Smolanczuk and A. Sobiczewski, in *Proceedings of XV Nuclear Physics Divisional Conference "Low Energy Nuclear Dynamics"*, St. Petersburg, Russia, 1995 (World Scientific, Singapore), p.313; R. Smolanñzuk, Phys. Rev. C **56**, 812 (1997); R. Smolanczuk, J. Skalski and A. Sobiczewski, Phys. Rev. C **52**, 1871 (1995).

[13] I. Muntian *et al.*, Acta Phys. Pol. B **34**, 2073 (2003); Phys. At. Nucl. **66**, 1015 (2003).

[14] S. Cwiok, W. Nazarewicz and P.H. Heenen, Phys. Rev. Lett. **83**, 1108 (1999); J.F. Berger, D. Hirata and M. Girod, Acta Phys. Pol. B **34**, 1909 (2003).

[15] S.Typel and B.A. Brown, Phys. Rev. C **67**, 034313 (2003).

[16] M. Bender, Phys. Rev. C **61**, 031302 (2000); P.-G. Reinhard *et al.*, in *Proc. Tours Symposium on Nuclear Physics IV*, Tours, France, 2000 (AIP, New York, 2001), p.377; Z. Ren, Phys. Rev. C **65**, 051304(R) (2002).

[17] Yu.Ts. Oganessian *et al.*, JINR Communication D7-2002-287 (2002); Lawrence Livermore National Laboratory Report, UCRL-ID-151619 (2003).

[18] Yu.Ts. Oganessian *et al.*, JINR Communication E7-2003-178 (2003); Phys. Rev. C, submitted.

[19] D.C. Hoffman and M.R. Lane, Radiochim. Acta **70/71**, 135 (1995).

[20] K. Morita, *Proceedings of the VIII International Conference on Nucleus-Nucleus Collisions (NN2003)*, Moscow, Russia, 2003, to be published.

[21] Yu.Ts. Oganessian *et al.*, Phys. Rev. Lett. **83**, 3154 (1999).

[22] Yu.Ts. Oganessian *et al.*, Nature **400**, 242 (1999).

[23] V. Ninov *et al.*, Phys. Rev. Lett. **83**, 1104 (1999); Phys. Rev. Lett. **89**, 039901 (2002).

[24] M.G. Itkis *et al.* J. Nucl. Radiochem. Sci. **3**, 57 (2002); in *Proceedings of the Second International Conference on Fission and Properties of Neutron-Rich Nuclei*, St. Andrews, Scotland, 1999 (World Scientific, Singapore), p.268.

DESCRIPTION OF PROPERTIES OF SUPERHEAVY NUCLEI

A. Sobiczewski[1], I. Muntian[1], O. Parkhomenko[1]

[1] *Soltan Institute for Nuclear Studies, PL-00-681 Warsaw, Poland*

Abstract Role of shell structure of heaviest nuclei in their properties is discussed. Quality of theoretical description of these properties is also considered in the paper. This quality is illustrated on the example of the α-decay energy which is one of the best experimentally studied quantities in the region of superheavy nuclei.

1. Introduction

The objective of this paper is to discuss the role of shell structure in properties of heaviest nuclei and to illustrate the quality of their theoretical description. This quality is illustrated on the example of the α-decay energy, which is one of the most important and, simultaneously, the best studied properties of superheavy nuclei.

2. Role of shell effects

Effects of shell structure play a basic role in the problem of superheavy nuclei. Without these effects, these nuclei would not simply exist. A quantitative analysis [1] has shown that these effects decrease mass (i.e. increase binding energy) of heaviest nuclei by up to about 5 MeV. The α-decay half-lives of them are increased by these effects by up to about 5 orders of magnitude. Especially large shell effects are found in the spontaneous-fission half-lives. This is illustrated in Fig. 1, where experimental (exp.) and theoretical (Y) values of the logarithm of $T_{\rm sf}$ are given. The theoretical values are calculated by a macroscopic (Yukawa-plus-exponential [2]) model, which does not include any shell structure of a nucleus. Thus, the difference between the two values is just shell effect in the spontaneous-fission half-life $T_{\rm sf}$. One can see that the effect delays the fission process in all considered nuclei, except only few lightest ones (isotopes of uranium). The delay increases from few orders (Pu isotopes) to about 15 orders of magnitude for the nucleus $^{260}106$ (^{260}Sg). For such heavy

W. Greiner et al. (eds.), Structure and Dynamics of Elementary Matter, 377–386.

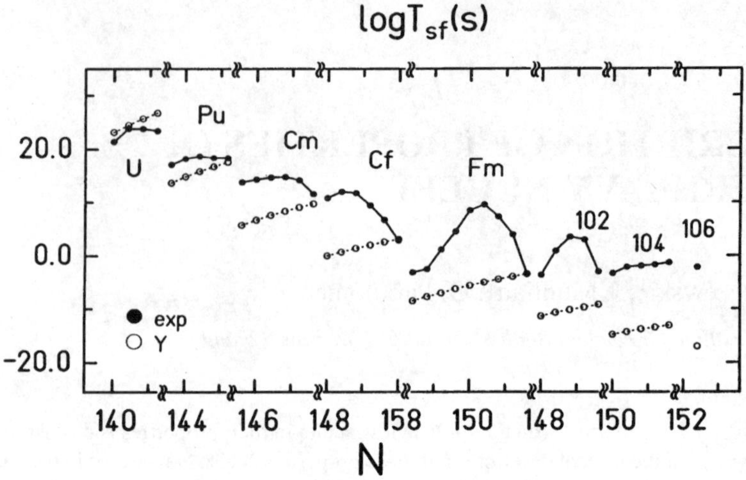

Figure 1. Logarithm of experimental (exp) and calculated within a macroscopic model (Y) spontaneous-fission half-lives T_{sf}, given in seconds.

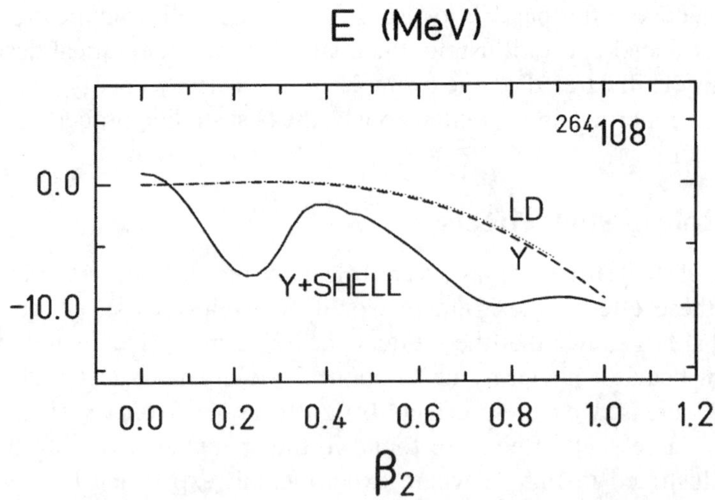

Figure 2. Static spontaneous-fission barrier of the nucleus $^{264}108$ (^{264}Hs) calculated in three cases: when only macroscopic part of energy is taken (Y and LD), and when shell correction to the energy is also included (Y+SHELL).

nuclei as $^{260}106$, with T_{sf} of the order of few milliseconds, this increase makes up practically the whole half-life of these nuclei. In other words, they would not exist without shell effects, as already mentioned above.

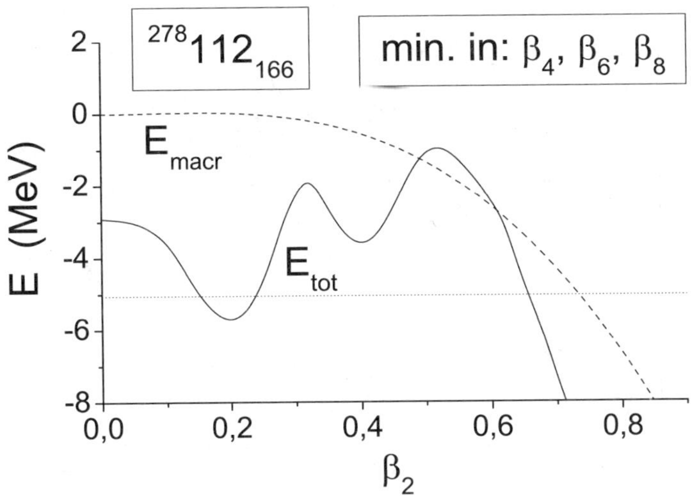

Figure 3. Static spontaneous-fission barrier calculated for the nucleus $^{278}112$ in two cases: when only the macroscopic ($E_{\rm macr}$) and when the total ($E_{\rm tot}$) energy of it is considered.

The mechanism by which practically the whole half-life of a very heavy nucleus is made up by shell effects is illustrated in Fig. 2. The figure illustrates the spontaneous-fission barrier of the nucleus $^{264}108$ (^{264}Hs), i.e. the dependence of the ground-state energy of this nucleus on its quadrupole-deformation parameter β_2. For each β_2, the energy is minimized with respect to the hexadecapole-deformation parameter β_4. The total fission barrier (Y+SHELL), including shell effects, is shown by solid line and its smooth, macroscopic part (obtained by the Yukawa-plus-exponential model (Y) [2]) by dashed line. The smooth barrier obtained by another macroscopic model (liquid drop, LD [3]) is also shown (dotted line), for comparison. One can see that a significant height (about 5 MeV) of the fission barrier is obtained only after the inclusion of shell effects. Without them, no fission barrier (Y and LD) appears.

Figure 3 shows the spontaneous-fission barrier for even heavier nucleus: $^{278}112$. The barrier is analyzed in a large, 4-dimensional deformation space $\{\beta_\lambda\}$, λ=2,4,6,8. One can see that shell correction fluctuates quite much as a function of deformation. Due to this, a two-hump barrier is generated. The correction is largest (about -5.8 MeV) at the equilibrium point, smaller (about -3.0 MeV) at the second minimum, even smaller (about -1,8 MeV) at the first saddle and the smallest (about +0.5 MeV) at the second saddle point. Significant shell correction at the saddle points are worth to be noticed as they are quite often neglected in various estimates of the fission-barriers heights of superheavy nuclei.

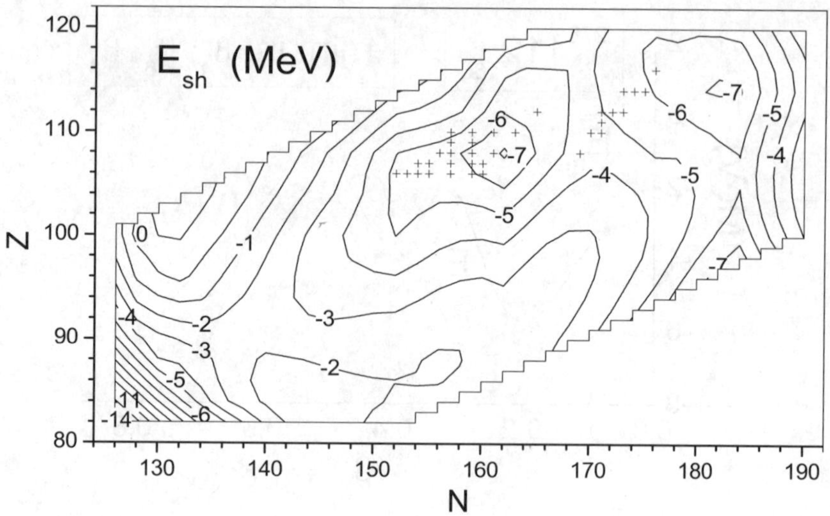

Figure 4. Contour map of the shell correction energy $E_{\rm sh}$. Crosses denote the heaviest nuclides synthesized up to now.

Important quantity is the global shell correction to energy (mass) of a nucleus in its ground state. It is defined as

$$E_{\rm sh} \equiv E(\beta_\lambda^0) - E_{\rm macr}(0), \tag{1}$$

where $E(\beta_\lambda^0)$ is the value of the total energy E of a nucleus at its equilibrium deformation β_λ^0 and $E_{\rm macr}(0)$ is the value of the macroscopic part of the energy at spherical shape of this nucleus. As the equilibrium deformation of a nucleus is zero (spherical shape) in a macroscopic model, $E_{\rm sh}$ is the gain in energy of a nucleus due to its shell structure.

The shell correction $E_{\rm sh}$ gives us a first orientation in the stability of a nucleus. Figure 4 [4] shows this quantity calculated for a large region of nuclei with proton number Z = 82 - 120 and neutron number N = 126 - 190. One can see that $E_{\rm sh}$ has three minima in this region. The first one, which is the deepest ($E_{\rm sh}$= –14.3 MeV), is obtained for the doubly magic spherical nucleus ^{208}Pb. The second one ($E_{\rm sh}$= –7.2 MeV) appears for the nucleus $^{270}108_{162}$, which is predicted [5, 6] to be a doubly magic deformed nucleus. The third minimum, of the same depth ($E_{\rm sh}$= –7.2 MeV) as that of the second minimum, is obtained for the nucleus $^{296}114_{182}$, which is close to the nucleus $^{298}114_{184}$ predicted [7–9] to be a doubly magic spherical nucleus, the next one to the last experimentally known ^{208}Pb. The problem of existence of spherical superheavy nuclei has been considered for already more than 30 years [10]. Besides these three minima, there appears a rather wide plateau around the nucleus ^{252}Fm, which, having a smaller (in absolute value) shell correction ($E_{\rm sh}$= –5.2 MeV) than the nucleus $^{270}108$ (^{270}Hs), may be considered as a nucleus with closed deformed subshells [5, 6].

One can also see in Fig. 4 that some of the already synthesized nuclei profit by 6-7 MeV in their binding energy from the shell correction. Without this profit they could not exist, as illustrated above.

The appearance of the region of nuclei around the second minimum (deformed superheavy nuclei [11]) constitutes the main change in our view of stability of heaviest nuclei in recent years. Before, it was believed for a long time that spherical superheavy nuclei, predicted to be situated around the third minimum, would constitute an island, separated from the usual peninsula of relatively long-lived nuclei by an "ocean" of full instability. After the appearance of deformed superheavy nuclei, however, the peninsula is expected to be extended to include also the spherical superheavy nuclei [12].

A wider discussion of the problem of superheavy nuclei may be found e.g. in the reviews [13–16]. Recently, there is a large activity in studying superheavy nuclei within self-consistent microscopic approaches (e.g. [17–24]).

3. Alpha-decay energy

In this section, we illustrate quality of description of α-decay energy Q_α of heavy and superheavy nuclei within a macroscopic-microscopic model in its recent version [25] and also within a few other approaches. The energy Q_α is one of the most important quantities for superheavy nuclei. The energy is directly obtained from the calculated masses of these nuclei [26]. Difference between calculated values $Q_\alpha^{\rm th}$ and experimental ones $Q_\alpha^{\rm exp}$ is shown in Fig. 5 for nuclei with proton number Z=84-94 and in Fig. 6 for Z=95-108. Experimental values $Q_\alpha^{\rm exp}$ are obtained from experimental masses [27].

One can see in Figs. 5 and 6 that:

(1) The discrepancy between $Q_\alpha^{\rm th}$ and $Q_\alpha^{\rm exp}$ is contained within the range $\pm\,0.5$ MeV, for almost all of the nuclei. Only for some isotopes of the elements with Z=86-92, calculated Q_α is smaller than measured one by more than 0.5 McV.

(2) Theoretical description of Q_α is better for heavier nuclei than for lighter ones. Probably one can state a more general rule that better description is obtained for better deformed (i.e. with larger deformation energy) nuclei. In other words, more distant is a nucleus from (spherical) magic numbers, better is the description of Q_α.

(3) Calculated Q_α show a good isotopic dependence. This is especially well seen in a long chain of Q_α of well deformed plutonium nuclei.

(4) Decay energies $Q_\alpha^{\rm th}$ of odd-A and odd-odd nuclei, in the calculations of which no additional adjustable parameters have been used, are reproduced similarly well as those of even-even nuclei.

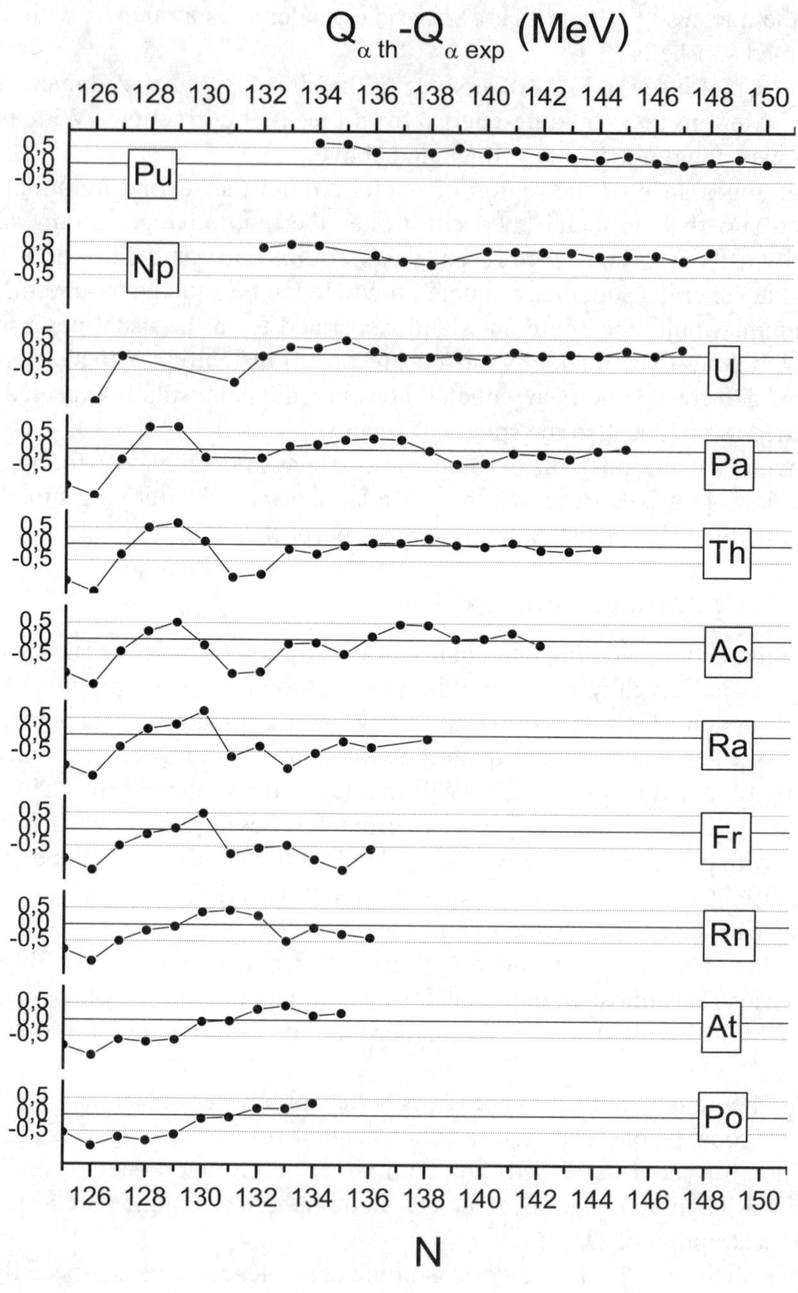

Figure 5. Difference between calculated and experimental values of the α-decay energy Q_α, shown for nuclei with proton number Z = 84 - 94.

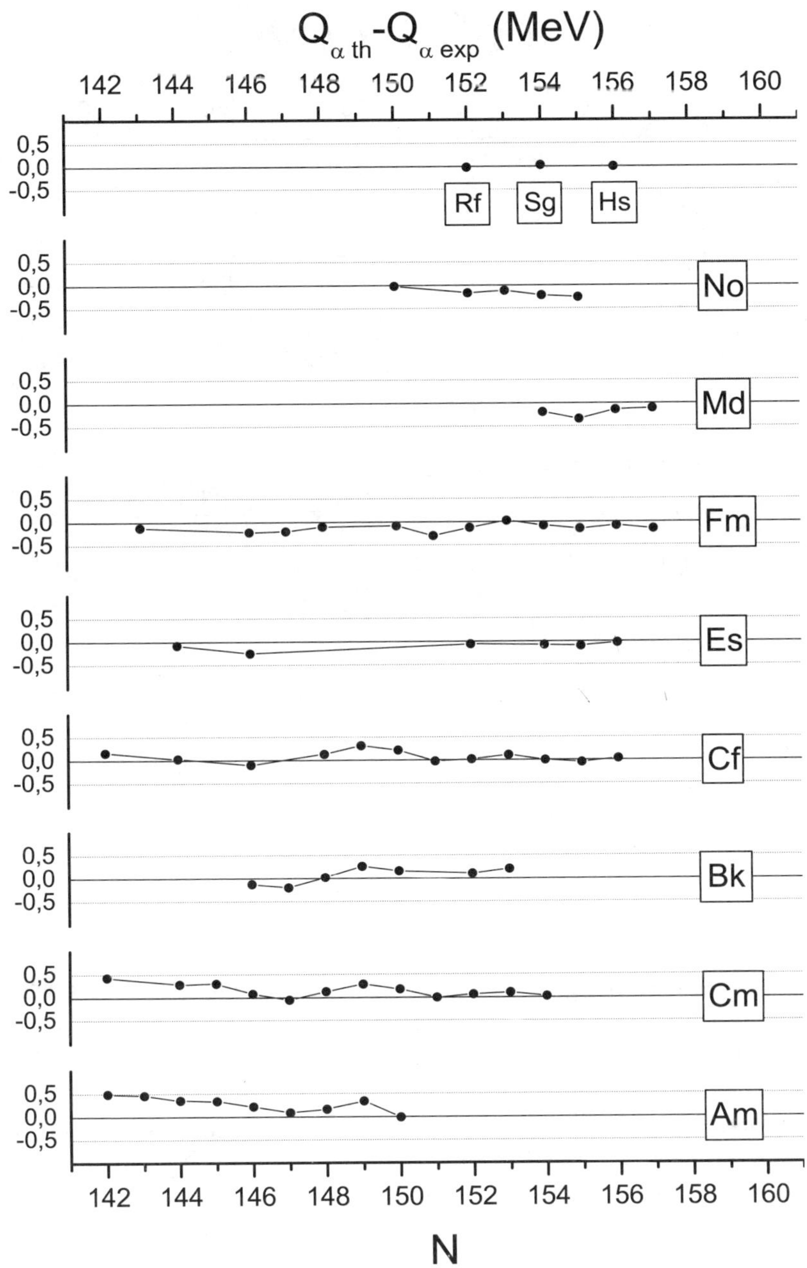

Figure 6. Same as in Fig. 5, but for $Z = 95$ - 108.

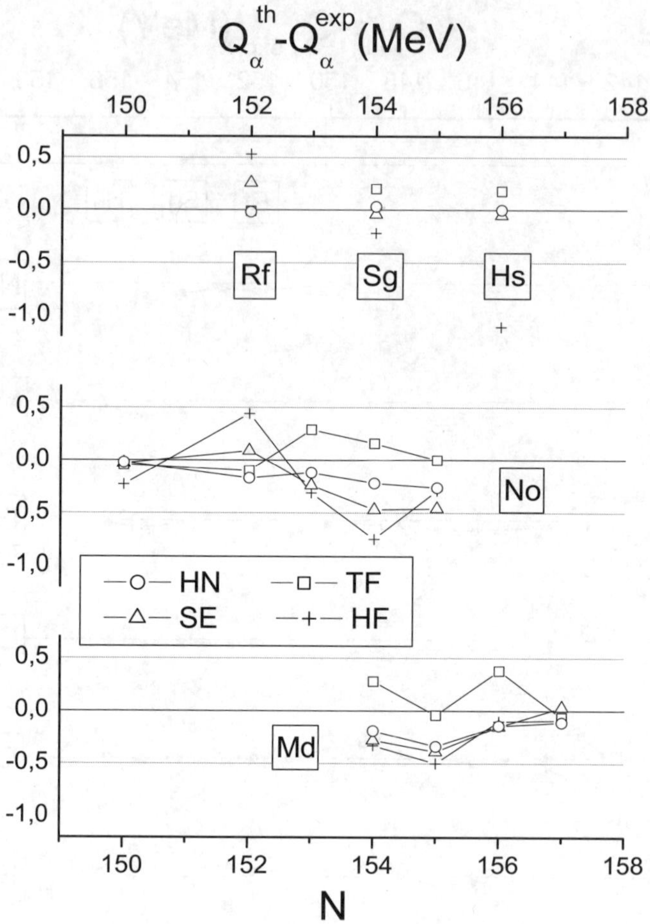

Figure 7. Same as in Fig. 5, but for Z = 101 - 108 and for 4 variants of theoretical approaches.

It is interesting to compare the quality of descriptions of Q_α obtained in various approaches. Figure 7 shows the results for $Q_\alpha^{\rm th}$ obtained within a semi-empirical (SE) shell model [28], macroscopic-microscopic model specially adapted to heaviest nuclei (HN) [25], macroscopic-microscopic model with the macroscopic part of the Thomas-Fermi type (TF) [29] and a purely microscopic model using the Hartree-Fock-BCS approximation (HF) [30]. The comparison is made for transfermium nuclei with Z=101-108. One can see that the discrepancy is contained within $\pm$0.5 MeV for three of the considered models. Only for the HF approach, it is larger (up to about 1 MeV). The best agreement is obtained within the HN approach.

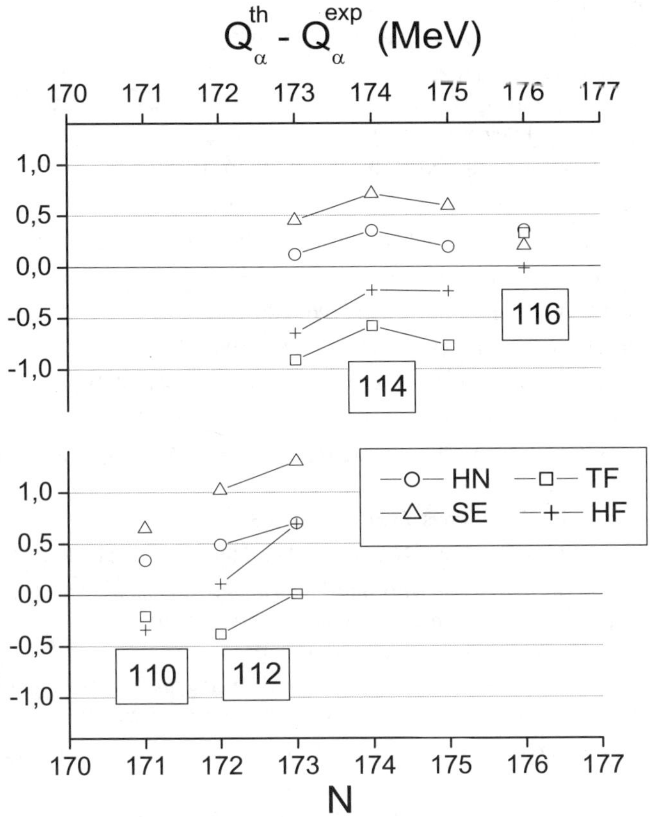

Figure 8. Same as in Fig. 7, but for $Z = 110 - 116$.

Finally, Fig. 8 (taken from Ref. [31]) shows the comparison for heaviest nuclei obtained recently. One can see that the discrepancy with experiment is larger than for lighter nuclei of Fig. 7. Also the discrepancy between the considered approaches, themselves, is larger. The TF results are relatively good for these nuclei.

Acknowledgements

Support by the Polish State Committee for Scientific Research (KBN), grant no. 2 P03B 039 22, and the Polish-JINR (Dubna) Cooperation Programme is gratefully acknowledged.

References

[1] Z. Patyk, A. Sobiczewski, P. Armbruster, and K.-H. Schmidt, Nucl. Phys. A **491** (1989) 267.

[2] H.J. Krappe, J.R. Nix, and A.J. Sierk, Phys. Rev. C **20** (1979) 992.

[3] W.D. Myers and W.J. Swiatecki, Ark. Fys. **36** (1967) 343.

[4] R. Smolanczuk and A. Sobiczewski, Proc. XV Nucl. Phys. Conf.:*Low Energy Nuclear Dynamics*, St. Petersburg (Russia) 1995, eds. Yu.Ts. Oganessian, W. von Oertzen, R. Kalpakchieva (World Scientific, Singapore 1995) p. 313.

[5] Z. Patyk and A. Sobiczewski, Nucl. Phys. A **533** (1991) 132.

[6] Z. Patyk and A. Sobiczewski, Phys. Lett. B **256** (1991) 307.

[7] A. Sobiczewski, F.A. Gareev, and B.N. Kalinkin, Phys. Lett. **22** (1966) 500.

[8] H. Meldner, Ark. Fys. **36** (1967) 593.

[9] U. Mosel and W. Greiner, Z. Phys. **222** (1969) 261.

[10] W.D. Myers and W.J. Swiatecki, Nucl. Phys. **81** (1966) 1.

[11] A. Sobiczewski, Z. Patyk, and S. Cwiok, Phys. Lett. B **224** (1989) 1.

[12] A. Sobiczewski, Z. Patyk, and S. Cwiok, Phys. Lett. B **186** (1987) 6.

[13] A. Sobiczewski, Fiz. Elem. Chastits At. Yadra **25** (1994) 295; Phys. Part. Nucl. **25** (1994) 119.

[14] P. Moeller and J.R. Nix, J. Phys. G **20** (1994) 1681.

[15] A. Sobiczewski, Usp. Fiz. Nauk **166** (1996) 943; Physics–Uspekhi **39** (1996) 885.

[16] A. Sobiczewski, in *Heavy Elements and Related New Fenomena*, vol. 1, eds. W. Greiner and R.K. Gupta (World Scientific, Singapore, 1999) chap. 9.

[17] J.F. Berger, L. Bitaud, J. Decharge, M. Girod, and S. Peru-Desenfants, in *Proceedings of the 24th International Workshop: "Extremes of Nuclear Stucture"*, Hirschegg (Austria) 1996, eds. H. Feldmeier, J. Knoll, and W. Noerenberg (GSI, Darmstadt 1996) p. 43.

[18] S. Cwiok, J. Dobaczewski, P.-H. Heenen, P. Magierski, and W. Nazarewicz, Nucl. Phys. A **611** (1996) 211.

[19] G.A. Lalazissis, M.M. Sharma, P. Ring, and Y.K. Gambhir, Nucl. Phys. A **608** (1996) 202.

[20] K. Rutz, M. Bender, T. Buervenich, T. Schilling, P.-G. Reinhard, J.A. Maruhn, and W. Greiner, Phys. Rev. C **56** (1997) 238.

[21] W. Greiner, J. Nucl. Radiochem. Sci. **3** (2002) 159.

[22] T. Buervenich, D.G. Madland, J.A. Maruhn, and P.G. Reinhard, J. Nucl. Radiochem. Sci. **3** (2002) 191.

[23] Zhongzhou Ren, J. Nucl. Radiochem. Sci. **3** (2002) 195.

[24] A.V. Afanasjev, T.L. Khoo, S. Frauendorf, G.A. Lalazissis, and I. Ahmad, Phys. Rev. C **67** (2003) 024309.

[25] I. Muntian, Z. Patyk, and A. Sobiczewski, Acta Phys. Pol. B **32** (2001) 691.

[26] I. Muntian, Z. Patyk, and A. Sobiczewski, Yad. Fiz. **66** (2003) 1051; Phys. At. Nucl. **66** (2003) 1015.

[27] G. Audi, O. Bersillon, J. Blachot, and A.H. Wapstra, Nucl. Phys. A **624** (1997) 1.

[28] S. Liran, A. Marinov, and N. Zeldes, Phys. Rev. C **62** (2000) 047301.

[29] W.D. Myers and W.J. Swiatecki, Nucl. Phys. A **601** (1996) 141.

[30] F. Tondeur, S. Goriely, J.M. Pearson, and M. Onsi, Phys. Rev. C **62** (2000) 024308.

[31] I. Muntian, S. Hofmann, Z. Patyk, and A. Sobiczewski, Acta Phys. Pol. B **34** (2003) 2073.

DISSIPATION–FLUCTUATION DYNAMICAL APPROACH TO FUSION LEADING TO THE SUPERHEAVY ELEMENTS

Yasuhisa Abe
Yukawa Institute for Theoretical Physics, Kyoto University,Kyoto 606-8502, Japan

Bertrand Bouriquet
Yukawa Institute for Theoretical Physics, Kyoto University,Kyoto 606-8502, Japan

Grigori Kosenko
Department of Physics, University of Omsk, RU-644077 Omsk, Russia

Abstract The two-step model and the statistical decay code KEWPIE is applied to the $1n$ reactions producing Z=110, Z=111 and Z=112. The experiments are well reproduced with a single scaling parameter to the shell correction energy. Thus, an extension of the recipe is expected to predict reliable values of the residue cross-sections of the elements Z=113 and Z=114 which will be a useful guide for planning of experiments.

Keywords: Superheavy elements, Fusion hindrance, Two-step model

1. Introduction

The production and the identification of so called super-heavy elements(SHE) have been made since the establishment of the nuclear shell model [1, 2]. The studies of shell structure predict an island of stability for those elements around the double closed shell next to ^{208}Pb. Several values for the magic number of Z have been predicted to be 114, 120 and 126, depending on the models used in the calculations, while that for N is 172 and 184 [3–5]. However, the major difficulty in the production of the SHE lies in the lack of knowledge about the reaction mechanism. In the region of the heavy nuclei, the so-called fusion-hindrance is known to exist experimentally [6]. But the mechanism is not yet understood well. As a consequence, it has been difficult to predict reliably which combination of target and projectile is promising and at what bombarding energy the residue cross section is optimum.

W. Greiner et al. (eds.), Structure and Dynamics of Elementary Matter, 387–394.

Let's assume the compound nucleus theory, though it might not be fully justified for such unstable systems. We can express the residue cross-section as

$$\sigma_{res} = \pi \bar{\lambda}^2 \sum (2J+1) P^J_{fusion}(E_{c.m.}).P^J_{surv}(E^*), \quad (1)$$

where J denotes a total spin of the system and $E_{c.m.}$ the incident energy in the center of mass. E^* is equal to $E_{c.m.} + Q$ with the fusion Q-value. As usual P^J_{fusion} and P^J_{surv} are the fusion and the survival probabilities for the spin J, respectively. The latter is obtained by the use of the statistical disintegration code KEWPIE [7], which is newly developed by solving the time-dependent Bateman equation. The former is the most unknown part. But recently, in the present authors *et al.* have proposed the two-step model, where the fusion process is divided into two phases: the approaching phase and the formation phase. They correspond respectively to the passing over the Coulomb barrier and to the shape evolution toward the spherical compound nucleus, starting from the pear-shaped sticking configuration of the incident system. Thus,

$$P^J_{fusion} = P^J_{sticking}.P^J_{formation} \quad (2)$$

Each probability can be obtained by solving dynamics of the process in each step. The amalgamated system is supposed to be exited, which means that the incident kinetic energy has been converted into the thermal energy. This conversion is presumed to start before the top of the Coulomb barrier in view of the deep-inelastic collisions(DIC) [8]. With this presumption we calculate the sticking probability with a Langevin equation including a friction force and its associated fluctuation with a time dependent temperature [9]. The formation probability is calculated by a multi-dimensional Langevin equation for the shape evolution with a constant temperature. The full description of this model has been given in the references [10, 11]. For other approaches, see Refs. [12].

We will first make a brief theoretical description of the two-step model. Then, we will check the ability of the model to reproduce known experimental data. Finally we will present predictions for the residue cross-sections for the reaction $^{70}Zn +^{209} Bi \rightarrow^{278} 113 + 1n$, $^{71}Ga +^{208} Pb \rightarrow^{278} 113 + 1n$ and $^{76}Ge +^{208} Pb \rightarrow^{283} 114 + 1n$ that have not yet been measured, but will be made in a near future.

2. The Two-Step Model for fusion

We briefly recapitulate here the theoretical framework of the two-step model used to calculate the fusion probability.

Approaching phase

To describe the approaching phase, a classical description of the relative motion between colliding nuclei is used with a friction force and its associated fluctuation force. In the original treatment [8, 13], the random force was not implemented, but is properly taken into account in the present model.

$$\begin{cases} \mu \frac{d^2 r}{dt^2} & = & -\frac{\partial V}{\partial r} - \frac{\partial}{\partial r}\frac{\hbar^2 L^2}{2\mu r^2} - C_r(r)\frac{dr}{dt} + \Gamma_r(t) \\ \frac{dL}{dt} & = & \frac{C_\phi(r)}{\mu}[L - \frac{5}{7}L_0] + \Gamma_\phi(t) \end{cases} \quad (3)$$

$$< \Gamma_i(t)\Gamma_j(t) >= 2T(t).C_i(r(t)).\delta_{ij}.\delta(t - t'), \quad (4)$$

where μ is the reduced mass and V is the sum of Coulomb and nuclear attractive potentials. $C_i(r)$ is the radial and the tangential friction, respectively. We notice here that the rolling friction is neglected. L_0 is the incident angular momentum and $\frac{5}{7}L_0$ so-called sliding limit. $\Gamma_i(t)$ denotes a Gaussian random force with zero mean value. The equation 4 is the dissipation-fluctuation theorem. In the case $i = j = \phi$, r^2 factor is necessary for $C_\phi(r)$ in *Eq.* 4. In the present description, the temperature increases as the energy is dissipated by the friction force.

This model, i.e., the surface fiction model (SFM)[8], succeeded in reproducing the main features of the deep inelastic collisions, but the spin of the outgoing fragments appears to require an inclusion of the rolling friction. Moreover the very different strengths between the radial and tangential friction forces may not be physical. And of course, its applicability is limited to the energy region above the Coulomb barrier, due to its classical nature. Nevertheless, it is simple and has no free parameters, so we employ it as a certain effective description of the approaching phase for the realisation of the two-step model [9, 10].

One of the characteristic results of the model is the Gaussian distribution of the radial momentum at the touching point. The origin of this distribution is inferred due to the Gaussian random force associated to the friction force. The width of this distribution is consistent with the temperature determined by the internal energy transferred from the incident kinetic energy. This distribution is used as an initial condition for the subsequent dynamics of the shape evolution from a pear shape toward the spherical shape. Thus, the connection between two phases is called “statistical”, and is completely new, different from adiabatic and diabatic processes.

Shape evolution phase

The pear-shaped system formed by the contact of the nuclei involved in the collision is located outside of the conditional saddle point [14, 15]. As a consequence, we have to solve the shape evolution toward the spherical shape under a friction force associated with a random force. The dynamics of this process is again treated by the multidimensional Langevin equation [16–19].

$$\begin{cases} \frac{dq_i}{dt} &= (m^{-1})_{ij}.p_j \\ \frac{dp_i}{dt} &= -\frac{\partial V^J}{\partial q_i} - \frac{1}{2}\frac{\partial}{\partial q_i}(m^{-1})_{jk}.p_j.p_k \\ & -\gamma_{ij}.(m^{-1})_{jk}.p_k + g_{ij}.\Gamma_j(t) \end{cases} \tag{5}$$

$$< g_{ik}g_{jk} >= \gamma_{ij}.T^J, \tag{6}$$

where V^J denotes the liquid drop model (LDM) potential energy surface plus the rotational energy for spin J [20, 21]. The equation 6 is again the dissipation-fluctuation theorem with temperature T^J, which is assumed constant for the simplicity. The friction tensor is calculated by using the so-called one-body model (OBM) [22], that is, a one-body wall-and-window formula with the two-center parametrisation of nuclear shape.

The ratio between the number of trajectories reaching the spherical configuration and the total number of trajectories gives the formation probability. Mostly, the trajectories do not go over the ridge line but go back to re-separation. We calculate this probability for various initial momenta at a touching point of a given system. Then, the total formation probability is obtain by the convolution of those probabilities with the initial distribution of momenta obtained in the analyses of the approaching phase, according to the statistical connection mentioned above.

3. Comparison with experiment

As the fusion probability is obtained, we can calculate fusion cross sections. Actually, the measured fusion excitation function for $^{58}Fe +^{208} Pb$ is reproduced almost perfectly, which will be discussed elsewhere [23]. This indicates that the fusion probabilities calculated with the two-step model are realistic. Now we calculate residue cross-sections according to *Eq.* 1. The decay calculations are made for each partial cross-section for a given J by the code KEWPIE which has only one free parameter, that is, the reduced friction coefficient (β) used for the Kramers factor [24]. This value is set to $\beta = 5.10^{+21}s^{-1}$ for all the results presented here in consistence with OBM. The essential ambiguities that remain are the masses and shell correction energies of the SHE. To know their influence we have made the calculations employing three different predictions of mass : Møller *et al.* [4], Koura [25], Myers & Swiatecki [26].

First, we will take $^{64}Ni +^{208} Pb \rightarrow^{271} 110 + 1n$ system as an example for which a precise excitation function is recently measured at RIKEN, in addition to the previous GSI data. The calculated peak positions in $E_{c.m.}$ are not far from the data, though there are small differences among those with the different masses, as in seen in Fig. 1. This is mainly due to the slight differences in the macroscopic part of the masses among the predictions, which is consistent with the qualitative explanation by Swiatecki *et al.* [27]. Absolute values of the peak heights of $1n$ cross-sections are larger than the experiment by a few to one order of magnitude for Møller *et al.*'s predictions, for example. Nevertheless, it would be remarkable that the calculations of the present model result in rather good reproductions of the experiments without using any free parameter. For more precise reproduction and thus accurate predictions, however, we might have to introduce a phenomenological scaling factor, for example, a constant reduction of the cross-sections. However, in view of a variety of predictions of the shell correction energy, we insteads introduce a scaling factor for the shell correction energy which is to be used all through each mass predictions employed, keeping its own characteristic tendency predicted over the elements and the isotopes under investigation. In order to reproduce the peak height of the experimental measurements for the reaction $^{64}Ni+^{208}Pb \rightarrow ^{271}110+1n$, the factor is 0.4 , 0.36 and 0.4 for the shell correction energy of Møller & Nix, Koura, Myers & Swiatecki, respectively.

In Fig. 1 is shown the comparisons between the calculations and the experiments [28, 29]. The energy of the experimental data plotted in this figure corresponds to the energy in the center of mass minus the energy loss at the middle of the target. The associated error bars in energy correspond to the

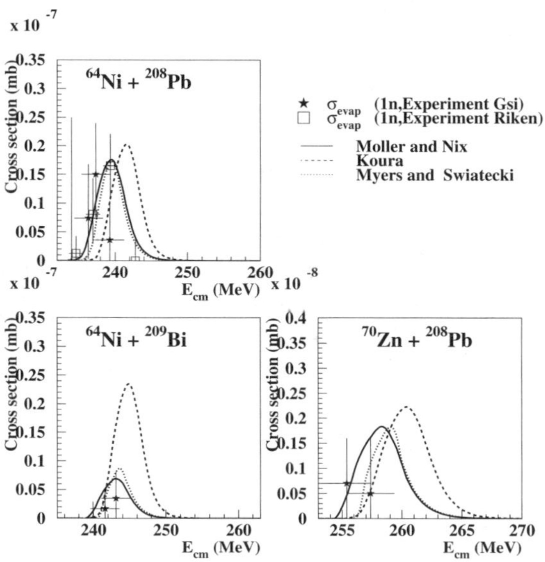

Figure 1. Calculated results with three different mass predictions are compared with the available data.

energy losses in the target. In the upper left panels of Fig. 1 we notice that the position of the peak in $E_{c.m.}$ is well reproduced, especially for the masses of Møller *et al.* and Myers & Swiatecki. For a justification of the scaling factor for of the shell correction we have made calculations for the systems $^{64}Ni + ^{209}Bi \rightarrow ^{272} 111 + 1n$ and $^{70}Zn + ^{208}Pb \rightarrow ^{277} 112 + 1n$, keeping the same value of the factor for the shell correction energy. The lower panels of Fig. 1 show an overall (calculated curves within the experimental error bars) good agreement between the experiments and the theory. Moreover, the global phenomenological decreasing trend of the peak heights of the cross-sections is well reproduced, assuming the Møller *et al.* and Myers & Swiatecki. (Note that the scales of the abscissa and the ordinate are different among the panels.) For those two mass predictions a remarkably good agreement between the experiment and theory is obtained. We should remind that the same model has been already applied to the hot fusion path (^{48}Ca + actinide reactions) [30] and resulted in a fairly good reproduction of the data [10, 11].

4. Predictions of the residue cross section for elements Z=113 and Z=114

As we have reproduced very well the residue cross-sections for the above known systems, we extend the present recipe with the same value of the scaling factor to the systems where measurements are not yet made. We will focus on the yet undiscovered element Z=113 and a new isotope of the Z=114 produced by the cold fusion path.

For the production of the Z=113, the reaction $^{70}Zn + ^{209}Bi \rightarrow ^{278} 113 + 1n$ or the reaction $^{71}Ga + ^{208}Pb \rightarrow ^{278} 113 + 1n$ is under consideration. As we see on the upper panels of Fig. 2, both reactions are predicted to lead to roughly the same residue cross-sections for all the mass predictions employed. This similarity comes from the fact that the reaction $^{71}Ga + ^{208}Pb \rightarrow ^{278} 113 + 1n$ is, in comparison with the reaction $^{70}Zn + ^{209}Bi \rightarrow ^{278} 113 + 1n$, unfavourable with respect to the higher fusion barrier, while it is favourable with respect to the survival probability due to the lower excitation energy (the higher Q-value).

Next, we predict an excitation function of $1n$ reaction for Z=114. This element has been already produced by the hot fusion path but the experiment with the reaction $^{76}Ge + ^{208}Pb \rightarrow ^{283} 114 + 1n$ will be the first measurement of the production of Z=114 by the cold fusion path. The results of the calculation are shown in the lower panel of Fig. 2. The predicted peak height is around a few tenth of pico-barn. To measure the residue with so low cross-sections is a challenge in experiment.

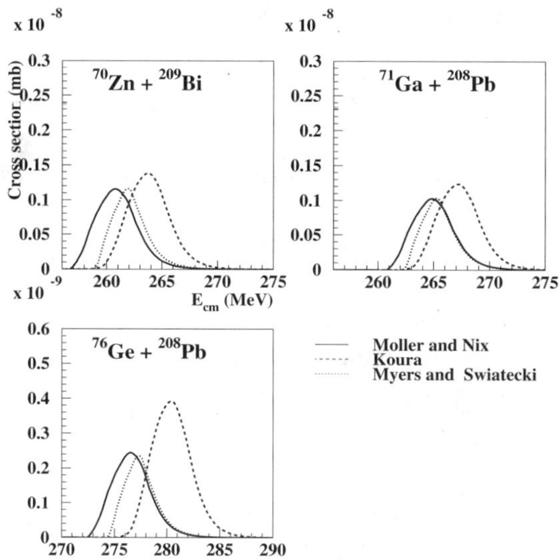

Figure 2. Predictions of the excitation functions of $1n$ reactions for the superheavy elements with Z=113 and 114.

5. Conclusion

As we are able to reproduce rather well both the positions and the absolute values of the peaks of measured $1n$ residue cross-sections by introducing only one scaling parameter for the predicted shell correction energies, the two step model for the fusion reaction appears to be quite promising. And the statistical code KEWPIE is accurate for SHE. Based on the success, we have made the predictions on the $1n$ cross-sections in ${}^{70}Zn +^{209} Bi \rightarrow^{278} 113 + 1n$, ${}^{71}Ga +^{208} Pb \rightarrow^{278} 113 + 1n$ and ${}^{76}Ge +^{208} Pb \rightarrow^{283} 114 + 1n$ without adjusting anything more. This is the first predictions of the excitation function of residue cross section for Z=113 and 114 by the calculation with the dynamical model of reactions. Of course, the ultimate test of the model is the future experiment performed on the above systems. The model is expected to provide a reliable guideline for future experiments for productions of super-heavy elements. Hopefully the present model provides an overview of the most suitable experiments and save a lot of experimental time.

Acknowledgments

The authors thank W.J.Swiatecki for fruitful discussions and K.Morita for the access to the data before publication. One of us (B.B) thanks the Yukawa

Institute for its warm hospitality. The authors acknowledge JSPS for supports (contracts P-01741, 1340278).

References

[1] M.G. Mayer and J.H.D. Jensen, *Elementary Theory of Nuclear Shell Structure*, 1955, Wiley, New York.

[2] see for a review, S. Hofmann and G. Münzenberg, *Rev. Mod. Phys.* **72**, 733 (2000).

[3] A. Sobiczewski, Lectures in the present Institutes and references therein.

[4] P. Møller et al., *Atomic Data and Nuclear Data Tables* **59**, 185 (1995).

[5] T. Buervenich, talk in the present Institutes and references therein.

[6] A.B. Quint et al., *Z. Phys.* **A346**, 119 (1993).

[7] B. Bouriquet et al., Submited to *Comp. Phys. Com.*

[8] D.H.E. Gross and H. Kalinowski, *Phys. Rept.* **45**, 175 (1978).

[9] G. Kosenko et al., *J. Nucl. and Radiochem. Sci.***3**, 19 (2002).

[10] C.W. Shen et al., *Phys. Rev. C66* 061602(R) (2002).

[11] Y. Abe et al., *Prog. Theor. Phys. Suppl.* 146, 104 (2002).

[12] W. Scheid et al., Lectures in the present Institutes, V. Yu Denisov and S. Hofmann, *Phys. Rev.* **C61**, 034606 (2001), V.I.Zagrebav, *Phys. Rev.* **C64**, 034606 (2001).

[13] D. Bangert and H. Freiesleben. *Nucl. Phys.* **A340**, 205 (1980).

[14] W.J. Swiatecki, *Physica Scripta* **24**, 113 (1981).

[15] Y. Abe et al. *Eur. Phys. J.* **A13**, 143 (2002)

[16] Y. Abe et al., *J. de Phys.* **47**, C4-329 (1986).

[17] Y. Abe et al., *Phys. Rept.* **275**, Nos. 2 and 3 (1996).

[18] T. Wada et al., *Phys. Rev. Lett.* **70**, 3538 (1993).

[19] Abe et al. *Phys. Rev.* **E61**, 1125-1133 (2000)

[20] A. Iwamoto et al., *Prog. Theor. Phys.* **55**, 115 (1976).

[21] K. Sato et al., *Z. Phys.* **A290**, 145 (1979).

[22] J. Blocki et al., *Ann. Phys. (NY)* **113**, 330 (1978).

[23] B. Bouriquet et al., to be published.

[24] H.A. Kramers, *Physica* **VII 4**, 284 (1940).

[25] H. Koura et al., *Nuc. Phys.* **A674**, 47 (2000).

[26] W.D. Myers and W.J. Swiatecki, *Ann. of Phys.* 204, 401 (1990).

[27] W.J. Swiatecki *Acta Physica Polonica*, **B 34**, 2049 (2003)

[28] S. Hofmann et al., *Rep. Prog. Phys.* 61, 639(1998).

[29] K. Morita et al *Private communication*

[30] Yu. Oganessian et al., *Phys. Rev. Lett.* **83**, 3154 (1999).

FUSION-FISSION DYNAMICS OF SUPERHEAVY NUCLEI BY THE FLUCTUATION-DISSIPATION MODEL

Y. Aritomo
Flerov Laboratory of Nuclear Reactions, JINR, Dubna, Russia
Department of Physics, University of Tokyo, Hongo, Bunkyo-ku, Tokyo, Japan,
E-mail: aritomo@jinr.ru

M. Ohta
Department of Physics, Konan University, 8-9-1 Okamoto, Kobe, Japan

T. Materna, F. Hanappe
Universite Libre de Bruxelles, 1050 Bruxelles, Belgium

L. Stuttgé
Institut de Recherches Subatomiques, F-67037 Strasbourg Cedex, France

Abstract

We discuss the fusion-fission process in superheavy mass region. By analyzing the mass distribution of fission fragments, we can distinguish between fusion-fission process and quasi-fission process. We investigate the process using the fluctuation-dissipation model and employ the three-dimensional Langevin equation. We find a lot of mass symmetric fission events come from the quasi-fission process in the superheavy mass region. In order to classify the fusion-fission paths and compare with the experimental data directly, we analyze the pre-scission neutron emission in the correlation with the fission fragments. The neutron multiplicity depends on the travelling time of the trajectory. It is useful to investigate the fusion-fission process.

1. Introduction

In the superheavy mass region, the existence of the Island Stability was predicted by the nuclear shell model [1]. The most recent experiments on the

W. Greiner et al. (eds.), Structure and Dynamics of Elementary Matter, 395–405.

synthesis of superheavy elements are approaching to the region and supports this hypothesis. The production of new superheavy elements are reported at Dubna and GSI [2–4]. Riken group succeeded to reproduce Z=110 and 111 in cold fusion reactions and measured the excitation function with high accuracy [5].

On the other hand, in the theoretical study, we are still at the stage of refining the model and investigating the property of undecided parameters. First of all, we have to clear up the mechanism of fusion-fission process. We do not directly discuss the evaporation residue cross section here, because there are many uncertainty parameters and it is very sensitive.

In fusion-fission process, a lot of relevant experimental data become available recently. For example, the mass and total kinetic energy distributions of fission fragments were extensively measured at Dubna [6]. The emission of neutrons in coincidence with fission fragments has been studied by DeMon group [7]. In this paper, using the concomitant data we try to investigate this process dynamically for aiming the more precise determining of the fusion-fission cross section in superheavy mass region.

We treat this process using the fluctuation-dissipation model. We employ the Langevin equation [8–12]. By analyzing the mass distribution of fission fragments, we can distinguish between fusion-fission process (FF) and quasi-fission process (QF).

In order to compare the fusion-fission path with the experimental data directly, we analyze the pre-scission neutron emission in the correlation with fission fragments. The neutron multiplicity depends on the travelling time of the trajectory. It is useful to investigate the fusion-fission process. We introduce the effect of the neutron emission with the Langevin calculation.

The main purpose of the present paper is to elucidate the mechanism of fusion-fission process. In section 2, we explain our framework and model. We discuss the mass distribution of fission fragment in section 3. By analyzing the Langevin trajectories in three dimensional coordinate space, we discuss the fusion-fission mechanism. We present the results for the excitation function of fusion-fission cross section in the reactions ^{48}Ca+^{244}Pu. In section 4, we present the neutron multiplicity in correlation with fission fragments.

2. Model

Usually, the fusion-fission cross section σ_{ER} is expressed as;

$$\sigma_{ER} = \frac{\pi\hbar^2}{2\mu_0 E_{cm}} \sum_{l=0}^{\infty}(2l+1)T_l(E_{cm},l)P_{CN}(E^*,l)W(E^*,l), \qquad (1)$$

where μ_0 denotes the reduced mass in the entrance channel. E_{cm} and E^* denote the incident energy in the center-of-mass frame and the excitation energy of the composite system, respectively. $P_{CN}(E^*, l)$ is the probability of forming a compound nucleus in competition with quasi-fission. $T_l(E_{cm}, l)$ is the capture probability of the lth partial wave. $T_l(E_{cm}, l)$ is calculated with the empirical coupled channel model which is suggested by Zagrebaev [13]. $W(E^*, l)$ denotes the survival probability of compound nuclei during deexitation, which is calculated by statistical model.

Our final goal is to estimate σ_{ER}. However, it is very difficult now because there are many unknown parameters at each stages of the whole reaction and it is very sensitive. Here, we focus our attention to the fusion-fission process which is rather complicated. We discuss on the fusion-fission probability $T_l P_{CN}(1 - W)$.

In order to calculate P_{CN}, we introduce the fluctuation-dissipation model and employ the Langevin equation. We adopt a three-dimensional nuclear deformation space with two-center parametrization [14, 15]. The neck parameter ϵ is fixed to 1.0 in the present calculation. The three collective parameters to be described by the Langevin equation are treated as follows: z_0 (distance between two potential centers), δ (deformation of fragments) and α (mass asymmetry of the colliding partner); $\alpha = (A_1 - A_2)/(A_1 + A_2)$, where A_1 and A_2 denote the mass number of target and projectile, respectively.

The multi-dimensional Langevin equation is given in the following form,

$$\begin{aligned}\frac{dq_i}{dt} &= \left(m^{-1}\right)_{ij} p_j, \\ \frac{dp_i}{dt} &= -\frac{\partial V}{dq_i} - \frac{1}{2}\frac{\partial}{\partial q_i}\left(m^{-1}\right)_{jk} p_j p_k - \gamma_{ij}\left(m^{-1}\right)_{ik} p_k + g_{ij} R_j(t), \end{aligned} \quad (2)$$

where summation over repeated indicates is tacitly assumed. V is the potential energy, m_{ij} and γ_{ij} are the shape-dependent collective inertia and dissipation tensors, respectively. The normalized random force $R_i(t)$ is assumed to be a white noise, *i.e.*, $\langle R_i(t)\rangle$=0 and $\langle R_i(t_1)R_j(t_2)\rangle = 2\delta_{ij}\delta(t_1 - t_2)$. The strength of random force g_{ij} is given by $\gamma_{ij}T = \sum_k g_{ij}g_{jk}$, where T is the temperature of the compound nucleus calculated from the intrinsic energy of the composite system as $E_{int} = aT^2$ with a denoting the level density parameter. Intrinsic energy of the composite system E_{int} is calculated for each trajectory as,

$$E_{int} = E^* - \frac{1}{2}\left(m^{-1}\right)_{ij} p_i p_j - V(q, l, T), \quad (3)$$

where E^* is the excitation energy of the composite system which is given as $E^* = E_{cm} - Q$ with Q denoting the $Q-$value of the reaction. At $t = 0$, each trajectory starts from the contact configuration with the initial velocity in the z_0 direction.

A hydrodynamical inertia tensor is adopted with the Werner-Wheeler approximation for the velocity field, and the wall-and-window one-body dissipation is adopted for the dissipation tensor [16–18].

3. Results

3.1 Mass distribution of fission fragments and fusion-fission cross section

First, we try to reproduce the experimental data of the mass and the total kinetic energy (TKE) distribution of fission fragment [6]. By analyzing the mass distribution, we can distinguish between FF process and QF process. The mechanism of FF and QF are clarified by analyzing the trajectory on the three-dimensional potential energy surface.

Figure 1 show the potential energy surface of liquid drop model with shell correction energy in nuclear deformation space for $^{256}102$, which is calculated using the code [19, 20]. In order to save a great deal of computation time, we use the scaling and employ the coordinate z, which is described in the reference [21, 22]. The position at $z = \alpha = 0$ corresponds to a spherical compound nucleus. Due to the shell structure of Pb and Sn, we can see the valley on the potential surface. The black arrow denotes the injection point. We can see the FF process is dominant and mass symmetric fission events are detected. The white arrows indicate the fusion-fission paths.

However, in the case of ^{48}Ca+^{244}Pu the situation changes. The potential energy surface of $^{292}114$ is shown in Fig. 2. In this case, QF process is dominant. Only few probability enters the spherical region.

At Dubna the experiments on the fission of superheavy nuclei in the reaction ^{48}Ca+^{244}Pu carried out and they present the fusion-fission cross section of compound nuclei which is derived from the mass symmetric fission fragments ($A/2 \pm 20$) [6]. The subsequent important question is whether all of the mass symmetric fission fragments come from the compound nuclei or not. As the final results of the experiments the mass symmetric fission fragments are detected, but there exists two possibilities where it comes from. One is that the mass symmetric fission fragments come from the compound nuclei and the other is that they come from QF.

We try to check them by using three-dimensional trajectory calculation with Langevin equation. In the results, about 90$\sim$99 % of mass symmetric fission fragments come from QF process, which we call *"deep quasi-fission process* (DQF)". We can see it more precisely in the reference [21, 22].

In heavy mass region, due to the large Coulomb repulsion force, potential energy surface has very steep slope in the direction of $+\delta$. Therefore, even if trajectories overcome the fusion barrier, almost all of them flow down to the $+\delta$ direction and can not reach the spherical region. Figure. 3 shows the samples

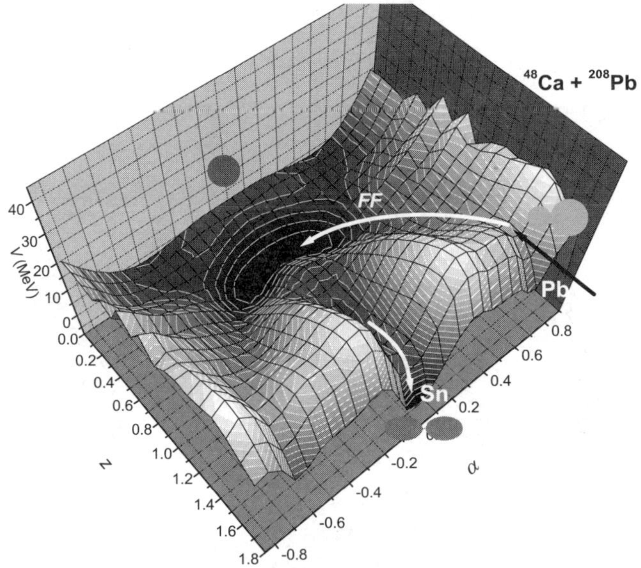

Figure 1. The potential energy surface of liquid drop model with shell correction energy in nuclear deformation space for $^{256}102$. z and α denote the separation between two potential center and the mass asymmetry, respectively. The white arrows indicate the fusion-fission paths.

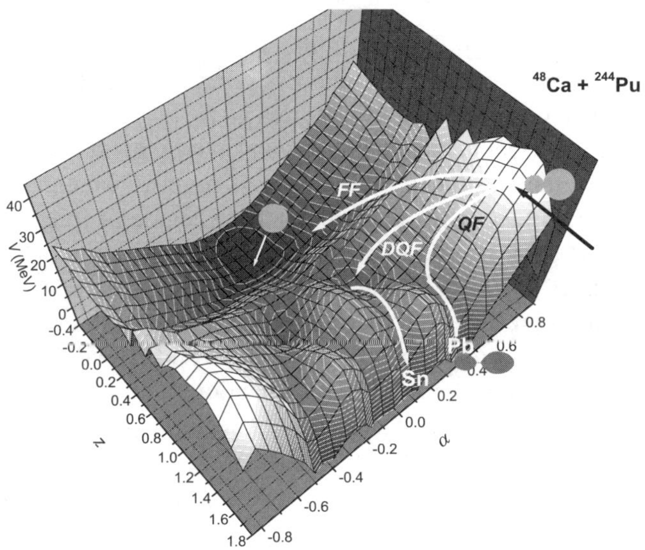

Figure 2. The potential energy surface of $^{292}114$. The QF, DQF and FF are denoted by white arrows.

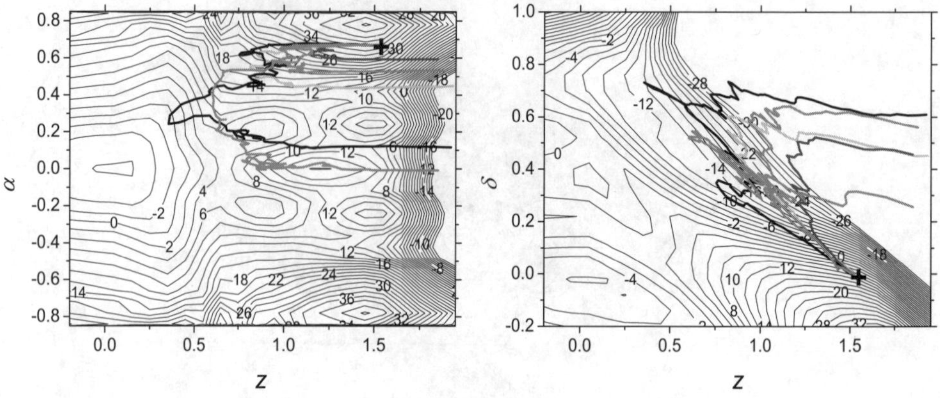

Figure 3. Samples of the trajectory which are projected on $z-\alpha$ plane (a) and $z-\delta$ plane, at $E^*=33$ MeV in the reaction ^{48}Ca+^{244}Pu. The trajectories go to $+\delta$ direction, and can not enter the region of spherical nucleus.

of trajectory which are projected on the $z-\alpha$ and $z-\delta$ plane, in the reaction ^{48}Ca+^{244}Pu at $E^*=33$ MeV. The cross point (+) denotes the touching point of the system. We can see that trajectories go to $+\delta$ direction, and can not enter the region of spherical nucleus. That is reason why the formation of compound nucleus is extremely reduced in heavy mass region. We can say that fusion is hindered by the flow of the trajectory to the δ degree of freedom.

Figure 4 (a) shows mass distribution of fission fragments in the reactions ^{48}Ca+^{244}Pu at E^*= 37 MeV. The mass asymmetric fission events is dominant. The calculation agrees with experimental data. The excitation function of the cross section are shown in Fig. 4 (b). The open and closed diamonds denote the capture cross section σ_{cap} and the cross section $\sigma_{A/2\pm20}$ which derived by the yield of the mass symmetric fission fragments with $\frac{A}{2}\pm 20$ in the experiments, respectively [6]. The calculated $\sigma_{A/2\pm20}$ is denoted by the solid line. We can see good agreement with the experimental data and our calculations. The calculated fusion-fission cross section σ_{CN} is denoted by the dashed line. This cross section σ_{CN} is derived from the trajectory crossing the three-dimensional fusion box. The fusion-fission cross section σ_{CN} is one or two order magnitude smaller than the cross section $\sigma_{A/2\pm20}$. We see that the cross section $\sigma_{A/2\pm20}$ includes the deep quasi-fission events. Such an information is very important to estimate the evaporation residue cross section.

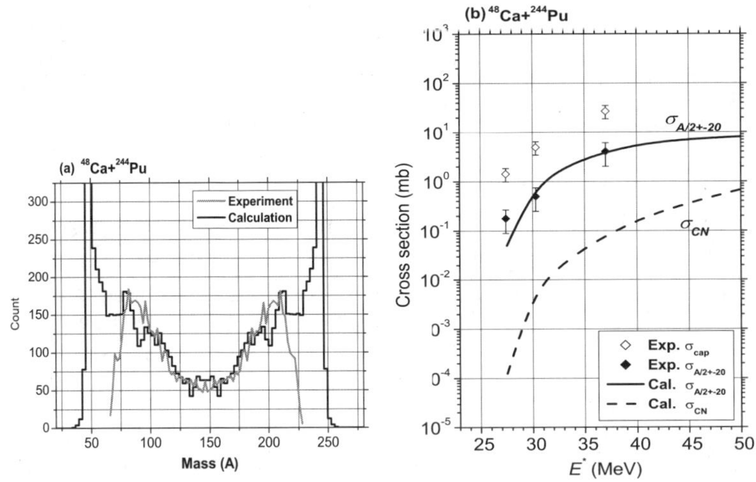

Figure 4. (a) Mass distribution of fission fragments in the reaction ^{48}Ca+^{244}Pu at E^*=37 MeV. The black line and gray line denote the calculation and experiment, respectively. (b) The cross section by the calculation and experiment. Lines and diamonds are given in the text.

3.2 Pre-scission neutron multiplicity

In order to classify the fusion-fission process more precisely, we analyze the pre-scission neutron multiplicity [6, 7]. The neutron multiplicity directly depends on the time scale of trajectory. So using the data, we can investigate the different class of the dynamical process and compare with experiment directly. We introduce the effect of neutron emission in Langevin code [23].

Demon group measured the neutron multiplicity in the reaction ^{58}Ni+^{208}Pb at $E^* = 185.9$ MeV [7]. The neutron multiplicity has two peaks (near 4 and 8 neutron emission), which is shown in Fig. 5 (a). We suppose that the first peak is connected with QF process and second one is connected with FF process. The neutron multiplicity depends on the travelling time of trajectories from the contact point to the scission one. Figure 6 shows the sample trajectories of QF process and FF process on $z-\alpha$-plane, which are denoted by gray and black lines, respectively. On the FF process, the trajectory is trapped in the pocket around spherical region. It takes a long time to stay in the pocket and it has a large chance to emit neutrons. The calculation is shown in Fig. 5 (b). We can see the two peaks which come from QF process and FF process. Actually the travelling time scale of each process from contact point to scission point are

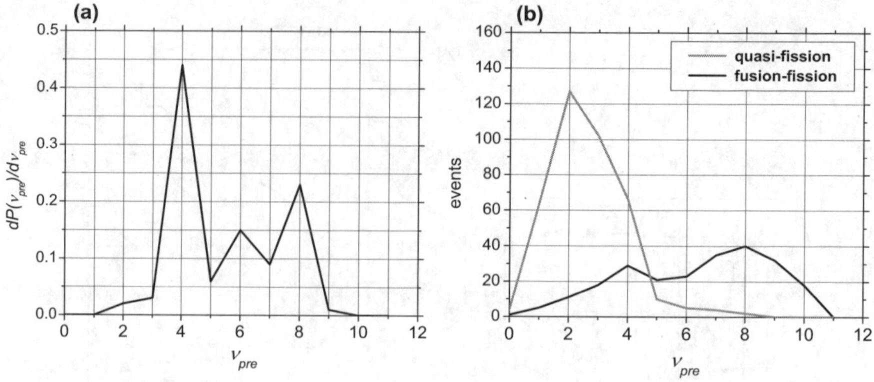

Figure 5. Pre-scission neutron multiplicity in the reaction ^{58}Ni+^{208}Pb at $*E^{*}$=189.5 MeV. (a) The experimental data. (b) Calculation.

clearly distinguished like as Fig. 7 (a). The time scale of FF process is longer than that of QF process.

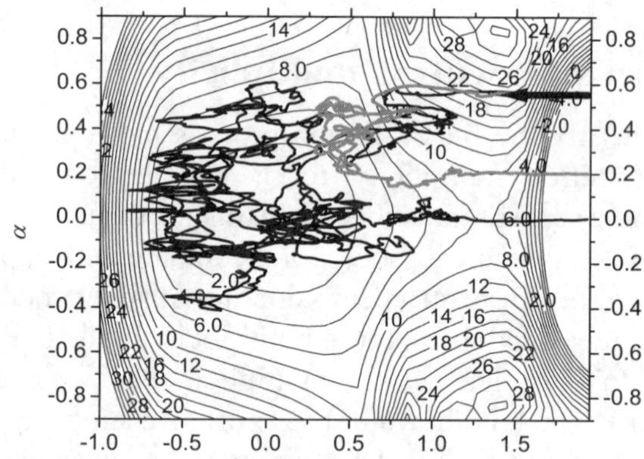

Figure 6. Samples of the trajectory projected on $z-\alpha$ plane at $E^{*} = 189.5$ MeV in the reaction ^{58}Ni+^{208}Pb. The trajectory of QF and FF are denoted by gray and black lines, respectively. The arrow denotes the injection point.

We investigate the neutron multiplicity in the reactions ^{48}Ca+^{208}Pb and ^{48}Ca+^{244}Pu [24]. The calculation and experimental data are shown in Fig. 8,

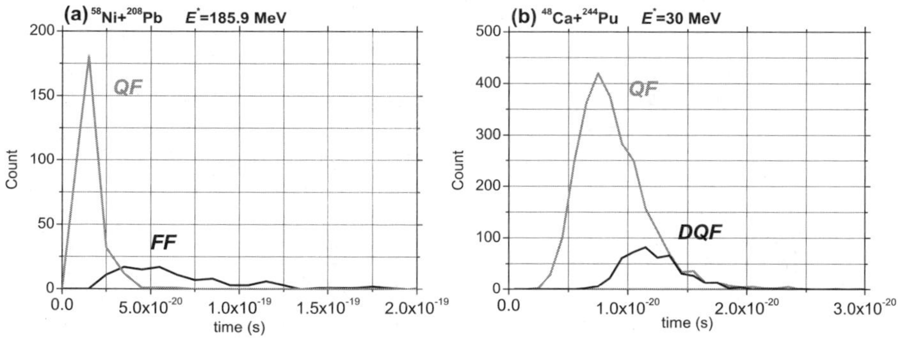

Figure 7. The travelling time from the contact point to the scission point in the reaction (a) ^{58}Ni+^{208}Pb, and (b) ^{48}Ca+^{244}Pu.

which excitation energy is about 40 MeV. In the reaction ^{48}Ca+^{208}Pb, the neutron multiplicity shows the single peak near 2 neutron emission, which comes from FF process. In the three dimensional Langevin calculation, we can see that the trajectory of FF process is dominant and single peak is reproduced.

In the ^{48}Ca+^{244}Pu case, in the Langevin calculation, we found mass symmetric fission events come from not only FF process but also DQF process. The experiment of neutron multiplicity shows the two peak near 2 and 4 neutron emission. Such two peaks are originated by two different fusion-fission mechanism. It looks that the first peak comes from QF process and the second one comes from mainly DQF or FF. However, the calculation can not reproduce the two peaks. We check the travelling time of both case, and they overlap each other, which is shown in Fig. 7 (b). In the calculation, the lifetime of neutron emission is rather short, because the excitation energy is rather low. The time scale of trajectory fluctuation is the same order of the life time of neutron emission. So we can not distinguish QF and DQF process clearly. Also the event of FF process is quit low. We see the only few FF events in the fusion-fission process. We have to investigate the parameters, for example friction tensor, level density parameter and neutron binding energy near di-nucleus configuration etc. We try to use the friction tensor which is derived from liner response theory [25–27]

Our analysis is useful for identifying the detail of the fusion-fission process in connection with the neutron multiplicity. The time scale of fusion-fission process depends on the strength of friction tensor. Also the time scale of neutron emission depends on the value of level density parameter and neutron binding energy at each nuclear shapes. The measured neutron multiplicity can be used for the determination of such values.

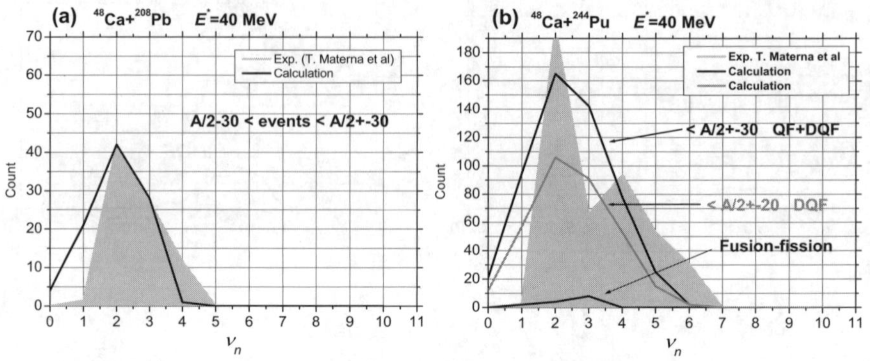

Figure 8. Pre-scission neutron multiplicity in the reaction (a) ^{48}Ca+^{208}Pb and (b) ^{48}Ca+^{244}Pu at E^*=40 MeV. Experiments are carried out by T. Materna. The absolute value of calculations are normalized.

References

[1] W.D. Myers and W.J. Swiatecki, Nucl. Phys. 81 (1966) 1; A. Sobiczewski *et. al.*, Phys. Lett. 22 (1966), 500.

[2] Yu.Ts. Oganessian et al., Nature **400** (1999), 242 ; Yu.Ts. Oganessian et al., Phys. Rev. Lett. **83**(1999), 3154.

[3] Yu.Ts. Oganessian et al., Phys. Rev. C **63** (2001), 011301(R).

[4] S. Hofmann and G. Munzenberg, Rev. Mod. Phys. 72 (2000) 733; S. Hofmann et al., Eur. Phys. J. A 14 (2002) 147.

[5] K. Morita, Proceedings of the VIII International Conference on Nucleus-Nucleus Collisions (NN2003), Moscow, Russia, 2003, to be published.

[6] M.G. Itkis et al., Proceeding of Fusion Dynamics at the Extremes (World Scientific, Singapore, 2001) p93.

[7] L. Donadiile et al, Nucl Phys, A **656** (1999), 259.

[8] C.E. Aguiar *et al.*, Nucl. Phys. A **491** (1989), 2031; C.E. Aguiar et al. Nucl. Phys. A **514** (1990), 205.

[9] T. Wada, Y. Abe and N. Carjan, Phys. Rev. Lett. **70**, 3538 (1993).

[10] Y. Aritomo, T. Wada, M. Ohta and Y. Abe, Phys. Rev. C **55** (1997), R1011; T. Wada, T. Tokuda, K. Okazaki, M. Ohta, Y. Aritomo and Y. Abe, Proceeding of DANF98, Slovakia 1998 (World Seientific)pp. 77; Y. Aritomo, T. Wada, M. Ohta and Y. Abe, Phys. Rev. C **59** (1999), 796;

[11] T. Tokuda, T. Wada and M. Ohta, Prog. Theor. Phys. **101** (1999), 607.

[12] C. Schmitt, J. Bartel, A. Surowiec and K. Pomorski, Acta Phy. Polon. B **32** (2001), 841

[13] V.I. Zagrebaev, Phys. Rev. C **64** (2001) ,034606.

[14] J. Maruhn and W. Greiner, Z. Phys. **251** (1972), 431.

[15] K. Sato, A. Iwamoto, K. Harada, S. Yamaji, and S. Yoshida, Z. Phys. A **288** (1978), 383.

[16] J. Blocki, Y. Boneh, J.R. Nix, J. Randrup, M. Robel, A.J. Sierk and W.J. Swiatecki, Ann. Phys. **113** (1978), 330.

[17] J.R. Nix and A.J. Sierk, Nucl. Phys. A **428** (1984), 161c.

[18] H. Feldmeier, Rep. Prog. Phys. **50** (1987), 915.

[19] S. Suekane, A. Iwamoto, S. Yamaji and K. Harada, JAERI-memo, 5918 (1974).

[20] A. Iwamoto, S. Yamaji, S. Suekane and K. Harada, Prog. Theor. Phys. **55** (1976), 115.

[21] Y. Aritomo, Proceedings of Int. Conf. on Nuclear Physics at Border Lines, Lipari, Italy, 2001 (World Scientific, Singapore, 2002) p38.

[22] Y. Aritomo, Proceedings of International Conference on Exotic Nuclei, Baikal, Russia, 2001 (World Scientific, Singapore, 2002)

[23] N.D. Mavlitov, P. Frobrich and SI.I. Gonchar, Z. Phys. A **342** (1992), 195.

[24] T. Materna et. al, Proceedings of the VIII International Conference on Nucleus-Nucleus Collisions (NN2003), Moscow, Russia, 2003, to be published.

[25] H. Hofmann, Phys. Rev. **284** (1997), 137.

[26] S. Yamaji, F.A. Ivanyuk, H. Hofmann, Nucl. Phys. A **612** (1997), 1.

[27] F.A. Ivanyuk, H. Hofmann, V.V. Pashkevich, S. Yamaji, Phys. Rev. C **55** (1997), 1730.

SELF-CONSISTENT MEAN-FIELD MODELS AND THEIR APPLICATION TO SUPERHEAVY NUCLEI

T. J. Bürvenich
Theoretical Division, Los Alamos National Laboratory, Los Alamos, New Mexico 87545, USA,
E-mail: tbuerven@lanl.gov

P.–G. Reinhard
Institut für Theoretische Physik II, Universität Erlangen–Nürnberg, Staudtstrasse 7, D–91058 Erlangen, Germany

J. A. Maruhn and W. Greiner
Institut für Theoretische Physik der J. W. Goethe Universität, D-60054, Frankfurt am Main, Germany

Abstract We review some predictions of modern self-consistent mean-field models concerning the shell structure and fission barriers of superheavy elements. Several parametrizations for each mean-field model are chosen to obtain a measure for the uncertainties in the extrapolations. The various forces agree on global trends but differ in detailed predictions of magic numbers and inner fission barrier heights.

1. Introduction

The quest for superheavy elements (SHE) has inspired nuclear structure physics for a long time. These elements can be defined as those nuclear species for which the liquid drop barrier vanishes, i.e., they owe their mere existence to quantum mechanical shell effects. Recent experimental progress at GSI [1] and FLNR in Dubna [2] leading to the synthesis of elements $Z = 112, 114$ and $Z = 116$ has triggered new theoretical work.

Selfconsistent mean-field models provide powerful tools for the study of SHE. Starting with an effective interaction – or, in modern terms, an energy functional – they do not contain any prejudice concerning the shape or structure of superheavy elements. They have reached a high quality for the description of nuclear ground-state properties of nuclei close to and away from the valley of β stability. Since (yet) their parameters cannot be determined from low-

W. Greiner et al. (eds.), Structure and Dynamics of Elementary Matter, 407–414.

momentum QCD, an adjustment to nuclear ground-state observables becomes necessary. The adjustment of these forces is biased and their predictive power is related to the fitting strategy. When extrapolating with these models and forces, it is thus still mandatory to consider a selection of them.

2. Framework

The models chosen for these studies are the Relativistic Mean-Field (RMF) model (see [4] for a review) and the Skyrme-Hartree-Fock (SHF) model [5].

As mentioned in the Introduction, it is important to take into account different forces since yet a *best force* does not exist. Each force has its own strongholds and shortcomings, and only a selection of forces can help to illuminate the question which features are model-like and which are only related to a special force. Among the many parametrizations on the market, we take a selection that proved to be successful in former studies [6–9, 11].

We consider 4 Skyrme-forces and 2 forces for the RMF model to obtain an overview of the generic features of the models. We use the Skyrme interactions SkP [12], SLy6 [13], SkI3, and SkI4 [14]. In the relativistic calculations, we employ NL3 [15] and NL-Z2 [19] parametrizations (for the analysis of shell structure, we also consider the relativistic forces NL-Z [16] and NL-SH [17] as well as the Skyrme force SkM* [18]). Please consult Ref. [3] for a characterization of these parameterizations. Pairing is treated within the BCS approximation using a zero-range delta force with the strength adjusted for each mean-field parameterization as described in Ref. [20]. The center-of-mass correction is performed as in the fit of the various forces. The actual recipe used in the fit influences the surface tension of the effective interactions which becomes visible at large deformations along the fission barrier, see [21] for a detailed discussion. The coupled mean-field equations for both SHF and RMF models are solved on a mesh in coordinate space using a Fourier representation of the derivatives and the damped gradient iteration method [22]. Calculations within both models share the same basic numerical routines which allows a direct comparison of the results. For an estimate of the remaining uncertainties of the deformation energy see [8].

3. Shell structure

Since superheavy elements get stabilized solely by shell effects, it is interesting to try to identify – in a given model – the magic numbers and doubly-magic candidates. To this end, K. Rutz et al. [6] have performed a scan of a large region of SHE with various mean-field forces. Figure 1 shows as grey-scale plots the two-proton and two-neutron shell gaps, defined as

$$\delta_{2\mathrm{p}}(N, Z) \;=\; 2B(N, Z) - B(N, Z-2) - B(N, Z+2)$$

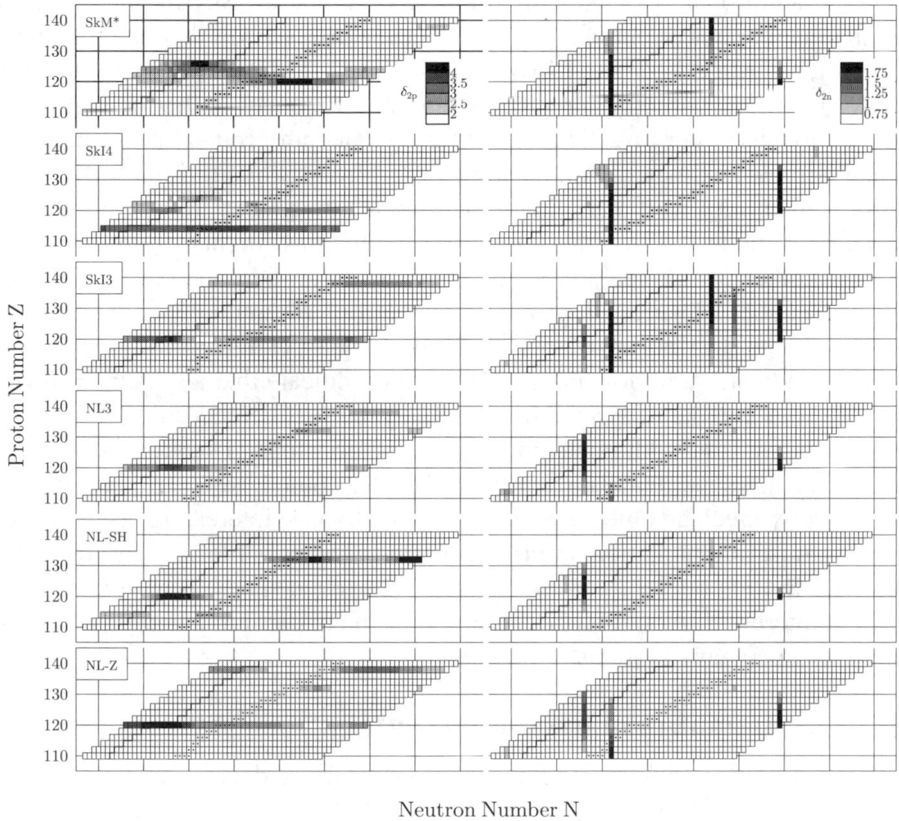

Figure 1. Grey scale plots of proton gaps (left column) and neutron gaps (right column) in the N-Z plane for spherical calculations with the forces as indicated. The assignment of scales differs for protons and neutrons, see the uppermost boxes where the scales are indicated in units of MeV. Nuclei that are stable with respect to β decay and the two-proton dripline are emphasized.

$$\begin{aligned} &= S_{2\mathrm{p}}(N,Z) - S_{2\mathrm{p}}(Z,Z+2), \\ \delta_{2\mathrm{n}}(N,Z) &= 2B(N,Z) - B(N-2,Z) - B(N+2,Z) \\ &= S_{2\mathrm{n}}(N,Z) - S_{2\mathrm{n}}(N+2,Z), \end{aligned}$$

where $S_{2\mathrm{p}}$ and $S_{2\mathrm{n}}$ denote the two-proton and two-neutron separation energies, respectively. Crossing a shell closure for a sort of nuclons, the separation energy drops down resulting in a peak of the shell gap. Thus, large shell gaps indicate large shell stabilization and hence magic numbers. From figure 1 it becomes apparant that the prediction of magic numbers is both model and force dependent. The relativistic models NL-Z, NL3 and NL-SH predict $^{292}120_{172}$ to be the doubly-magic superheavy nucleus. For NL-Z, at smaller proton numbers, an additional $N = 184$ shell appears. While all Skyrme forces agree on the magic neutron number $N = 184$, they disagree on the magic number for protons, namely $Z = 114$ (SkI4), $Z = 120$ (SkI3, SkM*), or $Z = 126$ (SkM*). These conflicting predictions hint at differences in the shell structure of the models. Later investigations revealed that spin-orbit splittings in heavy nuclei are overestimated with these Skyrme-forces, while the spin-orbit force works rather very well and without additional adjustments for the relativistic mean-field model. Thus, the relativistic predictions might be more realistic [19]. Another typical feature of selfconsistent mean-field models is visible in the figure: shell closures for protons and neutrons are mutually dependent.

The charge distribution of the $^{292}120_{172}$ nucleus, calulated with mean-field models, indicates a hollow inside. This could indicate that a system with 120 protons and 180 neutrons might essentially be a fullerene consisting of 60 α-particles and one additional binding neutron per alpha. The protons and neutrons of such a superheavy nucleus would be distributed over 60 α particles and 60 neutrons. Such an object could be expected to have interesting decay modes such as multifragmentation, multi-α emission. The possible condensation of α particles in light nuclei (in ground states and excited states) is a topic of great current interest. It would be fascinating if such condensation could occur also in these superheavy systems.

4. Systematics of fission barriers

For SHE, spontanous fission becomes a competing decay mode besides radioactive decay. Thus, the structure of fission barriers is of paramount interest. A first exploration of asymmetric shapes within self-consistent models, both SHF and RMF, was presented in [8].

In this contribution to the proceedings, we will focus especially on the predictions for the heights of the axial fission barriers of a selection of superheavy nuclei and their deformation properties. We will calculate them with two self-consistent mean-field models employing the parameterizations listed above. For a complete analysis that also includes results for actinides see [3].

The fission barriers for the RMF force NL-Z2 and the Skyrme force SLy6 for the superheavy systems under consideration are shown in figure 2. The global trends are common for both forces (and also the others employed in this study): There is a gradual transition from well-deformed nuclei with $\beta_2 \approx 0.3$ around the deformed $Z = 108$ and $N = 162$ shell closures to spherical shapes approaching $N = 184$. Note that earlier studies suggest that the neutron number is more important than the proton number to determine the ground-state shape. Intermediate systems around $Z = 114$, $N = 174$ have two distinct prolate and oblate minima at small deformation. While the ground-state deformation moves to smaller values, also the saddle point is shifted to smaller deformations. This

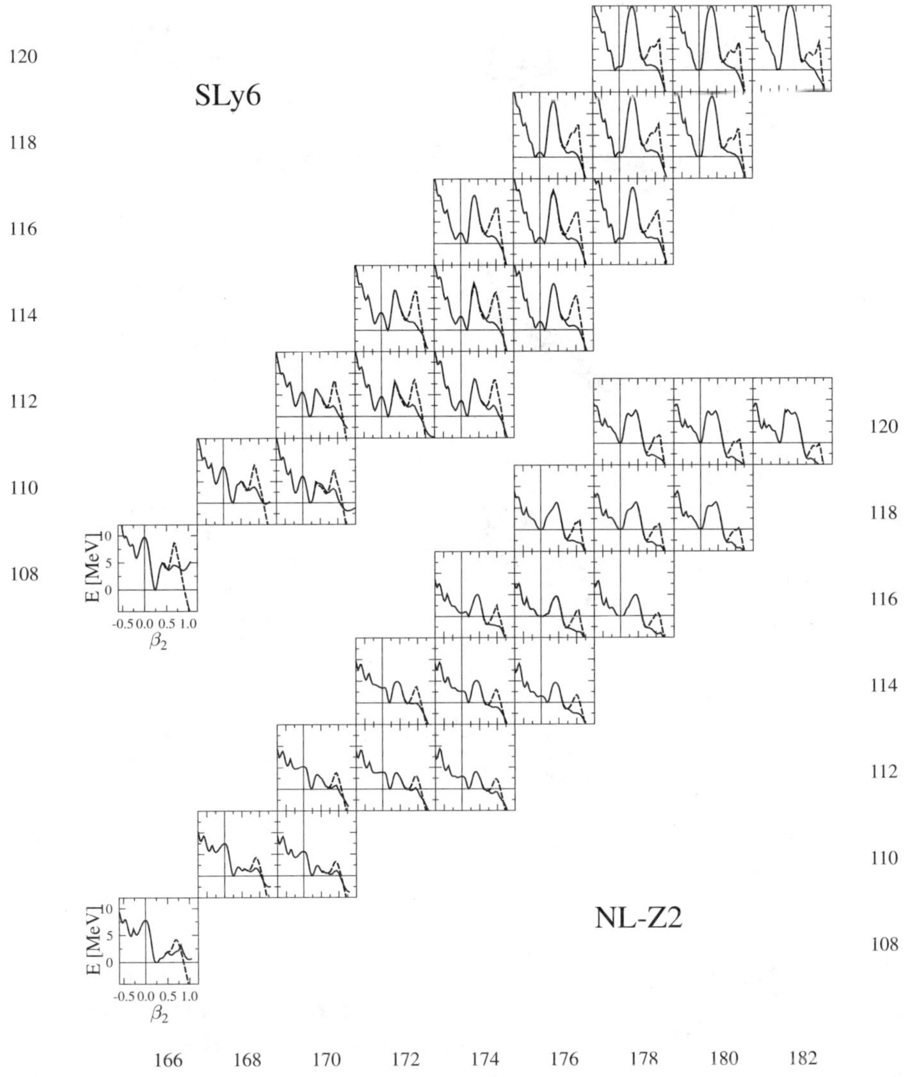

Figure 2. Axial fission barriers for the Skyrme force SLy6 (top) and the relativistic force NL-Z2 (bottom). Solid (dashed) lines denote the reflection-asymmetric (reflection-symmetric) path.

means that the width of the fission barrier is not necessarily larger for nuclei with spherical shell closures. The transitional nuclei in between belong to a regime of low fission barriers. This is consistent with the current (still sparse) experimental knowledge. The recent data from Dubna interpreted as the α-decay chain of $^{292}116$ indicate that fission is the preferred decay channel of

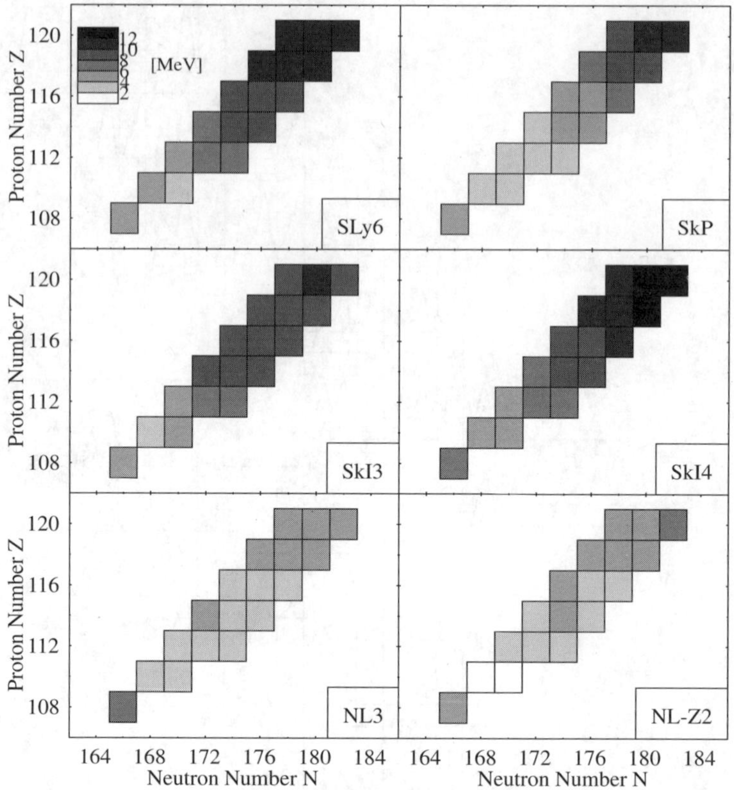

Figure 3. The height of the symmetric (inner) barrier calculated in axial symmetry for the forces as indicated.

$^{280}110_{170}$, but not for the heavier nuclides in this chain. Above $Z = 108$, the static fission path switches from symmetric to asymmetric fission. For most of the nuclei above $Z \geq 110$ considered here, reflection-asymmetric shape degrees of freedom remove completely the second barrier that is well-known from actinide nuclei, leading to a single-humped fission barrier only.

The axial fission barriers heights for all nuclei and forces under consideration are shown in figure 3. The striking result from this analysis is that there is a huge discrepancy in the prediction of barrier heights which is essentially a property of the model, not of a special force. The SHF model predicts barrier heights which are up to a factor of 2 larger than the corresponding predictions with the RMF model. There are differences in detail depending on the force, but the global trends are basically determined by the model, RMF or SHF. Our results confirm the earlier finding [10, 8] that barriers from RMF are systematically lower than those predicted by SHF. This effect is even more pronounced for the superheavy systems investigated here. It can be expected that the RMF gives lifetimes smaller by many orders of magnitude compared to SHF. These conclusions might, however, depend on the adjustment procedures for these forces. This needs still to be checked.

To make quantitative predictions, additional degrees of freedom like triaxiality get important which could change the picture again drastically. But it can be expected that the systematic differences among the models persist when adding non-axially symmetric degrees of freedom. Therefore, before performing such demanding calculations systematically, the reasons for the differences among the models will have to be understood.

5. Conclusions

For superheavy nuclei all models and forces agree on the systematic trends concerning the fission barriers, ground state deformations, and fission modes. There are differences in detail. For larger nuclei, there emerge large and systematic discrepancies between SHF and RMF concerning the barrier heights, reaching a factor two for $Z = 120$. The factor two in fission barriers around $Z = 120$ means in absolute numbers that SHF fission barriers are about 5 MeV larger and this amounts to many orders of magnitude longer fission lifetimes. Moreover, the SHF forces employed here predict also larger barriers than the more phenomenological mic-mac models. The reason for the systematic difference between SHF and RMF have yet to be found out. We suspect that the difference is caused by a different shell structure in the models, which is supported by the different predictions of magic numbers. This point deserves more investigation.

The authors would like to thank M. Bender, D. G. Madland, Y. T. Oganessian, and V. I. Zagrebaev for inspiring discussions as well as S. Schramm for help with coding problems. This work was supported in part by Bundesministerium für Bildung und Forschung (BMBF), Project No. 06 ER 808.

References

[1] S. Hofmann, Rep. Prog. Phys. **61** (1998) 639; S. Hofmann and G. Münzenberg, Rev. Mod. Phys. **72** (2000) 733; P. Armbruster, Ann. Rev. Nucl. Part. Sci. **50** (2000) 411.

[2] Yu. Ts. Oganessian it et al, Eur. Phys. J. A **5** (1999) 68; Phys. Rev. Lett. **83** (1999) 3154; Nature **400** (1999) 209; Phys. Rev. C **62** (2000) 041604(R); Phys. Rev. C **63** (2001) 011301(R).

[3] T. Bürvenich, M. Bender, P.-G. Reinhard and J. A. Maruhn, accepted for publication in Phys. Rev. C, nucl-th/0302056

[4] P.–G. Reinhard, Rep. Prog. Phys. **52**, (1989) 439.

[5] P. Quentin and H. Flocard Ann. Rev. Nucl. Part. Sci. 21 (1978) 523

[6] K. Rutz, M. Bender, T. Bürvenich, T. Schilling, P.–G. Reinhard, J. A. Maruhn and W. Greiner, Phys. Rev. C **56** (1997) 238

[7] T. Bürvenich, K. Rutz, M. Bender, P.-G. Reinhard, J. A . Maruhn, and W. Greiner, Eur. Phys. J. A **3** (1998) 139

[8] M. Bender, K. Rutz, P.-G. Reinhard, J. A. Maruhn, and W. Greiner, Phys. Rev. C **58** (1998) 2126

[9] M. Bender, K. Rutz, P.-G. Reinhard, J. A. Maruhn, and W. Greiner, Phys.Rev. C **60** (1999) 34304

[10] K. Rutz, J. A. Maruhn, P.–G. Reinhard, W. Greiner, Nucl. Phys. **A590** (1995) 680

[11] A.T. Kruppa, M. Bender, W. Nazarewicz, P.-G. Reinhard, T. Vertse, S. Cwiok, Phys.Rev C **61** (2000) 034313; M. Bender, W. Nazarewicz, P.-G. Reinhard, Phys.Lett. B **515** (2001) 42

[12] J. Dobaczewski, H. Flocard, and J. Treiner, Nucl. Phys. **A422** (1984) 103.

[13] E. Chabanat, P. Bonche, P. Haensel, J. Meyer, and R. Schaeffer, Nucl. Phys. **A635** (1998) 231, Nucl. Phys. **A643** (1998) 441(E).

[14] P.-G. Reinhard and H. Flocard, Nucl. Phys. **A584** (1995) 467.

[15] G. A. Lalazissis, J. König, and P. Ring, Phys. Rev. C **55** (1997) 540.

[16] M. Rufa, P.–G. Reinhard, J. A. Maruhn, W. Greiner und R. Strayer, Phys. Rev. **C38** (1988) 390

[17] M. M. Sharma, P. Ring, Phys. Rev. **C45** (1992) 2514

[18] J. Bartel, P. Quentin, M. Brack, C. Guet und H.–B. Hakansson, Nucl. Phys. **A386** (1982) 79

[19] M. Bender, K. Rutz, P.–G. Reinhard, J. A. Maruhn, W. Greiner, Phys. Rev. C **60** (1999) 034304.

[20] M. Bender, K. Rutz, P.–G. Reinhard, J. A. Maruhn, Eur. Phys. J. **A8** (2000) 59.

[21] M. Bender, K. Rutz, P.-G. Reinhard, J. A. Maruhn, Eur. Phys. J. A7 (2000) 467.

[22] V. Blum, G. Lauritsch, J. A. Maruhn, P.–G. Reinhard, J. Comp. Phys. **100**, (1992) 364.

INFLUENCE OF DEFORMATION ON THE SYNTHESIS OF SUPERHEAVY ELEMENTS

R. A. Gherghescu
Horia Hulubei National Institute for Physics and Nuclear Engineering, RO-76900, Bucharest-Magurele, Romania, E-mail: rgherg@ifin.nipne.ro

W. Greiner
Institut für Theoretische Physik der J. W. Goethe Universität, D-90054, Frankfurt am Main, Germany

Abstract A deformed two center shell model specialized for fusion and fission phenomena is presented. A neck dependent potential is introduced as a link between the two non-overlapped partners in fission configuration. The influence of the necking region and the two spheroidal deformations of the fragments on the shell correction energy is studied. Applications are presented for fusion processes of light and superheavy synthesis.

1. Introduction

Two-center shell models have been introduced in nuclear physics in Frankfurt by Greiner and co-workers [1–3]. The model presented in this work is an improved version which takes into account spheroidal deformations of the partners. As a new feature, a potential which describes the neck between two fission fragments is introduced. Such a model allows the study of microsopic changes beyond the touching point, where the two level schemes overlap. Spin-orbit operators are calculated using the deformed two center potential, hence it becomes a shape dependent operator. Finally the level diagram is obtained as a function of the distance between the two centers, the semiaxes ratios of the two spheroids and the neck radius. The level scheme is an input for the Strutinsky method to calculate the shell corrections. Variations of the shell effects with the above free parameters are calculated for light nuclei and superheavy elements synthesis within fusion like configuration. Influence of different neck radii on the shell corrections is also presented for the fission processes.

W. Greiner et al. (eds.), Structure and Dynamics of Elementary Matter, 415–422.

2. Deformed two-center shell model

A fission like configuration with two spheroidally deformed fragments and a necking region in-between is presented in Fig. 1. The free geometrical

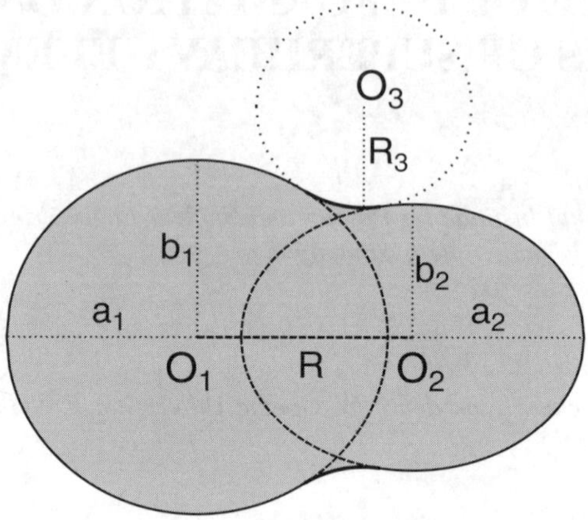

Figure 1. Fission like shape configuration for the deformed two-center shell model.

parameters of the model are: the distance between the centers R, the two pairs of semiaxes (a_1, b_1) and (a_2, b_2) and the radius R_3 of a sphere which generates the necking region when rolling around the symmetry axis. The nuclear surface must be equipotential. An oscillator type potential which describes such a binary system configuration reads:

$$V_{DTCSM}(\rho, z) = \begin{cases} V_1(\rho, z) & , (\rho, z) \in v_1 \\ V_g(\rho, z) & , (\rho, z) \in v_{g1} + v_{g2} \\ V_2(\rho, z) & , (\rho, z) \in v_2 \end{cases} \quad (1)$$

where

$$V_1(\rho, z) = \frac{1}{2} m_0 \omega_{\rho_1}^2 \rho^2 + \frac{1}{2} m_0 \omega_{z1}^2 (z + z_1)^2$$

$$V_g(\rho, z) = \begin{cases} V_{g1} = 2V_0 - [\frac{1}{2} m_0 \omega_g^2 (\rho - \rho_3)^2 + \frac{1}{2} m_0 \omega_g^2 (z - z_3)^2], (\rho, z) \in v_{g1} \\ V_{g2} = V_0 \quad , (\rho, z) \in v_{g2} \end{cases}$$

$$V_2(\rho, z) = \frac{1}{2} m_0 \omega_{\rho_2}^2 \rho^2 + \frac{1}{2} m_0 \omega_{z_2}^2 (z - z_2)^2 \quad (2)$$

where V_1 and V_2 are active in the space regions v_1 and v_2 and the neck potential is active in v_g. These regions are revealed by the equations:

$$V_1 = V_2$$

$$\begin{aligned} V_1 &= V_{g1} \\ V_2 &= V_g \end{aligned} \tag{3}$$

where the second equation defines v_{g1} and v_{g2} is the region between the inner surface of the fragments and the neck surface.

The angular momentum dependent operators are calculated using the above deformed oscillator potential. Their expression is a function of the fragment deformation and the neck geometry:

$$\begin{aligned} \mathbf{\Omega}^+(v_1) &= -e^{i\varphi}\left[\frac{\partial V_1(\rho,z)}{\partial\rho}\frac{\partial}{\partial z} - \frac{\partial V_1(\rho,z)}{\partial z}\frac{\partial}{\partial\rho} - \frac{i}{\rho}\frac{\partial V_1(\rho,z)}{\partial z}\frac{\partial}{\partial\varphi}\right] \\ &= -e^{i\varphi}\left[m_0\omega_{\rho_1}^2\rho\frac{\partial}{\partial z} - m_0\omega_{z_1}^2(z+z_1)\frac{\partial}{\partial\rho} - \frac{i}{\rho}m_0\omega_{z_1}^2(z+z_1)\frac{\partial}{\partial\varphi}\right] \end{aligned}$$

$$\begin{aligned} \mathbf{\Omega}^-(v_1) &= e^{-i\varphi}\left[\frac{\partial V_1(\rho,z)}{\partial\rho}\frac{\partial}{\partial z} - \frac{\partial V_1(\rho,z)}{\partial z}\frac{\partial}{\partial\rho} + \frac{i}{\rho}\frac{\partial V_1(\rho,z)}{\partial z}\frac{\partial}{\partial\varphi}\right] \\ &= e^{-i\varphi}\left[m_0\omega_{\rho_1}^2\rho\frac{\partial}{\partial z} - m_0\omega_{z_1}^2(z+z_1)\frac{\partial}{\partial\rho} + \frac{i}{\rho}m_0\omega_{z_1}^2(z+z_1)\frac{\partial}{\partial\varphi}\right] \end{aligned}$$

$$\begin{aligned} \mathbf{\Omega}_z(v_1) &= -\frac{i}{\rho}\frac{\partial V_1}{\partial\rho}\frac{\partial}{\partial\varphi} \\ &= -im_0\omega_{\rho_1}^2\frac{\partial}{\partial\varphi} \end{aligned} \tag{4}$$

with

$$\mathbf{\Omega}\mathbf{s}(v_1) = \frac{1}{2}(\mathbf{\Omega}^+(v_1)\mathbf{s}^- + \mathbf{\Omega}^-\mathbf{s}^+) + \mathbf{\Omega}_z(v_1)\mathbf{s}_z \tag{5}$$

The same applies for (v_2) and (v_g) spin-orbit operators terms, and the expressions are:

$$\begin{aligned} \mathbf{\Omega}^+(v_2) &= -e^{i\varphi}\left[m_0\omega_{\rho_2}^2\rho\frac{\partial}{\partial z} - m_0\omega_{z_2}^2(z-z_2)\frac{\partial}{\partial\rho} - \frac{i}{\rho}m_0\omega_{z_2}^2(z-z_2)\frac{\partial}{\partial\varphi}\right] \\ \mathbf{\Omega}^-(v_2) &= e^{-i\varphi}\left[m_0\omega_{\rho_2}^2\rho\frac{\partial}{\partial z} - m_0\omega_{z_2}^2(z-z_2)\frac{\partial}{\partial\rho} + \frac{i}{\rho}m_0\omega_{z_2}^2(z-z_2)\frac{\partial}{\partial\varphi}\right] \\ \mathbf{\Omega}_z(v_2) &= -im_0\omega_{\rho_2}^2\frac{\partial}{\partial\varphi} \end{aligned} \tag{6}$$

and:

$$\mathbf{\Omega}^+(v_g) = -e^{i\varphi}\left[m_0\omega_g^2(\rho-\rho_3)\frac{\partial}{\partial z} - m_0\omega_g^2(z-z_3)\frac{\partial}{\partial\rho} - \frac{i}{\rho}m_0\omega_g^2(z-z_3)\frac{\partial}{\partial\varphi}\right]$$

$$\mathbf{\Omega}^{-}(v_g) = e^{-i\varphi}\left[m_0\omega_g^2(\rho-\rho_3)\frac{\partial}{\partial z} - m_0\omega_g^2(z-z_3)\frac{\partial}{\partial \rho} + \frac{i}{\rho}m_0\omega_g^2(z-z_3)\frac{\partial}{\partial \varphi}\right]$$

$$\mathbf{\Omega}_z(v_g) = im_0\omega_g^2\frac{\rho-\rho_3}{\rho}\frac{\partial}{\partial\varphi} \qquad (7)$$

Finally the matrix elements are calculated as :

$$\begin{aligned}\langle i|DTCSM|j\rangle &= E_{osc}^{(d)}(n_\rho, |m|, \mu) + \langle i|\Delta V_1|j\rangle + \langle i|\Delta V_2|j\rangle \\ &+\langle i|V_g|j\rangle + \langle i|V_{\mathbf{\Omega s}}|j\rangle + \langle i|V_{\mathbf{\Omega}^2}|j\rangle \qquad (8)\end{aligned}$$

3. Results

Five level schemes are drawn for different neck radii in figure 2.

The plots are drawn for the fission of the superheavy element along the channel $^{306}122 \rightarrow ^{198}W + ^{108}Cd$. Level density disturbances are more evident in the high energy level zone. At intermediary stages of splitting, for $(R-R_i)/(R_f-R_i)$ around 0.4 fm there is an increase in the level energy. This is the result of the fact that the space region where the neck potential V_g is active becomes larger in volume when R_3 increases. The influence of the neck radius variation on the shell corrections E_{shell} is depicted in the lower plot of figure 2. A two-bump curve is common. One has to notice the lower values in the last part of the fission process for large (R_3).

Mass asymmetry influence is first presented for the light nuclei fusion reaction Mg+O →Ca considering different isotopes of the magnesium target in figure 3.

The Mg-target isotopes cover a range of deformation converted in semiaxis ratio $\chi = b_1/a_1$ as follows: $\chi(^{20}\text{Mg})$=0.814, $\chi(^{22}\text{Mg})$=0.690 and $\chi(^{24}\text{Mg})$ = 0.602. The calcium projectile is spheric. The lower plot shows the proton shell correction values whereas the upper plot displays the total E_{shell}. One can see that in the first half of the process with $(R-R_f)/(R_t-R_f)$ larger than 0.5, where R_t is the tangent point distance between centers, proton shell corrections for ^{24}Mg - reaction (the most deformed target) takes lower values. In the last part of the fusion ^{20}Mg - reaction has the lowest proton shell corrections. However, for the total value of E_{shell}, in the last part of the fusion process it is the ^{24}Mg - reaction which has the lower values. This can also be connected to the double magicity of ^{40}Ca. The levels of this double magic nucleus already manifest themselves for $(R-R_f)/(R_t-R_f) < 0.25$.

In figure 4 proton $E_{shell}^{(p)}$ and total shell corrections E_{shell} have been calculated in the synthesis of ^{266}Hs within the isotopic-isobaric reaction Fe+Pb →^{266}Hs.

Both proton and total curves have lower values for ^{56}Fe - projectile reaction. One has to mention in this case that ^{56}Fe is the only spherical projectile, whereas

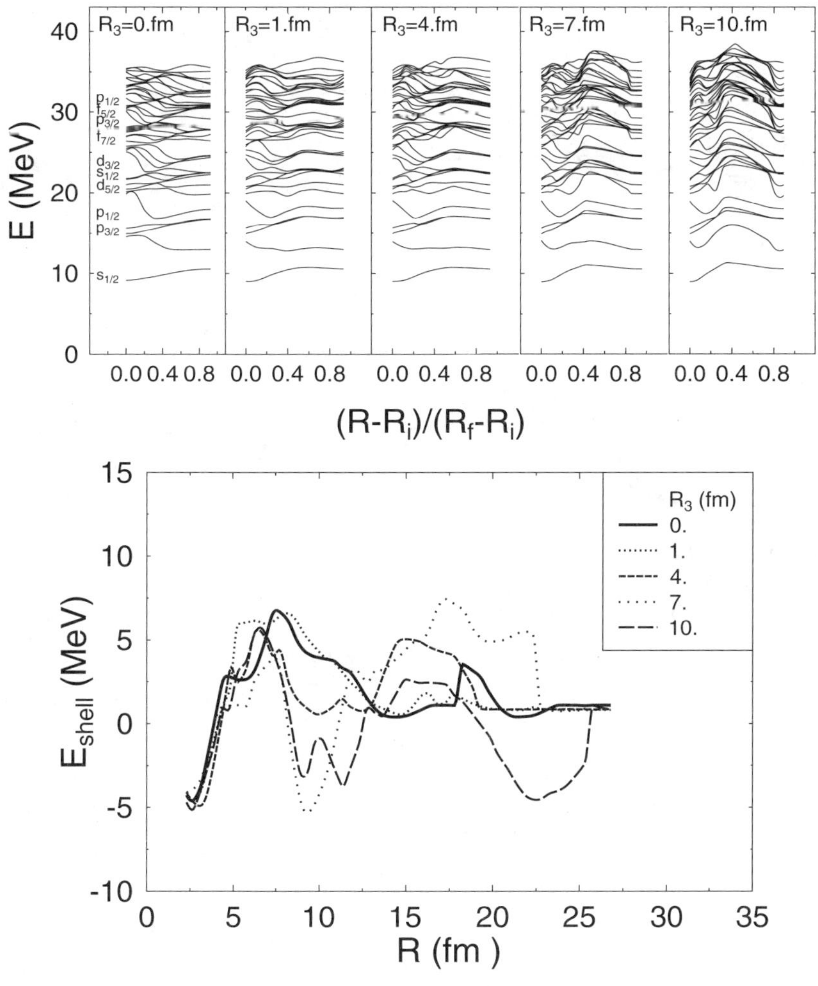

Figure 2. Level schemes (upper plot) and corresponding shell corrections for different neck radii.

the other iron isotopes are deformed. Eventhough the proton shell corrections are lower for ^{56}Fe - reaction, the negative values of the total shell corrections E_{shell} come from the neutron contribution. ^{56}Fe has the closest to magicity neutron number.

The last example treats mass asymmetry influence on the shell corrections in the synthesis of the superheavy nucleus $^{296}116$. Shell corrections are displayed in figure 5.

Calculations have been performed for four possible fusion reactions. Everyone of them contains two magic numbers. In this way one takes advantage

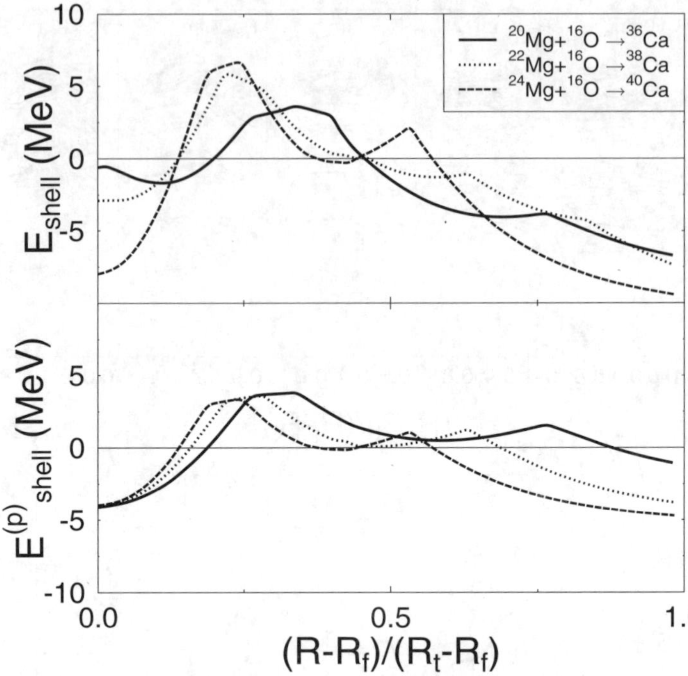

Figure 3. Mass asymmetry influence on shell corrections for the isotopic fusion reaction Mg+O →Ca.

of the strong negative shell correction due to proton and neutron closed shells. The lowest values of E_{shell} are obtained for ^{88}Se+^{208}Pb and ^{16}+^{280}Hs. Though the ^{208}Pb - reaction has a higher peak of the shell correction, one can see that it has the narrower width. Since both width and height are important in penetrability calculation and due to the exponential dependence of the action integral, a definite result could be obtained only after the macroscopic part is added and the total deformation energy is computed as their sum. The influence of the lead double magicity in this case goes beyond the touching point, up to $(R-R_f)/(R_t-R_f)$=0.65. The other two reactions show a much broader microscopic barrier. Also the higher values at the beginning of the overlap make them less suitable to be used in the synhtesis of ^{296}116.

4. Conclusion

This work presented a deformed two center shell model specialized for binary processes like fusion and fission reactions. The model is able to take into account partner deformations and necking together with the mass asymmetry degree of freedom. The evolution of the level scheme for the overlapping configuration can be followed from two separated independent systems up to

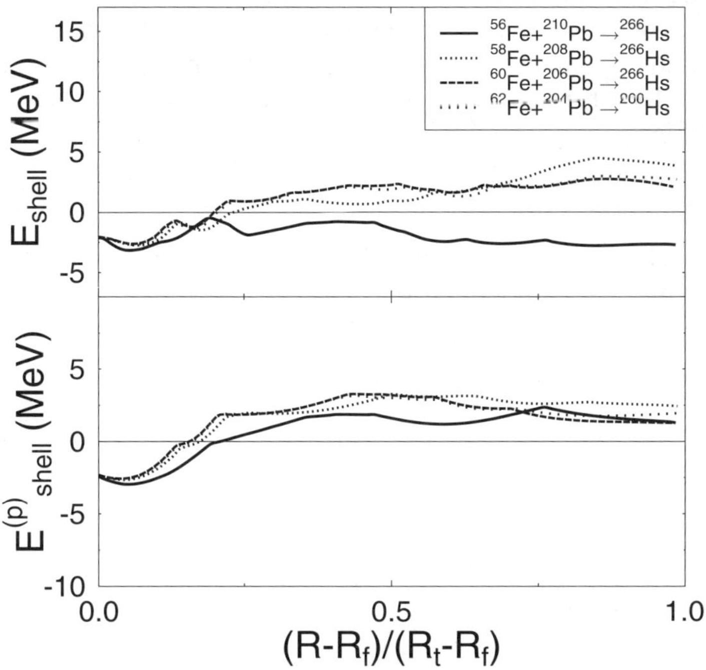

Figure 4. Proton and total shell corrections for the synthesis of ^{266}Hs.

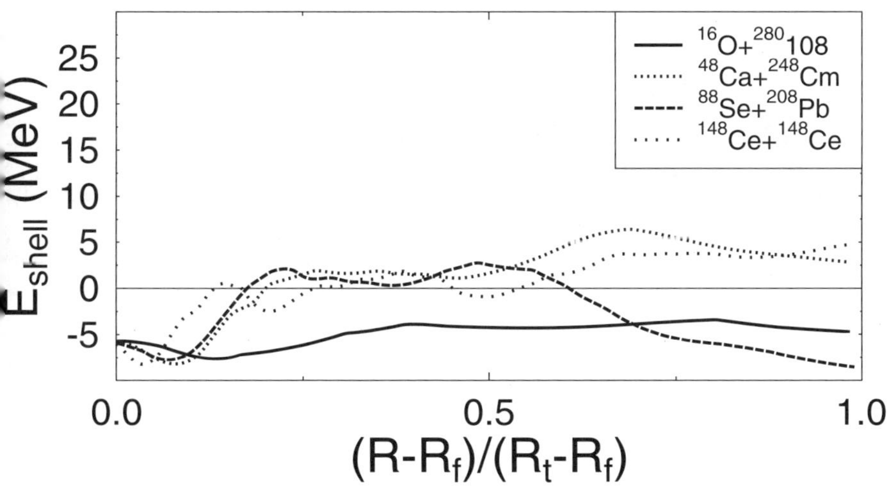

Figure 5. Mass asymmetry influence on shell corrections for the synthesis of ^{296}116.

the compound nucleus. In fusion reactions target and projectile can change their initial deformation. A complex path in the spheroidal space of deformation is obtained in fusion reactions. For fission configuration the neck radius plays an important role. Its value can change along the same fission channel as to lower the fission barrier.

References

[1] P. Holzer, U. Mosel, and W. Greiner, Nucl. Phys. A **138** (1969), 241.

[2] D. Scharnweber, U. Mosel, and W. Greiner, Nucl. Phys. A **164** (1971), 257.

[3] J. Maruhn and W. Greiner, Zeit. Phys. **251** (1972), 431.

SHELL EFFECTS IN FISSION AND QUASI-FISSION OF HEAVY AND SUPERHEAVY NUCLEI

M. G. Itkis
Flerov Laboratory of Nuclear Reaction
Joint Institute for Nuclear Research
141980 Dubna, Moscow region
itkis@flnr.jinr.ru

Abstract Results of the experiments aimed at the study of fission and quasi-fission processes in the reactions ^{12}C+^{204}Pb, ^{48}Ca+^{144,154}Sm, ^{168}Er, ^{208}Pb, ^{238}U, ^{244}Pu, ^{248}Cm; ^{58}Fe+^{208}Pb, ^{244}Pu, ^{248}Cm, and ^{64}Ni+^{186}W, ^{242}Pu are presented in the work. The choice of the above-mentioned reactions was inspired by recent experiments on the production of the isotopes $^{283}112$, $^{289}114$ and $^{283}116$ at Dubna [1],[2] using the same reactions. The ^{58}Fe and ^{64}Ni projectiles were chosen since the corresponding projectile-target combinations lead to the synthesis of even heavier elements. The experiments were carried out at the U-400 accelerator of the Flerov Laboratory of Nuclear Reactions (JINR, Russia), the XTU Tandem accelerator of the National Laboratory of Legnaro (LNL, Italy) and the Accelerator of the Laboratory of University of Jyvaskyla (JYFL, Finland) using the time-of-flight spectrometer of fission fragments CORSET[3] and the neutron multi-detector DEMON[4],[5]. The role of shell effects and the influence of the entrance channel on the mechanism of the compound nucleus fusion-fission and the competitive process of quasi-fission are discussed.

1. MED characteristics of heavy element fission fragments

The reactions ^{48}Ca+^{144,154}Sm,^{168,170}Er were investigated in order to understand the nature of the competition between the processes of fission and quasi-fission in dependence on the reaction entrance channel. For this purpose the fission cross-section was measured together with the evaporation residue cross-section in a wide energy range of ^{48}Ca ions. At the same time, mass-energy distributions (MED) of fission fragments, taken alone, are also of great interest. In the recent years our efforts were concentrated on the investigation of the multi-modal structure of the fission fragment MED in the region of transition nuclei $213 < A_{\rm CN} < 226$, which so far has been poorly studied. Thus, MED of the ^{216,218,220}Ra compound nuclei produced in reactions with ^{12}C were studied. It was found that the contribution of the asymmetric fission mode in

W. Greiner et al. (eds.), Structure and Dynamics of Elementary Matter, 423–438.

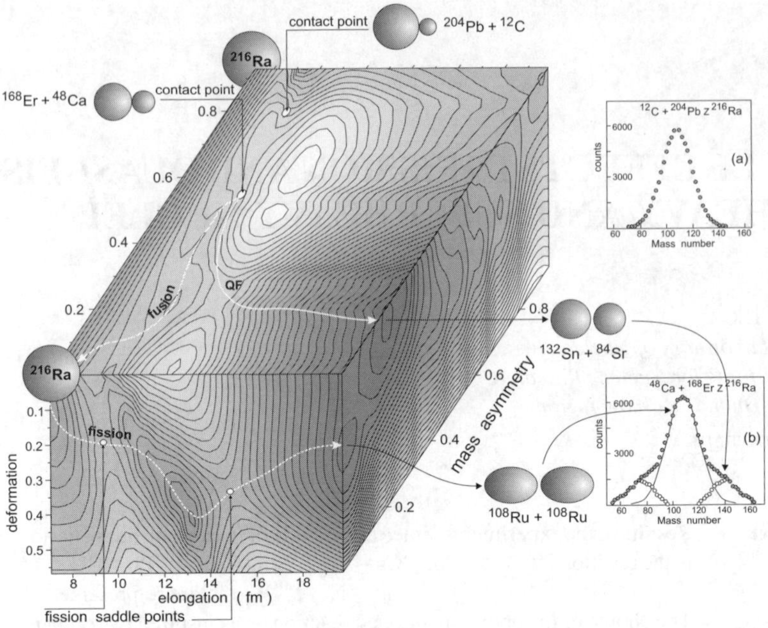

Figure 1. Potential energy surface as a function of mass asymmetry of the entrance channels, elongation and deformation for ^{216}Ra. The dark color corresponds to the lower potential energy.

the case of the ^{12}C+^{204}Pb reaction is 1.5% [6], and it is 30% in the case of the ^{48}Ca+^{168}Er reaction [7]. We connect such a sharp increase in the yield of asymmetric products in the reaction ^{48}Ca+^{168}Er with the quasi-fission process (QF), MED of which have a clearly expressed shell structure.

The properties of the fission fragment MED are such that they allow their interpretation by analogy with the low-energy fission of heavy nuclei as a manifestation of an independent mode of nuclear decay which competes with the classical fusion-fission (FF) process. To explain such a difference in the MED a theoretical calculation of the potential energy was made within the three-dimensional version of the "nucleon collectivization" model [8] based on the two-center shell model idea [9]. Figure 1 shows the result of this calculation. One can see that in the case of ^{12}C+^{204}Pb, mass distribution of which is shown in insert panel (a), the process of fusion-fission is the dominating one. In the case of the ^{48}Ca+^{168}Er reaction, there are two optimal paths at the potential energy surface. One of these paths (the white line marked as QF) leads to the ^{132}Sn+^{84}Sr exit channel without the compound nucleus formation. This potential energy minimum is determined by shell effects of the doubly magic ^{132}Sn. The second path corresponds to the classical fusion-fission process and leads to the symmetrical exit channel ^{108}Ru+^{108}Ru, the same as in the reaction with ^{12}C ions. That is why mass distribution of the ^{48}Ca+^{168}Er reaction products (insert panel (b)) is the superposition of both processes – the symmetrical classical fission and quasi-fission "shoulders".

Angular distributions for different fragment masses were measured in order to find experimental evidence of the quasi-fission nature of the "shoulders". Fig. 2 shows the angular distributions for the symmetrical mass range ($M = A/2 \pm 10$, solid circles) and for the "shoulders" ($M = 78 \pm 10$, open circles). The solid curve shows the results of calculations made in the framework of the Transition State Model [10]. As one can see from Fig. 2, the solid line fits well into the data for the symmetrical part of mass distribution, whereas the angular distribution for the "shoulders" has a pronounced forward-backward asymmetry, which is one of the distinctive features of the quasi-fission process.

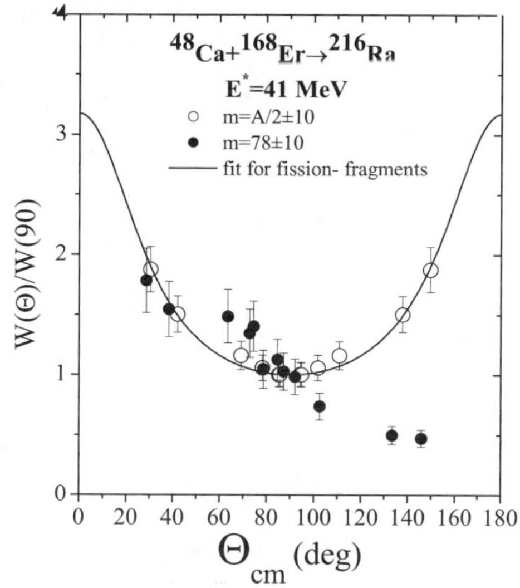

Figure 2. Experimental angular distributions of fission fragments from the reaction ^{48}Ca+^{168}Er at $E_{lab} = 195$ MeV.

For a further study of the influence of the entrance channel and deformations of the colliding nuclei on the competition between the fusion-fission and quasi-fission processes the reaction ^{48}Ca+^{144,154}Sm was chosen. Samarium is attractive as a target due to its numerous stable isotopes. Mass-energy distributions for the reaction ^{48}Ca+^{144,154}Sm are presented in Fig. 3. The left-hand panels show the data for the reaction with the ^{144,154}Sm targets at approximately the same excitation energy $E^* = 49$ MeV, and the right-hand panels show MEDs for nearly the same angular momentum $\langle l^2 \rangle = 900\hbar^2$. It is clearly seen that no quasi-fission "shoulders" are observed for ^{144}Sm, although in the case of ^{154}Sm the quasi-fission contribution reaches 30 % (see the mass distribution and $\langle TKE \rangle(M)$ dependence in Fig. 3). There are two reasons for such different course of the reactions with Sm isotopes: the first one is that ^{144}Sm is a spherical nucleus, whereas ^{154}Sm is a deformed one. The second reason is that in the case of ^{154}Sm the magic shells in both heavy and light "shoulders" ($Z = 28, N = 50$ in the light one and $Z = 50, N = 82$ in the heavy fragments) come into play. Since the compound nucleus has 10 neutrons less for the reaction with ^{144}Sm-target, only shells in the light fragments influence on the MED.

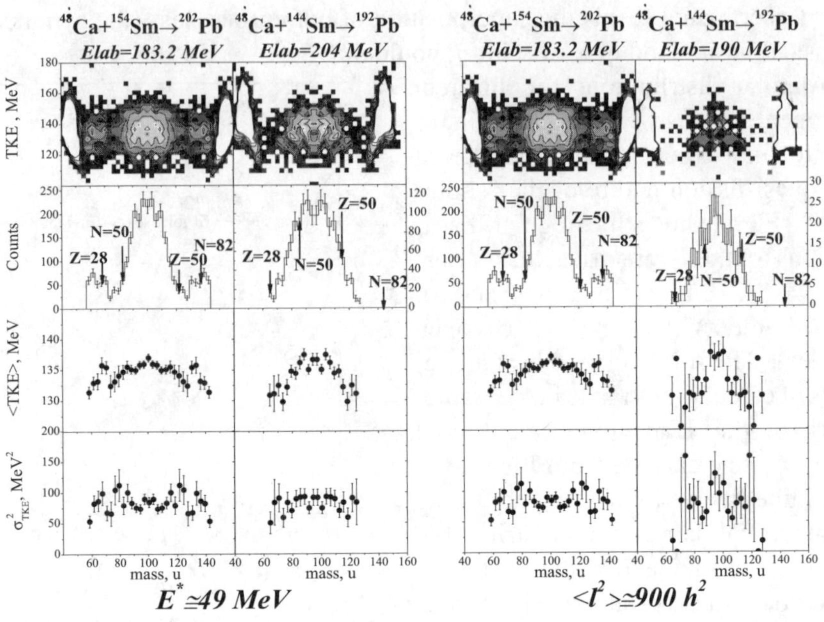

Figure 3. Two-dimensional TKE-Mass matrix, mass yield, average TKE and the variance $\sigma^2_{\mathrm{TKE}}(M)$ as a function of the mass of fragments for the reactions ^{48}Ca+^{144,154}Sm at the excitation energy $E^* = 49$ MeV (columns 1 and 2), and at angular momentum (columns 3,4) $\langle l^2 \rangle = 900\hbar^2$ (columns 3,4).

2. MED characteristics of fission fragments of super heavy elements

Figure 4 shows the data on MED of fission fragments of the ^{256}No, ^{286}112, ^{292}114 and ^{296}116 nuclei produced in the reactions with ^{48}Ca-ions at the same excitation energy $E^* \approx 33$ MeV. The main peculiarity of the data is the sharp transition from the predominant compound nucleus fission in the case of ^{256}No to the quasi-fission mechanism of decay in the case of the ^{286}112 and heavier nuclei. It is very important to note that despite the dominating contribution of the quasi-fission process into the total mass yield in the case of nuclei with $Z = 112-116$, in the symmetric region of fission fragment masses $(A/2 \pm 20)$ the process of fusion-fission of compound nuclei, in our opinion, prevails. It is demonstrated in the framings (see the bottom panels of Fig. 4), where one can see that the mass distribution of fission fragments of compound nuclei is asymmetric in shape with the light fission fragment mass at about $132-134$ a.m.u. Thus in the case of super heavy elements (SHE) with mass $132-134$ plays a stabilizing role, in contrast to the region of actinide nuclei, where the heavy fragment is the spherical one.

An experimental two-dimensional TKE-mass plot for the ^{48}Ca+^{248}Cm fusion-fission reaction is shown in Fig. 5 along with the corresponding potential en-

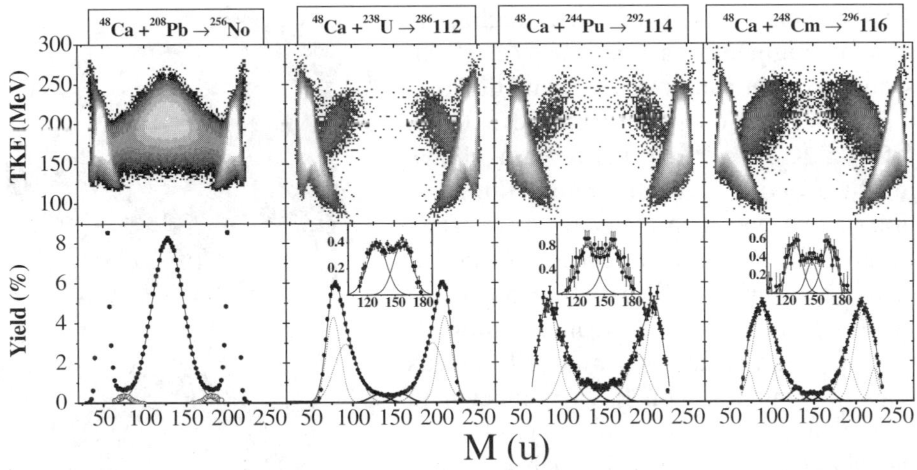

Figure 4. Two-dimensional matrices TKE-Mass (top panels) and mass yields (bottom panels) of fission fragments of ^{256}No, ^{286}112, ^{292}114 and ^{296}116 nuclei produced in the reactions with ^{48}Ca at the excitation energy $E^* = 33$ MeV.

ergy surface [8] determining evolution of the nuclear system. After the nuclei have come into contact, the nuclear system typically evolves in the asymmetric quasi-fission channels: path number 1 in Figure 5(a), which populates the area of fragment masses near $A = 208$ in Figure 5(b). The asymmetric quasi-fission channels are closer to the initial state in the configuration space of collective degrees of freedom as compared with the configurations through which the system has to pass on the way to the compound nucleus: trajectory number 3 in Figure 5(a). As a result,only a small part of the incoming flux reaches a compound nucleus configuration, and the fusion cross section turns out to be far less than the capture cross section. The distinction between $\sigma_{\rm fus}(E^*)$ and $\sigma_{\rm capt}(E^*)$ becomes still more evident at low excitation energies. Approximating the mass distribution of the quasi-fission fragments (area 1 in Figure 5(b)) by a Gaussian shape, we may easily single out the events in the symmetric region of fission fragments ($A_{\rm CN}/2 \pm 20$, area of dashed quadrangle in Figure 5(b)) which correspond to a regular fission. For the ^{48}Ca+^{248}Cm fusion-fission reaction, the tail of Gaussian gives no more than 20% of all the events in this region (see the details in [11]). A mass distribution corresponding to the symmetric region of fission fragments ($A_{\rm CN}/2 \pm 20$; area of dashed quadrangle in Fig. 5(b)) is shown in Fig. 5(c), compared with the typical mass yield of ^{238}U fission fragments [12]. As it can be seen from the figure, the mass spectra in the overlapped region practically coincide within the error bars. In the case of ^{238}U a two-humped fission mass distribution is regulated mainly by a doubly magic heavier fragment ^{132}Sn, which plays the role of a lighter fragment in the case of fission of a ^{296}116 nucleus at low excitation energies.

This means, that symmetric region of fission fragment masses ($A_{\rm CN}/2 \pm 20$) seems to originate mainly from the regular fusion-fission process in the reaction ^{48}Ca+^{248}Cm→^{296}116. However, as shown in [8], evolving from the initial configuration of two nuclei in contact into the state of spherical or near-spherical compound nucleus (path number 3 in Figure 5(a)), the system goes through the same configurations through which a compound nucleus goes in regular fission (path number 4 in Figure 4(a)), i.e., configurations close to the saddle point. When in such a configuration and in a state of complete thermodynamic equilibrium, the nuclear system is much likely to go into the fission channel (path number 2 in Figure 5(a)), without overcoming the saddle point, and producing a spherically symmetric compound nucleus. A process of this kind results in fragments that are practically not different from regular fission fragments, since in both cases the system follows the same path from the saddle point to the scission point. This means that among all the events resulting in the system going in regular near-symmetric fission channels, there are such events in which the system does not produce a true spherically symmetric compound nucleus.

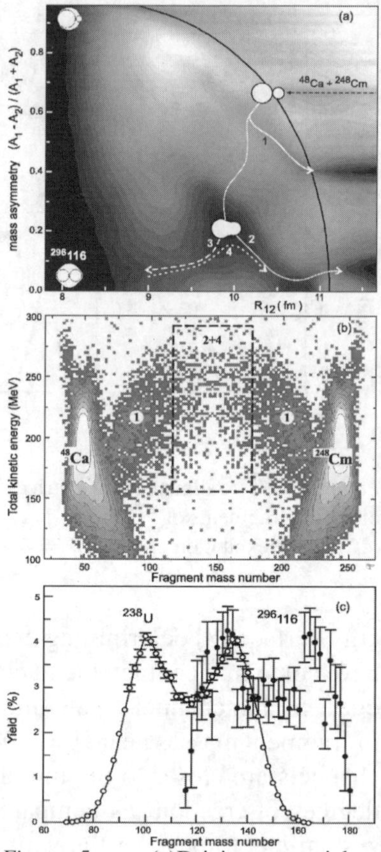

Figure 5. (a)Driving potential as function of mass asymmetry and distance between centers of two nuclei with zero deformation. (b) Two dimensional TKE-Mass matrix. (c) Mass spectra of fission fragments of ^{238}U at the excitation energy $E^* = 30$ MeV (solid circles) and those of the reaction ^{48}Ca+^{248}Cm (open circles).

Formerly emission of neutrons and γ-rays in correlation with fission fragments in the decay of superheavy compound systems at excitation energies near or below the Coulomb barrier had not been extensively studied. At the same time such investigations may be extremely useful for an additional identification of the fusion-fission and quasi-fission processes and consequently for a more precise determination of the cross-sections of these processes in the total yield of fragments. On the other hand, the knowledge of the value of the fission fragment neutron multiplicity can be used in the identification of SHE in experiments aimed at their synthesis.

The results of such investigations are presented in Table 1. As seen from the table, in all cases the total neutron multiplicity is considerably lower (by

Table 1. The measured values of neutron and γ-multiplicities

Reactions	$E_{\rm lab}$ (MeV)	E^* (MeV)	$M_n^{A/2\pm20}$	M_n^{QF}	$M_\gamma^{A/2\pm20}$	M_γ^{QF}
^{48}Ca+^{208}Pb	230	33	5.2 ± 0.7	3.1 ± 0.5	14.0 ± 1.3	9.3 ± 1.8
^{48}Ca+^{238}U	232	33	8.4 ± 1.2	4.9 ± 0.9	17.5 ± 1.6	12.6 ± 1.3
^{48}Ca+^{244}Pu	238	37	9.0 ± 1.2	4.6 ± 0.9	15.2 ± 1.8	10.4 ± 1.6
^{48}Ca+^{248}Cm	245	37	9.9 ± 1.4	5.6 ± 1.0	14.8 ± 1.8	12.5 ± 1.6

more than twice) for the region of fragment masses where the mechanism of quasi-fission predominates as compared with the symmetric region of fragment masses where, in our opinion, the process of fusion-fission prevails. Mass yields of fragments, the neutron and γ-ray multiplicities for different mass regions are presented in Fig. 6.

Summing up all the data, we plot the dependence of the quasi-fission relative yield $\sigma_{\rm QF}/\sigma_{\rm capt}$ as a function of the mass of compound nucleus, produced in reactions with ^{48}Ca ions at the excitation energy $E^* \approx 33 - 40$ MeV (Fig. 7). Solid circles are the measured reactions; the question signs are the combinations to be measured. Our prediction on the behavior of this curve is shown with line. The most possible explanation of such a nontrivial behavior of the

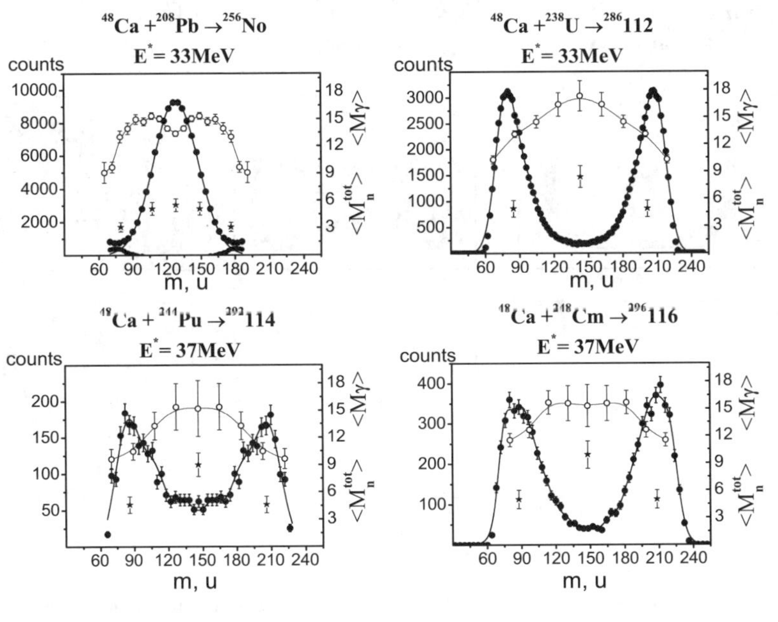

Figure 6. Mass yields of the fragments (solid circles), average neutron (solid stars) and γ-ray (open circles) multiplicities for ^{48}Ca-induced reactions with targets ^{208}Pb, ^{238}U, ^{244}Pu, ^{248}Cm.

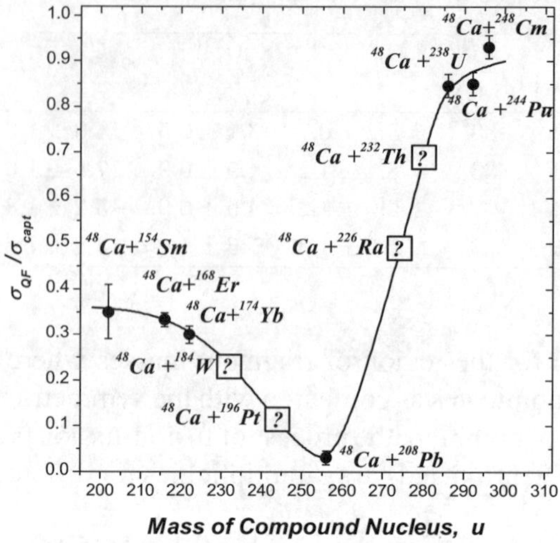

Figure 7. The QF cross section ($\sigma_{\rm QF}$) to the capture cross-section ($\sigma_{\rm tot}$) ratio dependence on the mass number of compound nuclei, produced in the reactions with ^{48}Ca ions at the same excitation energy $E^* \approx 33 - 40$ MeV.

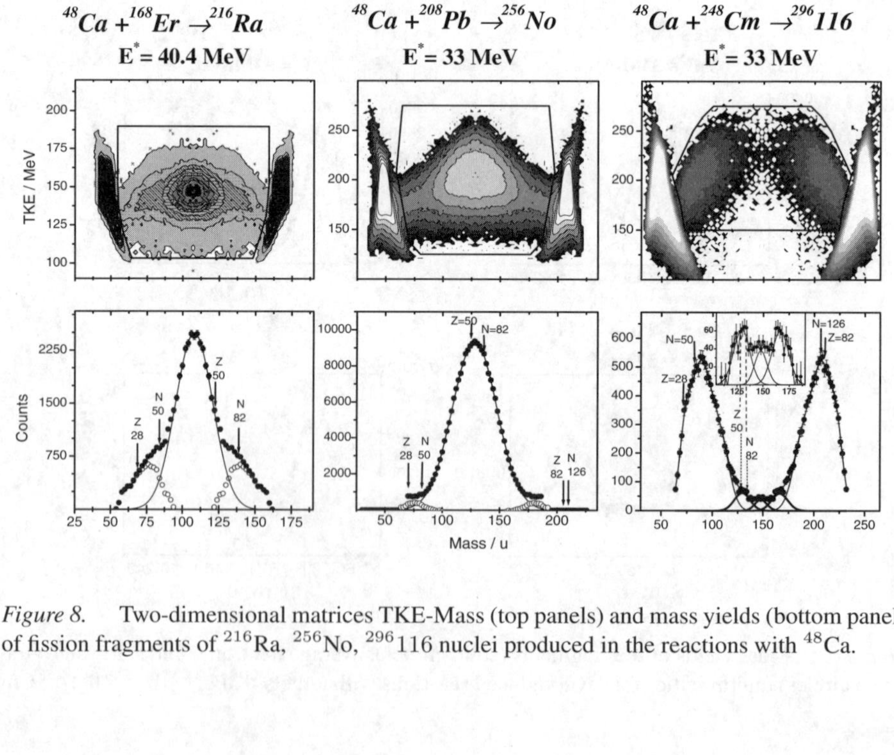

Figure 8. Two-dimensional matrices TKE-Mass (top panels) and mass yields (bottom panels) of fission fragments of ^{216}Ra, ^{256}No, ^{296}116 nuclei produced in the reactions with ^{48}Ca.

ratio $\sigma_{\rm QF}/\sigma_{\rm capt}$ is the corresponding probability of formation of the different spherical shells in the nascent fragments. The graphic example of the shell influence on the MED of the fission fragments is shown on Figure 8.

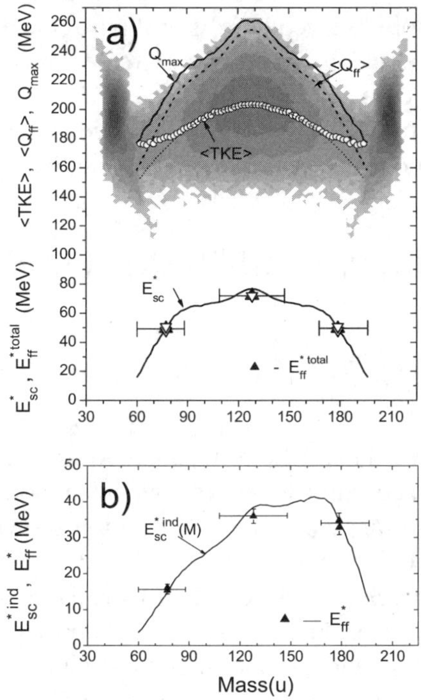

Figure 9. Energy balance for ^{48}Ca+^{208}Pb→^{256}No at the excitation energy $E^* = 33$ MeV (see text).

The knowledge of the total kinetic energy of fission fragments, mean neutron and γ-multiplicities allows us to calculate the energy balance of the reaction. The results of such calculations for the reaction ^{48}Ca+^{208}Pb ($E^* = 33$ MeV) are shown in Fig. 9. The calculations were made for three mass regions, i.e., for the light ($\Delta M_L = 60-88$), heavy ($\Delta M_H = 168-196$), and symmetric fission fragments ($\Delta M_S = 108-148$), since the neutron multiplicities were determined for these mass regions. Fig. 9(a) shows a two-dimensional matrix ($TKE-M$), together with the experimental dependence $\langle TKE\rangle(M)$ (the open circles) and its approximation by the parabola [13], as well as the average $\langle Q_{\rm ff}\rangle$ (the dashed line) and maximal $Q_{\rm max}$ (the solid line) reaction energies for the ground state of the compound nucleus.

It is clearly seen that the events belonging to the area of the QF shoulders ($M_L = 60-70$) are close in their energy to the maximal reaction energy $Q_{\rm max}$, and the fission fragments can be formed in this mass region only due to the excitation energy ($E^* \approx 33$ MeV). So, the fragments are weakly heated in this mass region which leads to the small multiplicities of the emitted neutrons and γ-quanta. The excitation energy of the nuclei at the scission point can be written as [14] :

$$E^*_{\rm sc} = \langle Q_{\rm ff}\rangle(M) + E^* - \langle TKE\rangle(M) - \langle M_{\rm pre}\rangle\langle E_{\rm pre}\rangle(M), \qquad (1)$$

where $\langle Q_{\rm ff}\rangle(M)$ is the average Q-value of the fission process for a given M; $E^* \approx 33$ MeV; $\langle TKE\rangle(M)$ is the experimental dependence; $\langle M_{\rm pre}\rangle\langle E_{\rm pre}\rangle$ is the mean energy released by a pre-fission neutron, calculated according to [6]. The excitation energy of the fragments is:

$$E^*_{\rm ff}(M) = E_\gamma(M) + \sum_{i=0}^{\langle M_{\rm post}\rangle(M)} [\langle B_n\rangle(M) + \langle E^i_n\rangle(M)], \qquad (2)$$

where $\langle E_\gamma \rangle$ is the average γ-quanta energy, $\langle B_n \rangle$ is the average neutron binding energy and $\langle E_n^i \rangle(M)$ is the average energy released by the i-th neutron. The value of $\langle E_n^i \rangle(M)$ for the fragments from the symmetric region ΔM_s was calculated using the approximation for the mean level density taken as $a = 0.093M$ [15] because shell effects were not observed here. For the fragments from the asymmetric mass regions $\Delta M_{L,H}$ we analyzed two extreme cases – the "heated" and "cold" ones. In the first case (the heated nucleus) $\langle E_n^i \rangle(M)$ was calculated using the same level density $a = 0.093M$. In the second case (the cold nucleus) we took the mean level density as $a = 0.07M$. Ideally, the reaction energy balance should look like this:

$$E^*_{\mathrm{sc}}(\Delta M_{\mathrm{H,L}}) = E^*_{\mathrm{ff}}(\Delta M_{\mathrm{H}}) + E^*_{\mathrm{ff}}(\Delta M_{\mathrm{L}}) = E^{*\mathrm{total}}_{\mathrm{ff}}(\Delta M_{\mathrm{L,H}}) \tag{3}$$

$$E^*_{\mathrm{sc}}(\Delta M_S) = 2E^*_{\mathrm{ff}}(\Delta M_S) = E^{*\mathrm{total}}_{\mathrm{ff}}(\Delta M_S) \tag{4}$$

The results of the analysis are demonstrated in the bottom panel of Fig. 9(a), where the excitation energy E^*_{sc} calculated according to (1) is shown by the solid line. Black triangles are the E_{sc} values, averaged over the mass region shown by horizontal lines, open triangles are the $E^{\mathrm{total}}_{\mathrm{ff}}$ value for the "cold"(upper triangles) and "heated" (lower triangles) nuclei. It is seen that there is good agreement between the experimental data and calculations. It is also possible to find from experimental data on the neutron and γ-multiplicities how the excitation energy is distributed between the heavy and light fragments. Indicated by triangles in Fig. 9(b) are the actual excitation energies E^*_{ff} which we found for the light, symmetric and heavy fragments. The solid curve shows the excitation energy for individual fragments at the scission point $E^{*\mathrm{ind}}_{\mathrm{sc}}(M)$, which was obtained in the assumption that the total energy E^*_{sc} (Eq. (2)) is divided in proportion to their masses: $E^{*\mathrm{ind}}_{\mathrm{sc}}(M) = (M/A_{\mathrm{CN}})E^*_{\mathrm{sc}}(M)$. One can see that $E^*_{\mathrm{ff}}(\Delta M_i)$ and $E^{*\mathrm{ind}}_{\mathrm{sc}}(M)$ coincide within the error bars. Therefore, it can be said with confidence that the total excitation energy of the fragment nuclei is divided in proportion to their masses. It means that the fissioning nucleus undergoes complete thermalization at any mass ratio of the fragments, including the quasi-fission region.

Columns (a),(b),(c) of Figure 10 show the data for the reactions of ^{58}Fe projectile on ^{232}Th, ^{244}Pu and ^{248}Cm targets, leading to the formation of the compound system ^{290}116 and the heaviest compound systems ^{302}120 and ^{306}122 (where $N = 182 - 184$), i.e. to the formation of the spherical compound nucleus, predicted by theory [16]. As seen from Figure 10, in these cases we observe an even stronger manifestation of the asymmetric mass distributions of ^{306}122 and ^{302}120 fission fragments with the light fragment mass at about 132. The corresponding structures are seen well in the $\langle TKE \rangle(M)$ distribution

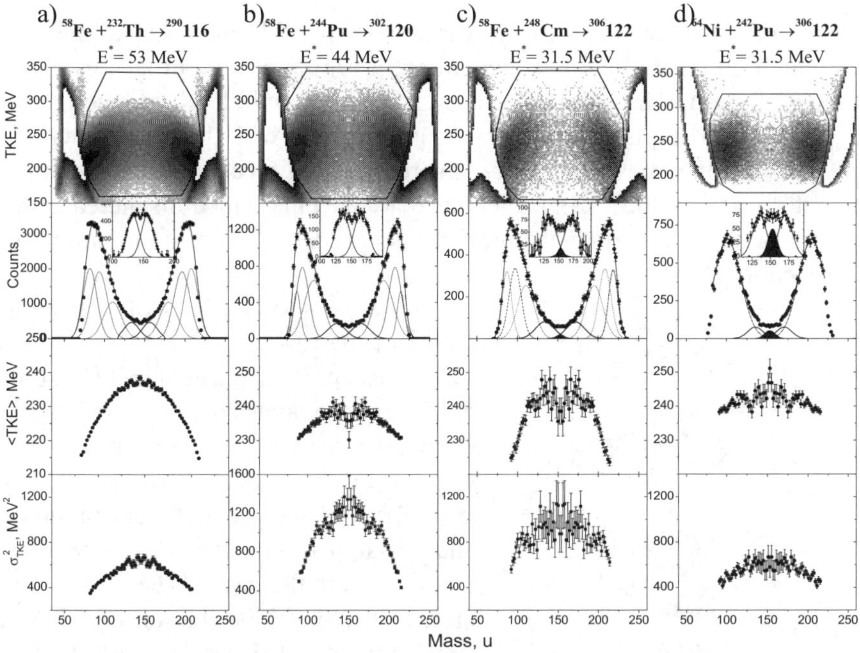

Figure 10. Two-dimensional TKE-Mass matrices, the mass yields, average TKE and the variances $\sigma^2_{\rm TKE}$ as a function of the fission fragment mass for $^{290}116$, $^{302}120$, $^{306}122$, produced in the reactions with ^{58}Fe and ^{64}Ni projectiles.

as a function of fragment mass. Only for the reaction ^{58}Fe+^{232}Th→^{290} 116 ($E^* = 53$ MeV) the valley in the region of $M = A/2$ disappears – this is seen from the mass yield distribution as well as from the $\langle TKE\rangle(M)$ dependence. This fact is connected with a reducing of the shell effects influence at so high excitation.

Figure 10(d) shows the data for the reaction ^{64}Ni+^{242}Pu→^{306}122. In this reaction the same nucleus ^{306}122 is formed at the same excitation energy as in the previous case (Fig. 10(c)), and the asymmetry of the entrance channel changes only slightly. Mass distributions look similar in both cases, with characteristic quasi-fission peaks in the region of doubly magic lead, however, energy characteristics of the fission fragments differ.

In the case of the reaction with ^{58}Fe-projectile the dependence $\langle TKE\rangle(M)$ has a valley in the symmetric part and peaks corresponding to mass 132 and complimentary masses. The ^{64}Ni-induced reaction yields a totally different dependence: $\langle TKE\rangle(M)$ changes only slightly, by about 5 MeV within the entire range of masses. For the reaction with ^{58}Fe ions, $\langle TKE\rangle(M)$ varies in the limit of 25 MeV.

For the reaction with ^{64}Ni ions the dependence $\sigma^2_{\rm TKE}(M)$ is parabolic and is smaller in amplitude than that in the reaction with ^{58}Fe ions, where it is seen despite low statistics that the variance $\sigma^2_{\rm TKE}(M)$ increases in the central part, which points to the co-existence in this region of both processes – fission and quasi-fission. Thus, it indirectly suggests that in the ^{64}Ni-induced reaction the contribution of quasi-fission grows as compared with the reaction with ^{58}Fe ions. This fact may possibly explain the reasons of failed attempts to synthesize element $^{293}118$ in the ^{86}Kr projectile reaction [17], since an increase in the asymmetry of the entrance channel in this case leads to the domination of the quasi-fission process.

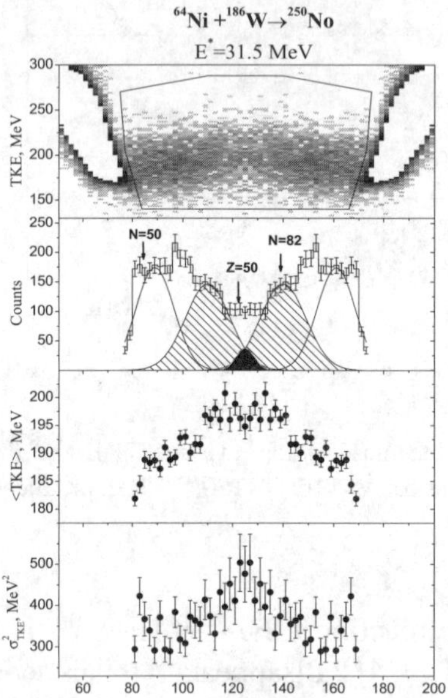

Figure 11. Two-dimensional TKE-Mass matrix, mass yield, average TKE and the variance $\sigma^2_{\rm TKE}$ as a function of the fragment mass for the reation ^{64}Ni+^{186}W→^{250}No at the exitation energy E$^* \approx 31.5$ MeV.

Figure 11 shows MED of fission fragments of the reaction ^{64}Ni+^{186}W→^{250}No. This reaction is of particular interest since the form of the mass distribution has a pronounced structure. In contrast to ^{256}No, formed in the reactions with ^{48}Ca ions (Fig. 8), it is seen that the contribution of quasi-fission fragments into the total mass distribution in the case of the ^{250}No greatly increases. Mass distribution of ^{250}No fission fragments also has a complicated structure – the asymmetric fission mode Standard II [18], connected with the formation of the deformed shell near the heavy fission fragment mass 140, is manifested in this case. Such situation is inherent in the elements from U to Cf. Besides, in the ^{250}No mass distribution, along with the asymmetric fission, there is a symmetric fission component S (Superlong), determined by the influence of the closed proton shell $Z = 50$. (see Fig 9.(b)). The manifestation of these processes is seen in the energy characteristics – dependences of $\langle TKE \rangle (M)$ and $\sigma^2_{\rm TKE}(M)$. The curve $\sigma^2_{\rm TKE}(M)$ has humps in the center ($M = 125$) and in the mass range $160 - 170$ which confirms the co-existence in these regions of several processes.

The obtained yields of ^{250,256}No fission fragments are included into the diagram of mass distributions by D. Hofmann [19] for the fission of trans-berkelium isotopes (Fig. 12). In the spontaneous and low-energy ($E^* \leq 33$ MeV) fission of No isotopes one can see a change in the form of the fission fragment yield, from the superposition of symmetric (S) and asymmetric (Standard-II) modes for ^{252}No, determined by the shells of fission fragments $N = 82$ and $Z = 50$, to the asymmetric form for ^{252}No, and a further transition to symmetric fission

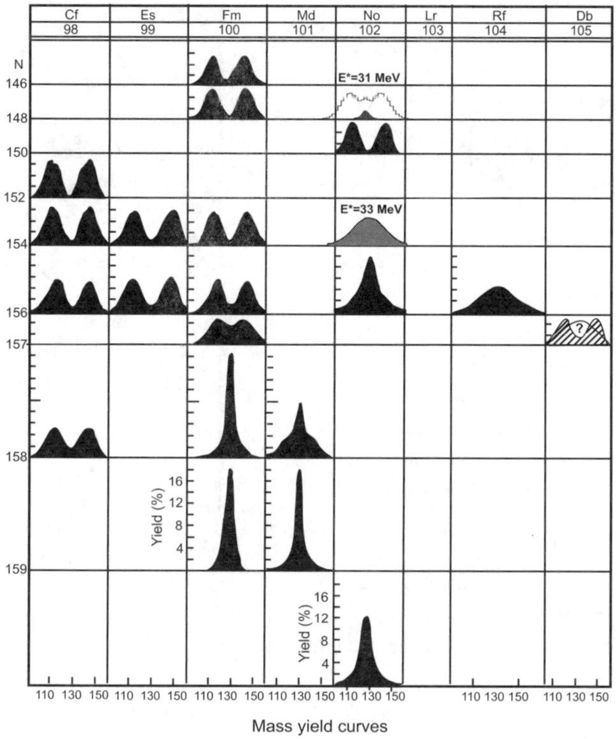

Figure 12. Schematic of mass-yield distributions, normalized to 200% fission fragments yield for SP of trans-Bk isotopes (from [19]) and induced fission (our work)

in the case of ^{256,258,272}No. Note that in the case of induced fission symmetric and asymmetric components become wider, and besides, the mass distribution width for spontaneous fission of No isotopes becomes smaller with increasing the neutron number at approaching $N = 164$, i.e. when in symmetric fission both fragments are close to magic ones by the proton and neutron numbers $(Z = 50, N = 82)$.

3. Capture and fusion-fission cross sections

Figure 13 shows the results of measurements of the capture $\sigma_{\rm cap}$ and the fusion-fission $\sigma_{\rm ff}$ cross sections (defined as $\sigma_{A/2\pm20}$) for a few studied reactions as a function of the initial excitation energy of the compound systems. Comparing the data on the cross sections $\sigma_{\rm ff}$ for the warm fusion (the reactions from ^{48}Ca+^{238}U to ^{58}Fe+^{248}Cm at E$^* \approx 33$ MeV) and for the cold fusion (the reactions ^{58}Fe+^{208}Pb and ^{86}Kr+^{208}Pb at E$^* \approx 14 - 15$ MeV [11]), one can obtain the following ratios: $\sigma_{\rm ff}(112)/\sigma_{\rm ff}(122) \approx 4 - 5$ in the first case and $\sigma_{\rm ff}(108)/\sigma_{\rm ff}(118) \geq 10^2$ for the second one, however the value of Z of

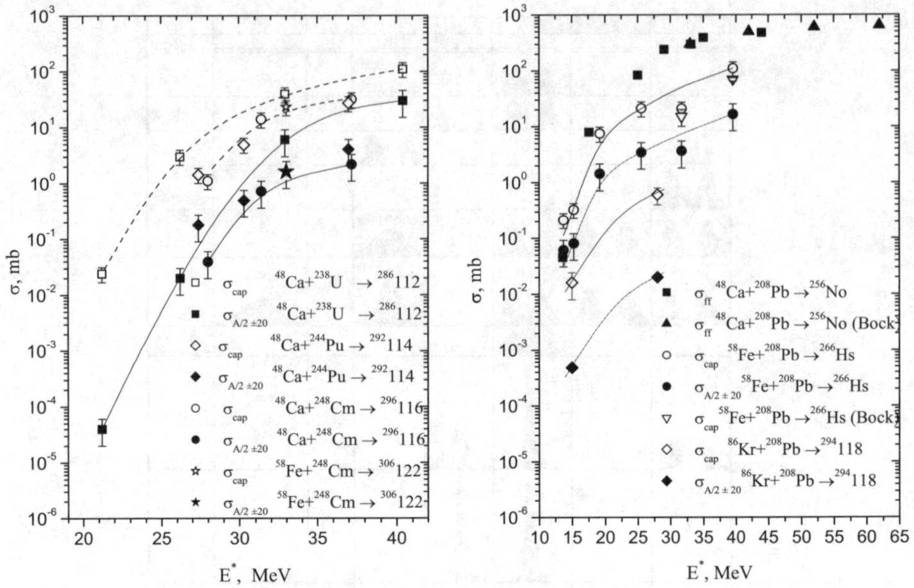

Figure 13. The capture cross section σ_{cap} and the fusion-fission cross section σ_{ff} for the indicated reactions as a function of the excitation energy.

the compound nucleus changes by the same 10 units in both cases. That makes the use of asymmetric warm fusion reactions for the synthesis of spherical superheavy nuclei quite promising.

Another interesting result is connected with the fact that the values of σ_{ff} for $^{256}102$ and $^{266}108$ at $E^* \approx 14-15$ MeV are quite close to each other, whereas the evaporation residue cross sections σ_{xn} [20] differ by almost three orders of magnitude $\sigma_{\text{ff}}/\sigma_{\text{xn}}$ which is evidently caused by a change in the Γ_f/Γ_n value for the above mentioned nuclei. At the same time, for the $^{294}118$ nucleus formed in the reaction $^{208}\text{Pb}+^{86}\text{Kr}$, the compound nucleus formation cross section is decreasing at an excitation energy of 14 MeV by more than two orders of magnitude according to our estimations ($\sigma_{\text{ff}} \approx 500$ nb is the upper limit) as compared with σ_{ff} for $^{256}102$ and $^{268}108$ produced in the reactions $^{208}\text{Pb}+^{48}\text{Ca}$ and $^{208}\text{Pb}+^{58}\text{Fe}$ at the same excitation energy. But when using the value of 2.2 pb for the cross section $\sigma_{\text{ER}}(1\text{n})$ from work [17], one obtains the ratio $\sigma_{\text{xn}}/\sigma_{\text{ff}} \approx 4 \cdot 10^{-6}$ for $^{294}118$, whereas for $^{266}108$ the ratio is $\sigma_{\text{xn}}/\sigma_{\text{ff}} \approx 10^{-6}$.

In one of recent works [21] it has been proposed that such unexpected increase in the survival probability for the $^{294}118$ nucleus is connected with the sinking of the Coulomb barrier below the level of the projectile's energy and, as a consequence, leads to an increase in the fusion cross section. However, our data do not confirm this assumption.

4. Conclusion and Outlook

Mass and energy distributions of fragments, fission and quasi-fission cross sections, multiplicities of neutrons and gamma-quanta have been studied for a wide range of nuclei with $Z = 82 - 122$ produced in reactions with ^{12}C, ^{22}Ne, ^{26}Mg, ^{48}Ca, ^{58}Fe, ^{64}Ni and ^{86}Kr ions at energies close and below the Coulomb barrier. In the case of the fission process as well as in the case of quasi-fission, the observed peculiarities of mass and energy distributions of the fragments, the ratio between the fission and quasi-fission cross sections, in dependence of the nucleon composition and other factors, are determined by the shell structure of the formed fragments. It is important to note that in the case of the quasi-fission process the influence of the shell effects on the observed characteristics is much stronger than in the case of classical fission of compound nuclei.

A further progress in the field of synthesis of superheavy nuclei can be achieved using hot fusion reactions between actinide nuclei and ^{48}Ca ions as well as actinide nuclei and ^{58}Fe or ^{64}Ni ions. Of course, for planning the experiments on the synthesis of superheavy nuclei of up to $Z = 122$, new research and more precise quantitative data obtained in the processes of fusion-fission and quasi-fission of these nuclei in reactions with ^{58}Fe and ^{64}Ni ions are required.

References

[1] Yu. Ts. Oganessian et al., Eur. Phys. J, A5 (1999) 63.

[2] Yu. Ts. Oganessian et. al., Nature 400 (1999) 242.

[3] N. A. Kondratiev et al., Proceedings of International Conference DANF'98, (World Scientific, Singapore, 2000) p. 431.

[4] M. Moszinski et al., Nucl. Instr.Meth, A350 (1994) 226.

[5] I. Tilquin et al., Nucl. Instr. Meth, A365 (1995) 446.

[6] I. V. Pokrovsky et al., Phys. Rev. C60 (1999) 041304.

[7] A. Yu. Chizhov et. al., Phys. Rev. C67 (2003) 011603.

[8] V. I. Zagrebaev, Phys. Rev. C64 (2001) 034606; J.Nucl. Radiochem.Sci., 3, No.1(2002)13.

[9] U. Mosel et al., Phys. Lett. B34 (1971) 587;J. Maruhn and W. Greiner, Z.Physik, 251 (1972) 431.

[10] B. B. Back et al., Phys. Rev. C32 (1985)195.

[11] M. G. Itkis et al., in Proceedings on Fusion Dynamics at the Extremes, (Dubna,2000), World Scientific, Singapore (2001) 93.

[12] A. A. Goverdovsky (private communication)

[13] M. G. Itkis et al., Sov. J. Part. Nucl. 19 (1988) 301.

[14] D. J. Hinde et al., Nucl. Phys, A452 (1986)550.

[15] A. V. Ignatyuk, Statistical Properties of Excited Atomic Nuclei (Energoatomizdat, Moscow, 1983) (in Russian) [translated as IAEA report, INDC(CCP)-233/L, Vienna, 1985].

[16] Z. Patyk, A. Sobiczewski, Nucl. Phys., A533 (1991) 132.

[17] V. Ninov et al., Phys. Rev. Lett 83 (1999) 1104.

[18] U. Brosa et al., Phys. Reports 197 (1990) 167.

[19] D. C. Hoffman et al., Phys. Rev. C41 (1990) 631.

[20] S. Hofmann, G. Münzenberg, Reviews of Modern Physics, 72 (2000) 3.

[21] W. D. Myers and W. J. Swiatecki, Phys. Rev. C62 (2000) 044610.

FISSION BARRIERS OF HEAVY AND SUPERHEAVY NUCLEI

I. Muntian[1] and A. Sobiczewski[1]

[1] *Soltan Institute for Nuclear Studies, PL-00-681 Warsaw, Poland*

Abstract Height of static spontaneous-fission barrier is calculated within a macroscopic-microscopic model. A multidimensional deformation space is used. Axial symmetry of a nucleus is assumed. Even-even nuclei with proton number Z = 98 - 102 and neutron number N = 140 - 164 are considered. For some of them, non-axial shapes are also taken into account, illustrating their importance.

1. Introduction

There is a big activity in experimental studies of excitation functions and cross sections for synthesis of heavy and superheavy nuclei (e.g. [1–6]). It is accompanied by theoretical studies of these quantities (e.g. [7–12]). For the latter studies, height of the fission barrier is one of the basic quantities. It has been recently analyzed within various approaches: macroscopic-microscopic [13–16], extended Thomas-Fermi plus Strutinski integral (ETFSI) [17, 18], relativistic mean field (RMF) [19] and Hartree-Fock with effective Skyrme interactions (SHF) [19].

The objective of the present paper is to extend our previous macroscopic-microscopic calculations [13–15] to even-even nuclei with Z = 98 - 102, i.e. lighter ones than considered previously. For these nuclei, there recently appeared experimental studies of excitation functions and cross sections [5, 6]. For some of them, there also exist experimental data for the height of the fission barrier which can be used as a test of the theoretical model.

2. Method of the calculations

The calculations are done within a macroscopic-microscopic approach, with the Yukawa-plus-exponential model used for the macroscopic part of the potential energy and the Strutinski shell correction, based on the Woods-Saxon single-particle potential, taken for the microscopic part. Details of the approach

W. Greiner et al. (eds.), Structure and Dynamics of Elementary Matter, 439–446.

are specified in [20]. A large deformation space is used in the analysis. The space depends on the assumed symmetry of the shape of a nucleus.

2.1 Axial symmetry

In the case of axial symmetry of a nucleus, the 4-dimensional space $\{\beta_\lambda\}$, λ = 2,4,6,8, describing reflection symmetric shapes, has been finally used. It has been checked, however, that the reflection asymmetric shapes of a nucleus have no contribution to its energy at the saddle point. For this check, the 7-dimensional space $\{\beta_\lambda\}$, λ = 2,3,4,...,8, has been taken.

2.2 Axial asymmetry

In this case, the 4-dimensional deformation space $\{\beta_\lambda, \gamma\}$, λ=2,4,6, is taken. The non-axiality parameter γ is introduced in the following way [21]:

$$
\begin{aligned}
R(\theta,\phi) = R_0 \{1 &+\beta_2 \left[\cos\gamma \mathrm{Y}_{20} + \frac{1}{\sqrt{2}} \sin\gamma \,(\mathrm{Y}_{22} + \mathrm{Y}_{2-2})\right] \\
&+\beta_4 \left[\frac{1}{12}(7 + 5\cos 2\gamma)\mathrm{Y}_{40} + \frac{1}{2}\sqrt{\frac{5}{6}} \sin 2\gamma \,(\mathrm{Y}_{42} + \mathrm{Y}_{4-2})\right. \\
&+ \left.\frac{1}{12}\sqrt{\frac{35}{2}}(1 - \cos 2\gamma)\,(\mathrm{Y}_{44} + \mathrm{Y}_{4-4})\right] \\
&+ \beta_6 \mathrm{Y}_{60}\},
\end{aligned} \tag{1}
$$

where radius of a nucleus $R(\theta, \phi)$ is expressed (in the intrinsic frame of reference) by spherical harmonics $\mathrm{Y}_{\lambda\mu}(\theta, \phi)$. Here, besides the quadrupole non-axiality (characterized by the parameter γ), also the hexadecapole non-axiality γ_4 is introduced, according to [21]. To avoid, however, too large deformation space, only a particular hexadecapole non-axiality is taken (with $\gamma_4 = -2\gamma$ in the notation of [21]). The potential energy is analyzed in the full 3-dimensional deformation space: $\{\beta_2, \gamma, \beta_4\}$. In each point of the space, however, the energy is minimized in the fourth, β_6, degree of freedom.

In both cases of axial and non-axial shapes of a nucleus, the saddle point and the energy at this point are found in the following way. For each analyzed nucleus, we choose two points in the deformation space: one around the equilibrium point of the nucleus, and the other in its fission valley. Then, we connect them by all possible paths passing through dense grid points inside the used space. For each path L, we find the point P_L, which corresponds to the largest energy along this path, $V(P_L)$. Saddle point P_{L_0} is the point, at which the energy $V(P_{L_0})$ is smallest among all the values $V(P_L)$.

The static spontaneous-fission barrier height $B_{\mathrm{f}}^{\mathrm{st}}$ is:

$$
B_{\mathrm{f}}^{\mathrm{st}} = V(P_{L_0}) - E_0, \tag{2}
$$

where the ground-state energy E_0 (corresponding to the equilibrium point P_{eq}) is

$$E_0 = V(P_{\text{eq}}) + E_{\text{zp}}. \tag{3}$$

For the zero-point energy E_{zp}, we take 0.7 MeV [22].

It is worth mentioning that the method of finding saddle point, used by us, avoids problems appearing when saddle point is searched by minimization of energy, at each elongation of a nucleus, in other degrees of freedom. In the latter case, one can easily come to a wrong solution for the saddle point, as discussed e.g. in [23].

3. Results

Figure 1 shows the results for B_{f}^{st} for even-even nuclei with proton number Z = 98 - 106. For the lowest three values, Z = 98, 100 and 102, the results are obtained in the present paper. One can see a clear effect, on B_{f}^{st}, of the closed neutron shell at N = 152 (large B_{f}^{st}), especially for lighter elements. A smaller effect of the closed shell at N = 162 is also seen. The effects of the closed shells are much larger in the spontaneous-fission half-lives than in the barriers B_{f}^{st}, as discussed and illustrated in [24, 22].

Figure 2 shows a comparison between the macroscopic-microscopic and ETFSI results for elements with various Z, from Z = 98 up to Z = 120. One can see that for Cf, the lightest element considered in Fig. 2, the difference between the two results is rather small, less, as a rule, than 0.5 MeV. For heavier elements, however, it becomes larger, up to about 3 MeV for some nuclei of darmstadtium (Ds, Z = 110). One can also see that the dependence of B_{f}^{st} on both Z and N, obtained within the two models, is different.

Figure 3 shows contour map of the potential energy of the nucleus ^{254}No projected on the plane ($\beta_2 \cos\gamma$, $\beta_2 \sin\gamma$). The line γ=0° corresponds to the axially symmetric (with respect to Oz axis) prolate shapes, and the line γ=60° is corresponding to the axially symmetric (with respect to the Oy axis) oblate shapes of a nucleus. The line γ=30° corresponds to shapes with maximal non-axiality. One can see that the saddle point in the case of axial symmetry (denoted by the symbol "+") has energy 2.9 MeV, while non-axiality shifts the saddle to the point denoted by the symbol "$\times$" and decreases its energy to 1.2 MeV, i.e. by 1.7 MeV. (The energy is normalized so, that its macroscopic part is zero at spherical shape of a nucleus). Thus, the reduction of the saddle-point energy of this nucleus by non-axiality is quite large.

Finally, Fig. 4 gives contour map for the nucleus ^{250}Cf. One can see that the inclusion of non-axial deformations significantly changes the shape and the energy of the nucleus at its saddle point, similarly as it does for the nucleus ^{254}No.

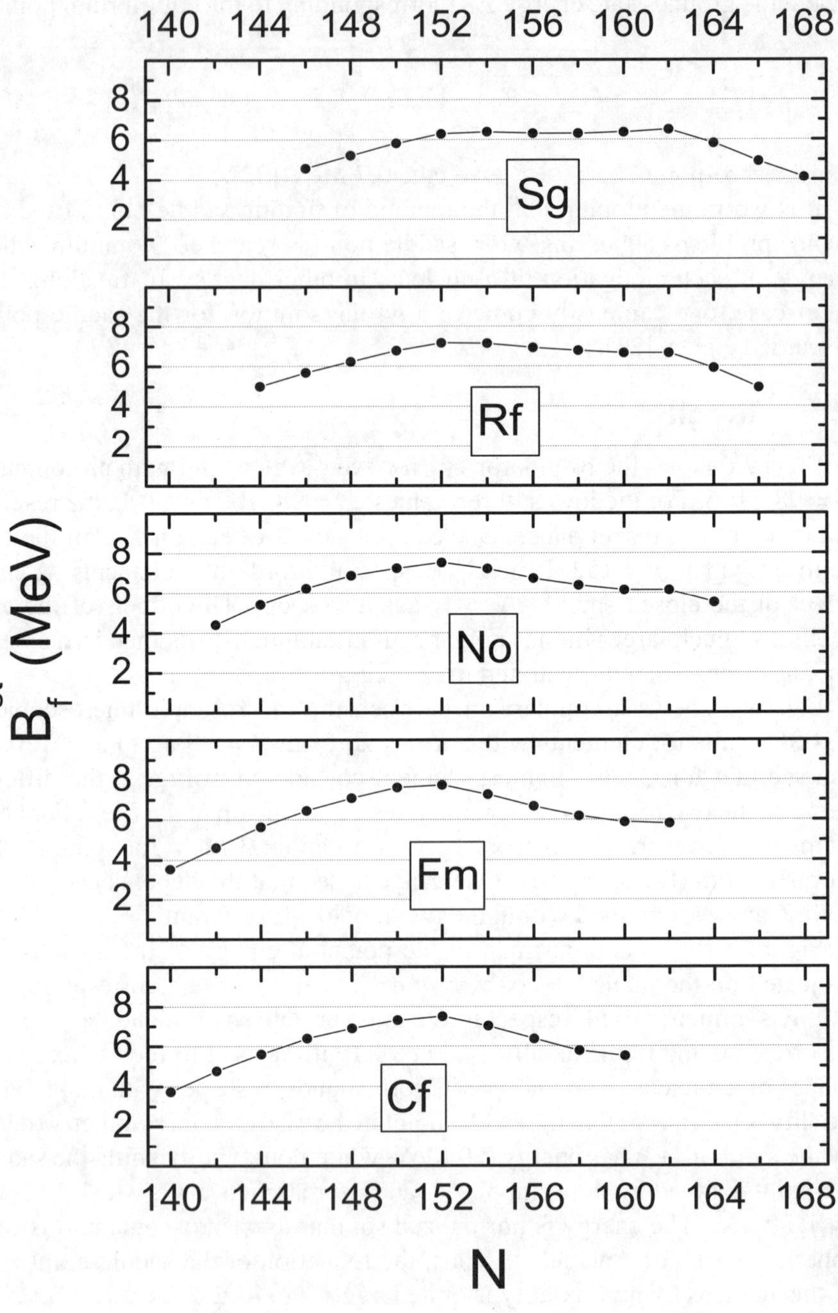

Figure 1. Fission barriers calculated for even-even nuclei with proton number Z=98-106.

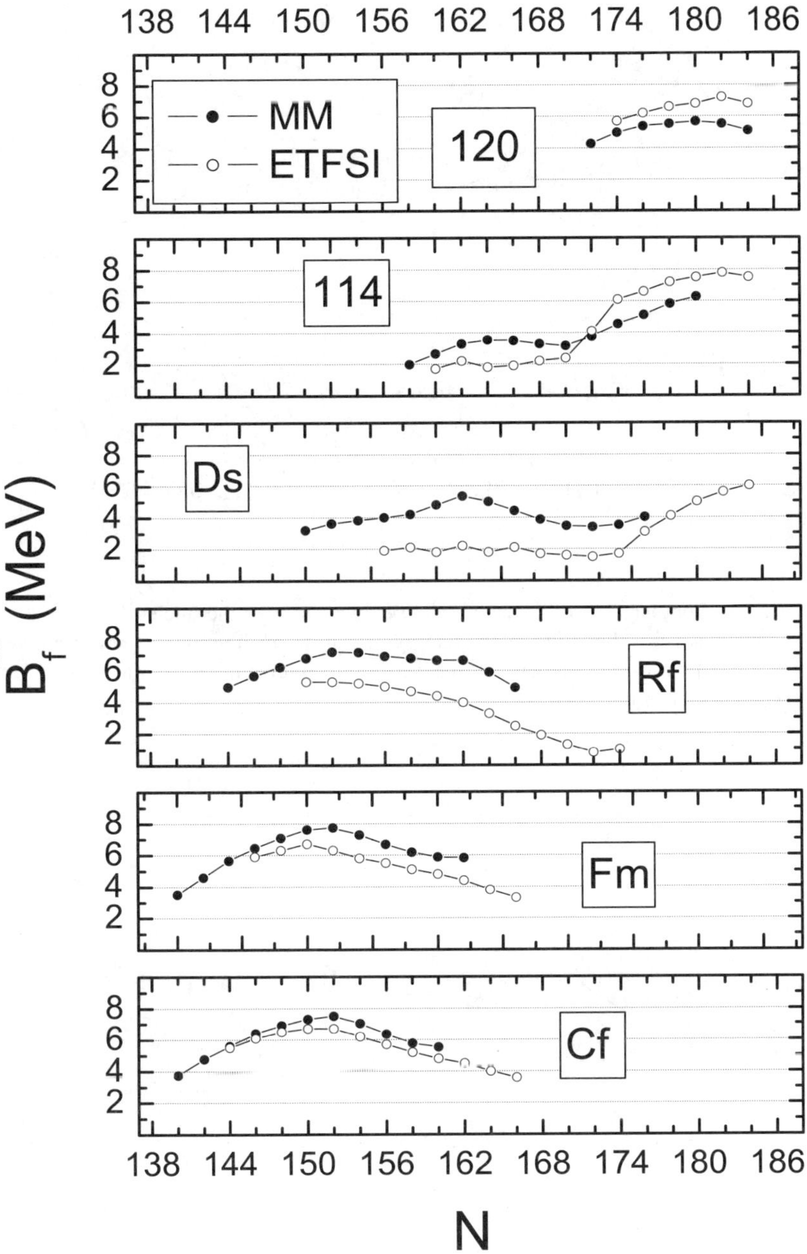

Figure 2. Comparison between barriers calculated within a macroscopic-microscopic (MM) and the ETFSI models.

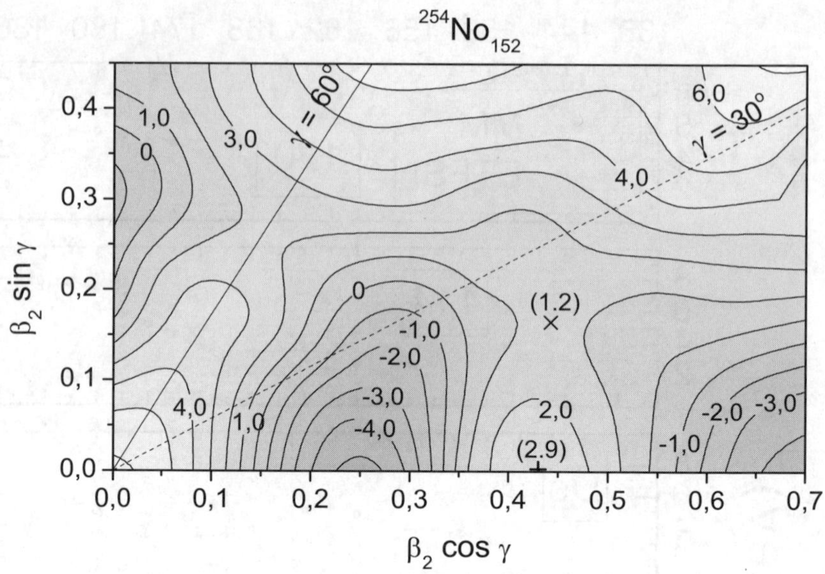

Figure 3. Contour map of potential energy calculated for the nucleus ^{254}No. Numbers at contour lines specify the value of the energy. Position of the saddle point is marked by the symbol "+", when axial symmetry of the nucleus is assumed, and by the symbol "×", when non-axiality is taken into account.

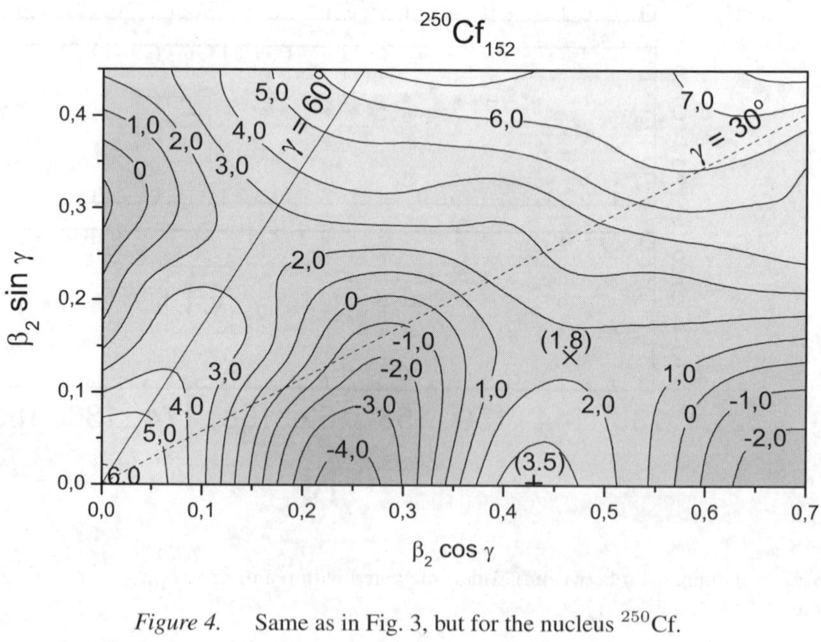

Figure 4. Same as in Fig. 3, but for the nucleus ^{250}Cf.

To compare our results for B_f^{st} with experimental values, we take the latter ones from the Obninsk evaluation [25] updated by V. Maslov in 1997, and from [26], quoted also in [17]. They are: 5.6 MeV, 6.2 MeV, 5.3 MeV, 5.4 MeV and 5.7 MeV for ^{250}Cf, ^{251}Cf, ^{252}Cf, ^{253}Cf and ^{255}Fm, respectively. Our values, calculated with the assumption of axial symmetry of these nuclei are: 7.5 MeV, 7.6 MeV, 7.0 MeV, 7.0 MeV and 7.3 MeV, respectively. Thus, the theoretical values are by 1.4 - 1.9 MeV too large. However, the inclusion of non-axial shapes decreases the barriers. For ^{250}Cf, it decreases B_f^{st} by 1.7 MeV, reducing the discrepancy from 1.9 MeV to 0.2 MeV. This stresses the importance of non-axial shapes in calculations of the barrier heights.

Acknowledgements

Support by the Polish State Committee for Scientific Research (KBN), grant no. 2 P03B 039 22, and the Polish-JINR (Dubna) Cooperation Programme is gratefully acknowledged.

References

[1] A. Tuerler, H.W. Gaeggeler, D.T. Jost *et al.*, Z. Phys. A **331** (1988) 363.

[2] Yu.A. Lazarev, Yu.V. Lobanov, R.N. Sagaidak *et al.*, Phys. Scr. **39** (1989) 422.

[3] H.W. Gaeggeler, D.T. Jost, A. Tuerler *et al.*, Nucl. Phys. A **502** (1989) 561c.

[4] S. Hofmann and G. Muenzenberg, Rev. Mod. Phys. **72** (2000) 733.

[5] Yu.Ts. Oganessian, V.K. Utyonkov, Yu.V. Lobanov *et al.*, Phys. Rev. C **64** (2001) 054606.

[6] A.V. Belozerov, M.L. Chelnokov, V.I. Chepigin *et al.*, Eur Phys. J. A **16** (2003) 447.

[7] G. Giardina, S. Hofmann, A.I. Muminov, and A.K. Nasirov, Eur. Phys. J. A **8** (2000) 205.

[8] V.I. Zagrebaev, Y. Aritomo, M.G. Itkis, Yu.Ts. Oganessian, and M. Ohta, Phys. Rev. C **65** (2001) 014607.

[9] M.G. Itkis, Yu.Ts. Oganessian, and V.I. Zagrebaev, Phys. Rev. C **65** (2002) 044602.

[10] Y. Abe and D. Bouriquet, Acta Phys. Pol. B **34** (2003) 1927.

[11] W.J. Swiatecki, K. Siwek-Wilczynska, and J. Wilczynski, Acta Phys. Pol. B **34** (2003) 2049.

[12] G.G. Adamian, N.V. Antonenko, and W. Scheid, Phys. Rev. C **69** (2004) 011601 (R).

[13] I. Muntian, Z. Patyk, and A. Sobiczewski, Acta Phys. Pol. B **34** (2003) 2141.

[14] A. Sobiczewski and I. Muntian, Nucl. Phys. A (2004), in press.

[15] I. Muntian, O. Parkhomenko, and A. Sobiczewski, Proc. 5th Tours Symposium, Tours (France) 2003 (AIP Conf. Series 2004), in press.

[16] P. Moeller, A.J. Sierk, and A. Iwamoto, Phys. Rev. Lett. (2004), in press.

[17] A. Mamdouh, J.M. Pearson, M. Rayet, and F. Tondeur, Nucl. Phys. A **644** (1998) 389.

[18] A. Mamdouh, J.M. Pearson, M. Rayet, and F. Tondeur, Nucl. Phys. A **679** (2001) 337.

[19] T. Buervenich, M. Bender, J.A. Maruhn, and P.-G. Reinhard, Phys. Rev. C **69** (2004) 014307.

[20] I. Muntian, Z. Patyk, and A. Sobiczewski, Acta Phys. Pol. B **32** (2001) 691.

[21] S.G. Rohozinski and A. Sobiczewski, Acta Phys. Pol. B **12** (1981) 1001.

[22] R. Smolanczuk, J. Skalski, and A. Sobiczewski, Phys. Rev. C **52** (1995) 1871.

[23] P. Moeller and A. Iwamoto, Phys. Rev. C **61** (2000) 047602.

[24] Z. Patyk, A. Sobiczewski, P. Armbruster, and K.-H. Schmidt, Nucl. Phys. A **491** (1989) 267.

[25] http://iaeand.iaea.or.at/ripl/

[26] H. Weigmann, *in* The Nuclear Fission Process, ed. C. Wagemans (CRC Press, Boca Raton, 1991) p. 30.

FUSION AND QUASIFISSION IN A MOLECULAR MODEL

G. G. Adamian[1,2], N. V. Antonenko[1], S. P. Ivanova[1], W. Scheid[3], and A. Zubov[1,3]

[1] *Joint Institute for Nuclear Research, 141980 Dubna, Russia*

[2] *Institute of Nuclear Physics, Tashkent 702132, Uzbekistan*

[3] *Institut für Theoretische Physik, Justus-Liebig-Universität, 35392 Giessen, Germany*

Abstract Fusion and quasifission are described with a molecular, so-called dinuclear system model. This model assumes a dynamics in the mass and charge asymmetry degrees of freedom via the exchange of nucleons between the clusters.

Keywords: Fusion, quasifission, superheavy nuclei, dinuclear model

1. Introduction

Superheavy elements are produced in fusion reactions with heavy nuclei, e.g. $^{48}\mathrm{Ca}+{}^{238}\mathrm{U}\rightarrow{}^{283}112+3\mathrm{n}$. These reactions can be described by the concept of a molecular model which assumes two touching nuclei (clusters) which exchange nucleons up to the moment when the system crosses the inner fusion barrier and an excited compound nucleus is formed. This description was introduced as the dinuclear system (DNS) concept by V. V. Volkov [1]. The dinuclear system has two main degrees of freedom: the nucleon transfer between the touching nuclei affecting the mass and charge asymmetries of the system and the decay of the system into two fragments what we denote as quasifission since not the compound nucleus but the dinuclear system is decaying. These degrees of freedom are described by the mass and charge asymmetry coordinates $\eta = (A_1 - A_2)/(A_1 + A_2)$ and $\eta_Z = (Z_1 - Z_2)/(Z_1 + Z_2)$ where A_1, A_2 and Z_1, Z_2 are the mass and charge numbers, respectively, and by the relative distance R between the centers of the clusters.

In Sect. 2 of this article we discuss the basic assumption of the DNS model that the touching nuclei are hindered by a repulsive potential to amalgamate directly into the compound nucleus. Sect. 3 presents the theory and results for the evaporation residue cross sections for the production of superheavy

W. Greiner et al. (eds.), Structure and Dynamics of Elementary Matter, 447–462.

nuclei in cold and hot fusion reactions. The competing quasifission process is separately treated and explained in Sect. 4 with a comparison of the calculated mass distributions and variances of the kinetic energy of the fragments with experimental data.

2. Diabatic Potentials for the Fusion Dynamics

The potential of the DNS model depends on the internuclear distance R, on the mass and charge asymmetry coordinates η and η_Z and on the deformations β_1 and β_2 of the clusters assumed in a pole-to-pole orientation. It is given by [2]

$$U(R,\eta,\eta_Z,\beta_1,\beta_2,J) = B_1(\beta_1^{gs}) + B_2(\beta_2^{gs}) + B_{def}(\beta_1 - \beta_1^{gs}, \beta_2 - \beta_2^{gs}) - B_{CN} + V(R,\eta,\eta_Z,\beta_1,\beta_2,J), \quad (1)$$

where B_1 and B_2 are the mass excesses (negative binding energies) of the fragments at their ground states with deformation parameters β_1^{gs} and β_2^{gs}, B_{def} is the energy of the additional dynamical deformation of the DNS nuclei and B_{CN} the mass excess of the compound nucleus. The nucleus-nucleus potential

$$V = V_C(R,\eta_Z,\beta_1,\beta_2) + V_N(R,\eta,\beta_1,\beta_2) + V_{rot}(R,\eta,\beta_1,\beta_2,J) \quad (2)$$

is the sum of the Coulomb potential, the nuclear potential and the centrifugal potential. For the nuclear part we use a double folding formalism with an effective density-dependent nucleon-nucleon interaction known from the theory of finite Fermi systems. The potential has a minimum at the distance $R_m = R_1(1+\beta_1\sqrt{5/(4\pi)})+R_2(1+\beta_2\sqrt{5/(4\pi)})+0.5$ fm where the DNS nuclei are in contact and evolve in the mass asymmetry coordinate. Towards the outside there is situated the Coulomb barrier at $R_b = R_m + 1.5$ fm which is denoted as the quasifission barrier of height B_{qf} measured with respect to the DNS minimum. Towards the inside the potential raises repulsively from the DNS minimum and forbids a direct transition of the nuclei to the compound system.

The proof of the steep rising of the potential for $R < R_m$ can be done semi-microscopically with a potential calculated with the macroscopic-microscopic method by using the Strutinsky formalism. The basic single-particle model for the calculation of the shell correction energy to the liquid drop and internuclear potentials is the two-center shell model (TCSM). We discriminate between adiabatic and diabatic two-center shell models. In the adiabatic model the nucleons follow the energetically deepest possible levels as a function of the distance and minimize the shell correction energy. The resulting adiabatic potentials are unsuitable for the calculation of fusion cross sections for superheavy nuclei. We found fusion probabilities [3], e.g. for the ^{110}Pd + ^{110}Pd system, obtained with an adiabatic two-center shell model which are much larger than the values deduced from the experimental data. In contrast the diabatic model assumes that the nucleons remain in their states during the internuclear motion due to a finite relaxation time of the mean field and, therefore, a highly excited dinuclear system is formed for smaller internuclear distances.

In Fig. 1 we show adiabatic and diabatic potentials for the ^{110}Pd + ^{110}Pd system as a function of the elongation $\lambda = \ell/(2R_0)$ of the dinuclear system where ℓ is the length of the system and $2R_0$ the diameter of the compound system. The diabatic potential rises to a strongly repulsive potential which confines the dinuclear system to the touching distance. This potential survives partly during the reaction time t_0 as found with time-dependent dynamical diabatic occupation numbers [4]:

$$\frac{dn_\alpha^{diab}(\lambda,t)}{dt} = -\frac{1}{\tau}(n_\alpha^{diab}(\lambda,t) - n_\alpha^{adiab}(\lambda)). \tag{3}$$

Here, n_α^{adiab} and n_α^{diab} are the adiabatic and diabatic occupation numbers of

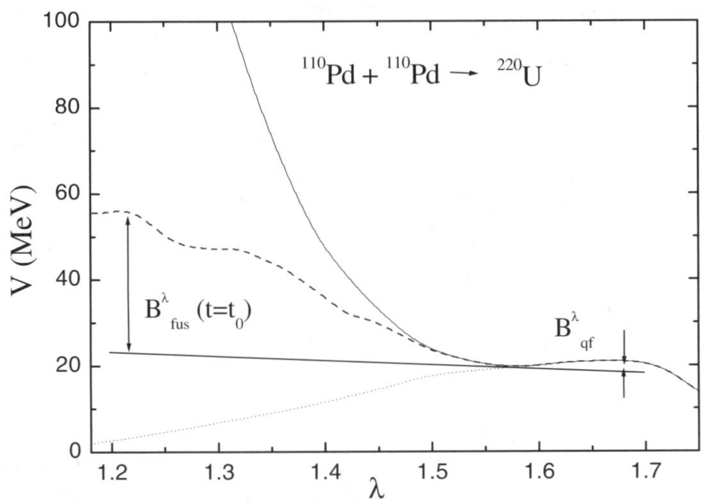

Figure 1. The diabatic time-dependent potentials (solid and dashed curves) and the adiabatic potential (dotted curve) for ^{110}Pd+^{110}Pd as a function of λ.

the two-center shell model levels and τ a relaxation time determined by a mean single-particle width depending on the diabatic occupation numbers. The decrease of the diabatic potential in the reaction time $t_0 = (3-4)\cdot 10^{-20}$s is not so strong that the barrier to the inside vanishes and that the adiabatic potential is already reached as shown in Fig. 1. Hence, the dinuclear system can not directly amalgamate or fuse together and, therefore, the basic assumption about the dinuclear system that it survives in a touching configuration can be seen as justified with this calculation. The results are also supported by further microscopical investigations [5], based on the SU(3) - symmetry with harmonic oscillator wave functions for the nucleons, which showed a structural forbiddenness for the motion to smaller relative distances.

3. Fusion to Superheavy Nuclei

In accordance with the DNS concept the evaporation residue cross section can be written as a sum over partial contributions:

$$\sigma_{ER}(E_{\text{c.m.}}) = \sum_J \sigma_{cap}(E_{\text{c.m.}}, J) P_{CN}(E_{\text{c.m.}}, J) W_{sur}(E_{\text{c.m.}}, J). \tag{4}$$

The factors are the partial capture cross sections, the fusion and survival probabilities. The contributing angular momenta in σ_{ER} are limited by the survival probability W_{sur} with $J_{max} \approx 10 - 20$ when highly fissile superheavy nuclei are produced for energies $E_{\text{c.m.}}$ above the Coulomb barrier. In this case the evaporation residue cross section can be approximated as

$$\sigma_{ER}(E_{\text{c.m.}}) = \sigma_{cap}^{eff}(E_{\text{c.m.}}) P_{CN}(E_{\text{c.m.}}, J=0) W_{sur}(E_{\text{c.m.}}, J=0) \tag{5}$$

with an effective capture cross section $\sigma_{cap}^{eff} = \pi \lambda\!\!\!^{-2} (J_{max}+1)^2 T(E_{\text{c.m.}}, J = 0)$. For reactions leading to optimal cross sections for superheavy nuclei, the bombarding energy $E_{\text{c.m.}}$ is above the outer Coulomb barrier, and we set $T(E_{\text{c.m.}}, J = 0) = 0.5$ near this barrier. Then the effective capture cross section results in the order of a few mb.

3.1 Fusion Probability

After the system is captured in a dinuclear configuration, the relative kinetic energy is transferred into potential and excitation energies. The dinuclear system develops in time by diffusion in the mass asymmetry and relative coordinates. The fusion probability P_{CN} gives the probability that the dinuclear system crosses the inner fusion barrier B_η^{fus} in η and forms the compound nucleus. This barrier is measured with respect to the potential $U(R_m, \eta_i)$ of the initial dinuclear configuration with the mass fragmentation η_i at the touching radius R_m. The fusion probability can be calculated by solving diffusion equations like the Fokker-Planck and master equations in the coordinates η and R or by using the Kramers approximation. The mean value $\bar{\eta}(t)$ tends to the symmetric fragmentation $\eta = 0$ with an increasing probability for quasifission. The quasifission is determined by the quasifission barrier $B_{qf}(\eta)$ measured with respect to the minimum of the potential $U(R, \eta)$ at $R = R_m$.

The minima in the potential $U(R_m, \eta)$ give a sensitive effect on the fusion probability and can be used in selecting optimum target and projectile combinations for producing superheavy elements. This idea goes back to Sandulescu *et al.* [6] who argued that the nuclei can fuse with higher probabilities along the valleys in an adiabatic potential in the R coordinate and pointed to the experimentally successful choice of target-projectile combinations with the Pb-nucleus as target as proof for their hypothesis. These ideas can simply be taken over into the DNS model. A certain initial system in a minimum is hindered by the barrier B_η^{sym} of the potential in η to change fast to more symmetric dinuclear systems which have a higher probability for quasifission because of the smaller quasifission barrier. Therefore, an asymmetric dinuclear system in a potential minimum has a longer lifetime with respect to its decay by quasifission than outside of a minimum and hence a larger chance to fuse by diffusion via nucleon transfer into the compound nucleus.

If the initial configuration is situated in a minimum of the driving potential $U(R_m, \eta)$, then the probability of the complete fusion depends on the quasistationary rate λ_η^{fus} for fusion, on λ_η^{sym} for going to a symmetric DNS and then decaying and on λ_R^{qf} for the decay of the initial DNS where the decays constitute the quasifission [7]:

$$P_{CN} = \lambda_\eta^{fus}/(\lambda_\eta^{fus} + \lambda_\eta^{sym} + \lambda_R^{qf}). \tag{6}$$

These rates can be expressed by two-dimensional Kramers-type formulas falling off exponentially with the fusion barrier B_η^{fus} in η, with the barrier B_η^{sym} in η in the direction to more symmetric configurations and with the quasifission barrier B_{qf}, respectively. The probability P_{CN} to overcome B_η^{fus} is approximately given by

$$P_{CN} \sim \exp(-(B_\eta^{fus} - \min[B_\eta^{sym}, B_{qf}])/T). \tag{7}$$

The barriers are taken from the potential $U(R, \eta)$; their heights are strongly influenced by shell and deformation effects. The temperature T is the local temperature of the initial DNS and is calculated as $T = \sqrt{E^*/a}$, where $a = A_{tot}/12$ MeV^{-1} ($A_{tot} = A_1 + A_2$) and E^* is the excitation energy of the DNS. The evolution of the initial DNS to more symmetric configurations and its decay are the main hindrance for complete fusion. In cold fusion reactions the quasifission mainly originates from the initial dinuclear system because of $B_{qf} < B_\eta^{sym}$, whereas in hot fusion reactions more symmetric dinuclear systems decay. In the later case, B_{qf} is initially larger than B_η^{sym}.

3.2 Survival Probability

The survival probability under the evaporation of x neutrons is treated as

$$W_{sur} = P_{xn}(E_{CN}^*) \prod_{i=1}^{x} \frac{\Gamma_n((E_{CN}^*)_i)}{\Gamma_n((E_{CN}^*)_i) + \Gamma_f((E_{CN}^*)_i)}. \tag{8}$$

Here, P_{xn} is the probability for the realization of the xn channel at the excitation energy $E_{CN}^* = E_{\rm c.m.} + Q_{fus}$ of the compound nucleus and the index i denotes the evaporation step. $(E^*_{CN})_i$ is the mean excitation energy of the compound nucleus at the beginning of step i with $(E^*_{CN})_1 = E^*_{CN}$. One can use an analytical expression for the ratio of the partial widths Γ_n and Γ_f of the neutron emission and fission, respectively. Also it is possible to calculate W_{sur} with statistical codes, e.g. GROGIF. In Fig. 2 we show the survival probability and the ratio Γ_n/Γ_f calculated for the neutron-evaporation in the fusion reactions ${}^{A}Z+{}^{208}$Pb for $J = 0$ with the GROGIF programme [8].

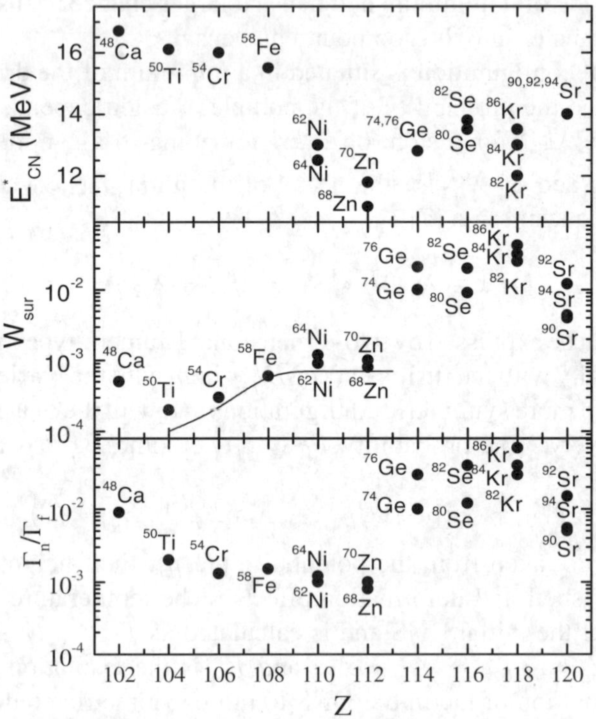

Figure 2. The upper part shows excitation energies used in the calculations. The middle and lower parts give W_{sur} and Γ_n/Γ_f, respectively, for the neutron-evaporation in the fusion reactions ${}^{A}Z$ + ^{208}Pb for $J = 0$ calculated with the code GROGIF.

3.3 Results for the Isotopic Trends

Various evaporation residue cross sections for cold and hot fusion reactions were obtained with the DNS model and published in journals and conference proceedings [9]. Here we discuss some recent results on the isotopic variation of fusion cross sections [7]. Fig. 3 shows that the production of the superheavy nucleus with Z=112 in the cold fusion reaction ^{70}Zn + ^{208}Pb with the emission of one neutron does not profit from the higher isospin. One can expect quite large cross sections in the ^{208}Pb-based reactions with ^{67}Zn and ^{68}Zn projectiles. The odd nucleus ^{275}112 has a larger P_{1n}, a larger fission barrier and a smaller neutron binding energy than the neighbouring even-even nuclei which leads to a larger W_{sur}. The fission barrier of the isotopes with Z=112 slightly increases with decreasing mass number from A=278 to 274 due to a large level spacing at N=162 for deformed nuclei. Going from ^{66}Zn to ^{68}Zn projectiles, we find a small decrease of P_{CN} but a stronger increase of the survival probability.

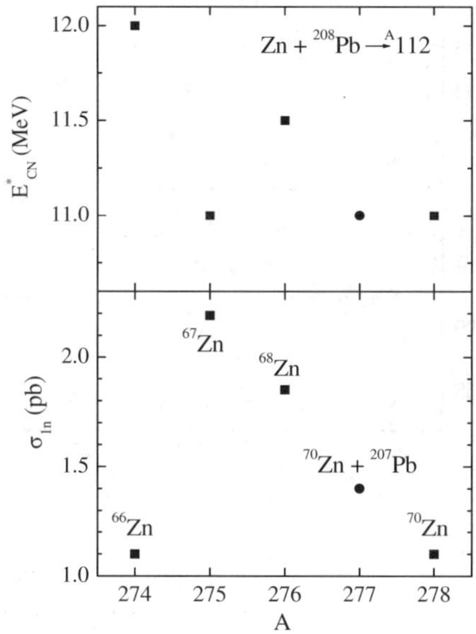

Figure 3. The calculated maximal evaporation residue cross sections in the $1n$ channel (lower part) at the corresponding excitation energies of the compound nuclei (upper part) for the fusion reactions Zn + ^{208}Pb $\rightarrow$ $^{(A-1)}$112 + n and ^{70}Zn + ^{207}Pb $\rightarrow$ ^{276}112 + n. Ref. [10] was used in the calculations.

The increase of the production cross section of element 112 with decreasing isospin was first suggested by Hofmann [11]. There was an experimental attempt to prove this effect in the reaction ^{68}Zn + ^{208}Pb. The lowest excitation energy in this experiment was about 1.5 MeV higher than the optimal one. We obtain $\sigma_{1n} = 0.3$ pb at this bombarding energy with our model.

In Fig. 4 we show the calculated maximal evaporation residue cross sections, the corresponding excitation energies of the compound nuclei and the Q-values in the 3 and $4n$ evaporation channels for the hot fusion reactions ^{48}Ca + ^{A}Pu. In this interval of the mass number the value of P_{CN} becomes larger with decreasing A in most cases. The larger P_{3n} and fission barrier and smaller neutron binding energy give a larger W_{sur} for the odd nucleus ^{241}Pu in comparison to the neighbouring even-even nuclei. The available experimental data [12] are well described. The calculated $\sigma_{4n} = 1$ pb for element 114 was reported before

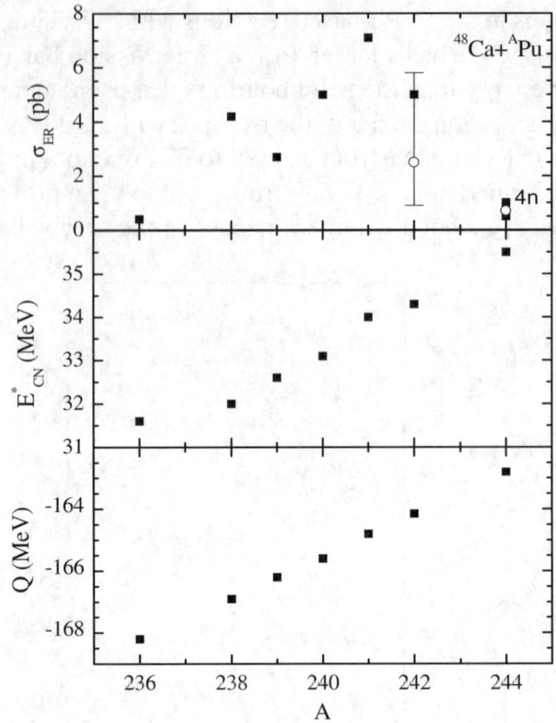

Figure 4. The calculated maximal evaporation residue cross sections for the $3n$ channel (upper part) at the corresponding excitation energies of the compound nuclei (middle part) and Q-values (lower part) for the fusion reactions ^{48}Ca + ^{A}Pu. For the reaction ^{48}Ca + ^{244}Pu with the same E^*_{CN} the cross section in the $4n$ evaporation channel is indicated. The experimental data of ^{48}Ca + ^{244}Pu $\rightarrow$ ^{288}114 + $4n$ ($E^*_{CN} \approx 35 - 37$ MeV) and ^{48}Ca + ^{242}Pu $\rightarrow$ ^{287}114 + $3n$ ($E^*_{CN} \approx 31 - 33$ MeV) are presented by open circles with error bars, respectively. Ref. [10] was used in the calculations.

the experiment. We like to mention that the calculations for all reactions were performed with the same parameters and assumptions.

4. The Quasifission Process

4.1 Master Equations for Nucleon Transfer

For the description of the quasifission process we choose master equations for the transfer of nucleons between the nuclei of the DNS. As coordinates we take the nucleon numbers of the light cluster, namely $Z_1 = Z$, $N_1 = N$ and $A_1 = A = Z + N$. The corresponding numbers of the heavy cluster are $Z_2 = Z_{tot} - Z_1$, $N_2 = N_{tot} - N_1$ and $A_2 = A_{tot} - A_1$, where Z_{tot}, N_{tot} and A_{tot} are the total charge, neutron and mass numbers of the DNS, respectively.

The starting point for the derivation of the master equations is the shell-model Hamiltonian of the DNS, which can be written in second quantization with the single-particle quantum numbers n_1 and n_2 of the clusters 1 and 2, respectively:

$$H = H_0 + V_{int}, \tag{9}$$

$$H_0 = \sum_{n_1} \epsilon_{n_1} a^+_{n_1} a_{n_1} + \sum_{n_2} \epsilon_{n_2} a^+_{n_2} a_{n_2} \tag{10}$$

$$V_{int} = \sum_{n_1,n_2} (g_{n_1,n_2}(R_m) a^+_{n_1} a_{n_2} + h.c.) \tag{11}$$

with $g_{n_1,n_2}(R_m) = \frac{1}{2}\langle n_1|U_1 + U_2|n_2\rangle$ where R_m is the distance of the nuclei at the minimum of the internuclear potential. Here, U_1 and U_2 are the single-particle potentials of the light and heavy nucleus, respectively. Since we assume a thermal equilibrium in the DNS we disregard the excitation of the light fragment by the heavy one and vice versa. Denoting the fragmentation by Z, N and $Z_{tot} - Z, N_{tot} - N$, we introduce the unperturbed states of the DNS, $|Z, N, n\rangle$, solving

$$H_0|Z, N, n\rangle = E^{Z,N}_n |Z, N, n\rangle \tag{12}$$

with the eigenvalues $E^{Z,N}_n$. Then a master equation for the probability $P_{Z,N}(n,t)$ to find the DNS in the state $|Z, N, n\rangle$ at time t can be formulated:

$$\begin{aligned} \frac{d}{dt} P_{Z,N}(n,t) &= \sum_{Z',N',n'} \lambda(Z,N,n|Z',N',n')[P_{Z',N'}(n',t) - P_{Z,N}(n,t)] \\ &- [\Lambda^{qf}_{Z,N}(n) + \Lambda^{fis}_{Z,N}(n)] P_{Z,N}(n,t). \end{aligned} \tag{13}$$

The transition rate λ ($\lambda(Z,N,n|Z',N',n') = \lambda(Z',N',n'|Z,N,n)$) can be calculated in time-dependent first order perturbation theory:

$$\begin{aligned} \lambda(Z,N,n|Z',N',n') &= \frac{1}{\Delta t} |\langle Z,N,n|V_{int}|Z',N',n'\rangle|^2 \\ &\times \frac{\sin^2[\Delta t(E^{Z,N}_n - E^{Z',N'}_{n'})/2\hbar]}{(E^{Z,N}_n - E^{Z',N'}_{n'})^2/4}, \end{aligned} \tag{14}$$

where the time interval $\Delta t = 10^{-22}$s is larger than the relaxation time of the mean field but considerably smaller than the reaction time. The quantities $\Lambda^{qf}_{Z,N}(n)$ and $\Lambda^{fis}_{Z,N}(n)$ are the rates for quasifission and for the fission of the heavy nucleus with $Z_{tot} - Z$ and $N_{tot} - N$ in the DNS, respectively.

Because of the single-particle character of the interaction energy (11), the transition rates (14) are only non-zero between states which differ by one particle-hole pair. In order to simplify the system of the differential equations, we assume that the DNS is in thermal equilibrium, and factorize $P_{Z,N}(n,t)$ in the form

$$P_{Z,N}(n,t) = P_{Z,N}(t)\Phi_{Z,N}(n,T). \tag{15}$$

Here, $\Phi_{Z,N}(n,T)$ is the probability for finding the DNS in the state n at a local temperature $T(Z,N)$ and is normalized to unity. Using Fermi occupation numbers for the single-particle states as functions of $T(Z,N)$ and summing over the DNS states n, we finally obtain the master equations used in the calculations [13]:

$$\begin{aligned}
\frac{d}{dt}P_{Z,N}(t) &= \Delta^{(-,0)}_{Z+1,N}\, P_{Z+1,N}(t) + \Delta^{(+,0)}_{Z-1,N}\, P_{Z-1,N}(t) \\
&+ \Delta^{(0,-)}_{Z,N+1}\, P_{Z,N+1}(t) + \Delta^{(0,+)}_{Z,N-1}\, P_{Z,N-1}(t) \\
&- (\Delta^{(-,0)}_{Z,N} + \Delta^{(+,0)}_{Z,N} + \Delta^{(0,-)}_{Z,N} \\
&+ \Delta^{(0,+)}_{Z,N} + \Lambda^{qf}_{Z,N} + \Lambda^{fis}_{Z,N})\, P_{Z,N}(t)
\end{aligned} \tag{16}$$

with ($n_1(T) = n_{n_1}(T), n_2(T) = n_{n_2}(T)$)

$$\Delta^{(\pm,0)}_{Z,N}(T) = \frac{1}{\Delta t}\sum_{n_1,n_2}^{Z} |g_{n_1,n_2}|^2\, n_{\substack{2\\1}}(T)(1 - n_{\substack{1\\2}}(T))\frac{\sin^2[\Delta t(\epsilon_{n_1} - \epsilon_{n_2})/2\hbar]}{(\epsilon_{n_1} - \epsilon_{n_2})^2/4},$$

$$\Delta^{(0,\pm)}_{Z,N}(T) = \frac{1}{\Delta t}\sum_{n_1,n_2}^{N} |g_{n_1,n_2}|^2\, n_{\substack{2\\1}}(T)(1 - n_{\substack{1\\2}}(T))\frac{\sin^2[\Delta t(\epsilon_{n_1} - \epsilon_{n_2})/2\hbar]}{(\epsilon_{n_1} - \epsilon_{n_2})^2/4},$$

$$\Lambda^{qf}_{Z,N}(T) = \sum_n \Lambda^{qf}_{Z,N}(n)\Phi_{Z,N}(n,T),\ \Lambda^{fis}_{Z,N}(T) = \sum_n \Lambda^{fis}_{Z,N}(n)\Phi_{Z,N}(n,T).$$

This system of equations has to be solved with the initial condition $P_{Z,N}(0) = \delta_{Z,Z_i}\cdot\delta_{N,N_i}$. The rates for the transfer of a proton or neutron from the heavy nucleus to the light one ($\Delta^{(+,0)}_{Z,N}$, $\Delta^{(0,+)}_{Z,N}$) and in opposite direction ($\Delta^{(-,0)}_{Z,N}$, $\Delta^{(0,-)}_{Z,N}$) depend on the temperature-dependent Fermi occupation numbers $n_{n_1}(T)$ and $n_{n_2}(T)$ of the single-particle states. We used single-particle states in spherical Woods-Saxon potentials with spin-orbit and Coulomb interactions. Also the rotation of the DNS is phenomenologically taken into account in the single-particle energies.

The decay rate $\Lambda^{qf}_{Z,N}(T)$ of the DNS for quasifission depends on the height B_{qf} of the outer potential barrier at $R_b = R_m + 1.5$ fm. The barrier is nearly independent of the angular momentum for $J < 70$ because the DNS has a large moment of inertia. The height of the barrier is about 4.5 MeV at $Z = 20$ and less than 0.5 MeV for $Z = Z_{tot}/2 \pm 10$. The decay rate can be treated with the

Kramers formula [13]

$$\Lambda^{qf}_{Z,N}(T) = \frac{\omega}{2\pi\omega^{B_{qf}}}\left(\sqrt{\left(\frac{\Gamma}{2\hbar}\right)^2 + (\omega^{B_{qf}})^2} - \frac{\Gamma}{2\hbar}\right) \times \exp\left(-\frac{B_{qf}(Z,N)}{T(Z,N)}\right). \quad (17)$$

The temperature $T(Z,N)$ is calculated with the Fermi-gas expression $T = (E^*/a)^{1/2}$ with the excitation energy $E^*(Z,N)$ of the DNS. For a nearly symmetric DNS we find about $T = 1.5$ MeV. An inverted harmonic oscillator with the frequency $\omega^{B_{qf}}$ approximates the potential around the top of the quasifission barrier, and a harmonic oscillator with the frequency ω gives the potential at the pocket. We use constant values for these quantities: $\hbar\omega^{B_{qf}} = 1.0$ MeV and $\hbar\omega = 2$ MeV, and set the width $\Gamma = 2.8$ MeV in (17).

4.2 Charge and Mass Yields

The charge and mass yields for quasifission can be calculated by the expression

$$Y_{Z,N}(t_0) = \Lambda^{qf}_{Z,N} \int_0^{t_0} P_{Z,N}(t)dt, \quad (18)$$

where $t_0 \approx (3-4)\cdot 10^{-20}$s is the reaction time which is about ten times larger than the time of deep-inelastic collisions. This time is determined by considering the balance equation for the probabilities:

$$\sum_{Z,N}(\Lambda^{qf}_{Z,N} + \Lambda^{fis}_{Z_{tot}-Z,N_{tot}-N}) \int_0^{t_0} P_{Z,N}(t)dt = 1 - \sum_{Z,N} P_{Z,N}(t_0). \quad (19)$$

The sum on the right hand side of (19) contains the fusion probability $P_{CN} = \sum_{Z<Z_{BG},N<N_{BG}} P_{Z,N}(t_0)$, defined by the fraction of probability existing for $Z < Z_{BG}$ and $N < N_{BG}$ at time t_0, where Z_{BG} and N_{BG} determine the barrier B^{fus}_{η} for fusion in the asymmetry coordinates. The DNS with $Z < Z_{BG}$ and $N < N_{BG}$ evolves to the compound nucleus in a time of 10^{-21}s which is short compared with the decay time of the compound nucleus.

The mass and charge yields of quasifission products are

$$Y(A) = \sum_Z Y_{Z,A-Z}(t_0), \qquad Y(Z) = \sum_N Y_{Z,N}(t_0). \quad (20)$$

The cross setion for quasifission can be calculated as

$$\sigma_{qf}(E_{\rm c.m.}) \approx (1 - P_{CN}(E_{\rm c.m.}) - P_{fis}(E_{\rm c.m.}))\sigma_{cap}(E_{\rm c.m.}) \quad (21)$$

with the fission probability $P_{fis}(E_{\rm c.m.})$ of the heavier fragment and the capture cross section $\sigma_{cap}(E_{\rm c.m.})$ responsible for the formation of the DNS:

$$\sigma_{cap}(E_{\rm c.m.}) = \frac{\pi\hbar^2}{2\mu E_{\rm c.m.}} J_{cap}(J_{cap}+1). \tag{22}$$

Here, $\hbar J_{cap} \leq (2\mu R_b^2[E_{\rm c.m.} - V(R_b)])^{1/2}$, and J_{cap} is smaller than the critical angular momentum J_{crit}. Trajectories with $J \geq J_{crit}$ contribute to deep-inelastic and quasi-elastic collisions.

4.3 Total Kinetic Energy Distribution

The average total kinetic energy (TKE) of the quasifission products and its dispersion depend strongly on the deformation of the fragments. For nearly symmetric dinuclear systems with $A = A_{tot}/2 \pm 20$, we found deformations which are about 3-4 times larger than the deformations of the nuclei in their ground states. These large polarisations of the DNS nuclei have to be regarded to explain the experimental TKEs of the quasifission fragments.

We assume that the distribution of the fragments as a function of Z, N and the deformations can be written as

$$W = W(Z,N,\beta_1,\beta_2) = Y_{Z,N}(t_0) w_{\beta_1}(Z,N) w_{\beta_2}(Z_{tot}-Z, N_{tot}-N). \tag{23}$$

The distributions of the deformations β_1 and β_2 are chosen as Gaussian distributions at fixed values of Z and N:

$$w_\beta(Z,N) = \frac{1}{\sqrt{2\pi\sigma_\beta^2}} \exp(-(\beta - \langle\beta\rangle)^2/(2\sigma_\beta^2)), \tag{24}$$

where $\sigma_\beta^2 = (\hbar\omega_{vib}/(2C_{vib}))\cdot\coth(\hbar\omega_{vib}/(2T))$ with the frequency ω_{vib} and the stiffness parameter C_{vib} of the quadrupole vibrations of the clusters. The determination of these quantities from the experimental spectra is described in detail in Ref. [13]. Using the distribution W, we get the average TKE as a function of the mass number A of the light fragment:

$$\langle TKE(A)\rangle = \int\int d\beta_1 d\beta_2 \sum_{\substack{Z,N\\ Z+N=A}} TKE\cdot W/(\int\int d\beta_1 d\beta_2 \sum_{\substack{Z,N\\ Z+N=A}} W), \tag{25}$$

where we set $TKE = V_C(R_b) + V_N(R_b)$ with the radius R_b at the position of the quasifission barrier. The variance of the TKE is

$$\begin{aligned}\sigma_{TKE}^2(A) \approx & \sum_Z (TKE)^2|_{\substack{\beta_1=\langle\beta_1\rangle\\ \beta_2=\langle\beta_2\rangle}} Y_{Z,A-Z}(t_0)/\sum_Z Y_{Z,A-Z}(t_0)\\ & - \langle TKE(A)\rangle^2 + (\sigma_{TKE}^{def}(A))_1^2 + (\sigma_{TKE}^{def}(A))_2^2\end{aligned} \tag{26}$$

with $(j = 1, 2)$

$$(\sigma_{TKE}^{def}(A))_j^2 = \sum_Z \left(\frac{\partial TKE}{\partial \beta_j} \right)^2 \Bigg|_{\substack{\beta_1=\langle\beta_1\rangle \\ \beta_2=\langle\beta_2\rangle}} \sigma_{\beta_j}^2 Y_{Z,A-Z}(t_0) / \sum_Z Y_{Z,A-Z}(t_0). \quad (27)$$

4.4 Results for Hot Fusion Reactions

The data for the mass yields and for the variances of the TKE of the fragments in the reactions with ^{48}Ca projectiles incident on actinide targets were measured by Itkis *et al.* [14] in Dubna. Fig. 5 shows the calculated mass yield $Y(A)$ and the variance of the TKE of the fragments as functions of the mass number for the hot fusion reaction $^{48}\mathrm{Ca} + {}^{238}\mathrm{U} \rightarrow {}^{286}112$ in comparison with experimental data. The incident energy corresponds to an excitation energy of the compound nucleus of 33.4 MeV. The calculated data in the figure are related to the primary fragments before neutron emission. Therefore, the maxima and minima in the calculated functions $Y(A)$ and $\sigma_{TKE}^2(A)$ are pronounced. Near the initial mass number A=48 the quasifission events overlap with the products of deep-inelastic collisions and were taken out in the experimental analysis since it is difficult to discriminate them from deep-inelastic events. The calculated peak near the initial mass number contains only quasifission events because we disregard angular momenta larger than the critical one which belong to deep-inelastic and quasi-elastic collisions.

Maxima in the mass and charge yields arise due to minima in the driving potential $U(R_m, Z, N, \beta_1, \beta_2, J)$ which are caused by shell effects in the dinuclear system. For $A > 48$, the maximum yield of the quasifission products appears around the nucleus ^{208}Pb for the heavy fragments ($A = 78$) where the driving potential has several minima. The height of the peak around $A = 80$ is 4.5 times larger than the height of the peaks in the symmetric mass region.

With our calculations we predict mass yields for light fragments with $A < 48$. Complementary to the heavier fragments with $A > 48$, the light fragments would give significant information about the dynamics of the dinuclear system on its way to the fused system. The yield of light products is known to be larger for higher beam energies. It would be a challenge for experimentalists to measure this region of the mass yield.

The minima in the dependence of σ_{TKE}^2 on A in Figs. 5 and 6 arise due to stiff nuclei in the DNS like Zr, Sn and Pb. The absolute height of the calculated σ_{TKE}^2 agrees well with the experimental data. The fluctuations due to the deformations contribute much more to the variance of the TKE than the fluctuations by the transfer of nucleons which come only into play for very asymmetric dinuclear systems.

In Fig. 6 we show the results of calculations in comparison with experimental data for the collision of ^{48}Ca with ^{248}Cm at an incident energy corresponding

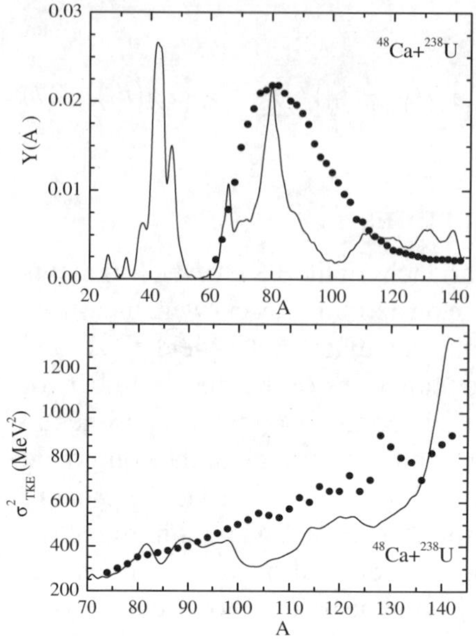

Figure 5. The calculated (solid lines) mass yield (upper part) and variance of the TKE (lower part) of the quasifission products as a function of mass number A of the light fragment for the hot fusion reaction ^{48}Ca+^{238}U $\rightarrow^{286}$112 at a bombarding energy corresponding to an excitation energy of the compound nucleus of 33.4 MeV. The experimental data [14] are shown by solid points.

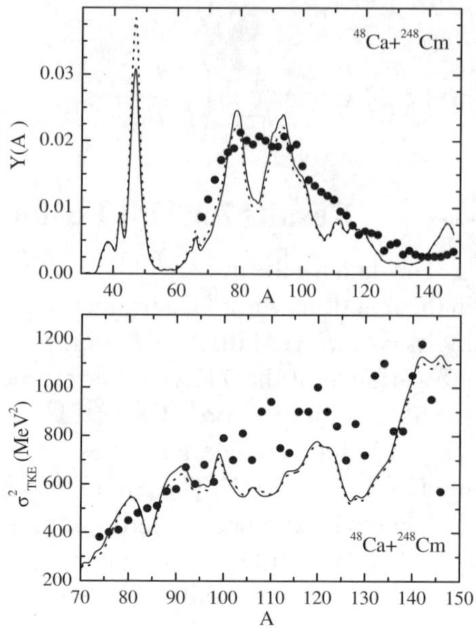

Figure 6. The same as in Fig. 5, but for the hot fusion reaction ^{48}Ca+^{248}Cm$\rightarrow^{296}$116 at the bombarding energy corresponding to an excitation energy of the compound nucleus of 37 MeV. The results calculated for $J = 0$ and 70 are presented by solid and dotted curves, respectively. The experimental data [14] are shown by solid points.

to an excitation of 37 MeV, respectively. As indicated by the solid and dotted curves in Fig. 6, we only find a weak dependence of the results on the angular momentum, chosen as $J = 0$ and 70, respectively, because the DNS has a large moment of inertia.

Beside the calculation for hot fusion reactions we also carried out calculations of quasifission yields and TKE variances for reactions with a ^{58}Fe beam, for cold fusion reactions with Pb targets and for reactions with lighter nuclei, e.g. ^{40}Ar + ^{165}Ho (for details see Ref. [13]). In Table 1 we list calculated relative contributions of the fusion-fission process with respect to the quasifission and the calculated total numbers of emitted neutrons for nearly symmetric quasifission splitting for examples of various types of reactions. The contribution of fusion-fission is mainly determined by the fusion probability P_{CN} since the survival probability of the excited compound nucleus is much less than unity. One recognizes that this contribution increases with incident en-

Table 1. The calculated results for the average variance σ^2_{TKE} of the TKE for the nearly symmetric quasifission products, the fraction of the fusion-fission events with respect to the quasifission events and the calculated average total number $M_n^{tot-sym}$ of emitted neutrons for nearly symmetric quasifission splitting in the mass region with $A_{tot}/2 - 20 \leq A \leq A_{tot}/2$. The reactions and the energies of the corresponding compound nuclei are indicated.

Reactions	E^*_{CN} (MeV)	σ^2_{TKE} (MeV^2)	$P_{CN}/\sum_{A=A_{tot}/2-20}^{A_{tot}/2} Y(A)$	$M_n^{tot-sym}$
^{40}Ar+^{165}Ho	89	119	1.1	5.5
	120	143	0.7	7.3
^{48}Ca+^{244}Pu	34.8	805	1.4×10^{-2}	7.5
	50	893	1.1×10^{-1}	8.5
^{86}Kr+^{208}Pb	17	738	2.1×10^{-7}	4.8
	30	813	2.0×10^{-5}	7.0

ergy, but remains small in the reactions considered. Therefore, the quasifission process mainly gives the yield of nearly symmetric products. For example, for the reaction ^{48}Ca ($E_{\text{c.m.}} = 193$ MeV) + ^{238}U (see Fig. 5) the calculated cross section of the yield of quasifission fragments with mass numbers $A_{tot}/2 \pm 20$ is about 4.5 mb at $J_{cap} = 25$ in good agreement with the measured value of about 5 mb.

5. Summary and Conclusions

The dinuclear system concept assumes two touching nuclei which exchange nucleons by transfer. With this molecular concept one describes the fusion to superheavy nuclei, the competing quasifission and also nuclear structure phenomena connected with cluster structures [15]. The available experimental data for fusion and quasifission of superheavy systems formed in lead- and actinide-based reactions can be well reproduced. This set of data gives an essential experimental support for the usefulness of the dinuclear system model. Future studies will settle down the basic principles of this concept.

References

[1] V.V. Volkov, Izv. AN SSSR ser. fiz **50** (1986) 1879

[2] G.G. Adamian, N.V. Antonenko, W. Scheid, Nucl. Phys. A **618** (1997) 176

[3] G.G. Adamian, N.V. Antonenko, S.P. Ivanova. W. Scheid, Nucl.Phys. A **646** (1999) 29

[4] A. Diaz-Torres, G.G. Adamian, N.V. Antonenko, W. Scheid, Phys. Lett B **481** (2000) 228

[5] G.G. Adamian, N.V. Antonenko, Yu. M. Tchuvil'sky, Phys. Lett. B **451** (1999) 289

[6] A. Sandulescu, R.K. Gupta, W. Scheid, W. Greiner, Phys. Lett B **60** (1976) 225; W. Greiner, Nuov. Cim. **110** A (1997) 1237

[7] G.G. Adamian, N.V. Antonenko, W. Scheid, *Unexpected isotopic trends in synthesis of superheavy nuclei*, preprint (2003); *Isotopic trends in production of superheavy nuclei in cold fusion reactions*, preprint (2003)

[8] G.G. Adamian, N.V. Antonenko, S.P. Ivanova, W. Scheid, Phys. Rev. **62** (2000) 064303; A.S. Zubov *et al.*, Phys. Rev. C **68** (2003) 014616

[9] *Proceedings of the Symposium on Nuclear Clusters: from Light Exotic to Superheavy Nuclei* (Rauischholzhausen, Germany, 2002), ed. R.V. Jolos and W. Scheid (EP Systema, Debrecen, 2003)

[10] P. Möller, J.R. Nix, At. Data Nucl. Data Tables **39** (1988) 213; LANL Preprint LA-UR-86-3983 (1986)

[11] S. Hofmann, Rep. Prog. Phys. **61** (1998) 636

[12] Yu. Ts. Oganessian *et al.*, Eur. Phys. J. A **13** (2002) 135; *ibidem* A **15** (2002) 201, and these Proceedings

[13] G.G. Adamian, N.V. Antonenko, W. Scheid, Phys. Rev. C **68** (2003) 034601

[14] M.G. Itkis *et al.* in Ref. [9], p. 315, and these Proceedings

[15] G.G. Adamian *et al.*, Phys. Rev. C **67** (2003) 054303; T.M Shneidman *et al.*, Phys. Rev. C **65** (2002) 024308

EXOTIC MATTER AT THE LOW DENSITY LIMIT: EXPLORING BOUND AND CONTINUUM STRUCTURES OF BORROMEAN HALO NUCLEI

J.S. Vaagen[1], G. Hagen[1], B.V. Danilin[2], S.N. Ershov[3] and I.J. Thompson[4]
[1] *SENTEF, Department of Physics, University of Bergen, N-5007 Bergen, Norway*
[2] *Russian Research Center "The Kurchatov Institute", 123182 Moscow, Russia*
[3] *Joint Institute for Nuclear Research, 141980 Dubna, Russia*
[4] *Department of Physics, University of Surrey, Guildford GU2 5XH, U.K.*

Abstract Studies of lighter dripline nuclei have revealed dramatic deviations from the matter distribution of ordinary stable nuclei and from shell model ordering. New stability patterns have been discovered, such as halo nuclei with co-existence of normal and low-density nucleonic matter, those of the Borromean nuclei being the most outstanding. Borromean nuclei provide the focus of these lectures, and also ongoing research on the halo continua.

1. Introduction

The nucleus is a self-organized system. The complexity of the nuclear many-body problem has made nuclear physics a field driven by discoveries of outstanding phenomena. With access to secondary exotic nuclear beams, the edges of the nuclear landscape itself are being explored, *i.e.* the very limits of nuclear existence. At these limits, the so-called neutron (proton) *driplines*, additional neutrons(protons) literally drip out of the nucleus. The exact locations of the boundaries are far from clear, and their complete determination awaits a better quantitative understanding of the nuclear system. Nuclei far from stability allow us to amplify and isolate particular aspects of the nuclear interaction and dynamics. Using what we learn from these nuclei we can then return to the nuclei of the world around us and understand them far better than ever before. Progress in nuclear structure is made at the various levels where one attempts to understand nuclear phenomena; (i) Experimental data and phenomenology, (ii) 'Local models' (effective models with few emergent degrees of freedom), (iii) 'Global nuclear models', (iv) Ab initio NN-interaction based procedures, (v) QCD.

W. Greiner et al. (eds.), Structure and Dynamics of Elementary Matter, 463–477.

There is an overall effort to explain higher levels in terms of lower ones, but a more reductionist "deductive approach" is likely to have as precursor an approach which tries to isolate and understand the characteristic degrees of freedom. These lectures highlight some characteristic findings in a field in rapid development, the physics of Borromean halo nuclei.

2. Halo Nuclei; Clusters and a Veil of extremely Dilute and Extended Neutron Matter

The granularity of a halo nucleus, say that of the cardinal case of ^{11}Li, is not trivial, see Fig. 1. The spatial extension is huge, the rms matter radius of ^{11}Li is as large as that of ^{48}Ca, and the radius of the halo neutrons is as large as for the outermost neutrons in ^{208}Pb. It is a true challenge for theory to account for both extension and granularity.

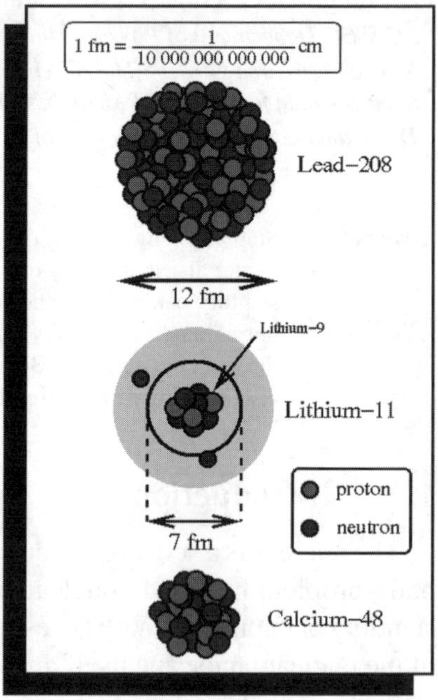

Figure 1. The matter size of ^{11}Li is compared to that of ^{48}Ca and ^{208}Pb

Light nuclei constitute so far the part of the nuclear landscape where the neutron dripline has been reached. Triggered by Tanihata's discovery (1985) [1] of vastly spatially extended nuclei (^{6,8}He; ^{11}Li; ^{11}Be) at the neutron dripline, the initial idea of (binary) halos was suggested by B. Jonson and P.G. Hansen [2]. Subsequent developments have deepened and enriched the picture of halos as outstanding structural dripline phenomena with extreme clusterization into an ordinary core nucleus and veil of halo nucleons – forming exceptionally dilute neutron matter. The origin of the stratification is purely quantum mechanical in nature, and only partly understood, but a prerequisite is low angular momentum motion for the halo particles and few-body dynamics such as in Borromean nuclei[1] character-

[1]The three Borromean rings, the heraldic symbol of the Italian Borromeo family, are interlocked in such a way that if any of them were removed, the other two would also fall apart. The three interwined Borromean rings are now widely used as the logo of the halo field.

ized by pairwise constituents with no bound states. In the limit of vanishing binding, extremely large halos may occur.

In a remarkable paper in atomic physics in 1988 [3], "Loosely bound states of three particles", Zhen and Macek write; "Although there is no corresponding experimental data to confirm these results at present, the theory is nevertheless of great interest because it predicts some new physical properties of three-body systems, among which the most remarkable one is the existence of loosely bound states in such systems, even though the interaction potential between the particles has no wells strong enough to bind any two of the three particles separately. Because of this, and since the three-body system is the simplest model in many-body problems, it seems desirable to use the three-body system (instead of two-body) as the basic block of the many-body system".

The dream had already come true, three years earlier – but in another field, at the neutron dripline when Isao Tanihata and his team made their discoveries in 1985. In the following decade ^{6}He and ^{11}Li became understood as realizations of exactly the mechanisms Zhen and Macek were speaking about, and were in 1993 baptized Borromean halo nuclei [4]. At that time we were, however, not aware of the Zhen/Macek paper.

The Borromean property is clearly exhibited by the hadronic stability of various isotope chains, see Fig. 2. Thus, while ^{5}He is unbound ^{6}He is bound, ^{7}He is unbound but ^{8}He is again bound. A recent review [5] looks at all the dripline nuclei.

Pioneering activities usually employ some kind of scaffolding and the discoveries are highlighted in terms of heuristic interpretations. These emphasize the new findings, and halo physics is no exception. The binary core-(point)dineutron picture of Hansen and Jonson for two-neutron halos (1987), originally just meant to illustrate how large extension and weak binding may be related, turned out to be an oversimplification. Since 1990, halo models have been developed where the N-N degree of freedom is no longer frozen, but chosen in accordance with the free NN interaction inspired by the dilute character of the halo. Thus focus has been shifted to features genuinely related to the intrinsic character of the halo, and of the interplay between halo degrees of freedom.

The present picture of a 'normal' nucleus veiled in a neutron halo, i.e. co-existence of normal and low density nuclear material, has emerged from a concerted effort between subsequent dedicated measurements and theory. How this stratification takes place still needs to be better understood, but nature seems to have provided a laboratory where 'eigenproperties' of the low-density halo can be studied, subjected only to a mild influence from the intrinsic structure of the core object. Few-body dynamics plays the crucial role in any adequate description of halo properties, since for halo bound states most of the wave function is concentrated in the classically forbidden region. The chain of He

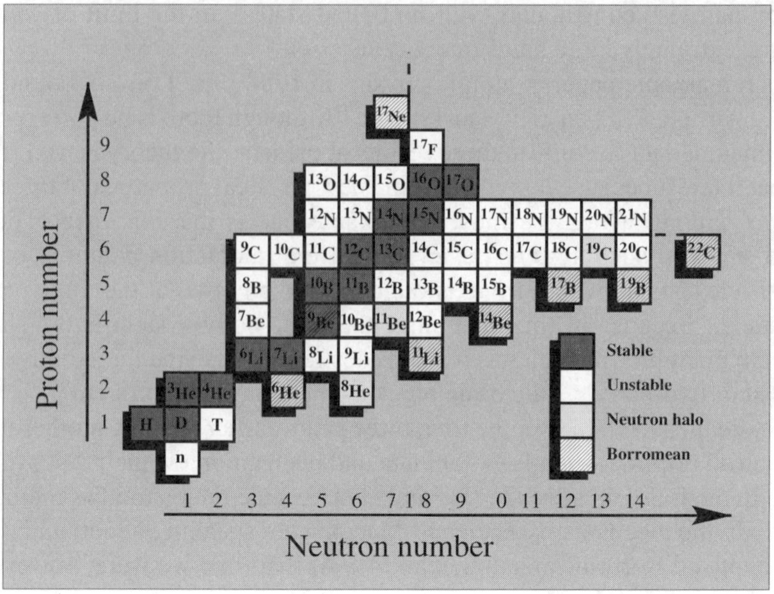

Figure 2. The nuclear chart exhibits the stability of Borromean nuclei among different isotope chains

isotopes with an alpha core has become particularly useful as set of benchmark systems.

In three-body models of halo nuclei such as ^{6}He, Pauli blocking is needed to remove components of the halo wave function that would disappear under full antisymmetrization. This, and also other aspects of the composite nature of the clusters make the challenge somewhat different from that of three-nucleon systems.

2.1 Enriching the Nuclear Paradigm

Thomas Kuhn, in his famous book "The Structure of Scientific Revolutions", claims that we are hypnotized by our paradigms. Kuhn's conclusions are questioned by Steven Weinberg in a recent article "The Revolution That Didn't Happen", in the New York Review [6]. In particular Weinberg argues that "He (Kuhn) was quite wrong in saying that it is no part of the work of normal science to find new sorts of phenomena". Here we side with Weinberg. New discoveries, also fundamental, do not as a rule require overthrowing paradigms, they rather extend and modify a ruling paradigm. What we are doing these days is to enrich and widen the nuclear paradigm. The lessons we have learned during the last decade or so by directing, with increasing success, our nuclear studies toward extreme conditions are truly remarkable.

The recent theory advances have influenced dripline physics conceptually, and also its methodology and logistics. Two-body recipes have been replaced by appropriate 3-body (few-body) theory. And most importantly: a theory framework with growing predictive power is emerging.

The Hoyle resonance in ^{12}C plays a crucial role in the synthesis of the elements; "It wouldn't be missed if it didn't exist" to paraphrase a statement by the Danish writer Piet Hein. *It is the doorway to our Universe.* The processes leading to this doorway *all proceed* from ^{4}He *via Borromean systems*: (i) in Red Giants via the triple α-process; (ii) in Supernovae via $^{9}_{4}Be_5$ also Borromean; and (iii) a less probable side-route to ^{9}Be via ^{6}He. Recall that neither $(\alpha\alpha)$, (αn) nor (nn) can form bound binary systems, only resonances.

3. Simplicity in Complexity

It is customary in fields with a rich history to try to fit the new findings within the ruling formalisms: in nuclear physics a many-particle shell model type picture. The shell model has, however, not turned out very suited for halo systems, and has to be fundamentally modified for neutrons so close to the continuum threshold. We will return to this. The concept of orbits in the sense of relative motions with respect to the core, is on the other hand still fruitful. Thus the recognition of the instrumental role of low angular momentum "intruder" orbits (s and p) has become essential to our understanding of exotic structure formation and the melting of shell structures at the neutron dripline. Borromean nuclei is most easily understood in a three-body cluster formulation, using the Hyper Harmonic method, which for bound states gives the leading physics in terms of only a few wave function components [4].

A dream dating back to the early years of nuclear physics is that of starting from the free binary interactions, for reactions as well as structure. The influence of the nuclear medium has complicated this approach. For loosely bound dripline nuclei this question has renewed interest. The nature of the halo seems sufficiently dilute to justify approaches where the starting point is the free nucleon-nucleon t-matrix.

3.1 Borromean Halos: Three-body Theory

Fig. 3 gives an impression of the average (r.m.s.) geometrical separations in ^{11}Li, indicating also the ranges of the interactions between the constituents. The geometry of the figure corresponds to three-body calculations, the gross structure is however insensitive to finer details of the calculations. For ^{6}He the relative geometry is quite similar, the core radius is now 1.5 fm. The abnormal size of the halo systems is obviously related to the unusually small separation energies ($\sim$ 1 MeV or less), and means that there is a large probability (often more than 50%) of the system being in classically forbidden configurations. The situation is truly one of 'residence in forbidden regions' [7].

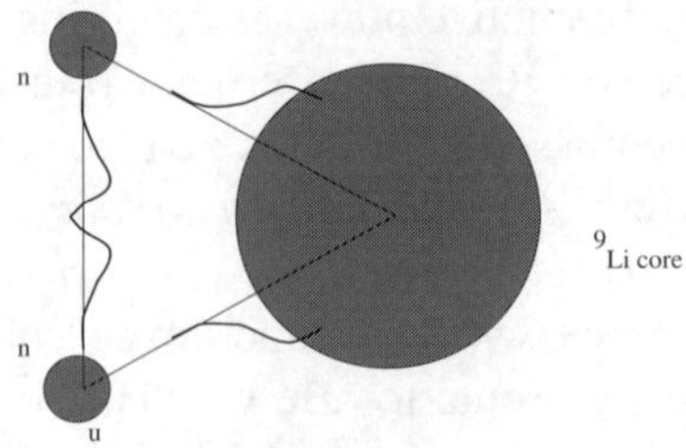

Figure 3. The geometrical structure of ^{11}Li and its core ^{9}Li with r.m.s radii 3.5 fm and 2.3 fm, exhibiting Borromean tunneling into classically forbidden regions. The ranges of the binary interactions are indicated. Distributing the 11 nucleons with a usual Fermi-type density distribution would give about 1/3 of normal nuclear density.

The physical building blocks for describing such systems are, just like for transfer reactions, appropriate overlap functions $\mathcal{O}_C^A(x) = \langle \Psi_C(C) \,|\, \Psi_A(C,x)\rangle$ (where x are the translationally invariant halo coordinates) between a given state A for the halo nucleus and a possible state C for the core. In a few-body model, the overlap coincides with the few-body wave function.

The asymptotic behaviour of a one-neutron overlap between bound states is simply $\exp(-\kappa_C^A r_{1C})$ where r_{1C} is the relative coordinate between the core and the valence neutron while the decay constant $\kappa_C^A = \sqrt{2\mu_{1C} S_C^A}/\hbar$ is governed by the separation energy between the two states being connected.

The asymptotic behaviour becomes essentially more complicated already for two halo neutrons. For Borromean nuclei [4], however, it simplifies to $\exp(-\kappa_C^A \rho)$ where ρ is the hyperradius, expressed in n-n and (nn)-C relative distances as

$$\rho^2 = \frac{1}{2} r_{nn}^2 + \frac{2C}{C+2} r_{(nn)C}^2$$

while the decay constant now is $\kappa_C^A = \sqrt{2MS_C^A}/\hbar$ and S_C^A the two-neutron separation energy, M is the nucleon mass. There is also a quantum number K, being the hypermoment quantum number associated with the 'angular' α conjugate of the hyperradius ρ. This is essentially a generalisation of the transformation to polar coordinates in the 2-body case.

If an initial expansion of the three-body wave function on hyperspherical harmonics (HH) is used [4], the three-body Schrödinger equation simply becomes a set of coupled equations for radial amplitude functions $\chi_{K\gamma}(\rho)$ in the hyperradius. The mathematical scheme bears a resemblance with that of a particle in a strongly deformed field $V_{K\gamma,K'\gamma'}(\rho)$:

$$\Big(-\frac{\hbar^2}{2M}\Big[\frac{d^2}{d\rho^2}-\frac{(K+3/2)(K+5/2)}{\rho^2}\Big]+V_{K\gamma,K\gamma}(\rho)-E\Big)\chi_{K\gamma}(\rho)$$
$$=-\sum_{K'\gamma'\neq K\gamma} V_{K\gamma,K'\gamma'}(\rho)\chi_{K'\gamma'}(\rho). \quad (1)$$

The coupling matrix elements are

$$V_{K'\gamma',K\gamma}(\rho)=\langle\mathcal{Y}_{K\gamma}(\Omega_5)\mid\sum_{j\neq i=1}^{3}V_{ij}\mid\mathcal{Y}_{K'\gamma'}(\Omega_5)\rangle \quad (2)$$

where V_{ij} are the binary potentials and $\gamma=\{l_x,l_y,L,S\}$. Here l_x and l_y are relative angular momenta, say $l_x=l_{nn}$, $l_y=l_{(nn)C}$ in the di-neutron set of Jacobi (T-)coordinates, while Ω_5 contains the hyperangle α in addition to the four spatial angles of a Jacobi set. The hypermoment $K=2n+l_x+l_y$ for some integer n = 0, 1, 2 ... For bound states ($E<0$) each $\chi_{K\gamma}(\rho)$ is regular for $\rho\rightarrow 0$ and falls off exponentially as $\exp(-\kappa_C^A\rho)$ for $\rho\rightarrow\infty$.

For a Borromean system the above HH equations can easily be solved for continuum states at positive energy $E>0$. The asymptotic boundary condition is again a generalisation of binary scattering to that of $3\rightarrow 3$ scattering and is given in the hyperradius by $\chi_K^{K'}(\rho)\rightarrow H_K^-(\rho)\delta_{KK'}-S_{KK'}H_K^+(\rho)$ for a unit-amplitude boundary condition in channel K'.

Borromean systems have no binary binding. So why is the system bound? Part of the explanation is related to the three-body HH centrifugal barrier, the term in eq.(1) which is proportional to $(K+3/2)(K+5/2)/\rho^2$. Even if $K=0$, implying that the total orbital angular momentum is zero, the three-body centrifugal barrier does not vanish, contrary to the two-body case. For sufficient binding between the bodies, this may provide a barrier inside which quasi-bound motion may be established, giving rise to three-body resonances in the continuum, even if there are only s-wave virtual states in the binary subsystems. Furthermore, as Efimov [8] first pointed out, the attraction in ρ contains components of asymptotic behaviour ρ^{-3} for the diagonal terms.

3.2 Two-neutron Borromean Ground States

The test-bench three-body Borromean halo nucleus has been ^{6}He. The ground-state properties of ^{6}He are quite well determined by the known $\alpha-n$ and $n-n$ pairwise potentials, and a variety of few-body calculations have been performed for the ground state wave function, and these are largely in agreement with each other. They support a picture in which the valence neutrons have a 90% probability of being in a $(0p_{3/2})^2$ configuration, with the $n-n$ attraction increasing the $S=0$ component of the wave function from an uncorrelated

67% to a correlated 89%. The structure of the 0^+ ground state of ^{6}He, with its spatial density dominated by a "di-neutron" and a "cigar" peak, is shown in fig. 4, and has been reproduced by a number of few-body procedures. 'Pure three-body' models using pairwise interactions underbind ^{6}He by around 0.5 MeV. This is almost certainly due to the neglect of the $t + t$ degree of freedom.

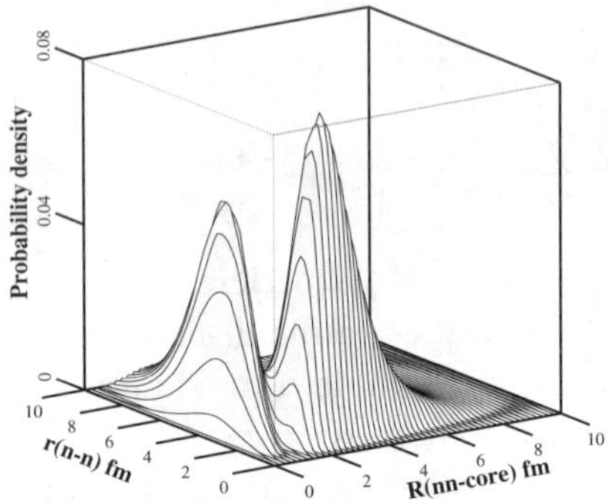

Figure 4. The joint radial probability of the halo neutrons in ^{6}He, averaged over all angles. The rightmost peak has the two neutrons closest together.

The ground-state properties of ^{11}Li are not so well determined, since the available information on the binary neutron-^{9}Li(core) channel is not sufficiently unambiguous and non-contradictory. It is now generally agreed, based on a variety of observables, that in a two-neutron halo system such as ^{11}Li, the intruder $(1s_{1/2})^2$ two-particle configurations can be mixed by the nn interaction with the $(0p_{1/2})^2$ configuration, so the ground state will be roughly an equal superposition of these two arrangements for ^{11}Li. The weights and wave functions are given by the solution of the three-body problem. We predicted [9] that the presence of a low-lying s virtual (antibound) state in the $n+^9$Li system, when included in three-body calculations, would have a pronounced effect on the structure of the ^{11}Li halo.

Low-energy two-neutron transfer reactions with Borromean nuclei have been shown to be an effective instrument for studying both the structure of such nuclei and the dynamics of nuclear reactions where they take part. A four-body model has been developed to describe such two-nucleon transfer processes within DWBA, and also extended to coupled reaction channels. Using a realistic 3-body bound state wave function of ^{6}He, the role of its separate components has been thoroughly studied. In particular, it was found that the "di-neutron" configuration of the ^{6}He nucleus gives the dominant contribution to the two-neutron transfer cross sections, measured at FLNR, Dubna, [10]. The possibilities of using multi-neutron transfer reactions for studying the structure of other radioactive nuclei, such as ^{8}He, have now become part of the schedule at leading international laboratories.

4. Exciting a Halo to the Borromean Continuum

Dripline physics, involving nuclei with only one or a few weakly bound states, is physics of threshold phenomena where structure and reaction theory merge. Recently the theoretical few-body approach has been successfully extended to also encompass three-body continuum structure, response functions and reactions involving Borromean systems. Borromean continua, for which the three-body threshold is the lowest-lying, are intriguing. Here the continuum pairing, and also the different decay channels via low-lying binary resonant states, can change shell-model expectations drastically. An example is the ^{6}He nucleus with only one experimentally confirmed narrow (2^{+}) resonance below the $t + t$ threshold at 13 MeV. The nature of experimentally observed strength concentration at 1–3 MeV (from three-body threshold) in ^{6}He, and at 1 MeV in ^{11}Li, is being clarified. The strength may contain three-body resonances, or a new kind of collective motion – an induced giant dipole moment due to oscillations of the inert core against the halo neutrons – that does not correspond to a pole in the S-matrix.

The intrinsic properties of Borromean continua are directly revealed in $3 \rightarrow 3$ scattering, which however is almost impossible to perform experimentally, but could be significant in neutron stars and supernovi. In ordinary binary reactions the continuum is usually explored via responses that induce transitions from the ground state to the continuum. A reliable way to study halo continuum properties is one-step transitions in nuclear reactions. A microscopic four-body reaction theory has recently been developed [11], a rather comprehensive task because of the intertwining of the ground state and continuum structures, influenced by reaction mechanisms.

In the vicinity of a 'true' three-body resonance the $3 \rightarrow 3$ scattering asymptotic amplitude has the analytical property

$$A(E_\kappa) \sim \frac{c}{E_\kappa - (E_0 - i\Gamma_0/2)} \tag{3}$$

where E_κ is the excitation energy calculated from the three-body threshold. E_0 and Γ_0 are the position and width of the resonance, while c depends on spatial and momentum angles in hyperspherical coordinates [4]. This coincides in the two-body case with the resonant structure of the wave function in the interior region, a property we have also observed (but not proved) for the Borromean three-body case :

$$\psi(\rho; E_\kappa) \sim C(E_\kappa)\ \psi^R(\rho) \tag{4}$$

with

$$|C(E_\kappa)|^2 = \frac{const}{(E_\kappa - E_0)^2 + \Gamma_0^2/4} \tag{5}$$

where $\psi^R(\rho)$ is the energy-independent form of the internal part of the scattering wave function. Using this resonance representation for the wave function

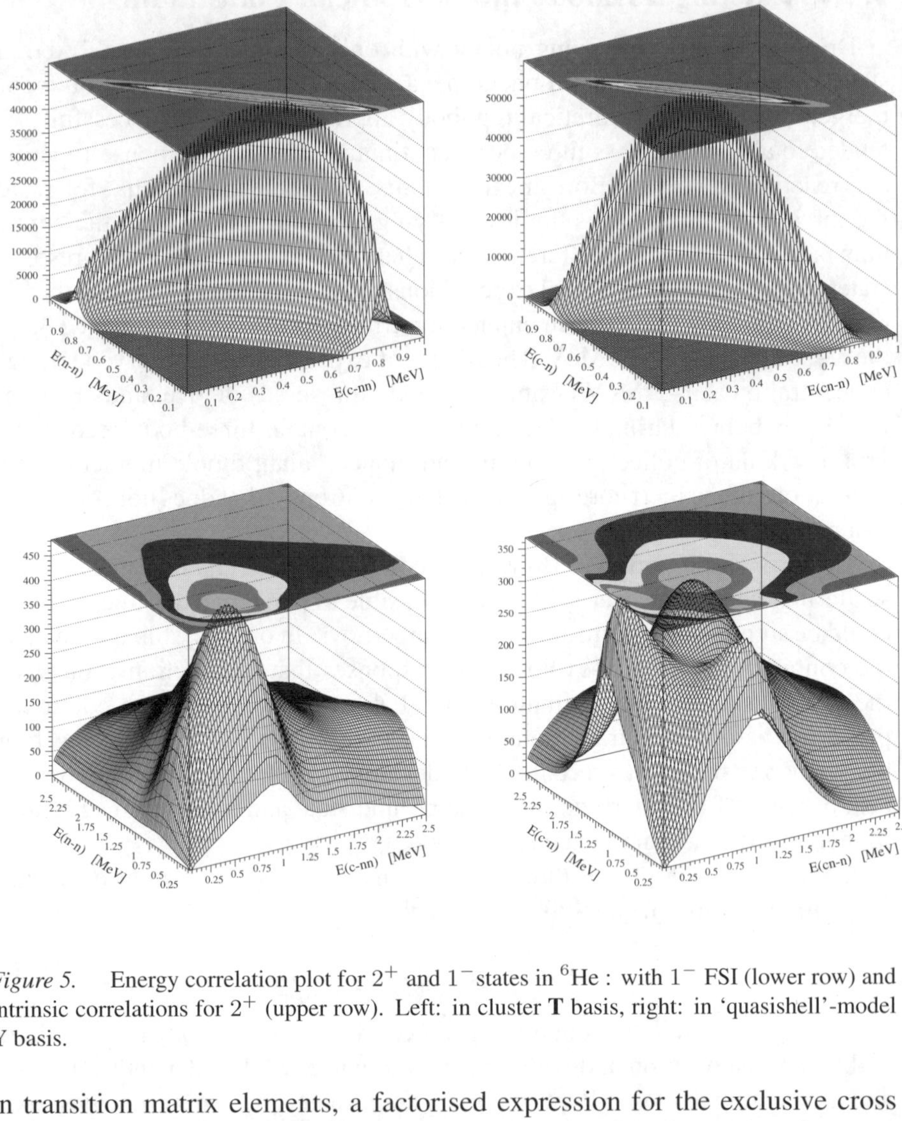

Figure 5. Energy correlation plot for 2^+ and 1^-states in ^{6}He : with 1^- FSI (lower row) and intrinsic correlations for 2^+ (upper row). Left: in cluster **T** basis, right: in 'quasishell'-model **Y** basis.

in transition matrix elements, a factorised expression for the exclusive cross section for three-body fragmentation can be obtained. After integration over the directions of momenta $\vec{k}_x, \vec{k}_y$ between the fragments, the double differential cross-section becomes proportional to :

$$\frac{d^2\sigma}{d\varepsilon_{x_i} d\varepsilon_{y_i}} \sim \sqrt{\varepsilon_{x_i}\varepsilon_{y_i}}\, |C(E_\kappa)|^2 \sim \frac{\sqrt{\varepsilon_{x_i}\varepsilon_{y_i}}}{(\varepsilon_{x_i} + \varepsilon_{y_i} - E_0)^2 + \Gamma_0^2/4} \tag{6}$$

where $\sqrt{\varepsilon_{x_i}\varepsilon_{y_i}}$ is the phase volume accessible to fragments in three-body decay.

Thus it is clear that a correlation plot of $d^2\sigma/d\varepsilon_{x_i} d\varepsilon_{y_i}$ should have maxima along a ridge which is the straight line $E_0 = \varepsilon_{x_i} + \varepsilon_{y_i}$ with sections in a contour plot of elliptical type with width $\sim \Gamma_0$. Halo neutrons and core can be combined in two different ways (T and Y) into binary Jacobian subsystems. One (T) is when the x-coordinate corresponds to the $n-n$ subsystem while y refers to $C-(nn)$. In the second case (Y) the x, y denote $C-n$ and $(Cn)-n$, respectively. The excitation energy E_κ and the phase volume $\sqrt{\varepsilon_x \varepsilon_y}$ are invariant for any binary Jacobian partitions. Then the ridge pattern of the energy correlation plot should be similar in T and Y partitions in the case of a true three-body resonance, and modified if not. This is a *decisive* tool for identification of a true three-body resonance directly from experimental data, in addition to calculated phase shifts.

The most extensive and reliable analysis has been carried out for ^{6}He where as a check point the known 2_1^+ resonance is reproduced. The calculations give 1^- strength concentrations at lower energies (soft modes) in the proximity of the three-body threshold and predict new 2^+, 1^+ (and possibly 0^+) resonances at slightly higher energies in ^{6}He. Fig. 5 shows how a correlation plot can be used to prove that while 2_1^+ is a 'true' three-body resonance, the much debated 1^- dipole strength is not!

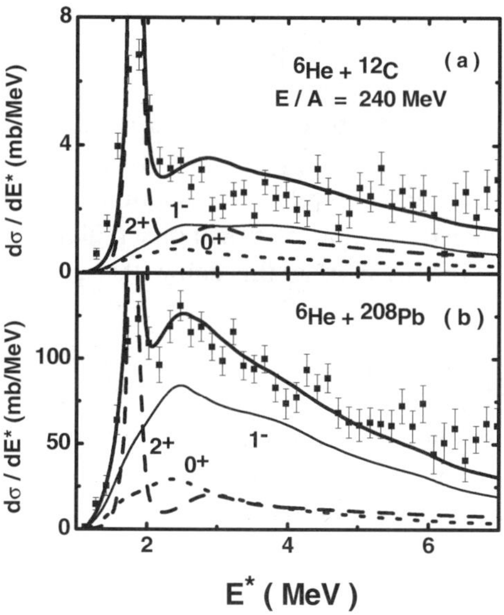

Figure 6. Spectrum of ^{6}He in collision with (a) ^{12}C and (b) ^{208}Pb at energy E/A 240 MeV. Dotted, thin solid and dashed lines are the monopole, dipole and quadrupole excitations, respectively. See [11] for further detail.

This correlated microscopic continuum structure has been used to gain new insight in halo fragmentation in diffractive breakup. Thus the crucial role of the low-lying three-body resonances and soft modes in determining distribution characteristics (width and shape) has been demonstrated. A surprising memory of the low-lying continuum structure prevails even in inclusive observables. Fig. 6 shows GSI data and calculations [11] for ^{6}He break-up to the halo continuum.

5. Ab initio Calculations of Halos

The most ambitious of these 'reductionist' attempts are Green's Function Monte Carlo (GFMC) calculations [12], extending previous variational Monte Carlo (VMC). According to the authors these are the first microscopic calculations that directly produce nuclear shell structure from realistic interactions that fit NN scattering data. Another line of approach are the large-basis no-core shell-model (NCSM) calculations of Barrett's group [13], where the shell model is combined with microscopic effective interactions derived from modern nucleon-nucleon potentials. Substantial progress has been made in recent years, but a three-body (Illinois) interaction has to be supplemented to make the GFMC able to bind loosely bound halo systems like ^{6}He and ^{8}He.

The size of the GFMC (and NCSM) calculations, regrettably, grows beyond what is manageable even before ^{11}Li is reached. The practitioners have, however, already had success with their pioneering attempts. Thus the GFMC calculations for A=6 do produce an alpha-like core object, and two alpha-particles for ^{8}Be, both promising features.

To check the information content in GFMC type wave functions, a joint venture with few-body modeling is needed to explore critical observables and to improve the asymptotic behavior of GFMC. Work to this end will require the best available computer support.

6. Expansions on a Berggren Basis

In nuclear physics, expansion of complex objects on an unperturbed single (two) particle basis, generated by a suitable potential, has been common practice. In the traditional shell model the harmonic oscillator basis has served as a paradigmatic tool for describing bound nuclei, providing a reasonable description for the more deeply bound single particle orbitals. As one approaches the nuclear driplines the binding reduces, and in the Borromean limit the core-n interaction is unable to uphold bound motion. Thus there will be no bound state part in the usual completeness relation, or *resolution of unity*, proven by Newton in Ref. [14],

$$\mathbf{1} = \sum_{nl} |\psi_{nl}\rangle\langle\psi_{nl}| + \sum_{l} \frac{1}{2} \int_{-\infty}^{\infty} dk k^2 |\psi_l(k)\rangle\langle\psi_l(k)|. \tag{7}$$

The exploration of the nuclear driplines has now pushed the traditional many body methods to their limits of applicability, and coupling to the binary continuum plays a vital role. A modification of the shell model has become known as the *Gamow shell model*, see Refs. [15, 16], or the *Berggren expansion*, where bound, resonant, antibound and continuum states are treated on equal footing, and has been under development for the last few years. As alluded to earlier, we had to include antibound (virtual) s motion in our three-body calculation of

^{11}Li to understand the physics findings of this remarkable system. Recently the role of antibound states in the Gamow shell model description of this system has also been discussed [16].

The study of two-body resonant structures has a long history in theoretical physics. There exists a variety of methods described in for example textbooks such as [14]. Among the more popular methods one finds the complex scaling method (CSM) and the method based on analytic continuation in the coupling constant (ACCC).

Recently we have considered an approach formulated for integral equations in momentum space. The method is based on deforming the contour integrals in momentum space, and known as the contour deformation or distortion method (CDM). It has been shown previously that a *contour rotation* in momentum space is equivalent to a rotation of the corresponding differential equation in coordinate space. The coordinate space analog is often referred to as the *dilation group transformation*, or *complex scaling*.

Complex scaling in coordinate space has for a long time been used extensively in atomic and molecular physics, see Ref. [17]. During the last decade it has also been applied in nuclear physics, as interest in loosely bound nuclear halo systems has grown, see for example Refs. [18].

There are several advantages in considering the contour deformation method in momentum space. First, most realistic potentials derived from field theoretical considerations are given explicitly in momentum space. Secondly, the boundary conditions are automatically built into the integral equations. Moreover, the Gamow states in momentum space are non-oscillating and rapidly decreasing, even for Gamow states with large widths far from the real energy axis, as opposed to the complex scaled coordinate space counterpart. These states are represented by strongly oscillating and exponentially decaying functions.

A revitalising of the contour deformation method in momentum space has therefore taken place. CDM is a method which allows for accurate and stable solutions of binary bound, anti-bound, capture, and decay states. We consider a generalized type of contour, allowing for an analytic continuation in the third quadrant of the complex k-plane. Antibound - and capture states near the scattering threshold may then be calculated at a specified accuracy. This choice of contour can be regarded as a special case of the *Berggren class* of contours [19]. Berggren studied various completeness relations derived by analytic continuation of the Newton completeness relation to the complex plane. The Berggren completeness includes discrete summation over resonant as well as bound states. Figure 7 gives an example of a Berggren type contour which also includes antibound states. To solve a many-body problem by expansions on a Berggren basis is no straightforward task. An important lesson so far for Borromean halos is that the continuum states on the complex contour give vi-

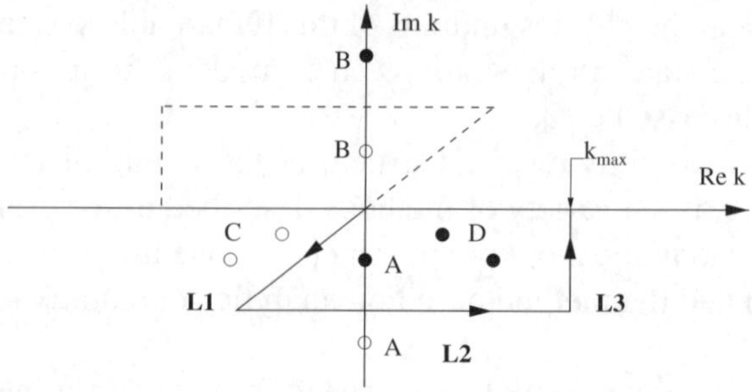

Figure 7. Contour $C^+_{R+T} = L_1 + L_2 + L_3$ is given by the solid line, while the contour C^-_{R+T} is given by the dashed line. The contour $C_{R+T} = C^+_{R+T} + C^-_{R+T}$ is clearly *inversion symmetric*. The two body spectrum which is exposed by this contour is marked by filled circles • and the excluded spectrum by open circles ◦. The full spectrum includes bound states (B), anti-bound (A), decay (D) and capture (C) resonant states.

tal contributions, in addition to those from physical resonances and antibound states.

7. Conclusions

After a decade of work, essential physics of Borromean bound states in two-nucleon halo nuclei is now understood in terms of a three-body cluster model formulation. The stratification itself has, however, to be understood at a deeper nucleonic level and encouraging developments to this end, such as within the Green's function Monte Carlo method, have been reported in recent years. These methods are also tools to understand multi-neutron halos, such as for ^{8}He, where a cluster formulation involves the challenges of the five-body problem. Loosely bound multi-neutron matter within halos provides a fascinating laboratory for low-density research. With few bound states close to multifragment break-up threshold, investigations of the physics of halo continua represent another direction of research in rapid development, involving also need for a better understanding of few-body reaction theory. We have given the status of the art for Borromean two-neutron halos.

8. Acknowledgments

We acknowledge contributions from other members of the RNBT (Russian-Nordic-British Theory) collaboration. Support from EPSRC Grants GR/R25514 and G/M82141 is acknowledged.

References

[1] I.Tanihata *et al*, Phys. Rev. Lett. **55** (1985) 2676; Phys. Lett. **B160** (1985) 380; I. Tanihata, J. Phys., **G22** (1996) 157.

[2] P.G. Hansen and B. Jonson, Europhys. Lett., **4** (1987) 409.

[3] Z. Zhen and J. Macek, Phys. Rev. **A38** (1988) 1193

[4] M.V. Zhukov, B.V. Danilin, D.V. Fedorov, J.M. Bang, I.J. Thompson and J.S. Vaagen, Physics Reports **231** (1993) 151; B.V. Danilin, I.J. Thompson, J.S. Vaagen and M.V. Zhukov, Nucl. Phys. A **632** (1998) 383. See also: E. Nielsen et al., Phys. Rep., **347** (2001) 373 and references therein.

[5] B. Jonson, Physics Reports **389** (2004) 1.

[6] Steven Weinberg, The New York Review (of Books), October 8, 1998, pp 48

[7] K. Riisager, Rev. Mod. Phys. **66** (1994) 1105.

[8] V.M. Efimov, Phys. Lett., **B33** (1970) 563; Nucl. Phys., **A210** (1973) 157; Comments Nucl. Part. Phys., **19** (1990) 271.

[9] I.J. Thompson and M.V. Zhukov, Phys. Rev., **C49** (1994) 1904.

[10] Yu.Ts. Oganessian, V.I. Zagrebaev and J.S. Vaagen, Phys. Rev. Lett. **82** (1999), 4996

[11] S.N. Ershov, B.V. Danilin, J.S. Vaagen, Phys. Rev., **C64** (2001) 064609; S.N. Ershov, B.V. Danilin, T. Rogde and J.S. Vaagen, Phys. Rev. Lett. **82** (1999), 908.

[12] S.C. Pieper and R.B. Wiringa, Ann. Rev. Nucl. Part. Sci. **51** (2001) 53

[13] P. Navratil and B.R. Barrett, Phys. Rev. **C57** (1998) 3119

[14] R. G. Newton *Scattering Theory of Waves and Particles* (Springer-Verlag, New York, 1982); V. I. Kukulin, V. M. Krasnopol'sky, and J. Horáček, *Theory of Resonances* (Kluwer Academic publishers, Amsterdam, 1989); A. I. Baz', Ya. B. Zel'dovich, and A. M. Perelomov *Scattering, Reactions and Decay in Nonrelativistic Quantum Mechanics* (Fiziko-Matematicheskoi Literatury Moskva 1966).

[15] N. Michel, W. Nazarewicz, M. Płoszajczak, and K. Bennaceur, Phys. Rev. Lett. **89**, 042502 (2002); N. Michel, W. Nazarewicz, M. Płoszajczak, and J. Okołowicz, nucl-th/0302060.

[16] R. Id Betan, R. J. Liotta, N. Sandulescu, and T. Vertse, Phys. Rev. C **67**, 014322 (2003); R. Id Betan, R. J. Liotta, N. Sandulescu, and T. Vertse, nucl-th/0307060

[17] N. Moiseyev, Phys. Rep. **302**, 211 (1998).

[18] A. Csoto, Phys. Rev. C **49**, 2244 (1994); T. Myo, K. Kato, S. Aoyama and K. Ikeda, Phys. Rev. C **63**, 054313 (1998); E. Garrido, D. V. Fedorov, and A. S. Jensen, Nucl. Phys. A **708**, 277 (2002); I. Raskinyte, *Resonances in few-body systems*, PhD Thesis University of Bergen 2002, unpublished.

[19] T. Berggren, Nucl. Phys. A **109**, 265 (1968).

FORMATION OF THE COMPOUND NUCLEUS IN FUSION OF HEAVY NUCLEI

A. Diaz-Torres
Institut für Theoretische Physik der Johann Wolfgang Goethe–Universität, Robert Mayer Str. 10, D–60054 Frankfurt am Main, Germany, E-mail: diaz@th.physik.uni-frankfurt.de

Abstract A new model that includes the time-dependent dynamics of the single-particle (s.p.) motion in conjunction with the macroscopic evolution of the system is proposed for describing the compound nucleus (CN) formation in fusion of heavy nuclei. The diabaticity initially keeps the entrance system around its contact configuration, but the gradual transition from the diabatic to the adiabatic potential energy surface (PES) leads to fusion or quasifission. Examples are given for some (near) symmetric central collisions.

1. Introduction

The competition between fusion and quasifission (reseparation before CN formation) can inhibit the fusion of massive systems by many orders of magnitude, e.g. [1]. Understanding this inhibition may be the key to forming more superheavy elements (SHE). Nowadays, there is no consensus [2] for the mechanism of the CN formation in fusion of heavy nuclei near the Coulomb barrier. Depending on the main coordinate for fusion, two sorts of models can be distinguished. In the first type [3–11], the fusion occurs along the radial coordinate using adiabatic PES obtained either with the liquid drop model or Strutinsky's macroscopic-microscopic method. The competition between fusion and quasifission, which depends on the fluctuactions [12], has only been included in recent models, e.g. [9–11]. However, the experimental data are not always explained. In the second type [13] (dinuclear system (DNS) model), the fusion happens in the mass asymmetry coordinate $\eta = (A_1 - A_2)/(A_1 + A_2)$ where A_1 and A_2 are the mass numbers of the nuclei. The DNS nuclei remain at the touching configuration and exchange nucleons until either all nucleons have been transferred from the lighter to the heavier fragment (complete fusion), or the DNS decays before the CN formation (quasifission). The model assumes a sudden (double-folding in frozen density approximation) PES in the radial coordinate, while the PES behaves adiabatically along the fusion path in the η

W. Greiner et al. (eds.), *Structure and Dynamics of Elementary Matter*, 479–486.

coordinate. Although the model has been used to explain many experimental evaporation residues (ER) cross sections, its theoretical foundation is not clear enough yet.

The present model is based on the following general ideas (which are well-established but have up to now not been used in combination in any of the current models of fusion): (i) Once the two nuclei are at the contact point, the system moves in a multidimensional space of collective coordinates **q** (shape parameters), (ii) this motion is governed by the master equation, and (iii) the nature of the s.p. motion is time-dependent, it is initially diabatic and then approaches the adiabatic limit due to residual two-body collisions. Consequently, the system moves in a time-dependent PES, $V(\mathbf{q}, t)$, which is initially diabatic and gradually becomes adiabatic. In the diabatic limit [14] (elastic nuclear matter), the nucleons do not occupy the lowest free s.p. energy levels as in the adiabatic case (plastic nuclear matter), but keep their quantum numbers and remain in the diabatic states during a collective motion of the system. This approach is realistic in the initial stage of collisions near the Coulomb barrier where the total excitation energy per nucleon $E^* \geq 0.03$ MeV [15, 16]. In contrast to the current models for CN formation, the present approach explicitly includes for the first time the time-dependent dynamics of the s.p. motion in conjunction with the macroscopic evolution of the system into the CN. The scenario presented here for the CN formation shows that following contact the diabaticity forms a valley in the PES where the system remains trapped around its touching configuration, but the gradual transition from the diabatic to the adiabatic PES allows the system to evolve in shapes leading to fusion or quasi-fission. The calculations are based on the master equation [17] and the diabatic two-center shell model (TCSM) [18] developed using the asymmetric TCSM (ATCSM) [19]. The critical ingredients of the model are (i) temperature, (ii) diabatic PES, (iii) adiabatic PES, (iv) transition from diabatic to adiabatic PES, and (v) shape parameters. The evaluation of each will first be described, then the results from the model will be presented.

2. Model

The time-dependent population probabilities p for the different configurations **q** (shapes) of the system are solutions of the master equation

$$\dot{p}(\mathbf{q}, t) = \sum_{\mathbf{q}'} [\Lambda(\mathbf{q}, \mathbf{q}', t) p(\mathbf{q}', t) - \Lambda(\mathbf{q}', \mathbf{q}, t) p(\mathbf{q}, t)], \tag{1}$$

where the macroscopic transition probabilities according to Ref. [20] are $\Lambda(\mathbf{q}, \mathbf{q}', t) = \kappa_0 exp[V(\mathbf{q}', t)/2T(\mathbf{q}', t) - V(\mathbf{q}, t)/2T(\mathbf{q}, t)]$, justified by the assumption that only the level density of the system determines the transition. The strengh constant κ_0 characterizes the global time scale and has a realistic value of $\sim 10^{22}$ s^{-1} [21]. The sum in Eq. (1) is extended only to the nearest configurations $\mathbf{q}'$ (the collective coordinate space is discretized). It is assumed that the system is initially at the contact configuration $\mathbf{q}_0$ where the s.p. occupation numbers obey a Fermi-distribution $n_\alpha^F(\mathbf{q}_0, T_0)$ for a temperature T_0. Configurations other than $\mathbf{q}_0$ are not populated at this time, and hence the initial condition for Eq. (1) is $p(\mathbf{q}, 0) = \delta_{\mathbf{q},\mathbf{q}_0}$. The temperature T_0 is related to the excitation of the system immediately after the capture process. This temperature can be estimated either as $T_0 \approx \sqrt{[E_{c.m.} - V(\mathbf{q}_0, 0)]/a}$, where $E_{c.m.}$ is the total incident energy in the center of mass frame and $a = A/12$ MeV^{-1} (A is the total mass number of the system), if the initial radial kinetic energy is dissipated when the nuclei reach the contact point, or using a frictional model, e.g. [10], for the capture process. The local (at fixed $\mathbf{q}$) temperature $T = \sqrt{E_{exc}/a}$ is defined by means of the local excitation energy

$$E_{exc}(\mathbf{q}, t) = a\, T_0^2 + \int_0^t \left(-\frac{d\Delta V_{diab}(\mathbf{q}, t')}{dt'}\right) p(\mathbf{q}, t')dt', \qquad (2)$$

which results from the decay of the diabatic part ΔV_{diab} of the potential V. ΔV_{diab} represents an energetic hindrance for the initial system to reach a configuration $\mathbf{q}$, if the nucleons follow diabatic levels during this process (elastic response). The local excitation of the system is caused by the loss of its elasticity [14]. ΔV_{diab} is calculated as

$$\Delta V_{diab}(\mathbf{q}, t) \approx \sum_\alpha \varepsilon_\alpha^{diab}(\mathbf{q})[n_\alpha(\mathbf{q}, t) - n_\alpha^F(\mathbf{q}, T)], \qquad (3)$$

where $\varepsilon_\alpha^{diab}$ are the diabatic levels with occupations $n_\alpha(\mathbf{q}, t)$ and α denotes the quantum numbers of these states. The diabatic levels $\varepsilon_\alpha^{diab}$ and their wave functions $\phi_\alpha^{diab}(\mathbf{r})$ are obtained with the maximum symmetry method [18]. The dynamical PES is defined as $V(\mathbf{q}, t) = V_{adiab}(\mathbf{q}, T) + \Delta V_{diab}(\mathbf{q}, t)$, where V_{adiab} is the adiabatic PES which is calculated using Strutinsky's macroscopic-microscopic method [19]. The nuclear part of the macroscopic energy is obtained with the Yukawa-plus-exponential method [22]. The diabaticity destroys the Fermi-distribution of the occupations, but the residual two-body collisions gradually recover it. This process is locally described by the relaxation equation [23]

$$\dot{n}_\alpha(\mathbf{q}, t) = -\tau^{-1}(\mathbf{q}, t)[n_\alpha(\mathbf{q}, t) - n_\alpha^F(\mathbf{q}, T)], \qquad (4)$$

where τ is an average relaxation time (in order to conserve the number of particles). The initial occupations $n_\alpha(\mathbf{q}, 0)$ are the diabatic ones obtained from $n_\alpha^F(\mathbf{q}_0, T_0)$. τ is defined as

$$\tau^{-1}(\mathbf{q}, t) = \frac{\sum_\alpha [n_\alpha(\mathbf{q}, t) - n_\alpha^F(\mathbf{q}, T)]\, \Gamma_\alpha(\mathbf{q}, T)}{N_{coll}\, \hbar \sum_\alpha n_\alpha(\mathbf{q}, t)}, \qquad (5)$$

where Γ_α are the widths of the s.p. levels. The factor N_{coll} is the average number of two-body collisions per nucleon to establish the equilibrium occupations n_α^F. The value $N_{coll} = 3$ [15] will be used. The expression (5) follows the idea that the relaxation process becomes slower when the occupations approach n_α^F. If the equilibrium was reached, the relaxation time would be infinite, i.e. the occupations would remain the same. The widths Γ_α are obtained with the parametrization given in Ref. [24]. Since the diabatic s.p. excitations occur around the Fermi surface, the values $\Gamma_0^{-1} = 0.061$ MeV^{-1} for half saturation density and $c = 20$ MeV will be used [24].

The collective coordinates **q** are the shape parameters of the ATCSM [19]: (i) the elongation $\lambda = l/2R_0$, which measures the length l of the system in units of the diameter $2R_0$ of the spherical CN and describes the relative motion, (ii) the deformation $\beta_i = a_i/b_i$ of the fragments, defined by the ratio of their semiaxes, (iii) the neck coordinate $\epsilon = E_0/E'$, defined by the ratio of the actual barrier height E_0 to the barrier height E' of the two-center oscillator, and (iv) the volume asymmetry of the nuclear shapes (equipotential shapes) $\xi = (V_1 - V_2)/(V_1 + V_2)$, where V_1 and V_2 are the volumes of the left and right regions divided by a plane at the necks between the fragments. The collective coordinate space is divided into three regions: (i) compact shapes around the spherical shape (fusion region), (ii) elongated shapes outside the initial contact configuration and beyond the Coulomb barrier (quasifission region), and (iii) intermediate shapes which could lead to fusion or quasifission. In the fusion region the physical mass asymmetry η (defined like ξ but in terms of the masses which are calculated using the microscopic density distribution $\rho(\mathbf{r}) = \sum_\alpha n_\alpha |\phi_\alpha^{diab}(\mathbf{r})|^2$) reaches a minimal plateau, i.e. a maximal number of nucleons move (their wave-functions spread) in the whole volume of the system due to the decrease of the barrier E_0 between the fragments. The fusion (P_{CN}) and quasifission (P_{QF}) probabilities are defined as the sum of the population probabilities p in the fusion and quasifission regions, respectively. In addition the CN excitation energy E_{CN} is defined as the average of the excitation energies (2) of the different shapes in the fusion region.

In the following the fragments are considered as spherical (β_i=1) and the neck coordinate is fixed at $\epsilon = 0.75$. With this value of ϵ the Coulomb barrier of the diabatic potential for an initial system with $T_0 = 0$ MeV is close to the barrier of the double-folding potential [16]. For computational reasons the calculations are done using the coordinates λ and ξ that are the relevant ones in the current models for CN formation. The non-linear set of equations (1-5) are solved by sucessive iterations using a small time step $\Delta t = 10^{-23}$ s. The master equation (1) is solved on a grid ($1 \leq \lambda \leq 1.8$, $|\xi| \leq 0.7$ where $\Delta\lambda = 0.02$ and $\Delta\xi = 0.1$) with appropriate boundary conditions to follow the continuous split of the initial population probabilities $p(\mathbf{q}, 0)$ into fusion and quasifission. Values of $|\xi| > 0.7$ are not included in the calculation because the TCSM used [19] is not appropriate for large ξ. The fusion and quasifission processes are determined at the timescale t_0 when $P_{CN}(t_0) + P_{QF}(t_0) \simeq 1$. The model will be applied to some (near) symmetric central ($l = 0$) collisions. In these calculations the excitation energy of the initial system at the contact configuration is 40 MeV.

3. Results

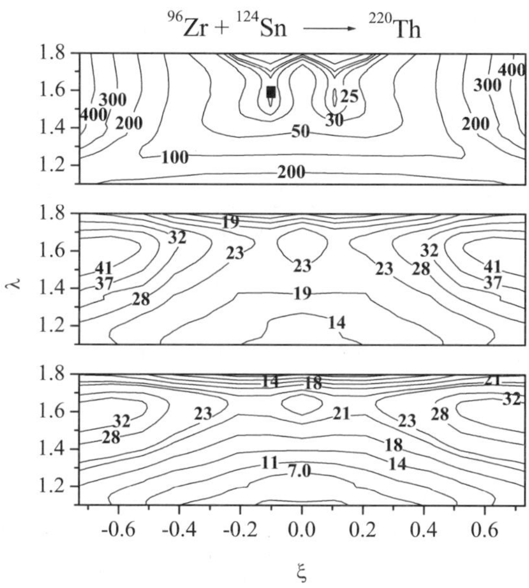

Figure 1. Dynamical PES for ^{96}Zr+^{124}Sn: (upper part) at t=0 s (diabatic PES), the square denotes the contact configuration; (middle part) at $t_0 = 4 \times 10^{-20}$ s; and (lower part) the adiabatic PES. See text for further details.

Fig. 1 shows the dynamical PES for ^{96}Zr+^{124}Zn (i) at the initial moment (diabatic PES, upper part) when the nuclei are at the contact configuration (square), (ii) at $t_0 = 4 \times 10^{-20}$ s (middle part), and (iii) the adiabatic PES (lower part). The PES is normalized with the macroscopic energy of the spherical CN. In Fig. 2, the distribution of the population probabilities $p(\lambda, \xi, t_0)$ on the PES of Fig. 1 (middle part) is presented. From these figures, it is observed that the diabaticity initially forms a valley confining the entrance system around its touching configuration as in the DNS model [13]. Nevertheless, the gradual transition to the adiabatic PES leads to fusion mainly in the elongation λ (relative distance) or the population of neighbouring configurations ending in the quasifission channel. In contrast to the DNS model, the diffusion along the ξ coordinate, while the nuclei remain at the contact point ($\lambda = 1.5 - 1.6$), is strongly suppressed

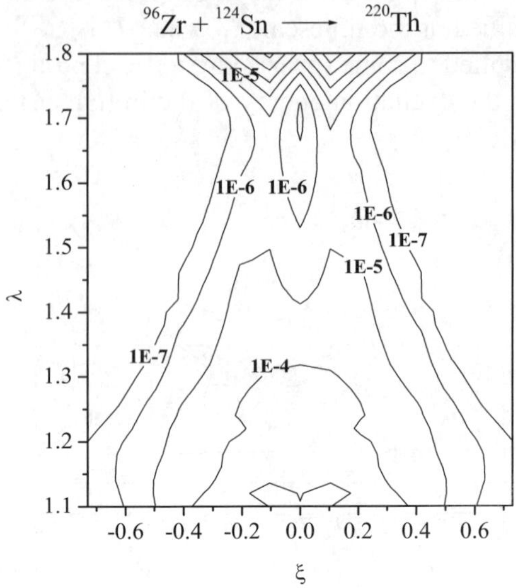

Figure 2. Distribution of the population probabilities $p(\lambda, \xi, t_0)$ for ^{96}Zr+^{124}Sn on the PES of Fig. 1 (middle part). The fusion region is $\lambda \leq 1.3$. See text for further details.

by a large diabatic hindrance. P_{CN} and P_{QF} are determined before the dynamical PES reaches the adiabatic one (comparing middle and lower parts in Fig. 1). The competition between fusion and quasifission is regulated by the dynamical PES. The probability of configurations which stay close to that of the entrance channel and reseparate is very large, as can be seen from Fig. 2. This causes the fusion hindrance for ^{96}Zr+^{124}Sn which is consistent with the experimental conclusions drawn in Ref. [1]. The final population probabilities in the fusion region ($\lambda \leq 1.3$ in Fig. 2) are very small, and as a result of the expression (2), E_{CN} is practically the same as the excitation energy of the initial system at the contact point, i.e. 40 MeV. The scenario seen so far is the same for other systems studied. Table I shows P_{CN}, t_0 and E_{CN} calculated for some (near) symmetric reactions. The values of the model parameters are the same for all the reactions. The decrease of the P_{CN} values from ^{90}Zr+^{90}Zr to ^{110}Pd+^{110}Pd is observed along with an increase of the quasifission probability beyond 90%. This is because (i) the repulsive character of the diabatic PES increases with increasing mass number A of the combined system and (ii) the valley of the diabatic PES becomes more shallow with increasing A. Consequently the timescale t_0 of the fusion and quasifission processes decreases with increasing A. The P_{CN} for ^{96}Zr+^{124}Sn agrees very well with the recent direct experimental measurement [25]. The $P_{CN}^{\mathrm{exp.}}$ includes the contribution of higher

Table 1. P_{CN}, t_0 (10^{-20} s) and E_{CN} (MeV) for some (near) symmetric central collisions. The P_{CN} values are compared to direct experimental $P_{CN}^{exp.}$ [25].

Reaction	P_{CN}	t_0	E_{CN}	$P_{CN}^{exp.}$
$^{90}Zr+^{90}Zr \rightarrow ^{180}Hg$	3×10^{-1}	5	40	
$^{100}Mo+^{100}Mo \rightarrow ^{200}Po$	1.4×10^{-1}	4.5	40	
$^{110}Pd+^{110}Pd \rightarrow ^{220}U$	1.7×10^{-2}	4	40	
$^{100}Mo+^{110}Pd \rightarrow ^{210}Ra$	7.5×10^{-2}	4.3	40	
$^{96}Zr+^{124}Sn \rightarrow ^{220}Th$	4.5×10^{-2}	4	40	5×10^{-2}

partial waves l, i.e. $l > 0$, although the main contribution results from low l for heavy systems at this energy. The P_{CN} obtained with the DNS model for this reaction [13] also agrees well with that experiment, while P_{CN} for, e.g. ^{110}Pd+^{110}Pd, is about one order of magnitude smaller than the present P_{CN} (see Table I). The fluctuation-dissipation model in Ref. [7] predicts P_{CN} values that are much larger than the present ones, i.e. about one order of magnitude larger for ^{110}Pd+^{110}Pd and ^{100}Mo+^{110}Pd. Direct experimental measurements of P_{CN} like in Ref. [25] are crucial to discriminate between the current models for CN formation. The dependence of P_{CN} on the parameter Γ_0^{-1} is strong for the heaviest systems studied, i.e. changing the latter from 0.061 MeV^{-1} to 0.03 MeV^{-1}, causes the P_{CN} decrease by about one order of magnitude due to the faster transition to the adiabatic PES, whereas the effect on the timescale t_0 is rather weak. The timescale t_0 is mainly determined by the quasifission process that occurs near the contact configuration of the initial system, where the diabatic effects are small. The dependence of P_{CN} on the excitation energy of the initial system at the touching point is rather weak, i.e. a saturation of the P_{CN} practically occurs from $\sim$ 20 MeV upwards.

4. Summary

A new realistic model for the CN formation in fusion of massive nuclei has been developed. It incorporates for the first time important physical effects which critically affect the evolution of the fusing system. The diabaticity initially keeps the entrance system around its touching point, but the gradual transition from the diabatic to adiabatic PES leads to fusion (mainly in the relative distance) or quasifission. The dynamical PES regulates the competition between fusion and quasifission. The probabilities for CN formation in some (near) symmetric central collisions have been obtained and found to agree very well with the recent direct experimental determination for ^{96}Zr+^{124}Sn. Direct measurements of the P_{CN} like in Ref. [25] along with distributions of the quasifission fragments are crucial to discriminate between the current models for CN formation. To calculate ER cross sections the present approach should be combined with other models that describe the initial capture stage and the survival of the CN against fission. A TCSM more appropriate for reactions with large mass-asymmetry is being developed which should be useful in predicting production cross sections of SHE.

Acknowledgments

The author thanks the Alexander von Humboldt Foundation for financial support.

References

[1] W. Reisdorf et al., Z. Phys. A **342** (1992), 411.

[2] *Proceedings of the Symposium on Nuclear Clusters* edited by Rostislav Jolos and Werner Scheid (EP Systema, Debrecen, Hungary, 2003).

[3] A. Sandulescu et al., Phys. Lett. B **60** (1976), 225.

[4] W.J. Swiatecki, Phys. Scr. **24** (1981), 113.

[5] C.E. Aguiar et al., Nucl. Phys. A **517** (1990), 205.

[6] Y. Aritomo et al., Phys. Rev. C **59** (1999), 796.

[7] T. Tokuda, T. Wada and M. Ohta, Prog. Theor. Phys. **101** (1999), 607.

[8] V. Yu. Denisov and S. Hofmann, Phys. Rev. C **61** (2000), 034606.

[9] V.I. Zagrebaev, Phys. Rev. C **64** (2001), 034606.

[10] C. Shen, G. Kosenko and Y. Abe, Phys. Rev. C **66** (2002), 061602.

[11] Y. Abe et al., Acta Physica Polonica B **34** (2003), 2091.

[12] Y. Abe et al., J. Phys. **47** (1986), C4-329.

[13] G.G. Adamian, N.V. Antonenko and W. Scheid, Nucl. Phys. A **618** (1997), 176; **627** (1997), 361; **678** (2000), 24; Phys. Rev. C **68** (2003), 034601.

[14] W. Nörenberg, Phys. Lett. B **104** (1981), 107.

[15] W. Cassing and W. Nörenberg, Nucl. Phys. A **433** (1985), 467.

[16] A. Diaz-Torres et al., Nucl. Phys. **652** (1999), 61; Phys. Lett. B **481** (2000), 228.

[17] W. Nörenberg, Phys. Lett. B **53** (1974), 289.

[18] A. Lukasiak, W. Cassing and W. Nörenberg, Nucl. Phys. A **426** (1984), 181.

[19] J. Maruhn and W. Greiner, Z. Phys. **251** (1972), 431.

[20] L.G. Moretto and J.S. Sventek, Phys. Lett. B **58** (1975), 26.

[21] P. Gippner et al., Phys. Lett. B **252** (1990), 198.

[22] H.J. Krappe, J.R. Nix and A.J. Sierk, Phys. Rev. C **20** (1979), 992.

[23] W. Cassing and W. Nörenberg, Nucl. Phys. A **401** (1983), 467.

[24] H. Hofmann, Phys. Rep. **284** (1997), 139.

[25] D.J. Hinde, M. Dasgupta and A. Mukherjee , Phys. Rev. Lett. **89** (2002), 282701.

ANISOTROPIC α - DECAY

D. S. Delion
Institute of Physics and Nuclear Engineering, Bucharest, Romania
E-mail: delion@theory.nipne.ro

A. Insolia
Dept. of Physics and Astronomy, Univ. of Catania and INFN, Catania, Italy
E-mail: Antonio.Insolia@ct.infn.it

R. J. Liotta
KTH - Institute of Physics, Stockholm, Sweden
E-mail: Liotta@msi.se

Abstract A miscroscopic description of the alpha decay of the odd mass nuclei is given for axially deformed nuclei. Realistic mean field + pairing residual interaction in a very large single particle basis is used. Results for ^{221}Fr, ^{241}Am and Pa - isotopes are reported. It is found that the approach gives predictions in good agreement with experimental data for well deformed nuclei. New calculations for $^{241-243}$Am, $^{253-255}$Es and $^{255-257}$Fm isotopes will be presented.

1. Introduction

It has been shown long time ago that in odd-mass actinides at very low temperature the alpha particles are emitted preferentially with respect to the direction of the total nuclear spin [1 5]. Recently, new experiments [6] have renewed the interest in this problem by reporting anisotropic emission in some At, near spherical isotopes, in connection with several theoretical descriptions of this effect.

Preferential emission of the alpha particles from deformed nuclei was first explained by Hill and Wheeler [7] and then by Bohr, Fröman and Mottelson [8] in terms of the penetration of the alpha cluster through the deformed Coulomb barrier. It was thus found that since for a prolate nucleus the barrier at the poles is thinner than at the equator, the probability to penetrate the barrier is larger along the nuclear symmetry axis. More recently, in order to explain observed anisotropies for almost spherical At isotopes, Berggren [9] proposed an alpha + core model. A quadrupole-quadrupole interaction

W. Greiner et al. (eds.), Structure and Dynamics of Elementary Matter, 487–494.

acting between the already existing structureless alpha cluster and an odd-mass core was diagonalised in a weak couplig scheme. The strength of the interaction was adjusted to obtain the energy of the emitted alpha particle. Using this model several solutions with pronounced anisotropy were obtained [10]. No good comparison with the available data was obtained. Buck et al. [11] describe alpha decay from odd mass nuclei in a similar model, in which the depth of the alpha-core potential (taken as a square well), the alpha formation probability and the number of nodes in the radial wave function are fitted to the experimental data. Rowley et al. [12] followed the same philosophy, diagonalising the quadrupole-quadrupole interaction in an extreme cluster model basis. Stewart et al. [13], using either semiclassical or coupled - channels transmission matrices without any formation mechanism, calculated also anisotropic α emission.

In a series of papers [14,15] we followed the traditional Hill and Wheeler line, but using a realistic deformed mean field with a large configuration space + pairing residual interaction in computing the preformation amplitude of the alpha cluster inside the nucleus. We estimated the penetration through the deformed Coulomb barrier within the framework of the WKB approximation. The anisotropy was explained mainly by the effect of the deformed barrier (see ref. [16] for an overview on the microscopic approach to the alpha decay problem).

The aim of the present talk is to give a short account about our work on the anisotropic alpha particle emission from odd-mass nuclei at low temperature. We will discuss some predictions of ref.s [14,15] as well as some more recent calculations, in connection with the experimental results obtained by Schuurmans et al. on α anisotropy in Fr and Pa isotopes [17,18]. We will show that our predictions have been very well confirmed by the recent experimental findings in well deformed nuclei.

Finally, new experimental data on α-decay anisotropy should be soon available [19]. In particular, new measurements have been planned for $^{241-243}$Am, $^{253-255}$Es and $^{255-257}$Fm isotopes. Calculations for such cases will be reported [20].

2. Microscopic description of the anisotropy

The mechanism describing the emission of the alpha particle used in refs. [14,15] consists in the classical two step process [21]: first the four nucleons cluster on a point at the nuclear surface with a given formation amplitude and afterwards this object penetrates the Coubomb barrier.

Let us consider the decay

$$B(I_i, K_i, M_i) \rightarrow A(I_f, K_f, M_f) + \alpha \tag{1}$$

where K_i, K_f and M_i, M_f are the projections of the initial and final total angular momenta in the intrinsic and laboratory frame, respectively [21].

We describe the mother and daughter nuclei within the BCS approximation. That is, the wave function of the nucleus X (A or B) is

$$|\phi^X\rangle = a^\dagger_{k\Omega}|(BCS)\rangle^X_\pi \otimes |(BCS)\rangle^X_\nu \tag{2}$$

where $\pi(\nu)$ label proton (neutron) degrees of freedom. The operator $a^+_{k\Omega}$ is the creation operator of the unpaired nucleon with projection Ω on the intrinsic symmetry axis and k denotes the other quantum numbers. In the case of favoured transitions of odd mass nucleus the quantum numbers $k\Omega$ of the odd nucleon are unchanged during the decay.

The formation amplitude can be written as [14,15]

$$F(R,\vartheta,\varphi) = \sum_L F_L(R,\vartheta,\varphi)$$

$$= \sum_L \int d\xi_\alpha d\xi_A [\phi_\alpha(\xi_\alpha)\phi^A(\xi_A)Y_L(R,\vartheta,\varphi)]^*_{J_B M_B}\phi^B(\xi_B) \tag{3}$$

where R is the distance between cluster and daughter nucleus and ξ are the internal coordinates. The intrinsic wave function of the alpha particle has a standard Gaussian form. Additional details on the evaluation of the multidimensional integral in eq. (3) can be found in ref.s [14,15].

Experimentally, nuclei are tipically first produced, then separated, implanted on a foil of ferromagnet material (cooled down to few $10^{-2}K$) and, eventually, oriented by applying a strong magnetic field. Anisotropy is thus measured with respect to the direction of the applied magnetic field [17,18].

If full alignement is not achieved in the orientation process of the implanted isotopes, the conditions are such that one has to average on the initial distribution of the angular momentum projections M_i. The total width is given by

$$\Gamma(\vartheta,\varphi) = \frac{\hbar v}{4\pi}\left(\frac{R}{G_0(E,R)}\right)^2 \sum_l F_l{}^2 W(\vartheta) \tag{4}$$

where F_l is the partial formation amplitude of the emitted alpha particle, i.e.

$$F_l = \exp\left(-\frac{2l(l+1)}{\chi}\sqrt{\frac{\chi}{kR}-1}\right)$$

$$\times \sum_\Omega{}' (-1)^{\Omega'} < I_i K_i l - \Omega' | I_f K_f > \sum_{l'} K^{\Omega'}_{ll'} a_{l'\Omega'}(R) \tag{5}$$

The matrix element $K_{ll'}$, as well as the quantities χ and $G_0(E,R)$ are defined as in ref. [14] (see also ref. [8] for additional detail). The microscopic formation amplitude enters into the calculation through the amplitude $a_{l'\Omega'}(R)$ in eq.(5).

The function W in eq. (4) determines the angular distribution of the emitted particle. After recoupling l and l' to the angular momentum L of the emitted alpha particle and assuming an axially symmetric nucleus, one gets

$$W(\vartheta) = \sum_L A_L P_L(\cos\vartheta) \tag{6}$$

where the amplitudes A_L are given in terms of the F_l amplitudes of eq. (5) [15].

3. Numerical calculations and discussion

We will apply the previous formalism to a few selected cases of anisotropic α decay from odd - even nuclei. In particular, we will discuss the ^{241}Am, the ^{221}Fr, and the ^{227,229}Pt - isotopes.

3.1 Application to the alpha-decay of ^{241}Am

We present in this section an application of the formalism developed above to the case of the favoured transition [15]

$$^{241}Am \rightarrow ^{237}Np + \alpha \tag{7}$$

for which $K_i = K_f = I_i$ and the Nilsson quantum numbers of both the mother and daughter nuclei are $\frac{5}{2}^-[523]$. The diagonalization of the deformed Woods-Saxon potential is done using 18 major shells [22]. The deformation parameters were chosen as $\beta_2 = 0.22$, $\beta_3 = 0$ and $\beta_4 = 0.08$. The total width was computed according to presented formalism. We obtained $\Gamma_{th} = 2.09 \times 10^{-34}$ MeV which is quite close to the experimental value $\Gamma_{exp} = 3.34 \times 10^{-34}$ MeV. As for the case of even - even nuclei [15,16], the absolute alpha decay widths, for odd nuclei, are given within the right order of magnitude. For instance, in the case of $^{243}Am \rightarrow ^{239} Np + \alpha$, we obtained $\Gamma_{th} = 1.17 \times 10^{-33}$ MeV, to be compared with the experimantal value $\Gamma_{exp} = 1.96 \times 10^{-33}$ MeV. However, the main goal of our analysis is to determine the influence of deformation on the angular distributions of the emitted alpha particles.

The model predicts a large enhancement of the anisotropy versus quadrupole deformation. The role of deformations with multipolarities higher than the quadrupole one is much less important [15]. The influence of the intrinsic structure of the mother and daughter wave functions can be estimated by studying the dependence of the angular distribution as a function of the angular momentum transfer L. We found that including only the $L = 0$ component, the total width Γ versus θ shows a variation which is 10% smaller with respect to the case in which all L components are included. This result puts in evidence the important role of the barrier deformation. As a matter of fact, the $L = 0$ part of the formation amplitude is isotropic and in such a case the calculated anisotropy has to be ascribed entirely to the barrier.

The function $W(\vartheta)$ is the relevant quantity regarding the anisotropy in alpha decay processes. Once the microscopic formation amplitude is calculated one can easily expand the function W in terms of even order Legendre polynomials, as shown in eq. (6)

Applying the reduction procedure of ref. [3], the calculated W values can be compared with the experimental data [3] at $1/T = 90.5^0 K^{-1}$. Our results are presented in table I.

ϑ	W_{th}	W_{exp}
0	1.500	1.610
$\pi/2$	0.736	0.714

Table I Function $W(\vartheta)$ at $\vartheta = 0$ and $\vartheta = \pi/2$ compared with the measured anisotropy for the favored transition $^{241}Am \rightarrow ^{237} Np + \alpha(\Omega^{\pi} = 5/2^{-})$.

The agreement can be considered remarkable, specially if one considers that the absolute normalization is given by the formation amplitude entering in the evaluation of the A_L coefficients.

3.2 The case of ^{221}Fr and Pa - isotopes

It is of great interest to refer the most recent results by Schuurmaans et al. [18] for well deformed nuclei. For the ^{221}Fr, a ground state $K = 1/2$ was assumed with a prolate deformation [18], while a previous calculation [14,15] used an oblate ground state with $K = I = 5/2$ (referred in the first line of Table II as a theoretical predictions for the A_2^{th}).

	β_2	β_3	A_2^{th}	A_2^{exp}
$^{221}_{87}Fr$	-0.069	0.0	-0.215	-0.375
	0.120	0.15	-0.373	
$^{227}_{91}Pa$	0.168	0.0	0.649	0.696
	0.168	0.1	0.748	
$^{229}_{91}Pa$	0.185	0.0	0.733	1.13
	0.1185	0.08	0.808	

Table II Deformations, experimental and computed coefficients A_2, for the $^{221}_{87}Fr$ and the indicated Pa isotopes (adapted from ref. [18]).

For the Pa isotopes a prolate ground state was taken for the favoured $5/2^- \rightarrow 5/2^-$ ($^{227}_{91}Pa \rightarrow \alpha + ^{223}_{89}Ac$) and $5/2^+ \rightarrow 5/2^+$ ($^{229}_{91}Pa \rightarrow \alpha + ^{225}_{89}Ac$) transitions.

The results are reported in the Table II. The agreement between the microscopic model and the experimental data is excellent. The calculation for the total widths gives $\Gamma = 0.64 \times 10^{-25}$ MeV and $\Gamma = 0.29 \times 10^{-28}$ MeV the $^{227}_{91}Pa$ and $^{229}_{91}Pa$, respectively. As already mentioned, we found that the $L = 0$ part in the formation amplitude is the largest contribution. The neglect of the higher multipoles in the formation amplitude produces anisotropy only slightly smaller $(10 - 20\%)$ in comparison with the case in which all multipolarities are included. This shows that the main role is played by the penetration through the deformed barrier. The effect is therefore expected to be larger for very well deformed nuclei.

4. Numerical results for $^{241-243}Am$, $^{253-255}Es$ and $^{255-257}Fm$ isotopes.

We have applied this formalism to a few selected cases of anisotropic α-decay from axially deformed odd - even nuclei. In particular, we will discuss the $^{241-243}Am$, $^{253-255}Es$, $^{255-257}Fm$ isotopes.

In the present version of the model we used the new single particle basis with two harmonic oscillator (ho) parameters initially applied to favoured α decay of even-even spherical and deformed nuclei in ref. [16]. The first part of the basis describes spectroscopic properties, while the second one is connected with the α-clustering properties [16]. The new basis was tested in a systematic calculation of the total α-decay widths for many nuclei, ranging from spherical to well deformed systems. A very good agreement with the experimental data was obtained [16].

For the nuclei we analyzed here, we used the same parametrisation as in ref. [2], i.e. the universal mean field diagonalized in a basis with two ho parameters, $\lambda = m\omega/\hbar$ and 0.7λ. The number of single particle states needed to reproduce the order of magnitude of the decay width turns out to be equal to 50. It is important to remark that the method of ref. [16] uses a spherical mean field to calculate the formation amplitude, i.e. only the a_{00} term is left in eq. (5).

Therefore, within the present model calculation, any anisotropy has to be ascribed entirely to the barrier penetration. On the other end a large configuration mixing together with a realistic mean field are essential ingredients to get not only the absolute value of the total α width but also the correct values for the A_L amplitudes of eq. (6). This is the main issue in our approach. The a_{00} formation amplitude provides the correct absolute normalization within the present model without any *ad hoc* adjustable parameter.

The deformation parameters used in the present calculations as well as the K^π quantum numbers of the considered odd nuclei can be found in ref. [20].

The function $W(\vartheta)$ of eq. (6) is the relevant quantity regarding the anisotropy in α emission processes. Once the microscopic formation amplitude is calculated one can easily expand the function W in terms of even order Legendre polynomials, as shown in eq. (6). In our calculations we considered favoured decays with $I_i = I_f = K_i = K_f$ and, therefore, in eq. (5) only a single term with $\Omega' = 0$ is left. It turns out, actually, that A_L amplitudes in eq. (6) practically do not depend upon the radius R [20]. We remind that $A_0 = 1$. The A_L amplitudes and the resulting anisotropy $W(\vartheta)$ [see eq. (6)] are reported in Table III, for the six nuclei considered in this paragraph [20]. When comparing the results of Table III with the future experimental data one should keep in mind that the calculations do assume angles equal to 0° and 90° as detection angles and that all nuclei are assumed fully aligned with the maximum projection of the total angular momentum in the laboratory frame. Additional experimental correction factors should be also in order for a proper comparison with the measurements.

^{A}X	A_2	A_4	$W(0^o)$	$W(90^o)$	Γ_{th}/Γ_{exp}
^{241}Am	1.083	0.059	2.143	0.481	0.26
^{243}Am	1.092	0.060	2.153	0.476	0.28
^{253}Es	1.467	0.180	2.647	0.334	0.81
^{255}Es	1.437	0.172	2.609	0.346	0.51
^{255}Fm	1.425	0.169	2.595	0.351	0.85
^{257}Fm	1.680	0.296	2.976	0.271	0.74

Table III Theoretical predictions for the α anisotropy for the indicated isotopes. In the last column we report the ratio between the calculated width and the experimental value for a given isotopes.

The results show anyway a large anisotropy with a preferred polar emission, as expected for such large quadrupole deformation in axially symmetric prolate nuclei.

The expected anisotropy shows also a measurable dependence on the overall deformation parameters relevant for the axially deformed ground state description [20].

In the last column one can see that the total decay width (integrated on angles) reproduces the experimental value within a factor of three (or four) in the worst cases ($^{241-243}Am$) and, actually, much better for all the other considered cases.

5. Concluding Remarks

In conclusion we have presented a realistic microscopic approach for the calculation of the formation amplitude for the alpha decay problem in axially symmetric deformed nuclei. The predicted absolute values of the total alpha widths are reproduced within $10-30\%$. [16]. In the model we performed a systematic microscopic calculation of quantities related to alpha particle emission from oriented odd-mass nuclei and, in particular, for $_{87}$Fr, $_{91}$Pa as well as Am and Es isotopes. The probability of emitting an alpha particle in the polar direction (or in the equatorial direction) is strongly dependent on the emission angle. For prolate (oblate) deformations polar (equatorial) emission is preferred. We also found that deformations higher than quadrupole can play an important role in some cases. In addition, we have found that, in well deformed nuclei, the main role in the observed anisotropy is played the deformed barrier penetration. This has been recently confirmed experimentally [18].

Acknowledgments

One of the author (A. I.) would like to thank the Organizers for the kind hospitality in Kemer during the NATO Advanced Study Institute.

References

[1] S.H.Hanauer, J.W.T.Dabbs, L.D.Roberts and G. W. Parker, Phys. Rev. **124** (1961)1512
Q.O.Navarro, J. O Rasmussen ans D.A.Shirley, Phys. Lett. **2** (1962)353

[2] A.J.Soinski, R.B.Frankel, Q.O.Navarro and D.D.Shirley, Phys. Rev. **C2** (1970) 2379

[3] A.J.Soinski and D.D.Shirley, Phys. Rev. **C10** (1974) 1488

[4] D. Vandeplassche, E. van Walle, C. Nuytten and L.Vanneste, Phys. Rev. Lett. **49** (1982) 1390

[5] F. A. Dilmanian et al., Phys. Rev. Lett. **49** (1982) 1909

[6] J. Wouters et al., Phys. Rev. Lett. **56** (1986) 1901
J. Wouters et al., Nucl. Instr. and Meth. **B26** (1987) 463;
N. G. Nicolis et al., Phys. Rev. **C41** (1990) 2118

[7] D.L. Hill and J. D. Wheeler, Phys. Rev. **89** (1953) 1102

[8] P. O. Fröman, Mat. Fys. Skr. Dan. Vid. Selsk. **1** (1957) no. 3;
A. Bohr, P. O. Fröman abd B. Mottelson, Dan. Mat. Fys. Medd. **29** (1955) no. 10

[9] T. Berggren, Phys. Lett. **197B** (1987) 1;
T. Berggren, Hyperfine Interactions **43** (1988) 407

[10] T. Berggren, Phys. Rev. **C50** (1994) 2494

[11] B. Buck, A.C. Merchant and S.M. Perez, J. Phys. **G18** (1992) 143

[12] N. Rowley, G.D. Jones and M.W. Kermode, J. Phys. **G18** (1992) 165

[13] T.L. Stewart et al., Phys. Rev. Lett. **77** (1996) 36; T.L. Stewart et al., Nucl. Phys. **A611** (1996) 332

[14] A. Insolia, P. Curutchet, R. J. Liotta and D. S. Delion, Phys. Rev. **C44** (1991) 545
D.S.Delion, A.Insolia and R.J.Liotta, Phys. Rev **C46** (1992) 1346

[15] D.S.Delion, A.Insolia and R.J.Liotta, Phys. Rev **C46** (1992) 884;
D.S.Delion, A.Insolia and R.J.Liotta, Phys. Rev **C49** (1994) 3024

[16] R. G. Lovas, R. J. Liotta, A. Insolia, K. Varga and D. S. Delion, Phys. Rep. **294** (1998) 265-362;
D.S.Delion, A.Insolia and R.J.Liotta, Phys. Rev. **C 54** (1996) 292

[17] P. Schuurmans et. al., (Nicole and Isolde Collaboration), Phys. Rev. Lett. **77** (1996) 4720

[18] P. Schuurmans et. al., (Nicole and Isolde Collaboration), Phys. Rev. Lett. **82** (1999) 4787

[19] N. Severijns, private communication;

[20] D.S.Delion, A.Insolia and R.J.Liotta, Phys. Rev. C **67** (2003) 054317

[21] R.G. Thomas, Progr.Theor.Phys. **12** (1954) 253
H. J. Mang, Ann. Rev. Nucl. Sci. **14** (1964) 1

[22] R. Bengtsson et al., Phys. Scrip. **39** (1989) 196;
S. Cwiok et al., Comp. Phys. Comm. **46** (1987) 379.

EXOTIC NUCLEI: WHERE DO WE GO!

S. Galès
IPN Orsay, IN2P3/CNRS & Université Paris-Sud, 91406 Orsay Cedex, France

Abstract The exploration of nuclei far of stability, in particular, the neutron rich side of the nuclear mass chart, raised new and challenging physics issues. The structure of neutron rich light nuclei, the coupling of low-lying excitations with the continuum, shells quenching and the interplay between nuclear properties far of stability and astrophysics, prompted the study of a new generation of "Radioactive Ion Beam Facilities" with high luminosity and the development of associated new experimental tools.
The performances of existing facilities will be reviewed and recent proposals using nuclear fission as a reaction process to produce heavier neutron rich nuclei (A=60-140) to enlarge the spectrum of available exotic nuclei will be presented. An ambitious European proposal based on the ISOL method "EURISOL" is being elaborated by a network of ten laboratories. The status of ongoing studies will be discussed.

1. Introduction

Fig.1 shows a map of nuclei in which the number of neutrons N and protons Z are plotted on the horizontal and vertical scales, respectively.

The black dots are associated with the ashes of the primordial explosion, e.g. the 328 nuclei, which are sufficiently stable to form our universe. The dotted area indicates the limits of known nuclei and the large region or continent in gray is called "Terra Incognita" where more than 4000 nuclei are predicted to exist and their properties are essentially unknown.

The territory to explore is enormous. We do not the main land, neither the shore where instability occurs. Questions of fundamental importance are under discussions. What are the proton and neutron magic numbers and how the effective n-n interaction is modified for nuclei with extreme N/Z ratios? Is a new form of organization of nucleons in nuclei (multi-neutron or helium like clusters) is likely to appear? Is there new opportunities to reach the island of stability for super-heavy systems (Z=114 to 120). There are also related issues in the field of nuclear astrophysics, since many of these very unstable species are believed to be produced in astrophysical sites.

W. Greiner et al. (eds.), Structure and Dynamics of Elementary Matter, 495–505.

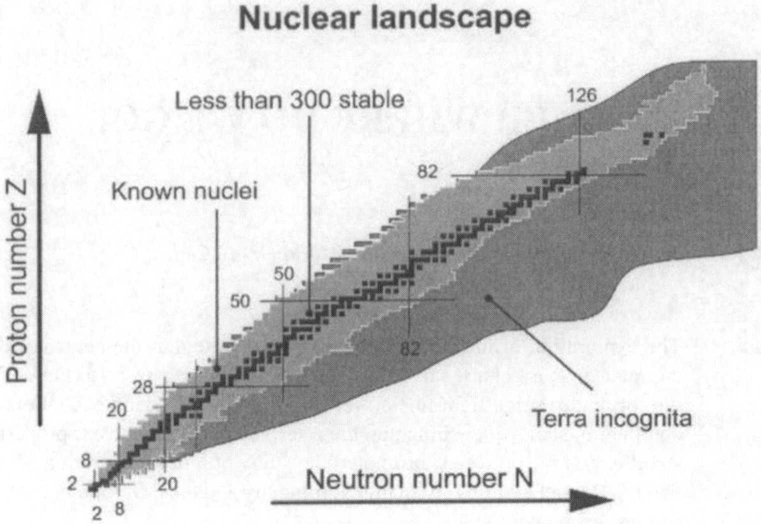

Figure 1. Nuclear mass chart showing the stability line (black dots), the newly discovered nuclei (light gray) and the large continent of unknown nuclei "Terra incognita" (dark gray)

2. Production of Radioactive Ion Beams

The new tools needed to explore this continent are intense secondary beams of Radioactive Ion Beams (RIB).There are two methods to produce secondary RIB, the fragmentation method or " IN FLIGTH" which takes advantage of the high incident velocity of high energy (50-1000 MeV/n) heavy ion stable beams incident on a thin target. The subsequent fragmentation process produces strongly forward focused RIB with roughly the same velocity as the incident beam.

The so-called " ISOL" method , Isotopic Separation On Line, is characterized by an intense primary beam (protons, fast or slow neutrons, photons or heavy ions) hitting a thick target. The exotic nuclei produced by the reaction diffuse from the target and are ionized on line before being sent to an electromagnetic separator for isotopic selection. The selected species can be further accelerated by injection into a booster linac or cyclotron.

The motivations for revisiting nuclear structure far off stability are now widely spread around the world. Various projects of first generation RIB facilities are operating around the world.

3. Status of first generation Radioactive Beam Facilities

3.1 Post acceleration of ISOL beams

The **ISOL** method was born in Europe [OKH51]. Following this historical attempt, the ISOLDE facility at CERN, based to day on the 1 GeV, 2μA proton driver, have produced during the last 25 years, more than one hundred of secondary radioactive ions, ranging from He, Li , Be up to isotopes with charge and mass Z=90, A=240. The "exotic" beams are accelerated up to 60KeV, using 1^+charge state and mass separated with high resolution mass spectrometers [HLR90]. Thus ISOL technique provides high-quality low energy (few hundred of KeV) radioactive beams, ideally suited for various types of experiments with device like COMPLIS [JS00] or high precision mass measurements by means of the instruments ISOLTRAP [GB98] and MISTRAL [DL98].

"Classical" ISOL separators can be found at many places in the world, e.g. at CERN, Jyvaskyla (Finland), Louvain-la-Neuve (Belgium), TRIUMPF (Canada), Orsay (France), Warsaw (Poland), Oak-Ridge (USA), CIAE Beijing (China), see the reviews made in [GM96].

At Louvain–La-Neuve, a pioneering work have been made during the last ten years [JV97] in order to achieve re-acceleration of low energy ISOL beams. The intense low energy light ion beams (protons, 30 MeV ,500 μA) delivered by the "Cyclone30" cyclotron are used to produce secondary RIB of ^{6,8}He, ^{7}Be, ^{13}N, ^{18}Ne, etc ... These beams are further ionized by an ECR source and injected into the K=110 machine "Cyclone" in order to provide post-accelerated RIB in the range 0.65-10 MeV/n.

This development has given rise to several projects based on this technique (for overviews and main references for Asian, European and North-America projects see [IT98-YEP02], ACM98-SG02] and [JAN98], respectively). To day, the HRIBF (Oak-Ridge, USA), which produce post-accelerated beams by a 25 MV tandem Van de Graff, has started its experimental program.

After successful commissioning of its post-accelerator (K=265 CIME cyclotron), the SPIRAL facility at GANIL has started with its first experiment made in September 2001. A wide energy range, from 2 to 25 MeV/n, is covered by SPIRAL for various secondary exotic beams like ^{6,8}He, ^{18,19}Ne , ^{22}O, ^{46}Ar, ^{72}Kr [ML03].

The REX-ISOLDE experiment at CERN has also started last year. The project was designed to achieve charge breeding of the 1^+ion for light secondary isotopes (He to Ca) produce by 1GeV, 2.5μA proton beam and further acceleration up to 2.2 MeV/n using an RFQ and a LINAC. Upgrade of the LINAC up to 3 and 5 MeV/n is planned.

The ISAC project at TRIUMF (Canada) make use of an existing medium energy proton accelerator (500MeV) with high intensity (from 10 to 100μA) as a driver associated to a dedicated linac booster to reach RIB in the energy range from 0.5 to 5MeV/n .The ISOL beams are presently available and post acceleration have been completed.

The EXCYT ISOL facility at Catania (Italy) is scheduled to start in 2004 and a few others ISOL projects: ALTO at Orsay (France), in Brazil (Sao Paulo), in China (CIAE Beijing), in India (VEC Calcutta) are either in the planning or construction phase.

3.2 Fragmentation. "In-Flight" facilities

In Western Europe, two in-flight RIB facilities are operational at GANIL (France) and at GSI (Germany). The GANIL cyclotrons produce heavy ions beams up to 95 MeV/n, with a present maximum beam power of 2KW. The beams are fragmented in a high power thin target preceded and followed by the super-conducting focusing lenses (SISSI). Under these conditions, the fragments are efficiently collected and identified by a doubly achromatic momentum selecting spectrometer LISE. This method pioneered by LISE has stimulated the development of a number of large devices for fragment separators all over the world (see YEP02).

At GSI, heavy ions beams up to Uranium are accelerated at energies up to 1 GeV/n by the UNILAC and the SIS synchrotron [HG02]. These beams are fragmented or fission is induced in a thin target and the fragments are collected and identified in the Fragment Recoil Separator FRS. The collected exotic species can be either studied at the focal plane of the FRS or cooled in the ESR storage ring where precise mass measurements, for example, can be performed. At present an intensity upgrade is taking place at GSI aiming at 10^{10}ions/pulse for Uranium beams in the SIS synchrotron.

In Eastern Europe, the Flerov laboratory at Dubna (Russia) has put into operation an "in-flight" facility based on its heavy ion cyclotrons (U400 and U400M) and the fragment separators ACCULINNA and COMBAS. Intense beams of ^{6,8}He (10^{4-7} pps) have been used to study neutron rich hydrogen isotopes [AAK02-YTO02].

In North America, the NCSL (USA) operates the K1200 super-conduc- ting cyclotron, which accelerates heavy ion beams up to 150 MeV/n. The facility includes the A1200 fragment separator. An upgrade has been recently achieved by coupling the K500 and K1200 super-conducting cyclotrons. Up to 1pμA light, N=Z ions at 200 MeV/n, and up to 1pnA Uranium ion beam at 100MeV/n are foreseen. A new powerful A1900 fragment separator has been built and the combination leads to about 10^3 increase of the RIB intensities [JAN98].

In Japan, the RIKEN laboratory includes a heavy ion ring cyclotron RRC, producing beams up to 135 MeV/n, and a fragment separator RIPS. A major upgrade of this facility is under construction. More details on the RIKEN Radioactive Beam Factory can be found in sect.4 of this paper and in [YEP02-

IT98]. To be complete, in China, the IMP laboratory in Lanzhou operates a K=450 sector separated cyclotron (heavy ions up to 80 MeV/n) and a fragment separator RIBLL [IT98]. A new project "à la GSI" based on a large booster synchrotron associated with storage and cooler rings for the production of high energy RIB is under construction at IMP Lanzhou.

4. Long term perspectives

For the long term future (5-10 years), working groups both in Europe [RNBF00], USA [RIA00] and Japan [IT98,YEP02] are actively looking into the production of very intense secondary RIB, characterized by higher intensities (a factor 100 to 1000), as compared to the RIB facilities discussed in subsection 3.1 and 3.2 of this paper.

4.1 Optimizing the production of RIB

Before going to a review of these long-term projects, I would like to discuss what are the key factors which determine the expected several order of magnitude increase of the RIB intensities.

The secondary radioactive beam intensity **I**, available for an experiment, is determined by equation (1):

$$I = \sigma \times \phi \times N \times \epsilon_1 \times \epsilon_2 \times \epsilon_3 \times \epsilon_4 \times \epsilon_5$$

Which contains the following factors: σ is the cross section for the production reactions, ϕ the primary beam intensity, N the thickness of the production target, ϵ_1 the efficiency of the release and the transfer to the ion source, ϵ_2 the efficiency of the latter, ϵ_3 the efficiency of the separator, ϵ_4 the delay transfer efficiency due to the radioactive decay losses, ϵ_5 the efficiency of the post-accelerator. The interdependence of these factors and their best combination depending on the production methods (ISOL or IN-FLIGHT) is subject to intense discussions among specialists.

4.2 Selected long-term projects for "In-Flight" RIB Facilities

For the "in-flight" method, the luminosity, i.e. the product $L = \phi \times N$ is definitively the way to progress. The other important factor is ε_3 the fragment separator efficiency which is of the order of 1% to 80%. Optimizing L implies to develop high-power heavy-ion accelerator and to go to high-energy (1 GeV/n) in order to maximize N (target thickness) and ϵ_3 the fragment separator efficiency (strong forward focusing of fragments allowing the use of small acceptance spectrometers).

With the exception of the upgrade of NSCL, already discussed in subsection 3.2, three major new fragmentation facilities are planned or in construction phase all around the world.

In Japan, at RIKEN, a new fragmentation facility is in the first phase construction.

The foreseen accelerator complex consists of a K510 fix-frequency ring cyclotrons (FRC), a K980 intermediate stage ring cyclotron (IRC) and a K2500 super conducting ring cyclotron (SRC). This new cyclotron cascade will be the post-accelerator of the existing K540 ring machine (RRC). The final beam output energy will be 400 MeV/n for light ions (p,d,..) and 350 MeV/n for heavy ions. The goal for the beam intensity is to reach 1pμA. The heavy ions SRC beams will be converted to RIB by a separator complex called "Big RIPS". This complex is formed by three distinct fragment separator, two of them used in time sharing mode to deliver RIB to experiments and the middle one (RIPS-M) send the selected RIB to a Multi-Use-Experimental Storage rings (MUSES).

MUSES have four rings, which have different functions, namely: Accumulation and Cooler (ACR), Booster Synchrotron (BSR) and Double Storage (DSR). The first phase of the new RIB factory (cyclotrons cascade and Big RIPS) is planned to be completed by the end of 2006, the second phase (MUSES rings) will be reviewed soon. Details can be found in [RAP00].

In North-America, NSAC, in its new long-range plan discussed and adopted in 2001, has given a high priority to the construction of an advanced **RIB facility called RIA**. The main physics opportunities has been defined in the Columbus White Paper [RIA00] and the present stage of design of RIA can be find in the report of the "ISOL task force" [RIA01].

Following the ISOL task force, the concept of RIA is a combination of both ISOL and a fragmentation facility. RIA is based on a flexible super conducting Linac driver capable of providing 400KW beam power, 900MeV for protons and 400 MeV/n beams of any stable isotope from He to Uranium. Rare short–lived isotopes will be produced by using the best techniques (projectile fragmentation, target fragmentation, fission and spallation).

An additional opportunity will be provided by fast in flight separated beams of rare isotopes. RIB at rest will be produced by conventional or new ISOL techniques and will be post accelerated by an RFQ plus Super conducting Linac up to 15 MeV/n but also fast beams of rare isotopes can be used directly after separation in a high-resolution fragment separator .The maximum yield available for post-acceleration will exceed 10^{12}pps for isotopes best produced by ISOL techniques whereas it may reach 10^{9}pps for rather exotic nuclei like ^{68}Ni or ^{132}Sn .

In Europe, the next generation of **"in flight" facility** has been discussed by working groups set up by NuPECC and the results of this work can be found in [RNBF00] and are summarized as follows . The new in-flight facility should provide:

Primary beams intensities of 2.10^{12}/s for all elements from hydrogen to uranium with energies up to 1.5 GeV/n. The associated instrumentation should include a large acceptance fragment separator with efficiency in excess to 50%, a new set up for reaction studies with 4π detector, a storage cooler ring system and an electron–heavy ion collider with a luminosity in excess of $10^{26}\text{cm}^{-2}\text{s}^{-1}$.

The GSI laboratory at Darmstadt has proposed to built such a facility as part of the GSI future project. The new accelerator complex builds on the existing facility as an injector. The key component of the new facility is a rapidly cycling, super-conducting double ring synchrotron accelerator with 100Tm and 300Tm magnetic rigidity, respectively. This double powerful ring synchrotron is coupled to smaller rings for storage, beam cooling, and in ring experimental instrumentation. In addition fixed target fragmentation are coupled to large fragment separator devices providing clean and intense secondary drip-line nuclei such as ^{48}Ni or ^{100}Sn with rates around a few per second (an increase of 10^6 compared to present capabilities) and intense neutron rich secondary beams of ^{132}Sn or ^{144}Xe with intensities of the order of 10^9 pps. This new large scale accelerator complex has been recently approved by the German government (see G. Muzenberg, contribution to this conference) .

4.3 Next generation of ISOL facilities at the highest power levels

According to equation (I), the step towards for a significant increase of RIB produced by the ISOL method depends on the driver power and on the efficiency factors of the target release and ion-source ionization.

To avoid long release time and chemical selectivity of ISOL thick solid target techniques, the RIA project has developed an improved version of the gas collector method used in the so-called IGISOL technique [JA87]. Here the main advantage of the fragment separator-ion guide method is the universality (no chemical selectivity) and the very short delay times (about 1 ms, drip-line nuclei). Considerable R&D is necessary to reach high release efficiencies with this novel method.

A more general limitation is given by the maximum heat deposition in the target due to the energy loss of the primary charged particle beams. In order to minimize the energy loss two different approaches could be used.

One may use directly an energetic (1GeV), low Z (protons, deuterons) charged particles Within this "one stage" approach, extensive R&D is under way at various laboratories (CERN-ISOLDE, RAL, IPN Orsay, GANIL) for the design of the target with a commonly accepted power limitation of 100 KW.

In the two-stage method, the source of intense neutral particles is either a powerful multi Megawatts proton driver (spallation source) or a reactor (high flux of thermal neutrons).

Another option is to use an intense electron accelerator associated to a tungsten converter, producing a high flux of energetic photons. In all cases, the neutral particles are used to induce hot or cold fission on a uranium target leading to the production of very neutron rich nuclei.

For cold fission, the MAFF project at Munich will make use of the very high cross-section (586 barns) of thermal neutrons on uranium (^{235}U) using the newly built high flux reactor FRM-II where a local flux of 10^{14} n/cm^2/s is expected [DH97].

For fission induced by hadrons, the PARRNe project was initiated by IPN Orsay to investigate the feasibility of using thick ^{238}U targets and by experimentally optimizing the release and transfer coefficient ϵ_1 and the radioactive decay losses ϵ_4 for the produced neutron-rich fission fragments.

A deuteron beam from the 15MV Tandem at Orsay was used to produce a neutron flux of 10^9/s with an energy of about 10 MeV originating from the breakup of the deuterons on a Be/C converter. The produced fission fragments are extracted and ionized by a ISOL MK5 ion source, analyzed by a mass spectrometer and identified in a background free gamma station [SM99,CL00,SKR00]. More than one hundred isotopes have been observed with yields ranging from 10^2 to 10^7 pps from Cu, Zn to Ba. A typical result is the production of 3.10^5 ^{132}Sn/s/μA [FI02]. If, as projected neutron fluxes of many orders of magnitude higher will become available for the next generation of proton/deuteron ISOL driver accelerator (see sec 4.4), a ^{132}Sn (1^+) beam of about 10^{11}pps could be injected in a post-accelerator.

The other possibility for a fission based RIB facility is the use of a low-energy (25-50 MeV) electron driver beam. The generated bremsstrahlung would induce photo-fission in a ^{238}U target. The Flerov laboratory has included this option in its upgrade program [YTO00].

To test the photo-fission scheme, an experiment has been recently achieved using for the electron driver the 50 MeV injector Linac of the LEP accelerator complex at CERN. A movable version of the PARRNe device [FC99,SE03] has been used to extract and collect noble gases. The target was identical (mass, geometry, temperature) to the one employed with deuteron beams. Therefore, the obtained results are directly comparable and the outcome of this test experiment, in agreement with the predictions of Oganessian et al, is that a fission yield of 0.06% per incident electrons [YTO00]. A new ISOL facility, called ALTO, based on the photo-fission concept is being built at IPN Orsay coupling a 50 MeV, 10μA, electron linear accelerator and the PARNNe set-up.

The fast neutron concept has also been recently adopted for an upgrade of the GANIL facility. A new dedicated high power linac driver (deuterons 40 MeV,

5mA) will produce up to 10^{13}fissions/s and therefore high intensity neutron-rich fission fragments using a similar target-ion source set-up. In this planned new project, called SPIRAL II [ML03], the neutron-rich nuclei will be further post-accelerated by the CIME cyclotron to energies between 5 and 9 MeV/n. Another important initiative, based on the same fast neutron concept, has been taken by Legnaro (Italy) with the proposal SPES to built a 100 MeV high power proton driver [AB02].

4.4 New ultra intense proton driver and the EURISOL project

Since the last few years considerable efforts are devoted by accelerators teams from several countries to develop HPPA (High Power Proton Accelerator) technology. The aim here is to reach multi-megawatt class. Such accelerators are expected to permit breakthroughs in several scien-tific domains (see Proceedings of EPAC 2000 [EP00]).

As a short summary one may list the **Condensed Matter Physics** (new generation of pulsed neutron beam), **Sub-critical Hybrid Reactor Systems** for the incineration of radioactive waste by transmutation, **Neutron Irradiation Accelerator** or **Neutrino beams** with fluxes of 10^{20}/year generated by HPPA.

For radioactive beams, not only the quest for intense charged particle beam directly (with most probably a limit at 100KW heat deposition in the production target), is the driving motivation, but also, as explained in the previous section, intense spallation neutron fluxes are considered. The NuPPEC study group on radioactive beams recommended, among others items to consider a next generation ISOL facility. From this work, the EURISOL design study funded by EU [EU01] was launched three years ago. This study is carried out by a network of ten European laboratories.

The EURISOL preliminary design report will be published by within a few months (January 2004). The main features of the proposed facility are summarized below.

A proton driver accelerator with variable energy from 200 MeV to 2 GeV, with a maximum average beam power of one to 5MW. Time structure as close to DC as possible. Option to trigger high intensity bunches of 2 to 100ms lengths. Option for light heavy ion beams is also consider (A/Z=2). This driver is based on a linac structure, consisting of three different sections, low energy RFQ, and intermediate and high-energy sections with super conducting RF cavities. Both SPIRAL II and SPES project, based on High Power Particle Accelerator (HPPA) may be seen as full prototype of future HPPA machines.

Exotic beams are produced from fission by fast neutrons from a **converter target** (mercury jet targets) designed for very high power (MW class). Note that limits to the production of fission products only. Neutron rich beams from He to Ni are produced by high-energy fragmentation or highly asymmetric fission. Proton rich nuclei will be produced with ISOLDE or SPIRAL type targets with maximum power limited to 100 KW.

Two separators with separately accessible front-end areas will be used to purify the radioactive beams, one with high acceptance (M/ΔM = 2000) and one with high resolution (M/ΔM higher than 20000). Typical secondary "exotic "nuclear beam intensities after mass separation and before post-acceleration will range from 10-100 pps for the most exotic species (like ^{14}Be, ^{50}Ar, ^{78}Ni) to 10^{9-12}pps for neutron rich fission fragments like $^{90-94}$Kr, $^{132-134}$Sn.

Post acceleration up to about 100 MeV/n for mass A=150 will be available through a heavy ion SC linac. It will cover the energy range from astrophysical region 0.2-1 MeV/n to the Coulomb barrier (5 MeV/n) and further the Fermi energy (30 MeV/n) regime for reaction studies including fragmentation of secondary beams (like ^{132}Sn) leading to reach the neutron drip line close to the Zr region.

The heavy ion post accelerator can be also use as a driver for the production of neutron and proton rich light exotic nuclei. Associated with the different energy regimes, the exotic beams will be delivered to four experimental halls with adequate instruments.

The estimated investments costs of the EURISOL facility (proton driver, target-ion sources and radioactivity handling laboratories, mass-separators, superconducting post accelerator, beams lines and four experimental halls with complete instrumentations) is of the order of 400 Meuros (excluding salaries).

For this project, the next major step will be the funding through the 6th Framework Program of EU of a "detailed design study" where major R&D on accelerators, high power targets, and instrumentation will be included.

Acknowledgments

I would like to thank my colleagues F.Azayez, D.Beaumel, Y.Blumenfeld, S.Fortier, D.Guillemaud-Mueller, F.Ibrahim, O.Sorlin, A.C.Mueller, D.Verney and all the members of the EURISOL working groups for the use of documents prior to publications and for stimulating discussions.

References

OKH51 O. Kofoed-Hansen and K.O. Nielsen , Dan. Mat. Fys. Medd. **26**, n ° 7 (1951)

HLR90, H.L. Rahvn, Nucl. Inst. Methods B**70** (1990) 107

JS00, J. Sauvage et al., Hyperfine Interactions **129** (2000) 303

GB98, G. Bollen et al., Proceedings of the International Conference on Exotic Nuclei and Atomic Masses 1998, edited by B.M. Sherill, D.J. Morissey and C.N. Davids, AIP Conference Proceedings 455, Woodbury NY, 1998, p.3

DL98, D. Lunney et al. in GB98 p.997

GM96, Proceedings of the 13th International Conference on Electromagnetic Isotope Separator and their applications, edited. by G. Münzenberg, Nucl. Inst. Methods B**126** (1996)
JV97, J. Vervier, Nucl. Phys. A**616** (1997) 97c
IT98, I. Tanihata in GB98, p.943
YEP02, M. Ishihara in Proceedings of the International Symposium on Exotic Nuclei, EXON 2001, 2edited by Yu.E. Penionzhkevich and E.A. Cherepanov ,World Scientific, 2002, p.596
ACM98, A.C. Mueller in GB98, p.933
SG02, S. Gales in YEP02, p.547
JAN98, J.A. Nolen in GB98, p.953
ML03, M. Lewitowitz, Nucl. Phys A**722** (2003) 67c
HG02, H. Geissel and G. Münzenberg in YEP02, p.567
YTO02, Yu.Ts. Oganessian and G.G. Gulbekian in YEP02, p.609
AAK02, A.A. Korsheninnikov in YEP02 p.317
RNBF00, Radioactive Nuclear Beam Facilities, NuPECC report, April 2000 and contribution to this conference by G. Münzenberg
RIA00, Scientific Opportunities with an advanced ISOL facility, http://www.phys.anl.gov
RAP00, Riken accelerator progress report 2000, vol (34), p.301
RIA01, http://srfsrv.jlab.org/isol/ISOLTaskForceReport.pdf
JA87, J. Arje et al., NIM B**26** (1987) 384
DH97, D. Habs et al., Nucl. Phys. A**616** (1997) 39c and http://www.ha.physik.uni-muenchen.de/maff/
SM99, S. Menard et al., Phys. ReV. ST-Acc & Beams 2 (1999) 033501
CL00, C. Lau, Thesis, Denis Diderot University, Paris 7, (2000)
SKR00, S. Kandri-Rody et al., NIM B**160** (2000) 1
FI02, F. Ibrahim et al., EPJA**15** (2002) 357
YTO00, Yu.Ts. Oganessian in Proceedings of the International Conference on RNB-5, 2000, 1-6 April 2000, Divonne les Bains, France
FC99, F. Clapier, A.C. Mueller, J. Obert et al., Phys. Rev. ST-Acc & Beams 1 (1999) 013501
SE03, S. Essabaa et al., NIM B**204** (2003) 780
AB02 A. Bracco and A. Pisent, SPES Technical design for an advanced EXOTIC ion beam facility at LNL, Internal Report (REP181/02/June 2002)
EP00, Proceedings of the European Accelerator Conference EPAC 2000, Vienna, Austria, June 2000, http://www.accelconf.cern.ch/accelconf/e00
EU01, EURISOL server, htpp://www.ganil.fr/eurisol/

STORAGE-RING EXPERIMENTS WITH EXOTIC NUCLEI: FROM MASS MEASUREMENTS TO THE FUTURE

G. Münzenberg
Gesellschaft für Schwerionenforschung (GSI) mbH, Planckstr. 1, 64291 Darmstadt, Germany and Johannes Gutenberg - Universität Mainz, D-55128 Mainz, Germany
E-mail: g.muenzenberg@gsi.de

Abstract Direct mass measurements with electron-cooled ions coasting in a heavy-ion storage ring allow the mapping of large areas of the nuclear chart, well suited to explore new regions and to investigate isospin effects as well as structure changes far-off stability. In this contribution first results will be discussed, experimental masses will be compared to theoretical predictions. A study of the isospin dependence of shell strength and pairing will be made.

The storage of atomic nuclei fully stripped or as few electron systems allows to study them under interstellar conditions, specific decay modes can be switched-off.

The prospects for precision experiments in storage and cooler rings including reaction studies at an internal target will be given.

The new GSI project of a new synchrotron facility with new instrumentation will have new impacts on structure research extending our knowledge into far unknown regions and providing new probes for structure research with unstable nuclei.

1. Introduction

The exploration of the limits of nuclear stability in isospin and charge including exotic decay modes and the evolution of shell structure are the central topics of modern structure research. The aim is to understand correlations and pairing, in-medium modification of the nucleon-nucleon interaction, and the behavior of nuclear matter with extreme neutron-to-proton ratios.

GSI research is based on a system of heavy ion accelerators which can accelerate heavy ions from all chemical elements of the periodic table ranging from protons to uranium, including rare isotopes with low abundances in nature[1]. The UNIversal Linear ACcelerator (UNILAC) is used for near Coulomb-barrier energies and as injector for the heavy-ion synchrotron (SIS) which can accel-

W. Greiner et al. (eds.), Structure and Dynamics of Elementary Matter, 507–514.

erate uranium ions to more than 1 AGeV. Unique is the Experimental Storage Ring (ESR) for storage and cooling of heavy ions.

Storage rings are ion traps with large acceptance. They can accept cocktail beams of a large number of different nuclides and at the same time in different ionic charge states such as bare, hydrogen, and helium-like[2]. Masses and half-lives reflect gross properties of nuclear matter. Direct mass measurements in storage rings are the ideal tool for systematic studies e.g. mapping the nuclear mass surface to get a first overview on the evolution of nuclear structure towards the limits of nuclear stability, best suited to explore isospin effects. Irregularities in the nuclear mass surface are first indications for structure changes.

Decay studies in storage rings allow to explore nuclear lifetime under interstellar conditions. Atomic nuclei can be stored as bare, hydrogen- or helium-like systems. Beta decay into bound nuclear states or the switching-off of specific decay modes such as electron capture or internal conversion may change nuclear decay rates and thus influence the calibration of cosmologic clocks[3].

Cooled beams of heavy ions in combination with thin targets located inside the storage ring are the ideal tool for precision experiments as atomic interactions such as energyloss or angular scattering can be eliminated. The effective target thickness is enhanced as the beam passes the target on each revolution.

The success of the first generation experiments with unstable nuclear beams motivated a number of second generation projects with improved accelerators and new experimental equipment. The new GSI facility will play a premier role as it will provide high beam intensities in combination with the highest energies available for unstable nuclear beams. I will be the only facility which uses a storage-ring system for structure research[4]. A heavy-ion storage ring combined with small intersecting electron storage ring will provide electrons as a new probe.

2. Mass measurements

The nuclei to be stored in the ESR are produced by fragmentation of relativistic heavy ions in peripheral collisions or by fission of relativistic ^{238}U. Typical projectile energies for storage-ring experiments are 600 AMeV to 900 AMeV. The nuclides of interest are separated in-flight by the FRS and injected into the experimental storage ring with their full energy. Optionally beam cocktails or isotopic beams can be provided[5].

The nuclei stored in the ring are electron cooled to momentum spreads of the order of 10^{-6} which finally determines the mass resolution[6]. The masses (or more precise, the mass over charge ratios) are obtained from the Fourier - transformed Schottky noise signal of the ions coasting in the ESR. The cooling time is about 10 s, it limits this method for long-lived species[2]. Operating the storage ring in a different ion optical mode, so that the beam trajectories become

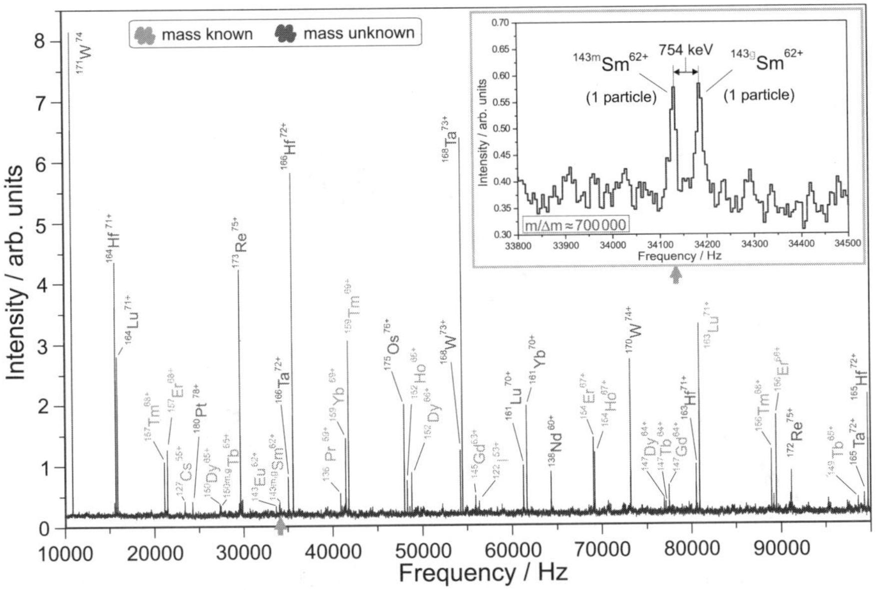

Figure 1. Portion of a mass spectrum of electron cooled projectile fragments stored in the ESR. The insert demonstrates the resolving power and sensitivity of this method.

isochronuous, permits to use the ring as a large time-of-flight spectrometer. Halflives of the order of microseconds become accessible. The flight time of the ions which make 100 to 1000 turns is measured with pickups using thin carbon foils and secondary-electron detection[14] .

Fig. 1 shows as an example a portion of a mass spectrum obtained with electron cooled fragments produced by fragmentation of ^{209}Bi. Known and unknown masses are well mixed which guarantees a precise and reliable calibration. This method permits the measurement even of single ions, as shown in the insert. The precision is of the order of 30 keV. Already in our first experiments 104 proton rich nuclides from tellurium to plutonium were measured[2]. In the meantime masses of 350 nuclides have been determined with a precision of 30 keV for the first time.

First experiments with the time-of flight method gave resolutions which are lower by about one order of magnitude. An overview over the nuclides with known masses is displayed in Fig. 2. Nuclei which have been measured for the first time at GSI are indicated[14].

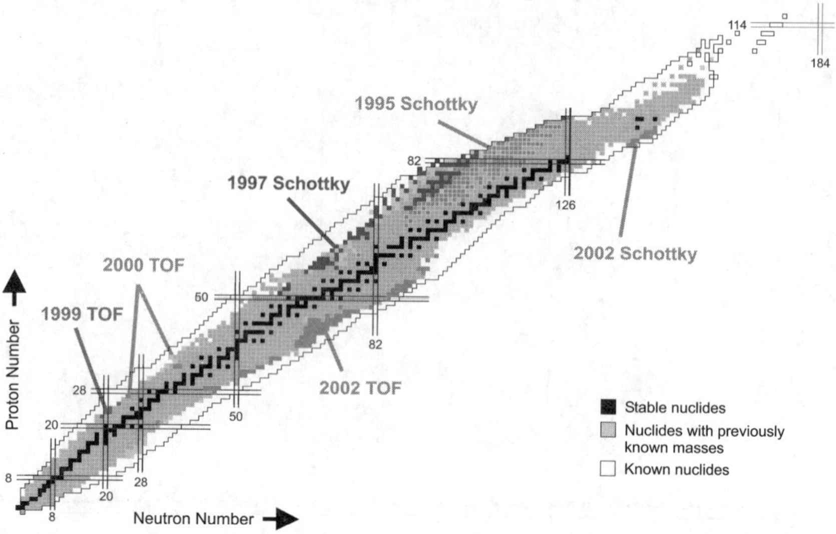

Figure 2. The nuclear chart with the new masses obtained from the FRS-ESR experiments.

3. Comparison to predictions

Systematic studies are ideally suited to test theoretical predictions. One fundamental characteristic of nuclei is the location of the driplines, the limits for the existence of matter in terms of isospin. Fig. 3 shows the comparison of experiment to theory for odd-mass nuclei with $Z > 71$. The relativistic mean-field calculation with the NL3 parametrization predicts the dripline precisely except for tantalum ($Z = 73$) and iridium ($Z = 79$) where it fails by one isotope [8].

A test for a limited and selected number of nuclei is always risky, as theory may by chance reproduce the data. A systematic investigation is much more reliable. This has been performed for the lead isotope chain, Fig 4. To extend this comparison beyond the limit of experimental data the Thomas - Fermi model has been taken as reference. The data marked by squares show excellent agreement to this model which is also true for the extended Thomas Fermi model with Strutinski integral and for the finite range droplet model. The latter however indicates an increasing deviation towards the neutron rich isotopes beyond the N = 126 neutron shell[9].

A basic question of structure research is the nature of closed shells and their persistence in regions off stability. Fig. 5 shows an investigation on the strength of the lead proton shell from beyond the neutron shell closure at $N = 126$ towards the proton dripline. The shell strength has been obtained from the double differential two-proton separation energies δ_{2p} according to Koopman's rule. The value for δ_{2p} drops by a factor of two from more than 6 MeV at the

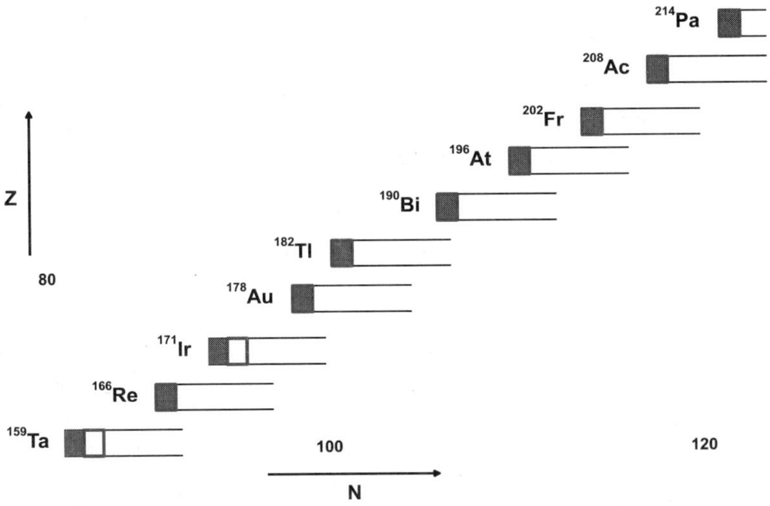

Figure 3. The proton drip line. Comparison between experiment and prediction.

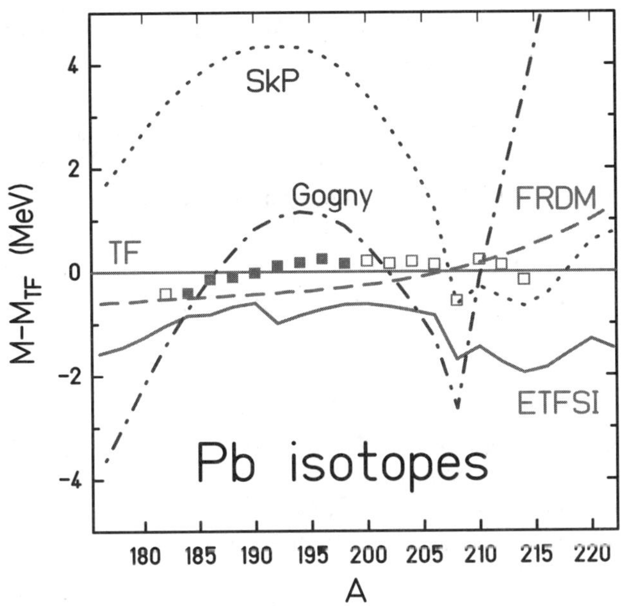

Figure 4. Experimental data for the lead isotopes compared to theoretical predictions[9].

closed neutron shell $N = 126$ to below 3 MeV for $N = 100$, which suggests a shell quenching[10]. A theoretical investgation[11] however shows that the data can be reproduced with a constant shell gap of close to -10 MeV as seen in the figure. The reason for the decrease in separation energies in presence of an almost constant shell gap is that the nuclear structure in this region changes fast, the neighbored elements of lead are deformed. This result may have a consequence for SHE predictions with only spherical calculations.

Very recently the isospin dependence of the neutron and proton pairing strength has been investigated. The pairing strength for neutrons as well as for protons increases strongly towards the proton dripline, an effect not predicted by theoretical models[12].

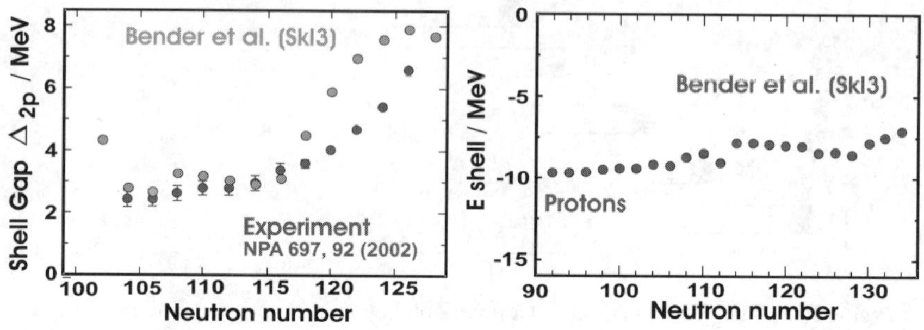

Figure 5. The lead proton shell. Left panel: Comparison of experiment and theory. Right panel: Shell corrections of Pb isotopes.

For the short-lived species a time-of-flight method has been developed operating the storage ring in the isochronuous mode where the revolution time of the stored ions is only determined by their mass. Nuclei with half-lives down to microseconds can be measured. First experiments for the masses of light nuclei and in the vicinity of the r-process path have been performed successfully with a precision of about 200 keV to 300 KeV at present. The data evaluation is in progress[13].

Mass measurements with cocktail beams stored in the ESR using Schottky spectroscopy or the time-of-flight method will become important for the systematic exploration of the large area of neutron rich nuclei which become accessible with the new generation of radioactive beam facilities. In forthcoming experiments regions around closed shell as well as along the cosmic nucleosynthesis paths including the rp- the r- process will be measured.

3.1 Decay studies of bare nuclei

A new experimental field opened up with heavy-ion storage rings are decay studies of bare nuclei or few-electron systems. A first result is the discovery of the "bound beta decay". This is the beta decay to bound electron orbitals which is only possible when these are empty[3]. This decay mode is specific to β emitters close to stability with small Q-values not sufficient for the decay electron to escape into the continuum. Such nuclei are stable in our natural environment but become unstable when bare. The first case was ^{163}Dy, later

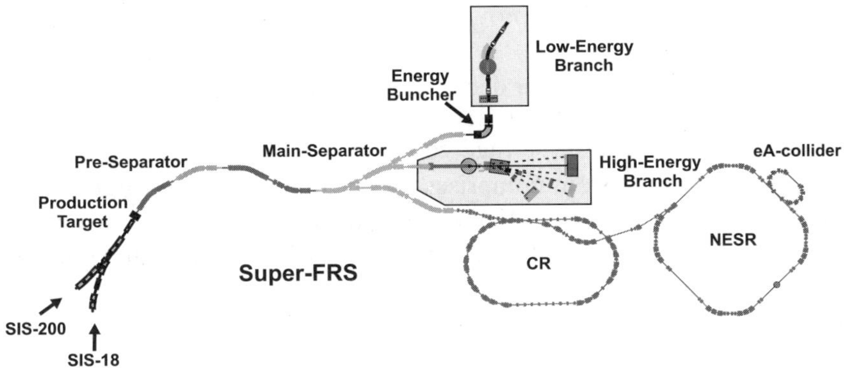

Figure 6. The proposed Next Generation Exotic Beam Facility at GSI with the two-staged superconducting fragment separator SuperFRS and its three experimental areas for high energies, used for reaction studies, low energy, used for structure physics, an the storage-ring complex

^{187}Re was studied. This nucleus serves as an astrophysical clock. Half-live changes in its history need to be corrected[3] for precise age estimates.

Electron capture is switched-off in bare nuclei. Such cases are of interest for instance in the observation of radiation from cosmic processes, e.g. ^{56}Ni originating from supernovae.

4. The next generation facility

The prerequisite to access more exotic species and to approach the limits of nuclear stability is a new generation of powerful heavy-ion accelerators which can deliver intense beams of all stable isotopes - including the rare ones - of all natural chemical elements.

The future GSI facility will be a multipurpose facility, using the experience in parallel operation for several experiments gained with the existing SIS18[16]. The accelerator system is based on a superconducting double-ring synchrotron system of 100 Tm and 200 Tm[1, 16]. The 100 Tm synchrotron is fast cycling to achieve a high duty factor providing intense beams for the production of unstable nuclei for structure research. The 200 Tm machine can be used as stretcher ring to provide large duty factors for slow extraction.

For exotic beam production the new facility will provide beam intensities of the order of 10^{12}/s for all natural elements of the periodic table at energies around 1 AGeV[16]. The high primary beam intensity is required to access exotic new regions, to allow for detailed structure studies of sufficiently exotic species, and to open the new field of reaction studies with unstable nuclei in storage rings. The high beam energy is necessary for clean separation of heavy elements and fission products. An improved multi-stage super conducting fragment separator of large acceptance adapted to the large phase-space of fission fragments, the "Super FRS", provides beams for three experimental areas[15] as shown in Fig. 6:

- A low energy branch for decay and in-beam studies,
- A high energy branch for reaction studies at full energy,
- A ring branch for the storage-cooler ring system.

The storage-ring system comprises an accumulator for in-flight separated projectile- and fission fragments with fast stochastic cooling, an experimental storage ring, an internal target for reactions at with low momentum transfer and highest precision, and a small intersecting electron ring for structure studies. First generation experiments are elastic electron scattering at low momentum transfer.

This next generation facility will allow to proceed further to the limits of the existence of nuclear matter and open up the possibilities to apply new experimental methods for structure research to investigate the many facets of nuclear structure. New probes will give new access to hitherto unknown and new facets of the atomic nucleus. Systematic studies of selected and most relevant features of the atomic nucleus, starting from the light few-body systems to the heavy multi-nucleon systems will help us to understand it as the building block of elementary matter.

References

[1] www.gsi.de

[2] T. Radon, H. Geissel, G. Münzenberg et al., Nucl. Phys. A **677** (2000), 75

[3] F. Bosch et al., Phys. Rev. Lett. **77**(1996), 5190

[4] Conceptual design report for "An International Accelerator Facility for Ions and Antoprotons", GSI, Darmstadt 2002, WWW.gsi.de available as CD-rom

[5] H. Geissel, G. Münzenberg and K. Riisager, Ann. Rev. Nucl. Part. Sci. **45** (1995), 163

[6] M. Steck et al., Phys. Rev. C **77** (1996), 3808

[7] M. Matos, Thesis Univ. Giessen, under preparation

[8] P. Ring, Priv. comm. 2003

[9] Z. Patyk et al., Phys. Rev. Lett. C **59** (1999), 704

[10] Yu. A. Novikov et al., Nucl. Phys. A **697** (2002), 92

[11] M. Bender et al., Eur. Phys. J. A **14** (2002), 23

[12] Yu. A. Novikov, priv. comm 2003

[13] M. Matos, Thesis Univ. Giessen, under preparation

[14] Yu. A. Litvinov, thesis Univ Giessen and GSI report, 2003

[15] H. Geissel et al., Nucl. Instr. Meth.

[16] H. Gutbrod, contribution to this conference

[17] S. Hofmann et al., contribution to this conference

NEW CANDIDATES FOR CLUSTER DECAY EXPERIMENTS

D. N. Poenaru[1,2], W. Greiner[2]

[1] *Horia Hulubei National Institute of Physics and Nuclear Engineering, PO Box MG-6, RO-077125, Bucharest-Magurele, Romania*

[2] *Institut für Theoretische Physik der J. W. Goethe Universität, Pf 111932, D-60054, Frankfurt am Main, Germany*

Abstract Heavy particle radioactivity was predicted in 1980. Starting in 1984 with ^{14}C radioactivity of ^{223}Ra, the following cluster emitters have been experimentally identified up to now: ^{221}Fr, $^{221-224,226}$Ra, ^{225}Ac, ^{228,230}Th, ^{231}Pa, $^{230,232-236}$U, ^{236,238}Pu, and ^{242}Cm. The detected new kinds of radioactivities are: ^{14}C, ^{20}O, $^{22,24-26}$Ne, ^{28,30}Mg, and ^{32,34}Si. Only upper limits were determined for ^{18}O radioactivity of ^{226}Th, ^{24,26}Ne emission from ^{232}Th, ^{23}F radioactivity of ^{231}Pa, ^{28}Mg emission from ^{233,235}U, ^{24,26}Ne emission from ^{234}U, ^{30}Mg emission from ^{237}Np, ^{34}Si radioactivity of ^{240}Pu and of ^{241}Am. The measured halflives are in good agreement with predicted values within analytical superasymmetric fission model. The strong shell effects observed in the present systematics was not fully exploited; new candidates for successful experiments are suggested.

1. Introduction

Heavy particle radioactivity was predicted [1] four years before the first experimental confirmation [2] of ^{14}C spontaneous emission from ^{223}Ra. Misleading conclusions including speculations since 1924 concerning the abundance of some gases (like N and Ne) in uranium ores were published earlier. One of the plotted examples we gave in 1980 to illustrate the new decay modes intermediate between fission and α-decay was the calculated penetrability versus the mass number ot the emitted cluster for different heavy ions emitted from ^{222}Ra and ^{224}Ra, from which it was clear that ^{14}C would have the greater chance to be emitted. Before any other model was published, we calculated, within our analytical superasymmetric fission (ASAF) model (e.g. [3]) the half-lives and branching ratios relative to α-decay for more than 150 decay modes, including all cases experimentally confirmed up to now. Comprehen-

W. Greiner et al. (eds.), Structure and Dynamics of Elementary Matter, 515–522.

sive tables were updated several times [4]. The state-of-the-art of theoretical and experimental achievements in the field was reviewed by Blendowske et al., Poenaru and Greiner, Hourany, Bonetti and Guglielmetti in various chapters of the multi-authored book [5].

The half-lives of ^{221}Fr, $^{221-224,226}$Ra, ^{225}Ac, ^{228,230}Th, ^{231}Pa, $^{230,232-236}$U, ^{236,238}Pu, and ^{242}Cm against spontaneous emission of ^{14}C, ^{18,20}O, ^{23}F, $^{22,24-26}$Ne, ^{28,30}Mg and ^{32,34}Si measured up to now are in good agreement with predicted values within ASAF model. Also the upper limits determined for ^{18}O radioactivity of ^{226}Th, ^{24,26}Ne emission from ^{232}Th, ^{23}F radioactivity of ^{231}Pa, ^{28}Mg emission from ^{233,235}U, ^{24,26}Ne emission from ^{234}U, ^{30}Mg emission from ^{237}Np, ^{34}Si radioactivity of ^{240}Pu and of ^{241}Am are not in contradiction with ASAF values.

Below we present a systematics of experimental data compared to calculations, from which one may suggest [6] possible new candidates for future experiments.

2. Measurable and model-dependent quantities

A parent nucleus with atomic number Z and mass number A decays into an emitted cluster A_e, Z_e and a daughter A_d, Z_d, i.e.

$$^{A}Z \rightarrow {}^{A_e}Z_e + {}^{A_d}Z_d$$

with conservation of hadron numbers (neutrons and protons): $N = N_e + N_d$ and $Z = Z_e + Z_d$, where $A = N + Z$, and similar relationships hold for A_e and A_d. The process is energetically allowed if the released energy,

$$Q = M(A, Z) - [M_e(A_e, Z_e) + M_d(A_d, Z_d)] \quad (1)$$

is a positive quantity, $Q > 0$. The atomic masses M, M_e, and M_d in units of energy, are taken from tables of experimental values evaluated and updated from time to time [7]. As in a cold fission process or in α-decay, the measured kinetic energy of the emitted particle is given by

$$E_k = QA_d/A \quad (2)$$

Another important measurable quantity is the partial decay half-life T related to the decay constant $\lambda = \ln 2/T$. This is expressed as a product of three (model-dependent) quantities:

$$\lambda = \nu SP \quad (3)$$

the frequency of assaults (ν), the preformation probability of the emitted cluster into the parent nucleus (S), and the (external) barrier penetrability (P). The frequency of assaults is usually considered to be in the range 10^{20}–10^{22} per second. For emitted clusters heavier or equal to an α-particle, the three (model-dependent) quantities within our ASAF model, in the equation of the half-life, $\log T(s) = \log[(\ln 2)/\nu(s^{-1})] - \log S - \log P$, take numerical values of the order -21, > 2, and > 12, respectively, leading to $T > 10^{-7}$ s.

In a fission-like model let us assume a potential barrier $E(R)$ dependent on a single parameter of deformation R which can be the distance between centres of two fragments supposed to have spherical shapes. The penetrability of the external part of the barrier may be calculated by using the one-dimensional Wentzel–Kramers–Brillouin (WKB) approximation

$$P = \exp(-K_s); \quad K_s = \frac{2}{\hbar} \int_{R_t}^{R_b} \sqrt{2B(R)E(R)} dR \tag{4}$$

In the same way, we gave a practical method to estimate the preformation probability as the penetrability of the internal part of the barrier

$$S = \exp(-K_{ov}); \quad K_{ov} = \frac{2}{\hbar} \int_{R_a}^{R_t} \sqrt{2B(R)E(R)} dR \tag{5}$$

where B is the nuclear inertia, equal to the reduced mass $\mu = mA_eA_d/A$ for separated fragments, m is the nucleon mass, E is the potential energy from which the Q-value has been subtracted out, R_a and R_b are the classical turning points. There is an internal barrier (the overlapping stage of two fragments) from the initial $R = R_a$ to the touching point separation distance R_t and an external one (separated fragments) from $R = R_t$ to the exit point R_b.

Since 1980, in order to make a systematic search for new cluster decay modes, we have had to take into account a large number of combinations of parent and emitted nuclei (at least 2000×250 = $5{\cdot}10^5$). The corresponding large number of computations can be performed in a reasonable time by using an analytical relationship for the half-life. Such a formula has been obtained on the basis of the Myers–Swiatecki's [8] liquid drop model (LDM). We substitute in the WKB eq. $B(R) = \mu$, and replace $E(R)$ by $[E(R) - E_{corr}] - Q$. E_{corr} is a phenomenological correction energy similar to the Strutinsky shell correction, also taking into account the fact that LDM overestimates fission barrier heights. In order to obtain an analytical relationship for K_{ov}, we use a parabolic approximation of the potential barrier:

$$E(R) = Q + (E_i - Q)[(R - R_i)/(R_t - R_i)]^2; \quad R \leq R_t \tag{6}$$

where the interaction energy at the top of the barrier, in the presence of a non-negligible angular momentum, $l\hbar$, is given by the Coulomb and centrifugal terms:

$$E_i = E_C + E_l = e^2 Z_e Z_d / R_t + \hbar^2 l(l+1)/(2\mu R_t^2) \tag{7}$$

After substitution into the above integral we obtain a closed formula for K_{ov}

$$0.2196 \left(\frac{E_b^0 A_e A_d}{A} \right)^{1/2} (R_t - R_i) \left[\sqrt{1-b^2} - b^2 \ln \frac{1+\sqrt{1-b^2}}{b} \right] \quad (8)$$

where the parameter $b^2 = (E_{corr} + E^*)/E_b^0$ appears as a consequence of the following expression of the first turning point (defined by $E(R_a) = Q$):

$$R_a = R_i + (R_t - R_i)[(E_{corr} + E^*)/E_b^0]^{1/2} \quad (9)$$

in which E^* is the excitation energy concentrated in the separation degree of freedom, and $E_b^0 = E_i - Q$ is the barrier height before correction.

For separated spherical fragments in the Coulomb-plus-centrifugal potential

$$E(R) = E_C R_t/R + E_l R_t^2/R^2; \quad R > R_t \quad (10)$$

it is not necessary to make any approximation; we get

$$K_s = 0.4392[(Q + E_{corr} + E^*)A_e A_d/A]^{1/2} R_b J_{rc} \quad (11)$$

with

$$\begin{aligned} J_{rc} \quad = \quad & (c) \arccos \sqrt{(1-c+r)/(2-c)} - [(1-r)(1-c+r)]^{1/2} + \\ & \sqrt{1-c} \ln \left[\frac{2\sqrt{(1-c)(1-r)(1-c+r)} + 2 - 2c + cr}{r(2-c)} \right] \end{aligned} \quad (12)$$

in which $r = R_t/R_b$ and $c = rE_C/(Q + E_{corr} + E^*)$. When the angular momentum $l = 0$, one has $c = 1$ and the above equations take a simpler form. The external turning point R_b [defined as $E(R_b) = Q$] is given by

$$R_t E_C\{1/2 + [1/4 + (Q + E_{corr} + E^*)E_l/E_C^2]^{1/2}\}/(Q + E_{corr} + E^*) \quad (13)$$

In order to reduce the number of fitting parameters, we took $E_v = E_{corr}$. Both shell and pairing effects are included in $E_{corr} = a_i(A_e)Q$ ($i = 1, 2, 3, 4$ for even–even, odd–even, even–odd, and odd–odd parent nuclei).

3. Predicted and confirmed half-lives

The calculations performed within ASAF model and reported before the experiments are compared with data in figure 1. The strong shell and pairing effects are better displayed in these plots by choosing the neutron and proton numbers of the daughter nuclei N_d and Z_d as independent variables. Minima of the halflives are obtained whenever N_d and Z_d are magic numbers (126 and 82, respectively). The doubly magic daughter ^{208}Pb has a very important

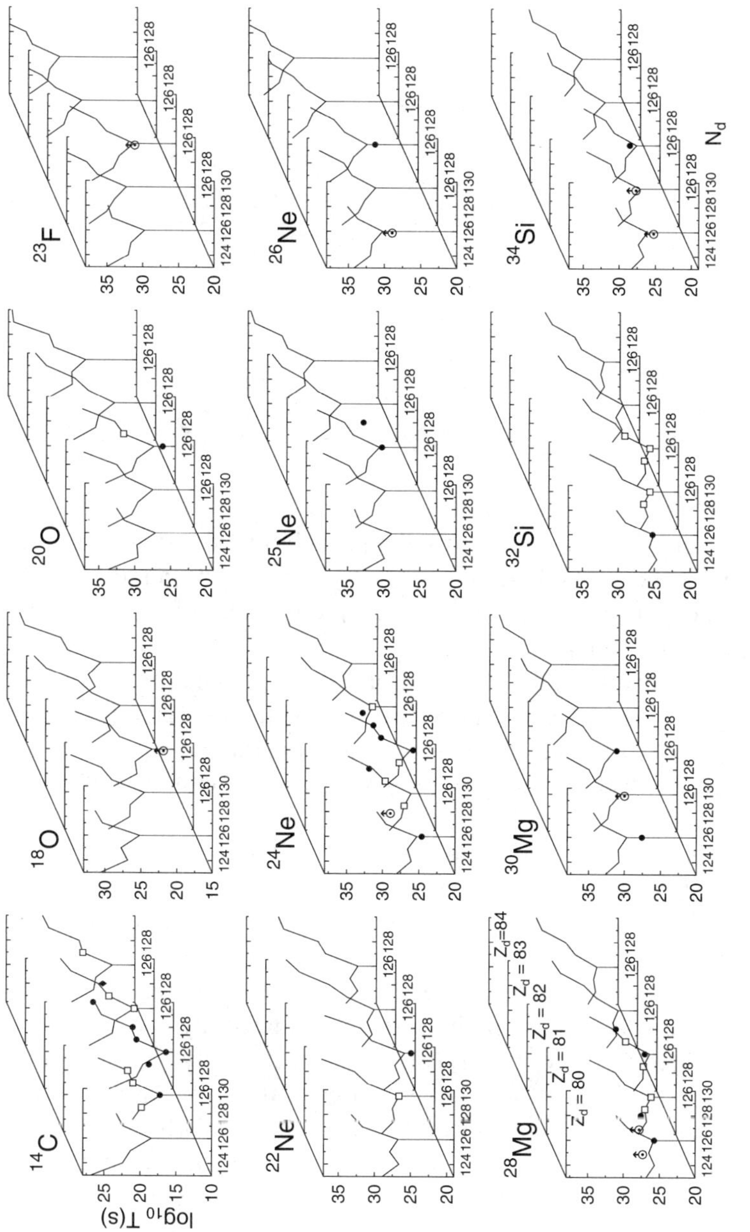

Figure 1. Predicted half-lives within ASAF model (lines) and measurements (points) for 12 kinds of cluster decay modes versus neutron number of the daughter nucleus. The calculations belonging to the same atomic number of the daughter are joined with a line. The experimentally determined upper limits are marked with a vertical arrow. The open squares are suggestions for new candidates.

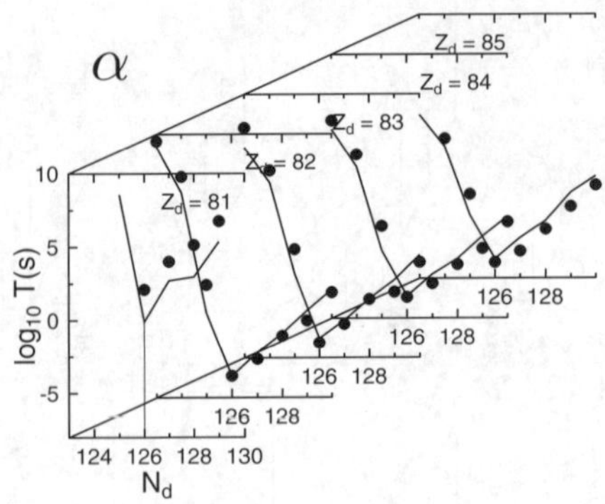

Figure 2. Similar to Figure 1 for α-decay.

contribution to enhance the emission probability of its partners. Other magic numbers of the emitted clusters, e.g. $N_e = 8$ for ^{14}C could be important too. This may explain why ^{14}C and not ^{12}C is preferentially emitted from Ra isotopes considered in this figure.

With the very strongly bound α-particle things are different, as may be seen from figure 2. Halflives as short as fractions of microseconds for α-decay should be compared with about 10^{11} s — the minimum for heavy particle decay modes. The largest branching ratio $b = T_\alpha / T_c$ measured up to now is of the order of 10^{-9}.

Experimental difficulties are mainly related to the low yield in the presence of a strong background of α-particles. A ΔE$\times$E telescope of two silicon detectors directly viewing the source was used in the first experiment, which was running six months in order to obtain 11 events. The solution adopted in the elegant experiments performed at Argonne [9] and Orsay [10] was to deflect the unwanted ^{4}He ions, simply or doubly ionized, by a strong magnetic field produced in a magentic spectrometer and to select only ^{14}C clusters to reach the detector in the focal plane. A source about 300 times stronger than that used by Rose and Jones was employed so that a run of only five days was necessary to obtain 11 events. With the SOLENO superconducting spectrometer it was possible to discover [11] the fine structure in cluster decay previously discussed [12] and to perform the most accurate experiment [10] by using high quality implanted sources of Ra isotopes made at the ISOLDE mass separator, CERN, Geneva.

Another method extensively used [13–15], is based on solid state nuclear track detectors (SSNTD) which are not sensitive to alphas and other low-Z particles, because they need a certain threshold of ionization. Very frequently such detectors are made from polyethylene terephtalate or from a phosphate

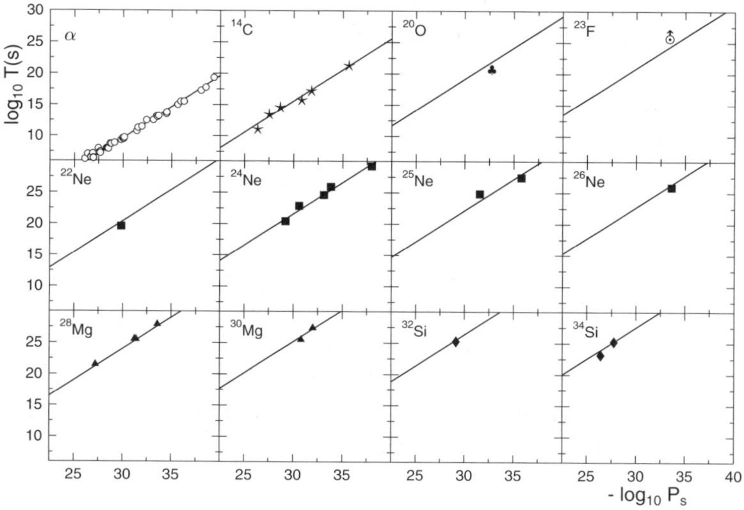

Figure 3. Universal curves and experimental points for heavy ion radioactivities and α-decay.

glass. They are cheap and handy, but like the photographic plates, do not deliver the information on-line; only after a suitable post-irradiation chemical etching are the tracks visible. The track can be seen, located and measured by manual or automatic scanning with a microscope, and one may derive the characteristics of the incoming particle from its shape and dimensions.

As may be seen from figure 1, until now only some of the most favourable cases were investigated (the points), leading to magic or almost magic proton and neutron numbers of daughter nuclei. The systematics of experimental results compared to calculations, are clearly showing other possible candidates for future experiments (open squares in this figure): ^{220,222,223}Fr, ^{223,224}Ac, and ^{225}Th as ^{14}C emitters; ^{229}Th for ^{20}O radioactivity; ^{229}Pa for ^{22}Ne decay mode; ^{230,232}Pa, ^{231}U, and ^{233}Np for ^{24}Ne radioactivity; ^{234}Pu for ^{26}Mg decay mode; ^{234,235}Np and ^{235,237}Pu as ^{28}Mg emitters, as well as ^{238,239}Am and $^{239-241}$Cm for ^{32}Si radioactivity. Also ^{33}Si decay of ^{241}Cm could be observed.

4. Universal curves

One may assume as a first approximation, that preformation probability only depends on the mass number of the emitted cluster, A_e, in the following simplified manner:

$$\log S = \frac{(A_e - 1)}{3} \log S_\alpha \tag{14}$$

where the preformation probability of the α particle S_α may be determined by fit to experimental data. The next assumption is that $\nu(A_e, Z_e, A_d, Z_d) =$ constant. From the fit one obtains: $S_\alpha = 0.0160694$ and $\nu = 10^{22.01}$ s^{-1}.

In this way one arrives at a single straight line *universal curve* on a double logarithmic scale

$$\log T = -\log P - 22.169 + 0.598(A_e - 1) \tag{15}$$

where

$$-\log P = c_{AZ}\left[\arccos\sqrt{r} - \sqrt{r(1-r)}\right] \tag{16}$$

with $c_{AZ} = 0.22873(\mu_A Z_d Z_e R_b)^{1/2}$, $r = R_t/R_b$, $R_t = 1.2249(A_d^{1/3} + A_e^{1/3})$, $R_b = 1.43998 Z_d Z_e/Q$, and $\mu_A = A_d A_e/A$. For all measurements performed until now the agreement is good, as can be seen in figure 3.

References

[1] A. Sandulescu, D. N. Poenaru, and W. Greiner, Sov. J. Part. Nucl. **11** (1980), 528.

[2] H. J. Rose and G. A. Jones, Nature **307** (1984), 245.

[3] D. N. Poenaru, M. Ivascu, A. Sandulescu, and W. Greiner, J. Phys. G: Nucl. Phys. **10** (1984), L183.

[4] D. N. Poenaru, D. Schnabel, W. Greiner, D. Mazilu, and R. Gherghescu, Atomic Data Nucl. Data Tables **48** (1991), 231.

[5] D. N. Poenaru and W. Greiner (eds.), *Nuclear decay modes*, Institute of Physics Publishing, Bristol, England, 1996.

[6] D. N. Poenaru, Y. Nagame, R. A. Gherghescu, and W. Greiner, Phys. Rev. C **65** (2002), 054308, Erratum: C66.049902.

[7] G. Audi and A. H. Wapstra, Nucl. Phys. A **595** (1995), 409.

[8] W. D. Myers and W. J. Swiatecki, Nucl. Phys. A **81** (1966), 1.

[9] W. Henning and W. Kutschera, Particle Emission from Nuclei, Vol. II, CRC Press, Boca Raton, Florida, 1989, p. 188.

[10] E. Hourany, Nuclear Decay Modes, IOP Publishing, Bristol, 1996, p. 350.

[11] L. Brillard, A. G. Elayi, E. Hourani, M. Hussonnois, J. F. Le Du, L. H. Rosier, and L. Stab, C. R. Acad. Sci. Paris **309** (1989), 1105.

[12] M. Greiner and W. Scheid, J. Phys. G: Nucl. Phys. **12** (1986), L229.

[13] P. B. Price and S. W. Barwick, Particle Emission from Nuclei, Vol. II, CRC Press, Boca Raton, Florida, 1989, p. 205.

[14] S. P. Tretyakova, Sov. J. Part. Nucl. **23** (1992), 156.

[15] R. Bonetti and A. Guglielmetti, Nuclear Decay Modes, Institute of Physics Publishing, Bristol, 1996, p. 370.

NUCLEAR FISSION AND STRUCTURE STUDIES WITH GAMMASPHERE

A.V. Ramayya[1], J.H. Hamilton[1], J.K. Hwang[1], P. Gore[1], E.F. Jones[1], D. Fong[1], G.M. Ter Akopian[2,3], A.V. Daniel[2,3], J.O. Rasmussen[4], S.-C. Wu[4], R. Donangelo[4,5], J. Kormicki[1], X.Q. Zhang[1], A. Rodin[2], A. Formichev[2], J. Kliman[2], L. Krupa[2], G. Chubaryan[6], and J.D. Cole[7]

[1]*Department of Physics, Vanderbilt University, Nashville, TN37235, USA*

[2]*Flerov Laboratory for Nuclear Reactions, Joint Institute for Nuclear Research, Dubna, Russia*

[3]*Joint Institute for Heavy Ion Research, Oak Ridge, TN37831, USA*

[4]*Lawrence Berkeley National Laboratory, Berkeley, CA 94720, USA*

[5]*Instituto de Física, Universidade Federal do Rio de Janeiro, C.P. 68528, 21945-970 Rio de Janeiro, Brazil*

[6]*Cyclotron Institute, Texas A and M University, Texas 77843-3366, USA*

[7]*Idaho National Engineering and Environmental Laboratory, Idaho Falls, ID83415, USA*

Abstract The γ-rays emitted in the spontaneous fission of ^{252}Cf has been studied with Gammasphere. The yields of correlated Mo-Ba pairs in binary fission with 0-10 neutron emission have been remesured with an uncompressed cube. The previous hot fission mode with 8-10 neutron emission seen in the Mo-Ba split is found to be smaller than earlier results but still present. New 0n binary SF yields are reported. By gating on the light charged particles detected with ΔE-E detectors mounted inside the Gammasphere chamber and $\gamma - \gamma$ coincidences with Gammasphere, the relative yields of correlated pairs in alpha ternary fission with zero to 6n emission are observed for the first time. Half-lives ($T_{1/2}$) of 7 states which decay by delayed γ transitions were determined from $\gamma - \gamma - \gamma$ coincidences. High spin states in neutron-rich ^{104,106}Mo nuclei have been investigated. The known ground state band, one-phonon and two-phonon γ-vibrational bands and two-quasiparticle bands are extended to high-spin. The one- and two-phonon γ-bands have remarkably close in energies. Several new quasiparticle bands are observed in both nuclei. In addition to their triaxial shape and large deformation, both high j two quasi-proton and two quasi-neutron orbitals close to the Fermi surface provide the necessary conditions for chiral bands. Two pairs of bands in ^{106}Mo are found to have all the characteristics of chiral doublets, the first such in an even-even nucleus.

W. Greiner et al. (eds.), Structure and Dynamics of Elementary Matter, 523–533.

1. Introduction

Studies of prompt gamma rays emitted in spontaneous fission with large detector arrays have given new insights into the fission process [1-5]. From $\gamma-\gamma-\gamma$ coincidence studies of the prompt γ-rays emitted in the spontaneous fission of ^{252}Cf with Gammasphere, yields of individual correlated pairs in binary and ternary fission were determined for the first time for 0 to 10 neutron emission[2-4]. Earlier we reported an ultra hot fission mode in the Mo-Ba split in the ^{252}Cf SF [2]. By using an uncompressed $\gamma-\gamma-\gamma$ cube, problems in the fission data analysis arising from complexity in the spectra in the 8-10 neutron emission yields were overcome. The new Mo-Ba yields show a reduced yield for the ultra hot mode. These data also allowed the extraction of more accurate zero neutron emission yields in 'cold' binary fission. A new experiment was carried out in which the light charged particles (LCP) involved in ternary fission were detected in a LCP-$\gamma-\gamma$ coincidence mode. From gating on the alpha particles and a gamma ray, the relative 0 to 6n yields associated with α ternary fission were extracted.

2. A new determination of the Ba-Mo yield matrix for ^{252}Cf

We carried out pioneering work on the quantitative determination of yield matrices, using $\gamma-\gamma$ and $\gamma-\gamma-\gamma$ coincidence data to extract yields of particular fragment pairs in the SF of ^{252}Cf [2,3,4]. One interesting finding was that $\approx 14\%$ of the ^{252}Cf barium-molybdenum split goes via a "hot fission" mode, where, as many as 10 neutrons are emitted [2,3]. This latter feature stimulated some theoretical speculations and also some skepticism, since the hot fission mode (called Mode 2) has been reported only in the Ba-Mo pairs in ^{252}Cf and not in ^{248}Cm spontaneous fission [6] . There have been some theoretical efforts to understand how this hot fission could arise [4,7,8]. In this work we used our 1995 Gammasphere data, taken by the GANDS95 collaboration [2]. The new analysis differs from the previous analyses where either uncompressed double coincidence spectra or compressed triple coincidence spectra were used. In both of these methods one faces problems, because of the vast number of γ-rays in the spectrum and particulary because of the degeneracy of several γ-rays in the 8-10 neutron emission yields for Mo-Ba. In this new analysis, using the uncompressed 3D data, we remeasured the pair-wise yields of barium (Z=56) with molybdenum (Z=42) partners. Because ^{104}Mo and ^{108}Mo have $2^+ \rightarrow 0^+$ transitions that are too close in energy to resolve and their $4^+ \rightarrow 2^+$ transitions are barely resolvable with peak-fitting routines, we have generally chosen to double-gate on the Ba fragments and measure the $2^+ \rightarrow 0^+$ intensities in the Mo partners. The barium double gates are on the $4 \rightarrow 2 \rightarrow 0$ cascade and the $3 \rightarrow 2 \rightarrow 0$ cascade, the latter being significant in the heavier bariums

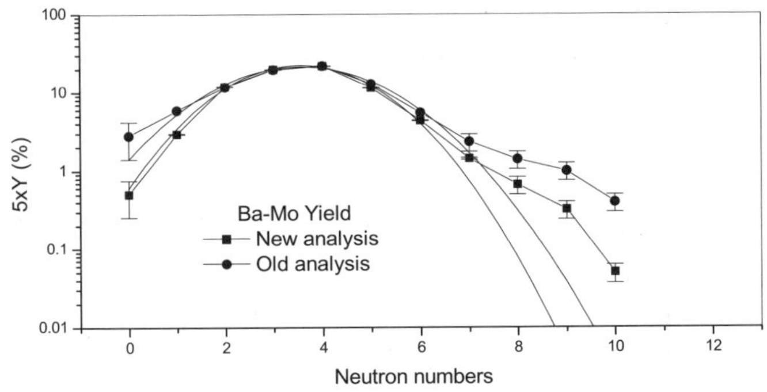

Figure 1. Mo-Ba yields vs. neutron numbers.

where octupole deformation is reported [9]. The odd-A nuclei are special cases. Their yields in our triple-coincidence analysis fall rather smoothly into the yield patterns of their even-even neighbors.

Fig. 1 shows semi-log plots of the summed Ba-Mo fission yields vs. neutron-emission number found in our work and in the previous work [2,3]. One sees that the hot fission mode is still present but its intensity is reduced by a factor of 3 from the 14% reported earlier [2]. Since our work was completed [10], Biswas *et al.* [11] also reported analogous data that show a similar small irregularity around 8 neutrons lost. They reported they could not observe a 10-neutron cell. We do report one such cell, ^{104}Mo-^{138}Ba, but with a large standard deviation as 0.003 ± 0.002 %.

3. Cold Binary Spontaneous Fission

Since the neutronless(clod) binary events are much smaller than those with neutrons emitted, double gating techniques have been employed to extract the yields for the cold binary fission. No direct measurements of yields of correlated pairs in cold binary fission had been made prior to our work. Earlier we reported the first results for the correlated pairs in cold binary fission in ^{252}Cf [1,3] and ^{242}Pu [12]. Subsequently we extracted additional and more accurate yields of cold binary fission [4] from Gammasphere data with 72 detectors. The cold binary fission yields are shown in Table 1 along with the theoretical values predicted by Sandulescu et al. [13]. In Table 1, the first reports of the cold binary fission of an odd-Z - odd-Z fragmentation is shown for the Tc and Cs pair and of an even Z- odd A pair. The over all agreement with theory is gen-

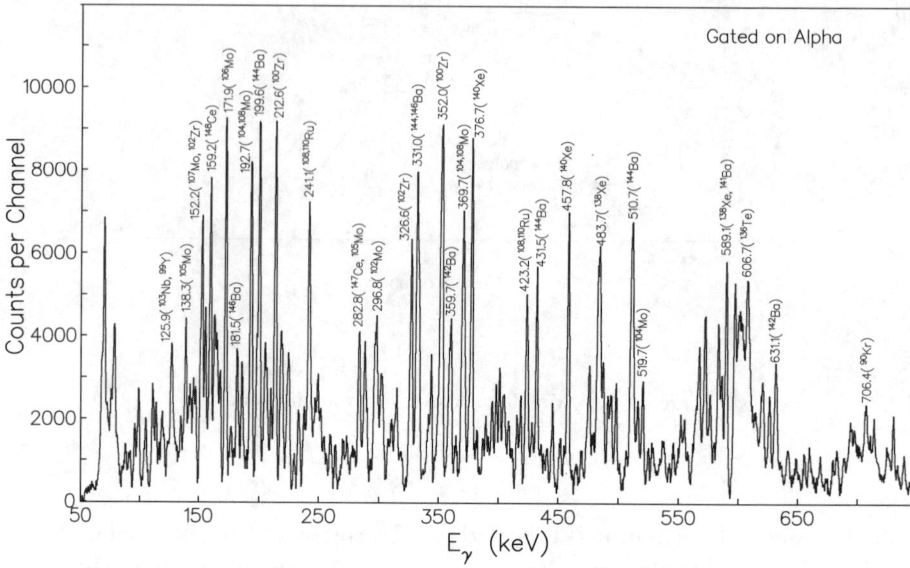

Figure 2. γ-ray spectrum in coincidence with α-particles.

erally good, including the predicted enhancement of the odd Z and odd A cases.

4. Light Charged Particle Ternary Fission

Another experiment was performed incorporating light charged particle(LCP) detectors to detect ternary particles in coincidence with γ rays in Gammasphere. The energy spectrum of LCPs emitted in the spontaneous fission of ^{252}Cf was measured by using two ΔE-E Si detector telescopes installed at the center of the Gammasphere array. With the position resolution of the strip detector (4 mm wide strips and 1 mm resolution along each strip), the ΔE-E telescopes provided unambiguous Z and A identification for all the LCPs of interest. The $\gamma-$ray spectrum in coincidence with ternary $\alpha-$particles is shown in Fig. 2. In this spectrum, one can easily see the $\gamma-$transitions for various partner nuclei where a ternary $\alpha-$particle is emitted. For example, Xe and Mo isotopes are partners where α and xn are emitted. Now, imposing an additional condition that the $\alpha-$gated $\gamma-$spectrum should be also in coincidence with the $2^+ \rightarrow 0^+$ transition in ^{142}Xe, one gets a very clean spectrum as shown in Fig. 3. From

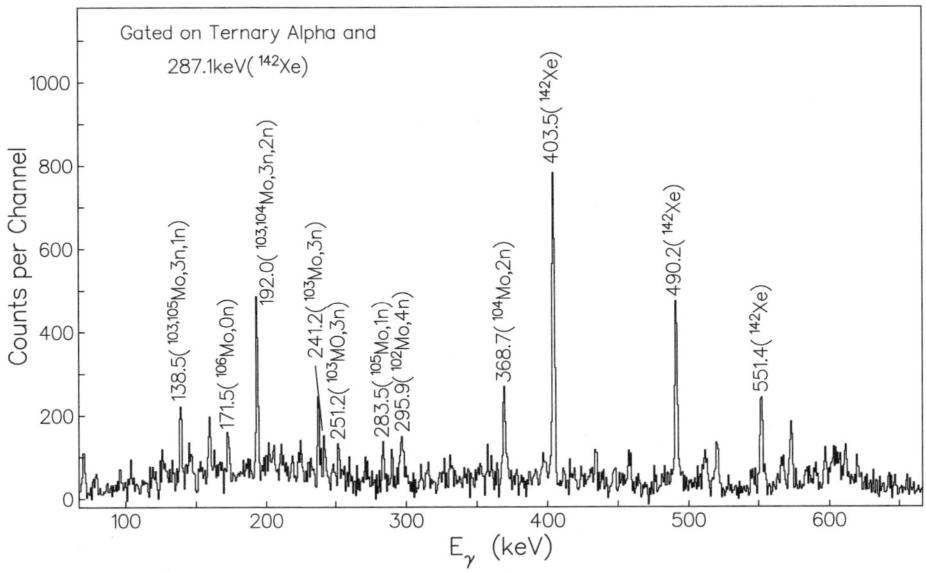

Figure 3. Coincident spectrum gated on α and 287.1 keV transition in ^{142}Xe

the analysis of the γ−ray intensities in these types of spectra, one can calculate the yield distributions. The yield distributions both for binary and ternary α−channel from 0 to 6n emission are shown Fig. 4 for two particular channels. These are the first relative 0-6n yields for any ternary α SF. Note the peak of the neutron emission yields for Ba-α-^{102}Zr is shifted up by about half an AMU from the Ba-^{106}Mo binary yield and so the average neutron emission in this α ternary SF channel is shfted down by about 0.4n. About 5-20 % of the α yield is from ^{5}He ternary fission in ^{252}Cf SF.

5. Half-life determination of several states of nutron-rich nuclei

Previously, half-lives of states in neutron-rich nuclei have been determined by single γ or $\gamma - \gamma$ coincidence relations for the delayed γ transitions emitted from the seperated isotopes produced in the fission of ^{235}U, ^{239}Pu, ^{248}Cm, and ^{252}Cf. Usually, more than 100 isotopes are produced in the fission of these heavy nuclei, with each isotope emitting many γ rays. With such complex

Table 1. Average cold binary fission yields from gates on two light fragment and two heavy fragment transitions

	A_L / A_H	Y_{exp}	Y_{the}	$Y_{the}^{(ren)}$
Zr/Ce	100/152	0.010(2)	0.38	0.004
	102/150	0.020(4)	2.82	0.033
	103/149	0.030(6)	4.21	0.049
	104/148	0.010(2)	1.03	0.012
Mo/Ba	104/148	0.010(2)	0.47	0.005
	106/146	0.040(8)	0.61	0.007
	107/145	0.070(14)	3.07	0.036
	108/144	0.030(6)	7.45	0.087
Tc/Cs	109/143	0.090(18)	11.03	0.128
Ru/Xe	110/142	0.060(12)	3.78	0.044
	111/141	0.10(2)	7.12	0.083
	112/140	0.020(4)	0.59	0.007
	114/138	0.020(4)	1.17	0.014
Pd/Te	116/136	0.050(20)	2.35	0.027

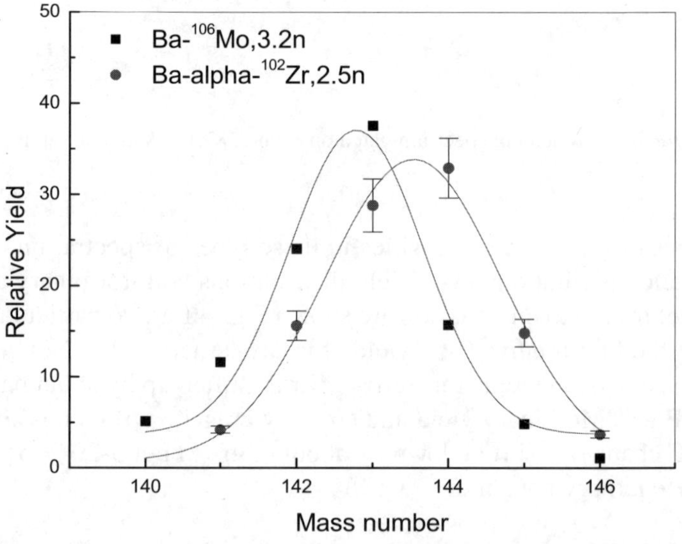

Figure 4. Yield spectrum for α

spectra, it is very difficult to isolate a single γ ray peak and measure the half-life of the state. The triple coincidence method outlined below can reduce the error associated with complexity of the γ ray spectra in SF.

Table 2. Half-lives ($T_{1/2}$ nsec) of several states (E_{IS}, keV). E(γ_1)/E(γ_3) are the double-gated transition energies. For ^{97}Sr, and ^{137}Xe, E(γ_2)/E(γ_3) and E(γ_1) are used instead.

Nuclei	E_{IS}	E(γ_1)/E(γ_3)	E(γ_2)	E(γ_0)	Present $T_{1/2}$
^{95}Sr	555.4	682.4/678.6	427.1	203.7	23.6(24)[14]
^{97}Sr	829.8	239.6/272.5	205.9	522.0	265(27)[14]
^{99}Zr	252.2	416.4/415.2	142.5	130.4	316(48)
^{108}Tc	330.6+x	123.4/341.6	125.7	154.0	94(10)
^{133}Te	1610.4	721.1/933.4	738.6	125.5	97(7)
^{134}Te	1691.8	2322.0/516.0	549.7	115.7	197(20)
^{137}Xe	1933.2	311.3/304.1	1046.4	314.1	10.1(9)

Consider a downward cascade consisting of $\gamma_3 - \gamma_2 - \gamma_1 - \gamma_0$ transitions where γ_0 is the outgoing transition from a state with long half-life and γ_1 is the incoming transition into the same state. Other higher states in this cascade have very short lifetimes. We set a double gate on E_{γ_3} and E_{γ_1} and compare the intensities of γ_0 and γ_2, N(γ_0) and N(γ_2) in the spectra. The coincidence time window (t_w) of interest is fixed for these triple γ coincidences and all others by requiring this condition on the $\gamma - \gamma - \gamma$ cube. In the present work, γ_1, γ_2, and γ_3, are in prompt coincidences. Therefore, the delay-time between γ_1, and γ_3 will be negligible. Since γ_0 is the ending transition in this cascade, the coincidence time window (t_w) limits the TDC time difference, t_{10}, between the γ_1 and γ_0 transitions and the intensity N(γ_0) observed from the state with the long lifetime. The N(γ_0) intensity determines the fraction of N(γ_2) counts observed from the state with the long half-life with decay constant, λ. Therefore, N(γ_0)/ N(γ_2) = $C(1 - e^{-\lambda t_w})$ can be applied in this case, where C is a constant.

The present time-gated triple γ coincidence method is useful for the half-life measurements of nuclear states in neutron-rich nuclei. We applied this triple γ coincidence method for the half-life measurements of 7 cases as shown in Table 2. As one example, coincidence spectra with double gate on 682.4 and 678.6 keV transitions in ^{95}Sr[14] is shown in Fig. 5. We are in the process of extending this technique to the 2^+ states with short half-lives in even-even nuclei. Several corrections have to be applied for this technique has to be useful because of the time-walks associated with low energy transitions.

6. Evidence for soft Chiral vibrations in ^{106}Mo

The neutron rich nucleus ^{106}Mo lies in a mass region where nuclei tend to take a triaxial shape. The observation of low-lying one and two phonon states of the gamma vibration indicates the softness of ^{106}Mo with respect to triaxial deformation. In such a soft nucleus the excitation of qp's will strongly modify the shape and may induce a stable triaxial shape [15,16]. A clear indication of

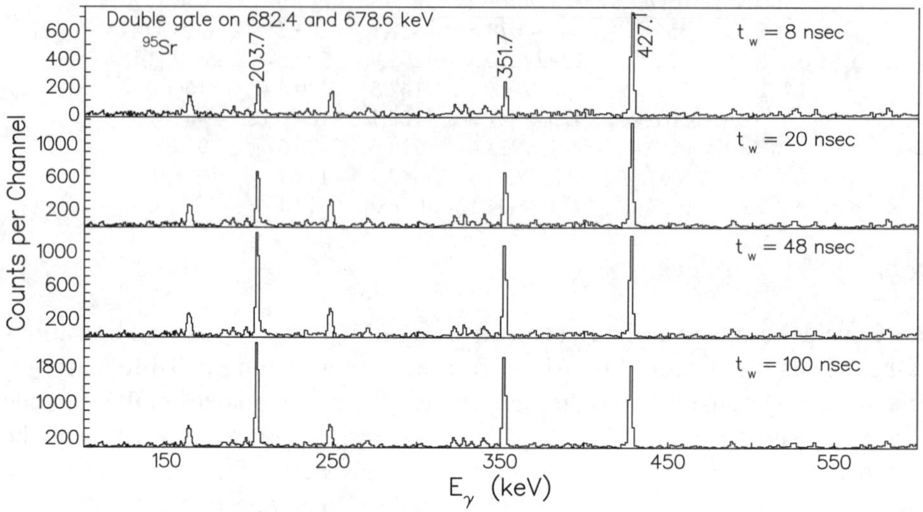

Figure 5. Coincidence spectra with double gate on 682.4 and 678.6 keV transitions in ^{95}Sr[14].

triaxial shapes is the observation of a pair of chiral doublet rotational bands, which consists of two sets of ΔI=1 sequences of states with the same parity and very close energies. Such chiral doubling emerges when the angular momentum has substantial components along all three principal axes of the triaxial density distribution. There are two energetically equivalent orientations of the angular momentum vector. In one case the short, intermediate and long axes form a right- handed system with respect to the angular momentum, in the other case a left-handed system. Examples for such chiral pairs of bands have been found in odd-odd nuclei around Z=59 and N=75[17]. In these structures, the angular momentum is composed of a component from the odd $h_{11/2}$ proton along the short axis, a component from the $h_{11/2}$ neutron hole along the long axis and a collective component along the intermediate axis. Chiral bands with an analogous composition have been predicted for odd-odd nuclei around Z=43 and N=65, where the odd $h_{11/2}$ neutron generates the angular momentum component along the short and the $h_{9/2}$ proton hole the component along the long axis[16]. The observation of a pair of bands with this structure in ^{104}Rb has been reported [18]. In ^{106}Mo(Fig. 6), the decays out of bands (4) and (5)

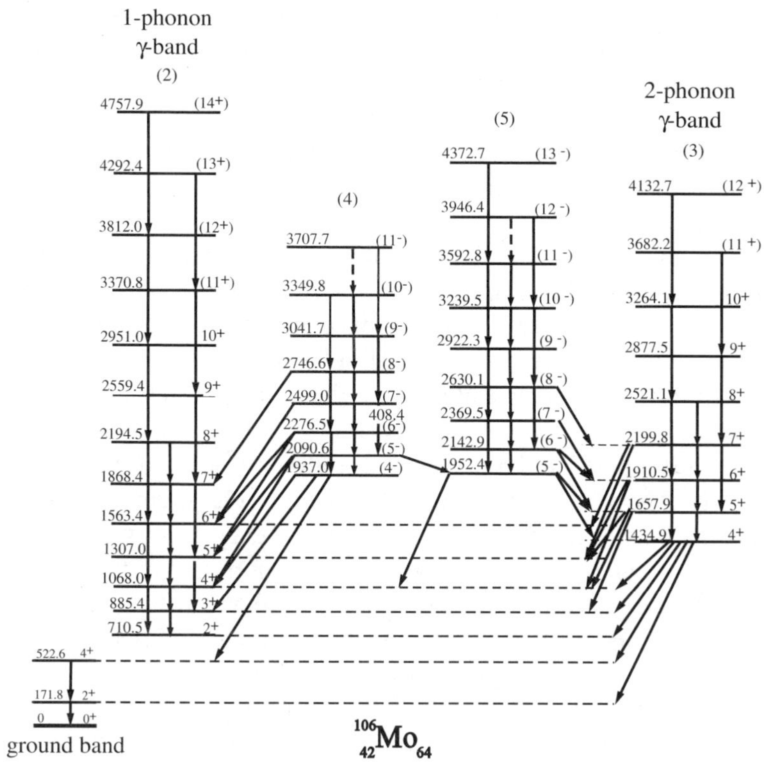

Figure 6. Level scheme of ^{106}Mo

are consistent with two-quasiparticle assignment with K = 4 and 5 for their band heads. Bands (4) and (5) have remarkably constant moments of inertia quite different from the MOI's of the ground and $\gamma-$vibrational bands. They have definite intertwined transitions between their even and odd spin members as expected for chiral doublets. Moreover as one goes up the bands the same spin level energies become closer together (ΔE become smaller) and the feeding between the doublets occurs only at the bottom. Both these characteristics are seen for chiral doublets (12). These data and the available proton and neutron orbitals lead us to propose bands (4) and (5) as chiral doublets with I^{π}=4^- and 5^- for their band heads. From our calculated Nilsson diagrams for protons and neutrons around N = 64 for $\beta_2\tilde{}0.3 - 0.4$, the neutron $g_{9/2}$ orbital is filled and the neutron $h_{11/2}$ partially filled. On the proton side, the $\pi g_{9/2}$orbital is partially filled and the strongly down- sloping $\pi h_{11/2}$ orbital is empty. The problem, as noted by Dimitrov and Frauendorf [19], is that the single particle levels in this region are not very well known so it is difficult to pick a definite assignment. We can only note that there are candidates near the Fermi surface to form chiral doublets for two quasi protons and for two quasi neutrons. These are the first chiral doublets proposed in an even-even nucleus. We emphasize that if these two doublet bands are not chiral doublets, then they are highly unusual structures not previously known.

One of the chiral bands decays into the one phonon gamma band and the other into the two phonon band. This pattern supports our interpretation in terms of a chiral vibration. Consider the symmetery of the probability amplitude of the angular momentum vector $\vec{J}$ with respect to the long axis, where ϕ is the angle of a rotation of $\vec{J}$ about this axis. For the one phonon state of the γ vibration on a prolate nucleus, this amplitude is $cos(2\phi)$ and $sin(2\phi)$ for even and odd I, respectively. In the case of the two-phonon state the amplitude is $cos(4\phi)$ and $sin(4\phi)$, respectively. These amplitudes are even or odd with respect to $\phi = \pi/2$. If the one-phonon state is even the two-phonon state is odd and vice verse. On the other hand, the two sectors $0 < \phi < \pi/2$ and $\pi/2 < \phi < \pi$ contain the left-handed and right-handed configurations of $\vec{J}$ with respect to the triaxial body. The zero and one phonon state of the chiral vibration are even and odd combinations of the two chiralities, i.e. the even zero-phonon state of the chiral vibration can only decay into the two-phonon state of the γ vibration, or the other way around. The experiment is in accordance with the first possibility.

The observations of bands -(4) and -(5) can be interpreted as a first chiral vibrator (nearly constant distance, difference of one unit in angular momentum in an even-even nucleus). The TAC calculations are consistent with this interpretation (very shallow in phi-direction). A different mechanism (as compared to the known cases) generates chirality, which proves the general nature of the concept.

In summary, study of SF isotopes with large detector arrays is an important tool for studying neutron rich nuclei produced in SF and also provides valuable information about fission dynamics and fission processes.

Acknowledgments

The work at Vanderbilt University, Lawrence Berkeley National Laboratory, and Idaho National Engineering and Environmental Laboratory are supported by U.S. Department of Energy under Grant No. DE-FG05-88ER40407 and Contract Nos. W-7405-ENG48, and DE-AC07-76ID01570, respectively. The Joint Institute for Heavy Ion Research is supported by U. of Tennessee, Vanderbilt University and U.S. DOE through contract No. DE-FG05-87ER40311 with U. of Tennessee.

References

[1] J.H. Hamilton *et al.*, J. Phys. G20 (1994) L85

[2] G.M. Ter-Akopian et al., Phys. Rev. Lett. 77 (1996) 32

[3] G.M.Ter-Akopian et al., Phys. Rev. C55 (1997) 1146

[4] A.V. Ramayya et al., *Heavy elements and related new phenomena, Volume I* World Scientific (1999) 477

[5] A.V. Ramayya et al., Phys. Rev. Lett. 81 (1998) 947

[6] N. Schultz, (Private communication).

[7] Yu.U. Pyatkov et al., Nucl. Phys. A624 (1997) 140

[8] R. Donangelo et al., Int. J. Mod. Phys. E7 (1998) 669

[9] J.H. Hamilton et al., Proc. Nuclear Structure 98, Gatlinburg (1999) 473

[10] S.-C. Wu et al., Phys. Rev. C62 (2000) 041601

[11] D.C. Biswas et al., Eur. Phys. J. A7 (2000) 189

[12] Y.X. Dardenne et al., Phys. Rev. C54 (1996) 206

[13] A. Sandulescu et al., Int. J. Modern Phy. E7 (1998) 625

[14] J.K. Hwang et al, Phys. Rev. **C67** 054304 (2003).

[15] S. Frauendorf and F.R. May, Phys. Lett. **B125**, 245 (1983).

[16] V. Dimitrov et al., Proceedings of the International Conference on Frontiers in Nuclear Sspectroscopy, Berkeley, 2002, APS Conference Proceedings, vol. 656, p. 151, 2003

[17] C. Petrache et al., Nucl. Phys. **A597**, 106 (1996).

[18] K. Koike et. al, Proceedings of the International Conference on Frontiers in Nuclear Sspectroscopy, Berkeley, 2002, APS Conference Proceedings, vol. 656, p 160, 2003.

[19] V. Dimitrov and S. Frauendorf, Proceedings of the Third International Conference on Fission and Properties of Neutron-Rich Nuclei, Sanibel Island, Florida, 2003, edited by J. H. Hamilton, A. V. Ramayyya and H. K. Carter (to be published in World Scientific, Singapore).

RELATIVISTIC DENSITY FUNCTIONAL THEORY: APPLICATIONS TO NUCLEI FAR FROM STABILITY

P. Ring
Physikdepartment, Technische Universität München, D-85748 Garching, Germany

Abstract Covariant density functional Theory (CDFT) is used to describe properties of the nuclear many-body system. Using only a very limited number of parameters it provides a universal self-consistent phenomenological method to reproduce and understand various phenomena in this strongly interacting mesoscopic system. It is based on the mean field idea. Correlations, vacuum polarization, exchange terms etc. are included in a phenomenological way. The time-dependent version allows the investigation of excited states such as rotational motion and collective vibrations. We give an overview over the basic concepts of density functional theory in nuclei with its various extensions and discuss recent applications for nuclei in and far from the valley of stability.

1. Introduction

The nucleus is a finite many-body system of strongly interacting particles. It is generally assumed, that Quantum Chromodynamics (QCD) is the fundamental theory, which determines the properties of such systems. However, at present and in foreseeable future, ab-initio calculations starting from the QCD Lagrangian are completely impossible and even attempts to derive the nuclear properties from models based on chiral symmetry in the spirit of QCD [1] require additional phenomenological input from experimental data in finite nuclear systems. On the other side density functional theory known in nuclear physics under the name of density dependent Hartree-Fock calculations [2, 3] provide a very successful phenomenological framework to describe and predict nuclear properties all over the periodic table. It addition it become clear that nuclei are relativistic systems even at low excitation energies. Lorentz dynamics determines basic nuclear properties and models including Lorentz invariance are able to reduce the number of phenomenological parameters considerably [4]. Essential progress for a realistic description of nuclear properties has been made by introducing a density dependence in the form of non-linear meson couplings [5]. The method has been extended for open shell nuclei considering

W. Greiner et al. (eds.), Structure and Dynamics of Elementary Matter, 535–542.

pairing correlations in the framework of relativistic Hartree-Bogoliubov theory (RHB) [6, 7]. This means density functional theory in the nuclear system has to be formulated in a covariant way and it has to be based not only on the normal density, but also on the pairing density of RHB theory. This theory has been applied successfully for the description of ground state properties of nuclei all over the periodic table.

In this talk we discuss recent applications of covariant density functional theory for the calculation of excited states, in particular for collective excitations such as vibrations and rotations.

2. Density functional theory in the rotating frame

In non-relativistic nuclear physics the cranking model plays an important role in the description of rotating nuclei. It has been introduced with phenomenological arguments already in the fifties by Inglis [8]. Later-on it has been shown, that it can be derived from a microscopic many-body theory as an approximate variation after angular momentum projection [9]. In the self-consistent version it allows to include alignment effects [10] as well as polarization effects induced by the rotation, such as Coriolis-anti-Pairing or changes of the deformation. But already in the simplified version of the *Rotating Shell Model* with fixed mean fields [11] it is able to describe successfully an extremely large amount of data in the high spin region of deformed nuclei.

The cranking idea can be used also for a relativistic description too [12]: one simply transforms the coordinate system to a frame rotating with constant angular velocity $\boldsymbol{\Omega}$ around a fixed axis in space assuming, as in non-relativistic nuclear physics, that this axis is perpendicular to the symmetry axis of the nucleus in its ground state. According to the cranking prescription the absolute value of the angular velocity $|\Omega|$ will be determined after the self-consistent solution of the equations of motion in the rotating frame by the Inglis condition [8]:

$$\frac{1}{|\Omega|}\langle \boldsymbol{\Omega}\boldsymbol{J}\rangle_{\Omega} \;=\; \sqrt{I(I+1)} \tag{1}$$

In this way we obtain the classical field equations in the rotating frame:

$$\{\boldsymbol{\alpha}(\boldsymbol{p}-\boldsymbol{V})+V+\beta(m-S)-\boldsymbol{\Omega}\boldsymbol{J}\}\psi_i \;=\epsilon_i\psi_i \tag{2}$$

$$\left\{-\Delta+(\boldsymbol{\Omega}\boldsymbol{L})^2\right\}\sigma \;+\; U'(\sigma) \;=-g_\sigma\rho_s \tag{3}$$

$$\left\{-\Delta+(\boldsymbol{\Omega}\boldsymbol{L})^2 \;+\; m_\omega^2\right\}\omega^0 \;= g_\omega\rho_v \tag{4}$$

$$\left\{-\Delta+(\boldsymbol{\Omega}\boldsymbol{J})^2 \;+\; m_\omega^2\right\}\boldsymbol{\omega} \;= g_\omega\boldsymbol{j} \tag{5}$$

and similar equations for ρ and A. These equations are very similar to the RMF equations in the non-rotating frame. There are only three essential differences:

(i) The Dirac equation (2) contains a Coriolis term $\boldsymbol{\Omega J}$ in full analogy to non-relativistic cranking. (ii) The Klein-Gordon equations for the mesons contain terms proportional to the square of the corresponding Coriolis terms. It turns out, however, that they can be neglected completely for all realistic cranking frequencies, because (i) they are quadratic in Ω and (ii) mesons being bosons are to a large extent in the lowest s-states with only small d-admixtures. (iii) The Coriolis operator in the Dirac equation breaks time-reversal invariance. Currents $\boldsymbol{j}$ are induced, which form the source of magnetic potentials in the Dirac equation (*nuclear magnetism*). In this way the charge current $\boldsymbol{j}_c$ is the source of the normal magnetic potential $\boldsymbol{A}$, the iso-scalar baryon current $\boldsymbol{j}$ is the source of the spatial components $\boldsymbol{\omega}$ of the ω-mesons and the iso-vector baryon current $\boldsymbol{j}_3$ is the source of the spatial component $\boldsymbol{\rho}_3$ of the ρ-mesons. In contrast to the Maxwellian magnetic field $\boldsymbol{A}$ having a small electromagnetic coupling, the large coupling constants of the strong interaction causes the fields $\boldsymbol{\omega}$ and $\boldsymbol{\rho}$ to be important in all cases, where they are not forbidden by symmetries, such as time reversal. They have a strong influence on the magnetic moments [13] in odd mass nuclei, where time reversal is broken by the odd particle, as well as on the moment of inertia in rotating nuclei, where time reversal is broken by the Coriolis field.

Shortly after the discovery of super-deformed bands in the $A \sim 150$ mass region, a second region of super-deformed bands has been found in the $A \sim 190$ mass region. Nuclei in this region have much larger proton numbers and therefore the deformation driving Coulomb energy plays a more important role as the surface energy favoring spherical shapes. At zero angular momentum the liquid drop energy increases therefore more smoothly with deformation than in the $A \sim 150$ mass region and we need less rotational energy to balance their effects on the nuclear shapes. Shell effects become important at much smaller angular velocities and thus we observe super-deformed minima already at relatively moderate spin values. If one goes in the periodic table to even higher charge numbers as in the Actinides, one does not need any rotational energy to balance Coulomb and surface energy. Super-deformed configurations at zero spin are well know in this region since the sixties under the name of *fission isomers*. Yrast super-deformed bands in ^{194}Pb and ^{194}Hg are linked by experimental observation of the transitions to the low-spin level scheme. In addition, there is a tentative linking of super-deformed band in ^{192}Pb. These data provide an opportunity to compare in a direct way not only calculated dynamic moments of inertia with experiment but also kinematical moments of inertia. At present the yrast super-deformed bands in ^{190,192}Hg and ^{196}Pb are not linked to the low-spin level scheme yet. Thus spin values consistent with the signature of the calculated yrast super-deformed configuration have to be assumed for the experimental bands when a comparison is made with respect to the kinematic moment of inertia $J^{(1)}$.

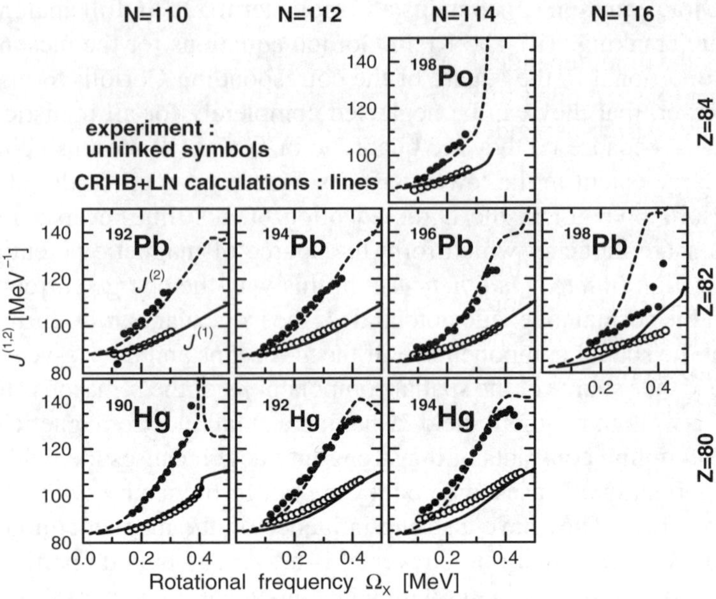

Figure 1. Experimental and calculated kinematic ($J^{(1)}$) and dynamic ($J^{(2)}$) moments of inertia of the yrast super-deformed bands in even-even nuclei of the $A \sim 190$ mass region. Experimental $J^{(1)}$ and $J^{(2)}$ moments of inertia are shown by open and solid circles, respectively. Solid and dashed lines are used for the $J^{(1)}$ and $J^{(2)}$ moments obtained in the cranked relativistic Hartree-Bogoliubov calculations with Lipkin-Nogmai projection of particle number (From Ref. [14])

In Fig. 1 we show theoretical results obtained within the Cranked Relativistic Hartree-Bogoliubov approach. The theoretical values for the dynamic and kinematic moments of inertia agree well with the experimental ones in the cases of the linked super-deformed bands in ^{194}Pb and ^{194}Hg.

Because of the relatively small angular velocities pairing correlations cannot be neglected in the super-deformed minima of the $A \sim 190$ mass region. As usual they are quenched with increasing angular velocities. This leads to increasing moments of inertia and therefore the experimentally observed super-deformed bands in this region are characterized by a dynamic moment of inertia $J^{(2)}$ increasing with increasing rotational frequency (see Fig. 1). Kinematic moments of inertia $J^{(1)}$ which are known experimentally from linked super-deformed bands in isotopes ^{192,194}Pb and ^{194}Hg show the same features and, in addition, we have the relation $J^{(2)} \geq J^{(1)}$. This forms an experimental indication for the fact that pairing correlations play a more important role in the super-deformed bands of the $A \sim 190$ mass region compared with those observed in the $A \sim 60$ and the $A \sim 150$ mass regions.

3. Relativistic random phase approximation

Using the time-dependent variational principle, relativistic density functional theory can be extended to the description of excited states such as collective vibrations by time-dependent RMF equations [15] . In the limit of small amplitudes one obtains from the time-dependent equations, as in the non-relativistic case, the random phase approximation (RPA). The derivation of relativistic RPA can be carried out in full analogy to the non-relativistic case, as it is described for instance in Ref. [16]. However, after the first developments of relativistic RPA codes [17–19] it took more than 10 years to find to obtain successful approximations to the time-dependent solutions [20]. Two essential points have to be taken into account: (i) The non-linear terms in the Lagrangian which describe an additional density dependence of the energy functional are crucial for any realistic description of nuclear surface properties [21]. (ii) The negative energy solutions of the static equation which, according to the no-sea approximation, are not taken into account in the calculation of the densities, have to be treated properly in the RPA (for details see Ref. [20]). For scalar excitations the RPA amplitudes connecting states in the Fermi-sea with those in the Dirac-sea play an important role. They have to be taken into account in order to have a consistent description of current conservation and the corresponding symmetries in relativistic RPA [22]. In practice this means, one has to take into account in addition to the usual particle-hole (ph)-configurations with particles above the Fermi-sea anti-particle hole (ah)-configurations, where the a sits in one of the orbits in the Dirac sea and h is a hole in the Fermi sea.

If one is interested in properties of exotic nuclei, one has to investigate the behavior of resonances as a function of the neutron or proton excess, i.e. one has to solve the corresponding problem in configurations with open shells, where pairing correlations play an important role. In this case one recovers the relativistic quasi-particle RPA equations (for details see Refs. [23, 24])

If one adds more and more neutrons to a closed shell nucleus such as ^{16}O a neutron skin develops and this has a strong influence on the spectrum of collective excitations of such a system. Since it does not require a large amount of energy to rearrange the neutrons in the skin, one has predicted low-lying collective dipole strength, which corresponds to the collective motion of the neutrons in the skin against the core of protons and neutrons (so-called pygmy resonances). The isovector dipole response in neutron-rich oxygen isotopes has recently attracted considerable interest because these nuclei are experimentally accessible and they might be good candidates for a possible identification of such low-lying collective soft modes [25].

The isovector strength function ($J^{\pi} = 1^{-}$) of the dipole operator for ^{22}O is displayed in the left panel of Fig. 2. Full solutions of the relativistic QRPA equations with pairing are compared with calculations without any pairing (simple

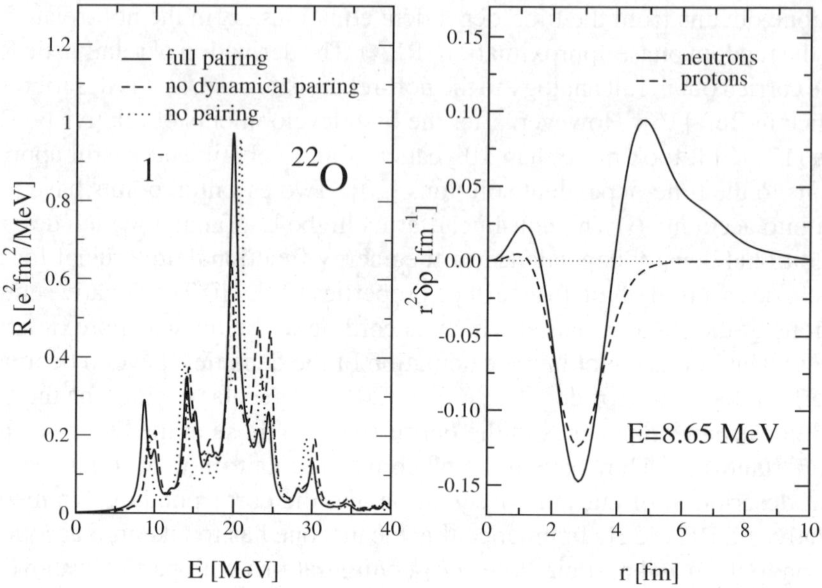

Figure 2. Left panel: The isovector strength function of the dipole operator in ^{22}O. The fully self-consistent RHB+RQRPA response (solid line) is compared with the RMF+RRPA calculation without pairing (dotted line), and with the RHB+RRPA calculation that includes pairing correlations only in the ground state (dashed line). Right panel: The proton and neutron transition densities for the peak at $E = 8.65$ MeV (from Ref. [23]).

RPA on a ground state without pairing) and with calculations without dynamical pairing, where pairing is only taken into account in the ground state with proper occupation probabilities, but the pairing interaction between the quasiparticle pairs is neglected. A large configuration space enables the separation of the zero-energy mode that corresponds to the spurious center of mass motion (Goldstone mode). In the present calculation for ^{22}O this mode is found at $E = 0.04$ MeV. The strength distribution shows a dominant peak at the Giant Dipole Resonance (IVGDR) in the region above 20 MeV and a smaller peak at low energies in the region below 10 MeV. These results for the low-lying dipole strength distribution in ^{22}O are in very good agreement with recent experimental data [25].

The inclusion of pairing correlations in the full RHB+RQRPA calculation enhances the low-energy dipole strength near the threshold. For the main peak in the low-energy region ($\approx$ 8.65 MeV), we display the proton and neutron transition densities in the right panel of Fig. 2. In contrast to the well known radial dependence of the IVGDR transition densities, where proton and neutron densities oscillate with opposite phases and where the amplitude of the isovector transition density is much larger than that of the isoscalar component, the proton and neutron transition densities for the main low-energy peak are in phase in the nuclear interior, there is no contribution from the protons in the surface region, the isoscalar transition density dominates over the isovector one in the interior, and the strong neutron transition density displays a long tail in the radial coordinate.

4. Conclusions and Outlook

We have seen that Covariant Density Functional Theory in the framework of relativistic Hartree-Bogoliubov theory provides a very powerful und successful scheme for a microscopic description of nuclear phenomena all over the periodic table, starting from very light systems to the region of super-heavy nuclei, from halo phenomena at the neutron drip line to proton emission for nuclei with large proton excess, from ground state properties in even even nuclei to collective excitations in super-deformed bands with extreme angular momenta and to collective dipole vibrations with extreme isospin. It is a universal theory, because it uses the same set of a hand full carefully adjusted parameters all over the periodic table and therefore it has a high predictive power.

Of course covariant density functional theory is not the ultimate description of all phenomena in nuclear structure. It is based on the mean field concept, which is connected with essential symmetries. Some of them are spontaneously broken, but they can be restored by projection onto the appropriate quantum numbers. The essential contrast to other mean field models such as density dependent Hartree-Fock theories is the fact, that Lorentz invariance is taken into account properly, which plays an important role in the nuclear many-body system. There are phase transitions and transitional nuclei, where simple mean field theory breaks down. In this cases fluctuations have to be taken into account by projection techniques or by the Generator Coordinate Method (GCM). The application of such methods is possible nowadays, but so far most of such investigations have been done within the non-relativistic framework. There is more work necessary to implement such methods also in the relativistic case.

So far the density functionals contain phenomenological parameters, which are determined by careful fits to a few well selected nuclei. New experimental data and more systematic fits will certainly improve those functionals even further. The ultimate goal, however, will be to derive these parameters from more fundamental theories. At the moment there is hope to get a qualitative understanding of those parameters from the bare nucleon-nucleon interaction or from QCD inspired models.

Acknowledgements

I would like to express my deep gratitude to my collaborators, in particular to A. V. Afanasjev, G. A. Lalazissis , T. Niksic, N. Paar, M. Serra, and D. Vretenar, for the active collaboration over many years on the topics addressed in this paper. This work has been supported in part by the Bundesministerium für Bildung und Forschung (BMBF), Project No. 06 TM193, and by the Gesellschaft für Schwerionenforschung (GSI).

References

[1] D. Vretenar and W. Weise, in *Extended Density Functionals in Nuclear Structure Physics*, Lecture Notes in Physics (G. Lalazissis, P. Ring, and D. Vretenar, eds.), vol. 641, Springer-Verlag, Heidelberg, 2003, p. 61.

[2] D. Vautherin and D. M. Brink, Phys. Rev. **C5** (1972), 626.

[3] J. Dechargé and D. Gogny, Nucl. Phys. **C21** (1980), 1568.

[4] B. B. Serot and J. D. Walecka, Adv. Nucl. Phys. **16** (1986), 1.

[5] J. Boguta and A. R. Bodmer, Nucl. Phys. **A292** (1977), 413.

[6] T. Gonzales-Llarena, J. L. Egido, G. A. Lalazissis, P. Ring, Phys. Lett. **B379** (1996), 13.

[7] P. Ring, Progr. Part. Nucl. Phys. **37** (1996), 193.

[8] D. R. Inglis, Phys. Rev. **96** (1954), 1059.

[9] R. Beck, H. J. Mang, and P. Ring, Z. Phys. **231** (1970), 26; P. Ring, R. Beck, and H. J. Mang, Z. Phys. **231** (1970), 10.

[10] B. Banerjee, H. J. Mang, and P. Ring, Nucl. Phys. **A215** (1973), 366.

[11] P. Ring and H. J. Mang, Phys. Rev. Lett. **33** (1974), 1174.

[12] W. Koepf and P. Ring, Nucl. Phys. **A493** (1989), 61.

[13] U. Hofmann and P. Ring, Phys. Lett. **B214** (1988), 307.

[14] A.V. Afanasjev, P. Ring, and J. König, Nucl. Phys. **A676** (2000), 196.

[15] D. Vretenar, H. Berghammer, and P. Ring, Nucl. Phys. **A581** (1995), 679.

[16] P. Ring and P. Schuck, *The nuclear many-body problem*, Springer, Heidelberg, 1980.

[17] R. J. Furnstahl, Phys. Lett. **B152** (1985), 313.

[18] M. L'Huillier and N. Van Giai, Phys. Rev. **C39** (1989), 2022.

[19] C. J. Horowitz and J. Piekarewicz, Nucl. Phys. **A511** (1990), 461.

[20] P Ring, Zhong-Yu Ma, Nguyen Van Giai, D. Vretenar, A. Wandelt, and Li-Gang Cao, Nucl. Phys. **A694** (2001), 249.

[21] Z.Y. Ma, N. Van Giai, H. Toki, and M. L'Huillier, Phys. Rev. **C42**(1997), 2385.

[22] J. F. Dawson and R. J. Furnstahl, Phys. Rev. **C42** (1990), 2009.

[23] N. Paar, P. Ring, T. Nikšić, and D. Vretenar, Phys. Rev. **C67** (2003), 034312.

[24] N. Paar, Phd thesis, Technical University of Munich (unpublished), 2003.

[25] A. Leistenschneider *et al.*, Phys. Rev. Lett. **86** (2001), 5442.

NUCLEI AND NUCLEAR MATTER IN A HADRONIC MODEL

S. Schramm
Institut für Theoretische Physik der J. W. Goethe Universität, D-60054, Frankfurt am Main, Germany

Abstract

I investigate the properties of nuclei within a hadronic chiral model approach. In a two-dimensional calculation I study deformation properties of all known even-even nuclei. I also discuss the influence of isospin forces on neutron distributions in heavy nuclei as well as on neutron star properties. In addition results for possible superheavy elements are presented. The model parameters are fitted by reproducing hadron masses and decay constants and properties of spherical nuclei.

1. Introduction

Hadronic models as effective descriptions of low-energy strong interaction physics have again become more popular in recent years, as one had to realize that the goal of deriving the properties of a nucleus, for instance, in terms of quark degrees of freedom would be unattainable any time in the near future.

In calculations close to the phase transition to a chirally restored phase a number of models constructed around the chiral σ model have been investigated. Typically, however, those models fail in their extrapolation back to normal matter and cannot reproduce basic groundstate properties of nuclear matter, especially a reasonable value for the compressibility.

From a purely nuclear structure point of view there have been some remarkable successes in describing finite nuclei within a relativistic meson field ansatz [1–4]. Therefore it is suggestive to explore models that can combine both of those regimes[5, 6]. One can demonstrate that it is possible to obtain a reasonable behavior of highly excited matter, a good description of nuclear saturation properties as well as a fair description of nuclei and hypernuclei within a single model and a single set of parameters.

In this article we follow this model approach studying finite nuclei across the whole chart of nuclides including deformation effects. To that end we optimize

W. Greiner et al. (eds.), Structure and Dynamics of Elementary Matter, 543–549.

the parameter set to reproduce spherical nuclei as well as hadron masses and decay constants via dynamical chiral symmetry breaking.

We check the quantitative value of the model and also look at possible superheavy nuclei.

2. Hadronic SU(3) Model

The model is constructed as a chirally invariant model incorporating a complete set of the lowest SU(3) multiplets of baryons and mesons. In addition also extensions to higher multiplets have been included [7]. Explicit symmetry breaking terms are introduced to simulate the effect of finite current quark masses within the hadronic approach. The initial scale symmetry of the model is incorporated with a scalar, flavor-singlet field χ. The potential for the field generates a nonzero vacuum expectation field that represents the gluon condensate in vacuum and its modification in the medium.

The model Lagrangian can be split into the following terms

$$\mathcal{L} = \mathcal{L}_{\text{kin}} + \sum_{W=X,Y,V,\mathcal{A},u} \mathcal{L}_{\text{BW}} + \mathcal{L}_{\text{VP}} + \mathcal{L}_{\text{vec}} + \mathcal{L}_0 + \mathcal{L}_{\text{SB}} \ . \tag{1}$$

$\mathcal{L}_{\text{kin}}$ is the kinetic energy term of the various fields, $\mathcal{L}_{\text{BW}}$ includes the interaction terms of the baryon octet with the spin-0 and spin-1 meson multiplets and with the photons. In $\mathcal{L}_{\text{VP}}$, the interaction terms of vector mesons with pseudoscalar mesons and with photons is summarized. $\mathcal{L}_{\text{vec}}$ generates the masses of the spin-1 mesons through interactions with spin-0 mesons. $\mathcal{L}_0$ contains the meson-meson interaction terms which induce the spontaneous breaking of chiral symmetry. It also includes the scale breaking logarithmic potential. Finally, $\mathcal{L}_{\text{SB}}$ introduces an explicit symmetry breaking of the $\text{U}(1)_A$, the $\text{SU}(3)_V$, and the chiral symmetry. A detailed discussion of all the terms can be found in [5].

Looking only at the terms relevant for the study of nuclei the total Lagrangian reduces to the following structure: The interaction terms of baryons, mesons, and the photon reads

$$\mathcal{L}_{int} = -\sum_i \bar{B}_i \left[g_{i_\omega}\omega_0\gamma_0 + g_{i\rho}\tau_3\rho_0^0\gamma_0 + \frac{1}{2}e(1+\tau_3)A_0\gamma_0 + m_i^*(\sigma,\zeta) \right] B_i \tag{2}$$

with B_i denoting protons and neutrons. ω_0 and $\rho_0 0$ are 0th components of the omega and rho meson, respectively, γ_0 is the photon. m_i* are the effective masses of the nucleons generated by the scalar fields discussed below. The various coupling constants of mesons and baryons result from the original $SU(3)$ structure [5]. The potential of the scalar fields follows as

$$\mathcal{L}_0 = -\frac{1}{2}k_0\chi^2(\sigma^2+\zeta^2+\delta^2) + k_1(\sigma^2+\zeta^2+\delta^2)^2 +$$

Table 1. Hadron masses (in MeV) for best-fit parameter set χ_m[10]. In brackets the experimental isospin-averaged values are shown [11]. Note that in the particle data book σ, ζ, δ correspond to $f_0(400-1200)$, $f_0(980)$, and $a_0(980)$, respectively. The pseudoscalar decay constants have values $f_\pi = 93$ MeV and $f_K = 122$ MeV.

m_N	939.2	(938.9)	m_Λ	1115.7	(1115.6)	m_Σ	1196.0	(1193.1)
m_Ξ	1331.3	(1318.1)	m_π	138.6	(138.0)	m_K	497.3	(495.6)
m_η	574.0	(547.5)	$m_{\eta'}$	897.3	(957.8)	m_σ	466.5	(400 - 1200)
m_ζ	1024.5	(980)	m_δ	973.26	(985)	m_ϕ	1020.0	(1019.4)
m_ω	780.6	(782.0)	m_ρ	761.1	(768.1)			
m_χ	960.0							

$$k_2\left(\frac{\sigma^4}{2}+\frac{\delta^4}{2}+3\sigma^2\delta^2+\zeta^4\right)$$
$$+k_3\chi\sigma^2\zeta - k_4\chi^4 - \frac{1}{4}\chi^4\ln\frac{\chi^4}{\chi_0^4} + \epsilon\chi^4\ln\frac{(\sigma^2-\delta^2)\zeta}{\sigma_0^2\zeta_0} \qquad (3)$$

In a quark language the scalar fields σ, δ, and ζ correspond to the non-strange isoscalar (σ), isovector (δ) quark-antiquark states, and the strange anti-strange state (ζ). The field χ is the scalar, isoscalar glueball field χ introduced in [8]. It has a vacuum expectation value χ_0 with $\chi_0^4 \sim\ < G^a_{\mu\nu}G^{a,\mu\nu} >$ representing the scalar gluon condensate[9]. σ_0 and ζ_0 are the respective vacuum expectation values of the meson fields that are generated by spontaneous symmetry breaking.

3. Results

By fitting the model parameters to hadronic masses and to properties of a set of 22 spherical nuclei we obtain a best fit parameter set χ_m [10]. The calculated masses of the hadrons are shown in the table. A good reproduction of the hadron spectra can be obtained. Also the $\eta - \eta'$ splitting arising from the $U_A(1)$ breaking can be reproduced.

Regarding the properties of infinite symmetric nuclear matter the required groundstate saturation properties can be obtained as is shown in Fig. 1. There the binding energy of nuclear matter as function of density is shown. From the curvature of this plot around saturation density one can read off the compressibility of nuclear matter, which is $\kappa = 215\,\mathrm{MeV}$. Studying the energy shift for varying the isospin content of the matter one gets the asymmetry energy $a_4 = 31.9\,\mathrm{MeV}$.

In a general survey of energies and deformations of even-even nuclei I computed axially deformed nuclei as solutions of the mean field equations of the Lagrangian (1) in a two-dimensional numerical calculation. The resulting agreement with experimental binding energies show an average error of 0.14% per-

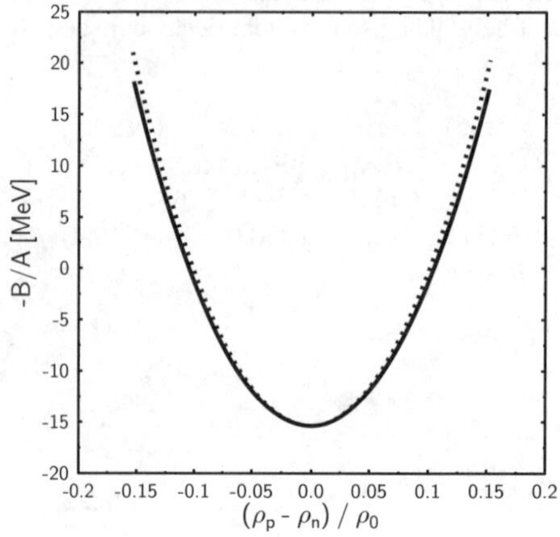

Figure 1. Binding energy of nuclear matter as function of density. density $\rho = .153\,\mathrm{GeV/fm^3}$. Results with (full line) and without (dotted curve) δ meson contributions are shown.

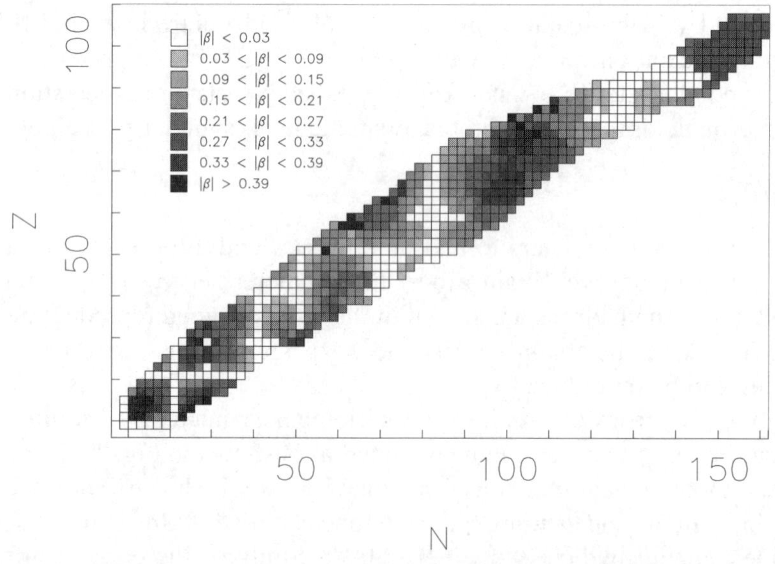

Figure 2. Groundstate quadrupole deformation $|\beta_2|$ in an axially symmetric 2-dimensional calculation.

cent for all even-even nuclei with A = 100, which is of the same quality as the best dedicated relativistic nuclear structure models achieve [10].

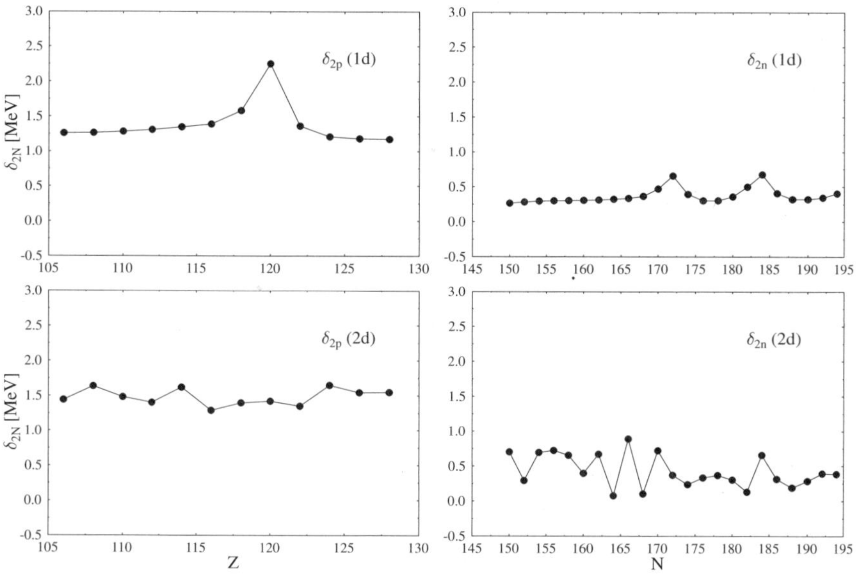

Figure 3. Two-nucleon gaps δ_{2p} (left panels) and δ_{2n} (right panels) in units of MeV in the range of hypothetical superheavy nuclei. The upper panels show the results of a calculation assuming spherical symmetry. The lower panels display the results for a fully 2-dimensional axially symmetric calculation.

The result is shown in Figure 2. In the figure the magic numbers can be clearly seen as regimes where the nuclear shapes are spherical. Studying a number of specific nuclei and comparing their deformations with measured values shows a very good agreement with experiment [10].

With the same parameter set I calculated nuclei in the superheavy regime. The results in Figure 3 show the proton and neutron gap energies for a spherical and a deformed calculation. One observes that in spherical approximation there is a signal for a shell closure with Z = 120 and rather weak ones for the neutrons at N=172 and 184. In a more realistic calculation of deformed nuclei those signals vanish and at least the gap energy cannot serve as a tool for pinning down possible superheavy elements.

In the general framework of this model one can in principle change the term structure in the Lagrangian influencing the isospin behaviour of the interactions. This was done in a calculation comparing the difference of the radii of neutron and proton distributions in ^{208}Pb Δr_{np} compared to the resulting radius of a

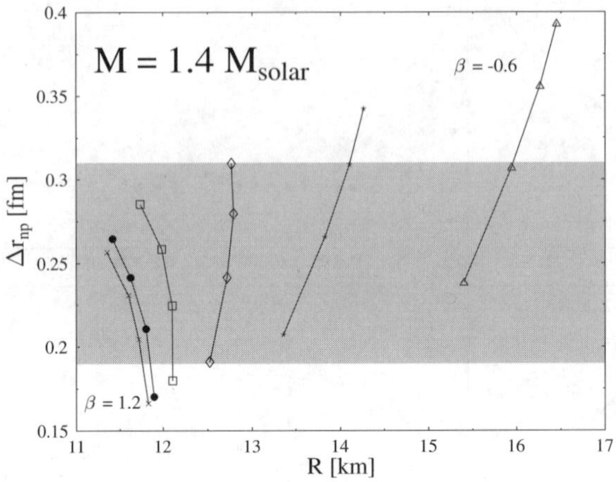

Figure 4. Relation of the difference of neutron and proton radius Δr_{np} of ^{208}Pb (full symbols) and the neutron star radius for a star with mass $M = 1.4\ M_{solar}$. The coupling strengths β and r_σ determine different additional isospin-dependent interactions [12]. Results for different values of β are marked with different symbols with values of 1.2, 1.0 0.6, 0.2, -0.2, -0.6 from the left to the right. For every value of β we varied r_σ between 0 and 0.3, with large values of r_σ yielding small values or Δr_{np}.

neutron star with a mass of 1.4 solar masses, calculated in the same model, as an extreme case of isospin asymmetry. Two additional isospin-dependent interaction terms are considered (see [12]), where the different strengths are regulated by the couplings β and r_σ. One can observe a correlation between neutron star radius and the nuclear neutron skin. A proposed 1 percent accurate measurement of the neutron radius in lead using parity violating electron scattering at JLAB [13, 14] (indicated by the shaded area in the plot), however, is not accurate enough to pin down the isospin-dependent parameters. A more accurate determination of neutron star radii below an accuracy of about 1 km could give important insight into the isospin dependence of the strong interactions.

4. Summary

I discussed a hadronic model based on chirally symmetric interactions within an flavor-SU(3) approach. In this framework I calculated nuclear properties using an improved parameter set obtained from fitting hadron masses and properties of spherical nuclei. The quality of the fit is as good as the results of other relativistic mean field models of nuclear structure [10]. The nuclear quadrupole deformations can be described very well. In a study of isospin effects one could see that better determinations of neutron star radii might help to pin down the isospin dependence of the strong interactions, measurements of radii of neutron distributions turn out to be too inaccurate for this purpose at the moment. In a calculation of superheavy elements with nuclear deformation the gap energies do not exhibit any pronounced magic numbers.

Acknowledgments

The author acknowledges the use of the computing resources of the Center for Scientific Computing of Frankfurt University.

References

[1] B. D. Serot and J. D. Walecka, Phys. Lett. **B87**, 172 (1979).

[2] A. R. Bodmer and C. E. Price, Nucl. Phys. **A505**, 123 (1989).

[3] P.-G. Reinhard, Rep. Prog. Phys. **52**, 439 (1989).

[4] G. A. Lalazissis, J. König and P. Ring, Phys. Rev. C **55**, 540 (1997).

[5] P. Papazoglou, D. Zschiesche, S. Schramm, J. Schaffner-Bielich, H. Stöcker, W. Greiner, Phys. Rev. **C59** (1998) 411

[6] Ch. Beckmann, P. Papazoglou, D. Zschiesche, S. Schramm, H. Stöcker, and W. Greiner, Phys. Rev. C **65**, 024301 (2002)

[7] D. Zschiesche, S. Schramm, J. Schaffner-Bielich, H. Stoecker, W. Greiner, Phys. Lett. B 547, 7 (2002).

[8] J. Schechter, Phys. Rev. D**21**, 3393 (1980)

[9] P. Papazoglou, S. Schramm, J. Schaffner-Bielich, H. Stöcker, and W. Greiner, Phys.Rev. C57, 2576 (1998).

[10] S. Schramm, Phys. Rev. C 66, 064310 (2002).

[11] D. E. Groom et al., Eur. Phys. J. **C15**, 1 (2000).

[12] S. Schramm, Phys. Lett. **B560**, 164 (2003).

[13] JLab experiment E-00-003, R. Michaels, P.A. Souder, G.M. Urciuoli (spokespersons)

[14] C. J. Horowitz and J. Piekarewicz, Phys.Rev.Lett. 86, 5647 (2001).

NEW EXPERIMENTAL APPROACH TO TERNARY FISSION RELYING ON THE γ-RAY SPECTROSCOPY

G. M. Ter-Akopian, A. V. Daniel, A. S. Fomichev, G. S. Popeko, and A. M. Rodin
Flerov Laboratory of Nuclear Reactions, Joint Institute for Nuclear Reserch
Dubna, 141980 Russia, E-mail: Gurgen.TerAkopian@jinr.ru

J. H. Hamilton, A. V. Ramayya, J. Kormicki, J. K. Hwang, D. Fong and P. Gore
Department of Physics, Vanderbilt University, Nashville, TN 37235, USA

J. O. Rasmussen, I. Y. Lee, A. O. Macchiavelli, and P. Fallon
Lawrence Berkeley National Laboratory, Berkeley, CA 94720, USA

J. D. Cole
Idaho National Engineering and Environmental Laboratory, Idaho Falls, ID 83415, USA

W. Greiner
Institut für Theoretische Physik, J.W. Goethe Universität, 60054 Frankfurt/M, Germany

Abstract Ternary fission of ^{252}Cf was studied at Gammasphere with using $\Delta E \times E$ particle telescopes. The abundance ratio of ^{10}Be LCP's emitted in the excited 2^+ state (E^*=3368 keV) and in the ground state was found to be $N(2^+)/N(0^+)=0.160\pm0.025$. No evidence was found for the 3368 keV γ rays emitted from a triple molecular state. For the first time charge distributions were obtained for fission fragments emitted with helium, beryllium and carbon LCP's.

1. Introduction

Spectroscopic γ ray measurements are known for their effectiveness in getting essential characteristics for low energy binary fission. Such are data on fragment independent yields [1] and the measured yields of individual fragment pairs obtained at particular charge splits of the fissile nucleus [2, 3]. Until the recent time, such measurements could not be done for the nuclear fission

W. Greiner et al. (eds.), Structure and Dynamics of Elementary Matter, 551–559.

accompanied by light charged particle (LCP) emission. The reason is that this process, often called ternary fission, is rare. In the case of the ^{252}Cf ternary fission, one ternary fission event comes about only one time per $\sim$260 binary fission events [4]. However, with the advent of contemporary 4π Ge detector arrays, such as the Gammasphere facility [5, 6], $\gamma - \gamma - \gamma$ coincidence technique could show its full potential for the study of such rare events. Using this technique the authors of Refs. [7, 8] for the first time obtained data on the yields of fragment pairs emitted in cold (neutron-less) ternary fission of ^{252}Cf associated with the ^{4}He and ^{10}Be LCP's.

It will be interesting to extend these data to the independent yields of other fragment pairs and/or individual fragments appearing in the ^{252}Cf ternary fission accompanied by different LCP's. Such data will be useful for specifying initial ternary fission configurations realized by the nuclear system just before the instant when the acceleration of the three bodies (the two main fragments and the LCP) begins in Coulomb field. The lack of experimental data characterizing these configurations led the authors of theory papers to make speculative assumptions when treating the problem of different LCP rates [9–12] and doing trajectory calculations for LPC's (see in Refs. [4, 9, 13]).

On the other hand, the spectroscopy of ternary fission γ rays is interesting because it could be helpful in filling the gap in our knowledge about the population probability of excited states in LCP's emitted in ternary fission. Existing data of this type are scarce. Only recently the population of a particle unstable level lying in ^{8}Li at E*=2.26 MeV was reported for the ternary fission of ^{252}Cf [14]. However one should not come to nothing more than the observation of particle unstable states because, apart from the large statistical error of the yield value reported in Ref. [14], the ambiguity is inherent to the population ratios measured for particle unstable states.

Among LCP's with Z>2 ^{10}Be has the largest yield. If its first excited state ($J^\pi = 2^+$) is populated, 3368 keV γ rays will be emitted as a result of $2^+ \to 0^+$ transition. The authors of Ref. [15, 16] reported on the observation of a 3.37 MeV peak in the γ spectrum recorded in coincidence with ^{10}Be LCP's emitted in the ^{252}Cf ternary fission. They also furnished evidences that this peak did not show Doppler broadening. Taking into consideration the short, 125 fs lifetime of ^{10}Be in its 2^+ state the authors of Ref. [15, 16] concluded that it could either mean an emission of the γ rays preferentially perpendicular to the ^{10}Be momentum vector or a substantial delay in the release of the excited ^{10}Be nuclei resulting in a fact that these nuclei stay at rest for an unusually long time before the beginning of their acceleration in the Coulomb field of the two main fragments. The result was open to questions because of the poor energy resolution of NaI detectors used in this experiment. To test this idea a γ−γ−γ coincidence experiment was made at Gammasphere with a high energy resolution [8, 17]. The data gave support to the observation reported in Refs.

[15, 16], but with a limited statistics and without a direct LCP identification. The idea that the ^{10}Be nucleus may stay between two fission fragments for a long time ($\sim 10^{-13}$ s) to create the so-called triple nuclear molecule opens up exciting possibility discussed in Ref. [18].

Thus, the observation of γ rays emitted by LCP's became important both for a deeper penetration into the dynamics of ternary fission and for examining this process for quasi-molecular states appearing presumably at the scission point.

2. Experiment

Keeping this in mind, we performed a new experiment at Gammasphere providing that events showing the coincidence of γ rays and LCP's originating from the spontaneous fission of ^{252}Cf were detected. Gammasphere was set to record γ rays with energy less than 5.4 MeV. The detection efficiency varied from a maximum value of 17% typical for the low energy part of the γ ray spectrum to 4.6% at 3368 keV. A ^{252}Cf sample giving $\sim 4 \times 10^4$ spontaneous fissions per second was installed in the center of a reaction chamber placed in a hollow sphere inside Gammasphere. The source was deposited on a 1.8 μ Ti foil and was tightly covered by gold foils on both sides. These foils had the least thickness that could stop the two main fragments originating in ternary fission. The fragment deceleration time was less than 1 ps ensuring one that γ rays were emitted by the fission fragments mainly when these fragments were at rest.

Eight ΔE$\times$E silicon detector telescopes were used for the LCP detection. Four telescopes were centered at polar angle $\theta = 30°$ ($\varphi = 45°$, 135°, 225° and 315°). Other four telescopes were at $\theta = 150°$ (azimuth angles were the same). The telescopes were inclined so that the lines connecting the centers of two opposite ΔE detectors and going through the center of the source were orthogonal to the detector surfaces. Each ΔE detector had an area of 10$\times$10 mm^2 and thickness varying between 9.0 and 10.5 μ. Each E detector was 400 μ thick and was 20$\times$20 mm^2 in area. All ΔE detectors were at a distance of 27 mm from the source. The distance between the ΔE and E detectors was 13 mm.

Data acquisition was triggered by ΔE or E signals with amplitudes above the threshold values, which were set to prevent the detection of two-fold pile-ups of α particles emitted in the ^{252}Cf radioactive decay. A ternary fission event was stored at a condition that at least one γ ray was detected within a 100 ns time interval. During two weeks, when measurements were made continuously, approximately 1.6×10^7 events were recorded.

3. Results and discussion

Resolution achieved for the $\Delta E \times E$ telescopes allowed us to well identify helium, beryllium, boron and carbon nuclei in each of the eight two-dimensional plots created from the data recorded by the telescopes. The lower primary energy of the recorded LCP's was 9, 21, 26, and 32, MeV for He, Be, B, and C LCP's, respectively. Thus only high energy parts of the full energy spectra were measured. Even under an usual assumption that the LCP energy distributions are Gaussians, it is impossible to estimate the mean values and widths of LCP energy distributions from these data. Therefore we used the mean values estimated in another experiment made in Dubna with an open ^{252}Cf source [19]. This allowed us to estimate the FWHM values by fitting the LCP energy distributions obtained with higher statistic in the Berkeley experiment. These fits gave the yields of LCP's occurring in the ^{252}Cf ternary fission. Obtained results are summarized in Table 1.

Table 1. Energies and yields of LCP's. The yields are normalized to 10^4 He LCP's.

	Be	B	C
$< E >$, MeV: this work/Ref. [20]	18.0(4) / 17.5(1.0)	20.5(1.0) / 21.2(1.0)	23.4(1.0) / 26(1)
FWHM, MeV: this work/Ref. [20]	17.03(8)/18(1)	19.8(2) / 19.3(1.0)	18.7(2) / –
Yield: this work/Ref. [20]	166(10) / 126(30)	16(3) / 6.3(4.0)	103(15) / –

Histogram shown in Fig. 1 with thin line presents the energy spectrum of prompt γ rays recorded in coincidence with Be LCP's. The thick solid line histogram demonstrates the same spectrum obtained after applying correction for Doppler shift. A distinct peak dominating in the last spectrum at 3368 keV corresponds to the $2^+ \rightarrow 0^+$ transition in ^{10}Be. The peak width, FWHM=65.7$\pm$4.5 keV, was estimated by using a Gaussian fit. This width is compatible with the value FWHM=62.1 keV obtained by the Monte Carlo simulation of the correction procedure made for γ rays emitted from moving ^{10}Be nuclei.

In the raw spectrum of γ rays recorded in coincidence with Be LCP's we do not see any distinct peak near 3368 keV, which could be associated with γ emission from stationary ^{10}Be nuclei. A thorough analysis made with the selection of γ rays emitted in about perpendicular direction to the trajectories of Be LCP's showed that there is a lack of any γ line which could be attributed to the emission from ^{10}Be standing in a triple nuclear molecule.

The number of counts obtained in the 3368 keV γ line is 263 ± 22. Taking into consideration the Gammasphere efficiency ($\simeq$4.6%), the total number of 3368 keV γ rays emitted by ^{10}Be nuclei becomes $I_\gamma = 5700 \pm 480$. After correction for the number of beryllium LCP's escaping the detection due the lack of recorded signals from γ rays, we estimated the total number of Be

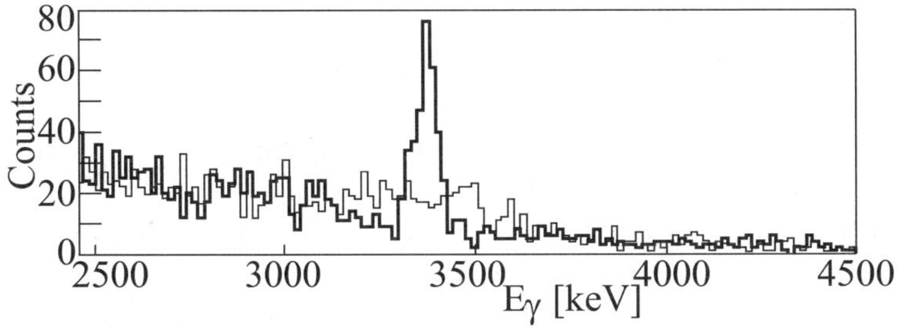

Figure 1. Histograms drawn with the thin and thick solid lines, respectively, show the spectra of γ rays coinciding with Be LCP's obtained before and after the Doppler shift correction.

LCP's recorded by the $\Delta E \times E$ telescopes as $N_{Be} = 51800 \pm 5000$. The energy resolution of the ΔE detectors did not allow us to directly separate the loci of beryllium isotopes in the identification ΔE×E plot. Comparing the loci obtained by Monte Carlo simulations for ^{9}Be, ^{10}Be and ^{11}Be with the experimental ΔE×E distributions we estimated the yield of ^{10}Be nuclei as $\sim$80% of all beryllium LCP's emitted in the ternary fission of ^{252}Cf. This observation is in agreement with the experimental data presented in Ref. [11].

Based on these results we estimated the population ratio of the excited 2^+ and the ground-state 0^+ levels in ^{10}Be as $N(2^+)/N(0^+)=0.160\pm0.025$. The upper limit for the probability that ^{10}Be emits its γ rays being at rest makes only 4% of this ratio. We note, however that these results are valid for ^{10}Be LCP's obtained at infinity with kinetic energy more than 21 MeV. In particular, one can not exclude that the hypothetical triple molecules decay in a way leading mainly to the low kinetic energy for the ^{10}Be LCP's which was beyond the reach of the present experiment. Since the earlier Gammasphere experiment [8, 17] relied solely on the observation of $\gamma - \gamma - \gamma$ coincidence events, a possibility still remains open that a narrow γ peak characteristic to the motionless ^{10}Be can be found if the low energy part of the energy distribution of ^{10}Be LCP's is detected. At the same time, the present result excludes any possibility that an effect of triple quasi molecular state, involving beryllium LCP's, could be observed in Refs. [15, 16] where the energy cut-off was 26 MeV for these clusters.

Using raw data we built a 4096×4096 channel two-dimensional spectrum for events involving two and higher fold γ ray signals recorded in coincidence with He ternary LCP's. This spectrum spanned energy range from 50 to 2700 keV along the two axes. We built also three linear spectra extending from 50 to 5500 keV separately for γ rays coinciding with helium, beryllium and carbon ternary LCP's.

In the two-dimensional spectrum we identified more than 170 $\gamma - \gamma$ coincidence peaks originating from fission fragments appearing in the He ternary fission of ^{252}Cf. Some of these peaks corresponded to the coincidences of ground state transitions occurring in complementary fragments entering into different fragment pairs. From these peak values, corrected for detection efficiency and electron conversion, we calculated independent yields for 63 fragment pairs corresponding to four charge splits of the main part of the ^{252}Cf nucleus remaining after the going out of the ternary He LCP. In addition, we estimated independent yields for two tin isotopes (^{130}Sn and ^{132}Sn) from their γ peak values obtained in the linear spectrum built for the He ternary fission events.

In the γ ray spectra observed in coincidence with Be LCP's we identified 72 peaks as due to transitions occurring in 38 fission fragments and estimated independent yields for these fission fragments. In the case of C LCP's, 70 γ peaks were identified, and independent yields were estimated for 35 fission fragments.

These results allowed us to present, for the first time, charge distributions obtained for fission fragments appearing in the ternary fission of ^{252}Cf in coincidence with helium, beryllium and carbon LCP's. The obtained charge distributions are presented in Fig. 2 (left panel). The fragment charge distribution known for the binary fission of ^{252}Cf is also presented in Fig. 2.

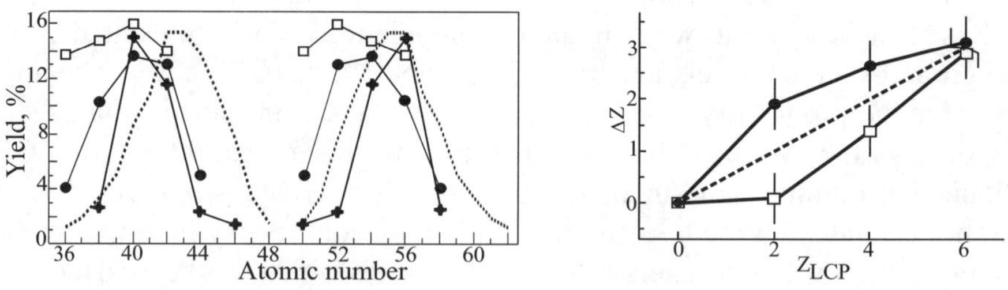

Figure 2. Left panel: Symbols, stars, closed circles and open squares, show the charge distributions obtained for main fragments emitted in the ternary fission of ^{252}Cf accompained, respectively, by He, Be and C LCP's. The dotted lines show the charge distribution known for the binary fission of ^{252}Cf. Right panel: Open squares and closed circles show the numbers of protons removed, respectively, from the heavy and light ternary fission fragmets which otherwise could be obtained as those emitted in binary fission.

From comparison made for the two charge distributions, one obtained for the He ternary fission and another one known for the binary fission of ^{252}Cf, we see that two protons entering the He LCP's come from the light fragments, otherwise obtained as those emitted in binary fission. Similar considerations show that Be nuclei take, on average, $\sim$2.7 protons from the light fragment with the rest of charge coming from the heavy fragment. Finally,we see that both fission fragments contribute about the same proton number in the formation of carbon LCP's. The average proton numbers removed from the light and heavy binary fission fragments by He, Be and C LCP's are presented in Fig. 2 (right panel).

It will be natural to say that, on average, the parts of nuclear matter contributing in these LCP's from the light and heavy binary fission fragment keep the same proportions as those established for protons. These observations seem to be more direct in their nature, and, hence, more correct, than analogous conclusions made on the basis of the energy and mass distributions of primary fragments (before neutron evaporation) deduced, in their turn, by iterative procedures from the measured energy rations of correlated fragments emitted in ternary fission [20].

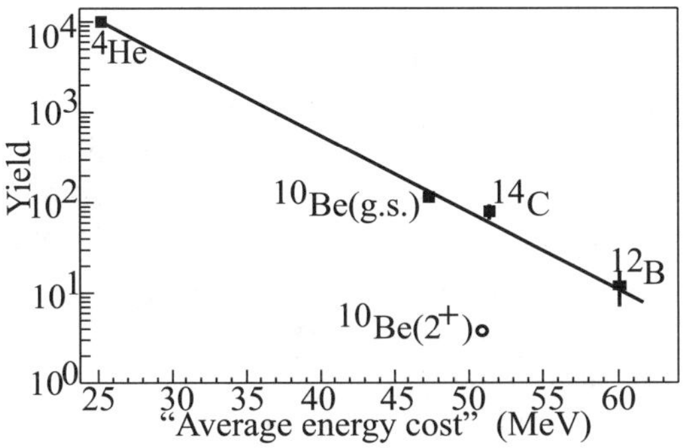

Figure 3. Correlation plot showing the relative yields of various LCP's in function of the "average energy cost. See text.

It is of interest to see how the measured LCP yields correlate with the "average energy cost" for the production of various third particles. The correlation plot presented in Fig. 3 was obtained using relative yield values presented in Table 1 for Be, B and C with respect to He LCP's. The "average energy cost" values were calculated following the prescription given in Ref. [9] and with the use of our results presented in Fig. 2. We assumed that ^{4}He, ^{12}B and ^{14}C exhaust the full yield values measured for the helium, boron and carbon LCP's. This approximation seems to be natural in view of relatively low rates reported for LCP's representing heavy helium isotopes [20] and abundance ratios measured for boron and carbon nuclei emitted in ternary fission [4, 9, 13]. As for Be LCP's, shown in Fig. 3 is the ground state yield of ^{10}Be obtained after correction made to the value given in Table 1. This correction takes into account the measured ratio N(2$^+$)/N(0$^+$)=0.160$\pm$0.025 and the 80 % proportion which we have estimated for ^{10}Be LCP's emitted in the ternary fission of ^{252}Cf. Fig. 3 demonstrates that the LPC yields fall down exponentially with the increase of "average energy cost".

Fig. 3 shows also one point representing the relative yield of ^{10}Be(2$^+$) nuclei emitted in the excited state. This estimate involves the measured ratio N(2$^+$)/N(0$^+$)=0.160$\pm$0.025 reduced by five times to take into account the spin factor which actually enhanced the 2$^+$ state yield. In the energy scale of Fig. 3, we just added the excitation energy of the ^{10}Be 2^+ state, 3368 keV, to the "average energy cost" estimated for ^{10}Be(g.s.). The two points representing ^{10}Be(g.s.) and ^{10}Be(2^+) show a remarkably more rapid decrease in the yield as compared with the exponential fitting the yields of other LCP's, taken with the exception of ^{10}Be(2^+).

4. Summary

In summary, He, Be, B and C LCP's emitted in the ^{252}Cf ternary fission with kinetic energy more than 9, 21, 26, and 32 MeV, respectively, were identified and their relative yields and energy distribution parameters were obtained. The extraordinary capability of Gammasphere in probing rare γ rays combined with ΔE$\times$E detector telescopes allowed us to measure for the first time the ratio of population probabilities, N(2^+)/N(0^+)=0.160$\pm$0.025, for ^{10}Be nuclei emitted in their excited (2^+) and ground (0^+) states. Assuming that a kind of statistical equilibrium is set at some stage in the ternary fission process (e.g. as discussed in the dynamical model [11]) and taking into account the statistical factor brought about by the spin, this ratio suggests a value of 1.0$\pm$0.1 MeV for the temperature parameter. Our results do not show any hint that the γ rays are emitted from ^{10}Be nuclei occurring at rest as a part of a triple nuclear molecule presumably formed at the scission point.

For the first time we presented charge distributions obtained for fission fragments appearing in the ternary fission of ^{252}Cf in coincidence with helium, beryllium and carbon LCP's. Based on the measured charge distributions we obtained data showing the average proton numbers removed from the light and heavy binary fission fragments by He, Be and C LCP's and presented a correlation plot showing how the measured LCP yields correlate with the "average energy cost" for the production of various third particles.

Acknowledgments

Work at Joint Institute for Nuclear Research was supported in part by the US Department of Energy contract #DE-AC011- 00NN4125, BBW1 Agreement No. 3498 (CRDF grant RPO-10301-INEEL) and by the joint RFBR-DFG grant (RFBR No. 02-02-04004, DFG No. 436RUS 113/673/0-1(R)). Work at Vanderbilt University, Lawrence Berkeley National Laboratory, Lawrence Livermore National Laboratory, and Idaho National Engineering and Environmental Laboratory are supported by U.S. Department of Energy under Grant No. DE-FG05-88ER40407 and Contract Nos. W-7405-ENG48, DE- AC03-76SF00098, and DE-AC07-76ID01570.

References

[1] E. Cheifets et al., Phys. Rev. C, **4**, 1913 (1971).

[2] J.H. Hamilton et al., Prog Part. Nucl. Phys., **35**, 635 (1995).

[3] G.M. Ter-Akopian et al., Phys. Rev. C, **55**, 1146 (1997).

[4] C. Wagemans, in *The Nuclear Fission Process*, ed. C. Wagemans (CRS Press, Boca Raton, FL, USA, 1991), Chapt. 12.

[5] P.J. Nolan, F.A. Beck, and D.B. Fossan, Ann. Rev. Nucl. Part. Sci. **45**, 561 (1994).

[6] R.V.F. Janssens and F.S. Stephens, Nucl. Phys. News, **6**, No.4, 9 (1996).

[7] A. Ramayya et al., Phys. Rev. C, **57**, 2370 (1998).

[8] A. Ramayya et al., Phys. Rev. Lett, **81**, 947 (1998).

[9] I. Halpern, Ann. Rev. Nucl Sci., **21**, 245 (1971).

[10] G.V. Valskii, Sov. J. Nucl. Phys., **24**, 140 (1976).

[11] V. A. Rubchenya and S. G. Yavshits, Z. Phys. **A329**, 217 (1988).

[12] G.A. Pik-Pichak, Phys. At. Nucl., **57**, 906 (1994).

[13] M. Mutterer and J. P. Theobald, in Nuclear Decay Modes, ed. D. Poenary, IOP Publ., Bristol, England, 1996, Chapt. 12.

[14] Yu. N. Kopach et al., Phys. Rev. C, **65**, 044614(16) (2002).

[15] P. Singer at al., in: *Proc. Int. 3rd Conf. DANF' 96, Casta Papernicka, Slovakia*, eds. J. Kliman and B. Pustylnik, (JINR, Dubna) 1996, p. 262.

[16] M. Mutterer at al., in: *Proc. Int. Conf. Fission and Properties of Neutron-Rich Nuclei*, eds. J.H. Hamilton and A.V. Ramayya (World Scientific, Singapore) 1998, p. 119.

[17] J. H. Hamilton et al., Acta Physica Slovaca, **49**, 31 (1999).

[18] W. Greiner, Acta Physica Slovaca, **49**, 9 (1999).

[19] G. M. Ter-Akopian et al., in: *Proc. Int. Conf. Fission and Properties of Neutron-Rich Nuclei*, Sanibel Island, Florida, 2002, (will be published by World Scientific, Singapore).

[20] M. Mutterer et al., Ibid as Ref. [15], p. 250.

HFB CALCULATIONS FOR NUCLEI FAR FROM STABILITY *

A.S. Umar,[1] V.E. Oberacker,[1] E. Teran,[2] A. Blazkiewicz[1]

[1] *Department of Physics and Astronomy, Vanderbilt University, Nashville, TN 37235, USA*

[2] *Physics Department, San Diego State University, San Diego, CA 92182-1233*

Abstract We have developed a new Hartree-Fock-Bogoliubov (HFB) code to study ground state and pairing properties of nuclei near the neutron and proton drip lines. The unique feature of our code is that it takes into account the strong coupling to high-energy continuum states. Using this code we investigate ground state properties of the sulfur isotopes from the line of stability up to the two-neutron dripline ($^{34-52}S$). In particular, we calculate two-neutron separation energies, quadrupole moments, and rms-radii for protons and neutrons. Evidence for shape coexistence is found in the very neutron-rich sulfur isotopes. We compare our calculations with results from relativistic mean field theory and with available experimental data.

1. Introduction

The study of structures and reactions of nuclei far from stability has been one of the most active fields of nuclear physics in the past decade[1]. The microscopic description of such nuclei will lead to a better understanding of the interplay among the strong, Coulomb, and the weak interactions as well as the enhanced correlations present in these many-body systems. It is generally acknowledged that an accurate treatment of the pairing interaction is essential for describing exotic nuclei[2, 3].

The large pairing correlations near the driplines can no longer be described by a small residual interaction. It becomes necessary to treat the mean field and the pairing field in a single self-consistent theory. Furthermore, the outermost

*This work is supported by U.S. DOE grant DE-FG02-96ER40963, and by the National Energy Research Scientific Computing Center (NERSC).

W. Greiner et al. (eds.), Structure and Dynamics of Elementary Matter, 561–567.

nucleons are weakly bound, which implies a large spatial extent, and they are strongly coupled to the particle continuum. Most HFB calculations to date are carried out in a truncated discrete harmonic oscillator basis[4–8]. This approach is quite appropriate for nuclei in the vicinity of the stability line. However, farther from stability, the continuum states become important and coordinate-based representations have numerous advantages: for example, the well-known 'French code' uses a truncated 3-D Hartree-Fock basis[9] which consists of both localized states and discretized continuum states; however, in this approach one can only include continuum states up to about 5 MeV of excitation energy. For nuclei in the vicinity of the driplines, continuum states with an equivalent single-particle energy of up to at least 60 MeV must be taken into account. One-dimensional calculations for spherical nuclei have been carried out in coordinate space for many years[2, 3], but only recently has our group succeeded in generalizing this approach to the case of deformed axially symmetric nuclei (HFB on a 2-D lattice)[10–12]. We utilize a novel computational technique, a Basis-Spline representation of wavefunctions and operators, which allows us to accurately describe high-energy continuum states in two space dimensions.

2. HFB Equations in Coordinate Space

For certain types of effective interactions (e.g. Skyrme mean field and pairing delta-interactions) the particle Hamiltonian h and the pairing Hamiltonian $\tilde{h}$ are diagonal in isospin space and local in position space, resulting in the following HFB equations with a 4x4 structure in spin space:

$$\begin{pmatrix} (h^q - \lambda) & \tilde{h}^q \\ \tilde{h}^q & -(h^q - \lambda) \end{pmatrix} \begin{pmatrix} \phi^q_{1,\alpha} \\ \phi^q_{2,\alpha} \end{pmatrix} = E_\alpha \begin{pmatrix} \phi^q_{1,\alpha} \\ \phi^q_{2,\alpha} \end{pmatrix} \tag{1}$$

with

$$h^q = \begin{pmatrix} h^q_{\uparrow\uparrow}(\mathbf{r}) & h^q_{\uparrow\downarrow}(\mathbf{r}) \\ h^q_{\downarrow\uparrow}(\mathbf{r}) & h^q_{\downarrow\downarrow}(\mathbf{r}) \end{pmatrix}, \quad \tilde{h}^q = \begin{pmatrix} \tilde{h}^q_{\uparrow\uparrow}(\mathbf{r}) & \tilde{h}^q_{\uparrow\downarrow}(\mathbf{r}) \\ \tilde{h}^q_{\downarrow\uparrow}(\mathbf{r}) & \tilde{h}^q_{\downarrow\downarrow}(\mathbf{r}) \end{pmatrix} . \tag{2}$$

The HFB equations have a mathematical structure that is similar to the Dirac equation: the spectrum of quasiparticle energies E is unbounded from above *and* below. The spectrum is discrete for $|E| < -\lambda$ and continuous for $|E| > -\lambda$. It is forbidden to choose positive and negative quasiparticle energies at the same time[4]. For even-even nuclei it is customary to solve the HFB equations with a positive quasiparticle energy spectrum $+E_\alpha$ and consider all negative energy states as occupied in the HFB ground state. The HFB mean field Hamiltonian has the same structure as the binding energy functional. Detailed expressions for the Skyrme mean fields and the Coulomb term, as well as the numerical methods and test comparisons to other calculations are given in reference[11].

3. Results

The sulfur isotopes ($Z = 16$) have been investigated several years ago by Werner et al. [13] using self-consistent mean field models: both Skyrme-HF and relativistic mean field (RMF) model calculations were carried out using a simple heuristic "constant pairing gap" approximation. Because of the well-known deficiencies of standard pairing theory in the exotic neutron-rich region, we have decided to re-investigate the sulfur isotope chain, starting at the line of stability ($N = 16$) up to the two-neutron dripline (which turns out to be $N = 36$ in our HFB calculations). Based on the above-mentioned earlier calculations, one may expect a wide range of ground state deformations, and in addition there has been some evidence for shape coexistence in this region [13]. Because our HFB code [11] has been specifically designed to describe deformed exotic nuclei, the sulfur isotopes are expected to provide a rich testing ground for our calculations.

Within the single-particle shell model, one would expect nuclei such as ${}^{44}_{16}S_{28}$ with "magic" neutron number $N = 28$ to be spherical. But the mean field theories predict, in fact, deformed intrinsic shapes as a result of "intruder" states. Furthermore, in some of these isotopes shape coexistence has been predicted [13]. All these phenomena depend strongly on the interplay between the mean field and the pairing field with is correctly described the the HFB theory. Furthermore, the neutron-rich $N \approx 28$ nuclei play a crucial role in astrophysics for the nucleosynthesis of the heavy Ca-Ti-Cr isotopes [14].

In radial (r) direction, our lattice extends from $0 - 12$ fm, and in symmetry axis (z) direction from $-12, ..., +12$ fm, with a lattice spacing of about 0.8 fm in the central region. Angular momentum projections $\Omega = 1/2, 3/2, ..., 21/2$ were taken into account.

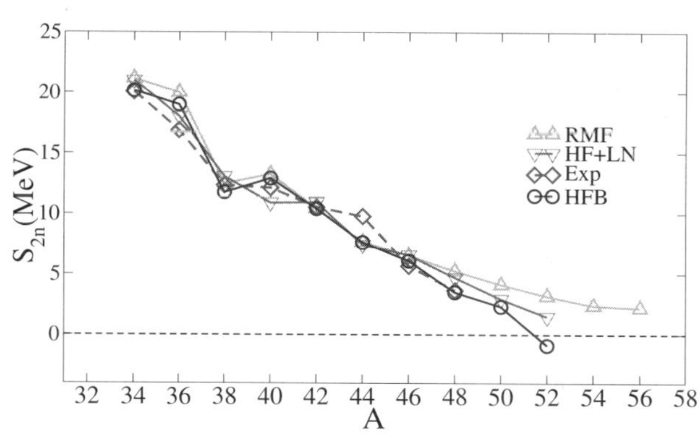

Figure 1. Two-neutron separation energies for sulfur isotopes. The dripline is located where the separation energy becomes zero.

Figure 1 shows the calculated two-neutron separation energies for the sulfur isotope chain. The two-neutron separation energy is defined as

$$S_{2n}(Z,N) = E_{bind}(Z,N) - E_{bind}(Z,N-2)\,. \tag{3}$$

Note that in using this equation, all binding energies must be entered with a *positive* sign. The position of the two-neutron dripline is defined by the condition $S_{2n}(Z,N) = 0$, and nuclei with negative two-neutron separation energy are unstable against the emission of two neutrons.

The two-neutron separation energies have been calculated using various methods: in addition to HFB calculations (i.e. selfconsistent mean field with pairing), we have also carried out Hartree-Fock calculations with added Lipkin/Nogami pairing (HF+LN), and we compare our results to the relativistic mean field with BCS pairing (RMF) calculations by Lalazissis et al. [15]. Experimental data based on measured binding energies [16] are available up to the isotope ^{48}S. Fig. 1 shows that both the HFB and RMF calculations are in good agreement with experiment where available but there are dramatic differences as we approach the two-neutron dripline: Our HFB calculations predict ^{50}S to be the last isotope that is stable against the emission of two neutrons. By contrast, the RMF approach predicts $S_{2n}(Z,N) > 0$ at least up to ^{56}S. Our HF+LN calculations also yield positive $2n$-separation energies in the mass region investigated here. It should be stressed that, on theoretical grounds, neither BCS-type nor Lipkin-Nogami type pairing is justified for the very neutron-rich isotopes. Furthermore, it is well-known that the HF+LN method breaks down at magic numbers (the pairing gap does not vanish).

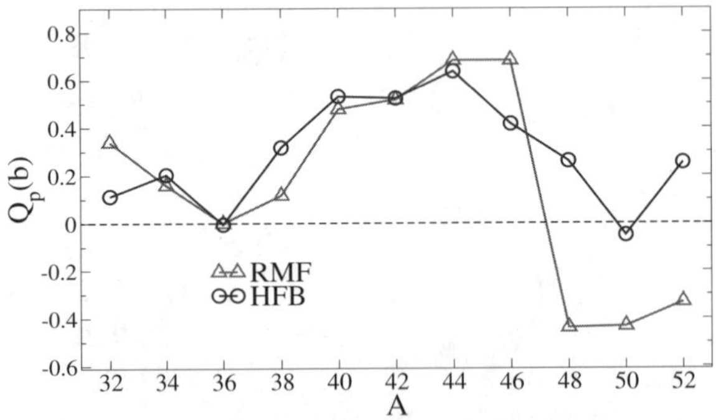

Figure 2. Quadrupole moment for protons (in units of barn) for even-even sulfur isotopes.

In Fig. 2 we show a comparison of the HFB and RMF results for the intrinsic electric quadrupole moments of the sulfur isotopes. In most cases, the HFB and RMF calculations show a similar trend: we observe a region with predominantly prolate deformation. Note, however, that for the most neutron-rich sulfur isotopes, our HFB theory predicts a prolate ground state whereas RMF theory yields an oblate shape. Direct measurements of electric quadrupole moments are only available for two of the sulfur isotopes. The data compilation of Stone [17] yields intrinsic electric quadrupole moments of $Q^{exp} = -0.15b$ for ^{32}S and $Q^{exp} = +0.04b$ for ^{34}S. Our HFB code yields $Q_{HFB} = +0.113b$ and $+0.203b$ for these two isotopes. The RMF calculations of Lalazissis et al. [15] give values of $Q_{RMF} = +0.339b$ and $+0.159b$, respectively.

In intermediate-energy Coulomb excitation experiments, the energies and $B(E2)$ values of the lowest excited 2^+ state were measured for $^{38,40,42}S$ [18] and for ^{44}S [19]. The analysis of the measured $BE(2)$ values in terms of the simple quadrupole-deformed rotor model yields the following experimental quadrupole deformations for the even sulfur isotopes $^{38-44}S$: $|\beta_2^{exp}| = 0.246(16),\ 0.284(16),\ 0.300(24),\ 0.258(36)$; our corresponding Skyrme-HFB theory results are $\beta_2^{HFB} = 0.16,\ 0.26,\ 0.25,\ 0.29$. Apparently, the HFB results for $^{40,42,44}S$ are in good agreement with experiment, but there is a discrepancy in the case of ^{38}S: Skyrme-HFB predicts a less deformed shape than the experimental value. There is an even larger discrepancy between experiment and the RMF calculations which yield an almost spherical shape, $\beta_2^{RMF} = 0.054$.

Both HFB and RMF calculations reveal shape coexistence in this region, with an energy difference between the ground state and the shape isomer that is usually quite small. For example, in the case of $^{48}_{16}S_{32}$ our HFB code yields a ground state binding energy of -362.56 MeV with a quadrupole deformation of $\beta_2 = 0.11$, and an oblate minimum at $\beta_2 = -0.15$ which is only 0.49 MeV higher than the ground state. The RMF predicts in this case a ground state binding energy of -362.97 MeV with oblate deformation of $\beta_2 = -0.25$, and a shape isomer with $\beta_2 = +0.179$ which is located 0.29 MeV above the ground state.

In Fig. 3 we show a comparison of the quadrupole moment for neutrons predicted by our HFB calculations and the RMF results of ref.[15]. In both cases, the general trend is very similar to the result obtained for protons.

In Fig. 4 we compare the root-mean-square radii of protons and neutrons predicted by our HFB calculations and the RMF calculations with BCS pairing [15]. Near the line of β-stability, the proton and neutron radii are almost identical, but as we approach the $2n$-dripline, we see clearly the development of a "neutron skin" as evidenced by the large difference between the neutron and proton rms radii. For example, in $^{50}_{16}S_{34}$ our HFB calculations yield $r_n = 3.935$ fm and $r_p = 3.364$ fm, respectively. In general, the RMF calculations predict larger neutron rms radii for mass numbers $A \geq 38$ than do our HFB calculations.

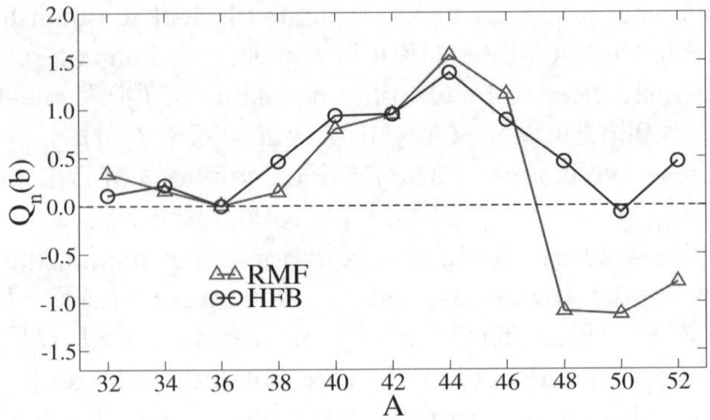

Figure 3. Quadrupole moment for neutrons (in units of barn) for even-even sulfur isotopes.

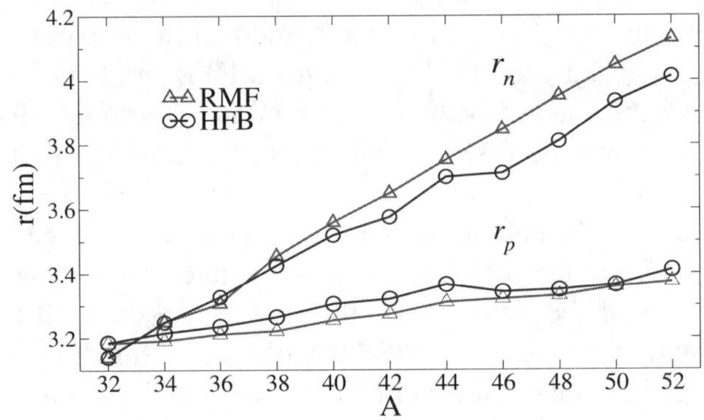

Figure 4. Root mean square radii of protons and neutrons for even-even sulfur isotopes.

4. Conclusions

We have calculated the ground state properties of the sulfur isotope chain ($Z = 16$), starting at the line of stability ($N/Z = 1$) up to the two-neutron dripline with a neutron-to-proton ratio of $N/Z \approx 2.2$. In particular, we have calculated two-neutron separation energies, quadrupole moments and rms radii for protons and neutrons. In comparing our HFB calculations with other theoretical methods (RMF with BCS pairing and HF+Lipkin/Nogami) we find similar results near stability but dramatic differences near the 2n-dripline (see Figures 1 - 4). For example, our HFB calculations predict ^{50}S to be the last isotope that is stable against the emission of two neutrons whereas the RMF approach predicts $S_{2n}(Z, N) > 0$ at least up to ^{56}S. For ^{48}S, the last even-even isotope for which experimental binding energies are available (Fig. 1), the experimental value for the two-neutron separation energy is 3.64 MeV, as compared to our HFB calculation result of 3.49 MeV and the RMF result of 5.24 MeV. Both our HFB calculations and the RMF calculations of ref.[15] predict the existence of shape isomeric states in the neutron-rich sulfur isotopes. In

Fig. 2 we compare calculated electric quadrupole moments. In most cases, the HFB and RMF calculations yield prolate quadrupole deformations. However, for the most neutron-rich sulfur isotopes, our HFB theory predicts a prolate ground state whereas RMF theory yields an oblate shape. Shape coexistence is found both in the HFB and RMF calculations, with fairly small energy difference between the ground state and the shape isomer. Specific results are given for ${}^{48}_{16}S_{32}$. A comparison between the root-mean-square radii of protons and neutrons clearly exhibits the development of a "neutron skin" in the neutron rich sulfur isotopes: for example, in ${}^{50}_{16}S_{34}$ our HFB calculations yield neutron and proton rms radii of $r_n = 3.935$ fm and $r_p = 3.364$ fm, respectively.

References

[1] *RIA Physics White Paper*, RIA 2000 Workshop, Raleigh-Durham, NC (July 2000), distributed by NSCL, Michigan State University.

[2] J. Dobaczewski, H. Flocard and J. Treiner, Nucl. Phys. **A422**, 103 (1984).

[3] J. Dobaczewski, W. Nazarewicz, T.R. Werner, J.F. Berger, C.R. Chinn and J. Dechargé, Phys. Rev. **C53**, 2809 (1996).

[4] P. Ring and P. Schuck, *The Nuclear Many-Body Problem*, (Springer Verlag, New York, 1980).

[5] A. Petrovici, K.W. Schmid, F. Grümmer, and A. Faessler, Nucl. Phys. **A517**, 108 (1990).

[6] C.R. Chinn, J.-F. Berger, D. Gogny and M.S. Weiss, Phys. Rev. **C45**, 1700 (1992).

[7] J.L. Egido and L.M. Robledo, Phys. Rev. Lett. **70**, 2876 (1993).

[8] M.V. Stoitsov, J. Dobaczewski, P. Ring, and S. Pittel, Phys. Rev. C **61**, 034311 (2000).

[9] J. Terasaki, P.-H. Heenen, H. Flocard, and P. Bonche, Nucl. Phys. **A600**, 371 (1996).

[10] E. Teran, V.E. Oberacker, and A.S. Umar, Heavy Ion Physics, Vol. 16/1-4, 437 (2002).

[11] E. Teran, V.E. Oberacker, and A.S. Umar, Phys. Rev. **C67**, 064314-1 (2003).

[12] V.E. Oberacker, A.S. Umar, E. Teran, and A. Blazkiewicz, (submitted to Phys. Rev. C, 2003).

[13] T.R. Werner, J.A. Sheikh, W. Nazarewicz, M.R. Strayer, A.S. Umar, and M. Misu, Phys. Lett. **B 333**, 303 (1994).

[14] O. Sorlin et al., Phys. Rev. C **47**, 2941 (1993).

[15] G.A. Lalazissis, S. Raman, and P. Ring, Atomic Data and Nuclear Data Tables **71**, 1 (1999).

[16] G. Audi and A.H. Wapstra, Nucl. Phys. **A595**, 409 (1995).

[17] N. Stone, "Table of Nuclear Moments (2001 Preprint)", Oxford University, United Kingdom.

[18] H. Scheit *et al.*, Phys. Rev. Lett. **77**, 3967 (1996).

[19] T. Glasmacher *et al.*, Phys. Lett. **B 395**, 163 (1997).

HOW LOW IS THE THERMODYNAMICAL LIMIT?

Tamás S. Biró, Gábor Purcsel
KFKI Research Institute for Particle and Nuclear Physics
H-1525 Budapest, Pf 49, Hungary
tsbiro@sunserv.kfki.hu

Berndt Müller
Physics Department, Duke University
Durham, NC 27708, USA

Abstract We analyze how can some dynamical process lead to exponential (Boltzmannian) distribution without instantaneous thermal equilibrium with a heat bath. We present a model for parton dressing which re-combines the exponential from cut power law minijet distributions fitted to pQCD as a limiting distribution. Thermal models and power law tails are interpreted within the framework of the more general Tsallis distribution.

Keywords: Thermal spectra, long tail, Tsallis

Introduction

Not only in heavy ion collisions, but also in pp and even in e^+e^- collisions exponential transverse mass spectra has been observed besides a power law tail at relatively high transverse momenta. The exponential shape at medium p_T was often interpreted as an indication that the source of the observed hadrons can be described by a temperature related to the slope of these spectra. This view is most prominent in the thermal models[1] which assume a thermally equilibrated state, well described by the familiar equilibrium thermodynamics.

The experimental data from CERN SPS and RHIC show an exponential shape in the transverse mass spectrum best fitted by $\exp((m - m_T)/T)$, where $m_T = \sqrt{m^2 + p_T^2}$ is the transverse mass of a particle with mass m[2]. The stiffness of these spectra, T, has still to be corrected for collective flow effects: for high momenta a relativistic blue-shift factor, for low momenta a mass dependent kinetic flow energy addition occurs in this parameter. The conjectured temperature, T_0, of a thermally equilibrated state is therefore usually higher than the directly observed slope parameter, T. Characteristic values at RHIC

W. Greiner et al. (eds.), Structure and Dynamics of Elementary Matter, 569–576.

are $T \approx 140$ MeV and $T_0 \approx 210$ MeV, so the assumed temperature is well above the color deconfinement temperature predicted by recent lattice QCD calculations.

There is, however, doubt whether a thermally equilibrated state can be formed in such a short time, during which the final state fireball at these high bombarding energies develops. There is no known hadronic process which would lead to a thermal state from a far off-equilibrium initial situation on a time scale of $1 - 2$ fm/c. It is strongly tempting to consider quark level mechanisms as reason for the observed exponential spectra. Several such mechanisms has already been suggested on a subhadronic level. Ref.[3] considered the Schwinger mechanism with a Gaussian fluctuating string tension, which leads to exponential m_T spectra. Having color ropes in mind, the effective string tension may have a Gaussian distribution due to a random color chargation process in the limit of high color. Another idea is to consider the fluctuation of temperature, distributed according to a Gamma distribution[4]. In this case the average of exponentials leads to a Tsallis distribution, which is still well approximated by an exponential at soft momenta, while ends in a power-law tail at high momenta. In this paper we shall have a closer look on a recently suggested hadronization mechanism, on the parton recombination[5], as a possible source of the experimentally observed exponential spectra. In this case we begin with cut power-law minijet distributions[6], having a form of the canonical Tsallis distribution[7], and combine their limiting distribution for many-fold recombination [8].

According to this goal first we review how the Gaussian distribution emerges as a limiting distribution and recall the central limit theorem of statistics. Then we consider a simple example not falling under the conditions of this renown theorem, but still leading to a non-trivial limiting distribution. Finally we relate this result to parton recombination and to the canonical Tsallis distribution, which gives an improved quantitative description of experimental findings compared to the naive thermal model.

1. The central limit theorem

In order to demonstrate, how fast a Gaussian limiting distribution may develop, let us first consider the distribution of the scaled sum of uniform random deviates. Let x_i be a uniform random deviate in $(-1, 1)$ and $P_n(x)$ the distribution of $x = \sqrt{\frac{3}{n}} \sum_{i=1}^{n} x_i$. The statement of the cited theorem is that in the limit of infinite n the distribution of the variable x exactly becomes the normal distribution.

$$P_\infty(x) = \frac{1}{\sqrt{2\pi}} e^{-x^2/2}. \tag{1}$$

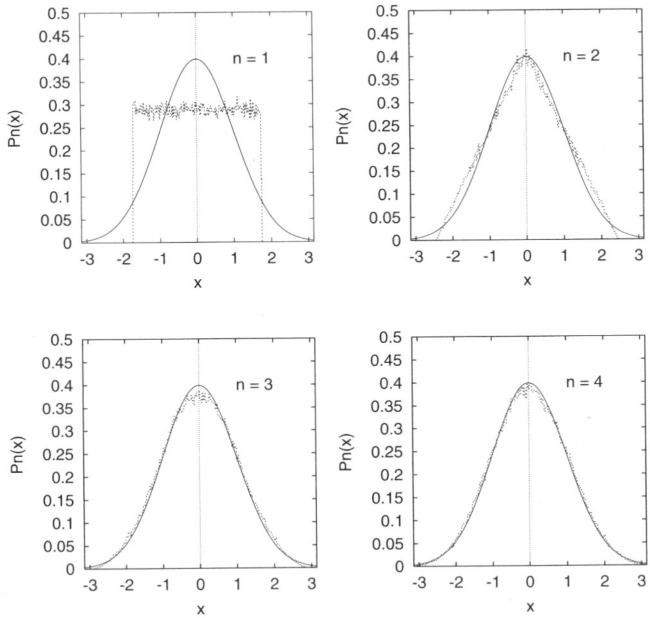

Figure 1. Comparison of histograms of $m = 200000$ sums of n uniform random deviates in $(-1, 1)$ scaled with $\sqrt{3/n}$ and the limiting Gauss distribution $\frac{1}{\sqrt{2\pi}} \exp(-x^2/2)$.

Before we review how and why this happens it is educating to make some guesses how large n actually should be in order to mistake such a distribution for a Gaussian. The answer is surprisingly low (cf. Fig.1). The $n = 1$ case is the uniform distribution, the $n = 2$ has a triangular shape. The finiteness of the sample ($m = 200000$) smoothes the differences out, it is therefore not easy to distinguish already the $n = 3$ or $n = 4$ case from the Gaussian in a numerical experiment.

Now we turn to the proof of the limiting distribution being a Gaussian. Consider the Fourier transform of $P_n(x)$:

$$\tilde{P}_n(k) = \int_{-\infty}^{\infty} dx\, e^{ikx} \prod_{i=1}^{n} \left(\frac{1}{2} \int_{-1}^{1} dx_i \right) \delta(x - \sqrt{\frac{3}{n}} \sum_{j=1}^{n} x_j). \tag{2}$$

After x-integration it turns to be a product of n equal factors:

$$\tilde{P}_n(k) = \left(\frac{\sin(k\sqrt{3/n})}{k\sqrt{3/n}} \right)^n . \tag{3}$$

The large n limit of this expression is a Gaussian:

$$\tilde{P}_\infty(k) = \lim_{n\to\infty} \left(1 - \frac{k^2}{2n} + \dots \right)^n \longrightarrow e^{-k^2/2}. \tag{4}$$

Fourier transforming back to x-space gives the required statement. The phenomenon is more general. Let x be a weighted sum of n independent random variables:

$$P_n(x) = \int (\prod_{i=1}^{n} w(x_i)dx_i)\, \delta(x - a_n \sum_{j=1}^{n} x_j). \tag{5}$$

each distributed according to the same probability density $w(x)$. Then the so called central moments, which describe correlations,

$$c_j(n) = \left(\frac{\partial}{i\partial k}\right)^j \log \tilde{P}_n(k)\Big|_{k=0}, \tag{6}$$

can be obtained easily from the central moments $c_j(1)$ of the $P_1(x) = w(x)$ distributions using $\log \tilde{P}_n(k) = n \log \tilde{w}(ka_n)$: $c_j(n) = na_n^j c_j(1)$. Whenever $a_n \propto 1/\sqrt{n}$, $c_j(n) \propto n^{1-j/2} c_j(1) \longrightarrow 0$ as $n \to \infty$ for $j > 2$. This completes the proff of the Central Limit Theorem.

There are exemptions to this theorem. Whenever $\log \tilde{P}_\infty(k)$ is not differentiable arbirary times in the vicinity of $k = 0$, or not the proper scaling factor a_n is used, or the base distribution $w(x)$ has a long tail not approaching zero fast enough. In the last case there can be a limiting distribution, it is just not Gaussian. As an interesting example we study the convolution of normalized Lorentzians

$$w(x) = \frac{R}{\pi} \frac{1}{x^2 + R^2} \tag{7}$$

The logarithm of the Fourier transform is not differentiable at $k = 0$: $\log \tilde{w}(k) = -|k|R$. The n-fold convolution's Fourier transform also has a non-differentiable logarithm, $\log \tilde{P}_n(k) = -nR|k|a_n$. It can, however, be found a finite limiting distribution for $(n \to \infty)$ if $a_n = 1/n$. In this case the Lorentzian is a fixed point of the convolution and scaling. The limiting Lorentzian distribution is also long-tailed, so the central moments diverge. Aiming at physical applications we have to regularize. We use a small parameter, m, for this purpose and at the end take the nontrivial double limit $m \to 0$ and $n \to \infty$. Starting with regularized Lorentzians, the Fourier transform has the form of the experimentally observed $m - m_T$ exponential,

$$\tilde{w}(k) = \exp\left(\frac{m - \sqrt{k^2 + m^2}}{T}\right) \tag{8}$$

with $T = 1/R$. Now $\log \tilde{w}(k) = R(m - \sqrt{k^2 + m^2})$ with vanishing odd central moments, and with regular even central moments, $c_{2j}(1) \propto \frac{R}{m^{2j-1}}$. The limiting distribution is obtained by keeping $M = nm$ constant and scaling with $a_n = 1/n$ again. It has the following Fourier spectrum

$$\tilde{P}_\infty(k) = e^{(M - \sqrt{k^2 + M^2})/T}. \tag{9}$$

This resembles the original regularized distribution, but with an eventually finite mass M. The low and high momentum limits are Gaussian and pure exponential, respectively.

2. Parton dressing to exponential spectra

Combining hadrons from many partons also results in a limiting distribution, which is not Gaussian. We consider an additive quark model where the hadron mass is made of n approximatevly equal partons: $M = \sum E_i = nm$. The variable, $x = \frac{\sum E_i x_i}{\sum E_i} = \frac{1}{n}\sum x_i$ is the center of energy coordinate. Its limiting distribution, $P_\infty(x)$ can be viewed as the hadron wave function squared. Its Fourier transform, $\tilde{P}(k)$ is proportional to the hadron spectrum.

While constructing from "equal-right" partons, we have $\tilde{P}_n(k) = \tilde{w}(k/n)^n$. The particular cases $n = 2$ and $n = 3$ of this formula have been used in parton recombination models of hadronization recently. The partons before hadronization show a cut power law transverse momentum distribution, as fitted to minijets obtained in the framework of pQCD based calculations:

$$\tilde{w}(k) = a\left(1 + \sqrt{k^2 + m^2}/b\right)^{-c} \tag{10}$$

Here b and c are parameters of the parton distribution, in no way connected to a thermal state, m is the transverse mass at zero rapidity, and finally a can be obtained from the normalization condition, $\tilde{w}(0) = 1$. Note that this distribution is a power law for large transverse momenta $\lim_{|k|\gg m,b} \tilde{w}(k) = ab^c\,|k|^{-c}$.

We construct a constituent quark and eventually hadrons from n such partons, keeping the total mass $M = nm$ fixed. We arrive at

$$\tilde{P}_n(k) = a^n\left(1 + \frac{1}{nb}\sqrt{k^2 + M^2}\right)^{-nc} \tag{11}$$

In the $n \to \infty$ limit this gives the exponential distribution in the transverse mass

$$\tilde{P}_\infty(k) \;=\; \exp\left(\frac{M - \sqrt{k^2 + M^2}}{b/c}\right). \tag{12}$$

This is exponential of the reduced transverse mass $M_T - M = \sqrt{k^2 + M^2} - M$ has a slope $T = b/c$. This is not a temperature, it is not related to any energy exchange with a heat reservoir. The large parameter by which one approaches the limiting distribution is the foldness of the recombination; in the present model the number of independent primordial partons dressing to a hadron constituent.

3. Tsallis distribution and thermal model

The cut power law used in minijet fits and modified as a function of energy in the hadronization model presented above accidentally coincides with the canonical Tsallis distribution. This distribution can be derived from a generalization of the Boltzmann entropy formula first suggested by Tsallis:

$$S_q = \frac{1}{1-q} \sum_i (w_i^q - w_i) \tag{13}$$

In order to relate to our model we use $q = 1 - 1/n$. We get

$$S_q = n \sum_i w_i (w_i^{-1/n} - 1) \longrightarrow - \sum_i w_i \ln w_i. \tag{14}$$

The canonical distribution maximizes the entropy with the following constraints: $\sum_i w_i = 1$ and $\sum_i w_i \varepsilon_i = E$. The use of a lagrange multipliers β leads to the following distribution which minimizes $S_q - \beta E$:

$$w_i = \frac{1}{Z} \left(1 + \frac{\beta \varepsilon_i}{n} \right)^{-n} \tag{15}$$

with $Z = \sum_i (1 + \beta\varepsilon_i/n)^{-n}$ for the proper normalization.

Based on this distribution the free energy can be derived as well as relations reminding to rhe familiar thermodynamical relations. The Tsallis distribution can hence be regarded as a generalization of equilibrium thermodynamics. Among its physical realizations it is particularly interesting the case of finite heat reservoirs.

In view of this formalism we interpret the one parton distribution as a canonical Tsallis distribution,

$$w_1 = (1 + \frac{E}{b})^{-c} = w_{\text{Tsallis}}(E; T, 1 - 1/c) \tag{16}$$

with $T = b/c$. Recombination of n such partons leads to another Tsallis distribution with the same "temperature", but closer to the equilibrium case $q = 1$:

$$w_n = w_1(E/n)^n = (1 + \frac{E}{nb})^{-nc} = w_{\text{Tsallis}}(E; T, 1 - 1/(nc)). \tag{17}$$

Arriving at the finally observed hadrons with $c \approx 8$ and $n = 2 \ldots 6$ one gets really close to the limiting exponential

$$w_\infty = w_{\text{Tsallis}}(E; T, 1) = e^{-E/T}. \tag{18}$$

This property may explain the relative, and so far unexplained, success of the thermal model.

Finally it is interesting to explore the potential of the Tsallis distribution with respect to particle numbers and the energy per particle, since these are the most profound predictions of the thermal model. For the sake of simplicity we consider here the analytically solvable case of massless particles.

The number of a particle in a volume V with internal degeneracy factor d is given by the integral of the canonical Tsallis distribution,

$$N = \frac{d}{2\pi^2} V \int_0^\infty p^2 dp \left(1 + \frac{E}{b}\right)^{-c}. \tag{19}$$

In the massless case, $E = |p|$, and we get

$$N = \frac{d}{2\pi^2} V b^3 \frac{2}{(c-1)(c-2)(c-3)}. \tag{20}$$

In general this is a bigger number for finite c than for the Boltzmann distribution at $nc \to \infty$: $N(c)/N(\infty) > 1$. This ratio is $2.4, 1.5, 1.3$ for $c = 8, 16, 24$.

The energy is given by another integral,

$$E = \frac{d}{2\pi^2} V \int_0^\infty p^2 dp\, E \left(1 + \frac{E}{b}\right)^{-c} \tag{21}$$

which can as well be calculated in a closed form for massless particles, $E = |p|$:

$$E = \frac{d}{2\pi^2} V b^4 \frac{6}{(c-1)(c-2)(c-3)(c-4)} \tag{22}$$

The energy per particle becomes $E/N = \frac{3b}{c-4}$ resulting in $E/N \approx 1.07$ GeV for $b = 1.39$ GeV and $c = 7.9$, and at the same time predicting a slope $T \approx 176$ MeV for the m_T exponential at low momenta $p_T \ll bc \approx 10$ GeV.

4. Conclusion

We reviewed limiting distributions of familiar short tailed and exceptional long tailed distributions. We have seen, that already a few, $n = 3 - 4$ independent random comopnents may lead to a distribution which is hard to distinguish from the limiting case on finite, fluctuating data samples.

Non Gaussian limiting distributions exist with altered (non $1/\sqrt{n}$) scaling laws. For the high energy hadronization process it is particularly interesting to consider cut power law distributions, which in the limiting case combine to exponential spectra. This happens with the $1/n$ scaling, pointing out a possible interpretation as hadron spectra showing the Fourier transform of the distribution of the center of energy of the recombining partons.

Minijet partons obtained using pQCD processes, and actually also experimentally observed spectra in pp collisions are Tsallis distributed at $T \approx$

170 − 200 MeV, and with $E/N \approx 1$ GeV. These two numbers can be simultanously described by a Tsallis distribution with the parameters $b = 1.39$ GeV and $c = 7.9$, when considering massless particles. The same simple consideration fails for the familiar Boltzmann distribution of the thermal equilibrium with a factor of 2. At finite mass these integrals can be calculated only numerically. Considering massive particles with a mass around the pion mass, our results do not change dramatically. At the mass of 1 GeV, however, the Boltzmann distribution reaches $E/N \approx 1$ GeV and the Tsallis distribution overshoots with a factor of 2.

Acknowledgments

This work has been supported by the Goethe University, Frankfurt and by the Hungarian National Research Fund OTKA (T034269). Special thanks to Professor Walter Greiner for organizing this NATO Advanced Study Institute meeting at Kemer, Turkey, which location created a perfectly relaxing and inspiring atmosphere.

References

[1] J.Letessier, J.Rafelski, A.Tounsi, Phys.Lett.B **328**, 499, (1999); J.Cleymans, H.Oeschler, K.Redlich, Phys.Lett.B **485**, 27, (2000); K.Redlich, K.Werner, Phys.Rev.Lett. **88**, 202501, (2001); W.Broniowski, A.Baran, W.Florkowski, nucl-th/0212053, nucl-th/0212052; P.Braun-Münziger, K.Redlich, J.Stachel, in Quark Gluon Plasma 3, eds.: R.Hwa, X-N.Wang, World Scientific Publ.Co., (2003), nucl-th/0304013; F.Becattini, G.Pettini, Nucl.Phys.A **715**, 557, (2003).

[2] PHENIX collaboration, nucl-ex/0307022, nucl-ex/0307010, nucl-ex/0304022, nucl-ex/0212014, Acta Phys.Hung. Heavy Ion Physics **15**, 291, (2002); STAR collaboration, hep-ex/0306056, nucl-ex/0306024; NA49 collaboration, nucl-ex/0306022.

[3] A.Bialas, Phys.Lett.B **466**, 301, (1999).

[4] G.Wilk, Z.Wlodarczyk, Phys.Rev.Lett. **84**, 2770, (2000); hep-ph/0002145; hep-ph/0004250.

[5] R.J.Fries, B.Müller, C.Nonaka, S.A.Bass, nucl-th/0301087, nucl-th/0305079, nucl-th/0306027.

[6] R.J.Fries, B.Müller, D.K.Srivastava, Phys.Rev.Lett. **90**, 132301, (2003); D.K.Srivastava, C.Gale, R.J.Fries, Phys.Rev.C **67**, 034903, (2003).

[7] C.Tsallis, J.Stat.Phys. **52**, 479, (1988), Physica A **221**, 277, (1995); A.Plastino, A.R.Plastino, Brazilian J. Phys. **29**, 50, (1999).

[8] T.S.Biro, B.Müller, hep-ph/0309052.

SIGNATURES OF LARGE EXTRA DIMENSIONS

S. Hossenfelder, M. Bleicher and H. Stöcker
Institut für Theoretische Physik
J. W. Goethe Universität
D-60054, Frankfurt am Main, Germany
hossi@th.physik.uni-frankfurt.de

Abstract String theory suggests modifications of our spacetime such as extra dimensions and the existence of a mininal length scale. In models with addidional dimensions, the Planck scale can be lowered to values accessible by future colliders. Effective theories which extend beyond the standart-model by including extra dimensions and a minimal length allow computation of observables and can be used to make testable predictions. Expected effects that arise within these models are the production of gravitons and black holes. Furthermore, the Planck-length is a lower bound to the possible resolution of spacetime which might be reached soon.

This is a summary of a talk given at the NATO Advanced Study Institute in Kemer, Turkey, Oct. 2003.

1. Introduction

The standard model (SM) of particle physics yields an extremely precise theory for the electroweak and strong interaction. It allowed us to improve our view of nature in many ways but leaves us with several unsolved problems. E.g. we still have to understand the large number of free parameters in the SM, the question of the fermion families, the mechanism of electroweak symmetry breaking, CP violation, and of course the puzzle of quantum gravity.

Gravity, if it is quantized as a spin-2 field in the canonical way, is non-renormizable. The only reason why the SM yields such high accuracy without concerning gravity is that gravity is much weaker than the other interactions. The effects of quantum gravity get as important as the effects of the SM only at the so-called Planck-scale which is reached at energies near the Planck-mass of $m_{\rm p} \approx 10^{16}$ GeV or at distances near the Planck-length $l_{\rm p} \approx 10^{-20}$ fm, resp. This comparable weakness of the gravitational interaction, also known als the "hierarchy-problem", is another yet unexplained fact which singles out gravity and has to be understood for achieving a successful unifying theory.

W. Greiner et al. (eds.), Structure and Dynamics of Elementary Matter, 577–584.

All these problems have in common that they can not be explained within the SM itself. The SM, it seems, is a low energy limit of a more general theory. It is one of the most exciting and challenging tasks on physicists in the 21st century to go on and look beyond the SM.

On the one hand, huge steps in this direction have been undergone by string theorists in the last decades. Supersymmetric string theory is – up to current knowledge – todays most promising candidate of a Grand Unified Theory (GUT). In addition to the SM symmetries, it provides naturally the existence of a spin-2 particle. Moreover, this canditate for quantum gravity is finite (at least in pertubation theory).

On the other hand, the need to look beyond infected many experimental groups which search for SM - violating processes.

Unfortunately, there is a gap between theory and experiment as the most obvious predictions of string theory – excited particles, one loop corrections – have to be hidden by broken symmetry at low enery scales and thus, there is no way so far to verify string theory. Whenever there is a theory in its full mathematical beauty that can not be applied to computate observables, one needs to make approximations. In a certain way, theoretical physics is the art of approximation; it is the art of model building; it is the art of simplification.

"Science may be described as the art of systematic over-simplification."
— Karl Popper, The Observer, August 1982

The recently proposed models of extra dimensions are models that can be used to fill the gap between theoretical conclusions and experimental possibilities. These models are motivated by string theory but do not have to cope with all the stringy implications. They are kind of an effective model for a theory beyond the SM. The main idea taken from string theory are the notions of open and closed strings which provide us naturally with two different kinds of particles. The closed strings describe the graviton, the open strings the other interacting particles. So, there is a natural reason why gravity is different. Further, supersymmetric string theory does only work properly (no anomalies) in spacetimes with extra dimensions. These extra dimensions are compactified to finite extension by empiric means: we have not seen them. When compactifying the extra dimensions we have to confine the open strings to be attached on a three-dimensional submanifold: our 3-brane, the universe we are used to.

There are different ways to build a model of extra dimensional space-time. Here, we want to mention only the most common ones:

1 The ADD-model proposed by N. Arkani-Hamed, S. Dimopoulos and G. Dvali [1] adds d extra spacelike dimensions without curvature, in general each of them compactified to the same radius R. All SM particles are

confined to our brane, while gravitons are allowed to propagate freely in the bulk.

2 The setting of the model from L. Randall and R. Sundrum [2, 3] is a 5-dimensional spacetime with an non-factorizable geometry. The solution for the metric is found by analyzing the solution of Einsteins field equations with an energy density on our brane, where the SM particles live. In the type I model [2] the extra dimension is compactified, in the type II model [3] it is infinite.

3 Within the model of universal extra dimensions [4] all particles (or in some extensions, only bosons) can propagate in the whole multi-dimensional spacetime. The extra dimensions are compactified on an orbifold to reproduce standard model gauge degrees of freedom.

In the following we will focus on the ADD-model. For a more general review the reader is refered to [5].

The ADD-model explains the hierarchy between the electroweak and the Planck-scale with the large volume of the extra dimensions. Consider the Poisson-equation for a point particle of mass m in $d+3$ spacelike dimensions. The coupling constant will have dimension mass^{-d-2}. This new mass-scale is the new higher dimensional Planck-mass and will be denoted by $M_{\rm f}$, the new Planck-length is $L_{\rm f} = 1/M_{\rm f}$.

The power law for the potential $V(r)$ goes with the distance r from the source as r^{-d-1}. This holds in the compactified scenario for distances much smaller than the radius R of the extra dimensions. For large distances $r \gg R$, d of the powers factorize and we have to match this to the usual $1/r$ power law in three dimensions with the familiar coupling $m_{\rm p}^2$. This yields

$$\frac{V}{m} = \frac{1}{M_{\rm f}^{d+2}} \frac{1}{r^{d+1}} \to \frac{1}{M_{\rm f}^{d+2}} \frac{1}{R^d} \frac{1}{r} = \frac{1}{m_{\rm p}^2} \frac{1}{r} \quad , \tag{1}$$

and so one obtains the relation

$$m_p^2 = M_{\rm f}^{2+d} R^d \quad . \tag{2}$$

With the assumption $M_{\rm f} \sim$TeV, the compactification scale R ranges from $1/10$ mm to 10^2 fm, resp. $1/R$ from 10^{-2} eV to 10 MeV, if d runs from 2 to 7. d=1 is excluded since this would imply an extra dimension of the size of the solar system. It was shown, that this setting with large[1] extra dimensions (LXDs) can be motivated by string theory and it indeed lowers the unification scale to values $\approx M_{\rm f}$ [6].

[1]Large when compared to the Planck-scale.

The lowered Planck-scale leads to a vast number of observable effects. The most obvious one is a modification of Newtons law at small distances, which is in todays measurable range for $d = 2$. There are several groups working on this sub-mm gravity measurements [7]. So far, $R > 0.18$ mm can be excluded. Next generation experiments are expected to yield precise measurements up to μm-distances.

This increase of the gravitational force at small distances enables the production of black holes at energy scales $\approx M_{\mathrm{f}}$ which can be reached at the LHC. Further, with the lowered scale, the production of gravitons becomes significant at energies $\approx M_{\mathrm{f}}$. Also, there are contributions to cross sections from the virtual graviton exchange. In the following, we will briefly discuss possible consequences for high energy physics, observable effects and the extension of the model to include minimal-scale effects.

2. Gravitons

Gravitons are treated as pertubations[2] h_{IJ} of the higher dimensional metric tensor $g_{IJ} = \eta_{IJ} + h_{IJ}$. For an effective description on our brane, this pertubation tensor can be decomposed in a spin-2 tensor, which describes the graviton, vector fields and scalar fields. From this ansatz one obtains the Lagrangian by minimal coupling to SM fields. The analysis shows that only the spin-2 field couples to the energy-momentum tensor.[3] With the interaction terms, one can then derive the Feynmann rules [8] for the quantized fields. This enables us to compute graviton cross sections at least at tree-level.

In absence of matter fields, the fields obey an $d+3$ dimensional wave equation. Due to the periodicity of the extra dimensions, we can expand the solutions in a fourier series with n_i/R, where $n = (n_1, .., n_d)$ is the number-vector of the excitation level. The quantized momentum in the direction of the extra dimensions yields an apparent mass term for the graviton if described effectively on our brane:

$$\eta^{IJ}\partial_I\partial_J = \Box - \sum_d \frac{n_i^2}{R^2} \quad .$$

Thus, we have a tower of massive gravitons. Since the level spacing $1/R$ is tiny compared to collider energies, the number $N(\sqrt{s})$ of excited graviton levels, that can be occupied with an energy $\sqrt{s}$ is $N(\sqrt{s}) \propto (\sqrt{s}R)^d$. This large phase-space of the gravitons is in the effective description responsible for the importance of the effect. Consider e.g. the process $e^+e^- \to G + \gamma$. With the

---

[2]Capital indices run from 0 to $d+1$, small latin indices run over the extra dimensions $4..4+d$ only, small greek indices run over the non-compactified dimensions, from 0 to 3.

[3]The trace of the scalar fields, also known as dilaton or radion, couples to the trace of the energy-momentum tensor.

estimation for the total cross-section[4]

$$\sigma(e^+e^- \to G+\gamma) \propto \frac{\alpha}{m_{\rm p}^2} N(\sqrt{s}) = \frac{\alpha}{s}\left(\frac{\sqrt{s}}{M_{\rm f}}\right)^{d+2} \quad ,$$

where we have used eq. (2), we see, that the graviton processes get as important as the SM processes at energies $\sqrt{s} \approx M_{\rm f}$.

The primary observable effect for real graviton production is an apparent non-conservation of momentum on our brane, since the graviton leaves our brane and is not detected. Therefore, production of a single jet at high transverse momentum is a promising signal to look for, the main contribution for LHC arises from the subprocess $qg \to qG$.

Analysis of present data yields contraints on real and virtual graviton processes and so gives a lower bound on the new scale $M_{\rm f}$ in the range 1 TeV, the exact value depending on d. For more details see e.g. [9] and references therein.

3. Black Holes

In general relativity a (non-charged, non-rotating) black hole of mass M is described by the Schwarzschild-metric. This metric is diagnonal, spherical symmetric and has $g_{tt} = -1/g_{rr} = 1 + 2\phi$, with ϕ the Newtonial potential. It can be shown, that this holds too (up to factors of order 1) for the higher dimensional Schwarzschild-solution [10]. Here, of course we have to use the $d+3$ dimensional potential. That is, the zero R_H of the metric coefficients, which gives the horizon radius is (again up to factors of order 1)

$$\frac{M}{M_{\rm f}^{d+2}} \frac{1}{R_H^{d+1}} \approx 1 \quad .$$

It is not surprising, that a black hole of mass $\approx M_{\rm f}$ has a horizon-radius of $R_H \approx L_{\rm f}$. Due to the lowered Planck-scale this horizon-radius is in the range of distances that can be reached at LHC-energies. Note, that we can use the spherical symmetric solution since the black-hole radius for collider-energies is much smaller than the radius of the extra dimensions and the periodic boundary conditions can be neglected.

¿From general relativistic arguments[5], the cross-section for black hole production is $\sigma = \pi R_H^2$. It has been discussed whether this cross-section is reliable and it has been shown in several approaches that it holds at least up to energies $\approx 10 M_{\rm f}$ [11].

[4] Here, G denotes the graviton.

[5] This is known as Thorn's hoop conjecture.

Using the QCD parton-distribution functions it is then possible to compute differential and total cross-sections for the black-hole production at LHC. The total number of black holes depends only weakly on d and is of order 10^9 per year, that is ≈ 30 black holes per second [12]!

The produced black holes will undergo evaporation which is more a decay because of their high temperature [13]. Unfortunately, the precise description of this process falls into the regime of quantum gravity and is unknown, the main question being whether the black hole evaporates completely or whether a stable relic is left [14].

There are several observables for the black hole detection. First, there will be a sharp drop in the jet-spectrum at high transverse momentum $> M_\mathrm{f}$. High energetic jets can not be produced any more since their energy will create a black hole. The black hole decay then yields multi-jet events with energies $< M_\mathrm{f}$ or, in the scenario with relics resp., mono-jets. Since the black hole radiates thermally, this processes are flavour-blind. The detection of black holes would not only allow us to test the large extra dimension model but it would be an enormous exciting possibility to examine the properties of an truly extreme state of matter on the junction between general relativity, thermodynamics and quantum field theory.

For more details on this subject see e.g. [12].

4. Minmal Length

Even if a full description of quantum gravity is not yet available, there are some general features that seem to go hand in hand with all promising candidates for such a theory. Besides the the need for a higher dimensional spacetime, there is the existence of a minimal length scale.

In perturbative string theory, the feature of a fundamental minimal length scale arises from the fact that strings can not probe distances smaller than the string scale. If the energy of a string reaches the Planc-mass m_p, excitations of the string can occur and cause a non-zero extension [15]. Due to this, uncertainty in position measurement can never become smaller than l_p. This can also be understood in a heuristic way. Consider an experiment to test a spacetime-structure from about Planck-length. The Compton-wavelength a particle must have in order to resolve Planck-length is just Planck-mass as follows from the uncertainty principle. Since the particle has Planck-mass its perturbation of the spacetime-metric can not be neglected any longer and thus causes an additional uncertainty at high energies. For a review, see [16].

In order to implement the notion of a minimal length into the model of LXDs, let us now suppose that one can increase the momentum p arbitrarily, but that the wave-vector k has an upper bound. This effect will show up when p approaches a certain scale M_f. The physical interpretation of this is that particles can not possess arbitrarily small Compton-wavelengths $\lambda = 2\pi/k$ and that arbitrarily small scales can not be resolved anymore.

To incorporate this behaviour, we assume a relation $k = k(p)$ between p and k which is an uneven function (because of parity) and which asymptotically approaches $1/L_{\rm f}$.[6] There are several approaches how to deal with this generalisation, see e.g. [17]. We will use the analysis of [18]. The modified commutator algebra in quantum mechanics then reads

$$[\hat{x}, \hat{p}] = +\mathrm{i}\frac{\partial p}{\partial k} \Rightarrow \Delta p \Delta x \geq \frac{1}{2}\left|\left\langle \frac{\partial p}{\partial k} \right\rangle\right| \quad . \tag{3}$$

which results in a generalised uncertainty relation. In a power series expansion of the function $k(p)$, we find an additional term in the uncertainty, which is quadratic in p. A further consequence of the existence of the minimal lenght is a dropping of the momentum space measure with the functional determinante $\partial k/\partial p$.

This generalized uncertainty relation influences observables of high energy physics. Mainly, it states that effects at high energies are much less then expected from the SM only. Consider e.g. the black hole formation: the beam energy needed with the generalized uncertainty to focus enough energy-density in a small region of space-time will increase. We want to point out that the minimal length effects do strongly modify predictions of the LXD-scenario and have to be included for completeness.

5. Summary

"Truth in science can best be defined as the working hypothesis best suited to open the way to the next better one." — *Konrad Lorenz*

The models with large extra dimensions do not claim to be a theory of everything but provide a useful basis to make predictions beyond the standard model. Observables, like graviton or black-hole production, can be used to test general features of our spacetime, such as the number and the size of the extra dimensions or the existence of a minimal length scale.

Acknowledgments

This work was supported by GSI, DFG, BMBF and DESY. The authors thank all members of the LXD-Group at Frankfurt University for fruitful discussions. We want to apologize for every contribution on the topic of Large Extra Dimension that has not been mentioned here.

[6]Note that this is similar to introducing an energy dependence of Planck's constant $\hbar \to \hbar(p)$.

References

[1] N. Arkani-Hamed, S. Dimopoulos & G. Dvali, Phys. Lett. B **429**, 263-272 (1998); I. Antoniadis, N. Arkani-Hamed, S. Dimopoulos & G. Dvali, Phys. Lett. B **436**, 257-263 (1998); N. Arkani-Hamed, S. Dimopoulos & G. Dvali, Phys. Rev. D **59**, 086004 (1999).

[2] L. Randall & R. Sundrum, Phys. Rev. Lett. **83** 3370-3373 (1999) [hep-ph/9905221].

[3] L. Randall & R. Sundrum, Phys. Rev. Lett. **83** 4690-4693 (1999) [hep-th/9906064].

[4] T. Appelquist, H. C. Cheng and B. A. Dobrescu, Phys. Rev. D **64**, 035002 (2001); C. Macesanu, C. D. McMullen and S. Nandi, Phys. Rev. D **66**, 015009 (2002); T. G. Rizzo, Phys. Rev. D **64**, 095010 (2001).

[5] V.A. Rubakov, Phys. Usb. **44**, 871 (2001); Y.A. Kubyshin, *Lectures given at the XI. School "Particles and Cosmology"*, Baksam, Russia, April 2001.

[6] I. Antoniadis, Phys. Lett. **B 246**, 377-384 (1990); I. Antoniadis & M. Quiros, Phys. Lett. **B 392**, 61 (1997); K.R. Dienes, E. Dudas & T. Gherghetta, Nucl. Phys. **B537** 47 (1999) .

[7] J. C. Long and J. C. Price, arXiv:hep-ph/0303057; C. D. Hoyle *et al*, Phys. Rev. Lett. **86**, 1418 (2001); J. Chiaverini *et al*, Phys. Rev. Lett. **90**, 151101 (2003).

[8] Tao Han, J. D. Lykken & Ren-Jie Zhang Phys. Rev. **D59** 105006 (1999); J. L. Hewett, Phys. Rev. Lett. **82**, 4765 (1999); G. F. Giudice, R. Rattazzi and J. D. Wells, Nucl. Phys. B **544**, 3 (1999).

[9] J. Hewett & M. Spiropulu, Ann. Rev. Nucl. Part. Sci. 52 , 397-424, (2002); Y. Uehara, Mod. Phys. Lett. A **17**, 1551 (2002); E. A. Mirabelli, M. Perelstein & M. E. Peskin, Phys. Rev. Let. **82** 2236-2239 (1999).

[10] R. C. Myers & M. J. Perry, Ann. Phys. **172**, 304-347 (1986).

[11] R. Emparan, Phys. Rev. D **64**, 024025 (2001); S. N. Solodukhin, Phys.Lett. **B 533**, 153-161, (2002); A. Jevicki and J. Thaler, Phys.Rev. **D 66**, 024041 (2002); D. M. Eardley and S. B. Giddings, Phys. Rev. **D 66**, 044011 (2002).

[12] P.C. Argyres, S. Dimopoulos & J. March-Russell, Phys. Lett. **B 441** 96 (1998) [hep-th/9808138]; S. B. Giddings, Gen. Rel. Grav. **34**, 1775 (2002) S. Hossenfelder, S. Hofmann, M. Bleicher & H. Stöcker, Phys. Rev. **D 66** 101502 (2002);

[13] R. Casadio & B. Harms, Phys.Rev. **D 64**, 024016 (2001); R. Casadio & B. Harms, Phys.Lett. **B 487** 209-214 (2000).

[14] Y. Aharonov, A. Casher and S. Nussinov, Phys. Lett. **191 B**, 51 (1987); T. Banks and M. O'Loughlin, Phys. Rev. **D 47**, 540 (1993); T. Banks, M. O'Loughlin and A. Strominger, Phys. Rev. **D 47**, 4476 (1993); J. D. Barrow, E. J. Copeland and A. R. Liddle, Phys. Rev. **D 46**, 645 (1992); S. Alexeyev *et al* Class. Quant. Grav. **19**, 4431-4444 (2002).

[15] D. J. Gross and P. F. Mende, Nucl. Phys. B **303** (1988) 407. D. Amati, M. Ciafaloni and G. Veneziano, Phys. Lett. B **216** (1989) 41; E. Witten, Phys. Today, **49** (1996) 24.

[16] L. J. Garay, Int. J. Mod. Phys. A **10** (1995) 145; A. Kempf, [arXiv:hep-th/9810215] .

[17] A. Kempf and G. Mangano, Phys. Rev. D **55** (1997) 7909; I. Dadic, L. Jonke and S. Meljanac, Phys. Rev. D **67** (2003) 087701; J. Martin and R. H. Brandenberger, Phys. Rev. D **63** (2001) 123501; F. Brau, J. Phys. A **32** (1999) 7691; R. Akhoury and Y. P. Yao, Phys. Lett. B **572**, 37 (2003).

[18] S. Hossenfelder *et al*, Phys. Lett. B 575 (2003), 84.

BARYON-BARYON INTERACTIONS IN FREE SPACE AND MATTER

Christoph Keil[1], Horst Lenske[1], Carsten Greiner[2]

[1] *Univ. Giessen, Heinrich-Buff-Ring 16, 35392 Giessen, Germany*

[2] *Univ. Frankfurt, Robert-Mayer-Str. 10, 60054 Frankfurt/M., Germany*

Abstract We study the interactions of the octet baryons in a relativistic meson exchange approach connecting free-space and in-medium interactions with those in finite (hyper)nuclei. A short sketch of the model including the relativistic T- and G-matrix and the mapping of the G-matrix onto density dependent vertex functionals in the density dependent relativistic hadron field theory (DDRH) is given. From new spectroscopic data on medium mass hypernuclei we extract the single particle spectrum including spin-orbit splitting. An extraction of the $\Lambda\omega$ vector- and tensor vertex as well as the $\Lambda\sigma$-vertex shows a strong breaking of SU(3) symmetry. As an application of our hypernuclear structure calculations we present Auger neutron rates in excited hypernuclei calculated within DDRH. The $\Lambda\Lambda$ HBT is discussed as an additional approach to constrain the Λ-meson couplings.

1. Introduction

The presented approach aims at a microscopic link between interactions for free baryon-baryon scattering, their in-medium interaction in infinite matter and in finite nuclei. Such a comprehensive description is essential for extrapolations to not yet experimentally observed or even never observable regions of the (hyper)nuclear chart, infinite hypermatter in neutron stars and baryon-baryon interactions in heavy ion collisions. In the nuclear sector there is abundant information available, i.e., the coupling constant can be fixed from free NN scattering and then propagated to finite nuclei where calculations of their properties and comparison to experimental data yield an important check of consistency. In the hyperon sector the model can be used then to extract from single particle spectra of hypernuclei free hyperon coupling constants which are unaccessible otherwise.

W. Greiner et al. (eds.), Structure and Dynamics of Elementary Matter, 585–592.

2. The density dependent relativistic hadron field theory

Our approach is based on the covariant meson exchange framework. To generate the effective microscopic interaction in finite nuclei an approach using T-matrix theory with a three dimensionally reduced two-nucleon propagator is applied. In this way the same mesons with the same bare couplings are used throughout the model. The in-medium interaction is calculated in Dirac-Brueckner theory, i.e., a T-matrix calculation in infinite matter where the intermediate two-baryon propagators are Pauli-blocked and dressed by Hartree-Fock self-energies. Since a G-matrix calculation for finite nuclei is technically not feasible DDRH theory is used to apply the G-matrix interactions in calculations of finite nuclei.

DDRH theory [1] is a Walecka type model designed for relativistic mean-field (RMF) calculations of finite nuclei using G-matrix interactions. Contrary to the conventional RMF models density dependent vertex functionals $\Gamma(\hat{\rho})$ are used instead of constant couplings and meson self-interactions to account for finite density effects. $\hat{\rho}$ is a bilinear of the baryonic field operators. Using the self-energies, in-medium effects of the G-matrix interaction are mapped onto the vertex functionals [1, 2]. Deriving the Dirac equation from the DDRH Lagrangian yields additional *rearrangement self-energies*, coming from the variation of the vertex functionals with respect to the baryonic field operators and describing static polarizations of the nuclear medium [1]. They do not affect the bulk properties of a nucleus but modify the single particle structure only. This treatment of the density dependence leads to a covariant field theoretically and thermodynamically consistent theory (see [1] for the proofs). DDRH theory was applied successfully to magic and exotic nuclei in RMF approximation [1, 2].

3. Hypernuclear structure

Hypernuclear spectroscopy has made important progress in accuracy in the recent years, allowing now to extract information about the interaction systematics between hyperons and nucleons within a microscopic model. With this understanding of the interaction and upcoming even more precise γ-spectroscopy experiments on hypernuclei at KEK, JLab, GSI and MAMI-C it will even be possible to study more subtle interaction effects beyond the mean-field which are not accessible in such a clean way using nucleon spectroscopy and conventional nuclear structure.

3.1 DDRH for hypernuclei

DDRH theory has been extended from the use in pure isospin nuclei to hypernuclei involving the octet baryons [3] and to hyperon star matter [4].

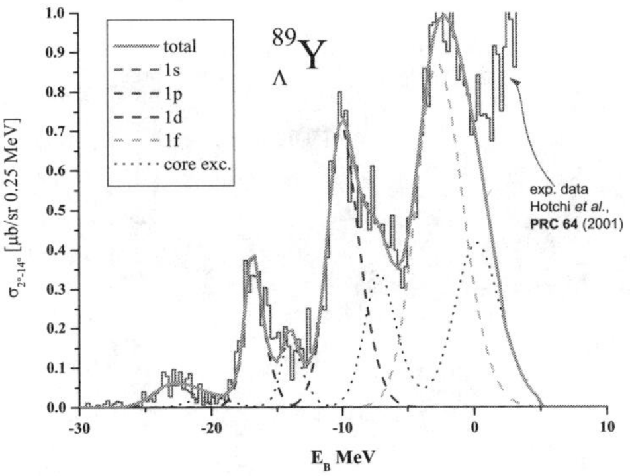

$1s_{1/2}$	-20.94 ± 0.64 MeV
$1p_{3/2}$	-17.02 ± 0.07 MeV
$1p_{1/2}$	-16.68 ± 0.07 MeV
$1d_{5/2}$	-10.26 ± 0.07 MeV
$1d_{3/2}$	-9.71 ± 0.07 MeV
$1f_{7/2}$	-3.04 ± 0.11 MeV
$1f_{5/2}$	-2.26 ± 0.11 MeV

Figure 1. The figure shows the refined fit to extract the single particle spectrum from the experimental data, taking also into account jj interactions. In the table the extracted single particle spectrum of ${}^{89}_{\Lambda}Y$ is shown.

Lacking a G-matrix calculation for the complete baryon octet – work is in progress on that – an approximate treatment of the hyperon-meson vertices was obtained from a diagramatic analysis of the Dirac-Brueckner equations. It was shown [3] that the functional shape of the vertices should be approximately the same as that of the nuclear ones, depending on the density of the respective hyperon iso-multiplet instead of the nuclear density. The strength has to be rescaled by a function $R_{Y\alpha}$ which turns out to be approximately constant and can in first order be related to the free coupling constants as $R_{Y\alpha} = \frac{g_{Y\alpha}}{g_{N\alpha}}$.

Single Λ hypernuclei are the only well known hypernuclear systems. The Λ being an isoscalar and electrically neutral baryon just couples to the σ and the ω meson simplifies its theoretical description tremendously. The two corresponding scaling factors relate directly to the free couplings $g_{\Lambda\sigma}$ and $g_{\Lambda\omega}$ by the approximate relations given above. Using spectral information of hypernuclear spectroscopy including resolved spin-orbit doublets these two constants can be extracted unambiguously from DDRH RMF calculations for the respective hypernucleus – or incompatibilities of the calculated and measured spectral structures will give definite hints to deformations and dynamics beyond mean field.

3.2 Evaluation of experimental data

Recent experimental data measured at KEK on ${}^{89}_{\Lambda}Y$ and ${}^{51}_{\Lambda}V$ [5] show a significant broadening of the single particle peaks as compared to the experimental resolution. The simple analysis with a two uncorrelated gaussian fit of each single particle peak - motivated by a spin-orbit (s.o.) splitting assumption - which was performed in [5] lead to a single particle spectrum which could not

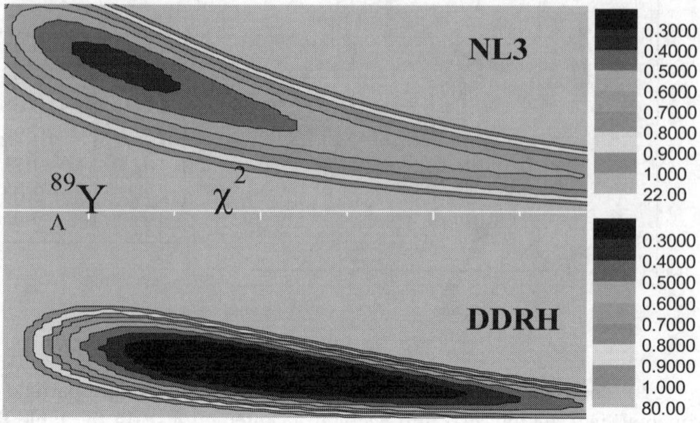

Figure 2. χ^2 plane of fitting the Λ meson coupling constants to match the experimental single particle spectrum.

be reproduced by DDRH RMF calculations. Taking into account also the high ground state spin of the ^{88}Y and ^{50}V cores with $I^{\pi} = 4^+, 6^+$, respectively, we performed a refit of the data. The broad single particle peaks were modeled by two gaussians of which the individual strength was constrained by the relative multiplicities of their s.o. states. The width of each gaussian and the relative distance of their means were fixed by assuming the spin dependent part of the single particle energies to be $E_s = E_{ls} \langle l \cdot s \rangle + E_{jj} \langle I \cdot j \rangle$ (I being the core spin and j the total angular momentum of the Λ state). The s.o. and jj matrix elements are the same for the whole fitting procedure. In contrast to the fit with uncorrelated gaussians our fit gets significantly more stable. A strong reduction of the s.o. strength compared to the simpler fit is observed being consistent with high precision measurements of the s.o. splitting in $^{13}_{\Lambda}C$ [6]. The s.o. and jj matrix elements are obtained as $E_{ls} = 223 \pm 153\ keV$ and $E_{jj} = 61 \pm 3\ keV$.

3.3 Hypernuclear structure calculation with DDRH

The fairly small s.o. splitting brings into the game the formerly already discussed $\Lambda\omega$ tensor interaction (see, e.g., [7]). This tensor interaction introduces terms proportional to $\overline{\psi}\sigma^{\mu\nu}\psi\partial_{\mu}\omega_{\nu}$ in the interaction Lagrangian being of the same Dirac structure as the s.o. interaction in the Dirac equation. According to spin-flavor SU(6) the tensor vertex "f" is minus that of the vector interaction "g", leading to an almost vanishing s.o. splitting.

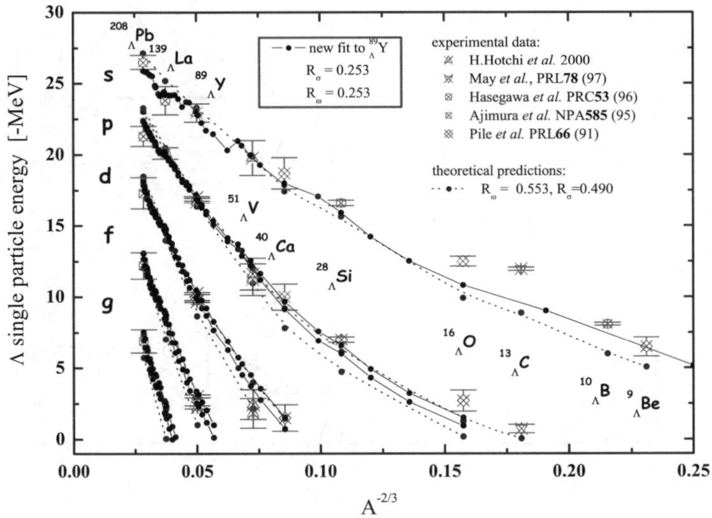

Figure 3. DDRH Λ single particle spectra compared to the world data.

Including the tensor interaction we performed a least squares fit of calculated DDRH RMF Λ single particle energies to the experimental ones with respect to the variables R_σ, R_ω and f/g (which should be $2/3$, $2/3$ and -1, respectively, according to SU(3)). A very strong linear correlation between R_σ and R_ω is found, induced by the spectral gross structure (known already from earlier hypernuclear RMF calculations not considering s.o. splitting [3]). The favored region with $\chi^2 < 1$ is shown in fig. 2. It is centered around $R_\sigma(=R_\omega) = 0.25$ and $f/g = 0$. This indicates a substantial violation of SU(3) symmetry. Effects which are still under investigation and might lead to minor changes are for example the influence of deformations of the nuclear core. Calculations for the world data set on hypernuclei with an even-even core is shown in fig. 3. The calculation with our new parametrization (dots linked by the solid line) shows clear improvements especially in the low mass region compared to our previous parameters.

Concerning model building the strong violation of SU(3) symmetry is bothering since SU(3) relations used in the derivation of, e.g., hyperon-hyperon interactions could no longer be applied. From the nuclear structure point of view the very weak Λ couplings are in contrast very encouraging. Since the strange sector seems to almost decouple from the isospin sector it might be possible in future high accuracy spectroscopic experiments to measure nuclear bulk properties like mass radii and deformations through Λ hypernuclear spectroscopy.

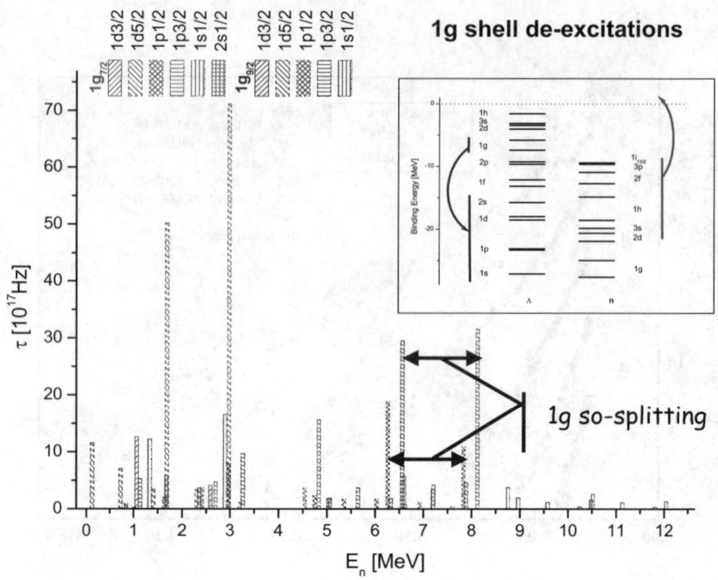

Figure 4. The partial life times of the 1g Λ shell in ${}^{209}_{\Lambda}$Pb due to Auger neutron emission are plotted versus the kinetic energy of the outgoing neutron. The insert shows the Λ and neutron level schemes for this transition where the involved states are marked.

3.4 Hypernuclear Auger effect

In preparation of a proposed experiment at JLab Auger rates for the capture of a Λ and the de-excitation of heavy hypernuclei are calculated within the DDRH model [8]. Different to a previous attempt [9] our approach treats the binding mean-field, wave functions and interactions in a self-consistent manner. The process is described by the production of a neutron-particle–neutron-hole state due to the de-excitation of the Λ, where the particle is unbound. A schematical illustration of the process for the 1g shell is shown in the insert of fig. 4. The details of the calculation will be given elsewhere. Fig. 4 shows the partial life times for the Λ transitions starting from the 1g shell plotted against the kinetic energy of the emitted neutron. The manifestation of the s.o. splitting is marked. Due to the complexity of the full spectrum it will be important to tag the Λ initial state in the measurements of Auger spectra. A further analysis of the global spectral features of the hypernuclear Auger effect is in progress.

4. ΛΛ HBT correlations

A method complementing the one presented in sec. 1.3.3 for achieving information on baryon-baryon interactions is the measurement of Hanbury-Brown-Twiss (HBT) two particle correlations in very high energetic collisions of heavy

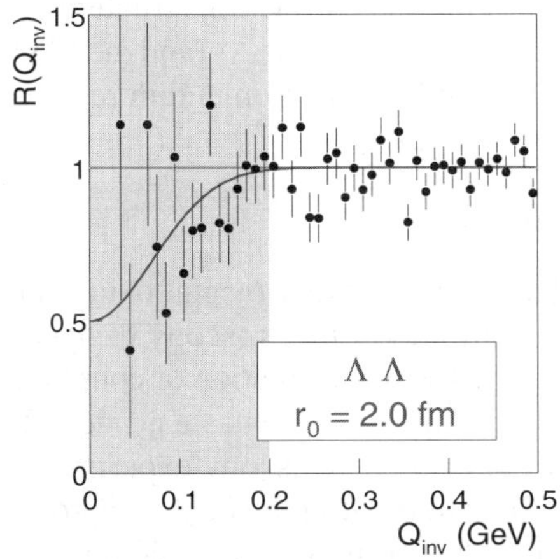

Figure 5. HBT correlations in the $\Lambda\Lambda$ channel measured by NA49 [12]. The shaded area marks the relevant relative momenta influenced by strong interaction between the Λs

ions. This was studied first for hyperons ($\Lambda\Lambda$) in [10]. It was shown on the example of a s-wave resonance in the $\Lambda\Lambda$ channel that interactions between the two particles modify quite significantly the standard correlations solely due to fermion quantum statistics, namely resulting in a negative correlation of -0.5 when the relative momentum q_{rel} approaches 0. The HBT correlation function R measures the ratio of the product of the two differential single particle cross sections to the differential two particle cross section, each normalized to the respective total cross section:

$$\frac{1}{\sigma_{12}} \frac{d^6\sigma_{12}}{d\vec{p}_1 d\vec{p}_2} = (1 + R(q_{rel})) \frac{1}{\sigma_1} \frac{d^3\sigma_1}{d\vec{p}_1} \frac{1}{\sigma_2} \frac{d^3\sigma_2}{d\vec{p}_2} \tag{1}$$

First measurements for $\Lambda\Lambda$ correlations have been performed by WA97 which, however, were of still rather poor statistics [11]. More encouraging preliminary results were obtained in a more recent analysis by NA49 [12], see fig. 5. The error bars, especially in the relevant region below 200 MeV relative momentum, are still too large due to insufficient statistics.

Within the same meson exchange model as described in the above section these correlations can be computed by solving the T-matrix equation $\mathcal{T}(q', q, s) = \mathcal{V}(q', q, s) + \int d^4k \mathcal{V}(q', k, s) G^{(2)}(k, s) \mathcal{T}(k, q, s)$ and then compared to data. In the case of sufficently high statistics this will give detailed information on the interactions of the two particles. More accurate data is likely to be taken at RHIC in the near future. However, there are still some “impurities” in the extraction of interaction informations in that way. First, the experimental spectra include secondary Λs from the electro-magnetic decay of initially produced Σ^0 hyperons which thus have to be accounted for in the calculated correlations.

They will somehow wash out the correlation structure of the pure $\Lambda\Lambda$ channel [10]. Secondly, the source function of the Λs (and the Σ^0s) needs to be modelled as this will influence the width of the momentum region in which correlations are pronounced.

5. Conclusions

The accuracy of the discussed experimental data is already at the limits of what can be expected from meson spectroscopy of hypernuclei. Still, looking at the large uncertainties in the determination of coupling constants from these data, see fig. 2, more precise measurements are needed. These will be provided by the beginning new era of γ spectroscopy experiments on hypernuclei. On the theory side it will be essential to replace the approximate treatment of the in-medium Λ interactions by a fully microscopic one. Work on such a Dirac-Brueckner G-matrix calculation is in progress.

Acknowledgments

This work was supported by the European Graduate School Giessen -Copenhagen, DFG contract Le 439/5 and GSI.

References

[1] C. Fuchs, H. Lenske and H. H. Wolter, Phys. Rev. C **52** (1995) 3043 [arXiv:nucl-th/9507044].

[2] F. Hofmann, C. M. Keil and H. Lenske, Phys. Rev. C **64** (2001) 034314 [arXiv:nucl-th/0007050].

[3] C. M. Keil, F. Hofmann and H. Lenske, Phys. Rev. C **61** (2000) 064309 [arXiv:nucl-th/9911014].

[4] F. Hofmann, C. M. Keil and H. Lenske, Phys. Rev. C **64** (2001) 025804 [arXiv:nucl-th/0008038].

[5] H. Hotchi *et al.*, Phys. Rev. C **64** (2001) 044302.

[6] H. Kohri *et al.* [AGS-E929 Collaboration], Phys. Rev. C **65** (2002) 034607 [arXiv:nucl-ex/0110007].

[7] J. Mares and B. K. Jennings, Phys. Rev. C **49** (1994) 2472.

[8] C. Keil and H. Lenske, Phys. Rev. C **66**, 054307 (2002) [arXiv:nucl-th/0207084].

[9] A. Likar, M. Rosina and B. Povh, Z. Phys. A **324**, 35 (1986).

[10] C. Greiner and B. Müller, Phys. Lett. B **219** (1989) 199.

[11] E. Andersen *et al.* [WA97 Collaboration], J. Phys. G **25** (1999) 423.

[12] C. Blume for NA49 at Quark Matter 2001, Brookhaven, USA;
http://na49info.cern.ch/na49/Archives/Images/Publications/QM2001:Blume/

NO PIONS IN CHIRAL NAMBU-JONA-LASINIO MODEL WITH LESS THAN THREE QUARK COLORS

H. Kleinert
Institut für Theoretische Physik,
Freie Universität Berlin, Arnimallee 14, 14195 Berlin, Germany,
E-mail: kleinert@physik.fu-berlin.de URL: http://www.physik.fu-berlin.de/~kleinert

Abstract We point out a severe problem in the presently popular use of the Nambu-Jona-Lasinio Model in nuclear physics to describe the chiral properties of quarks. Although the model does explain the spontaneous generation of quark masses and thus the difference between current and constituent quarks, it does not allow for Goldstone bosons to be identified with pions. This is prevented by violent fluctuations of the direction vector in the (σ, π) field space if the number of quark colors is less than 5. The phenomenon is first illustrated by a solvable version of the model in $2+\epsilon$ dimensions with a large number N_c of quarks which is exactly solvable to lowest order in ϵ and $1/N_c$.

1. Introduction

In the past four decades, the Nambu-Jona-Lasinio (NJL) model [1] of zero-mass fermions with quartic chirally-symmetric interaction has provided us with considerable insight into the mechanisms of spontaneous symmetry breakdown. By functional integral techniques it can formally be turned into a pure boson theory. In an SU(3)× SU(3)-symmetric version, the NJL model has been shown to be equivalent to a chirally SU(3)×SU(3) invariant σ-model which reproduces all well known relations of current algebra [2].

The NJL model is exactly solvable in the limit of an infinite number of quarks $N \to \infty$. For an attractive sign of the interaction, a *collective quantum field* [3] with two components representing the fermion-antifermion excitations $\sigma \propto \bar{\psi}\psi$, $\pi \propto i\bar{\psi}\gamma_5\psi$ acquires a nonzero vacuum expectation giving rise to a nonzero quark mass and thus to a spontaneous breakdown of the chiral symmetry of the action. Choosing the expectation value to point in the direction of σ, say $\Sigma = \langle\sigma\rangle$, the pion field is massless. It is the Goldstone particle of the

W. Greiner et al. (eds.), Structure and Dynamics of Elementary Matter, 593–611.

spontaneously broken chiral symmetry of the model. This scenario has so far been the favored model explanation for existence of low-mass pions in nature.

The purpose of this lecture is to point out that the actual number of quark colors $N_c = 3$ is too small to ensure the validity of this important $N \to \infty$ -property of the model. A factor N_c enters the next-neighbor coupling of the two-dimensional field vectors (σ, π) which for small N_c undergo phase transition to a disordered state just as a magnetic Heisenberg model above the Curie point. As a consequence, the model does *not* possess the desired zero-mass Goldstone bosons to be identified with the pions in nature.

To make the argument convincing, we shall first develop it in a simplified version of the model describing a large number N of massless fermions in $D = 2 + \epsilon$ dimensions. This model is known as chiral Gross-Neveu model, and its properties are known exactly to orders ϵ and $1/N$. After that it will be easy to transfer the conclusions to the NJL model.

The crucial observation will be that in $2 + \epsilon$ dimensions, the chiral Gross-Neveu Model exhibits *two* phase transitions, one in which pairs of fermions form bound pair states analogous to the Cooper pairs in a superconductor, and another one in which the paired phase becomes incoherent in a Kosterlitz-Thouless-like transition [4]. The latter transition occurs if the number of field components N drops below a certain critical value $N_c \lesssim 8$. The latter phase does not possess massless Goldstone bosons, and this situation is shared by the NJL model for only three quark colors.

2. Original Gross-Neveu Model

The original Gross-Neveu model has the following O(N)-symmetric Lagrange density

$$\mathcal{L} = \bar{\psi}_a i \partial\!\!\!/ \, \psi_a + \frac{g_0}{2N} \left(\bar{\psi}_a \psi_a \right)^2 . \tag{1}$$

where the index a runs from 1 to N. At the mean-field level, the effective action is equal to the initial action

$$\Gamma \left[\Psi, \bar{\Psi} \right] = \mathcal{A} \left[\Psi, \bar{\Psi} \right] = \int d^D x \left[\bar{\Psi} i \partial\!\!\!/ \, \Psi + \frac{g_0}{2N} \left(\bar{\Psi}_a \Psi_a \right)^2 \right] . \tag{2}$$

In general, we obtain all Green functions from the generating functional

$$Z[\eta, \bar{\eta}] = e^{iW[\eta, \bar{\eta}]} = \int \mathcal{D}\psi \mathcal{D}\bar{\psi} e^{i\mathcal{A}[\psi, \bar{\psi}] + i(\bar{\psi}\eta + \text{ç})}, \tag{3}$$

where $\eta(x)$ and $\bar{\eta}(x)$ are fermionic anticommuting sources. A collective quantum field [3] $\sigma \sim g\bar{\psi}\psi$ is introduced to rewrite (3) as

$$Z[\eta, \bar{\eta}] \;=\; \int \mathcal{D}\psi \mathcal{D}\bar{\psi} \mathcal{D}\sigma e^{i \int d^D x \left[\bar{\psi}_a (i\partial\!\!\!/ - \sigma) \psi_a + (\bar{\psi}\eta + \text{ç}) - N\sigma^2 / 2g_0 \right]} . \tag{4}$$

The fields $\psi(x)$ are integrated out according to the rule to yield a generating functional containing only the collective field $\sigma(x)$:

$$Z[\eta, \bar{\eta}] = \int \mathcal{D}\sigma e^{i\mathcal{A}_{\text{coll}}[\sigma] - \bar{\eta} G_\sigma \eta}, \tag{5}$$

with the collective action

$$\mathcal{A}_{\text{coll}}[\sigma] = N \left\{ -\frac{N}{2g_0}\sigma^2 - i\text{Tr}\log\left[i\partial\!\!\!/ - \sigma(x)\right] \right\}. \tag{6}$$

where Tr denotes the functional trace.

In the limit $N \to \infty$, the field σ is squeezed into the extremum of the action and we obtain the effective action

$$\frac{1}{N}\Gamma\left[\Sigma, \psi, \bar{\psi}\right] = -\frac{1}{2g_0}\Sigma^2(x) - i\text{Tr}\log\left[i\partial\!\!\!/ - \Sigma(x)\right] + \frac{1}{N}\Psi_a\left[i\partial\!\!\!/ - \Sigma(x)\right]\Psi_a. \tag{7}$$

The extremum of $\Gamma\left[\Sigma, \psi, \bar{\psi}\right]$ is given by the equations of motion,

$$\left[i\partial\!\!\!/ - \Sigma(x)\right]\Psi_a(x) = 0, \tag{8}$$

$$\Sigma(x) = g_0 \text{tr} G_\Sigma(x, x) - \frac{1}{N} g_0 \bar{\psi}_a \psi_a(x), \quad G_\Sigma(x, y) = \frac{i}{i\partial\!\!\!/ - \Sigma}, \tag{9}$$

where the trace symbol tr is restricted to the Dirac indices. The expectation Ψ_a of a fermionic field is always zero, so that we only must solve the *gap equation*

$$\Sigma(x) = g_0 \text{tr} G_\Sigma(x, x). \tag{10}$$

Thus, as far as the extremum is concerned, we may study only the purely collective part of the exact action

$$\frac{1}{N}\Gamma[\Sigma] = -\frac{1}{2g_0}\Sigma^2 - i\text{Tr}\log iG_\Sigma^{-1}. \tag{11}$$

The ground state is given by a constant gap field Σ_0, for which (10) yields either $\Sigma_0 = 0$ or

$$1 \;=\; g_0\,\text{tr}(1)\int \frac{d^D p_E}{(2\pi)^D}\frac{1}{p_E^2 + \Sigma_0^2}, \tag{12}$$

where we have performed a Wick rotation $p^0 \to ip^4$ to euclidean momenta $p_\mu \equiv (p^1, p^2, p^3, p^4)$ with the metric $p_E^2 = -p^2$. The Dirac matrices have dropped out, except for the unit matrix whose trace is $2^{D/2}$ for even D. This expression may be continued to any non-integer value of D.

For a constant Σ, the effective action gives rise to an effective potential

$$\frac{1}{N}v(\Sigma) = -\frac{1}{N}\Gamma[\Sigma] = \frac{1}{2g_0}\Sigma^2 - \mathrm{tr}(1)\frac{1}{2}\int \frac{d^D p_E}{(2\pi)^D} \log\left[p_E^2 + \Sigma^2\right]. \quad (13)$$

Performing the integral yields in $D = 2 + \epsilon$ dimensions with $\epsilon > 0$

$$\frac{1}{N}v(\Sigma) = \frac{\mu^\epsilon}{2}\left[\frac{\Sigma^2}{g_0\mu^\epsilon} - b_\epsilon \left(\frac{\Sigma}{\mu}\right)^{2+\epsilon} \mu^2\right], \quad (14)$$

where μ is an arbitrary mass scale, and the constant b_ϵ stands for

$$b_\epsilon = \frac{2}{D}2^{\epsilon/2} S_D \Gamma(D/2)\Gamma(1-D/2) = \frac{2}{D}\frac{1}{(2\pi)^{D/2}}\Gamma(1-D/2), \quad (15)$$

which has an ϵ-expansion $b_\epsilon \sim -\left[1 - (\epsilon/2)\log\left(2\pi e^{-\gamma}\right)\right]/\pi\epsilon + \mathcal{O}(\epsilon)$. A renormalized coupling constant g may be introduced by the equation

$$\frac{1}{g_0\mu^\epsilon} - b_\epsilon \equiv \frac{1}{g}, \quad (16)$$

so that

$$\frac{1}{N}v(\Sigma) = \frac{\mu^\epsilon}{2}\left\{\frac{\Sigma^2}{g} + b_\epsilon \Sigma^2 \left[1 - \left(\frac{\Sigma}{\mu}\right)^\epsilon\right]\right\}. \quad (17)$$

Extremizing this we obtain either $\Sigma_0 = 0$ or a nonzero Σ_0 solving the gap equation (12) in the form

$$1 - \frac{g^*}{g} = \frac{D}{2}\left(\frac{\Sigma_0}{\mu}\right)^\epsilon, \quad (18)$$

where $g^* = -1/b_\epsilon \approx \pi\epsilon$. A nontrivial solution of this is called *gap*. It specifies the mass which the fermions acquire from the attractive interactions, and will be denoted by M. The second derivative of $v(\Sigma)$ shows that the solutions $\Sigma_0 = 0$ and $\Sigma_0 \neq 0$ are stable for $g_0 > 0$ and $g_0 < 0$, respectively. Denoting the solution Σ_0 of (18) for $g = \infty$ by M_∞, we may write the g-dependence of the gap as

$$M(g) = M_\infty \left(1 - \frac{g^*}{g}\right)^{1/\epsilon}. \quad (19)$$

In terms of M, the effective potential (17) can be rewritten as

$$\frac{1}{N}v(\Sigma) = -\frac{1}{4\pi\epsilon}M^D\left[D\left(\frac{\Sigma}{M}\right)^2 - 2\left(\frac{\Sigma}{M}\right)^D\right]. \quad (20)$$

It has a minimum at $\Sigma = M$, where it yields the condensation energy $v(M) = -NM^D/4\pi$.

3. Correlation Functions of Pair Field

If N is no longer infinite, the pair field σ in the partition function (4) performs fluctuations around the extremal value $\Sigma_0 = M$. For large N, the correlation functions of $\sigma(x)$ can be extracted from the leading effective action (7) at $\Sigma_0 = M$. Setting $\Sigma(x) = M + \Sigma'(x)$, we expand

$$\delta^2\Gamma = -\frac{N}{2}\left[\frac{\Sigma'^2}{g_0} + i\,\mathrm{Tr}\left(\frac{i}{i\partial\!\!\!/ - M}\Sigma'\frac{i}{i\partial\!\!\!/ - M}\Sigma'\right)\right], \qquad (21)$$

implying a propagator of the σ'-field

$$G_{\sigma'\sigma'} = -\frac{1}{N}\frac{i}{1/g_0 + \Pi(q)}, \qquad (22)$$

where $\Pi(q)$ is given by the self-energy diagram

$$\Pi(q) \quad = \quad -2^{D/2}\int\frac{d^Dk_E}{(2\pi)^D}\frac{k_E(k-q)_E - M^2}{(k_E^2 + M^2)\left[(k-q)_E^2 + M^2\right]},$$

the integral being performed over euclidean D-dimensional energy-momenta. With standard Feynman methods, this can be transformed into a simple integral

$$\Pi(q) \quad = \quad -\frac{D(D-1)}{2}b_\epsilon M^\epsilon\int_0^{-1}dx\left[\frac{q_E^2}{M^2}x(1-x) + 1\right]^{\epsilon/2}. \qquad (23)$$

Inserting here Eq. (18), we obtain

$$\frac{1}{g_0} + \Pi(q) = \mu^\epsilon\left(\frac{1}{g^*} - \frac{1}{g}\right)$$
$$\times\left\{-1 + (D-1)\int_0^1 dx\left[\frac{q_E^2}{M^2}x(1-x) + 1\right]^{\epsilon/2}\right\}, \qquad (24)$$

which can be expanded for small q as

$$\frac{1}{g_0} + \Pi(q) = \epsilon\mu^\epsilon\left(\frac{1}{g^*} - \frac{1}{g}\right)\left[1 + (D-1)\frac{1}{12}\frac{q_E^2}{M^2} + \ldots\,.\right] \qquad (25)$$

This shows that the propagator (22) has a correlation length

$$M = \left(\frac{D-1}{12M^2}\right)^{1/2}. \qquad (26)$$

Inserting the g-dependence of M from (19), we see that

$$M = \frac{1}{M_\infty}\left(\frac{D-1}{12}\right)^{1/2}\left(1 - \frac{g^*}{g}\right)^{-1/\epsilon} \qquad (27)$$

so that the coherence length diverges for $g \to g^*$ with a critical exponent $\nu = 1/\epsilon$.

3.1 Chiral and Complex Pair Field Version of Model with Goldstone Bosons

We want to prove the existence of two phase transitions in the chiral version of the Gross-Neveu model, whose Lagrange density is

$$\mathcal{L} = \bar{\psi}_a i \partial\!\!\!/ \, \psi_a + \frac{g_0}{2N} \left[\left(\bar{\psi}_a \psi_a \right)^2 + \left(\bar{\psi}_a i \gamma_5 \psi_a \right)^2 \right]. \tag{28}$$

The collective field action (6) is then replaced by

$$\mathcal{A}_{\rm coll}[\sigma] = N \left\{ -\frac{N}{2g_0} (\sigma^2 + \pi^2) - i \mathrm{Tr} \log \left[i \partial\!\!\!/ - \sigma(x) - i \gamma_5 \pi \right] \right\}. \tag{29}$$

This model is invariant under the continuous set of chiral O(2) transformations which rotate σ and π fields into each other. This model is equivalent to yet another one which is closely related to the BCS model of superconductivity. Its Lagrange density is

$$\mathcal{L} = \bar{\psi}_a i \partial\!\!\!/ \, \psi_a + \frac{g_0}{2N} \left(\bar{\psi}_a C \bar{\psi}_a^T \right) \left(\psi_b^T C \psi_b \right). \tag{30}$$

Here C is the matrix of charge conjugation which is defined by

$$C \gamma^\mu C^{-1} = -\gamma^{\mu T}. \tag{31}$$

In two dimensions, we choose the γ-matrices as $\gamma^0 = \sigma^1$, $\gamma^1 = -i\sigma^2$, and $C = \gamma^1$. Note that $\left(\bar{\psi}_a C \bar{\psi}_a^T \right)^\dagger = \psi_a^T C \psi_a$, implying that $g_0 < 0$ corresponds to an attractive potential. The second model goes over into the first by replacing $\psi \to \frac{1}{2}(1-\gamma_5)\psi + \frac{1}{2}(1+\gamma_5) 2 C \bar{\psi}^T$, where superscript T denotes transposition. In the Lagrange density (30) we introduce a complex collective field by adding a term $(N/2g_0) \left| \Delta - \frac{g_0}{N} \psi_b^T C \psi_b \right|^2$, leading to the partition function

$$Z[\eta, \bar{\eta}] = \int \mathcal{D}\psi \mathcal{D}\bar{\psi} \mathcal{D}\Delta \exp \left\{ i \int d^D x \left[\bar{\psi}_a i \partial\!\!\!/ \, \psi_a \right.\right.$$
$$\left.\left. + \frac{1}{2} \left(\Delta^\dagger \psi_a^T C \psi_a^\dagger + \text{ç} \right) + \bar{\psi} \eta + \bar{\eta} \psi - \frac{N}{2g_0} |\Delta|^2 \right] \right\}. \tag{32}$$

The relation with the previous collective fields σ and π is $\Delta = \sigma + i\pi$. In order to integrate out the Fermi fields we rewrite the free part of the Lagrange density in the matrix form

$$\frac{1}{2} \left(\psi^T C, \bar{\psi} \right) \begin{pmatrix} 0 & i\partial\!\!\!/ \\ i\partial\!\!\!/ & 0 \end{pmatrix} \begin{pmatrix} \psi \\ C \bar{\psi}^T \end{pmatrix}, \tag{33}$$

which is the same as $\bar{\psi} i \partial\!\!\!/ \, \psi$, since $\psi^T C C \bar{\psi}^T = \bar{\psi}\psi$, $\psi^T C \overleftrightarrow{\partial\!\!\!/} C \bar{\psi}^T = \bar{\psi} \overleftrightarrow{\partial\!\!\!/} \psi$. But then the interaction with Δ can be combined with (33) in the form $\frac{1}{2} \phi_i^T G_\Delta^{-1} \phi$, where

$$\phi = \begin{pmatrix} \psi \\ C\bar{\psi}^T \end{pmatrix}, \quad \phi^T = \left(\psi^T, \bar{\psi}C^{-1}\right) \tag{34}$$

are doubled fermion fields, and

$$iG_\Delta^{-1} = \begin{pmatrix} C & 0 \\ 0 & C \end{pmatrix} \begin{pmatrix} \Delta & i\not{\partial} \\ i\not{\partial} & \Delta^\dagger \end{pmatrix} = -\left(iG_\Delta^{-1}\right)^T \tag{35}$$

is the inverse propagator in the presence of the external field Δ. Now we perform the functional integral over the fermion fields, and obtain

$$Z[j] = \int \mathcal{D}\Delta\mathcal{D}\Delta^\dagger e^{iN\mathcal{A}[\Delta]+\frac{1}{2}j_a^T G_\Delta j_a}, \tag{36}$$

where $\mathcal{A}[\Delta]$ is the collective action

$$\mathcal{A}[\Delta] = -\frac{1}{2}|\Delta|^2 - \frac{i}{2}\mathrm{Tr}\log iG_\Delta^{-1}, \tag{37}$$

and j_a is the doubled version of the external source

$$j = \begin{pmatrix} \bar{\eta}^T \\ C^{-1}\eta \end{pmatrix}. \tag{38}$$

This is chosen so that $\bar{\psi}\eta + \bar{\eta}\psi = \frac{1}{2}\left(j^T\phi - \phi^T j\right)$. In the limit $N \to \infty$, we obtain from (36) the effective action

$$\frac{1}{N}\Gamma[\Delta, \Psi] = \frac{1}{2g_0}|\Delta|^2 - \frac{i}{2}\mathrm{Tr}\log iG_\Delta^{-1} + \frac{1}{N}\bar{\Psi}_a iG_\Delta^{-1}\Psi_a \tag{39}$$

in the same way as in the last chapter for the simpler model with a real σ-field.

The ground state has $\Psi = 0$, so that the minimum of the effective action implies for Δ_0 either $\Delta_0 = 0$ or the gap equation

$$1 - \frac{g_0}{2}\mathrm{Tr}G_{\Delta_0}, \tag{40}$$

where we may assume Δ_0 to be real. With the Green function

$$G_{\Delta_0}(x, y) = \int \frac{d^D p}{(2\pi)^D} e^{-ip(x-y)} \frac{i}{p^2 - \Delta_0} \begin{pmatrix} \Delta_0 & \not{p} \\ \not{p} & -\Delta_0 \end{pmatrix} \begin{pmatrix} C^{-1} & 0 \\ 0 & C^{-1} \end{pmatrix}, \tag{41}$$

the gap equation (40) takes the same form as (12):

$$1 = g_0 \,\mathrm{tr}(1) \int \frac{d^D p}{(2\pi)^D} \frac{1}{p^2 + M^2}, \tag{42}$$

where we have again set $M \equiv \Delta_0$. The renormalization of the coupling constant and of the effective potential yields the same equation for $v(\Delta_0) = v(M)$ as before, so that the previous stability discussion for $g < g^*$ and $g > g^*$ holds also here.

Let us now study the propagator of the complete Δ-field. For small deviations $\Delta' \equiv \Delta - \Delta_0$ away from the ground state value we find from (39) the quadratic term

$$\frac{1}{N}\delta^2\Gamma = -\frac{1}{2}\left\{\frac{|\Delta|^2}{g_0} + \frac{i}{2}\mathrm{Tr}\left[\begin{pmatrix}\Delta'^\dagger & 0\\ 0 & \Delta'\end{pmatrix} G_M \begin{pmatrix}\Delta'^\dagger & 0\\ 0 & \Delta'\end{pmatrix} G_M\right]\right\}. \quad (43)$$

The second term in curly brackets may be written more explicitly as

$$\frac{i}{2}\left[M^2(\Delta'^2+\Delta'^{*2})2^{D/2}\int\frac{d^Dk}{(2\pi)^D}\frac{i}{k^2-M^2}\frac{i}{(k-q)^2-M^2}\right.$$
$$\left.+2|\Delta'|^2\int\frac{d^Dk}{(2\pi)^D}\frac{i}{k^2-M^2}\frac{i}{(k-q)^2-M^2}\mathrm{tr}[\not k(\not k-\not q)]\right],$$

and becomes

$$\frac{1}{2}\left\{M^2\left(\Delta'^2+\Delta'^{\dagger 2}\right)\tilde\Pi(q_E^2/M^2)+2|\Delta'|^2\left[\Pi(q_E^2/M^2)-M^2\tilde\Pi(q_E^2/M^2)\right]\right\},$$

where $\Pi(q_E^2/M^2)$ is the previous self-energy (22), and $\tilde\Pi\,(q_E^2/M^2)$ is the function

$$\begin{aligned}\tilde\Pi\left(q_E^2/M^2\right) &= i2^{D/2}\int\frac{d^Dk}{(2\pi)^D}\frac{i}{k^2-M^2}\frac{i}{(k-q)^2-M^2}\\ &= -\frac{D}{2}b_\epsilon(1-D/2)\int_0^1 d^Dx\left[q_E^2x(1-x)+M^2\right]^{D/2-2}. \quad (44)\end{aligned}$$

As a result, the action for the quadratic deviations from $\Delta_0 = M$ can be written as

$$\frac{1}{N}\delta^2\Gamma = -\frac{1}{2}\left[\left(\frac{1}{g_0}+A\right)|\Delta'|^2+\frac{1}{2}B\left(\Delta'^2+\Delta'^{*2}\right)\right], \quad (45)$$

with the coefficients

$$A = -\frac{D}{2}b_\epsilon M^\epsilon\left[(D-1)J_1^\epsilon\left(q_E^2/M^2\right)-(D/2-1)J_2^\epsilon(q_E^2/M^2)\right], \quad (46)$$
$$B = \frac{D}{2}b_\epsilon M^\epsilon(D/2-1)J_2^\epsilon\left(q_E^2/M^2\right), \quad (47)$$

and the integrals

$$J_1^\epsilon(z) = \int_0^1 d^Dx\,[zx(1-x)+1]^{D/2-1}, \quad (48)$$
$$J_2^\epsilon(z) = \int_0^1 d^Dx\,[zx(1-x)+1]^{D/2-2}. \quad (49)$$

Thus the propagators of real and imaginary parts of the field Δ' are

$$G_{\Delta'_{\rm re}\Delta'_{\rm re}} = -\frac{i}{N}\frac{1}{1/g_0 + A + B}, \quad G_{\Delta'_{\rm im}\Delta'_{\rm im}} = -\frac{i}{N}\frac{1}{1/g_0 + A - B}. \quad (50)$$

The excitation spectrum is given by the zeros of the denominator functions

$$\frac{1}{g_0}+A+B=\frac{D}{2}b_\epsilon M^\epsilon\left[1-(D-1)J_1^\epsilon\left(q_E^2/M^2\right)\right],$$
$$\frac{1}{g_0}+A-B=\frac{D}{2}b_\epsilon M^\epsilon\left\{\left[1-(D-1)J_1^\epsilon\left(q_E^2/M^2\right)\right]+(D-2)J_2^\epsilon\left(q_E^2/M^2\right)\right\}.$$

By expanding $J_1^\epsilon(z),\ J_2^\epsilon(z)$ in powers of $z = q_E^2/M^2 \approx 0$,

$$J_1^\epsilon(z) \sim 1 + \frac{D-2}{12}z + \mathcal{O}(z^2), \quad J_2^\epsilon(z) \sim 1 + \frac{D-4}{12}z + \mathcal{O}(z^2), \quad (51)$$

we find

$$\frac{1}{g_0} + A + B = -\frac{D}{2}b_\epsilon M^\epsilon (D-2)\left(1 + \frac{D-1}{12}z\right) + \mathcal{O}(z^2), \quad (52)$$
$$\frac{1}{g_0} + A - B = -\frac{D}{2}b_\epsilon M^\epsilon \frac{D-2}{12}3z + \mathcal{O}(z^2). \quad (53)$$

Inserting here the gap equation (19), we obtain

$$\frac{1}{g_0}+A+B = \epsilon\mu^\epsilon\left(\frac{1}{g^*} - \frac{1}{g}\right)\left(1 + \frac{D-1}{12}\frac{q_E^2}{M^2}\right) + \dots, \quad (54)$$
$$\frac{1}{g_0}+A-B = \epsilon\mu^\epsilon\left(\frac{1}{g^*} - \frac{1}{g}\right)\frac{1}{4}\frac{q_E^2}{M^2} + \dots. \quad (55)$$

Recalling (25) we see that the propagator of $\Delta'_{\rm re}$ coincides with that of Σ' in the standard Gross-Neveu model, so that the fluctuations of $\Delta'_{\rm re}$ have the same correlation length (26), with at a critical exponent ν in (27) as g approaches g^*.

In contrast, the propagator of the imaginary part of Δ' has now a pole at $q^2 = 0$:

$$G_{\Delta'_{\rm im}\Delta'_{\rm im}} = \frac{1}{N}\frac{4}{\epsilon}\left(\frac{1}{g^*} - \frac{1}{g}\right)^{-1} M^2\,\frac{i}{q^2} + \text{regular part at } q^2 = 0. \quad (56)$$

The sign of the pole term guarantees a positive norm of the corresponding particle state in the Hilbert space. The particle is a Nambu-Goldstone boson.

4. Second Phase transition

We are now prepared to show that the pair version of the chiral Gross-Neveu model in $2 + \epsilon$ dimensions has two phase transitions. Consider first the case

$\epsilon = 0$ where the collective field theory consists of complex field Δ with O(2)-symmetry $\Delta \to e^{i\phi}\Delta$. From the work of Kosterlitz and Thouless [4] we know that such a field system possesses macroscopic excitations of the form of vortices and antivortices. These attract each other by a logarithmic Coulomb potential, just like a gas of electrons and positrons in two dimensions. At low temperatures, the vortices and antivortices form bound pairs. The grand-canonical ensemble of pairs exhibits quasi-long-range correlations. At some temperature T_c, the vortex pairs break up, and the correlations becomes short-range. The phase transition is of infinite order.

This transition is most easily understood in a model field theory involving of a pure phase field $\theta(x)$, with a Lagrange density

$$\mathcal{L} = \frac{\beta}{2}[\partial\theta(x)]^2, \tag{57}$$

where β is the stiffness of the θ-fluctuations. The important feature of the phase field θ is that it is a cyclic field with $\theta = \theta + 2\pi$. In order to ensure that such jumps by 2π carry no energy, the gradient in the Lagrange density needs a modification which allows for the existence of vortices and antivortices. This will not be discussed here in detail — the reader may consult the ample literature on this subject [5, 6]. We only state here that after including vortices and antivortices at positions x_i, x_j, their partition function can be written as

$$Z = \sum_{\text{gas}} \exp\left\{ 4\pi^2\beta \sum_{i<j} q_i q_j \frac{1}{2\pi} \log(|x_i - x_j|/r_0) \right\}, \tag{58}$$

where r_0 is the size of the vortices. For a single vortex-antivortex pair, the average square distance r^2 is

$$< r^2 > \propto \int_{r_0}^{\infty} dr r\, r^2 e^{-2\pi\beta \log(r/r_0)} \propto \frac{1}{2\pi\beta - 4}. \tag{59}$$

This diverges as the stiffness falls below $\beta_{\text{KT}} = 2/\pi \approx 0.63662$. A more detailed study shows that this is an exact result for a very dilute system of vortices and antivortices [6].

The large-stiffness state with bound vortex pairs has a coherent phase field $\theta(x)$, the low-stiffness state with separated vortex pairs exhibits incoherent phase fluctuations. The same situation is found in three dimensions, only that the excitations are vortex lines. These become infinitely long and prolific in a second-order phase transition [6] at a critical point $\beta_c \approx 0.33$.

The result (56) for $\epsilon = 0$ can now be used to estimate a critical value of the number of field components $N = N_c$ below which the phase fluctuations of the complex field Δ' become so violent that the system has a phase transition. For

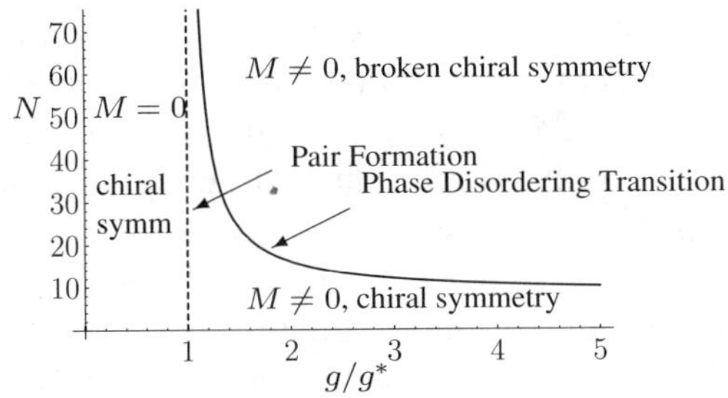

Figure 1. The two transition lines in the $N - g$-plane of the chiral Gross-Neveu model in $2+\epsilon$ dimensions. For $\epsilon = 0$, the vertical transition line coincides with the N-axis, and the solid hyperbola degenerates into a horizontal line at $N_c = 8$. The quark masses and chiral properties are indicated.

this we write $\Delta'_{\rm im} = M\theta$ and find from (56) a propagator of the θ-field

$$G_{\theta\theta} \approx \frac{i}{N}\frac{4\pi}{q^2} + \text{regular terms.} \tag{60}$$

Comparing this with the propagator for the model Lagrange density (57)

$$G_{\theta\theta} = \frac{1}{\beta}\frac{i}{q^2} \tag{61}$$

we identify the stiffness $\beta = N/4\pi$. The pair version of the chiral Gross-Neveu model has therefore a vortex-antivortex pair breaking transition if N falls below the critical value $N_c = 8$.

Consider now the model in $2+\epsilon$ dimensions where pairs form at $g = g^* \approx \pi\epsilon$. A comparison between the propagator (56) and (61) yields a stiffness of phase fluctuations

$$\beta = \frac{N}{4\pi}\left(1 - \frac{g^*}{g}\right). \tag{62}$$

The linear vanishing of the stiffness with the distance of the coupling constant g from the critical value g^* is in agreement with a general scaling relation, according to which the critical exponent of bending rigidity should be equal to $(D-2)\nu$. Since the model has $\nu = 1/\epsilon$ [see (27)], this yields $(D-2)\nu = 1$, which is precisely the exponent in Eq. (62).

The stiffness (62) implies the existence of a phase transition in the neighborhood of two and in three dimensions at roughly

$$N_c \approx 8\left(1 - \frac{g^*}{g}\right)^{-1}, \quad D \approx 2, \qquad N_c \approx 4.19\left(1 - \frac{g^*}{g}\right)^{-1}, \quad D = 3. \tag{63}$$

As N is lowered below these critical values, the phase fluctuations of the pair field Δ become incoherent and the pair condensate dissolves. The different phases are indicated in Fig. 1. In the chiral formulation of the same model, the intermediate phase has chiral symmetry in spite of a nonzero spontaneously generated "quark mass" $M \neq 0$. The reason why this is possible is that the "quark mass" depends only on $|\Delta_0|$, thus allowing for arbitrary phase fluctuations preserving chiral symmetry.

The sceptical reader may wonder whether the solid hyperbola in Fig. 1 is not simply the proper (albeit approximate) continuation of the vertical line for smaller N. There are two simple counterarguments. One is formal: For infinitesimal ϵ the first transition lies precisely at $g = g^* = \pi\epsilon$ for *all* N, so that the horizontal transition line is clearly distinguished from it. The other argument is physical. If N is lowered at some very large g, the binding energy of the pairs *increases with* $1/N$ {in two dimensions, the binding energy is $4M\sin^2[\pi/2(N-1)]$}. It is then impossible that the phase fluctuations on the horizontal branch of the transition line, which are low-energy excitations, unbind the strongly bound pairs. This will only happen in the limit $N \to \infty$ where the binding energy becomes zero and the two transition curves merge into a single curve. This is the situation in the BCS theory of superconductivity, where Cooper pair binding and pair condensation coincide.

In the ordinary Gross-Neveu model, the analog of the phase disordering transition is an Ising transition, in which the vacuum expectation value of σ jumps between $-\Sigma_0$ and Σ_0 in a disorderly fashion. In two dimensions, this occurs at some critical value N_c. In $2+\epsilon$ dimensions, this transition should again exist independently of the transition at which the system enters into a state of nonzero Σ_0. It would be interesting to see these two transitions in either model confirmed by Monte-Carlo simulations.

5. Transfer of Argument to NJL Model

What are the implication of these conclusions for the NJL model whose Lagrangian looks very similar to (28)? This question was answered in Ref. [7]. The model used in nuclear physics contains N_f quark fields $\psi(x)$, one for each flavor. Each of them may appear with N_c colors, such that the total number of quarks is $N = N_f \times N_c$. Thus the Lagrangian of the relevant part of the model is given by [8]

$$\mathcal{L} = \bar{\psi}\,(i\partial\!\!\!/ - m_0)\,\psi + \frac{g_0}{2N_c}\left[(\bar{\psi}\psi)^2 + (\bar{\psi}\lambda_a i\gamma_5\psi)^2\right], \tag{64}$$

where an implicit summation over the adjoint flavor labels $a = 1, 2, 3, \ldots$ is assumed. Since the fluctuation phenomenon to be discussed will be caused by the almost massless modes, we may restrict ourselves to the almost massless up and down quarks, so that a is an isovector label $a = 1, 2, 3$. A small diagonal quark mass matrix m_0 breaks slightly the $SU(2) \times SU(2)$ part of the chiral symmetry which lifts the mass of the pion to a small nonzero value which we shall ignore here.

At this point it must be said that this model is not really a well-defined quantum field theory in the continuum world in which it is supposed to be applied. Since it is not renormalizable, it requires a momentum space cutoff Λ in all Feynman diagrams involving the quark fields. Moreover, after having functionally transformed the system to σ and π fields, the calculation of corrections to the $N_c \to \infty$ limit requires the calculation of Feynman diagrams of these collective fields which are all divergent and require a further cutoff Λ_π. There exists no consistent procedure to go about this except by defining the NJL model on a lattice. But then the results will depend on the details of the lattice and calculations will be more complicated. Moreover, Lorentz invariance will be violated. Thus all results of the NJL model may only be considered as rough indications of what could happen in nature — no result is really reliable. It is within this rough description of nature that we formulate our no-pion conclusion.

In four dimensions, the critical stiffness of the directional fluctuations can no longer be deduced from the KT argument of a Coulomb gas. Instead, we may proceed as follows.

First we calculate as before the Goldstone propagator corresponding to (56), and find

$$G_\pi = \frac{(2\pi)^2}{N_c}\left[\ln\left(1+\frac{\Lambda^2}{M^2}\right) - \frac{\Lambda^2}{\Lambda^2+M^2}\right]^{-1} q_E^2 \equiv \frac{1}{Z_\pi^{(\Lambda)}}\frac{1}{q_E^2}. \tag{65}$$

It contains the above-discussed cutoff Λ in the quark momenta. The factor in front of q_E^2 is the usual wave function renormalization constants $Z_\pi^{(\Lambda)}$ of the pion field It determines the stiffness of the Goldstone field fluctuations:

$$\beta_\pi = Z_\pi^{(\Lambda)} = \frac{N_c}{(2\pi)^2}\left\{\ln\left[1+\left(\frac{\Lambda}{M}\right)^2\right] - \frac{\Lambda^2}{M^2+\Lambda^2}\right\}. \tag{66}$$

As a consequence of the spontaneous symmetry breakdown, the fluctuations of the pseudoscalar fields are massless. These fields appear in the x-space version of the Goldstone action the pure gradient form (57). Due to chiral symmetry, we can easily find a symmetric Lagrangian which has such gradients for its free Goldstone bosons:

$$\mathcal{A}_0[\pi] = \frac{\beta_\pi}{2}\int d^D x \left\{[\partial \pi_a(x)]^2\right\}, \tag{67}$$

where $\pi_a(x)$ is a two-dimensional vector $\pi_a \equiv M(\sigma, \pi) \equiv (\cos\theta, \sin\theta)$. The length of the field vector M is determined from the gap equation [compare

(42)], or equivalently from the minimum of the effective potential in the (σ, π)-plane. This is an O(2)-symmetric nonlinear σ-model. On a lattice, this model is well-known to have a phase transition. The action (67) has therefore the same form as (57) with a stiffness of phase fluctuations $\beta \equiv \beta_\pi M^2$. If this drops below a certain critical value β_c, the orientations of the unit vectors become uncorrelated. The same thing happens in a continuous theory with a momentum space cutoff in the π field, all it Λ_π. The critical value β_c is most easily estimated by extending artificially the O(2) unit vectors $u_a \equiv \pi_a/M$ to O(N_u) vectors and assuming N to be large. Thus we study the model

$$\mathcal{A}_0[\pi'] = \frac{\beta}{2} \int d^D x \left\{ [\partial n_a(x)]^2 \right\}, \tag{68}$$

which is the standard form of an O(N_n)-symmetric nonlinear σ-model. Although the limit of large N_n may be feared to yield a bad result for β_c for the small number $N_u = 2$, it is known to agree quite well with Monte Carlo simulations.

The estimate is derived as follows: We relax the unit vector constraint for the vectors n_a in (68) by introducing an additional field $\lambda(x)$ playing the role of a Lagrange multiplier. The n_a-fields can then be integrated out in the partition function, leading to an action

$$S = \frac{N_n}{2} \mathrm{Tr} \ln \left[-\partial^2 + \lambda(x) \right] - \beta \int d^D x \frac{\lambda(x)}{2}, \tag{69}$$

where Tr denotes the functional trace (the summation over the fields component has already been performed). For a large number N_n of components, the fluctuations are suppressed, and the field $\lambda(x)$ becomes a constant satisfying a second gap equation

$$\beta = N_n \int \frac{d^D k}{(2\pi)^D} \frac{1}{k^2 + \lambda}. \tag{70}$$

If there is a nonzero solution $\lambda \neq 0$, this will play the role of a square mass of the n_a-fluctuations, and represents an order parameter in the directional phase transition. The model has a phase transition at a critical stiffness

$$\beta_c = N_n \int \frac{d^D k}{(2\pi)^D} \frac{1}{k^2}. \tag{71}$$

For a smaller stiffness, the phase fluctuations are so violent that the system goes into a disordered phase with $\lambda \neq 0$ giving all fields n_a a nonzero square mass λ. Since the fields n_a are the normalized σ and π fields of the model, this determines an equal nonzero square mass of σ and π_i mesons, and thus a restoration of chiral symmetry.

Note that the quarks are still massive: their constituent mass is a consequence of the *formation* of the pairs, which are strongly bound for small N_c. The phase transition taking place at the critical value of the stiffness, on the other hand, is related to the Bose-Einstein condensation of the pairs. At small N_c, the two processes are widely separated. This separation of the two transitions (pair formation and pair condensation) can be judged by the simple fluctuation criterion in Ref. [9].

In our model, the number N_n is equal to four, which is not very large. Fortunately, Monte Carlo studies of the model [10–12] have shown that $N_n = 4$ is large enough to ensure the existence of the transition and the quantitative reliability of the theoretical estimate of the critical stiffness (71). From an evaluation of (71) *on a lattice*, and a comparison with Monte Carlo studies, we estimate that the critical stiffness obtained from (71) is correct to within less than 2% [11, 13] or 6% [10, 12, 13]. The same maximal error is expected if we work in the continuum using a momentum cutoff scheme.

For $N_n = 4$ and a cutoff Λ_π in the integral (71) over pion momenta, the critical stiffness is given by

$$\beta_c = \frac{4}{16\pi^2}\Lambda_\pi^2. \tag{72}$$

By comparing this with the stiffness of the model in (66), we find

$$N_c = \left(\frac{\Lambda_\pi}{\Lambda}\right)^2 \left(\frac{\Lambda}{M}\right)^2 \left\{ \ln\left[1 + \left(\frac{\Lambda}{M}\right)^2\right] - \frac{(\Lambda/M)^2}{1+(\Lambda/M)^2} \right\}^{-1}. \tag{73}$$

This equation determines the number N_c of identical quarks which is necessary to produce a large enough stiffness β to prevent the restoration of chiral symmetry. Only if the number of colors exceeds this critical value, will the model possess a phase in which the pion is a massless Goldstone boson, and σ a meson with a mass twice as large as that of the constituent quarks. The critical number (73) is plotted as the solid curve in Fig. 2 for $\Lambda_\pi = \Lambda$. We see that $N_c = 5$ would be the smallest allowed value. This number, however, is incompatible with color SU(3). This suggests that the Nambu–Jona-Lasinio model always remains in the symmetric phase, due to chiral fluctuations. It can therefore not be used to describe the chiral symmetry breakdown of quark physics, as has been claimed by many publications, which have appeared in particular in nuclear physics [8].

The choice $\Lambda_\pi = \Lambda$ would be natural in the only consistently defined versions of the NJL model which must live on a lattice. In many papers which use a continuum formulation of the model, which is necessarily inconsistent, different heuristic cutoffs Λ_π and Λ have been used to fit different data, but $\Lambda_\pi i$ cannot be made too small. The phase with broken symmetry for three colors would be reached only if the quark loop integration is cut off at $\Lambda^2 \gtrsim 11M^2$. Such a large value, however, is incompatible with the experimental value of the pion decay constant $f_\pi \approx 0.093$ which is given, in the large-N_c limit of the

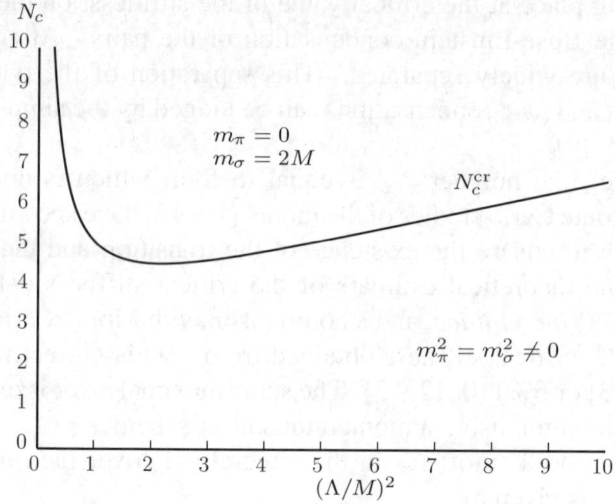

Figure 2. Solid curve shows critical number of colors $N_c^{\rm cr}$ as a function of Λ/M for $\Lambda_\pi = \Lambda$, which corresponds to the only consistent definition of the NJL model on a lattice.

model, by $f_\pi/M = Z_\pi^{1/2}$. For typical estimates of constituent quark masses $M \in (300, 400)$ MeV [2], we find that Λ^2/M^2 should lie in the range $(3.3, 7.3)$, the highest value corresponding to the lowest possible mass 300 MeV.

We must emphasize that our conclusions cannot be reached in the dimensional regularization scheme since, as explained at the end of Section 3, the integral in (71) determining the critical stiffness vanishes. due to the special property of dimensional regularization that all integrals over pure momentum powers vanish identically: $\int d^Dk\,(k)^\alpha = 0$ (Veltman's rule). Thus, arbitrary pure powers of the cutoff $\Lambda^{\alpha+D}$ have no counterpart in dimensional regularization. Only logarithmic divergences can be related to diverging pole terms $1/\epsilon \to \infty$ for $\epsilon \to 0$. It is therefore inconsistent to relate ϵ to Λ by setting $\Gamma(\epsilon/2-1) \approx \Lambda^2/M^2$, as proposed by Krewald and Nakayama [14]. Only the logarithmic divergences can be mapped to the small-ϵ divergences, relating $\Gamma(\epsilon/2-1) \approx -\ln(1+\Lambda^2/\rho_0^2)$. With their inconsistent identification, Krewald and Nakayama matched Λ by an $\epsilon > 2$ which lies in the wrong region $D < 2$, the physically relevant range being $D \in (2+\epsilon, 4-\epsilon)$. Note that the matching of the logarithm at the level of the effective potential leads to the properly matched gap equation, thus having circumvented the unphysical properties of the analytic regularization. The free use of this scheme in renormalizable field theories relies on the fact that all infinities are eventually absorbed in unobservable bare quantities, such that the artificial zeros of the integrals over pure powers of momenta cannot produce problems. In nonrenormalizable theories, on the other hand, only a cutoff (or a related Pauli-Villars regularization) is physical, and analytic regularization must be treated with caution. This is seen even more dramatically in integrals which do not have logarithmic infinities. For example the condensation energy of the model in $D = 3$ dimensions would be a finite negative number in analytic regularization, while being a linearly divergent positive function of the cutoff.

This is why dimensional regularization is inapplicable in the NJL model.

Our conclusions were derived from a study of only the σ, π collective fields. The inclusion of other flavors does not help preventing the restoration since the associated pseudoscalar mesons are quite massive, making their fluctuations irrelevant to the described phenomenon.

6. Conclusion

We have shown that within a certain nonperturbative approximation, the Nambu–Jona-Lasinio model does not really display the spontaneous symmetry breakdown for whose illustration it was constructed. The fluctuations of σ- and π-fields restore chiral symmetry and make σ and π equally massive. If our conclusion survives more refined approximations, this would invalidate a large number of publications, especially in nuclear physics, which have been based on the existence of a symmetry-broken ground state of the model. In particular, all studies of the temperature dependence of the symmetry-broken state [8] would deal with nonexisting objects, thus calling for further investigations. Finally, we note that our no-go result for the Nambu–Jona-Lasinio model does not imply problems with the effective-action approach to chiral dynamics. Certainly, there exists an effective chiral action for the meson sector of quantum chromodynamics which does contain almost massless pions for $N_c = 3$. It is only the Nambu–Jona-Lasinio model as it stands which is incapable of describing these for such a low number of colors. In fact, a recent paper [15] prompted by a first version of our preprint on the Los Alamos server points out that an extension of the Nambu–Jona-Lasinio model by interactions involving higher-dimensional operators is not subject to our no-go theorem. Another escape is possible by adding gradient and quartic interaction terms for σ- and π-fields to the initial action, thus extending the Nambu–Jona-Lasinio model to a linear sigma model.

Let us end by remarking that a criticism of the no-pion result for three quark colors was published in Ref. [16] which, however, misses completely the point. The author of that article claims that the conclusion is unreliable since it contains a momentum cutoff, or a lattice spacing, and is not ünuniversal", in contrast to the two-dimensional result. This, however, is a wrong argument since the NJL model for which the conclusion is derived can also be defined only via a cutoff or on a lattice. As a consequence, also the no-pion result must, of course, depend on this cutoff and cannot be universal. The *position* of a phase transition is never universal, except accidentally for the two-dimensional Kosterlitz-

Thouless transition in the zero-fugacity limit. In general, universal properties have been established only for certain properties of the immediate neighborhood of second-order transitions. Thus universal power laws in $N_c - N_c^{\rm cr}$ would be observable if an experimentalist could vary N_c continuously in the neighborhood of $N_c^{\rm cr}$. which is obviously impossible.

References

[1] Y. Nambu and G. Jona Lasinio, Phys. Rev. **122**, 345 (1961); **124**, 246 (1961).

[2] H. Kleinert, *On the Hadronization of Quark Theories*, Lectures presented at the Erice Summer Institute 1976, in *Understanding the Fundamental Constituents of Matter*, Plenum Press, New York, 1978, A. Zichichi ed., pp. 289-390. (http://www.physik.fu-berlin.de/˜kleinert/kleiner_re53/preprintg.html)

[3] H. Kleinert, *Collective Quantum Fields,* Lectures presented at the First Erice Summer School on Low-Temperature Physics, 1977, in Fortschr. Physik **26**, 565-671 (1978) (http://www.physik.fu-berlin.de/˜kleinert/55)

[4] V. L. Berezinskii, Zh. Eksp. Teor. Fiz., 1970, vol. 59, No 3, p.907 - 920;
J. Kosterlitz, D. Thouless. J. Phys., 1973, vol. C6, No 7, p.1181 - 1203.

[5] H. Kleinert, *Theory of Fluctuating Nonholonomic Fields and Applications: Statistical Mechanics of Vortices and Defects and New Physical Laws in Spaces with Curvature and Torsion*, in: *Proceedings of a NATO Advanced Study Institute on Formation and Interactions of Topological Defects* at the University of Cambridge, England (cond-mat/9503030) (http://www.physik.fu-berlin.de/˜kleinert/kleiner_re227/preprint.html)

[6] H. Kleinert, *Gauge Fields in Condensed Matter*, World Scientific, 1989 (http://www.physik.fu-berlin.de/˜kleinert/kleiner_re.html#b1).

[7] H. Kleinert and B. Van den Bossche, *No Spontaneous Breakdown of Chiral Symmetry in Nambu-Jona-Lasinio Model*
Phys. Lett. B **474** , 336 (2000) (hep-ph/9907274); *No Massless Pions in Nambu-Jona-Lasinio Model due to Chiral Fluctuations*, (hep-ph/9908284) °

[8] M. Lutz, S. Klimt, and W. Weise, Nucl. Phys. **A 542**, 521 (1992).
S. P. Klevansky, Rev. Mod. Phys. **64**, 649 (1992).
T. Hatsuda and T. Kunihiro, Phys. Rep. **247**, 221 (1994).
B. Van den Bossche, *A chiral Lagrangian with three flavors, axial and scale anomalies* (nucl-th/9807010).

[9] H. Kleinert, Phys. Rev. Lett. **84**, 286 (2000) cond-mat/9908239).

[10] A. Guha and S.-C. Lee, Nucl. Phys. B **240**, 141 (1984).

[11] A. Hasenfratz, K. Jansen, J. Jersak, H.A. Kastrup, C.B. Lang, H. Leutwyler, T. Neuhaus, Nucl. Phys. B **356** 332 (1991)

[12] A.R. Hoch and R.R. Horgan, Nucl. Phys. B **380**, 337 (1992).

[13] Simulations of the four-dimensional 0(4) model on a simple-cubic lattice gives $\beta^{\rm cr} \approx 0.6090$ [11]. This is to be compared with the approximation (71) calculated for a simple-cubic lattice, where $\beta^{\rm cr} = 4 \times 0.1549 \approx 0.6196$ which is thus correct to within less than 2%. (For references [10, 12], we find $\beta^{\rm cr} \approx 0.584$ (6%) and $\beta^{\rm cr} \approx 0.64$ (3%), respectively.) The value 0.1549 is taken from the list of lattice Coulomb potentials at the origin in Table 6.4 of the textbook [6] on p. 178.

[14] S. Krewald and K. Nakayama, Ann. Phys. **216**, 201 (1991). The developements made in this reference are correct. However, the application section 5 is wrong.

[15] T. Lee and Y. Oh, *Higher Dimension Operators and Chiral Symmetry Breaking in Nambu–Jona-Lasinio model*, nucl-th/9909078.

[16] E. Babaev, *Nonlinear sigma model approach for phase disorder transitions and the pseudogap phase in chiral Gross-Neveu, Nambu-Jona-Lasinio models and strong-coupling superconductors*, Int. J. Mod. Phys. A **16**, 1175 (2001) (hep-th/9909052)

SELF-CONSISTENT DESCRIPTION OF STRONGLY INTERACTING MATTER AT HIGH TEMPERATURES

J. Ruppert
Institut für Theoretische Physik der J. W. Goethe Universität, D-60054, Frankfurt am Main, Germany, email: ruppert@th.physik.uni-frankfurt.de

D. Röder
Institut für Theoretische Physik der J. W. Goethe Universität, D-60054, Frankfurt am Main, Germany, email: roeder@th.physik.uni-frankfurt.de

Abstract Chiral symmetry restoration at nonzero temperature is studied in the framework of the $O(4)$ linear sigma model and the $U(N_f)_r \times U(N_f)_\ell$ linear sigma model with $N_f = 2$, 3 and 4 quark flavors. We investigate the temperature dependence of the masses of the scalar and pseudoscalar mesons within the Hartree approximation as derived from the Cornwall-Jackiw-Tomboulis formalism. We find that the masses of the non-strange and strange mesons at nonzero temperature depend sensitively on the number of light quark flavors N_f. On the other hand, due to the large charm quark mass, neither do charmed mesons significantly affect the properties of the other mesons, nor do their masses change appreciably in the temperature range around the chiral symmetry restoration temperature.

1. Introduction

For N_f massless quark flavors, the QCD Lagrangian has a chiral $U(N_f)_r \times U(N_f)_\ell = SU(N_f)_r \times SU(N_f)_\ell \times U(1)_V \times U(1)_A$ symmetry. Here, $V = r + \ell$, while $A = r - \ell$. The $U(1)_V$ symmetry corresponds to baryon number conservation. It is always respected and thus plays no role in the symmetry breaking patterns considered in the following. In the vacuum, a non-vanishing expectation value of the quark condensate, $\langle \bar{q}_\ell \, q_r \rangle \neq 0$, spontaneously breaks the above symmetry to $SU(N_f)_V$. This gives rise to N_f^2 Goldstone bosons which dominate the low-energy dynamics of the theory. As shown by 't Hooft [1], instantons break the $U(1)_A$ symmetry explicitly to $Z(N_f)_A$ [2]. (For the low-energy dynamics of QCD, however, this discrete symmetry is irrelevant.)

W. Greiner et al. (eds.), Structure and Dynamics of Elementary Matter, 613–620.

Consequently, one of the N_f^2 Goldstone bosons becomes massive, leaving $N_f^2 - 1$ Goldstone bosons. The $SU(N_f)_r \times SU(N_f)_\ell \times U(1)_A$ symmetry of the QCD Lagrangian is also explicitly broken by nonzero quark masses. The $N_f^2 - 1$ low-energy degrees of freedom then become pseudo-Goldstone bosons. For nonzero quark masses, the chiral symmetry of QCD is explicitly broken.

As an alternative to lattice QCD calculations, one can also use linear sigma models to investigate the chiral symmetry restoring phase transition [3]. Studying the linear sigma model at nonzero temperature, however, requires many-body resummation schemes, because infrared divergences cause naive perturbation theory to break down [4]. In these resummation schemes, one necessarily has to make approximations by selecting certain subsets of diagrams. The most commonly and also in this work employed scheme is the Hartree approximation.

In the present work, we follow this line of arguments and investigate the change of meson masses and quark condensates with temperature in the framework of $U(N_f)_r \times U(N_f)_\ell$ linear sigma models. We focus in particular on the question how the number of quark flavors N_f changes the temperature dependence of the meson masses and the quark condensates.

In the present work we discuss the results obtained in the case of non-vanishing quark masses (explicit chiral symmetry breaking) in the presence of the $U(1)_A$-anomaly. For a detailed discussion of the results obtained in other possible symmetry breaking patterns with and without $U(1)_A$-anomaly and with explicit chiral symmetry breaking or in the chiral limit we refer the reader to the more extensive study [3].

Our calculations are done in the Hartree approximation which we derive within the Cornwall-Jackiw-Tomboulis (CJT) formalism [6]. The Hartree approximation has the advantage that the meson self-energies become independent of momentum and energy and one only has to solve gap equations for the meson masses as a function of temperature.

The remainder of this paper is organized as follows. In Sec. 2, we apply the CJT formalism *pars pro toto* to the $O(4)$ model to obtain the Hartree-Approximation. For the detailed - and somewhat lengthy formulas - in the $U(N_f)_r \times U(N_f)_\ell$ linear sigma models for $N_f = 2, 3$, and 4 we refer the reader to [3]. In Sec. 3, we discuss the resulting temperature dependence of the masses. We use the imaginary-time formalism to compute quantities at nonzero temperature. Our notation is

$$\int_k f(k) \equiv T \sum_{n=-\infty}^{\infty} \int \frac{d^3\mathbf{k}}{(2\pi)^3} f(2\pi i n T, \mathbf{k}) \quad . \tag{1}$$

Our units are $\hbar = c = k_B = 1$. The metric tensor is $g^{\mu\nu} = \mathrm{diag}(+,-,-,-)$.

2. Linear Sigma Models in the Hartree Approximation

In this subsection, we apply the CJT formalism to the $O(N)$ model. Our numerical results presented in Sec. 3 are exclusively for the case $N = 4$. As already discussed in the introduction we refer the reader to [3] for the detailed - and somewhat lengthy formulas - in the $U(N_f)_r \times U(N_f)_\ell$ linear sigma models for $N_f = 2, 3$, and 4 flavors. Here we concentrate exclusively on the simplest linear sigma model, namely the $O(N)$-model.

The CJT effective potential for the $O(N)$ model is [3]

$$\begin{aligned} V[\bar{\sigma}, \bar{S}, \bar{P}] &= U(\bar{\sigma}) + \frac{1}{2} \int_k \left[\ln \bar{S}^{-1}(k) + S^{-1}(k; \bar{\sigma})\, \bar{S}(k) - 1 \right] \\ &+ \frac{N-1}{2} \int_k \left[\ln \bar{P}^{-1}(k) + P^{-1}(k; \bar{\sigma})\, \bar{P}(k) - 1 \right] \\ &+ V_2[\bar{\sigma}, \bar{S}, \bar{P}] \,, \end{aligned} \tag{2}$$

where $\bar{\sigma}$ is the expectation value of the scalar field (in the presence of external sources). Because the vacuum of QCD has even parity, the expectation values of the pseudoscalar fields can be set to zero. The quantities $\bar{S}$ and $\bar{P}$ are the full propagators for scalar and pseudoscalar particles, while S^{-1} and P^{-1} are the corresponding inverse tree-level propagators,

$$S^{-1}(k; \bar{\sigma}) = -k^2 + m_\sigma^2(\bar{\sigma}) \,, \tag{3a}$$

$$P^{-1}(k; \bar{\sigma}) = -k^2 + m_\pi^2(\bar{\sigma}) \,, \tag{3b}$$

where the tree-level masses are

$$m_\sigma^2(\bar{\sigma}) = \mu^2 + \frac{12\,\lambda}{N}\, \bar{\sigma}^2 \,, \tag{4a}$$

$$m_\pi^2(\bar{\sigma}) = \mu^2 + \frac{4\,\lambda}{N}\, \bar{\sigma}^2 \,. \tag{4b}$$

The constant μ^2 is the bare mass term in the Lagrangian of the $O(N)$ model, while λ is the four-point coupling constant. For $\mu^2 < 0$, the $O(N)$ symmetry is spontaneously broken to $O(N-1)$, leading to $N-1$ Goldstone bosons. The tree-level potential is

$$U(\bar{\sigma}) = \frac{1}{2}\, \mu^2\, \bar{\sigma}^2 + \frac{\lambda}{N}\, \bar{\sigma}^4 - H\, \bar{\sigma} \,, \tag{5}$$

where H is a term which breaks the $O(N)$ symmetry explicitly to $O(N-1)$. $V_2[\bar{\sigma}, \bar{S}, \bar{P}]$ is the sum of all two-particle irreducible diagrams. In the following, we restrict ourselves to the Hartree approximation, i.e., we take into account only the double-bubble diagrams. These diagrams have no explicit $\bar{\sigma}$ dependence. Then, only tadpole diagrams (with resummed propagators) contribute to the self-energies, cf. Fig. 1.

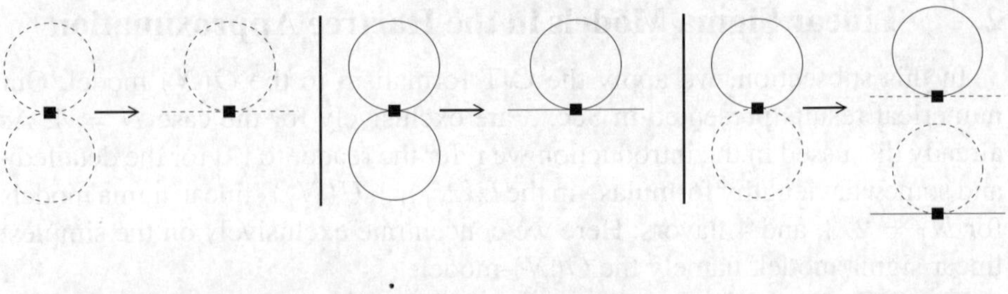

Figure 1. The double-bubble diagrams. Full lines are scalar particles, dashed lines are pseudoscalar particles.The picture shows different types of tadpole contributions to the self-energies obtained by cutting a line in the corresponding double-bubble diagrams.

In the Hartree approximation,

$$
\begin{aligned}
V_2[\bar{S},\bar{P}] &= 3\,\frac{\lambda}{N}\left[\int_k \bar{S}(k)\right]^2 + (N+1)(N-1)\,\frac{\lambda}{N}\left[\int_k \bar{P}(k)\right]^2 \\
&+ 2\,(N-1)\frac{\lambda}{N}\left[\int_k \bar{S}(k)\right]\left[\int_p \bar{P}(p)\right] . \qquad (6)
\end{aligned}
$$

The CJT stationarity conditions

$$\left.\frac{\delta V[\bar{\phi},\bar{S},\bar{P}]}{\delta\bar{\phi}}\right|_{\bar{\phi}=\phi,\ \bar{S}=\mathcal{S}\ \bar{P}=\mathcal{P}} = 0\ , \qquad (7a)$$

$$\left.\frac{\delta V[\bar{\phi},\bar{S},\bar{P}]}{\delta\bar{S}(k)}\right|_{\bar{\phi}=\phi,\ \bar{S}=\mathcal{S}\ \bar{P}=\mathcal{P}} = 0 \qquad (7b)$$

$$\left.\frac{\delta V[\bar{\phi},\bar{S},\bar{P}]}{\delta\bar{P}(k)}\right|_{\bar{\phi}=\phi,\ \bar{S}=\mathcal{S}\ \bar{P}=\mathcal{P}} = 0, \qquad (7c)$$

read

$$
\begin{aligned}
H &= \mu^2\,\sigma + \frac{4\lambda}{N}\,\sigma^3 + \frac{4\lambda}{N}\,\sigma\int_q\left[3\,\mathcal{S}(q) + (N-1)\,\mathcal{P}(q)\right]\ , && (8a)\\
\mathcal{S}^{-1}(k) &= -k^2 + M_\sigma^2\ , && (8b)\\
\mathcal{P}^{-1}(k) &= -k^2 + M_\pi^2\ , && (8c)
\end{aligned}
$$

where M_σ and M_π are the σ and pion masses dressed by the tadpole contributions,

$$
\begin{aligned}
M_\sigma^2 &= m_\sigma^2(\sigma) + \frac{4\lambda}{N}\left[3\int_q \mathcal{S}(q) + (N-1)\int_q \mathcal{P}(q)\right], && (9a)\\
M_\pi^2 &= m_\pi^2(\sigma) + \frac{4\lambda}{N}\left[\int_q \mathcal{S}(q) + (N+1)\int_q \mathcal{P}(q)\right]\ . && (9b)
\end{aligned}
$$

Using Eqs. (9a) and (9b), Eq. (8a) can be written in the compact form:

$$H = \sigma \left[M_\sigma^2 - \frac{8\lambda}{N} \sigma^2 \right] . \tag{9c}$$

Equations (9a), (9b), and (9c) are the stationarity conditions of the $O(N)$ model in the Hartree approximation.

In principle, the calculation of the tadpole integrals $\int_q \mathcal{S}(q)$ and $\int_q \mathcal{P}(q)$ requires renormalization. Renormalization of many-body approximation schemes is nontrivial, but does not change the results qualitatively [5]. We therefore simply omit the vacuum contributions to the loop integrals,

$$\int_q \mathcal{S}(q) = \int \frac{d^3\mathbf{q}}{(2\pi)^3} \frac{1}{\epsilon_\mathbf{q}[M_\sigma^2]} \left(\exp\left\{ \frac{\epsilon_\mathbf{q}[M_\sigma^2]}{T} \right\} - 1 \right)^{-1} , \tag{10a}$$

$$\int_q \mathcal{P}(q) = \int \frac{d^3\mathbf{q}}{(2\pi)^3} \frac{1}{\epsilon_\mathbf{q}[M_\pi^2]} \left(\exp\left\{ \frac{\epsilon_\mathbf{q}[M_\pi^2]}{T} \right\} - 1 \right)^{-1} . \tag{10b}$$

Here, $\epsilon_\mathbf{q}[M^2] = \sqrt{\mathbf{q}^2 + M^2}$ is the relativistic energy of a quasiparticle with mass M and momentum $\mathbf{q}$.

3. Results: Chiral Symmetry Restoration in Linear Sigma Models

In Fig. 2, we show the masses of the mesons as a function of temperature for explicit chiral symmetry breaking, including the $U(1)_A$ anomaly. This is the case where chiral symmetry breaking results in the smallest residual symmetry group, $SU(N_f)_V \times U(N_f)_A \to SU(2)_V$.

In Fig. 2(a) the masses of the σ meson and the pions are shown for all models. The σ meson and the pion become degenerate in mass in the chirally restored phase. Comparing the σ meson and pion masses in the $O(4)$ model with those in the $U(2)_r \times U(2)_\ell$ model, the difference is almost negligible up to temperatures of 150 MeV. In the chirally restored phase, the masses behave linearly with temperature, but grow faster in the $U(2)_r \times U(2)_\ell$ model than in the $O(4)$ model. The reason is that there are twice as many fields in the former model than in the latter, which results in twice as many tadpole-like contributions in the equations for the in-medium masses. These come with a positive sign and thus increase the masses.

Comparing the results of the $U(3)_r \times U(3)_\ell$ model with those of the $U(2)_r \times U(2)_\ell$ model, one observes differences already at a temperature of about 100 MeV. Furthermore, the masses become even larger in the chirally restored phase. The reason for this behavior are the strange degrees of freedom in the $N_f = 3$ case which lead to additional tadpole-like terms in the self-energies. As above, they lead to an increase of the in-medium masses.

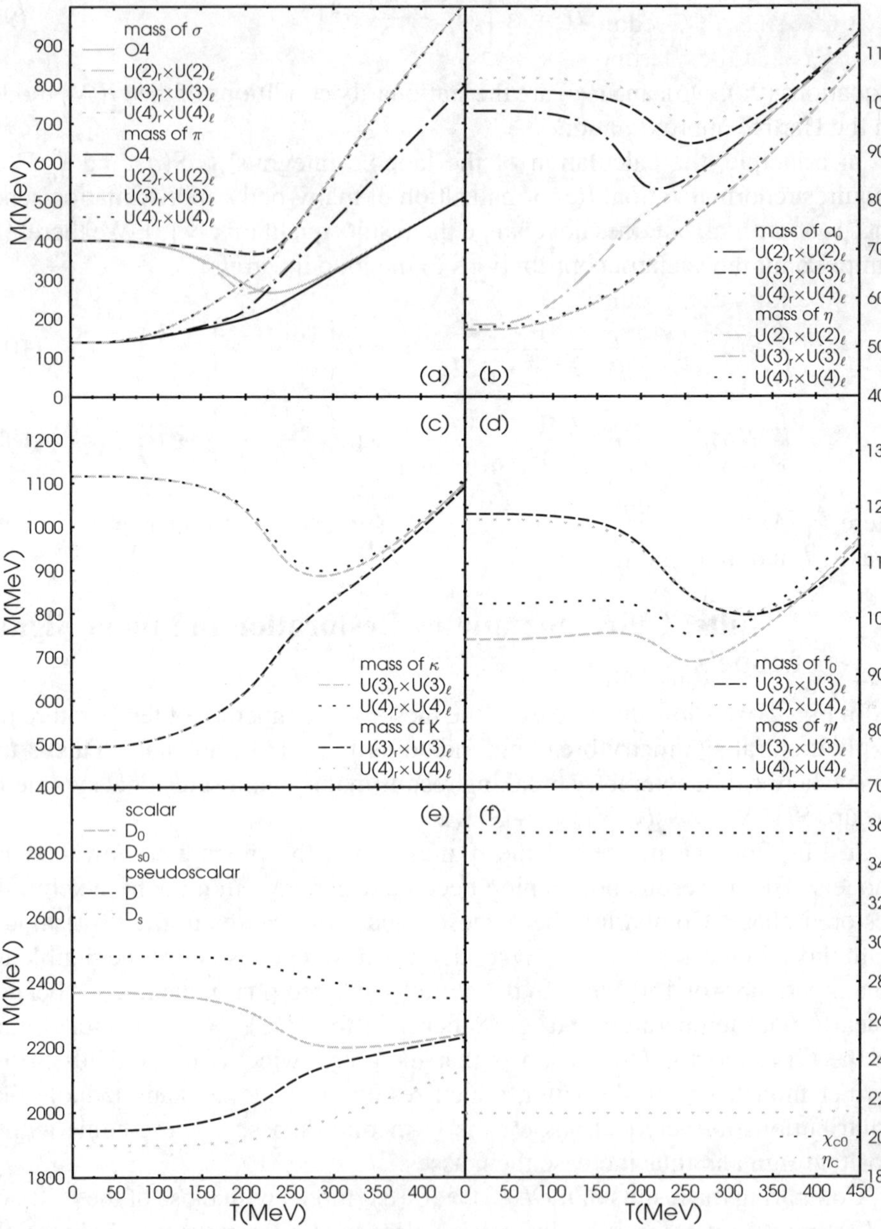

Figure 2. The meson masses as a function of temperature for the different models studied here, for the case with $U(1)_A$ anomaly and explicit chiral symmetry breaking.

Finally, one observes that virtually nothing changes in comparison with the $U(3)_r \times U(3)_\ell$ model in the temperature range of interest when including the charm degrees of freedom in the framework of the $U(4)_r \times U(4)_\ell$ model. The reason is that the charm quark is large compared to the temperature, $m_c \gg T$, and the contributions from charmed particles to the equations for the in-medium masses is suppressed. For two reasons, this is a non-trivial result. First, the equations for the in-medium masses are structurally different for the $U(4)_r \times U(4)_\ell$ model. Second, although the tadpole terms are strongly suppressed for particles with masses much larger than the temperature, the equation for the masses and the condensates form a *nonlinear* system of coupled equations, i.e., small perturbations could lead to large quantitative changes in the solution.

In Fig. 2(b) the masses of the a_0 and the η mesons are shown as functions of temperature. Qualitatively, the behavior of these masses is the same in all models. The a_0 meson mass is constant up to temperatures of 150 – 200 MeV. Then it decreases, before increasing again above temperatures of 200 – 250 MeV. The η meson mass is constant up to $T \simeq 50$ MeV, and then monotonously increases with temperature. At large temperatures, a_0 and η become degenerate in mass, indicating restoration of chiral symmetry. In the $U(2)_r \times U(2)_\ell$ model, this happens somewhat earlier, at about 250 MeV, than in the other two cases.

In the $U(2)_r \times U(2)_\ell$ model, the a_0 and η meson masses are used to determine the parameters of the model. Thus, at zero temperature, the masses coincide with their correct vacuum values. In the $U(3)_r \times U(3)_\ell$ model, the predicted η mass deviates only by 2% from its vacuum value. The a_0 mass is also rather close to the correct value; the predicted mass is about 4% too large. In the $U(4)_r \times U(4)_\ell$, the η mass is reproduced with excellent accuracy (the deviation is less than 1%), while the a_0 mass is within 7% of its vacuum value.

In Fig. 2(c) the κ meson (now referred to as $K_0^*(1430)$ [7]) and kaon masses are shown as a function of temperature. The results for the $U(3)_r \times U(3)_\ell$ and $U(4)_r \times U(4)_\ell$ model are almost identical. The κ meson and kaon become degenerate in mass at temperatures of the order of 400 MeV. In both models, the vacuum kaon mass is used as input, while the κ mass is predicted. The deviation to the vacuum value in nature is about 21%.

Figure 2(d) shows the masses of the f_0 and the η' meson as a function of temperature. These mesons also become degenerate in mass at temperatures of the order of 400 MeV. In the $U(3)_r \times U(3)_\ell$ model, the predicted η' mass is rather close to its value in nature; the deviation is 0.6%. In the $U(4)_r \times U(4)_\ell$ model, the η' mass deviates from its correct vacuum value by about 7%. The f_0 mass is predicted in both models. If we identify this state with the $f_0(1370)$, these predicted masses deviate by 14% from their correct values.

The masses of the D_{s0}, D_0, D_s, and D mesons are shown in Fig. 2(e). The masses of the pseudoscalar mesons are known but the scalar mesons and their masses have not yet been experimentally identified. It is somewhat peculiar that the charmed, strange D_s meson is lighter than the charmed, non-strange D meson. This is an artifact of the particular set of coupling constants chosen here. For a different choice, this unphysical ordering of the masses can be reversed. Then, however, the masses of the other mesons deviate by an unacceptably large extent from their physical values. Temperature has virtually no effect on the heavy charmed mesons: their mass changes at most by 10%, even in the chirally restored phase. Due to the non-linear nature of the coupled system, this is a non-trivial result, although not completely unexpected: we expect significant changes of the meson masses only when the temperature becomes of the order of the mass. For the charmed mesons with masses of the order of 2 GeV, this is never the case in the temperature range of interest.

Finally, in Fig. 2(f) we show the masses of the χ_{c0} and η_c meson. Their large masses do not change at all for temperatures below 450 MeV. The vacuum values for both meson masses are predicted. While the mass for the χ_{c0} is within 5% of its correct value, the deviation for the η_c is somewhat larger ($\simeq$ 30%).

To summarize, the masses of the scalar mesons remain approximately constant up to temperatures around 150 MeV, and then slightly decrease before they become degenerate with the masses of the pseudoscalar mesons. On the other hand, with the exception of the η' mass, the pseudoscalar masses in general increase monotonously with temperature.

Acknowledgments

J.R. acknowledges support from the Studienstiftung des deutschen Volkes (German National Merit Foundation) and wants to thank the organizers of the NATO Advanced Study Institute for their invitation.

References

[1] G. 't Hooft, Phys. Rev. D **14**, 3432 (1976) [Erratum-ibid. D **18**, 2199 (1978)].

[2] R. D. Pisarski and F. Wilczek, Phys. Rev. D **29** (1984) 338.

[3] D. Röder, J. Ruppert and D. H. Rischke, Phys. Rev. D **68**, 016003 (2003) [arXiv:nucl-th/0301085].

[4] L. Dolan and R. Jackiw, Phys. Rev. D **9** (1974) 3320.

[5] J. T. Lenaghan and D. H. Rischke, J. Phys. G **26** (2000) 431 [arXiv:nucl-th/9901049]. J. T. Lenaghan, D. H. Rischke and J. Schaffner-Bielich, Phys. Rev. D **62** (2000) 085008 [arXiv:nucl-th/0004006].

[6] J. M. Cornwall, R. Jackiw and E. Tomboulis, Phys. Rev. D **10**, 2428 (1974).

[7] K. Hagiwara *et al.* [Particle Data Group Collaboration], Phys. Rev. D **66**, 010001 (2002).

THE PHYSICS OF CLUSTERS

Andrey V. Solov'yov
A. F. Ioffe Physical-Technical Institute, Russian Academy of Sciences, Polytechnicheskaya 26, St. Petersburg 194021, Russia, E-mail: solovyov@th.physik.uni-frankfurt.de

Abstract A brief introduction to atomic cluster physics, the interdisciplinary field, which developed fairly successfully during last years, is presented. A review of recent achievements in the detailed *ab initio* description of structure and properties of atomic clusters and complex molecules is given. The main trends of development in the field are discussed and some of its new focuses are outlined. Particular attention is devoted to the role of quantum and many-body phenomena in the formation of complex muli-atomic systems and the methods of theoretical investigation of their specific properties. The role of the simplified model approaches accurately developed from the fundamental physical principles is stressed. Various illustrations are made for sodium, magnesium clusters, fullerenes and clusters of noble gas atoms.

1. Introduction

During the last decade it was recognized, both experimentally and theoretically, that complex molecules and atomic clusters (ACs) often possess unique properties, which make them a new object of physical research, rather different from both a single atom and from the solid state. The knowledge of the detailed electronic and ionic structure of single complex molecules and nano- clusters can be essential for various practical applications, such as the formation of new materials, nano-structures, in the design of drugs and biologically active spices as well as for the understanding the fundamental issues, such as functioning quantum and thermodynamic laws in the nano-scale systems or mechanisms for complex multi-atomic systems formation, self-assembly and functioning.

The demand in the understanding of principles of assembly and functioning of complex multi-atomic systems such as bio-molecules or nano-clusters is tremendous, because of the potential use of this knowledge for purposes of microelectronics, of biochemistry, the drug industry etc. The problems of self-organization, of self-assembly and of the functioning of complex multi-atomic aggregates and their interactions have been addressed both theoretically and

W. Greiner et al. (eds.), Structure and Dynamics of Elementary Matter, 621–634.

experimentally in a large number of papers from different perspectives. Often, these problems can be reduced to the problem of the interaction of a limited number of atoms within a complex molecule or even to the interaction of a single atom or ion with a certain fragment of a complex molecule (an active center responsible for a certain function) or a cluster structure. Thus, in order to achieve a real breakthrough in the field, one needs to learn how to handle (both theoretically and experimentally), i.e. to be able to manipulate experimentally and predict theoretically properties of multi-atomic systems containing about 100 atoms, or maybe a little less or a little more than this limit. With this knowledge in hand, one can then move towards a detailed *ab initio* understanding of the properties of larger multi-atomic systems, bio-molecules (proteins, DNA), which typically consist of rather small fragments (amino acids or bases) whose structure and interactions do not involve more than 100 atoms.

These structures are, nowadays, subjects of very intensive theoretical and experimental studies in many physical, chemical and biological laboratories and institutions worldwide. A variety of methods have been used to investigate these objects (see, e.g., [1–10] and references therein). Due to these efforts a vast amount of physical, chemical and biological data on the properties of complex multi-atomic systems and their interaction with the environment have been accumulated. However, it can be stated that until now there is no consistent theoretical approach, based solely on the fundamental principles of quantum physics, which might allow one, not only to explain systematically the known experimental data, but also, and this is quite essential, to predict new properties of the objects and new phenomena related to them. Nearly all theoretical approaches developed so far can be termed 'phenomenological' in the sense that each one of them substitutes the full quantum-mechanical description of the dynamics of constituents of a multi-atomic structure with a model theory which uses a set of parameters deduced from the experimental data. Each of these models is able to reproduce a limited number of particular properties of a complex multi-atomic system of a particular type, since the sets of the parameters involved are not of a universal nature. Thus, the model theories have severe restrictions: in each case they can explain but few of the experimental data. From them, often, one can hardly draw general conclusions or produce predictions on the properties of other structures. A more accurate description of the electronic and ionic structures, internal dynamics and interaction with external objects and fields has to be elaborated. The development of such an approach, which is multifaceted and includes not only theoretical investigations based on the first principles of the quantum many-body theory but also implies a great amount of experimental work and computing, is a subject of current joint efforts of specialists in various fields of physics and chemistry.

The complete theoretical description of nano-scale systems consisting of about 100 atoms is extremely difficult [5]. So far, an *ab initio* many-body quantum mechanical description accounting for all electrons in the system can be used effectively for systems of a few tens of atoms [6, 7] rather than hundreds. The computer power required for such calculations grows exponentially with increasing molecular or cluster size. Therefore, one needs to invoke various simplified model approaches in order to describe complex systems of sufficiently large size [5, 8, 9].

In this paper, the role of the careful choice of the model and the importance of accounting for many-body and quantum phenomena is demonstrated for the process of formation of the AC magic numbers within the cluster size range $N < 150$. Here N is the number of atoms in the system. This example shows that the high predictive power of a model can be achieved on the basis of detailed comparison of the predictions of the model and *ab initio* approaches with each other and with experiment for relatively small systems, consisting of tens of atoms, and by the extrapolation of the model postulates towards larger scale systems [5–7].

In order to illustrate these ideas, in the next sections, we briefly discuss fission, fusion and collision processes involving ACs as well as some general aspects of AC science.

2. Atomic cluster science

A group of atoms bound together by interatomic forces is called an atomic cluster. There is no qualitative distinction between small clusters and molecules. However, as the number of atoms in the system increases, ACs acquire more and more specific properties making them unique physical objects different from both single molecules and from the solid state.

In nature, there are many different types of AC: van der Waals clusters, metallic clusters, fullerenes, molecular, semiconductor, mixed clusters, and their shapes can depart considerably from the common spherical form: arborescent, linear, spirals, etc. Usually, one can distinguish between different types of clusters by the nature of the forces between the atoms, or by the principles of spatial organization within the clusters. Clusters can exist in all forms of matter: solid state, liquid, gases and plasmas.

In figure 1, we present images of a few clusters in order to show a big variety of the cluster forms existing in nature. We also show the structure of the $\alpha/\beta-$ triosophosphate-isomerase globule aiming to stress that complex molecules such as proteins can be treated as clusters of subunits and that each of the subunits is a cluster on its own.

The novelty of AC physics arises mostly from the fact that cluster properties explain the transition from single atoms or molecules to the solid state. Modern experimental techniques have made it possible to study this transition. By increasing the cluster size, one can observe the emergence of the physical

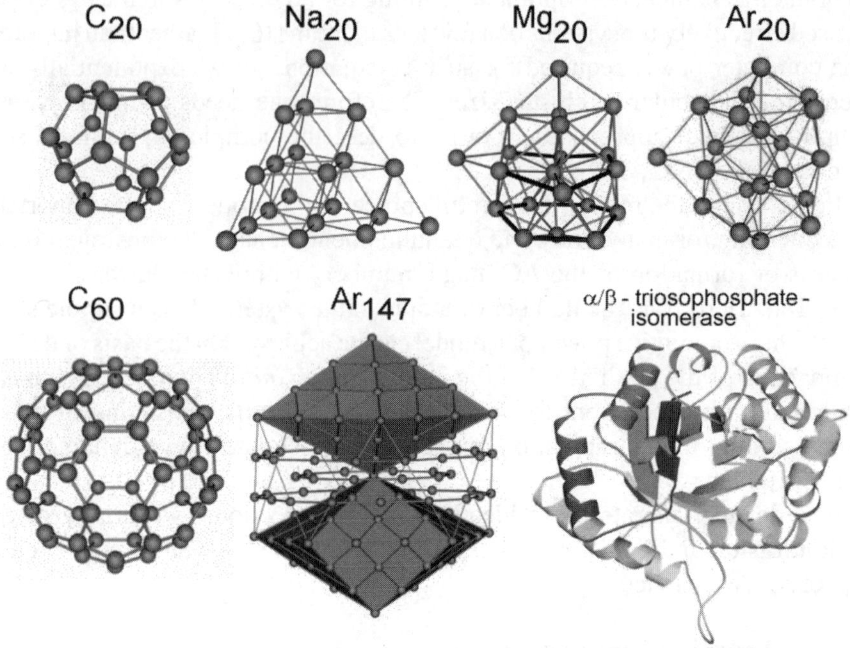

Figure 1. Different nature of forces between the atoms results in different principles of their organization within clusters and complex molecules. Geometries of the presented ACs have been calculated in [6, 7, 5, 10], structure of the protein globule ($\alpha/\beta-$ triosophosphate-isomerase) is taken from [4].

features in the system, such as plasmon excitations, electron conduction band formation, superconductivity and superfluidity, phase transitions, fission and many more. Most of these many-body phenomena exist in solid state but are absent for single atoms.

The science of clusters is a highly interdisciplinary field. ACs concern astrophysicists, atomic and molecular physicists, chemists, molecular biologists, solid-state physicists, nuclear physicists, plasma physicists, technologists all of whom see them as a branch of their subjects but cluster physics is a new subject in its own right.

Significant progress achieved in the field over the past two decades brought the understanding of ACs as new physical objects with their own distinctive properties. This became clear after such experimental successes as the discovery of the fullerene C_{60}, of the electronic shell structure in metal clusters, the observation of plasmon resonances in metal clusters and fullerenes, the observation of magic numbers for various other types of clusters, the formation of singly and doubly charged negative cluster ions and many more. A complete

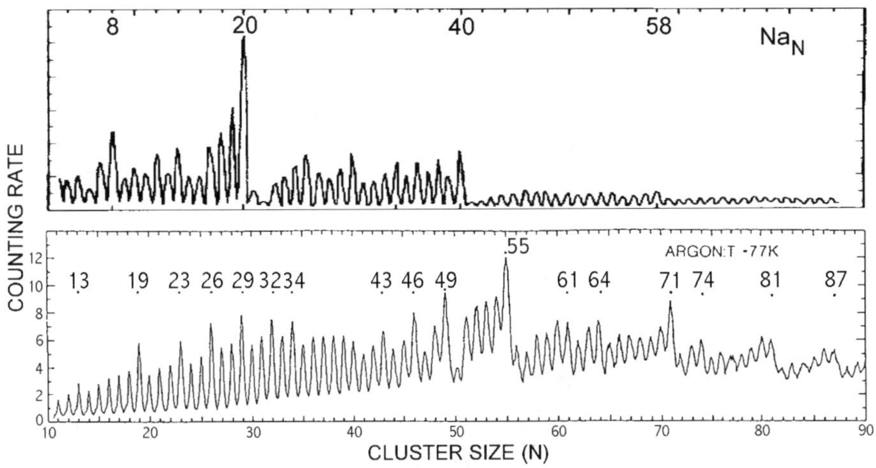

Figure 2. Mass spectra measured for Ar and Na clusters (see [11, 13, 16] and references therein). The intense peaks indicate enhanced stability.

review of this field can be found in review papers and books, see e.g. [1–3, 11–16].

3. Distinctive properties of atomic clusters: cluster magic numbers

ACs, as new physical objects, possess some properties, which are distinctive characteristics of these systems. The cluster geometry turns out to be an important feature of clusters, influencing their stability and vice-versa. The determination of the most stable cluster forms is not a trivial task and the solution of this problem is different for various types of cluster. The stability of clusters and the their transformations is a theme which does not exist at the atomic level and is not of great significance for solid state but is of crucial importance for AC systems. This problem is closely connected to the problem of cluster magic numbers.

The sequence of cluster magic numbers carries essential information about a cluster's electronic and ionic structure. Understanding the magic numbers of a cluster is pretty well equivalent to understanding its electronic and ionic structure [5]. A good example of this kind occurs for sodium clusters. In this case, the magic numbers arise from the formation of closed shells of delocalised electrons, one from each atom (see [11, 15] and references therein). Another example is the discovery of fullerenes, and in particular the C_{60} molecule [17], by means mass spectroscopy of carbon clusters.

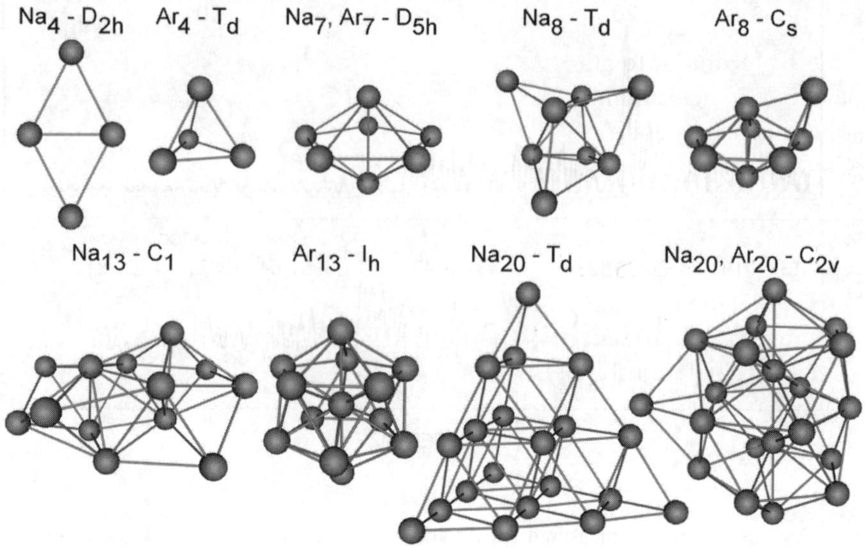

Figure 3. Geometries and the point symmetry groups of some Na and Ar clusters calculated in [5, 6].

In figure 2, we present the mass spectra measured for Ar and Na clusters (see [11, 13] and references therein), which clearly demonstrate the emergence of magic numbers. The forces binding atoms in these two different types of clusters are different. The argon (noble gas) clusters are formed by van der Waals forces, while atoms in the sodium (alkali) clusters are bound by the delocalized valence electrons moving in the entire cluster volume. The differences in the inter-atomic potentials and pairing forces lead to the significant differences in structure between Na and Ar clusters, their mass spectra and their magic numbers.

In figure 3, we present and compare the geometries of a few small Na and Ar clusters of the same size. It is clear from figure 3 that different principles of cluster organization result in different geometries of the alkali and noble gas cluster families.

Such differences can easily be explained. The van der Waals forces lead to enhanced stability of cluster geometries based on the most dense icosahedral packing. The most prominent peaks in mass spectra of argon clusters correspond to completed icosahedral shells of 13, 55, 147, 309 ets atoms. The origin of these magic numbers can be understood on the basis of the classical equations. The origin of the sodium cluster magic numbers is different and is based on the principles of quantum mechanics. In this case the cluster magic numbers 8, 20,

34, 40, 58, 92 ets correspond to the completed shells of the delocalised electrons: $1s^2 1p^6 1d^{10} 2s^2 1f^{14} 2p^6$ ets. This feature of small metal clusters make them qualitatively similar to atomic nuclei for which quantum shell effects play the crucial role in determining their properties [18].

The enhanced stability of cluster systems can be characterized by computing the second differences in cluster binding energies. In figure 4 we present Ar clusters binding energies and their second differences calculated in [5]. The correspondence of the peaks in figure 4 to those in the Ar clusters mass spectrum shown in figure 2 is readily established.

Finally, let us stress the obvious connection between AC physics and physics and chemistry of large molecules, such as proteins or DNA, which in fact can be treated as large clusters of amino acids or bases. The characteristic size of a fragment (amino acid or base) in such clusters is of the order of a few tens of atoms, i.e. the size of a small cluster. It is obvious that the knowledge gained from the AC studies is relevant for the bio-molecular investigations and vice versa. A bunch of interesting phenomena can arise at the juncture of the two fields. For example, fusion of ACs with bio-molecules can create new objects which can be handled as easy as ACs or possess some specific properties and characteristics of ACs, but at the same time carry all essential features of bio-molecules and participate in bio-processes.

4. Collisions involving atomic clusters

Properties of clusters can be studied by means of photon, electron and ion scattering (see Course 9 by A.V Solov'yov in [2, 3] and references therein). These methods are the traditional tools for probing properties and internal structure of various physical objects.

Interesting phenomena arise in elastic collisions of electrons with ACs. For example, the diffraction of fast electrons of the fullerene C_{60} molecule was predicted and later observed [19]. The diffraction pattern in the electron elastic scattering cross section caries important information on the electron density in the vicinity of the fullerene's surface.

Electron excitations in metal cluster systems have a profoundly collective nature (see [12] and references therein). They can be pictured as oscillations of electron density against ions, the so-called plasmon oscillations. This name is carried over from solid state physics where a similar phenomenon occurs. Collective electron excitations have also been studied for single atoms and molecules. In this case the effect is known under the name of the shape or the giant resonance. The name giant resonance came to atomic physics from nuclear physics, where the collective oscillations of neutrons against protons have been investigated [18].

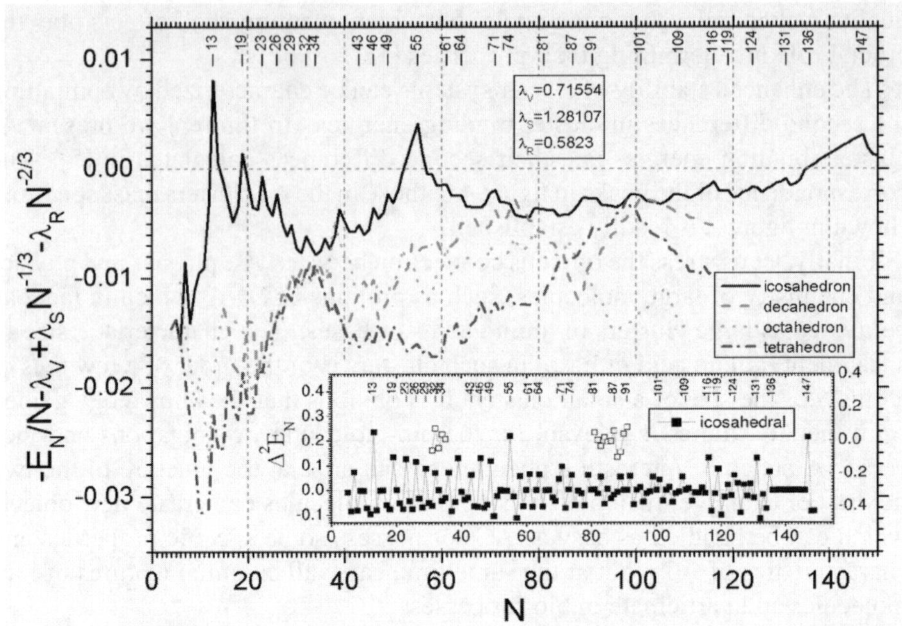

Figure 4. Binding energies and their second differences for Ar clusters calculated in [5].

The interest of plasmon excitations in small metal clusters is connected with the fact that the plasmon resonances carry a lot of useful information about cluster electronic and ionic structure. By observing plasmon excitations in clusters one can study, for example, the transition from the pure classical Mie picture of the plasmon oscillations to its quantum limit or to detect cluster deformations by the value of splitting of the plasmon resonance frequencies.

The plasmon resonances can be seen in the cross sections of various collision processes: photabsorption and photoionization, electron inelastic scattering, electron attachment, bremsstrahlung (see Course 9 by A.V Solov'yov in [2]). Both surface and volume plasmons can be excited. In electron collisions and in the multiphoton absorption regime, plasmons with large angular momenta play an important role in the formation the cross sections of these processes [20].

In figure 5, we present experimentally measured and theoretically calculated cross section for the photoabsoption of some Na and Mg clusters [21]. The cross sections are resonantly enhanced owing to the excitation of plasmon oscillations in the target cluster.

Plasmon excitations in clusters decay via the Landau damping mechanism, while the relaxation of single electron excitations in clusters occurs via the

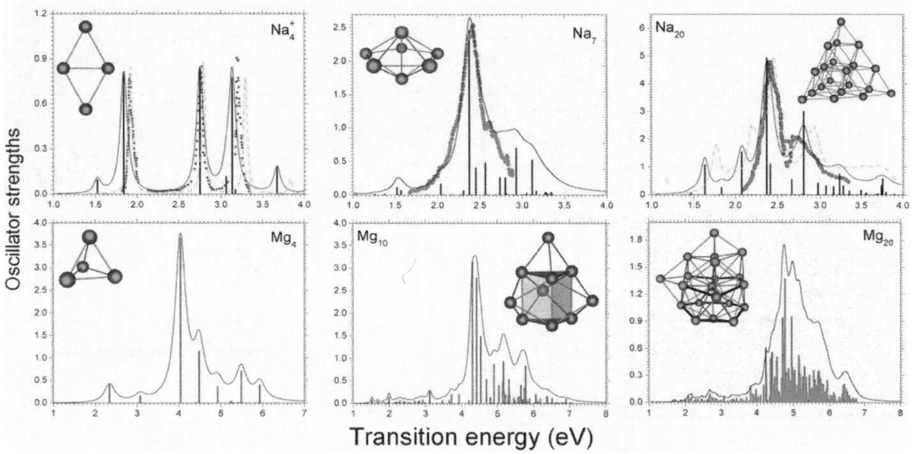

Figure 5. Photoabsortion spectra of some Na and Mg clusters [21].

interaction with the vibrations of ions, i.e. via the electron-phonon interaction (see Course 9 by A.V Solov'yov in [2]).

Collisions involving ACs rise many more interesting physical problem. For example, in collisions one can study phase transitions (solid-liquid or liquid-gas) in mesoscopic systems or the cluster multifragmentation process.

5. Fission instability of multiply charged clusters.

Multicharged ACs become unstable towards fission. The process of multicharged metal clusters fission is qualitatively analogous to nuclear fission. The fission instability of charged liquid droplets was first described by Lord Rayleigh in 1882 [22] within the framework of classical electrodynamics. The review of recent work on metallic cluster fission, one can find in [2, 8, 14, 15].

The fission process of ACs is interesting because it reveals the obvious parallel of AC studies with nuclear physics, where the fission process of nuclei has been studied for many decades [18]. The experiments on cluster fission provide a very good opportunity to test various concepts, approximations and AC models. It convincingly demonstrates the importance of the correct accounting for the quantum and many body phenomena in the description of multi-atomic systems. Dynamic aspects of the AC fission problem are also of great interest, because in the contrary to nuclear physics in the fission of ACs all the forces in the system are known and thus one can develop the full dynamic description of the process.

To illustrate the fission of charged metal clusters, we plot in figure 6 the fission barriers for the symmetric and asymmetric fission channels of Na_{18}^{2+}: $Na_{18}^{2+} \rightarrow$

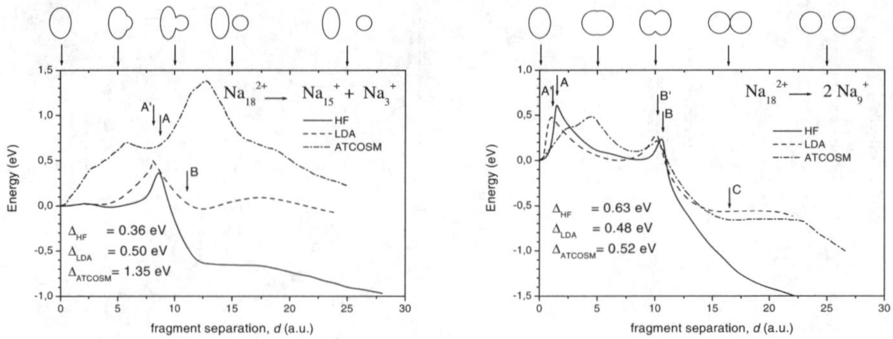

Figure 6. Fission barriers for the asymmetric and symmetric fission channels of Na_{18}^{2+}: $Na_{18}^{2+} \rightarrow Na_{15}^{+} + Na_{3}^{+}$ and $Na_{18}^{2+} \rightarrow 2Na_{9}^{+}$ calculated in [8]

$Na_{15}^{+} + Na_{3}^{+}$ and $Na_{18}^{2+} \rightarrow 2Na_{9}^{+}$. The bariers plotted in figure 6 have been calculated in [8] within the two-center LDA and Hartree-Fock jellium model and compared with the asymmetric two-center-oscillator shell model (ATCOSM). Figure 6 demonstrates the evolution of cluster shape during the fission process, the importance of cluster deformations, many-electron correlation and shell effects.

6. Fusion process of atomic clusters.

The formation of a sequence of cluster magic numbers should be closely connected to the mechanisms of cluster formation and growth. It is natural to expect that one can explain the magic numbers sequence and find the most stable cluster isomers by modelling mechanisms of cluster assembly and growth, i.e. the fusion process of ACs [5].

The problem of magic clusters is closely connected to the problem of searching for global minima on the cluster multidimentional potential energy surface. The number of local minima on the potential energy surface increases exponentially with the growth cluster size and is estimated to be of the order of 10^{43} for $N = 100$ [2, 5]. Thus, searching for global minima becomes increasingly difficult problem for large clusters. There are different algorithms and methods of the global minimisation, which have been employed for the global minimisation of AC systems (see [2, 5] and references therein). These techniques are often based on the Monte-Carlo simulations.

Alternatively, the algorithm based on the dynamic searching for the most stable isomers in the cluster fusion process has been recently proposed [5]. The calculations performed with this new algorithm demonstrated that this

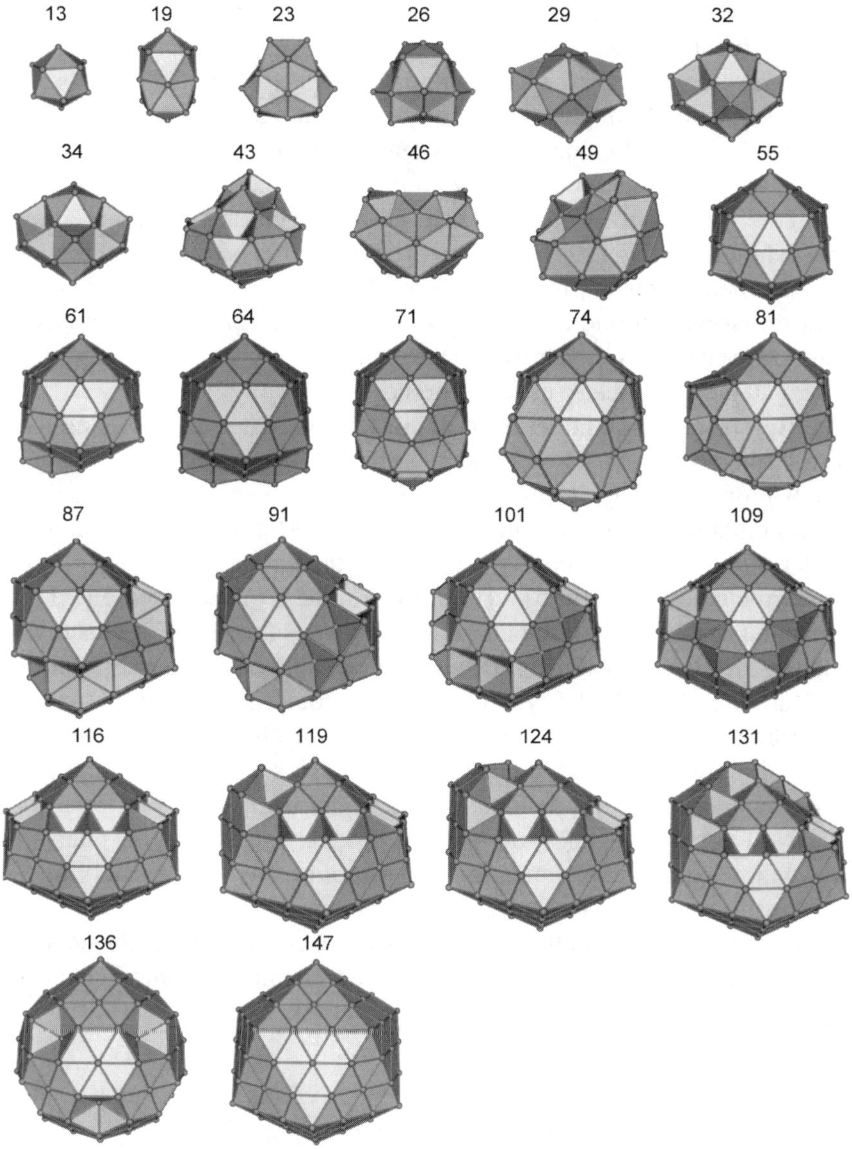

Figure 7. Images of the Lennard-Jonnes global energy minimum cluster isomers [5, 10]. The mass numbers of the pictured clusters correspond to the magic numbers of the noble gas (Ar, Kr, Xe) clusters.

approach is an efficient alternative to the known techniques of the cluster global minimisation. The big advantage of the fusion approach consists in the fact that it allows to study not just the optimized cluster geometries, but also their formation mechanisms.

In the recent work [5], the fusion algorithm was formulated in a most simple, but general form. In the most simple scenario, it was assumed that atoms in a cluster are bound by Lennard-Jones potentials and the cluster fusion takes place atom by atom. In this process, new atoms are placed on the cluster surface in the middle of the cluster faces. Then, all atoms in the system are allowed to move, while the energy of the system is decreased. The motion of the atoms is stopped, when the energy minimum is reached. The geometries and energies of all cluster isomers found in this way are stored and analysed. The most stable cluster configuration (cluster isomer) is then used as a starting configuration for the next step of the cluster growing process.

Starting from the initial tetrahedral cluster configuration and using the strategy described above, the cluster fusion paths have been analysed up to the cluster sizes of more than 150 atoms. We have found that in this way practically all known global energy minimum structures of the Lennard-Jonnes clusters can be determined. Figure 7 shows the images of the Lennard-Jonnes global energy minimum cluster isomers [5]. The mass numbers of the represented clusters correspond to the magic numbers of the noble gas (Ar, Kr, Xe) clusters.

So far, the cluster fusion algorithm has been applied to the noble gas clusters which are based on the LJ type of the inter-atomic interaction. However, the fusion process can be generated in a similar way for the systems, like metal clusters, holding together by quantum forces. This technique can also be used for the simulation of the fusion process of complex bio-molecules (proteins and DNA) or for the study of proteins folding. It would be interesting to see to which extent the parameters of inter-atomic interaction can influence the cluster fusion process, the corresponding sequence of magic numbers and also to study the clusterization in the nuclear matter. Studying cluster thermodynamic characteristics with the use of the developed technique is another interesting theme which is left open for future considerations.

7. Conclusions

In recent years, AC physics has made very significant progress, but a large number of problems in the field are still open. The transition of matter from the atomic to the solid state implies changes of organization which turn out to be a good deal more subtle and complex than was originally supposed. Different type of clusters, composite clusters, various size ranges, cluster geometries, complex molecules (including biological), clusters on a surface and in plasmas, all provide additional themes which make this field of science very rich and varied. Collisions involving ACs, mass spectroscopy and laser techniques provide tools for experimental studies of the cluster structure and properties.

However, what are the experimental limitations? Where should the theory go next? Where does the future lie? Could clusters one day become the smallest devices or be used to make the smallest devices? Could one manipulate cluster isomers for the production new materials and nano-structures? What is the difference between a cluster and a bio-molecule or a virus? Could the molecules as complex as proteins or DNA and their functions be understood on the basis of the classical mechanics or one ultimately needs to invoke the quantum theory? What are the principles of the matter self-organization, self-assembling and functioning on the nanoscale? We merely mention such intriguing questions in this paper, but we hope that at least some of them will be resolved during the future development of the physics of clusters.

Acknowledgement(s)

I would like to express my gratitude to Professor W.Greiner for invitating me to the NATO ASI "Structure and dynamics of elementary matter" and the excellent organization of this conference. I acknowledge the support of this work by the the Russian Foundation of Basic Research (grant No 03-02-16415-a), the Russian Academy of Sciences (grant No 44) and the INTAS.

References

[1] J.P. Connerade, A.V. Solov'yov and W. Greiner, *Europhysicsnews* **33** (2002) 200.

[2] C. Guet, P. Hobza, F. Spiegelman and F. David (eds.), *NATO Advanced Study Institute, Session LXXIII, Summer School "Atomic Clusters and Nanoparticles"*, Les Houches, France, July 2-28, 2000, EDP Sciences and Springer Verlag, Berlin, Heidelberg, New York, Hong Kong, London, Milan, Paris, Tokyo, 2001.

[3] J.P. Connerade and A.V. Solov'yov (eds.), *Latest Advances in Atomic Clusters Collision: Fission, Fusion, Electron, Ion and Photon Impact*, Imperial College Prass and World Scientific, London, 2003 (in print).

[4] A.V. Finkelshtein and O.B. Ptizin, *Physics of ptoteins*, University, Moscow, 2002.

[5] I.A. Solov'yov, A.V. Solov'yov, W. Greiner, A. Koshelev and A. Shutovich, *Phys.Rev.Lett.* **90** (2003) 053401; I.A. Solov'yov, A.V. Solov'yov and W. Greiner, *Journal of Chemical Physics* (2003) in print; http://xxx.lanl.gov/ [arXiv: physics/0306185, v.1, 26 Jun (2003)]

[6] I.A. Solov'yov, A.V. Solov'yov and W. Greiner, *Phys. Rev.* **A65** (2002) 053203.

[7] A.G.Lyalin, I.A. Solov'yov, A.V. Solov'yov and W. Greiner, *Phys. Rev.* **A67** (2003) 063203-(1-13).

[8] A.G. Lyalin, A.V. Solov'yov and W. Greiner, *Phys. Rev. A* **65** (2002) 043202.

[9] A. Matveentsev, A.G. Lyalin, I.A. Solov'yov, A.V. Solov'yov and W. Greiner, *Int.J.Mod.Phys.* **E12** (2003) 81.

[10] A. Koshelev, A. Shutovich, I.A. Solov'yov, A.V. Solov'yov, W. Greiner, *Proceedings of International Workshop "From Atomic to Nano-scale"*, Old Dominion University, Decem-

ber 12th-14 th, 2002, Norfolk, Virginia, USA (2002), editors Colm T. Whelan and Jim Mc Guire, Old Dominion University, 184-194 (2003)

[11] W.A. de Heer, *Rev. Mod. Phys.* **65** (1993) 611.

[12] C. Bréchignac, J.P. Connerade, *J.Phys.B: At. Mol. Opt. Phys.* **27** (1994) 3795.

[13] H. Haberland (ed.), *Clusters of Atoms and Molecules, Theory, Experiment and Clusters of Atoms*, Springer Series in Chemical Physics 52, Berlin, Heidelberg, New York, Springer, 1994.

[14] U. Näher, S. Bjørnholm, S. Frauendorf, F. Garcias and C. Guet, *Physics Reports* **285** (1997) 245.

[15] W. Ekardt (ed.), *Metal Clusters*, Wiley, New York, 1999.

[16] S.Sugano and H.Koizumi, *Microcluster Physics*, Second Edition, Springer, Berlin, Heidelberg, London, 1998.

[17] H.W. Kroto et al., *Nature*, **318** (1985) 163.

[18] J.M. Eisenberg and W. Greiner, *Nuclear Theory*, North Holland, Amsterdam, 1987.

[19] L.G. Gerchikov, P.V. Efimov, V.M. Mikoushkin and A.V. Solov'yov, *Phys.Rev.Lett.* **81** (1998) 2707.

[20] J.P. Connerade and A.V. Solov'yov, *Phys. Rev.* **A66** (2002) 013207.

[21] I.A. Solov'yov, A.V. Solov'yov and W. Greiner, *XXIII International Conference on Photonic, Electronic and Atomic Collisions, Abstracts of Contributed Papers*, Stockholm, Sweeden, 23-29 July 2003; in *Latest Advances in Atomic Clusters Collision: Fission, Fusion, Electron, Ion and Photon Impact*, Editors J.P. Connerade and A.V. Solov'yov, Imperial College Prass and World Scientific, London, 2003 (in print)

[22] Lord Rayleigh, *Philos. Mag* **14** (1882) 185.

PHASE TRANSITION IN SUPERCONDUCTIVITY AT TWO-PHONON EXCHANGE BETWEEN THE ELECTRONS

N.A.Enaki and V.V. Eremeev
Institute of Applied Physics, Academy of Sciences of Moldova, Academiei str. 5, Chisinau MD-2028, Moldova, E-mail: eremeev@cc.acad.md

Abstract The two-quanta cooperative effects can appear in superconductivity, where the Cooper pairs are created due to the simultaneous two-phonon exchange between electrons. It occurs when the one-phonon exchange integral between the band electrons is smaller than that of the two-phonon exchange. This is possible in many-band superconducting materials, in which the two-phonon exchange integral arises through the virtual bands of the material. In the two-phonon processes, a more complicated temperature dependence on the order parameter is expected. A rigorous study of this anomalous temperature dependence on the order parameter of superconductors is presented in this paper.

1. Introduction

The new cooperative emission phenomenon for dipole-forbidden transitions of an inverted system of radiators can be observed in two-photon spontaneous emission processes [1]. The condition under which a sample of inverted radiators goes into the regime of the two-photon cooperative emission was found. In the two-photon emission process the hydrogen- or helium-like atoms collectively generate pairs of entanglement photons (biphotons) relative to dipole-forbidden transitions $|2S\rangle - |1S\rangle$.

Let us study the electronic transitions between the states $|\mathbf{k}\rangle$ and $|\mathbf{k}'\rangle$ in the low band of a two-band superconductor with absorption or radiation of phonons (Fig. 1). In order to neglect the one-phonon transitions between these states, let us consider that the second band is situated at a distance energetically larger than k_BT (k_B is the Boltzmann constant, T is the temperature) on the Fermi level of the first band. In the case when the one-phonon transition between the first and the second bands $g_{21}(\mathbf{q})$ is larger than the matrix element of the one-phonon transition in the intrinsic first band $g_{11}(\mathbf{q})$, the two-phonon transition between the states $|\mathbf{k}\rangle$ and $|\mathbf{k}'\rangle$ becomes larger than the one-phonon transition. This becomes possible when the first and the second bands are formed from the local atomic levels with different symmetries. The anomalous behavior of the

W. Greiner et al. (eds.), Structure and Dynamics of Elementary Matter, 635–638.

order parameter as a function of temperature is observed due to the fact that the two-phonon exchange constant between the electrons increases as a function of temperature.

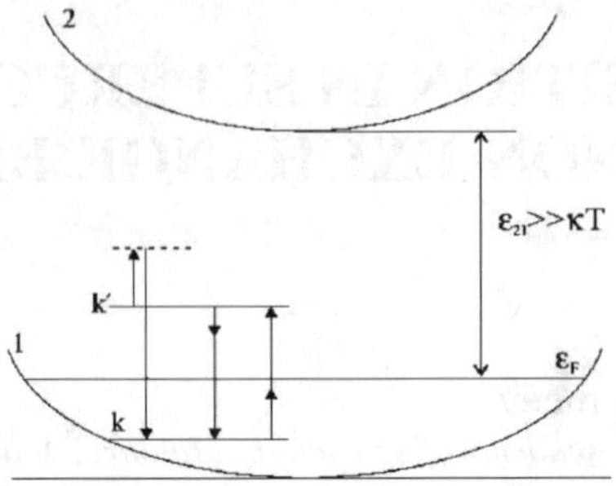

Figure 1. The band structure of a superconductor

2. Phase-transition in two-phonon model of superconductivity

In this section, the two-band electronic system, which interacts with the phononic field, is proposed for theoretical discussion. For the simplicity, the Columbian interaction between electrons is not taken into consideration. Therefore the Hamiltonian of such a system can be represented though the two-phonon interaction between the electrons of lower band in the following form [2]

$$\begin{aligned} H &= \sum_k \epsilon_1'(\mathbf{k}) a_k^\dagger a_k \\ &- \frac{1}{V}\Bigg\{\sum_{k_1}\sum_{q,q_1} g_{2,1}(\mathbf{q}) g_{2,1}^*(\mathbf{q}_1) \frac{a_{1,k+q-q_1}^\dagger a_{1,k} b_{q_1}^\dagger b_q}{\epsilon_2(\mathbf{k}+\mathbf{q}) - \epsilon_1(\mathbf{k}+\mathbf{q}-\mathbf{q}_1) - \hbar\omega_{q_1}} \\ &+ g_{2,1}(\mathbf{q}) g_{2,1}(\mathbf{q}_1) \frac{a_{1,k+q+q_1}^\dagger a_{1,k} b_{q_1} b_q}{\epsilon_2(\mathbf{k}+\mathbf{q}) - \epsilon_1(\mathbf{k}+\mathbf{q}+\mathbf{q}_1) + \hbar\omega_{q_1}} + \text{h.c.}\Bigg\} \quad (1) \end{aligned}$$

Here V is the volume of the crystal; $b_{\mathbf{q}}^\dagger$ ($b_{\mathbf{q}}$) is the boson creation (annihilation) operator with $\mathbf{q}$ wave-vector, and $a_{n,\mathbf{k}}^\dagger$ ($a_{n,\mathbf{k}}$) is the electron creation (annihilation) operator in m-band ($m = 1, 2$) with the quasi-wave vector $-\mathbf{k}$ and spin $-\sigma$; $\epsilon_m(k)$ is the energy of the electron, which can be calculated from the level of the chemical potential. The Hamiltonian (1) was obtained with the consideration that interband matrix element of the electron-phonon interaction g_{12} is larger than intraband matrix elements g_{11}, g_{22} ($g_{12} \gg g_{11}$). After the elimination of bi-boson operators one obtains an effective Hamiltonian which describes the two-quanta interaction between the electrons. Its detailed form is given in ref. [2].

The influence of the second virtual band position on the creation of the two-phonon Cooper effect in the first band plays an important role in the proposed model. After a simple modification of the effective Hamiltonian the following approximate interaction Hamiltonian between the electrons is obtained

$$H = \sum_k \epsilon_1'(\mathbf{k}) a_k^\dagger a_k - \frac{G(T)}{V} \sum_{k,k_1} a_{k_1}^\dagger a_{-k_1}^\dagger a_{-k} a_k \,, \tag{2}$$

where $\pm k = \pm(\mathbf{k}, \frac{1}{2})$ and

$$G(T) = \frac{4}{V}\sum_{q_1} \frac{|g_{21}(\mathbf{q}-\mathbf{q}_1)|^2 |g_{21}(\mathbf{q}_1)|^2}{\epsilon_{21}^2} \left[\frac{1}{\hbar\omega_{q-q_1}} + 4N_{q_1} \frac{\hbar\omega_{q-q_1}}{(\hbar\omega_{q-q_1})^2 - (\hbar\omega_{q_1})^2} \right] \tag{3}$$

Introducing the superconductivity order parameter $\Delta(T) = \frac{G(T)}{2V}$ $\sum_{\mathbf{k}_1} \left\langle a_{\mathbf{k}_1}^\dagger a_{-\mathbf{k}_1}^\dagger \right\rangle$ we can approximate the Hamiltonian (3) in the following form

$$H = \sum_k \epsilon_1'(k) a_k^\dagger a_k - \sum_k (\Delta(T) a_{-k} a_k + \text{h.c.}) \tag{4}$$

After intermediary transformation in (3) one can obtain the final expression for $G(T)$

$$G(T) = G^{(0)}(1+\gamma) \quad ; \quad G^{(0)} = \frac{|M_{21}|^4}{\pi^2 \epsilon_{21}^2} \frac{1}{\hbar c} \ln \left| \frac{q_{\max}}{k_F} \right|$$

$$\gamma = \left(\alpha \ln \left| \frac{q_{\max}}{k_F} \right| \right)^{-1} \sum_{n=1}^{\infty} \frac{1}{n} \left[e^{-n\alpha} \mathrm{Ei}(n\alpha) - e^{n\alpha} \mathrm{Ei}(-n\alpha) \right] \tag{5}$$

Here $\alpha = \hbar c k_F/(k_B T)$, while c is the sound velocity and k_F is the Fermi quasi-wave vector. After the diagonalisation of (4), using Bogoliubov transformations [3] the following relation between the order parameters at zero-temperature and non zero-temperature is obtained

$$\ln \frac{\Delta(0)}{\Delta(T)} = 2 \sum_{n=1}^{\infty} (-1)^{n+1} \mathrm{K}_0 \left(n \frac{\Delta(T)}{k_B T} \right) - \frac{1}{\lambda} \frac{\gamma}{1+\gamma} \tag{6}$$

where K_0 is MacDonald's function; $\lambda = |G^{(0)}|\nu_F$; $\Delta(0) = 2\hbar\omega_{\max}\exp(-1/\lambda)$; $\nu_F = m^* k_F/(2\pi^2\hbar^2)$ is the density of states on the Fermi surface. The comparison of dependence of order parameter as function of temperature $\Delta(T)$ with BCS-like model with same $\Delta(0)$ is given in Fig. 2.

In order to estimate the power law dependence of order parameter near the critical temperature T_c one can consider the MacDonald function to integral

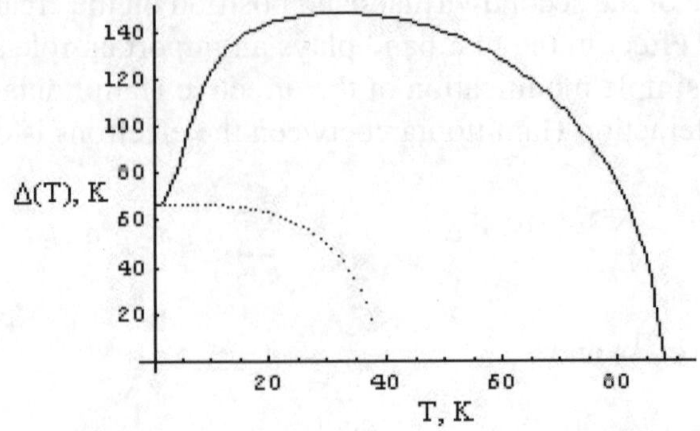

Figure 2. The temperature dependence of superconductivity order parameter $\Delta(T)$. Dotted line: BCS-like theory; full line: Two-phonon superconductivity theory.

form and decompose this integral on the small parameter $\Delta(T)/(k_B T)$. Taking into account the expression for T_c one obtains the following temperature dependence of the order parameter

$$\Delta(T) = \frac{4\pi k_B T}{\sqrt{14\zeta(3)}} \left[\ln \frac{T}{T_c} + \frac{1}{\lambda} \left(\frac{\sigma T^2}{1 + \sigma T^2} - \frac{\sigma T_c^2}{1 + \sigma T_c^2} \right) \right]^{1/2} \tag{7}$$

where $\zeta(3)$ is the Riemann zeta-function, $\zeta(3) \simeq 1.2$.

3. Results and Discussions

In this paper the two-phonon exchange between the electrons through the phonon field was discussed. It is shown that the exchange integral between electrons increases with temperature due to the fact that the two-quantum scattering effect stimulates the formation of Cooper-pairs in the superconductor. Such temperature dependence is very important for the formation of the superconducting state. This effect opens a new concept on the formation of new correlation of phase transitions in the processes of multi-quantum exchange between the carriers.

References

[1] N.A. Enaki and M. Macovei, Phys. Rev. **A 56** (1997), 3274

[2] N.A.Enaki and V.V.Eremeev, New J. of Phys. **4** (2002), 80.1

[3] N.N. Bogoliubov Jr. et al., in: *The Mathematical Methods of Statistical Mechanics of Model Systems*, (Moscow: Nauka) 1989

FACTORIZATION OF CHARGE AND NUCLEAR FORMFACTORS FOR CLUSTERIZED NUCLEI

K. A. Gridnev[1,2], K. V. Ershov[1], V. G. Kartavenko[2,3] and W. Greiner[2]
[1] *Institute of Physics, St.Petersburg State University, 198904, Russia*
[2] *J. W. Goethe Universität, D-60054, Frankfurt am Main, Germany*
[3] *Joint Institute for Nuclear Research, Dubna, Moscow District, 141980, Russia*

Abstract We considered folded formfactors for the clusterized light nuclei. Having assumed a gaussian form for the alpha-particle formfactor, charge and nuclear formfactors were factorized in two parts: one for alpha-particle and another one for the alpha-particle distribution in the nucleus. Applications to electron and heavy ion scattering are considered.

1. Introduction

In the past years experimental technique like the one with inverse geometry [1] provided a new information on the excitation functions, as well as on spin and parity assignments. Previously elastic scattering of electrons as projectiles has been used to study the charge density distributions and charge radii. On the other hand, heavy ions were used to investigate nuclear matter and potential generating wave functions in the elastic channel. Today we know well that light nuclei up to calcium have a cluster structure, namely alpha-particle structure. At certain conditions the scattering amplitude can be expressed through the formfactor. In the case of electrons it is a charge formfactor, while in case of heavy ions it is a nuclear formfactor. In both cases these formfactors in the case of cluster structure can be factorized as two multipliers: one for distribution of clusters in nuclei and the second for the cluster formfactor by itself.

2. Charge formfactors

Considering high energy electrons as a probe it is worth noting previous attempts to extract information about cluster structure [2]. The role of cluster configurations in the nucleus was investigated in [3]. In [2] authors used nucleon gaussian trial wave functions in the s-state to calculate cluster wave function.

W. Greiner et al. (eds.), Structure and Dynamics of Elementary Matter, 639–643.

Only in this case, when momentum transfer is high it is possible to factorize the formfactor. In general formfactors cannot be factorized. However the first attempt to factorize charge formfactor was done by [4]. He used a folded charge distribution given by

$$\rho(r) = \int \rho_0(\vec{r}')\rho_1(\vec{r} - \vec{r}')d^3\vec{r}' \tag{1}$$

where

$$\int \rho(\vec{r})d^3\vec{r} = \int \rho_0(r)d^3\vec{r} = \int \rho_1(\vec{r})d^3\vec{r} = 1$$

When Eq. (1) is substituted in the expression of formfactor:

$$F(\vec{q}) = \int \rho(\vec{r})e^{i\vec{q}\vec{r}}d^3r \tag{2}$$

it follows that

$$F(\vec{q}) = F_0(q)F_1(\vec{q}) \tag{3}$$

where F_0 and F_1 result from the substitution of ρ_0 and ρ_1 respectively, in Eq. (2):

$$F_0(\vec{q}) = \int \rho_0(\vec{r})e^{i\vec{q}\vec{r}}d^3r, \qquad F_1(\vec{q}) = e^{-g^2q^2/2} \tag{4}$$

$\rho_1(r)$ had the gaussian form

$$\rho_1(r) = (2\pi g^2)^{3/2}e^{-r^2/2g^2} \tag{5}$$

In his review [5] considers the value in Eq. (5) as "smearing", because the transition charge density must be concentrated on a shell of about nuclear radius, in the form of delta-function smeared by convolution in Eq. (1). Helm's model represents the transition moments of $\rho_0(r)$ as spherical shells at a radius R,

$$\rho_L^{tr}(r) = \widehat{J}_0\beta_L^{J_0J}R^{-2}\delta(r - R) \tag{6}$$

with a strength parameter to be fitted by experiment. The reduced matrix element reads

$$\widehat{J}_0^{-1} < J||M_L(q)||J_0 >= \beta_L^{J_0J}F_1(q)j_L(qR) \tag{7}$$

If we consider a nucleus consisting of α-particles, the formfactor (spherical Bessel function) represents a spherical shell made of α-particles. This result has been obtained considering high energy electrons scattered on fullerenes [6]

$$F(q) = F_A(q)\sum_k \exp{(i\vec{q}\vec{r}_j)} = F_A(q)n(q) \tag{8}$$

where $F_A(q)$ is the formfactor of a single carbon atom and $n(q)$ is the formfactor of the carbon atomic concentration.

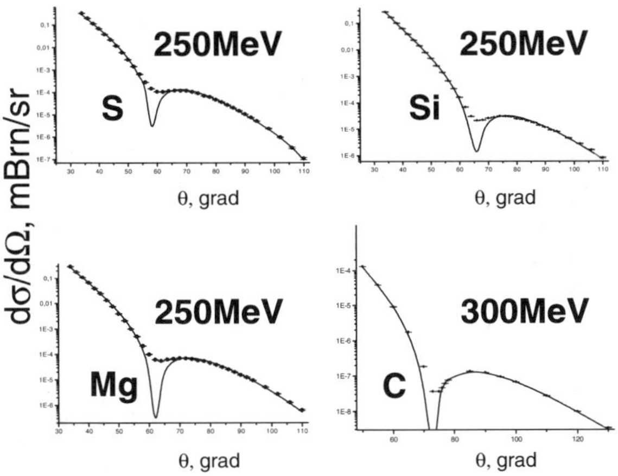

Figure 1. Electron scattering.

We considered the model when α-particles are whithin some surface shell. To certain extent this is a variant of the surface well model [7]. Assuming the quasicristaline structure of α-particles in nuclei we considered two possible types of α-particle density distribution, namely surface and volume distributions. In the first case the formfactor will be proportional to the Bessel function of order zero j_0 in the second case it will be the Bessel function of order one j_1. In Fig. 1 one can see scattering of electrons on ^{12}C, ^{16}O, ^{28}Si. At lower energies the surface distribution of alpha-particles in light nuclei is favorable. The Bessel function of order zero oscillates with the period $q = 2\pi/R$, where R is the nuclear radius.

3. Heavy ion scattering

[8] showed that elastic scattering of halo nuclei from a stable target can give simple direct evidence for the structure of the halo. The corresponding elastic scattering differential cross section is:

$$(d\sigma/d\Omega)_{el} = |F(q)|^2 (d\sigma/d\Omega)_{pt} \tag{9}$$

where $F(q) = \int d\vec{r} |\psi_0(\vec{r})|^2 e^{i\vec{q}\vec{r}}$ and $(d\sigma/d\Omega)_{pt}$ is the cross section for a point particle. The plane wave approximation for the elastic exchange at large angles gives the same result

$$\left(\frac{d\sigma}{d\Omega E_\alpha}\right) \approx \left|\frac{F(q)}{q^2} \int dr Y_{l,k_\alpha}(r) V_{int}(r) j_{l,q}(r)\right|^2 \tag{10}$$

where

$$Y_{l,k_\alpha} = \begin{cases} (\hbar^2 k^2 \Gamma \psi_l(r)/((4m_\alpha(E-E_0)+\Gamma^2/4) & \text{for r<R} \\ \exp(i(\sigma_l+\delta_l))\sin(\Phi_{\alpha l}) & \text{for r}\to\infty \end{cases}, \tag{11}$$

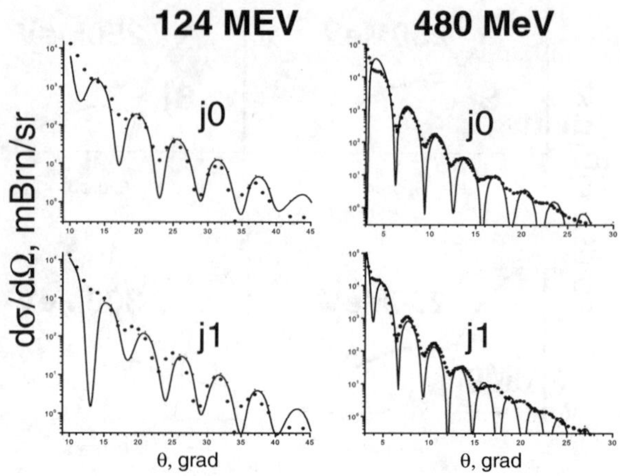

Figure 2. $^{16}O - ^{16}O$ scattering.

$$\Phi_{\alpha l} = k_\alpha r + \delta_l - l\pi/2 - \eta \ln(2k_\alpha r).$$

Whenever α-particles are scattered on α-particles or some nuclei possessing the alpha-particle structure applying the folding model is straightforward. The customary folding model is then reformulated for α-particles as if nucleons play the role of α-particles. The authors in [9] describe the elastic scattering $^{12}C + ^{12}C$ in terms of folding model for α-particles. Elastic scattering of $\alpha + ^{16}O$ finds the same treatment in [10, 11]. In both references they used the following formfactors for ^{12}C and ^{16}O respectively

$$h_\alpha(^{12}C) = (1 - 0.37q^2 + 0.0208q^4)e^{-0.167q^2} \tag{12}$$

$$h_\alpha(^{16}O) = (1 - 0.534q^2 + 0.0432q^4)e^{-0.27q^2} \tag{13}$$

Note that in both cases the formfactor for the alpha-particle takes the gaussian form. We applied the approach with factorized formfactors for the $^{16}O + ^{16}O$ scattering at 4 energies; 124 mev, 250 Mev, 350 Mev, 480 Mev. Experimental data has been taken from [12]. The results are presented in Fig. 2. The fitting was done according to the nodes of Bessel functions of order zero and one. In Fig. 2 one can see how the role of volume effects increases with the energy. Within this approach we are able to scan the cluster structure shell by shell. At higher energies the role of Fraunhoffer diffraction becomes more important.

4. Conclusion

We have used the factorized formfactors for clusterized light nuclei assuming the α-particle density having the gaussian form. It is necessary to stress that in the microscopic approach charge densities of ^{12}C were derived by folding the finite proton size determined by gaussian [13]. Factorized formfactors give us a possibility to scan nuclear surface and some intrinsic layers of a nucleus.

Acknowledgments

Work was supported in part by Deutsche Forschungsgemeinschaft (grant 436 RUS 113/24/0-4), Russian Foundation for Basic Research (grant 03-02-04021) and the Heisenberg–Landau Program (JINR).

References

[1] T. Lönroth, V. Z. Goldberg, M. Guttormsen, E. Indola, K.-M. Källman, R. Liotta, E. Melby, G. V. Rogatchev, M. V. Rozhkov, A. Schiller, T. Siskonen, S. Torilov, W. H. Trzaska, and R. Wolski, World Scientific, 2000, p. 166.

[2] E. V. Inopin, V. S. Kinchakov, V. K. Lukyanov, and Yu. S. Pol, Annals of Physics **118** (1979), 307.

[3] K. A. Gridnev, M. P. Kartamishev, J. S. Vaagen, V. K. Lukyanov, and G. S. Anagnastatos, Int. J. Mod. Phys. **11E** (2002), 359.

[4] R. H. Helm, Phys. Rev. **104** (1956), 1466.

[5] H. Überall, Springer Tracts Modern Physics **49** (1969), 1.

[6] L. G. Gershikov, P. V. Efimov, V. M. Mikoushkin, and A. V. Solov'yov, Phys. Rev. Lett. **81** (1998), 2707.

[7] A. I. Baz, V. Z. Goldberg, K. A. Gridnev, V. M. Semjonov, and E. Hefter, Z. Physik **28A** (1977), 171.

[8] R. C. Johnson, J. S. Al-Khalili, and J. A. Tostevin, Phys. Rev. Lett. **79** (1997), 2771.

[9] Li Qing-Run and Yang Yong Xu, Phys. Rev. **47C** (1993), 2393.

[10] ______, Nucl. Phys. **561A** (1993), 181.

[11] ______, Europhys. Lett. **21** (1993), 657.

[12] Dao T.Khoa, W. Von Oertzen, H.G. Bohlen, and V. M. Nuoffer, Nucl. Phys. **672A** (2000), 387.

[13] M. Kamimura, Nucl. Phys. **351A** (1981), 456.

ENVIRONMENTAL CONTAMINATIONS OF MOUNTAIN REGIONS: PROBLEMS AND DECISIONS

Kazimir Karimov and Razia Gainutdinova
Institute of Physics, National Academy of Sciences, Chui Prosp., 265-A 720071 Bishkek, Kyrgyz Republic

Abstract The problem of radioactive contamination of the environment in the mountainous areas of Kyrgyzstan is discussed. The main sources of radioactive and toxic contamination of an environment are classificated.

The problem of radiation safety is connected basically to conditions of uranium industry waste tailings storage. Is shown that these tailings render a negative influence on the environment. Ecological conditions in these mountain areas are deteriorating from year to year because of the activation of dangerous geological processes, such as earthquakes, landslides, landslips.

The possible consequences of environmental changes in zones of increased risk with radioactive and toxic contamination are considered.

Is shown, that the risk from the influence of technogenic and natural ecological accidents, in particular such as breaking of dams and waste tailings, is sharply increased because of transboundary influence on territory and population of the neighbouring states.

1. Introduction

One of the major problems of the environmental contamination in mountain regions of Kyrgyzstan is the assessment of radioactive and toxic contamination levels caused by nuclear tests, technogenic accidents which took place in the neighbouring countries and in connection with the breach of safe storage and destruction of uranium and hazardous waste tailings in the territory of Kyrgyzstan.

Long-term nuclear tests in various areas of the world and the accident at the NPS have caused the origin of stratospheric deposit of artificial radionuclides, the fallout of which have resulted in a global radioactive pollution of the biosphere. Maximum fall-outs have been observed in latitudinal zone 40-60^{0}N and in mountain regions.

Thus, the territory of Kyrgyzstan is constantly exposed to the pollution by radionuclides caused either by stratospheric fallouts or tropospheric and local fallouts as a result of regulary repeated nuclear tests. Until now no complex

W. Greiner et al. (eds.), Structure and Dynamics of Elementary Matter, 645–648.

ecologic estimation of the consequences of nuclear tests to the environment, both by regions and the world as a whole, has been conducted.

The other source of the radioactive pollution of Kyrgyzstan is the pollution by uranium tailings located on its territory. Waste products of uranium production present a real threat for the environment and public health.

2. Sources of Radioactive Contaminations

The Kyrgyz Republic is situated in the central part of Eurasia and therefore the atmospheric processes have several regional peculiarities. The contamination level of the atmosphere by technogenic sources essentially depends on the physical-geographical and meteorological conditions in the region. The influence of geophysical conditions on the territorial distribution of the contaminations is one of the main objectives in the researches of transfer mechanisms.

The troposphere in middle latitudes is characterized by a heightened level of inhomogeneities of the atmospheric parameters in horizontal and vertical directions. Characteristics of the latitudinal distribution of Hadley's vertical circulation cells and the structure of tropopause promote to high level of toxic contaminations fallouts in the latitudinal zone 40-60^0 N [1,2].

One of the external sources of the radioactive contaminations in Kyrgyzstan were atmospheric fallouts of technogenic radionuclides as a result of the nuclear tests in neighboring China. The distance from the eastern boundary of Kyrgyzstan to the Lop Nor polygon is about 900 km.

The chronology of nuclear tests conducted in China from 1964 till 1998 is as follows. From 1980 only underground nuclear tests took place. Subsurface and atmospheric nuclear tests conducted in China after 1964 are the main cause of the radioactive contamination of the environment of the USSR including Kyrgyzstan.

According to the data of the radiometric network of the USSR State Hydrometeorological Committee the widescale contamination of the lower layers of the atmosphere by the products of nuclear tests in China began in 1968. The consequences of the last atmospheric explosion in 1980 are still observed today.

The maximum contamination level (2-4)10^{-15} Cu/m^3 of the radionuclides Sr-90 and Cs-137 had been observed in 1971. In this period the concentration of Sr-90 and Cs-137 in the air above the whole territory of the USSR had been raised 100 times. The long-living radionuclides Sr-90 and Cs-137 made considerable contribution in the radiation situation in Kyrgyzstan. Thus, the radioactive contamination of the environment in 1966-1985 was solely caused by the Chinese nuclear tests.

In 1986 the consequences of the Chernobyl accident were been added.

In Kyrgyzstan radiometric observations have been carried out during the whole period of Chinese nuclear tests. The specific beta-activity of the atmospheric fallouts was determined. In addition the content of Sr-90 and Cs-137 were determined by radiochemical methods. The gamma-spectrometric measurements of the atmospheric fallouts and territorial contamination were carried out using five-channel spectrometers.

The total monthly beta-activity of the stratospheric fallouts in Kyrgyzstan increased 5-10 times after the nuclear tests conducted in China in 1964-1980. Long-term observations showed that after the atmospheric nuclear tests in China the stratospheric fallouts took place in Kyrgyzstan in 20-25 days and observed during some months until the change of circulation type [2].

These results show the contamination of the environment in Kyrgyzstan by technogenic radionuclides after conducting atmospheric nuclear tests in China during 1964-1980.

In May 1986 after the Chernobyl accident a sharp more than 80fold increase of the total monthly beta-activity of tropospheric fallouts took place.

The analysis of geophysical conditions in the atmosphere and results of radiometric observations in Kyrgyzstan in periods of conducting of underground nuclear tests in China in 1991-1995 showed the transfer of radioniclides to the territory of Kyrgyzstan. The effects from underground tests will take place only by eastern flows in the lower atmosphere [1]. The east direction flows in this period took place in the atmosphere from surface ground up to 3 km. The values of the wind velocity were 7-10 m/s. The west direction wind velocity occured at the altitudes higher than 3 km.

In Bishkek city after explosions in Lob-Nor gamma-spectrometric analysis of air samples was carried out. In air the gamma-active isotope Ruthenium-103, was found to exceed by 1,5-2,5 times the background concentrations [7]. Incidentally, the increase of concentration in air of these isotopes was synchronous with an increas of the general radiation background.

There are many uranium tailings in Kyrgyzstan which are the internal sources of radioactive contaminations. A lot of tailings and pit dumps remained after closing of uranium mines and applied factories. Most of the tailings are situated near densely populated areas and ground-water basins.

The background level of gamma-radiation varies between 20 up to 500 mcR/h on the surface of tailing dumps. The radiation background in the territory of Kyrgyzstan is within ecologically safe limits, 9-25 mcR/hours, except for local territories in areas with uranium tailings, natural anomalies and territories which have undergone "spotty" radionuclides pollution of an artificial origin as a result of atmospheric fallouts [2]. In the territory of the Kyrgyz Republic 44 waste tailings and 28 mountain mine dams are located. Among them there are 68% uranium tailings. The olume of collected wastes makes 56 million cubic meters consisting of radioactive substances, salts of heavy metals and toxic matters. These wastes make 34 million tons with total radioactive activity more than 88 thousand Cu.

On the territory of Kyrgyzstan (the southern slope of Fergansky range) there is a massive potential source of radioactive (uranium production wastes) and, possibly, chemical pollution threatening to the densely populated Fergansky valley (the territory of Kyrgyzstan and Usbekistan). A sharp activation of current engineering- and geological processes in 1993-1994 (earthquakes and huge landslides provoked by them) increased a probability of the break-down of radioactive pollution of a very big territory with hard to prognosticate consequences, that may have a global scale and may lead to great losses of material and resources.

3. Conclusions

The problem of radiation safety in Kyrgyzstan is connected with the pollution of an environment in mountain regions by radionuclides from external and internal sources, such as atmospheric fallouts and uranium tailings and mountain mine dams. The essential environmental pollution was observed after catastrophic accidents on objects of uranium tailings.

The present situation which exists on tailings and dams requires the development and realization of a complex of measures to ensure the safety of the population and protect the environment.

References

[1] Karimov, K.A. (1998) Programme on radiation safety of the population and the environment, liquidation of the consequences of the radiation impact, Nuclear Tests: Proceedings of the NATO ARW on Long-Term Consequences of Nuclear Tests for the Environment and Population Health (Semipalatinsk/Altai Case Study). Springer-Verlag, Berlin, Heidelberg, NATO ASI Series 2: Environment - Vol. 36, 2-5.

[2] Karimov, K.A. and Gainutdinova, R.D. (1999) Environmental changes within Kyrgyzstan, Proceedings of the NATO ARW on Environmental Change, Adaptation and Security. Dordrecht, The Netherlands: Kluwer Academic Publishers, NATO ASI Series 2: Environment - Vol. 65, 201-204.

USING CALORIMETRY TO INVESTIGATE THE AZIMUTHAL ANISOTROPY IN HEAVY-ION COLLISIONS

I.P. Lokhtin, S.V. Petrushanko, L.I. Sarycheva and A.M. Snigirev
D.V. Skobeltsyn Institute of Nuclear Physics, M.V. Lomonosov Moscow State University
sergant@lav01.sinp.msu.ru

Abstract The azimuthal anisotropy of energy flow and jet production in semi-central heavy ion collisions are among the most promising observables of partonic energy loss in quark-gluon plasma. A method to reconstruct the nuclear event plane with calorimetric information (instead of tracker ones) is presented. Detailed GEANT 3 simulation is done for calorimeters of CMS detector of future LHC [1].

Keywords: Ultrarelativistic heavy ion collisions, quark-gluon plasma, azimuthal anisotropy

The experimental investigation of ultra-relativistic nuclear collisions offers a unique possibility of studying the properties of strongly interacting matter at high energy density. In that regime, hadronic matter is expected to become deconfined, and a gas of asymptotically free quarks and gluons is formed, the so-called quark-gluon plasma (QGP), in which the colour interactions between partons are screened owing to collective effects [2]. One of the important tools to study QGP properties is azimuthal correlations. Recent anisotropic flow data from RHIC [3–5] can be described well by hydrodynamical models for semi-central collisions and transverse momentum, p_T, up to $\sim$ 2 GeV/c (the coefficient of elliptic flow v_2, which is defined as the coefficient of the second harmonic of the particle azimuthal distribution with respect to the reaction plane, appears to be monotonously growing with increasing p_T [6] in this case). On the other hand, the majority of microscopical Monte-Carlo models underestimate the flow effects (except [7]). Measurements of v_2 present one of the effective tools to test various models.

Azimuthal angle of reaction plane Ψ_n can be determined via n-th harmonic from the equation [8]

$$\tan n\Psi_n = \frac{\sum\limits_i \omega_i \sin n\varphi_i}{\sum\limits_i \omega_i \cos n\varphi_i}, \tag{1}$$

where φ_i is the azimuthal position of the i-th particle, ω_i is a weight, and the sum

W. Greiner et al. (eds.), Structure and Dynamics of Elementary Matter, 649–651.

runs over all particles. A procedure to optimize the weight is usually needed in order to achieve the best accuracy (e.g. selecting particles of a particular type or using the transverse momenta of particles as weights). Ψ_n is determined in the $-\pi/n < \Psi_n < \pi/n$ region.

When the azimuthal distribution of particles is described by the elliptic form, the event plane angle Ψ_0 can be determined via second harmonics, $n = 2$ in equation (1). The coefficient v_2 is $\langle\cos 2(\varphi - \Psi_0)\rangle$.

Instead of tracker (which is the common practice of v_2 determination) at LHC we can use calorimetry information [1]. In this case the weights ω_i can be introduced as energy deposition in calorimetric sector i of position φ_i, weight $\omega_i = E_i(\varphi_i)$.

To generate elliptic flow, a simple hydrodynamical model giving a hadron spectrum as a superposition of a thermal hadron distribution and collective flow was used [9, 10]. A sample of 100 Pb–Pb events at impact parameter $b = 6$ fm was utilized for the analysis. A full simulation (GEANT 3) and reconstruction of the calorimeter responses was used. Figures 1 illustrate the energy deposit as function of azimuthal angle for the hadronic (HCAl) and electromagnetic (ECAL) calorimeters of CMS.

The estimated resolution of reconstruction of azimuthal angle of reaction plane is $\sigma(\Psi_0) \approx 0.08 - 0.12$ rad. The ECAL is more suitable than the HCAL for event plane determination. This effect is mainly due to the significant distorting influence of the magnetic field on the energy flow in the HCAL (especially for low p_T-particles). Another important result is that although the anisotropic flow is maximal at mid-rapidity, the much larger values of energy deposition in the endcaps ($1.5 < |\eta| < 3$) result in reducing the relative fluctuations and, accordingly, in a much better event plane resolution. Moreover, energy flow measurements in the barrel ($|\eta| < 1.5$) are more sensitive to the magnetic field than those in the endcaps. This observation emphasizes the dominant role of the endcaps for the event plane determination, and shows the possibility of measuring the reaction plane with only the endcaps. Since the jet rapidity distribution is peaked at mid-rapidity, most jets are expected to be produced in the barrel region. A measurement of the reaction plane without barrel can minimize distortions due to the presence of jets. Such an approach will measure the energy flow (used for the analysis) and jets in different rapidity windows.

Acknowledgments

Author thanks the organizers for the warm welcome and the hospitality.

References

[1] I.P. Lokhtin, S.V. Petrushanko, L.I. Sarycheva and A.M. Snigirev , Azimuthal Anisotropy and Jet Quenching in Heavy-Ion Collitions with CMS Calorimetry, CMS Note 2003/019.

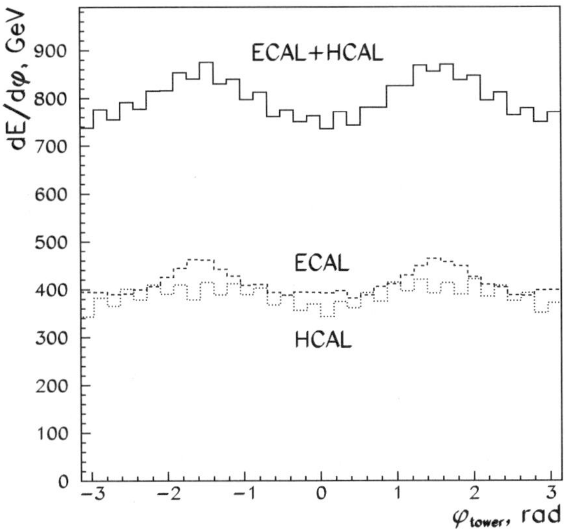

Figure 1. Energy deposition in CMS electromagnetic calorimeter (ECAL, dashed histogram), CMS hadron calorimeter (HCAL, dotted histogram), total deposition (HCAL+ECAL, solid histogram); Pb−Pb collisions at $b = 6$ fm (hydrodynamics, detailed CMS geometry simulation with GEANT 3).

[2] Proceedings of 16th International Conference on Ultrarelativistic Nucleus-Nucleus Collisions "Quark Matter'2002" (Nantes, France, July 18-24, 2002), Nucl. Phys. **A 715** (2003).

[3] K.H. Ackermann *et al.* (STAR Coll.), Phys. Rev. Lett. **86** (2001) 402; C. Adler *et al.* (STAR Coll.), Phys. Rev. Lett. **87** (2001) 182303; Phys. Rev. Lett. **90** (2003) 032301; R.J.M.Snellings *et al.* (STAR Coll.) Nucl. Phys. **A 698** (2002) 193.

[4] R.A. Lacey *et al.* (PHENIX Coll.), Nucl. Phys. **A 698** (2002) 559.

[5] I.C. Park *et al.* (PHOBOS Coll.), Nucl. Phys. **A 698** (2002) 564; B.B. Back *et al.* (PHOBOS Coll.), Phys. Rev. Lett. **89** (2002) 222301; [arXiv:nucl-ex/0210036].

[6] P.F. Kolb, J. Sollfrank, and U. Heinz, Phys. Rev. **C 62** (2000) 054909; P.F.Kolb *et al.*, Phys. Lett. **B 500** (2001) 232; Nucl. Phys. **A 696** (2001) 175.

[7] E.E. Zabrodin, C. Fuchs, L.V. Bravina, and A. Faessler, Phys. Lett. **B 508** (2001) 184.

[8] S.A. Voloshin and Y. Zhang, Z. Phys. **C 70** (1996) 665; A.M. Poskanzer and S.A. Voloshin, Phys. Rev. **C 58** (1998) 1671.

[9] N.A. Kruglov, I.P. Lokhtin, L.I. Sarycheva, and A.M. Snigirev, Z. Phys. **C 76** (1997) 99.

[10] I.P. Lokhtin, L.I. Sarycheva, and A.M. Snigirev, Phys. Lett. **B 537** (2002) 261; Nucl. Phys. **A 715** (2003) 633.

JET QUENCHING STUDIES IN HEAVY ION COLLISIONS AT LHC

I.P. Lokhtin, A.M. Snigirev
D.V. Skobeltsyn Institute of Nuclear Physics, M.V. Lomonosov Moscow State University
igor@lav01.sinp.msu.ru

Abstract The jet production processes, which can carry information about the earliest stages of quark-gluon system evolution undergoing final state medium-induced effects in heavy ion collisions at LHC, are analyzed.

Keywords: Ultrarelativistic heavy ion collisions, quark-gluon plasma, QCD, hard jets

Jets play one of the central roles as a promising tool to study properties of quark-gluon plasma (QGP) expected to be created in heavy ion collisions at RHIC and LHC. Medium-induced energy loss of energetic partons, the so-called jet quenching, has been proposed to be very different in cold nuclear matter and in QGP, resulting in many challenging observable phenomena (see for the review [1] and references therein). In particular, softening jet fragmentation function and, as a consequence, suppression of high-p_T hadron spectrum in nucleus-nucleus collisions relative to their production in independent nucleon-nucleon interactions, are considered [2–5]. Recent RHIC data on inclusive high-p_T charge and neutral hadron production from STAR [6], PHENIX [7], PHOBOS [8] and BRAHMS [9] experiments show such kind of suppression and are in agreement with the jet quenching hypothesis. However, since at the moment direct event-by-event reconstruction of jets and their characteristics is not available in RHIC experiments, the assumption that integrated yield of all high-p_T particles originates only from jet fragmentation is not fully clear. At the LHC, a new regime is reached where hard and semi-hard QCD multi-particle production can certainly dominate over underlying soft events. CMS experiment at LHC [10] will be able to provide adequate jet reconstruction using calorimetric measurements [11]. Thus identification of leading particle in a jet (i.e. particle carrying the maximal fraction of jet p_T) allows the measurement of jet fragmentation function (JFF) to be done. Comparison of JFF in AA and pp collisions (or in central and peripheral AA interactions) may give information about in-medium modification of JFF.

W. Greiner et al. (eds.), Structure and Dynamics of Elementary Matter, 653–655.

The crucial question here is: how much energy loss falls *outside* the typical jet cone and is truly lost to the jet? There are discussions on this subject in the literature [12–16]. In fact, since coherent Landau-Pomeranchuk-Migdal radiation induces a strong dependence of the radiative energy loss of a jet on the angular cone size, it will soften particle energy distributions inside the jet, increase the multiplicity of secondary particles, and to a lesser degree, affect the total jet energy. On the other hand, collisional energy loss turns out to be practically independent of jet cone size and causes the loss of total jet energy, because the bulk of "thermal" particles knocked out of the dense matter by elastic scatterings fly away in almost transverse direction relative to the jet axis [12]. Moreover, the total energy loss of a jet will be sensitive to the experimental capabilities for low-p_T particles, products of soft gluon fragmentation.

In our recent paper [17] we analyze relation between in-medium softening JFF and suppression of jet spectrum due to energy loss outside the jet cone at the LHC case. The method used are event-by-event Monte-Carlo simulation of rescattering and energy loss of partons in QGP (for model details one can refer to our previous papers [12, 18, 17]). The approach relies on an accumulative energy losses, when gluon radiation is associated with each scattering in expanding medium together including the interference effect by the modified radiation spectrum as a function of decreasing temperature $dE/dl(T)$. The basic kinetic integral equation for the energy loss ΔE as a function of initial energy E and path length L has the form

$$\Delta E(L,E) = \int\limits_0^L dl \frac{dP(l)}{dl} \lambda(l) \frac{dE(l,E)}{dl}, \quad \frac{dP(l)}{dl} = \frac{1}{\lambda(l)} \exp\left(-l/\lambda(l)\right), \quad (1)$$

where l is the current transverse coordinate of a parton, dP/dl is the scattering probability density, dE/dl is the energy loss per unit length, $\lambda = 1/(\sigma\rho)$ is in-medium mean free path, $\rho \propto T^3$ is medium density at temperature T, σ is the integral cross section of parton interaction in the medium. Such numerical simulation of free path of a hard jet in QGP allows us to obtain any kinematical characteristic distributions of jets in final state.

We examine here the fraction ε of partonic energy loss carried out of the jet cone as a phenomenological parameter. It has been found (see Fig.1 of [17]) that if ε close to 0 ("small-angular radiation" dominates), then the factor of jet suppression $Q^{\rm jet}$ is close to 1 (there is almost no jet rate suppression), and effect on JFF softening is maximal. Increasing ε (the contribution from "wide-angular radiation" and collisional loss grows) results in stronger jet suppression ($Q^{\rm jet}$ value decreases), but effect on JFF softening becomes smaller, especially for highest z (the p_T fraction of a jet carried by the leading hadron).

We have also found the distinctive anti-correlation between strengthening jet suppression and JFF softening determined mainly by fraction of partonic energy loss carried outside the jet cone (Figure 2 of [17]). The physical reason for the effect to be opposite in the jet suppression factor and the fragmentation function is it follows. Increasing ε results in decreasing final jet transverse momentum, $p_T^{\rm jet} = p_T^{\rm jet}(\Delta p_T^{\rm jet} = 0) - \varepsilon \cdot \Delta p_T^{\rm jet}$ (which is the denominator in definition of $z \equiv p_T^h / p_T^{\rm jet}$ in JFF) without any influence on the numerator of z and, as a consequence, in reducing effect on JFF softening, while the integral jet suppression factor becomes larger. The remarkable prediction here is that the effect on jet rate suppression becomes comparable with the effect on JFF softening at quite reasonable value $\varepsilon \sim 0.3$.

Finally, we considered the situation for b-quark induced jet which is tagged by leading muon. We have found the qualitatively similar results for anti-correlation between jet suppression and JFF softening, but with lesser absolute in-medium effect due to "dead cone" suppression of gluon radiation from heavy quark as compared with light parton [19].

Acknowledgments

I.P.L. thank the organizers of the Institute for the warm welcome, the hospitality and creating a stimulating atmosphere.

References

[1] R. Baier, D. Schiff, and B.G. Zakharov, Annual Rev. Nucl. Part. Sci. 50 (2000) 37.

[2] M. Gyulassy and X.-N. Wang, Phys. Lett. B 68 (1992) 1480.

[3] R. Baier, Yu. L. Dokshitzer, A.H. Mueller, and D. Schiff, JHEP 0109 (2001) 033.

[4] M. Gyulassy, P. Levai, and I. Vitev, Phys. Lett. B 538 (2002) 282.

[5] C.A. Salgado and U.A. Wiedemann, Phys. Rev. Lett. 89 (2002) 092303.

[6] C.Adler et al. (STAR Coll.), Nucl. Phys. A 698 (2002) 64.

[7] K.Adcox et al. (PHENIX Coll.), Nucl. Phys. A 698 (2002) 511.

[8] B.B. Back et al. (PHOBOS Coll.), Nucl. Phys. A 698 (2002) 655.

[9] I.G. Bearden et al. (BRAHMS Coll.), Nucl. Phys. A 698 (2002) 29.

[10] CMS Collaboration, Technical Proposal, CERN/LHCC 94-38.

[11] G. Baur et al., "Heavy Ion Physics Programme in CMS", CERN CMS Note 2000/060.

[12] I.P. Lokhtin and A.M. Snigirev, Phys. Lett. B 440 (1998) 163.

[13] R. Baier, Yu. L. Dokshitzer, A.H. Mueller, and D. Schiff, Phys. Rev. C 60 (1999) 064902.

[14] B.G. Zakharov, JETP Lett. 70 (1999) 176.

[15] U.A. Wiedemann and M. Gyulassy, Nucl.Phys. B 560 (1999) 345; U.A. Wiedemann, Nucl. Phys. B 588 (2000) 303; Nucl. Phys. A 690 (2001) 731.

[16] M. Gyulassy, P. Levai, and I. Vitev, Nucl. Phys. B 594 (2001) 371.

[17] I.P. Lokhtin and A.M. Snigirev, Phys. Lett. B 567 (2003) 39.

[18] I.P. Lokhtin and A.M. Snigirev, Eur. Phys. J. C 16 (2000) 527.

[19] Yu.L. Dokshitzer and D.E. Kharzeev, Phys. Lett. B 519 (2001) 199.

DECONFINEMENT AND FREEZE-OUT CONDITIONS FOR HADRONIC MATTER

V.K. Magas[1], H. Satz[1,2]

[1] *Center for Physics of Fundamental Interactions (CFIF), Instituto Superior Tecnico, Av. Rovisco Pais, 1049-001 Lisbon, Portugal, E-mail: vladimir@cfif.ist.utl.pt*

[2] *Fakultät für Physik, Universität Bielefeld, Postfach 100 131, D-33501 Bielefeld, Germany, Email: satz@physik.uni-bielefeld.de*

Matter implies the existence of a large-scale interconnected medium of uniform nature. When such a system breaks up into fragments much smaller than the size of the volume in which it is contained, it has undergone a change of state. Such a phase transition is similar to freeze-out (FO) in hadronic matter (HM). We would like to study this in a very simple geometric model of HM [1] - matter formed by hadrons, which can overlap; the volume of an individual hadron is $V_h = (4\pi/3)r_h^3$.

In such a model we can define freeze-out in geometric way: it occurs at the point at which the size of hadronic clusters falls below the size of the given overall spatial volume. This point is determined in percolation theory [2] and for three space dimensions becomes

$$n_f = 0.34/V_h, \tag{1}$$

where n is hadron density. This, however, does not mean that at $n = n_f$ clusters fill all the volume - $V_{empty}/V = \exp\{-0.34\} \simeq 0.71$. On the other hand, the vacuum, measured in terms of the hadronic scale r_h, also forms percolating clusters. >From percolation theory we know that at

$$n_c = 1.24/V_h \tag{2}$$

vacuum percolation stops and vacuum effectively disappears [3]. Since the disappearance of the physical vacuum is a basic feature of deconfinement, it seems natural to relate this threshold to the confinement/deconfinement transition.

On a more dynamical level, there is another approach to the problem. The confinement transition leads to interacting HM. It is known that if the interactions between the constituents of such a medium are dominated by resonance formation and decay, the system can effectively be considered as an ideal gas of all possible resonances [4]. In our context this means that if and when the interacting HM is resonance dominated, FO occurs at the point of confinement, i.e.

W. Greiner et al. (eds.), Structure and Dynamics of Elementary Matter, 657–659.

determined by condition (2). For the systems of vanishing or low baryon density, this appears quite well supported. Experimentally, the species abundances analysis indicates that these are indeed determined by a single FO temperature, $T_f \simeq 170-175$ MeV, obtained in e^+e^- annihilation, $p-p$, $p-\bar{p}$ interactions as well as in heavy ion collisions [5, 6]. This temperature is moreover completely in agreement with finite temperature lattice QCD results [7].

On the other limit, for dense nuclear matter at low temperature, the situation is quite different. The interaction between two nucleons does not lead to resonance formation; instead it is dominated by Fermi statistics and baryon repulsion. Hence here we expect FO to occur only when the nucleons no longer form interconnected matter, i.e. at the $n = n_f$ transition point. If we use the nucleon radius $r_n \simeq 0.8$ fm , Eq. (1) specifies $n_f \simeq 0.16$ fm^{-3} as the FO density, which is just the normal nuclear density. Thus, the correspondence between hadron percolation and freeze-out works well at $T = 0$ and defines nuclear matter as the most dilute system of nucleons which still forms connected matter.

In the relativistic, multi component gas, where the conservation of number of distinct particles is not the main interest, we associate the chemical potentials to conserved quantum numbers B, Q and S. The situation with strangeness is more complicated than with others due to large mass of strange quarks. Even at most energetic heavy ion collisions nowadays we can not expect to create such a hot and long lived system that strange quarks will also be in thermal equilibrium at hadronization time. In order to take into account the degree of equilibration of the strange quarks the parameter γ_s was introduced [9].

So far we have only specified a way to determine $T_f = T_c$ at $\mu = 0$ and μ_f at $T = 0$. To determine the entire freeze-out curve in the $T - \mu$ plane, we have to know how contributions from non-resonant baryon interactions modify the ideal resonance gas picture. We assume [1]: A) the sector of vanishing baryon density freezes out according to the resonance gas approximation and by vacuum percolation; B) the sector of finite baryon number freezes out according to baryon percolation; C) (A) and (B) processes go in such a way to give

$$n(T,\mu) = \frac{1.24}{V_h}\left[1 - \frac{n_B(T,\mu)}{n(T,\mu)}\right] + \frac{0.34}{V_h}\left[\frac{n_B(T,\mu)}{n(T,\mu)}\right] \qquad (3)$$

as the defining equation for the chemical FO curve in all $T - \mu$ plane (n_B is a baryon density). The Eq. (3) has correct $\mu = 0$ and $T = 0$ limits, discussed above, and moreover it determines FO fully in terms of geometric clustering, based on the intrinsic hadronic scale, with no adjustable parameters of any kind.

Figure 1 shows our curves and the FO points from the four different facilities [10] (see [1] for more results). The experimental results agree very well with our prediction, except the lowest energy point (SIS data). While we have no explanation for this, it should be noted that at such low energies, secondary hadron production is a rather rare process, making a FO determination from species abundances more difficult: only 4 particle ratios were measured at SIS and the fit with T and μ (γ_s was fixed to 1, although one could expect a lower value) through these 4 points was not so good - $\chi^2/d.o.f. \approx 14 - 15$ [6].

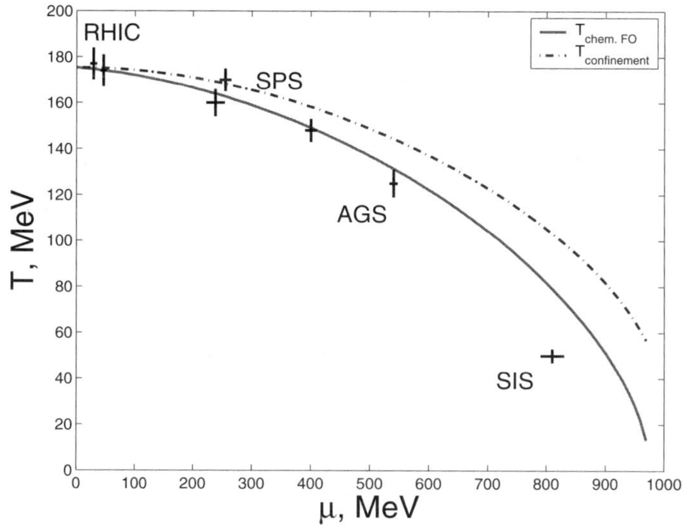

Figure 1. Confinement and freeze-out in hadronic matter. The chemical FO data for heavy ion collisions are from [10]. For the calculations we have used $\gamma_s = 0.5$; $r_n = 0.8$ fm for baryons, and $r_m = r_n/(2/3)^{3/4} \simeq r_n/1.1$ for mesons, as suggested by the bag model; all observed resonances [8] up to masses of $2.5\ GeV$ are taken into account. >From [1].

References

[1] V. Magas, H. Satz, hep-ph/0308155, to appear in Eur. Phys. J. C.

[2] See e.g., M. S. Isichenko, Rev. Mod. Phys. **64** (1992), 961.

[3] G. Baym, Physica A **96** (1979), 131; T. Çelik, F. Karsch and H. Satz, Phys. Lett. B **97** (1980), 128; H. Satz, Nucl. Phys. A **642** (1998), 130.

[4] E. Beth and G. E. Uhlenbeck, Physica **4** (1937), 915; R. Hagedorn, Nuovo Cim. Suppl. **3** (1965), 147; Nuovo Cim. A **56** (1968), 1027.

[5] F. Becattini, Z. Phys. C **69** (1996), 485; F. Becattini and U. Heinz, Z. Phys. C **76** (1997), 269; F. Becattini, Nucl. Phys. A **702** (2002), 336.

[6] F. Becattini et al., Phys. Rev. C **64** (2001), 024901.

[7] See e.g., F. Karsch, Lect. Notes Phys. **583** (2002), 209.

[8] K. Hagiwara et al., Phys. Rev. D **66** (2002), 010001.

[9] J. Rafelski, Phys. Lett. B **262** (1991), 333.

[10] For a recent compilation, see e.g. P. Braun-Munzinger et al., nucl-th/0304013.

RELATIVISTIC QRPA DESCRIPTION OF EXCITATIONS IN EXOTIC NUCLEI

N. Paar[1,2,3], T. Nikšić[2,3], D. Vretenar[2,3], and P. Ring [3]

[1]*Institut für Kernphysik, Technische Universität Darmstadt*
Schlossgartenstr. 9, D-64289 Darmstadt, Germany
nils.paar@physik.tu-darmstadt.de

[2]*Physics Department, Faculty of Science, University of Zagreb, Croatia*

[3]*Physik-Department der Technischen Universität München D-85748 Garching, Germany*

Abstract The fully self-consistent relativistic quasiparticle random-phase approximation (RQRPA) has been formulated in the the canonical basis of the relativistic Hartree-Bogoliubov (RHB) model and applied in the analysis of the isovector dipole response in nuclei with large neutron excess. The dipole response is characterized by the fragmentation of the strength distribution and its spreading into the low-energy region and by mixing of isoscalar and isovector modes. In light nuclei the onset of dipole strength in the low-energy region is due to single particle excitations of the loosely bound neutrons. In heavier nuclei, low-lying dipole states appear, which are characterized by a more distributed structure of the RQRPA amplitude. An analysis of the transition densities and velocity distributions in neutron rich N=82 isotones and tin isotopes reveals the dynamics of the dipole pygmy resonance: the vibration of the excess neutrons against the inert core composed of equal numbers of protons and neutrons.

Keywords: dipole excitations, pygmy resonances, exotic nuclei, neutron skin, relativistic mean field model, relativistic quasiparticle RPA

In recent years the multipole response of unstable nuclei far away from the valley of β-stability became a very active field of research, both experimentally and theoretically. These nuclei are characterized by unique structure properties: the weak binding of the outermost nucleons and the effects of the coupling between bound states and the particle continuum. On the neutron rich side, in particular, the modification of the effective nuclear potential leads to the formation of nuclei with very diffuse neutron densities, to the occurrence of the neutron skin and halo structures. These phenomena will also affect the multipole response of unstable nuclei, in particular the electric dipole and quadrupole excitations, and new collective modes of excitations might arise in nuclei near the drip

W. Greiner et al. (eds.), Structure and Dynamics of Elementary Matter, 661–663.

line. Non-resonant excitations into the continuum near the dissociation threshold may also provide information on the single-partice spectra of the ground state. In order to investigate multipole response in exotic nuclei, in particular the low-lying excited states, we have formulated the self-consistent relativistic quasiparticle random phase approximation (RQRPA). The single-particle spectra of the ground state for open-shell nuclei are determined from the Relativistic Hartree-Bogoliubov model (RHB), where for the interaction in the particle-hole channel effective Lagrangians with nonlinear meson self-interactions are used (NL3 effective force), and pairing correlations are described by the pairing part of the finite-range Gogny interaction (D1S parameterization). Furthermore, the RHB eigenspectrum is transformed into the canonical basis of single-particle states which diagonalizes the density matrix, where the RHB ground state has a BCS form. The canonical basis of the RHB model is used as the ground state in investigations of excitation phenomena in finite nuclei. The RQRPA equations are derived from the small amplitude limit of the time-dependent RHB model. The RQRPA configuration space is composed from two-quasiparticle configurations which include states with both nucleons in the discrete bound levels, states with one nucleon in the bound level and one nucleon in the continuum, and also states with both nucleons in the continuum. In addition, the RQRPA configuration space also contains pair configurations formed from the fully or partially occupied states and the empty negative-energy states from the Dirac sea. The present RHB+RQRPA model is fully self-consistent: for the interaction in the particle-hole channel effective Lagrangians with nonlinear meson self-interactions are used, and pairing correlations are described by the pairing part of the finite-range Gogny interaction. The RHB+RQRPA model has been employed to study the isovector dipole response, in particular to disclose the properties of the low-energy dipole excitations. Investigations of E1 strength in oxygen isotopes have recently attracted considerable interest since these nuclei might be good candidates for a possible identification of the low-lying collective soft mode (pygmy dipole resonance) which corresponds to the oscillations of excess neutrons out of phase with the core composed of an equal number of protons and neutrons. However, the present investigation indicates that in light nuclei, the low-lying strength is composed only from the neutron single-particle excitations. The self-consistent inclusion of the pairing correlations in the RHB+RQRPA model is of a particular importance, not only to decouple the spurious 1^- excitation as a zero-energy mode, but it also enhances the low-lying strength near the threshold. Motivated by very recent data on the concentration of electric dipole strength below the neutron separation energy, N=82 semimagic nuclei have also been studied within the present model.In medium-heavy neutron rich tin isotopes and N=82 isotones, among several dipole states in the low-energy region between 7 MeV and 9 MeV, a single state is found with a more distributed structure of the RQRPA amplitude, characterized by a coherent superposition of many neutron two-quasiparticle configurations. The evolution of isovector dipole strength functions for several tin isotopes are shown in Fig.1(a), and the corresponding proton and neutron transition densities for the main low-lying states are displayed in Fig.1(b). A relatively strong peak below 10 MeV is characterized by the dynamics of the pygmy resonance. Its proton and neutron transition densities are in phase in the nuclear interior, there are no contributions from the protons in the surface region, the isoscalar transition density dominates over the isovector one in the interior, and the strong neutron transition density displays a long tail in the radial coordinate. Very recently the LAND/FRS group at GSI has performed an experiment with a secondary beam of ^{132}Sn and neighboring isotopes and data are currently under analysis.

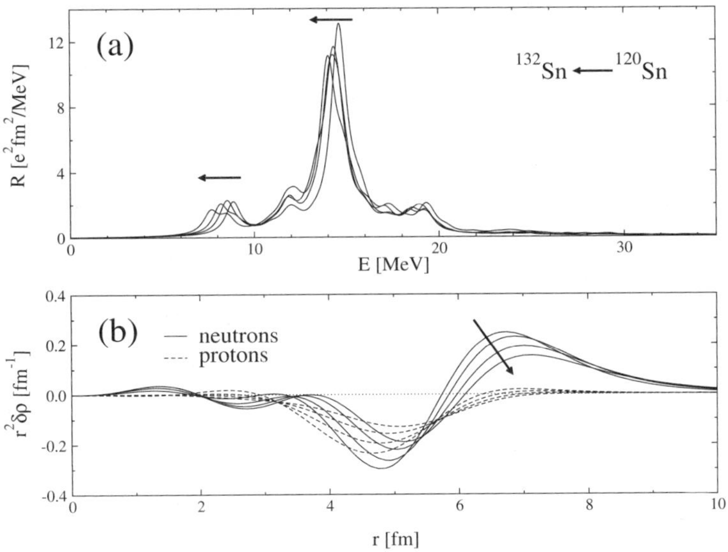

Figure 1. (a) Isovector dipole strength distributions in ^{120}Sn, ^{124}Sn, ^{128}Sn, and ^{132}Sn isotopes, (b) The corresponding proton and neutron transition densities for the main low-lying excitation energy below 10 MeV.

References

M. Matsuo, Nucl. Phys. A 696, 371 (2001).
D. Vretenar, N. Paar, P. Ring, and G. A. Lalazissis, Nucl. Phys. A 692, 496 (2001).
A. Leistenschneider et al., Phys. Rev. Lett. 86, 5442 (2001).
A. Zilges et. al. Phys. Lett. B 542, 43 (2002).
N. Paar, T. Nikšić, D. Vretenar and P. Ring, Phys. Rev. C 67, 034312 (2003).

$\Sigma(1385)$ RESONANCE STUDIES WITH STAR AT $\sqrt{S_{NN}} = 200$ GEV

Sevil Salur for the STAR Collaboration
Yale University, Physics Department,
272 Whitney Ave. New Haven, CT
sevil.salur@yale.edu

Abstract In p+p, d+Au, and Au+Au collisions at $\sqrt{s_{NN}} = 200$ GeV with the STAR detector at RHIC, $\Sigma(1385) \rightarrow \Lambda + \pi$ were measured using two techniques; three-particle mixing and a hybrid mixing technique. We present results from both of these methods and compare the invariant mass spectra and the backgrounds.

Keywords: Resonances, Relativistic Heavy Ion Collisions

Introduction

Due to the very short lifetime ($\sim$ few fm/c) of resonances a large fraction of the decays occur inside the reaction zone in the most central relativistic heavy ion collisions. These resonances have a high probability of not being reconstructed as their daughter particles are likely to re-scatter in the medium. Resonances with higher transverse momentum are more likely to be reconstructed because of their longer relative lifetimes. This means they are more likely to decay outside of the medium and hence their daughter particles would interact less with the medium[1]. This may give rise to effects such as changes in the physical properties of the resonances, width broadening and mass shifts, or changes in the transverse momentum (p_T) spectra due to re-scattering of the decay particles. Hence the study of resonances provides an additional tool to determine the hadronic expansion time between chemical and thermal freeze-out. Due to its strange quark content and high mass, $\Sigma(1385)$ may also give additional information about strangeness enhancement, one of the possible signatures of the Quark Gluon Plasma [2].

There are also several models currently available that attempt to describe hadronic matter with resonances. One is based on the thermal model and argues that the direct measurement of resonances can probe both the hadronization temperature and the lifetime of the interacting hadron gas phase [3]. In this model, the ratios of resonances to lighter particles with identical number of

W. Greiner et al. (eds.), Structure and Dynamics of Elementary Matter, 665–667.

valance quarks are sensitive to the freeze-out temperature because all of the chemical dependence cancels out in the ratio.

1. Analysis and Particle Identification

Charged daughter particles are identified by the momentum and the energy loss per unit length (dE/dx) measured in the Time Projection Chamber of STAR (TPC)[4]. For long-lived particles ($c\tau \sim$ few cm) such as the K_s^0, Λ, and Ξ, the decay vertex topology is used for identification. This technique cannot be used for resonances due to their very short lifetimes ($c\tau_{\Sigma(1385)} = 5fm/c$). However an alternative method, called the mixing technique, can be used and also applied to identify other short-lived particles such as pentaquarks and dibaryons.

The decay channel that we investigate is $\Sigma(1385) \rightarrow \Lambda + \pi$. We use two methods to identify Λ's. In the first method, the three particle mixing technique (TPM), every π is combined with every p to produce a Λ candidate. Then the Λ candidate is combined with all remaining π to get $\Sigma(1385)$. In the second method, the hybrid mixing technique (HM), $\Sigma(1385)$'s are identified by combining topologically reconstructed Λ's with π's. In both techniques the background is described by combining π 's from one event with the Λ's from another event.

2. Status of Current Studies

$\Sigma(1385) \rightarrow \Lambda + \pi_{bachelor}$ is identified in p+p, d+Au, and Au+Au collisions with the mixing techniques. Figure 1 is the invariant mass spectra of the $\Sigma(1385)$ for the various nuclear systems studied. With the TPM technique, in the lower kinematic limit of the invariant mass spectrum a background structure appears (Fig 1-a). This initial structure increases in the d+Au collisions and totally dominates the spectrum in Au+Au collisions, due to an increase in background combinatorial statistics. Monte Carlo studies show that a partial contribution of the background structure comes from the misidentification of the Λ's with the $\pi_{bachelor}$ and misidentification of K's as π's. A more detailed study with the simulations is needed to find the ratio of each contribution and the other possible sources. The HM technique can be used to identify $\Sigma(1385)$ with less efficiency but with a cleaner background (Fig 1-b,c,d). Since Λ's can be identified more cleanly with the decay topology technique, the initial structure disappears. Antiparticle to particle ratios of Ξ and $\Sigma(1385)$ are observed as 0.89 ± 0.04 and 0.90 ± 0.07 respectively in p+p collisions. These values are consistent with a net baryon-free collision environment.

With available events, it is possible to study the p_T spectra of the $\Sigma(1385)$ for all systems. In order to correct for STAR's efficiency and acceptance, a study using Monte Carlo tracks embedded into real events is necessary.

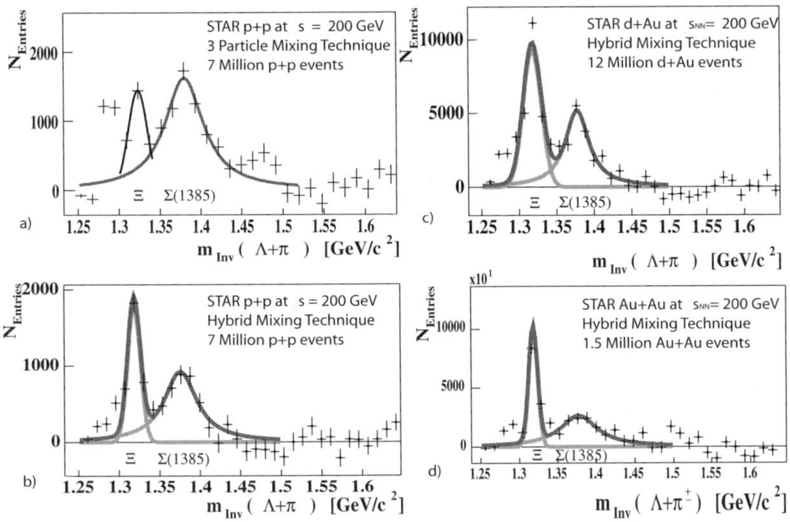

Figure 1. Invariant mass spectra of Ξ and Σ(1385) after mixed-event background subtraction: (a) Invariant mass spectrum of Σ(1385) in p+p using the three particle mixing technique. The Ξ has the same decay channel as the Σ(1385). The Ξ is fit with a gaussian distribution and the Σ(1385) is fit with a Breit-Wigner. (b) Invariant mass spectrum in p+p using the hybrid mixing technique. With respect to the TPM technique the structure at the low M_{inv} in the background disappears with cleanly identified Λ's. (c) Invariant mass spectrum in d+Au using the hybrid mixing technique. (d) Invariant mass spectrum in Au+Au using the hybrid mixing technique.

3. Summary

Using the mixing technique Σ(1385) and Ξ are constructed in $\sqrt{s_{NN}}$ = 200 GeV p+p, d+Au, and Au+Au data. Raw yields and ratios are presented. Corrections to these values are being investigated and will be resolved in the very near future. Fireball temperature of the collision can be studied using the p_T distributions after the corrections. Studying Au+Au collisions is essential to understand possible rescattering and medium effects on resonances. It is possible to study pentaquarks with STAR and with the upcoming run starting this January 2004 we will continue our investigation of exotic particles by using the STAR Silicon Vertex Tracker (and eventually a new Time of Flight detector system).

References

[1] Marcus Bleicher, hep-ph/0212378

[2] J.Rafelski and B. Muller, Phys. Rev. Lett. **48**, 1066 (1982)

[3] G.Torrieri and J.Rafelski, J. Phys. G **28**, 1911 (2002)

[4] STAR Collaboration, Nuc. Ins. Meth. **A 499**, 624 (2003)

THE INVESTIGATION OF MULTIPOLARITY OF THE ELECTROMAGNETIC TRANSITIONS IN SOME EVEN-EVEN HF ISOTOPES

Nurettin Türkan[1], Hüseyin Erdem[2], Davut Olgun[2], İhsan Uluer[2] and Sait İnan[3]

[1] *Erciyes University, Yozgat Faculty of Arts and Science*

[2] *Kirikkale University, Faculty of Arts and Science*

[3] *Celal Bayar University, Faculty of Education*

Abstract The aim of this work is to describe the results of calculation of multipole mixtures in some even-even Hf isotopes. Deformed nuclei have been studied for many years and an enormous quantity of detailed spectroscopy now exsts. But yet, our understanding of these nuclei remains suprisingly limited. The medium-to heavy- mass hafnium (Hf) isotopes are located in the rare-earth mass region where well-deformed nuclei can be populated to very high spin. The IBA model [1] has been applied to deformed nuclei [2], usually with considerable success [3, 4]. As a result of the phenomenological description of deformed nuclei in IBM , Hf isotopes has been occupied along the SU(3)-O(6) leg of the IBM triangle. In this work, we have carried out calculations for such isotopes and obtained reasonably good agreement with emprical systematics in the region. The aim of this work is to describe the results of calculation of multipole mixtures in ^{174,176,178,180}Hf.

The consistent Q formalism (CQF) [1] has been the standard choice for the description of deformed nuclei in the sd-IBM [2]. The Hamiltonian employed in the current approach is of the form of the dipole and quadrupole interactions [2],

$$H = k_1 L \cdot L - k_2 Q \cdot Q \tag{1}$$

Here L is the boson angular momentum operator and Q is the quadrupole operator which is given by

$$Q = [s^+\tilde{d} + d^+\tilde{s}]^{(2)} + \chi[d^+\tilde{d}]^{(2)} \tag{2}$$

where $\chi = \chi_Q/5^{1/2}$ and χ_Q denotes the parameter defining the ratio of the two terms in the operator [1, 4]. When $\chi_Q = -35^{1/2}/2$ the SU(3) limit of model comes to appearance (thus the ratio of the two terms in Q is $-7^{1/2}/2$). When $\chi_Q = 0$ then Q in the Eq. (2) becomes the generator of O(6) and a specific form of O(6) appears. The gross structural changes involved in the transition

W. Greiner et al. (eds.), Structure and Dynamics of Elementary Matter, 669–671.

between SU(3) and O(6)-like structure can be predicted simply as a function of χ_Q and the boson number N [4]. This feature comes from the particularly simple structure of the Hamiltonian of Eq. (1). The $L \cdot L$ term is always diagonal and so, it has no effect on the final wave functions. The relative sizes of all the matrix elements are specified by χ_Q (for a given boson number N). In addition, the relative energies of states of the same spin will also depend only on χ_Q and N. The k_1 is a parameter that is acting as a scaling factor, which determines the overall energy splitting of the states.

The calculation of E2 transition probabilities in the model involves the use of an operator,

$$T(E2) = e_2 Q \tag{3}$$

where e_2 represents the boson effective charge. The E2 quadrupole operator is defined to be identical to that in the Hamiltonian, so that its structure is also given by χ_Q. The eigenvalue expression corresponding to the terms $Q \cdot Q$ (with $\chi_Q \equiv 0$) and $L \cdot L$ has only two constants [4],

$$E = A'[(N+\sigma)(N+\sigma+4)+\tau(\tau+3)] + c'L(L+1) \tag{4}$$

where $A' = 2k_2$ and $c' = k_1$.

To calculate $\delta(E2/M1)$ values of some transitions in ^{174}Hf, ^{176}Hf, ^{178}Hf and ^{180}Hf isotopes k_2 parameters of 20.7, 22.0, 21.8 and 22.2 are used. Also, by the rules of IBM-1, when the proton and neutron are taken as identical particles the total boson numbers of these Hf isotopes are given by 15, 16, 15 and 14.

We mainly focused on the description of high-spin states, which has not been done before.

The comparison of calculated $\delta(E2/M1)$ multipole mixing ratios of some transitions in some even-even hafnium isotopes with some previous experimental results [5] are given in a table. At the result, the calculated values of $\delta(E2/M1)$ were seen in good agreement with that of some results of previous experiment.

The calculated $\delta(E2/M1)$ values for $4_2(765\text{keV})4_1, 2_2(1138\text{keV})2_1$, $2_4(1183\text{keV})2_1$ and $3_1(1197\text{keV})2_1$ transitions in ^{174}Hf, ^{176}Hf , ^{178}Hf and ^{180}Hf isotopes are almost the same as the previous experimental results. Altough the calculated $\delta(E2/M1)$ results for $2_2(809\text{keV})2_1$ and $6_2(699\text{keV})6_1$ in ^{174}Hf, $4_3(1301\text{keV})4_1$ in ^{176}Hf and $2_2(1082\text{keV})2_1$ and $2_6(1403\text{keV})2_1$ in ^{178}Hf are a bit different than the experimental value, they are within the error limits of experiment.

It is clearly seen from the table that Hafnium isotopes display a sudden change in the structure between the SU(3)-O(6) transition leg of the symmetric triangle by yielding a rapid approach towards the SU(3) limit.

Isotope	Spin-Parity $I_i \rightarrow I_f$	Transition Energy (keV)	Mixing Ratio $\delta(\mathrm{E2/M1})$	
			Present	Experimental [5]
^{174}Hf	$2_2 \rightarrow 2'_1$	809	5.06	11.00
	$4_2 \rightarrow 4'_1$	765	2.50	2.50
	$6_2 \rightarrow 6_1$	699	1.56	0.90
	$2_5 \rightarrow 2_2$	418	2.28	–
	$2_4 \rightarrow 2_1$	1225	5.64	–
	$3_2 \rightarrow 2_1$	1245	4.40	–
	$4_3 \rightarrow 4_1$	1097	3.82	–
^{176}Hf	$2_2 \rightarrow 2_1$	1138	3.99	4.00
	$2_3 \rightarrow 2_1$	1159	3.78	–
	$2_4 \rightarrow 2_1$	1252	3.14	–
	$2_5 \rightarrow 2_1$	1290	2.96	–
	$4_3 \rightarrow 4_1$	1301	2.02	0.70
^{178}Hf	$2_2 \rightarrow 2_1$	1082	0.47	11.00
	$2_4 \rightarrow 2_1$	1183	0.41	0.41
	$2_6 \rightarrow 2_1$	1403	0.36	0.77
	$2_8 \rightarrow 2_1$	1468	0.35	–
^{180}Hf	$3_1 \rightarrow 2_1$	1197	9.59	9.60
	$2_2 \rightarrow 2_1$	1089	8.79	–

Table: Comparison of calculated multipole mixing ratios of some transitions in some even-even Hafnium isotopes

References

[1] F. Iachello and A. Arima, The interacting boson model (Cambridge Univ. Press, Cambridge 1987)

[2] S.C. Li, S. Kuyucak, Nucl. Phys. A604 (1996) 305

[3] D.D. Warner, R.F. Casten and W.F. Davidson, Phys. Rev. C24 (1981) 1713

[4] D.D. Warner and R.F. Casten, Phys. Rev. C28 (1983) 1798

[5] J. Lange, K. Kumar, J.H. Hamilton, Rev. Mod. Phys. 54,(1982)161

QUANTUM DYNAMICS OF BRACHISTOCHRONE PROBLEM

Melike B. Yucel and Nuri Unal
Akdeniz University, Physics Department, Antalya Turkey, myucel@akdeniz.edu.tr and nuriunal@akdeniz.edu.tr

Abstract In this study we discuss the quantum dynamics of a particle, which moves classically on the brachistochrone curve corresponding to the minimization of the time functional, in a linear gravity potential. We derive the Lagrangian and the Hamiltonian of the particle, which moves also on the brachistochrone curve by the minimization of the action functional. The solutions of the Schrödinger's equation for this Hamiltonian give the energy spectrum, and the confluent hypergeometric functions as the wave functions. The problem combines the infinite-well and harmonic oscillator potentials. We also discuss the solutions of the Schrödinger's equation for the particle in the periodic extension of the original brachistochrone problem. We show that the band structure arised from Floquet theory and the problem is equivalent to the periodic δ-potential problem for the particle with positive energy in the limit of infinite potential.

1. Introduction

Brachistochrone is one of the oldest problems of physics and also one of the earliest problems proposed in the calculus of variations [1, 2], and it is known as "the shortest time problem of a particle moving between two points on a vertical plane". The period of oscillation for the brachistochrone curve depends on the amplitude.

The aim of this study is to discuss the quantum dynamics of a particle, which moves classically on the brachistochrone curve in a homogeneous force field and, the solution of the Schrödinger equation for the particle in the periodic extension of the original brachistochrone problem [3].

2. Classical system and quantization of the system

The minimization of the time function, t for a massive particle moving on a brachistochrone curve, so that the net force on it is the constant force of gravity, $-mgy$ gives the parametric solutions as $x = a(\theta - \sin\theta)$ and $y = a(1 - \cos\theta)$; where, the variable θ and the constant a are the angle of rotation

W. Greiner et al. (eds.), Structure and Dynamics of Elementary Matter, 673–675.

and the radius of the circle rolling on the x-axis, respectively. We classically write the Lagrangian of the particle by the transformation $u = a\cos(\theta/2)$:

$$L\left(u, \dot{u}\right) = 8m\left[\dot{u}^2 + \omega^2\left(a^2 - u^2\right)\right], \tag{1}$$

where $\omega^2 = g/4a$. The equation of motion is $\ddot{u} + \omega^2 u = 0$. So, we obtain the classical solutions as $\theta(t) = 2\cos^{-1}\left(\frac{A\sin\omega t + B\cos\omega t}{a}\right)$. The Hamiltonian of system:

$$H = \frac{p_u^2}{2\mu} - \frac{\mu\omega^2}{2}\left(a^2 - u^2\right) \tag{2}$$

where $\mu = 16m$. The energy eigenvalue equation is $\widehat{H}\Psi_E(\vec{r}) = E\Psi_E(\vec{r})$. For the sake of simplicity we introduce the dimensionless parameters, $\lambda = E'/\hbar\omega$, $\varsigma = \sqrt{2\mu\omega/\hbar}u$, where $E' = E + \mu\omega^2 a^2/2$. Here, we assume that $E \geq -\mu\omega^2 a^2/2$. The resulting unnormalized eigenfunctions and the energy eigenvalues are obtained as :

$$\begin{aligned}
\Psi_{p_n}^{(+)}(\varsigma) &= Ce^{-\frac{\varsigma^2}{4}}\,{}_1F_1\left(-\tfrac{p_n}{2}, \tfrac{1}{2}; \tfrac{\varsigma^2}{2}\right); n = \text{even integers} \\
\Psi_{p_n}^{(-)}(\varsigma) &= C'e^{-\frac{\varsigma^2}{4}}\,\varsigma\,{}_1F_1\left(\tfrac{1-p_n}{2}, \tfrac{3}{2}; \tfrac{\varsigma^2}{2}\right); n = \text{odd integers} \\
E_{p_n}^{(\pm)} &= -\tfrac{1}{2}\mu\omega^2 a^2 + \hbar\omega\left(p_n + \tfrac{1}{2}\right)
\end{aligned} \tag{3}$$

where $p = \lambda - 1/2$. C and C' are the constants. The geometry of the problem gives the boundary conditions as $\Psi^{(\pm)}\left(\pm\sqrt{2\mu\omega/\hbar}a\right) = 0$. We calculate the roots of it by using a graphical method and a process called curve fitting [4]:

$$p_n \cong n + \frac{(n+1)^2\,\hbar}{\mu\omega a^2}; \quad n = 0, 1, 2, ... \text{(integers)} \tag{4}$$

As it is given in Eq. (4), if $\sqrt{2\mu\omega/\hbar}a$ is very large, these solutions give the integer values for p. The eigenfunctions and the energy levels obtaining in the limit, $\omega \rightarrow \infty$ are the same as those of a harmonic oscillator solutions. The condition $\sqrt{2\mu\omega/\hbar}a << 1$ represents the case of an infinite well potential.

3. Periodic potential problem for brachistochrones

We consider an infinite array of the cycloid shaped potentials with the interval b in which the potential is zero in one dimension. The particle moves under the influence of the periodic potentials, $V = -\mu\omega^2\left(a^2 - \varsigma^2/\epsilon^2\right)/2$, where $\epsilon^2 = 2\mu\omega/2$. Thus the solutions of the Schrödinger equation are periodic according to Floquet theory [5].

The energies of the bound states are in the interval $-\mu\omega^2 a^2/2 < E < 0$. If we write the eigenfunctions in the period $-\epsilon a < \varsigma < 2b + 3\epsilon a$; and use the continuity conditions of the functions, Ψ and of the derivations of them, we obtained the following condition for allowed energy values in the limits, $b \to 0$, $2\epsilon a \to 0$, $V_0 = -\mu\omega^2 a^2/2 \to -\infty$,:

$$-1 \leq \cosh(\kappa b) - \Theta \frac{\sinh(\kappa b)}{\kappa b} \leq 1 \tag{5}$$

where, $\Theta = \lim_{\substack{2\epsilon a \to 0 \\ -1/2\mu\omega^2 a^2 \to -\infty}} \frac{2}{3}\mu\omega^2 \epsilon a^3$ =constant and $\kappa^2 = |E| / \hbar\omega$.

We derive the bound states with positive energy by substituting $k = -i\kappa$. In the same limits as above, the energy function $f(E)$ becomes

$$f(E) = \cos kb - \Theta \frac{\sin kb}{kb} \tag{6}$$

In this limit, the results are similar the band structure that has been derived with δ-potential problem.

4. Discussion

In this paper, we investigate the least time problem by the transformation $\varsigma = \sqrt{2\mu\omega/\hbar} a \cos(\theta/2)$, and obtained the solutions with this variable. We have also considered the classically allowed region as the interval $\left[-\sqrt{2\mu\omega/\hbar} a, \sqrt{2\mu\omega/\hbar} a\right]$, the potential is infinite at the outside of this interval. The problem combines the infinite-well and harmonic oscillator potentials and reduces to the harmonic oscillator for $\omega \to \infty$ and to the infinite-well potential for $\omega \to 0$. The solutions reflect the properties of the energy spectrum that is proportional to n and n^2. The eigenfunctions have even- and odd- parity. Although the p_n values in Eq. (4) are not the exact analytic expression for the energy eigenvalues of the system, they reflect the behaviour of the problem. Secondly, we have shown that the band structure arised from Floquet theory if the potential is periodic.

Acknowledgement
This research is partially supported by Akdeniz University (20.01.0105.12).

References

[1] H. Sussmann, J. C. Willems, in 35. Conference on Desicion and Control, Kobe, Japan, 1996

[2] H. Erlichson, Eur. J. Phys. 20, 299, 1999

[3] M. B. Yucel, N. Unal, Eur. Phys. J. D 25, 123, 2003

[4] P. Calter, Problem Solving with Computers, McGraw-Hill Book Company, USA, 1973, p. 125

[5] P. M. Morse, H. Feshbach, Methods of Theoretical Physics, McGraw-Hill Book Company, Inc., New York, 1953, vol. I